# REFERENCE CARD

**BEGINNING ALGEBRA Fifth Edition**
R. David Gustafson and Peter D. Frisk
ISBN 0-534-35832-2

**Brooks/Cole Publishing Company**
I**T**P® An International Thomson Publishing Company
Visit Brooks/Cole on the Internet
http://www.brookscole.com

## CHAPTER 1 REAL NUMBERS AND THEIR BASIC PROPERTIES

**Natural numbers:** 1, 2, 3, 4, 5, . . .

**Whole numbers:** 0, 1, 2, 3, 4, 5, . . .

**Integers:** . . . , $-4, -3, -2, -1, 0, 1, 2, 3, 4, \ldots$

**Rational numbers:** Fractions with integer numerators and nonzero integer denominators

**Real numbers:** Rational numbers and irrational numbers

**Prime numbers:** 2, 3, 5, 7, 11, 13, 17, . . .

**Composite numbers:** 4, 6, 8, 9, 10, 12, 14, 15, . . .

**Even integers:** . . . , $-6, -4, -2, 0, 2, 4, 6, \ldots$

**Odd integers:** . . . , $-5, -3, -1, 1, 3, 5, \ldots$

If $n$ is a natural number, then $x^n = \overbrace{x \cdot x \cdot x \cdot \cdots \cdot x}^{n \text{ factors of } x}$

**The closure properties:**

$x + y$ is a real number.

$x - y$ is a real number.

$xy$ is a real number.

$\dfrac{x}{y}$ is a real number ($y \neq 0$).

**The commutative properties:**

$x + y = y + x$

$xy = yx$

**The associative properties:**

$(x + y) + z = x + (y + z)$

$(xy)z = x(yz)$

**The distributive property:**

$x(y + z) = xy + xz$

## CHAPTER 2 EQUATIONS AND INEQUALITIES

Sale price = regular price − markdown

Retail price = wholesale cost + markup

Percentage = rate · base

## CHAPTER 4 POLYNOMIALS

If $m$ and $n$ are integers, then

$$x^m x^n = x^{m+n} \qquad (x^m)^n = x^{m \cdot n}$$

$$(xy)^n = x^n y^n \qquad \left(\frac{x}{y}\right)^n = \frac{x^n}{y^n} \ (y \neq 0)$$

$$\frac{x^m}{x^n} = x^{m-n} \ (x \neq 0) \quad x^0 = 1 \ (x \neq 0)$$

$$x^{-n} = \frac{1}{x^n} \ (x \neq 0)$$

**Special products:**

$$(x + y)^2 = x^2 + 2xy + y^2$$
$$(x - y)^2 = x^2 - 2xy + y^2$$
$$(x + y)(x - y) = x^2 - y^2$$

## CHAPTER 5 FACTORING POLYNOMIALS

**Zero-factor property:** If $a$ and $b$ are real numbers, then

If $ab = 0$, then $a = 0$ or $b = 0$.

**Factoring the difference of two squares:**

$$x^2 - y^2 = (x + y)(x - y)$$

**Factoring the sum and difference of two cubes:**

$$x^3 + y^3 = (x + y)(x^2 - xy + y^2)$$
$$x^3 - y^3 = (x - y)(x^2 + xy + y^2)$$

**Factoring perfect trinomial squares:**
$$a^2 + 2ab + b^2 = (a + b)^2$$
$$a^2 - 2ab + b^2 = (a - b)^2$$

## CHAPTER 6   PROPORTION AND RATIONAL EXPRESSIONS

If there are no divisions by 0, then

$$\frac{a}{b} = \frac{a \cdot c}{b \cdot c} \qquad \frac{a}{1} = a \qquad \frac{a}{0} \text{ is undefined}$$

$$\frac{a}{b} \cdot \frac{c}{d} = \frac{a \cdot c}{b \cdot d} \qquad \frac{a}{b} \div \frac{c}{d} = \frac{a}{b} \cdot \frac{d}{c}$$

$$\frac{a}{d} + \frac{b}{d} = \frac{a + b}{d} \qquad \frac{a}{d} - \frac{b}{d} = \frac{a - b}{d}$$

## CHAPTER 7   ROOTS AND RADICAL EXPRESSIONS

**The Pythagorean theorem:** If $c$ is the length of the hypotenuse of a right triangle and $a$ and $b$ are the lengths of its legs, then
$$a^2 + b^2 = c^2$$

**The distance formula:**
$$d = \sqrt{(x_2 - x_1)^2 + (y_2 - y_1)^2}$$

If $a$ and $b$ are not both negative, then

$$\sqrt{ab} = \sqrt{a}\sqrt{b} \qquad \sqrt{\frac{a}{b}} = \frac{\sqrt{a}}{\sqrt{b}} \quad (b \neq 0)$$

$$x^{1/n} = \sqrt[n]{x} \qquad x^{m/n} = \sqrt[n]{x^m} = \left(\sqrt[n]{x}\right)^m$$

## CHAPTER 8   WRITING EQUATIONS OF LINES; VARIATION

**Slope of a nonvertical line:**
$$m = \frac{y_2 - y_1}{x_2 - x_1} \quad (x_2 \neq x_1)$$

**Equations of a line:**

| | |
|---|---|
| $y - y_1 = m(x - x_1)$ | point–slope form |
| $y = mx + b$ | slope–intercept form |
| $ax + by = c$ | general form |
| $y = b$ | a horizontal line |
| $x = a$ | a vertical line |

**Variation:**

| | |
|---|---|
| $y = kx$ | direct variation |
| $y = \dfrac{k}{x}$ | inverse variation |
| $y = kxz$ | joint variation |

## CHAPTER 9   QUADRATIC EQUATIONS AND GRAPHING QUADRATIC FUNCTIONS

The solutions of $x^2 = c$ are $x = \sqrt{c}$ and $x = -\sqrt{c}$.

**The quadratic formula:**
$$x = \frac{-b \pm \sqrt{b^2 - 4ac}}{2a} \quad (a \neq 0)$$

The graph of the equation $y = ax^2 + bx + c$ is a parabola. It opens upward when $a > 0$ and downward when $a < 0$. The $x$-coordinate of the vertex of the parabolic graph is $x = -\frac{b}{2a}$.

The graph of the equation $y = (x - h)^2 + k$ is a parabola with vertex at $(h, k)$.

FIFTH EDITION

# *Beginning Algebra*

*To*
*Caitlin Mallory Barth*
*Nicholas Connor Barth*
*Prescott Alexander Heighton*
*Laurel Marie Heighton*
*Yasmeen Ann Lee*
*and Daniel Mark*
*Tyler Joseph*
*Spencer Roy*
*Skyler Roy*

## Books in the Gustafson/Frisk Series

■ ■ ■ ■ ■ ■ ■ ■ ■ ■ ■  *FIFTH EDITION*

# *Beginning Algebra* ■ ■ ■ ■ ■ ■ ■

R. David Gustafson

*Rock Valley College*

Peter D. Frisk

*Rock Valley College*

Brooks/Cole Publishing Company

I(T)P™ An International Thomson Publishing Company

Pacific Grove • Albany • Bonn • Boston • Cincinnati • Detroit • London • Madrid • Melbourne
Mexico City • New York • Paris • San Francisco • Singapore • Tokyo • Toronto • Washington

Publisher: *Robert W. Pirtle*
Marketing Team: *Jennifer Huber, Christine Davis, Debra Johnston*
Editorial Assistant: *Peggi Rodgers*
Production Editor: *Ellen Brownstein*
Production Service: *Hoyt Publishing Services*
Manuscript Editor: *David Hoyt*
Permissions Editor: *Carline Haga*
Interior Design: *E. Kelly Shoemaker, Vernon Boes*

Interior Illustration: *Lori Heckelman*
Photo Research: *Terry Powell*
Cover Design: *Roy Neuhaus*
Cover Photo: *Ed Young*
Art Coordinator: *David Hoyt*
Typesetting: *The Clarinda Company*
Cover Printing: *Phoenix Color Corp.*
Printing and Binding: *World Color Book Services (Taunton)*

*For more information, contact:*

BROOKS/COLE PUBLISHING COMPANY
511 Forest Lodge Road
Pacific Grove, CA 93950
USA

International Thomson Publishing Europe
Berkshire House 168-173
High Holborn
London WC1V 7AA
England

Thomas Nelson Australia
102 Dodds Street
South Melbourne, 3205
Victoria, Australia

Nelson Canada
1120 Birchmount Road
Scarborough, Ontario
Canada M1K 5G4

International Thomson Editores
Seneca 53
Col. Polanco
11560 México, D. F., México

International Thomson Publishing GmbH
Königswinterer Strasse 418
53227 Bonn
Germany

International Thomson Publishing Asia
221 Henderson Road
#05-10 Henderson Building
Singapore 0315

International Thomson Publishing Japan
Hirakawacho Kyowa Building, 3F
2-2-1 Hirakawacho
Chiyoda-ku, Tokyo 102
Japan

Printed in the United States of America

10  9  8  7  6  5  4  3  2  1

**Library of Congress Cataloging-in-Publication Data**

Gustafson, R. David (Roy David), [date]
  Beginning algebra / R. David Gustafson, Peter D. Frisk. — 5th ed.
    p.    cm.
  Includes index.
  ISBN 0-534-35832-2 (hardcover : alk. paper)
  1. Algebra.  I. Frisk, Peter D., [date].  II. Title.
QA152.2.G85  1998
512.9—dc21

98-27554
CIP

**Photo credits: p. 5,** The British Museum; **p. 190,** Courtesy of Texas Instruments; **p. 199,** Courtesy of IBM Corporation; **p. 261,** Martin Bond, Science Photo Library; **p. 339,** Ken Eward, Science Source; **p. 492,** Courtey of Princeton University.

## To the Instructor

*Beginning Algebra,* fifth edition, is written for students studying algebra for the first time and for those who need a review of basic algebra. It presents all of the topics associated with a first course in algebra, providing students with a thorough foundation in the basic skills of algebra and problem solving.

Our goal has been to write a book that

- Is enjoyable to read.
- Is easy to understand.
- Is relevant.
- Develops the skills necessary for success in future academic courses or on the job.

Although the material has been extensively revised, this fifth edition retains the basic philosophy of the highly successful previous editions. The revisions include several improvements in line with the NCTM standards, the AMATYC Crossroads, and the current trends in mathematics reform. For example, more emphasis has been placed on graphing and problem solving.

### ■ GENERAL CHANGES IN THE FIFTH EDITION

The overall effects of the changes made to the fifth edition are as follows:

- *To increase the emphasis on learning mathematics through graphing.* Although graphing calculators are incorporated throughout, their use is not required. All of the topics are fully discussed in traditional ways. Of course, we recommend that instructors use the graphing calculator material.

- *To increase the emphasis on problem solving through realistic applications.* The variety of application problems has been increased significantly, and all application problems are labeled with special titles.

- *To fine-tune the presentation of topics* for better flow of ideas and for clarity.

- *To increase the visual interest by the use of color.* Color is included not just as a design feature, but to highlight terms that instructors would point to in a classroom discussion.

## ■ SPECIFIC CHANGES IN THE FIFTH EDITION

To make the book more useful to students, we have:

- Included Self Checks following most examples.

- Included Vocabulary and Concepts problems in each exercise set. The Hints on Studying Algebra that appear in this preface recommend that students begin their study time with review, so Review problems are placed at the beginning of each exercise set. Most exercise sets follow this sequence:

  1. Review problems
  2. Vocabulary and Concepts problems
  3. Practice problems
  4. Application problems
  5. Writing problems
  6. Something to Think About problems

- The format of the chapter summaries has been changed. Now the important concepts are listed in the left-hand column, with relevant review exercises beside them in the right-hand column.

- In application problems, the book consistently uses this problem-solving technique:

  1. Analyze the problem.
  2. Form an equation.
  3. Solve the equation.
  4. State the conclusion.
  5. Check the result.

- The number of Cumulative Review Exercises has been increased. They now occur after every even-numbered chapter, with a Sample Final Exam given after Chapter 9.

- Numerous warnings about common errors have been added. They are marked with the easily recognizable symbol ▣.

- More geometry material has been included throughout the book.

- The emphasis on tables and graphs has been increased.

Specific changes to the chapters are as follows.

**Chapter 1**   When number lines are introduced, both open and closed circles and parentheses and brackets are used to designate intervals on the number line. After students understand both notations, the book uses parentheses and brackets. The work on algebraic expressions has been expanded.

**Chapter 2**   The basic work on solving equations has been divided into two sections. Graphics like Figure 2-1 help students understand the properties of equality.

Many application problems involve markdown, markup, and geometry. Through these exercises, students learn to solve simple equations by the addition and subtraction properties of equality. In applied percent problems, students solve equations by the division and multiplication properties of equality. Applications have been eliminated from Section 2.4 so that students can concentrate on equation-solving techniques. There are now fewer types of word problems in Section 2.5, so the section is more manageable. The section on formulas now follows the section on applications of equations.

**Chapter 3** now covers graphing linear equations and systems of equations and inequalities. A new section is devoted entirely to the coordinate system, emphasizing reading information from graphs. Section 3.2 discusses graphing linear equations. An application of graphing linear equations has been added to the section. Many more applications now illustrate the use of intersecting graphs. Work with graphing calculators has been expanded.

**Chapter 4**   now covers exponents and polynomials. In Section 4.2, finding present value has been included as an application of negative exponents. In Section 4.4, functions are introduced in an informal way. Function notation is used instead of polynomial notation. Several simple polynomial functions are graphed. The use of graphing calculators is covered. Application problems involving adding, subtracting, multiplying, and dividing polynomials are also included.

**Chapter 5**   now covers factoring. The method of factoring trinomials has been streamlined. This chapter remains a thorough treatment of traditional factoring techniques.

**Chapter 6**   now covers proportion and rational expressions. The material on ratio has been extensively revised to show that ratios can be interpreted as unit costs and rates. This edition includes many applications that involve comparison shopping. The material on proportions has also been extensively rewritten and features many more applications of proportions in everyday living. The skill work on manipulating rational expressions has been placed toward the end of the chapter.

**Chapter 7**   now covers roots and radical expressions. Again the practical material comes first, followed by the manipulation of radical expressions. Section 7.1 introduces square roots and the Pythagorean theorem, with many applications. The square root function is discussed. Section 7.2 introduces cube roots and higher-order roots. The cube root function is discussed. Section 7.3 covers radical equations and the distance formula, including applications.

The chapter concludes with a thorough treatment of the algebra of radical expressions.

**Chapter 8**   revisits the coordinate system and discusses slope, writing equations of lines, functions, and variation. The sections on slope and writing equations of lines have been reorganized for clarity. The work on functions summarizes the basic types of functions, reviews function notation, and introduces the vertical line test. Basic rational functions are discussed, and applications are given. More applications occur throughout.

**Chapter 9**   covers quadratic equations and quadratic functions. The material on complex numbers has been moved to an appendix.

## ◼ CALCULATORS

The use of calculators is assumed throughout the book. We believe that students should learn calculator skills in the mathematics classroom. They will then be prepared to use calculators in science and business classes and for nonacademic purposes. The directions within each exercise set indicate which exercises require calculators.

## ◼ ANCILLARIES FOR THE INSTRUCTOR

*Instructor's Edition*   In the Instructor's Edition, the answer is printed in blue next to each exercise.

*Test Manual*   The Test Manual contains four ready-to-use forms for every chapter test. Two of the tests are free-response, and two are multiple-choice.

*Complete Solutions Manual*   The Complete Solutions Manual contains worked-out solutions for each problem in the main text.

*Video Tutorial Series*   These text-specific videotapes feature examples from every chapter and section of the text. All examples that are worked on the videos are cross-referenced with a video icon in the text.

*Thomson World Class Learning$^{TM}$ Testing Tools*   This fully integrated suite of programs includes World Class Test 1.0, World Class Test On-Line 1.0, and World Class Manager 1.0. The program provides text-specific algorithmic testing options designed to offer instructors greater flexibility.

## ◼ ANCILLARIES FOR THE STUDENT

*Study Guide*   Written by George Grisham and Robert Eicken, the Study Guide is designed to help students master the material in the text. Each chapter of the Study Guide contains chapter objectives, additional explanations of worked examples, exercises involving student participation, warnings and hints for common student errors, and end-of-chapter tests. Available for sale at your college bookstore.

*Student Solutions Manual*  Compiled by Michael Welden, the Student Solutions Manual gives complete solutions for the odd-numbered exercises in the book. Available for sale at your college bookstore.

*InterActive Algebra*  This intuitive text-specific tutorial provides explanations of concepts along with carefully graded, algorithmically generated examples and exercises. The fifth section of each topic is a motivational game. An optional on-screen calculator is always available. Because a mathematics educator developed this software, the questions are all free-response. Students must work through the problem; they do not have the opportunity to "point and guess." Hints are provided when a student answers incorrectly. The management system gives the student a performance report upon completion of each unit. InterActive Algebra may be packaged with the text or sold as a stand-alone supplement.

*Student Video*  This "Greatest Hits" student video features concepts and skills students traditionally have the most difficulty comprehending. Examples—chosen by the authors—are drawn from every chapter of the text. Available for sale at your college bookstore, the student video may also be sold as a stand-alone supplement.

# ■ EXAMPLES OF FEATURES IN THE TEXT

■ ■ ■ ■ ■ ■ ■ ■ ■

**MATHEMATICS IN MEDICINE**    The red cells of our blood pick up oxygen in the lungs and carry it to all parts of the body. Each red cell is a tiny disc with an approximate radius of 0.00015 inch. Because the amount of oxygen carried depends on the surface area of the cells, and the cells are so tiny, a very great number is needed—25 trillion in an average adult.

   What is the total surface area of all the red blood cells in the body?

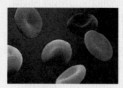

After reading this chapter, you will be able to answer this question.

◄ Each chapter begins with an application that sparks interest in the subject matter.

## 4.1  Natural-Number Exponents
■ EXPONENTS ■ POWERS OF EXPRESSIONS ■ THE PRODUCT RULE FOR EXPONENTS ■ THE POWER RULES FOR EXPONENTS ■ THE QUOTIENT RULE FOR EXPONENTS

◄ Section heads are easy to find, and sections are divided into subsections.

**Getting Ready**    *Evaluate each expression.*

**1.** $2^3$          **2.** $3^2$          **3.** $3(2)$          **4.** $2(3)$

**5.** $2^3 + 2^2$    **6.** $2^3 \cdot 2^2$    **7.** $3^3 - 3^2$    **8.** $\dfrac{3^3}{3^2}$

◄ Getting Ready exercises prepare students for the material in the section.

### ■ EXPONENTS

We have used natural-number exponent[s]... ample,

$$2^5 = 2 \cdot 2 \cdot 2 \cdot 2 \cdot 2 = 32$$
$$x^4 = x \cdot x \cdot x \cdot x$$

These examples suggest a definition f[or]

Perspectives are found ▶ throughout the book.

**172**    CHAPTER 3  GRAPHING AND SOLVING SYSTEMS OF EQUATIONS AND INEQUALITIES

■ ■ ■ ■ ■ ■ ■ ■ ■ ■   PERSPECTIVE

As a child, René Descartes was frail and often sick. To improve his health, eight-year-old René was sent to a Jesuit school. The headmaster encouraged him to sleep in the morning as long as he wished. As a young man, Descartes spent several years as a soldier and world traveler, but his interests included mathematics and philosophy, as well as science, literature, writing, and taking it easy. The habit of sleeping late continued throughout his life. He claimed that his most productive thinking occurred when he was lying in bed. According to one story, Descartes first thought of analytic geometry as he watched a fly walking on his bedroom ceiling.

Descartes might have lived longer if he had stayed in bed. In 1649, Queen Christina of Sweden decided that she needed a tutor in philosophy, and she requested the services of Descartes. Tutoring would not have been difficult, except that the queen scheduled her lessons before dawn in her library with her windows open. The cold Stockholm mornings were too much for a man who was used to sleeping past noon. Within a few months, Descartes developed a fever and died, probably of pneumonia.

### ■ GRAPHING MATHEMATICAL RELATIONSHIPS

Every day, we deal with quantities that are related.

- The distance that we travel depends on how fast we are going.
- Our weight depends on how much we eat.
- The amount of water in a tub depends on how long the water has been running.

   We can often use graphs to visualize relationships between two quantities. For example, suppose that we know the number of gallons of water that are in a tub at several time intervals after the water has been turned on. We can list that information in a **table of values** (see Figure 3-6).

Functions are presented using tables, ▶ graphs, and equations. Functions are discussed throughout the book.

Examples worked on videotape ▷
are marked with an icon.

Problems are worked using a
five-step problem-solving strategy: ▷
1. Analyze the problem.
2. Form an equation.
3. Solve the equation.
4. State the conclusion.
5. Check the result.

### ■ MOTION PROBLEMS

 **EXAMPLE 5**   Chicago and Green Bay are about 200 miles apart. A car leaves Chicago traveling toward Green Bay at 55 mph at the same time as a truck leaves Green Bay bound for Chicago at 45 mph. How long will it take them to meet?

*Analyze the problem*   Motion problems are based on the formula $d = rt$, where $d$ is the distance traveled, $r$ is the rate, and $t$ is the time. We can organize the information of this problem in chart form, as in Figure 2-9(a).

|  | $r$ · $t$ = $d$ |  |  |
|---|---|---|---|
| Car | 55 | $t$ | $55t$ |
| Truck | 45 | $t$ | $45t$ |

(a)

(b)

FIGURE 2-9

We know that the two vehicles travel for the same amount of time—say, $t$ hours. The faster car travels $55t$ miles, and the slower truck travels $45t$ miles. At the time they meet, the total distance can be expressed in two ways: as the sum $55t + 45t$ and as 200 miles.

*Form an equation*   Let $t$ represent the time that each vehicle travels until they meet. Then $55t$ represents the distance traveled by the car, and $45t$ represents the distance traveled by the truck. After referring to Figure 2-9(b), we form the equation

| The distance the car goes | plus | the distance the truck goes | equals | the total distance. |
|---|---|---|---|---|
| $55t$ | + | $45t$ | = | 200 |

*Solve the equation*   We can solve the equation as follows:

$$55t + 45t = 200 \qquad \text{The equation to solve.}$$
$$100t = 200 \qquad \text{Combine like terms.}$$
$$t = 2 \qquad \text{Divide both sides by 100.}$$

*State the conclusion*   The vehicles meet after 2 hours.

*Check the result*   During those 2 hours, the car travels $55 \cdot 2 = 110$ miles, while the truck travels $45 \cdot 2 = 90$ miles. The total distance traveled is $110 + 90 = 200$ miles. Since this is the total distance between Chicago and Green Bay, the answer checks.   ■

### ■ FUNCTION NOTATION

There is a special notation for function

**Function Notation**
The notation $y = f(x)$ denotes that th

⚠ **WARNING!**   The notation $f(x)$ does not mean "$f$ times $x$."

The notation $y = f(x)$ provides a way to denote the values of $y$ in a function that correspond to individual values of $x$. For example, if $y = f(x)$, the value of $y$ that is determined by $x = 3$ is denoted as $f(3)$. Similarly, $f(-1)$ represents the value of $y$ that corresponds to $x = -1$.

 **EXAMPLE 6**   Let $y = f(x) = 2x - 3$ and find   **a.** $f(3)$,   **b.** $f(-1)$,   **c.** $f(0)$,   and   **d.** $f(0.2)$.

*Solution*   **a.** We replace $x$ with 3.

$$f(x) = 2x - 3$$
$$f(3) = 2(3) - 3$$
$$= 6 - 3$$
$$= 3$$

**b.** We replace $x$ with $-1$.

$$f(x) = 2x - 3$$
$$f(-1) = 2(-1) - 3$$
$$= -2 - 3$$
$$= -5$$

**c.** We replace $x$ with 0.

$$f(x) = 2x - 3$$
$$f(0) = 2(0) - 3$$
$$= 0 - 3$$
$$= -3$$

**d.** We replace $x$ with 0.2.

$$f(x) = 2x - 3$$
$$f(0.2) = 2(0.2) - 3$$
$$= 0.4 - 3$$
$$= -2.6$$   ■

*Self Check*   Using the function of Example 6, find   **a.** $f(-2)$   and   **b.** $f\left(\frac{3}{2}\right)$.
*Answers*   **a.** $-7$,   **b.** $0$

◁ Warning symbols appear throughout.

◁ Color is used to show substitutions.

◁ Most examples are followed by a Self Check.

■ ■ ■ ■ ■ ■ ■ ■ ■ ■ **Making Tables and Graphs**

GRAPHING
CALCULATORS

So far, we have graphed equations by making tables of values and plotting points. This method is usually tedious and time-consuming. Fortunately, the task of making tables and graphing equations is much easier when we use a graphing calculator.

Several brands of calculators are available. Although we will use calculators to make tables and graph equations, we will not show complete keystrokes for any specific brand. For these details, please consult your owner's manual.

All graphing calculators have a **viewing window** that is used to display tables and graphs. We will first discuss how to make tables and then discuss how to draw graphs.

**MAKING TABLES**   To construct a table of values for the equation $y = x^2$, simply press the  Y =  key, enter the expression $x^2$, and press the  2nd  and  TABLE  keys to get a screen similar to Figure 3-17(a). You can use the up and down keys to scroll through the table to obtain a screen like Figure 3-17(b).

TI-83 graphing calculator
(Courtesy of Texas Instruments)

(a)

FIG

◀ Material involving the use of graphing calculators appears throughout the book.

Orals   *Solving by factoring.*

**1.** $x^2 - 25 = 0$                              **2.** $x^2 - 5x + 6 = 0$

*Solve by the square root method.*

**3.** $x^2 = 100$                                 **4.** $x^2 = 49$

Oral exercises help prepare ▶
students for the exercise set.

Review exercises are at the ▶
beginning of each exercise set.

Vocabulary and Concepts exercises ▶
are included in each exercise set.

EXERCISE 9.1

*REVIEW*   Write each expression without parentheses.

**1.** $(y - 1)^2$                **2.** $(z + 2)^2$                **3.** $(x + y)^2$
**4.** $(a - b)^2$                **5.** $(2r - s)^2$               **6.** $(m + 3n)^2$

*VOCABULARY AND CONCEPTS*   Fill in each blank to make a true statement.

**7.** Any equation that can be written in the form $ax^2 + bx + c = 0$ $(a \neq 0)$ is called a _____ equation.

**8.** In the equation $3x^2 - 4x + 5 = 0$, $a = \_\_$, $b = \_\_\_$, and $c = \_\_$.

**9.** If $ab = 0$, then $a = \_\_$ or $b = \_\_$.

**10.** In the quadratic equation $ax^2 + bx + c = 0$, a $\neq \_\_$.

**11.** The equation $x^2 = c$ $(c > 0)$ has _____ solutions.

**12.** The solutions of $x^2 = c$ $(c > 0)$ are _____ and _____.

Practice exercises provide ▶
plenty of drill.

*PRACTICE*   In Exercises 13–20, solve each equation.

**13.** $(x - 2)(x + 3) = 0$              **14.** $(x - 3)(x - 2) = 0$
**15.** $(x - 4)(x + 1) = 0$              **16.** $(x + 5)(x + 2) = 0$
**17.** $(2x - 5)(3x + 6) = 0$            **18.** $(3x - 4)(x + 1) = 0$
                                          **20.** $(x + 2)(x + 3)(x - 4) = 0$

*APPLICATIONS*   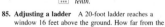   Use a calculator to help solve each problem. If an answer is not exact, give it to the nearest tenth.

**85. Adjusting a ladder**   A 20-foot ladder reaches a window 16 feet above the ground. How far from the wall is the base of the ladder?

**86. Length of guy wires**   A 20-foot-tall tower is secured by three guy wires fastened at the top and to anchors 15 feet from the base of the tower. How long is each guy wire?

**87. Height of a pole**   A 34-foot-long wire reaches from the top of a telephone pole to a point on the ground 16 feet from the base of the pole. Find the height of the pole.

**88. Length of a path**   A rectangular garden has sides of 28 and 45 feet. Find the length of a path that extends from one corner to the opposite corner.

**90. Television**   The size of the television screen shown in Illustration 3 is the diagonal measure of its rectangular screen. How large is the screen if it is 21 inches wide and 17 inches high?

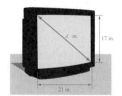

*d* in.

17 in.

21 in.

◀ Applications are plentiful throughout the book. Each application problem has a title.

**GEOMETRY**  *In Exercises 79–84, find x.*

**79.**

**80.**

**81.**

**82.**

**83.**

**84.**

**85.** Find the complement of 37°.

**86.** Find the supplement of 37°.

**87.** Find the supplement of the complement of 40°.

**88.** Find the complement of the supplement of 140°.

**WRITING**

**89.** Explain what it means for a number to satisfy an equation.

**90.** How can you tell whether a number is the solution of an equation?

**91.** Explain what Figure 2-1 is trying to show.

**92.** Explain what Figure 2-2 is trying to show.

**SOMETHING TO THINK ABOUT**

**93.** If two lines intersect as in Illustration 2, angles 1 and 2 and angles 3 and 4 are called **vertical angles.** Let the measure of angle 1 be various numbers and compute the values of the other three. What do you discover?

◄ Geometry is emphasized throughout.

◄ Writing exercises appear in every exercise set.

◄ Something to Think About exercises provide material for discussion.

Projects are included ▶
for group learning.

In its new format, the Chapter ▶
Summary is more accessible
to students.

548  CHAPTER 7  ROOTS AND RADICAL EXPRESSIONS

■ ■ ■ ■ ■ ■ ■ ■  **PROJECT**  *(continued)*

- The pendulum's period $t$ and length $l$ are related by the formula $t = a\sqrt{l}$ for some number $a$. From your experimental data, find the approximate value of $a$.
- Use your formula to predict the period of a pendulum 2 meters long.
- Time the period of a 2-meter pendulum. How close was your prediction?

## CHAPTER SUMMARY

**CONCEPTS**

**REVIEW EXERCISES**

**SECTION 7.1**  Square Roots and the Pythagorean Theorem

The number $b$ is a **square root** of $a$ if $b^2 = a$.

The **principal square root** of a positive number $a$, denoted by $\sqrt{a}$, is the positive square root of $a$.

**1.** Find each square root.

**a.** $\sqrt{25}$  **b.** $\sqrt{64}$  **c.** $-\sqrt{144}$  **d.** $-\sqrt{289}$

**e.** $\sqrt{256}$  **f.** $-\sqrt{64}$  **g.** $\sqrt{169}$  **h.** $-\sqrt{225}$

**2.** Use a calculator to find each root to three decimal places.

**a.** $\sqrt{21}$  **b.** $-\sqrt{15}$

   $-\sqrt{57.3}$  **d.** $\sqrt{751.9}$

aph each function.

$f(x) = \sqrt{x}$  **b.** $f(x) = 2 - \sqrt{x}$

606  CHAPTER 8  WRITING EQUATIONS OF LINES; VARIATION

### ■ Cumulative Review Exercises

*In Exercises 1–4, simplify each expression.*

**1.** $(3x^2 + 2x) + (6x^3 - 3x^2 + 1)$

**2.** $(5x^3 - 2x) - (2x^3 - 3x^2 - 3x - 1)$

**3.** $3(6x^2 - 3x + 3) + 2(-x^2 + 2x - 5)$

**4.** $5(3x^2 - 4x - 1) - 2(-2x^2 + 4x + 3)$

*In Exercises 5–8, do each multiplication.*

**5.** $(5x^4y^3)(-3x^2y^3)$

**6.** $-3x^2(-5x^3 - 3x^2 + 2)$

**7.** $(2x + 3)(3x + 4)$

**8.** $(4x - 3y)(3x - 2y)$

*In Exercises 9–10, do each division.*

**9.** $x + 3)\overline{x^2 - x - 12}$

**10.** $2x - 1)\overline{2x^3 + x^2 - x - 3}$

◄ Cumulative Review Exercises follow Chapters 2, 4, 6, and 8.

# To the Student

Congratulations. You now own a state-of-the-art textbook that has been written especially for you. We have tried to write a book that you can read and understand. The book includes carefully written narrative and an extensive number of worked examples with Self Checks.

To get the most out of this course, you must read and study the textbook properly. We recommend that you work the examples on paper first and then work the Self Checks. Only after you thoroughly understand the concepts taught in the examples should you attempt to work the exercises. A *Student Solutions Manual* contains the solutions to the odd-numbered exercises.

Since the material presented in *Beginning Algebra*, fifth edition, will be of value to you in later years, we suggest that you keep this book. It will be a good source of reference and will keep at your fingertips the material that you have learned here.

We wish you well.

## ■ HINTS ON STUDYING ALGEBRA

The phrase "Practice makes perfect" is not quite true. It is *perfect* practice that makes perfect. For this reason, it is important that you learn how to study algebra to get the most out of this course.

Although we all learn differently, there are some hints on how to study algebra that most students find useful. Here are some things you should consider as you work on the material in this course.

*Plan a strategy for success.* To get where you want to be, you need a goal and a plan. Your goal should be to pass this course with a grade of A or B. To earn one of these grades, you must have a plan to achieve it. A good plan involves several points:

- Getting ready for class
- Attending class
- Doing homework
- Arranging for special help when you need it
- Having a strategy for taking tests

*Getting ready for class.* To get the most out of every class period, you will need to prepare for class. One of the best things you can do is to preview the material in the text that your instructor will be discussing. Perhaps you will not understand all of what you read, but you will understand it better when the instructor discusses the material in class.

Be sure to do your work every day. If you get behind and attend class without understanding previous material, you will be lost and will become frustrated and discouraged. Make a promise that you will always prepare for class, and then keep that promise.

*Attending class.* The classroom experience is your opportunity to learn from your instructor. Make the most of it by attending every class. Sit near the front of the room, where you can easily see and hear. It is easy to be distracted and lose interest if you sit in the back of the room. Remember that it is your responsibility to follow the discussion, even though that takes concentration and hard work.

Pay attention to your instructor, and jot down the important things that he or she says. However, do not spend so much time taking notes that you fail to concentrate on what your instructor is explaining. It is much better to listen and understand the big picture than just to copy solutions to problems.

Don't be afraid to ask questions when your instructor asks for them. If something is unclear to you, it is probably unclear to many other students as well. They will appreciate your willingness to ask. Besides, asking questions will make you an active participant in class. This will help you pay attention and keep you alert and involved.

*Doing homework.* It requires practice to excel at tennis, master a musical instrument, or learn a foreign language. In the same way, it requires practice to learn mathematics. Since practice in mathematics is the homework, homework is your opportunity to practice your skills and experiment with ideas.

It is very important for you to pick a definite time to study and do homework. Set a formal schedule and stick to it. Try to study in a place that is comfortable and quiet. If you can, do some homework shortly after class, or at least before you forget what was discussed in class. This quick follow-up will help you remember the skills and concepts your instructor taught that day.

Each formal study session should include three parts:

**1.** Begin every study session with a review period. Look over previous chapters and see if you can do a few problems from previous sections, chosen randomly. Keeping old skills alive will greatly reduce the time you will need to prepare for tests.

**2.** After reviewing, read the assigned material. Resist the temptation of diving into the exercises without reading and understanding the examples. Instead, work the examples and Self Checks with pencil and paper. Only after you completely understand the principles behind them should you try to work the exercises.

Once you begin to work the exercises, check your answers with those printed in the back of the book. If one of your answers differs from the printed answer, see if the two can be reconciled. Sometimes answers can have more than one form. If you decide that your answer is incorrect, compare your work to the example in the text that most closely resembles the exercise, and try to find your mistake. If you cannot find an error, consult the *Student Solutions Manual*. If nothing works, mark the problem and ask about it in your next class meeting.

**3.** After completing the written assignment, preview the next section. This preview will be helpful when you hear that material discussed during the next class period.

You probably know the general rule of thumb for college homework: two hours of practice for every hour in class. If mathematics is hard for you, plan on spending even more time on homework.

To make homework more enjoyable, study with one or more friends. The interaction will clarify ideas and help you remember them. If you must study alone, try talking to yourself. A good study technique is to explain the material to yourself out loud.

*Arranging for special help.* Take advantage of any special help that is available from your instructor. Often, the instructor can clear up difficulties in a very short time.

Find out whether your college has a free tutoring program. Peer tutors can often be of great help.

*Taking tests.* Students often get nervous before a test, because they are afraid that they will not do well. There are many different reasons for this fear, but the most common one is that students are not confident that they know the material.

To build confidence in your ability to work tests, rework many of the problems in the exercise sets, work the exercises in the Chapter Summaries, and take the Chapter Tests. Check all answers with those printed at the back of the text.

Then guess what the instructor will ask, build your own tests, and work them. Once you know your instructor, you will be surprised at how good you can get at picking test questions. With this preparation, you will have some idea of what will be on the test. You will have more confidence in your ability to do well. You should notice that you are far less nervous before tests, and this will also help your performance.

When you take a test, work slowly and deliberately. Scan the test and work the easy problems first. This will build confidence. Tackle the hardest problems last.

## Acknowledgments

We are grateful to the following people, who reviewed the manuscript at various stages of its development. They all had valuable suggestions that have been incorporated into the text.

Helen Banes
*Kirkwood Community College*

Theresa Barrie
*Texas Southern University*

Robert Billups
*Citrus Community College*

Elaine D. Bouldin
*Middle Tennessee State University*

David Byrd
*Enterprise State Junior College*

Baruch Cahlon
*Oakland University*

Don Cohen
*SUNY-Cobleskill*

Patricia Cooper
*St. Louis Community College-Park Forest*

Sally Copeland
*Johnson County Community College*

Kay D. Crow
*Northeast Mississippi Community College*

Elwin Cutler
*Ferris State University*

Elias Deeba
*University of Houston-Downtown*

Edward Doran
*Front Range Community College*

Arthur Dull
*Diablo Valley College*

Robert Eicken
*Illinois Central College*

Marc Glucksman
*El Camino College*

George Grisham
*Bradley University*

Robert G. Hammond
*Utah State University*

Mitzy Johnson
*Northeast Mississippi Community College*

Robert Keicher
*Delta College*

Katherine McLain
*Consumnes River College*

Laurie McManus
*St. Louis Community College-Meramac*

Wayne Milloy
*Crafton Hills College*

Myrna Mitchell
*Pima Community College*

John Monroe
*University of Akron*

Carol M. Nessmith
*Georgia Southern University*

Kent Neuerburg
*Consumnes River College*

Paul Peck
*Glenville State College*

Mary Ann Petruska
*Pensacola Junior College*

Thea Prettyman
*Essex Community College*

Janet Ritchie
*SUNY-Old Westbury*

Michael Rosenthal
*Florida International University*

Jack W. Rotman
*Lansing Community College*

Irwin Schochetman
*Oakland University*

Erik A Schreiner
*Western Michigan University*

Kenneth Seydel
*Skyline College*

David Sicks
*Olympia College*

Willie Taylor
*Texas Southern University*

Douglas Tharp
*University of Houston-Downtown*

Lynn E. Tooley
*Bellevue Community College*

Gary VanVelsir
*Anne Arundel Community College*

Gerry C. Vidrine
*Louisiana State University*

Rosalyn Wells
*Georgia State University*

Clifton T. Whyburn
*University of Houston*

Hette Williams
*Broward Community College*

George J. Witt
*Glendale Community College*

We are grateful to Diane Koenig, who read the entire manuscript and worked every problem. We wish to thank George Witt for his assistance in the preparation of some of the exercises. We also wish to thank the staff at Brooks/Cole, especially Bob Pirtle, Ellen Brownstein, and Peggi Rodgers. We are also grateful to David Hoyt for coordinating production and to Lori Heckelman for fine artwork.

*R. David Gustafson*
*Peter D. Frisk*

# ■ ■ ■ ■ ■ ■ ■ ■ ■ ■ CONTENTS

*FIFTH EDITION*

# *Beginning*
# *Algebra*

# 1

# Real Numbers and Their Basic Properties

MATHEMATICS
FOR FUN

When three professors attending a convention in Las Vegas registered at the ho-tel, they were told that the room rate was $120. Each professor paid his $40 share.

Later the desk clerk realized that the cost of the room should have been $115. To fix the mistake, she sent a bellhop to the room to refund the $5 over-charge. Realizing that $5 could not be evenly divided among the three professors and not wanting to start a quarrel, the bellhop refunded only $3 and kept the other $2.

Since each professor received a $1 refund, each paid $39 for the room, and the bellhop kept $2. This gives $39 + $39 + $39 + $2, or $119. What happened to the other $1?

After you have read this chapter, you will be able to answer this question.

## 1.1 Real Numbers and Their Graphs

■ SETS OF NUMBERS ■ EQUALITY, INEQUALITY SYMBOLS, AND VARIABLES ■ THE NUMBER LINE
■ GRAPHING SUBSETS OF THE REAL NUMBERS ■ ABSOLUTE VALUE OF A NUMBER

Getting Ready

**1.** Give an example of a number that is used for counting.

**2.** Give an example of a number that is used when dividing a pizza.

**3.** Give an example of a number that is used for measuring very cold temperatures.

**4.** What other types of numbers can you think of?

### ■ SETS OF NUMBERS

A **set** is a collection of objects. For example, the set

{1, 2, 3, 4, 5}

contains the numbers 1, 2, 3, 4, and 5. The members, or **elements,** in a set are listed within braces { }.

Two basic sets of numbers are the set of **natural numbers** (often called the **positive integers**) and the **whole numbers.**

**Natural Numbers**
The **natural numbers** (or the **positive integers**) are the numbers

1, 2, 3, 4, 5, 6, 7, 8, 9, 10, . . .

**Whole Numbers**
The **whole numbers** are the numbers

0, 1, 2, 3, 4, 5, 6, 7, 8, 9, 10, . . .

The three dots in the previous definitions, called an **ellipsis,** indicate that the lists of numbers continue on forever.

We can use whole numbers to describe many real-life situations. For example, some cars get 30 miles per gallon of gas, and some students might pay $1,750 in tuition.

Numbers that show a loss or a downward direction are called **negative integers,** denoted as $-1$, $-2$, $-3$, and so on. For example, a debt of $1,500 can be denoted as $-\$1,500$, and a temperature of 20° below zero can be denoted as $-20°$.

The set of negative integers and the set of whole numbers together form the set of **integers.**

**Integers**
The **integers** are the numbers

$$\ldots, -5, -4, -3, -2, -1, 0, 1, 2, 3, 4, 5, \ldots$$

Because the set of natural numbers and the set of whole numbers are included within the set of integers, we say that these sets are **subsets** of the set of integers.

Integers cannot describe every real-life situation. For example, a student might study $3\frac{1}{2}$ hours, or a television set might cost $217.37. To describe these situations, we need fractions, more formally called **rational numbers.**

**Rational Numbers**
A **rational number** is any number that can be written as a fraction with an integer in its numerator and a nonzero integer in its denominator.

Some examples of rational numbers are

$$\frac{3}{2}, \quad \frac{17}{12}, \quad -\frac{43}{8}, \quad 0.25, \quad \text{and} \quad -0.66666\ldots$$

The decimals $0.25$ and $-0.66666\ldots$ are rational numbers, because $0.25$ can be written as the fraction $\frac{1}{4}$, and $-0.66666\ldots$ can be written as the fraction $-\frac{2}{3}$.

Since every integer can be written as a fraction with a denominator of 1, every integer is also a rational number. Since every integer is a rational number, the set of integers is a subset of the rational numbers.

**WARNING!** Because division by 0 is undefined, expressions such as $\frac{6}{0}$ and $\frac{0}{0}$ do not represent any number.

Numbers such as $\sqrt{2}$ and $\pi$ are not rational numbers, because they cannot be written as fractions with an integer numerator and a nonzero integer denominator. Such numbers are called **irrational numbers.** We can find decimal approximations for irrational numbers by using a calculator. For example,

$$\sqrt{2} \approx 1.414213562\ldots \qquad \text{Press } \boxed{\sqrt{\phantom{x}}}. \text{ Read} \approx \text{as ``is approximately equal to.''}$$
$$\pi \approx 3.141592654\ldots \qquad \text{Press } \boxed{\pi}.$$

If we combine the rational and the irrational numbers, we have the set of real numbers.

> **Real Numbers**
> A **real number** is any number that is either a rational number or an irrational number.

Figure 1-1 shows how the various sets of numbers are interrelated.

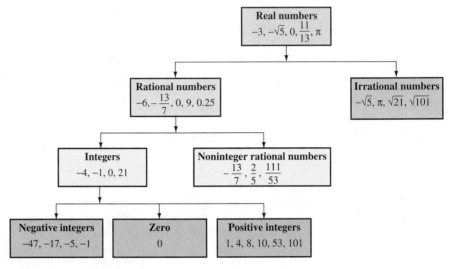

FIGURE 1-1

EXAMPLE 1    List the numbers in the set $\left\{-3, 0, \frac{1}{2}, 1.25, \sqrt{3}, 5\right\}$ that are   **a.** natural numbers,   **b.** whole numbers,   **c.** negative integers,   **d.** rational numbers,   **e.** irrational numbers,   and   **f.** real numbers.

*Solution*    **a.** The only natural number is 5.

**b.** The whole numbers are 0 and 5.

**c.** The only negative integer is $-3$.

**d.** The rational numbers are $-3, 0, \frac{1}{2}, 1.25$, and 5. $\left(1.25 \text{ is rational, because } 1.25 \text{ can be written in the form } \frac{5}{4}.\right)$

**e.** The only irrational number is $\sqrt{3}$.

**f.** All of the numbers are real numbers.    ∎

*Self Check*    List the numbers in the set $\left\{-2, 0, 1.5, \sqrt{5}, 7\right\}$ that are   **a.** positive integers   and   **b.** rational numbers.

*Answers*    **a.** 7,   **b.** $-2, 0, 1.5, 7$

The natural numbers greater than 1 that can be divided evenly only by 1 and themselves are called **prime numbers.** The nonprime natural numbers greater than 1 are called **composite numbers.**

**Prime numbers:**    $\{2, 3, 5, 7, 11, 13, 17, 19, 23, 29, \ldots\}$
**Composite numbers:**    $\{4, 6, 8, 9, 10, 12, 14, 15, 16, 18, 20, 21, 22, \ldots\}$

Integers that can be divided evenly by 2 are called **even integers.** Integers that cannot be divided evenly by 2 are called **odd integers.**

**Even integers:**    $\{\ldots, -10, -8, -6, -4, -2, 0, 2, 4, 6, 8, 10, \ldots\}$
**Odd integers:**    $\{\ldots, -9, -7, -5, -3, -1, 1, 3, 5, 7, 9, \ldots\}$

**EXAMPLE 2**    List the numbers in the set $\{-3, -2, 0, 1, 2, 3, 4, 5, 9\}$ that are    **a.** prime numbers, **b.** composite numbers,    **c.** even integers,    and    **d.** odd integers

*Solution*    **a.** The prime numbers are 2, 3, and 5.

**b.** The composite numbers and 4 and 9.

**c.** The even integers are $-2$, 0, 2, and 4.

**d.** The odd integers are $-3$, 1, 3, 5, and 9.    ■

### ◼ EQUALITY, INEQUALITY SYMBOLS, AND VARIABLES

To show that two expressions represent the same number, we use the **is equal to** sign (=). Since $4 + 5$ and 9 represent the same number, we can write

$4 + 5 = 9$        Read as "the sum of 4 and 5 is equal to 9."

Likewise, we can write

$5 - 3 = 2$        Read as "the difference between 5 and 3 equals 2," or "5 minus 3 equals 2."

$4 \cdot 5 = 20$        Read as "the product of 4 and 5 equals 20," or "4 times 5 equals 20."

and

$30 \div 6 = 5$        Read as "the quotient obtained when 30 is divided by 6 is 5," or "30 divided by 6 equals 5."

We can use **inequality symbols** to show that expressions are not equal.

| Symbol | Read as |
| --- | --- |
| $\neq$ | "is not equal to" |
| $<$ | "is less than" |
| $>$ | "is greater than" |
| $\leq$ | "is less than or equal to" |
| $\geq$ | "is greater than or equal to" |

**EXAMPLE 3**    **Inequality symbols**

**a.** $6 \neq 9$        Read as "6 is not equal to 9."

**b.** $8 < 10$        Read as "8 is less than 10."

**c.** $12 > 1$        Read as "12 is greater than 1."

**d.** $5 \leq 5$        Read as "5 is less than or equal to 5." (Since $5 = 5$, this is a true statement.)

**e.** $9 \geq 7$        Read as "9 is greater than or equal to 7." (Since $9 > 7$, this is a true statement.)    ◼

Tell whether each statement is true or false: **a.** $12 \neq 12$, **b.** $7 \geq 7$, and **c.** $125 < 137$.

**a.** false, **b.** true, **c.** true

Inequality statements can be written so that the inequality symbol points in the opposite direction. For example,

$$5 < 7 \quad \text{and} \quad 7 > 5$$

both indicate that 5 is a smaller number than 7. Likewise,

$$12 \geq 3 \quad \text{and} \quad 3 \leq 12$$

both indicate that 12 is greater than or equal to 3.

In algebra, we use letters, called **variables,** to represent real numbers. For example,

- If $x$ represents 4, then $x = 4$.
- If $y$ represents any number greater than 3, then $y > 3$.
- If $z$ represents any number less than or equal to $-4$, then $z \leq -4$.

**WARNING!** In algebra, we usually do not use the times sign ($\times$) to indicate multiplication. It might be mistaken for the variable $x$.

## ■ THE NUMBER LINE

We can use the **number line** shown in Figure 1-2 to represent sets of numbers. The number line continues forever to the left and to the right. Numbers to the left of 0 are negative, and numbers to the right of 0 are positive.

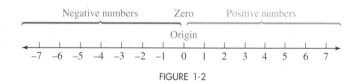

FIGURE 1-2

**WARNING!** The number 0 is neither positive nor negative.

The number that corresponds to a point on the number line is called the **coordinate** of that point. For example, the coordinate of the **origin** is 0.

Many points on the number line do not have integer coordinates. For example, the point midway between 0 and 1 has the coordinate $\frac{1}{2}$, and the point midway between $-3$ and $-2$ has the coordinate $-\frac{5}{2}$ (see Figure 1-3).

FIGURE 1-3

Numbers represented by points that lie on opposite sides of the origin and at equal distances from the origin are called **negatives** (or **opposites**) of each other. For example, 5 and $-5$ are negatives (or opposites). We need parentheses to express the opposite of a negative number. For example, $-(-5)$ represents the opposite of $-5$, which we know to be 5. Thus,

$$-(-5) = 5$$

This suggests the following rule.

> **Double Negative Rule**
> If $x$ represents a real number, then
> $$-(-x) = x$$

If one point lies to the *right* of a second point on a number line, its coordinate is the *greater*. Since the point with coordinate 1 lies to the right of the point with co-ordinate $-2$ (see Figure 1.4(a)), it follows that $1 > -2$.

If one point lies to the *left* of another, its coordinate is the *smaller* (see Figure 1-4(b)). The point with coordinate $-6$ lies to the left of the point with coordinate $-3$, so it follows that $-6 < -3$.

(a)                    (b)

FIGURE 1-4

## ■ GRAPHING SUBSETS OF THE REAL NUMBERS

Figure 1-5 shows the graph of the natural numbers from 2 to 8. The points on the line are called the **graphs** of their corresponding coordinates.

FIGURE 1-5

**EXAMPLE 4**    Graph the set of integers between $-3$ and 3.

*Solution*    The integers between $-3$ and 3 are $-2, -1, 0, 1,$ and 2. The graph is shown in Figure 1-6.

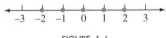

FIGURE 1-6

Self Check     Graph the set of integers between −4 and 0.

Answer
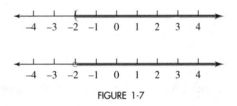

Graphs of many sets of real numbers are **intervals** on the number line. For example, two graphs of all real numbers $x$ such that $x > -2$ are shown in Figure 1-7. The parenthesis or the open circle at −2 shows that this point is not included in the graph. The arrow pointing to the right shows that all numbers to the right of −2 are included.

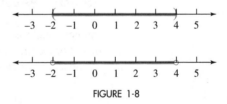

FIGURE 1-7

Figure 1-8 shows two graphs of the set of real numbers $x$ between −2 and 4. This is the graph of all real numbers $x$ such that $x > -2$ and $x < 4$. The parentheses or open circles at −2 and 4 show that these points are not included in the graph. However, all the numbers between −2 and 4 are included.

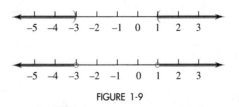

FIGURE 1-8

EXAMPLE 5     Graph all real numbers $x$ such that $x < -3$ or $x > 1$.

Solution     The graph of all real numbers less than −3 includes all points on the number line that are to the left of −3. The graph of all real numbers greater than 1 includes all points that are to the right of 1. The two graphs are shown in Figure 1-9.

FIGURE 1-9     ■

Self Check     Graph all real numbers $x$ such that $x < -1$ or $x > 0$. Use parentheses.

Answer

EXAMPLE 6

Graph the set of all real numbers from $-5$ to $-1$.

*Solution*

The set of all real numbers from $-5$ to $-1$ includes $-5$ and $-1$ and all the numbers in between. In the graphs shown in Figure 1-10, the brackets or the solid circles at $-5$ and $-1$ show that these points are included.

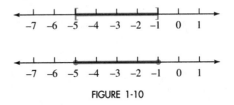

FIGURE 1-10

■

*Self Check*

*Answer*

Graph the set of real numbers from $-2$ to 1. Use brackets.

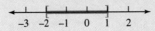

## ■ ABSOLUTE VALUE OF A NUMBER

On a number line, the distance between a number $x$ and 0 is called the **absolute value** of $x$. For example, the distance between 5 and 0 is 5 units (see Figure 1-11). Thus, the absolute value of 5 is 5:

$|5| = 5$     Read as "The absolute value of 5 is 5."

Since the distance between $-6$ and 0 is 6,

$|-6| = 6$     Read as "The absolute value of $-6$ is 6."

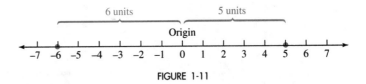

FIGURE 1-11

Because the absolute value of a real number represents that number's distance from 0 on the number line, the absolute value of every real number $x$ is either positive or 0. In symbols, we say

$|x| \geq 0$     for every real number $x$

**EXAMPLE 7**    Evaluate **a.** $|6|$, **b.** $|-3|$, **c.** $|0|$, and **d.** $-|2+3|$.

*Solution*    **a.** $|6| = 6$, because 6 is six units from 0.

**b.** $|-3| = 3$, because $-3$ is three units from 0.

**c.** $|0| = 0$, because 0 is zero units from 0.

**d.** $-|2 + 3| = -|5| = -5$    ■

**Self Check**    Evaluate **a.** $|8|$, **b.** $|-8|$, and **c.** $-|-8|$.
*Answers*     **a.** 8, **b.** 8, **c.** $-8$

Orals    *Describe each set of numbers in your own words.*

**1.** natural numbers            **2.** whole numbers

**3.** integers                   **4.** rational numbers

**5.** real numbers               **6.** prime numbers

**7.** composite numbers          **8.** even integers

**9.** odd integers               **10.** irrational numbers

*Find each value.*

**11.** $-|15|$                    **12.** $|-25|$

# EXERCISE 1.1

***VOCABULARY AND CONCEPTS***    *Fill in each blank to make a true statement.*

**1.** A ___ is a collection of objects.

**2.** The numbers 1, 2, 3, 4, 5, . . . form the set of _____ numbers.

**3.** The set of _____ numbers is the set {0, 1, 2, 3, 4, 5, . . .}.

**4.** The set of _____ is the set {. . . , $-3, -2, -1, 0, 1, 2, 3, . . .$}.

**5.** Since every whole number is also an integer, the set of whole numbers is called a _____ of the set of integers.

**6.** $\sqrt{2}$ is an example of an _____ number.

**7.** If a natural number is greater than 1 and can be divided exactly only by 1 and itself, it is called a _____ number.

**8.** A composite number is a _____ number that is greater than 1 and is not _____.

**9.** The symbol $\neq$ means _____.

**10.** The symbol ___ means "is less than."

**11.** The symbol $\geq$ means _____.

**12.** The opposite of $-7$ is ___.

**13.** The figure 

$$\xleftarrow{\quad} \underset{-3}{|} \ \underset{-2}{|} \ \underset{-1}{|} \ \underset{0}{|} \ \underset{1}{|} \ \underset{2}{|} \ \underset{3}{|} \xrightarrow{\quad}$$ 

is called a _____ line.

**14.** The distance between 8 and 0 on a number line is called the _____ of 8.

***PRACTICE*** *In Exercises 15–26, list the numbers in the set* $\left\{-3, -\frac{1}{2}, -1, 0, 1, 2, \frac{5}{3}, \sqrt{7}, 3.25, 6, 9\right\}$ *that are*

**15.** Natural numbers     **16.** Whole numbers     **17.** Positive integers     **18.** Negative integers

**19.** Integers     **20.** Rational numbers     **21.** Real numbers     **22.** Irrational numbers

**23.** Odd integers     **24.** Even integers     **25.** Composite numbers     **26.** Prime numbers

*In Exercises 27–34, simplify each expression. Then classify the result as a natural number, an even integer, an odd integer, a prime number, a composite number, and/or a whole number.*

**27.** $4 + 5$     **28.** $7 - 2$     **29.** $15 - 15$     **30.** $0 + 7$

**31.** $3 \cdot 8$     **32.** $8 \cdot 9$     **33.** $24 \div 8$     **34.** $3 \div 3$

*In Exercises 35–48, place one of the symbols* $=$, $<$, *and* $>$ *in each box to make a true statement.*

**35.** $5 \ \boxed{\phantom{x}} \ 3 + 2$     **36.** $9 \ \boxed{\phantom{x}} \ 7$     **37.** $25 \ \boxed{\phantom{x}} \ 32$     **38.** $2 + 3 \ \boxed{\phantom{x}} \ 17$

**39.** $5 + 7 \ \boxed{\phantom{x}} \ 10$     **40.** $3 + 3 \ \boxed{\phantom{x}} \ 9 - 3$     **41.** $3 + 9 \ \boxed{\phantom{x}} \ 20 - 8$     **42.** $19 - 3 \ \boxed{\phantom{x}} \ 8 + 6$

**43.** $4 \cdot 2 \ \boxed{\phantom{x}} \ 2 \cdot 4$     **44.** $7 \cdot 9 \ \boxed{\phantom{x}} \ 9 \cdot 6$     **45.** $8 \div 2 \ \boxed{\phantom{x}} \ 4 + 2$     **46.** $0 \div 7 \ \boxed{\phantom{x}} \ 1$

**47.** $3 + 2 + 5 \ \boxed{\phantom{x}} \ 5 + 2 + 3$     **48.** $8 + 5 + 2 \ \boxed{\phantom{x}} \ 5 + 2 + 8$

*In Exercises 49–54, write each statement as a mathematical expression.*

**49.** Seven is greater than three.

**50.** Five is less than thirty-two.

**51.** Eight is less than or equal to eight.

**52.** Twenty-five is not equal to twenty-three.

**53.** The result of adding three and four is equal to seven.

**54.** Thirty-seven is greater than or equal to the result of multiplying three and four.

*In Exercises 55–66, rewrite each inequality statement as an equivalent inequality in which the inequality symbol points in the opposite direction.*

**55.** $3 \le 7$     **56.** $5 > 2$     **57.** $6 > 0$     **58.** $34 \le 40$

**59.** $3 + 8 > 8$     **60.** $8 - 3 < 8$     **61.** $6 - 2 < 10 - 4$     **62.** $8 \cdot 2 \ge 8 \cdot 1$

**63.** $2 \cdot 3 < 3 \cdot 4$     **64.** $8 \div 2 \ge 9 \div 3$     **65.** $\dfrac{12}{4} < \dfrac{24}{6}$     **66.** $\dfrac{2}{3} \le \dfrac{3}{4}$

*In Exercises 67–74, graph each pair of numbers on a number line. In each pair, indicate which number is the greater and which number lies farther to the right.*

**67.** 3, 6

**68.** 4, 7

**69.** 11, 6

**70.** 12, 10

**71.** 0, 2

**72.** 4, 10

**73.** 8, 0

**74.** 20, 30

*In Exercises 75–86, graph each set of numbers on the number line.*

**75.** The natural numbers between 2 and 8

**76.** The prime numbers from 10 to 20

**77.** The even integers greater than 10 but less than 20

**78.** The even integers that are also prime numbers

**79.** The numbers that are whole numbers but not natural numbers

**80.** The prime numbers between 5 and 15

**81.** The natural numbers between 15 and 25 that are exactly divisible by 6

**82.** The odd integers between $-5$ and 5 that are exactly divisible by 3

**83.** The real numbers between 1 and 5

**84.** The real numbers greater than or equal to 8

**85.** The real numbers greater than or equal to 3 or less than or equal to $-3$

**86.** The real numbers greater than $-2$ and less than 3

*In Exercises 87–94, find each absolute value.*

**87.** $|36|$

**88.** $|-30|$

**89.** $|0|$

**90.** $|120|$

**91.** $|-230|$

**92.** $|18 - 12|$

**93.** $|12 - 20|$

**94.** $|100 - 100|$

**WRITING**

**95.** Explain why there is no greatest natural number.

**96.** Explain why 2 is the only even prime number.

**97.** Explain how to determine the absolute value of a number.

**98.** Explain why zero is an even integer.

***SOMETHING TO THINK ABOUT***   *Consider the following sets: the integers, natural numbers, even and odd integers, positive and negative numbers, prime and composite numbers, and rational numbers.*

**99.** Find a number that fits in as many of these categories as possible.

**100.** Find a number that fits in as few of these categories as possible.

## 1.2 Fractions

■ FRACTIONS ■ SIMPLIFYING FRACTIONS ■ MULTIPLYING FRACTIONS ■ DIVIDING FRACTIONS
■ ADDING FRACTIONS ■ SUBTRACTING FRACTIONS ■ MIXED NUMBERS ■ DECIMALS
■ ROUNDING DECIMALS ■ APPLICATIONS

Getting Ready

**1.** Add:   132
45
73

**2.** Subtract:   321
173

**3.** Multiply:   437
38

**4.** Divide:   $37\overline{)3,885}$

### ■ FRACTIONS

In the **fractions**

$$\frac{1}{2}, \quad \frac{3}{5}, \quad \frac{2}{17}, \quad \text{and} \quad \frac{37}{7}$$

the number above the bar is called the **numerator,** and the number below the bar is called the **denominator.**

We often use fractions to indicate parts of a whole. In Figure 1-12(a), a rectangle has been divided into 5 equal parts, and 3 of the parts are shaded. The fraction $\frac{3}{5}$ indicates how much of the figure is shaded. In Figure 1-12(b), $\frac{5}{7}$ of the rectangle is shaded. In either example, the denominator of the fraction shows the total number of equal parts into which the whole is divided, and the numerator shows the number of these equal parts that are being considered.

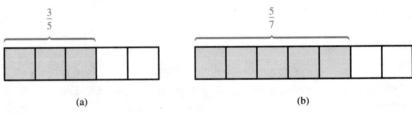

(a)                                     (b)

FIGURE 1-12

We can also use fractions to indicate division. For example, the fraction $\frac{8}{2}$ indicates that 8 is to be divided by 2:

$$\frac{8}{2} = 8 \div 2 = 4$$

**WARNING!**  Note that $\frac{8}{2} = 4$, because $4 \cdot 2 = 8$, and that $\frac{0}{7} = 0$, because $0 \cdot 7 = 0$. However, $\frac{6}{0}$ is undefined, because no number multiplied by 0 gives 6. Since every number multiplied by 0 gives 0, $\frac{0}{0}$ is indeterminate. Remember that the denominator of a fraction cannot be 0.

### ■ SIMPLIFYING FRACTIONS

A fraction is in **lowest terms** when no integer other than 1 will divide both its numerator and its denominator exactly. The fraction $\frac{6}{11}$ is in lowest terms, because only 1 divides both 6 and 11 exactly. The fraction $\frac{6}{8}$ is not in lowest terms, because 2 divides both 6 and 8 exactly.

We can **simplify** a fraction that is not in lowest terms by dividing both its numerator and its denominator by the same number. For example, to simplify $\frac{6}{8}$, we divide both numerator and denominator by 2.

$$\frac{6}{8} = \frac{6 \div 2}{8 \div 2} = \frac{3}{4}$$

From Figure 1-13, we see that $\frac{6}{8}$ and $\frac{3}{4}$ are equal fractions, because each one represents the same part of the rectangle.

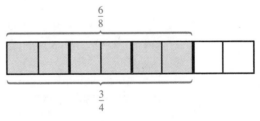

FIGURE 1-13

When a composite number has been written as the product of other natural numbers, we say that it has been **factored.** For example, 15 can be written as the product of 5 and 3.

$$15 = 5 \cdot 3$$

The numbers 5 and 3 are called **factors** of 15. When a composite number is written as the product of prime numbers, we say that it is written in **prime-factored form.**

**EXAMPLE 1**     Write 210 in prime-factored form.

*Solution*     We can write 210 as the product of 21 and 10 and proceed as follows:

$$210 = 21 \cdot 10$$
$$210 = 3 \cdot 7 \cdot 2 \cdot 5 \qquad \text{Factor 21 as } 3 \cdot 7 \text{ and factor 10 as } 2 \cdot 5.$$

Since 210 is now written as the product of prime numbers, its prime-factored form is $210 = 2 \cdot 3 \cdot 5 \cdot 7$. ∎

**Self Check**     Write 70 in prime-factored form.
*Answer*     $2 \cdot 5 \cdot 7$

To simplify a fraction, we factor its numerator and its denominator and divide out all common factors that appear in both the numerator and denominator. For example,

$$\frac{6}{8} = \frac{3 \cdot 2}{4 \cdot 2} = \frac{3 \cdot \cancel{2}^{1}}{4 \cdot \cancel{2}_{1}} = \frac{3}{4} \qquad \text{and} \qquad \frac{15}{18} = \frac{5 \cdot 3}{6 \cdot 3} = \frac{5 \cdot \cancel{3}^{1}}{6 \cdot \cancel{3}_{1}} = \frac{5}{6}$$

**WARNING!**     Remember that a fraction is in lowest terms only when its numerator and denominator have no common factors.

**EXAMPLE 2**     Simplify each fraction, if possible.

**a.** To simplify $\frac{6}{30}$, we factor the numerator and denominator and divide out the common factor of 6.

$$\frac{6}{30} = \frac{6 \cdot 1}{6 \cdot 5} = \frac{\cancel{6}^{1} \cdot 1}{\cancel{6}_{1} \cdot 5} = \frac{1}{5}$$

**b.** To attempt to simplify $\frac{33}{40}$, we factor the numerator and denominator and hope to divide out any common factors.

$$\frac{33}{40} = \frac{3 \cdot 11}{2 \cdot 2 \cdot 2 \cdot 5}$$

Since the numerator and denominator have no common factors, $\frac{33}{40}$ is in lowest terms. ∎

**Self Check**     Simplify $\frac{14}{35}$.
*Answer*     $\frac{2}{5}$

The preceding examples illustrate the **fundamental property of fractions.**

### The Fundamental Property of Fractions
If $a$, $b$, and $x$ are real numbers, then

$$\frac{a \cdot x}{b \cdot x} = \frac{a}{b} \quad (b \neq 0 \text{ and } x \neq 0)$$

## ■ MULTIPLYING FRACTIONS

### Multiplying Fractions
To multiply fractions, we multiply their numerators and multiply their denominators. In symbols, if $a$, $b$, $c$, and $d$ are real numbers, then

$$\frac{a}{b} \cdot \frac{c}{d} = \frac{a \cdot c}{b \cdot d} \quad (b \neq 0 \text{ and } d \neq 0)$$

For example,

$$\frac{4}{7} \cdot \frac{2}{3} = \frac{4 \cdot 2}{7 \cdot 3} \qquad \text{and} \qquad \frac{4}{5} \cdot \frac{13}{9} = \frac{4 \cdot 13}{5 \cdot 9}$$

$$= \frac{8}{21} \qquad\qquad\qquad\qquad = \frac{52}{45}$$

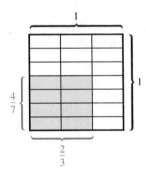

FIGURE 1-14

To justify the rule for multiplying fractions, we consider the square in Figure 1-14. Because the length of each side of the square is one unit and the area is the product of the lengths of two sides, the area is 1 square unit.

If this square is divided into 3 equal parts vertically and 7 equal parts horizontally, it is divided into 21 equal parts, and each represents $\frac{1}{21}$ of the total area. The area of the shaded rectangle in the square is $\frac{8}{21}$, because it contains 8 of the 21 parts. The width, $w$, of the shaded rectangle is $\frac{4}{7}$; its length, $l$, is $\frac{2}{3}$; and its area, $A$, is the product of $l$ and $w$:

$$A = l \cdot w$$
$$\frac{8}{21} = \frac{2}{3} \cdot \frac{4}{7}$$

This suggests that we can find the product of

$$\frac{4}{7} \quad \text{and} \quad \frac{2}{3}$$

by multiplying their numerators and multiplying their denominators.

Fractions such as $\frac{8}{21}$ whose numerators are less than their denominators are called **proper fractions**. Fractions such as $\frac{52}{45}$ whose numerators are greater than their denominators are called **improper fractions**.

**EXAMPLE 3** Do each multiplication.

**a.** $\dfrac{3}{7} \cdot \dfrac{13}{5} = \dfrac{3 \cdot 13}{7 \cdot 5}$   Multiply the numerators and multiply the denominators.

$\qquad = \dfrac{39}{35}$

**b.** $5 \cdot \dfrac{3}{15} = \dfrac{5}{1} \cdot \dfrac{3}{15}$   Write 5 as the improper fraction $\frac{5}{1}$.

$\qquad = \dfrac{5 \cdot 3}{1 \cdot 15}$   Multiply the numerators and multiply the denominators.

$\qquad = \dfrac{5 \cdot 3}{1 \cdot 5 \cdot 3}$   To attempt to simplify the fraction, factor the denominator.

$\qquad = \dfrac{\overset{1}{\cancel{5}} \cdot \overset{1}{\cancel{3}}}{1 \cdot \underset{1}{\cancel{5}} \cdot \underset{1}{\cancel{3}}}$   Divide out the common factors of 3 and 5.

$\qquad = 1$   $\qquad \frac{1 \cdot 1}{1 \cdot 1 \cdot 1} = 1.$ ∎

**Self Check** Multiply $\frac{5}{9} \cdot \frac{7}{10}$.

**Answer** $\frac{7}{18}$

**EXAMPLE 4** **European travel** Out of 36 students in a history class, three-fourths have signed up for a trip to Europe. If there are 30 places available on the flight, will there be room for one more student?

*Solution* We first find three-fourths of 36.

$\dfrac{3}{4} \cdot 36 = \dfrac{3}{4} \cdot \dfrac{36}{1}$   Write 36 as $\frac{36}{1}$.

$\qquad = \dfrac{3 \cdot 36}{4 \cdot 1}$   Multiply the numerators and multiply the denominators.

$\qquad = \dfrac{3 \cdot 4 \cdot 9}{4 \cdot 1}$   To simplify the fraction, factor the numerator.

$\qquad = \dfrac{3 \cdot \overset{1}{\cancel{4}} \cdot 9}{\underset{1}{\cancel{4}} \cdot 1}$   Divide out the common factor of 4.

$\qquad = \dfrac{27}{1}$

$\qquad = 27$

Twenty-seven students plan to go on the trip. Since there is room for 30 passengers, there is room for one more. ∎

## ■ DIVIDING FRACTIONS

One number is called the **reciprocal** of another if their product is 1. For example, $\frac{3}{5}$ is the reciprocal of $\frac{5}{3}$, because

$$\frac{3}{5} \cdot \frac{5}{3} = \frac{15}{15} = 1$$

**Dividing Fractions**
To divide two fractions, we multiply the first fraction by the reciprocal of the second fraction. In symbols, if $a$, $b$, $c$, and $d$ are real numbers, then

$$\frac{a}{b} \div \frac{c}{d} = \frac{a}{b} \cdot \frac{d}{c} = \frac{a \cdot d}{b \cdot c} \quad (b \neq 0, c \neq 0, \text{ and } d \neq 0)$$

**EXAMPLE 5**   Do each division.

**a.** $\dfrac{3}{5} \div \dfrac{6}{5} = \dfrac{3}{5} \cdot \dfrac{5}{6}$    Multiply $\frac{3}{5}$ by the reciprocal of $\frac{6}{5}$.

$= \dfrac{3 \cdot 5}{5 \cdot 6}$    Multiply the numerators and multiply the denominators.

$= \dfrac{3 \cdot 5}{5 \cdot 2 \cdot 3}$    Factor the denominator.

$= \dfrac{\overset{1}{\cancel{3}} \cdot \overset{1}{\cancel{5}}}{\underset{1}{\cancel{5}} \cdot 2 \cdot \underset{1}{\cancel{3}}}$    Divide out the common factors of 3 and 5.

$= \dfrac{1}{2}$

**b.** $\dfrac{15}{7} \div 10 = \dfrac{15}{7} \div \dfrac{10}{1}$    Write 10 as the improper fraction $\frac{10}{1}$.

$= \dfrac{15}{7} \cdot \dfrac{1}{10}$    Multiply $\frac{15}{7}$ by the reciprocal of $\frac{10}{1}$.

$= \dfrac{15 \cdot 1}{7 \cdot 10}$    Multiply the numerators and multiply the denominators.

$$= \frac{3 \cdot \overset{1}{\cancel{5}}}{7 \cdot 2 \cdot \underset{1}{\cancel{5}}} \qquad \text{Factor the numerator and the denominator, and divide out the common factor of 5.}$$

$$= \frac{3}{14}$$ ∎

## ■ ADDING FRACTIONS

**Adding Fractions with the Same Denominator**
To add fractions with the same denominator, we add the numerators and keep the common denominator. In symbols, if $a$, $b$, and $d$ are real numbers, then

$$\frac{a}{d} + \frac{b}{d} = \frac{a+b}{d} \quad (d \neq 0)$$

For example,

$$\frac{3}{7} + \frac{2}{7} = \frac{3+2}{7} \qquad \text{Add the numerators and keep the common denominator.}$$

$$= \frac{5}{7}$$

Figure 1-15 illustrates graphically why $\frac{3}{7} + \frac{2}{7} = \frac{5}{7}$.

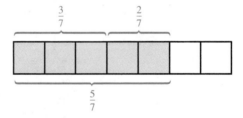

FIGURE 1-15

To add fractions with unlike denominators, we rewrite the fractions so that they have the same denominator. For example, we can multiply both the numerator and denominator of $\frac{1}{3}$ by 5 to obtain an equal fraction with a denominator of 15:

$$\frac{1}{3} = \frac{1 \cdot 5}{3 \cdot 5} = \frac{5}{15}$$

To rewrite $\frac{1}{5}$ as an equal fraction with a denominator of 15, we multiply the numerator and the denominator by 3:

$$\frac{1}{5} = \frac{1 \cdot 3}{5 \cdot 3} = \frac{3}{15}$$

Since 15 is the smallest number that can be used as a denominator for $\frac{1}{3}$ and $\frac{1}{5}$, it is called the **least** or **lowest common denominator** (the **LCD**).

To add the fractions $\frac{1}{3}$ and $\frac{1}{5}$, we rewrite each fraction as an equal fraction having a denominator of 15, and then we add the results:

$$\frac{1}{3} + \frac{1}{5} = \frac{1 \cdot 5}{3 \cdot 5} + \frac{1 \cdot 3}{5 \cdot 3}$$

$$= \frac{5}{15} + \frac{3}{15}$$

$$= \frac{5 + 3}{15}$$

$$= \frac{8}{15}$$

**EXAMPLE 6**   Add $\dfrac{3}{10} + \dfrac{5}{28}$.

*Solution*   To find the LCD, we find the prime factorization of each denominator and use each prime factor the greatest number of times it appears in either factorization:

$$\left.\begin{array}{l} 10 = 2 \cdot 5 \\ 28 = 2 \cdot 2 \cdot 7 \end{array}\right\} \text{LCD} = 2 \cdot 2 \cdot 5 \cdot 7 = 140$$

Since 140 is the smallest number that 10 and 28 divide exactly, we write both fractions as fractions with the LCD, 140.

$$\frac{3}{10} + \frac{5}{28} = \frac{3 \cdot 14}{10 \cdot 14} + \frac{5 \cdot 5}{28 \cdot 5} \qquad \text{Write each fraction as a fraction with a denominator of 140.}$$

$$= \frac{42}{140} + \frac{25}{140}$$

$$= \frac{42 + 25}{140} \qquad \text{Add the numerators and keep the denominator.}$$

$$= \frac{67}{140}$$

Since 67 is a prime number, it has no common factor with 140. Thus, $\frac{67}{140}$ is in lowest terms.   ∎

**Self Check**   Add $\frac{3}{8} + \frac{5}{12}$.
*Answer*   $\frac{19}{24}$

■ SUBTRACTING FRACTIONS

**Subtracting Fractions with the Same Denominator**
To subtract fractions with the same denominator, we subtract their numerators and keep their common denominator. In symbols, if $a$, $b$, and $d$ are real numbers, then

$$\frac{a}{d} - \frac{b}{d} = \frac{a - b}{d} \quad (d \neq 0)$$

For example,

$$\frac{7}{9} - \frac{2}{9} = \frac{7 - 2}{9} = \frac{5}{9}$$

To subtract fractions with unlike denominators, we write them as equivalent fractions with a common denominator. For example, to subtract $\frac{2}{5}$ from $\frac{3}{4}$, we write $\frac{3}{4} - \frac{2}{5}$, find the LCD of 20, and proceed as follows:

$$\frac{3}{4} - \frac{2}{5} = \frac{3 \cdot 5}{4 \cdot 5} - \frac{2 \cdot 4}{5 \cdot 4}$$

$$= \frac{15}{20} - \frac{8}{20}$$

$$= \frac{15 - 8}{20}$$

$$= \frac{7}{20}$$

**EXAMPLE 7**   Subtract 5 from $\dfrac{23}{3}$.

*Solution*

$$\frac{23}{3} - 5 = \frac{23}{3} - \frac{5}{1}$$   Write 5 as the improper fraction $\frac{5}{1}$.

$$= \frac{23}{3} - \frac{5 \cdot 3}{1 \cdot 3}$$   Write $\frac{5}{1}$ as a fraction with a denominator of 3.

$$= \frac{23}{3} - \frac{15}{3}$$

$$= \frac{23 - 15}{3}$$   Subtract the numerators and keep the denominator.

$$= \frac{8}{3}$$

■

*Self Check*   Subtract $\frac{5}{6} - \frac{3}{4}$.

*Answer*   $\frac{1}{12}$

## ■ MIXED NUMBERS

The **mixed number** $3\frac{1}{2}$ represents the sum of 3 and $\frac{1}{2}$. We can write $3\frac{1}{2}$ as an improper fraction as follows:

$$3\frac{1}{2} = 3 + \frac{1}{2}$$

$$= \frac{6}{2} + \frac{1}{2} \qquad 3 = \frac{6}{2}.$$

$$= \frac{6+1}{2} \qquad \text{Add the numerators and keep the denominator.}$$

$$= \frac{7}{2}$$

To write the fraction $\frac{19}{5}$ as a mixed number, we divide 19 by 5 to get 3, with a remainder of 4.

$$\frac{19}{5} = 3 + \frac{4}{5} = 3\frac{4}{5}$$

**EXAMPLE 8**    Add $2\frac{1}{4} + 1\frac{1}{3}$.

*Solution*    We first change each mixed number to an improper fraction.

$$2\frac{1}{4} = 2 + \frac{1}{4} \qquad\qquad 1\frac{1}{3} = 1 + \frac{1}{3}$$

$$= \frac{8}{4} + \frac{1}{4} \qquad\qquad\quad = \frac{3}{3} + \frac{1}{3}$$

$$= \frac{9}{4} \qquad\qquad\qquad\quad = \frac{4}{3}$$

Then we add the fractions.

$$2\frac{1}{4} + 1\frac{1}{3} = \frac{9}{4} + \frac{4}{3}$$

$$= \frac{9 \cdot 3}{4 \cdot 3} + \frac{4 \cdot 4}{3 \cdot 4} \qquad \text{Change each fraction into a fraction with the LCD of 12.}$$

$$= \frac{27}{12} + \frac{16}{12}$$

$$= \frac{43}{12}$$

Finally, we change $\frac{43}{12}$ to a mixed number.

$$\frac{43}{12} = 3 + \frac{7}{12} = 3\frac{7}{12}$$    ■

Self Check    Add $5\frac{1}{7} + 4\frac{2}{3}$.

Answer    $9\frac{17}{21}$

**EXAMPLE 9**    **Fencing land**    The three sides of a triangular lot measure $33\frac{1}{4}$, $57\frac{3}{4}$, and $72\frac{1}{2}$ meters. How much fencing will be needed to enclose the area?

Solution    We can find the sum of the lengths by adding the whole-number parts and the fractional parts of the dimensions separately:

$$33\frac{1}{4} + 57\frac{3}{4} + 72\frac{1}{2} = 33 + 57 + 72 + \frac{1}{4} + \frac{3}{4} + \frac{1}{2}$$

$$= 162 + \frac{1}{4} + \frac{3}{4} + \frac{2}{4}$$

Change $\frac{1}{2}$ to $\frac{2}{4}$ to obtain a common denominator.

$$= 162 + \frac{6}{4}$$

Add the fractions by adding the numerators and keeping the common denominator.

$$= 162 + \frac{3}{2}$$

$\frac{6}{4} = \frac{2 \cdot 3}{2 \cdot 2} = \frac{\cancel{2} \cdot 3}{\cancel{2} \cdot 2} = \frac{3}{2}$.

$$= 162 + 1\frac{1}{2}$$

Change $\frac{3}{2}$ to a mixed number.

$$= 163\frac{1}{2}$$

To enclose the area, $163\frac{1}{2}$ meters of fencing will be needed.    ■

Self Check    A rectangular lot is $85\frac{1}{2}$ feet wide and $140\frac{2}{3}$ feet deep. Find its perimeter.

Answer    $452\frac{1}{3}$ ft

## ■ DECIMALS

Rational numbers can always be changed to decimal form. For example, to write $\frac{1}{4}$ and $\frac{5}{22}$ as decimals, we use long division:

$$\begin{array}{r} 0.25 \\ 4)\overline{1.00} \\ \underline{8} \\ 20 \\ \underline{20} \end{array}$$

$$\begin{array}{r} 0.22727\ldots \\ 22)\overline{5.00000} \\ \underline{4\,4} \\ 60 \\ \underline{44} \\ 160 \\ \underline{154} \\ 60 \\ \underline{44} \\ 160 \end{array}$$

The decimal 0.25 is called a **terminating decimal.** The decimal 0.2272727 . . . (often written as 0.2$\overline{27}$) is called a **repeating decimal,** because it repeats the block of digits 27. Every rational number can be changed into either a **terminating** or a **repeating decimal.**

*Terminating Decimals*

$$\frac{1}{2} = 0.5$$

$$\frac{3}{4} = 0.75$$

$$\frac{5}{8} = 0.625$$

*Repeating Decimals*

$$\frac{1}{3} = 0.33333 \ldots \quad \text{or} \quad 0.\overline{3}$$

$$\frac{1}{6} = 0.16666 \ldots \quad \text{or} \quad 0.1\overline{6}$$

$$\frac{5}{22} = 0.2272727 \ldots \quad \text{or} \quad 0.2\overline{27}$$

The decimal 0.5 has one **decimal place,** because it has one digit to the right of the decimal point. The decimal 0.75 has two decimal places, and 0.625 has three.

To *add* or *subtract* decimal fractions, we first align their decimal points and then add or subtract.

```
  25.568          25.568
 +2.74           -2.74
 -------         -------
  28.308          22.828
```

To do the previous operations with a scientific calculator, we would press these keys:

25.568 + 2.74 =    and    25.568 − 2.74 =

To *multiply* decimal fractions, we multiply the numbers and then place the decimal point so that the number of decimal places in the answer is equal to the sum of the decimal places in the factors.

```
     3.453     Here there are three decimal places.
  ×  9.25      Here there are two decimal places.
  --------
    17265
     6906
    31 077
  --------
    31.94025   The product has 3 + 2 = 5 decimal places.
```

To do this multiplication with a scientific calculator, we would press these keys:

3.453 × 9.25 =

To *divide* decimals, we move the decimal point in the divisor to the right to make the divisor a whole number. We then move the decimal point in the dividend the same number of places to the right.

1.23)30.258   Move the decimal point in both the divisor and the dividend two places to the right.

We align the decimal point in the quotient with the repositioned decimal point in the dividend and then use long division.

$$
\begin{array}{r}
24.6 \\
123\overline{)3025.8} \\
\underline{246}\phantom{00.0} \\
565\phantom{.0} \\
\underline{492}\phantom{.0} \\
73\,8 \\
\underline{73\,8}
\end{array}
$$

To do the previous division with a scientific calculator, we would press these keys:

30.258    $\div$    1.23    $=$

## ■ ROUNDING DECIMALS

When decimal fractions are long, we often **round** them to a specific number of decimal places. For example, the decimal fraction 25.36124 rounded to one place (or to the nearest tenth) is 25.4. Rounded to two places (or to the nearest one-hundredth), the decimal is 25.36. To round decimals, we use the following rules.

> **Rounding Decimals**
> 1. Determine to how many decimal places you wish to round.
> 2. Look at the first digit to the right of that decimal place.
> 3. If that digit is 4 or less, drop it and all digits that follow. If it is 5 or greater, add 1 to the digit in the position to which you wish to round, and drop all of the digits that follow.

## ■ APPLICATIONS

A **percent** is the numerator of a fraction with a denominator of 100. For example, $6\frac{1}{4}$ percent, written $6\frac{1}{4}\%$, is the fraction $\frac{6.25}{100}$, or the decimal 0.0625. In problems involving percent, the word *of* often indicates multiplication. For example, $6\frac{1}{4}\%$ of 8,500 is the product 0.0625(8,500).

**EXAMPLE 10**    **Auto loans**    Juan signs a one-year note to borrow $8,500 to buy a car. If the rate of interest is $6\frac{1}{4}\%$, how much interest will he pay?

*Solution*    For the privilege of using the bank's money for one year, Juan must pay $6\frac{1}{4}\%$ of $8,500. We calculate the interest, $i$, as follows:

$$i = 6\tfrac{1}{4}\% \text{ of } 8,500$$
$$= 0.0625 \cdot 8,500 \qquad \text{The word } of \text{ means } times.$$
$$= 531.25$$

Juan will pay $531.25 interest.    ■

Self Check    In Example 10, how much interest will Juan pay if the rate is 9%?
Answer    $765

Orals   *Simplify each fraction.*

**1.** $\dfrac{3}{6}$    **2.** $\dfrac{5}{10}$    **3.** $\dfrac{10}{20}$    **4.** $\dfrac{25}{75}$

*Do each operation.*

**5.** $\dfrac{5}{6} \cdot \dfrac{1}{2}$    **6.** $\dfrac{3}{4} \cdot \dfrac{3}{5}$    **7.** $\dfrac{2}{3} \div \dfrac{3}{2}$    **8.** $\dfrac{3}{5} \div \dfrac{5}{2}$

**9.** $\dfrac{4}{9} + \dfrac{7}{9}$    **10.** $\dfrac{6}{7} - \dfrac{3}{7}$    **11.** $\dfrac{2}{3} - \dfrac{1}{2}$    **12.** $\dfrac{3}{4} + \dfrac{1}{2}$

**13.** $2.5 + 0.36$    **14.** $3.45 - 2.21$

**15.** $0.2 \cdot 2.5$    **16.** $0.3 \cdot 13$

*Round each decimal to two decimal places.*

**17.** $3.244993$    **18.** $3.24521$

## EXERCISE 1.2

**REVIEW**   *Decide whether the following statements are true or false.*

**1.** 6 is an integer.    **2.** $\dfrac{1}{2}$ is a natural number.    **3.** 21 is a prime number.    **4.** No prime number is an even number.

**5.** $8 > -2$    **6.** $-3 < -2$    **7.** $9 \leq |-9|$    **8.** $|-11| \geq 10$

*Place an appropriate symbol in each box to make the statement true.*

**9.** $3 + 7 \boxed{\phantom{x}} 10$    **10.** $\dfrac{3}{7} \boxed{\phantom{x}} \dfrac{2}{7} = \dfrac{1}{7}$    **11.** $|-2| \boxed{\phantom{x}} 2$    **12.** $4 + 8 \boxed{\phantom{x}} 11$

**VOCABULARY AND CONCEPTS**   *Fill in each blank to make a true statement.*

**13.** The number above the bar in a fraction is called the _____.

**14.** The number below the bar in a fraction is called the _____.

**15.** To _____ a fraction, we divide its numerator and denominator by the same number.

**16.** To write a number in prime-factored form, we write it as the product of _____ numbers.

**17.** If the numerator of a fraction is less than the denominator, the fraction is called a _____ fraction.

**18.** If the numerator of a fraction is greater than the denominator, the fraction is called an _____ fraction.

**19.** If the product of two numbers is ___, the numbers are called reciprocals.

**20.** $\dfrac{ax}{bx} =$ ___.

**21.** To multiply two fractions, _____ the numerators and multiply the denominators.

**22.** To divide two fractions, multiply the first fraction by the _____ of the second fraction.

**23.** To add fractions with a common denominator, add the _____ and keep the common _____.

**24.** To subtract fractions with a common denominator, _____ the numerators and keep the common _____.

**25.** $75\dfrac{2}{3}$ means $75$ ___ $\dfrac{2}{3}$.

**26.** 0.75 is an example of a _____ decimal.

**27.** $5.3\overline{27}$ is an example of a _____ decimal.

**28.** A _____ is the numerator of a fraction whose denominator is 100.

***PRACTICE***   *In Exercises 29–36, write each fraction in lowest terms. If the fraction is already in lowest terms, so indicate.*

**29.** $\dfrac{6}{12}$   **30.** $\dfrac{3}{9}$   **31.** $\dfrac{15}{20}$   **32.** $\dfrac{22}{77}$

**33.** $\dfrac{24}{18}$   **34.** $\dfrac{35}{14}$   **35.** $\dfrac{72}{64}$   **36.** $\dfrac{26}{21}$

*In Exercises 37–48, do each multiplication. Simplify each result when possible.*

**37.** $\dfrac{1}{2}\cdot\dfrac{3}{5}$   **38.** $\dfrac{3}{4}\cdot\dfrac{5}{7}$   **39.** $\dfrac{4}{3}\cdot\dfrac{6}{5}$   **40.** $\dfrac{7}{8}\cdot\dfrac{6}{15}$

**41.** $\dfrac{5}{12}\cdot\dfrac{18}{5}$   **42.** $\dfrac{5}{4}\cdot\dfrac{12}{10}$   **43.** $\dfrac{17}{34}\cdot\dfrac{3}{6}$   **44.** $\dfrac{21}{14}\cdot\dfrac{3}{6}$

**45.** $12\cdot\dfrac{5}{6}$   **46.** $9\cdot\dfrac{7}{12}$   **47.** $\dfrac{10}{21}\cdot 14$   **48.** $\dfrac{5}{24}\cdot 16$

*In Exercises 49–60, do each division. Simplify each result when possible.*

**49.** $\dfrac{3}{5}\div\dfrac{2}{3}$   **50.** $\dfrac{4}{5}\div\dfrac{3}{7}$   **51.** $\dfrac{3}{4}\div\dfrac{6}{5}$   **52.** $\dfrac{3}{8}\div\dfrac{15}{28}$

**53.** $\dfrac{2}{13}\div\dfrac{8}{13}$   **54.** $\dfrac{4}{7}\div\dfrac{20}{21}$   **55.** $\dfrac{21}{35}\div\dfrac{3}{14}$   **56.** $\dfrac{23}{25}\div\dfrac{46}{5}$

**57.** $6\div\dfrac{3}{14}$   **58.** $23\div\dfrac{46}{5}$   **59.** $\dfrac{42}{30}\div 7$   **60.** $\dfrac{34}{8}\div 17$

*In Exercises 61–84, do each addition or subtraction. Simplify each result when possible.*

**61.** $\dfrac{3}{5}+\dfrac{3}{5}$   **62.** $\dfrac{4}{7}-\dfrac{2}{7}$   **63.** $\dfrac{4}{13}-\dfrac{3}{13}$   **64.** $\dfrac{2}{11}+\dfrac{9}{11}$

**65.** $\dfrac{1}{6}+\dfrac{1}{24}$   **66.** $\dfrac{17}{25}-\dfrac{2}{5}$   **67.** $\dfrac{3}{5}+\dfrac{2}{3}$   **68.** $\dfrac{4}{3}+\dfrac{7}{2}$

**69.** $\dfrac{9}{4} - \dfrac{5}{6}$  **70.** $\dfrac{2}{15} + \dfrac{7}{9}$  **71.** $\dfrac{7}{10} - \dfrac{1}{14}$  **72.** $\dfrac{7}{25} + \dfrac{3}{10}$

**73.** $3 - \dfrac{3}{4}$  **74.** $5 + \dfrac{21}{5}$  **75.** $\dfrac{17}{3} + 4$  **76.** $\dfrac{13}{9} - 1$

**77.** $4\dfrac{3}{5} + \dfrac{3}{5}$  **78.** $2\dfrac{1}{8} + \dfrac{3}{8}$  **79.** $3\dfrac{1}{3} - 1\dfrac{2}{3}$  **80.** $5\dfrac{1}{7} - 3\dfrac{2}{7}$

**81.** $3\dfrac{3}{4} - 2\dfrac{1}{2}$  **82.** $15\dfrac{5}{6} + 11\dfrac{5}{8}$  **83.** $8\dfrac{2}{9} - 7\dfrac{2}{3}$  **84.** $3\dfrac{4}{5} - 3\dfrac{1}{10}$

*In Exercises 85–92, do each operation.*

**85.** $23.45 + 135.2$  **86.** $345.213 - 27.35$  **87.** $67.235 - 22.45$  **88.** $12.17 + 3.457$

**89.** $3.4 \cdot 13.2$  **90.** $4.21 \cdot 2.73$  **91.** $0.23\overline{)1.0465}$  **92.** $4.7\overline{)10.857}$

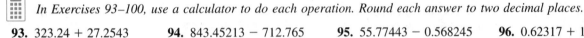 *In Exercises 93–100, use a calculator to do each operation. Round each answer to two decimal places.*

**93.** $323.24 + 27.2543$  **94.** $843.45213 - 712.765$  **95.** $55.77443 - 0.568245$  **96.** $0.62317 + 1.3316$

**97.** $25.25 \cdot 132.179$  **98.** $234.874 \cdot 242.46473$  **99.** $0.456\overline{)4.5694323}$  **100.** $43.225\overline{)32.465748}$

**APPLICATIONS**  *Solve each problem.*

**101. Buying fencing**  How many meters of fencing are needed to enclose the square field shown in Illustration 1?

$30\dfrac{2}{5}$ meters

ILLUSTRATION 1

**102. Spring plowing**  A farmer has plowed $12\frac{1}{3}$ acres of a $43\frac{1}{2}$-acre field. How much more needs to be plowed?

**103. Perimeter of a garden**  The four sides of a garden measure $7\frac{2}{3}$ feet, $15\frac{1}{4}$ feet, $19\frac{1}{2}$ feet, and $10\frac{3}{4}$ feet. Find the length of the fence needed to enclose the garden.

**104. Making clothes**  A designer needs $3\frac{1}{4}$ yards of material for each dress he makes. How much material will he need to make 14 dresses?

**105. Minority population**  22% of the 11,431,000 citizens of Illinois are nonwhite. How many are nonwhite?

**106. Quality control**  In the manufacture of active-matrix color LCD computer displays, many units must be rejected as defective. If 23% of a production run of 17,500 units is defective, how many units are acceptable?

**107. Freeze-drying**  Almost all of the water must be removed when food is preserved by freeze-drying. Find the weight of the water removed from 750 pounds of a food that is 36% water.

**108. Planning for growth**  This year, sales at Positronics Corporation totaled $18.7 million. If the projection of 12% annual growth is true, what will be next year's sales?

**109. Speed skating**  In tryouts for the Olympics, a speed skater had times of 44.47, 43.24, 42.77, and 42.05 seconds. Find the average time. (*Hint:* Add the numbers and divide by 4.)

110. **Cost of gasoline**  Otis drove his car 15,675.2 miles last year, averaging 25.5 miles per gallon of gasoline. If the average cost of gasoline was $1.27 per gallon, find the fuel cost to drive the car.

111. **Paying taxes**  A woman earns $48,712.32 in taxable income. She must pay 15% tax on the first $23,000 and 28% on the rest. In addition, she must pay a Social Security tax of 15.4% on the total amount. How much tax will she need to pay?

112. **Sealing asphalt**  A rectangular parking lot is 253.5 feet long and 178.5 feet wide. A 55-gallon drum of asphalt sealer covers 4,000 square feet and costs $97.50. Find the cost to seal the parking lot. (Sealer can only be purchased in full drums.)

113. **Installing carpet**  What will it cost to carpet the area shown in Illustration 2 with carpet that costs $29.79 per square yard? (One square yard is 9 square feet.)

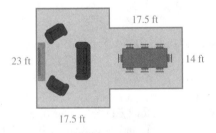

23 ft    17.5 ft    14 ft

17.5 ft

ILLUSTRATION 2

114. **Inventory costs**  Each television a retailer buys costs $3.25 per day for warehouse storage. What does it cost to store 37 television sets for three weeks?

## WRITING

121. Describe how you would find the common denominator of two fractions.

123. Explain how to convert a mixed number into an improper fraction.

115. **Manufacturing profits**  A manufacturer of computer memory boards has a profit of $37.50 on each standard-capacity memory board, and $57.35 on each high-capacity board. The sales department has orders for 2,530 standard boards and 1,670 high-capacity boards. Which order should production fill first, to receive the greater profit?

116. **Dairy production**  A Holstein cow will produce 7,600 pounds of milk each year, with a $3\frac{1}{2}\%$ butterfat content. Each year, a Guernsey cow will produce about 6,500 pounds of milk that is 5% butterfat. Which cow produces more butterfat?

117. **Feeding dairy cows**  Each year, a typical dairy cow will eat 12,000 pounds of food that is 57% silage. To feed 30 cows, how much silage will a farmer use in a year?

118. **Comparing bids**  Two contractors bid on a home remodeling project. The first bids $9,350 for the entire job. The second contractor will work for $27.50 per hour, plus $4,500 for materials. He estimates that the job will take 150 hours. Which contractor has the lower bid?

119. **Choosing a furnace**  A high-efficiency home heating system can be installed for $4,170, with an average monthly heating bill of $57.50. A regular furnace can be installed for $1,730, but monthly heating bills average $107.75. After three years, which system has cost more altogether?

120. **Choosing a furnace**  Refer to Exercise 119. Decide which furnace system will have cost more after five years.

122. Explain how to convert an improper fraction into a mixed number.

124. Explain how you would decide which of two decimal fractions is the larger.

*SOMETHING TO THINK ABOUT*

**125.** In what situations would it be better to leave an answer in the form of an improper fraction?

**126.** When would it be better to change an improper-fraction answer into a mixed number?

**127.** Can the product of two proper fractions be larger than either of the fractions?

**128.** How does the product of one proper and one improper fraction compare with the two factors?

## 1.3  Exponents and Order of Operations
■ EXPONENTS ■ ORDER OF OPERATIONS ■ GEOMETRY

**Getting Ready**  *Do the operations.*

**1.** $2 \cdot 2$

**2.** $3 \cdot 3$

**3.** $3 \cdot 3 \cdot 3$

**4.** $2 \cdot 2 \cdot 2$

**5.** $\dfrac{1}{2} \cdot \dfrac{1}{2}$

**6.** $\dfrac{1}{3} \cdot \dfrac{1}{3} \cdot \dfrac{1}{3}$

**7.** $\dfrac{2}{5} \cdot \dfrac{2}{5} \cdot \dfrac{2}{5}$

**8.** $\dfrac{3}{10} \cdot \dfrac{3}{10} \cdot \dfrac{3}{10}$

### ■ EXPONENTS

To show how many times a number is to be used as a factor in a product, we use *exponents*. In the expression $2^3$, 2 is called the **base** and 3 is called the **exponent.**

$$\text{Base} \longrightarrow 2^3 \longleftarrow \text{Exponent}$$

The exponent of 3 indicates that the base of 2 is to be used as a factor three times:

$$2^3 = \overbrace{2 \cdot 2 \cdot 2}^{3 \text{ factors of } 2} = 8$$

 **WARNING!**  Note that $2^3 = 8$. This is not the same as $2 \cdot 3 = 6$.

In the expression $x^5$ (called an **exponential expression** or a **power of** *x*), 5 is the **exponent** and *x* is the **base.** The exponent of 5 indicates that a base of *x* is to be used as a factor five times.

$$x^5 = \overbrace{x \cdot x \cdot x \cdot x \cdot x}^{5 \text{ factors of } x}$$

In expressions such as $x$ or $y$, the exponent is understood to be 1:

$$x = x^1 \qquad \text{and} \qquad y = y^1$$

In general, we have the following definition.

> **Natural-Number Exponents**
> If $n$ is a natural number, then
> $$\overset{n \text{ factors of } x}{\overbrace{x^n = x \cdot x \cdot x \cdot \cdots \cdot x}}$$

**EXAMPLE 1**    Write each expression without using exponents.

**a.** $4^2 = 4 \cdot 4 = 16$     Read $4^2$ as "4 squared" or as "4 to the second power."

**b.** $5^3 = 5 \cdot 5 \cdot 5 = 125$     Read $5^3$ as "5 cubed" or as "5 to the third power."

**c.** $6^4 = 6 \cdot 6 \cdot 6 \cdot 6 = 1{,}296$     Read $6^4$ as "6 to the fourth power."

**d.** $\left(\dfrac{2}{3}\right)^5 = \dfrac{2}{3} \cdot \dfrac{2}{3} \cdot \dfrac{2}{3} \cdot \dfrac{2}{3} \cdot \dfrac{2}{3} = \dfrac{32}{243}$     Read $\left(\dfrac{2}{3}\right)^5$ as "$\dfrac{2}{3}$ to the fifth power." ∎

**Self Check**    Evaluate   **a.** $7^2$   and   **b.** $\left(\frac{3}{4}\right)^3$.
**Answers**       **a.** 49,   **b.** $\frac{27}{64}$

We can find powers using a calculator. For example, to find $2.35^4$, we enter these numbers and press these keys:

$$2.35 \;\boxed{y^x}\; 4 \;\boxed{=}$$

The display will read $\boxed{\text{30.49800625}}$ . Some calculators have a $\boxed{x^y}$ key rather than a $\boxed{y^x}$ key.

In the next example, the base of an exponential expression is a variable.

**EXAMPLE 2**    Write each expression without using exponents.

**a.** $y^6 = y \cdot y \cdot y \cdot y \cdot y \cdot y$     Read $y^6$ as "y to the sixth power."

**b.** $x^3 = x \cdot x \cdot x$     Read $x^3$ as "x cubed" or as "x to the third power."

**c.** $z^2 = z \cdot z$     Read $z^2$ as "z squared" or as "z to the second power."

**d.** $a^1 = a$     Read $a^1$ as "a to the first power." ∎

*Self Check*

*Answers*

Write each expression without using an exponent:   **a.** $a^3$   and   **b.** $b^4$.
**a.** $a \cdot a \cdot a$,   **b.** $b \cdot b \cdot b \cdot b$

## ■ ORDER OF OPERATIONS

Suppose that you are asked to contact a friend if you see a Rolex watch for sale while traveling in Switzerland. After locating the watch, you send the following message to your friend.

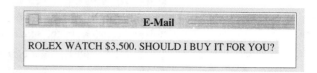

| E-Mail |
| --- |
| ROLEX WATCH $3,500. SHOULD I BUY IT FOR YOU? |

The next day, you receive this response.

| E-Mail |
| --- |
| NO PRICE TOO HIGH! REPEAT...NO! PRICE TOO HIGH. |

The first statement says to buy the watch at any price. The second says not to buy it, because it is too expensive. The placement of the exclamation point makes these statements read differently, resulting in different interpretations.

When reading a mathematical statement, the same kind of confusion is possible. To illustrate, we consider the expression $3 + 2 \cdot 4$, which contains the operations of addition and multiplication. We can calculate this expression in two different ways. We can do the addition first and then do the multiplication. Or we can do the multiplication first and then do the addition. However, we will get different results.

| *Method 1: Add First* | | *Method 2: Multiply First* | |
| --- | --- | --- | --- |
| $2 + 3 \cdot 4 = 5 \cdot 4$ | Add 2 and 3. | $2 + 3 \cdot 4 = 2 + 12$ | Multiply 3 and 4. |
| $= 20$ | Multiply 5 and 4. | $= 14$ | Add 2 and 12. |

Different results

To eliminate the possibility of getting different answers, we will agree to do multiplications before additions. The correct calculation of $2 + 3 \cdot 4$ is

$$2 + 3 \cdot 4 = 2 + 12$$
$$= 14$$

To indicate that additions should be done before multiplications, we must use **grouping symbols** such as parentheses (  ), brackets [  ], or braces {  }. In the expression $(2 + 3)4$, the parentheses indicate that the addition is to be done first:

$$(2 + 3)4 = 5 \cdot 4$$
$$= 20$$

To guarantee that calculations will have one correct result, we will always do calculations in the following order.

---

## Rules for the Order of Operations

Use the following steps to do all calculations within each pair of grouping symbols, working from the innermost pair to the outermost pair.

**1.** Find the values of any exponential expressions.

**2.** Do all multiplications and divisions, working from left to right.

**3.** Do all additions and subtractions, working from left to right.

When all grouping symbols have been removed, repeat the rules above to finish the calculation.

In a fraction, simplify the numerator and the denominator separately. Then simplify the fraction, whenever possible.

---

**WARNING!**   Note that $4(2)^3 \neq (4 \cdot 2)^3$:

$$4(2)^3 = 4 \cdot 2 \cdot 2 \cdot 2 = 4(8) = 32 \quad \text{and} \quad (4 \cdot 2)^3 = 8^3 = 8 \cdot 8 \cdot 8 = 512$$

Likewise, $4x^3 \neq (4x)^3$, because

$$4x^3 = 4xxx \quad \text{and} \quad (4x)^3 = (4x)(4x)(4x) = 64x^3$$

**EXAMPLE 3**    Evaluate $5^3 + 2(8 - 3 \cdot 2)$.

*Solution*    We do the work within the parentheses first and then simplify.

| | |
|---|---|
| $5^3 + 2(8 - 3 \cdot 2) = 5^3 + 2(8 - 6)$ | Do the multiplication within the parentheses. |
| $= 5^3 + 2(2)$ | Do the subtraction within the parentheses. |
| $= 125 + 2(2)$ | Find the value of the exponential expression. |
| $= 125 + 4$ | Do the multiplication. |
| $= 129$ | Do the addition. |

Self Check    Evaluate $5 + 4 \cdot 3^2$.

Answer    41

---

**EXAMPLE 4**    Evaluate $\dfrac{3(3 + 2) + 5}{17 - 3(4)}$.

Solution    We simplify the numerator and the denominator separately and then simplify the fraction.

$$\frac{3(3 + 2) + 5}{17 - 3(4)} = \frac{3(5) + 5}{17 - 3(4)} \qquad \text{Do the addition within the parentheses.}$$

$$= \frac{15 + 5}{17 - 12} \qquad \text{Do the multiplications.}$$

$$= \frac{20}{5} \qquad \text{Do the addition and the subtraction.}$$

$$= 4 \qquad \text{Do the division.} \qquad \blacksquare$$

Self Check    Evaluate $\dfrac{4 + 2(5 - 3)}{2 + 3(2)}$.

Answer    1

---

**EXAMPLE 5**    If $x = 3$ and $y = 4$, evaluate   **a.** $3y + x^2$   and   **b.** $3(y + x^2)$.

Solution    **a.** $3y + x^2 = 3(4) + 3^2$      Substitute 3 for $x$ and 4 for $y$.

$\qquad\qquad\quad = 3(4) + 9$      Evaluate the exponential expression.

$\qquad\qquad\quad = 12 + 9$      Do the multiplication.

$\qquad\qquad\quad = 21$      Do the addition.

**b.** $3(y + x^2) = 3(4 + 3^2)$      Substitute 3 for $x$ and 4 for $y$.

$\qquad\qquad\quad = 3(4 + 9)$      Evaluate the exponential expression.

$\qquad\qquad\quad = 3(13)$      Do the addition in the parentheses.

$\qquad\qquad\quad = 39$      Do the multiplication.    $\blacksquare$

Self Check    If $x = 2$ and $y = 3$, evaluate $2(x^2 + y^3)$.

Answer    62

**EXAMPLE 6**    If $x = 4$ and $y = 3$, evaluate $\dfrac{3x^2 - 2y}{2(x + y)}$.

*Solution*    $\dfrac{3x^2 - 2y}{2(x + y)} = \dfrac{3(4^2) - 2(3)}{2(4 + 3)}$     Substitute 4 for $x$ and 3 for $y$.

$= \dfrac{3(16) - 2(3)}{2(7)}$     Find the value of $4^2$ in the numerator and do the addition in the denominator.

$= \dfrac{48 - 6}{14}$     Do the multiplications.

$= \dfrac{42}{14}$     Do the subtraction.

$= 3$     Do the division.    ■

**Self Check**    If $x = 2$ and $y = 5$, evaluate $\dfrac{x^2 + 6y}{2(x + y) + 3}$.

*Answer*    2

## ■ GEOMETRY

To find perimeters and areas of geometric figures, substituting numbers for variables is often required. The **perimeter** of a geometric figure is the distance around it, and the **area** of a figure is the amount of surface that it encloses. The perimeter of a circle is called its **circumference.**

**EXAMPLE 7**    **Circles**  Find  **a.** the circumference  and  **b.** the area  of the circle shown in Figure 1-16.

14 cm

FIGURE 1-16

*Solution*    **a.** The formula for the circumference of a circle is

$$C = \pi D$$

where $C$ is the circumference, $\pi$ is approximately $\frac{22}{7}$, and $D$ is the diameter — the distance through the center of the circle. We can approximate the circumference by substituting $\frac{22}{7}$ for $\pi$ and 14 for $D$ in the formula and simplifying.

$$C = \pi D$$

$$C \approx \frac{22}{7} \cdot 14 \qquad \text{Read } \approx \text{ as "is approximately equal to."}$$

$$C \approx \frac{22 \cdot \overset{2}{14}}{\underset{1}{7} \cdot 1} \qquad \text{Multiply the fractions and simplify.}$$

$$C \approx 44$$

The circumference is approximately 44 centimeters. To use a calculator, we would press these keys:

$$\boxed{\pi} \quad \times \quad 14 \quad =$$

The display will read 43.98229715. . . . The result is not 44, because a calculator uses a better approximation for $\pi$ than $\frac{22}{7}$.

**b.** The formula for the area of a circle is

$$A = \pi r^2$$

where $A$ is the area, $\pi \approx \frac{22}{7}$, and $r$ is the **radius** of the circle. (The radius is one-half of the diameter.) We can approximate the area by substituting $\frac{22}{7}$ for $\pi$ and 7 for $r$ in the formula and simplifying.

$$A = \pi r^2$$

$$A \approx \frac{22}{7} \cdot 7^2$$

$$A \approx \frac{22}{7} \cdot \frac{49}{1} \qquad \text{Evaluate the exponential expression.}$$

$$A \approx \frac{22 \cdot \overset{7}{49}}{\underset{1}{7} \cdot 1} \qquad \text{Multiply the fractions and simplify.}$$

$$A \approx 154$$

The area is approximately 154 square centimeters. To use a calculator, we would press these keys:

$$\boxed{\pi} \quad \times \quad 7 \quad x^2 \quad =$$

The display will read 153.93804.   ∎

*Self Check*   Find  **a.** the circumference   and  **b.** the area of a circle with a diameter of 28 meters. $\left(\text{Use } \frac{22}{7} \text{ as an estimate for } \pi.\right)$ Check your results with a calculator.

*Answers*   **a.** 88 m,   **b.** 616 m$^2$

Table 1-1 shows the formulas for the perimeter and area of several geometric figures.

TABLE 1–1

| Figure | Name | Perimeter | Area |
|---|---|---|---|
| | Square | $P = 4s$ | $A = s^2$ |
| | Rectangle | $P = 2l + 2w$ | $A = lw$ |
| | Triangle | $P = a + b + c$ | $A = \dfrac{1}{2}bh$ |
| | Trapezoid | $P = a + b + c + d$ | $A = \dfrac{1}{2}h(b + d)$ |
| | Circle | $C = 2\pi r = \pi D$ | $A = \pi r^2$ |

The **volume** of a three-dimensional geometric solid is the amount of space it encloses. Table 1-2 shows the formulas for the volume of several solids.

TABLE 1-2

| Figure | Name | Volume |
|---|---|---|
| | Rectangular solid | $V = lwh$ |
| | Cylinder | $V = Bh$, where $B$ is the area of the base |
| | Pyramid | $V = \frac{1}{3}Bh$, where $B$ is the area of the base |
| | Cone | $V = \frac{1}{3}Bh$, where $B$ is the area of the base (If the base is a circle, then $B = \pi r^2$.) |
| | Sphere | $V = \frac{4}{3}\pi r^3$ |

**EXAMPLE 8**

**Winter driving**   Find the number of cubic feet of road salt in the conical pile shown in Figure 1-17. Round the answer to two decimal places.

*Solution*   We can find the area of the circular base by substituting $\frac{22}{7}$ for $\pi$ and 14.30 for the radius.

$$A = \pi r^2$$

$$\approx \frac{22}{7}(14.3)^2$$

$$\approx 642.6828571 \qquad \text{Use a calculator.}$$

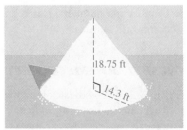

18.75 ft

14.3 ft

FIGURE 1-17

We then substitute 642.6828571 for $B$ and 18.75 for $h$ in the formula for the volume of a cone.

$$V = \frac{1}{3}Bh$$

$$\approx \frac{1}{3}(642.6828571)(18.75)$$

$$\approx 4{,}016.767857 \qquad \text{Use a calculator.}$$

To two decimal places, there are 4,016.77 cubic feet of salt in the pile. ∎

**Self Check**
To the nearest hundredth, find the number of cubic feet of water that can be contained in a spherical tank that has a radius of 9 feet. $\left(\text{Use } \pi \approx \frac{22}{7}.\right)$

**Answer** 3,054.86 ft$^3$

**Orals** *Find the value of each expression.*

**1.** $2^5$      **2.** $3^4$      **3.** $4^3$      **4.** $5^3$

*Simplify each expression.*

**5.** $3(2)^3$      **6.** $(3 \cdot 2)^2$      **7.** $3 + 2 \cdot 4$

**8.** $10 - 3^2$      **9.** $4 + 2^2 \cdot 3$      **10.** $2 \cdot 3 + 2 \cdot 3^2$

## EXERCISE 1.3

**REVIEW**

**1.** On the number line, graph the prime numbers between 10 and 20.

**2.** Write the inequality $7 \le 12$ as an inequality using the symbol $\ge$.

**3.** Classify the number 17 as a prime number or a composite number.

**4.** Evaluate $\dfrac{3}{5} - \dfrac{1}{2}$

**VOCABULARY AND CONCEPTS** *Fill in each blank to make a true statement.*

**5.** An _____ indicates how many times a base is to be used as a factor in a product.

**6.** In the expression $x^5$, $x$ is called the _____ and 5 is called an _____.

**7.** In the expression $3 + 4 \cdot 5$, the _____ should be done first.

**8.** Parentheses, brackets, and braces are called _____ symbols.

*Write the appropriate formula to find each quantity.*

**9.** The perimeter of a square _____

**10.** The area of a square _____

**11.** The perimeter of a rectangle _____

**12.** The area of a rectangle _____

**13.** The perimeter of a triangle _____

**14.** The area of a triangle _____

**15.** The perimeter of a trapezoid _____

**16.** The area of a trapezoid _____

**17.** The circumference of a circle _____

**18.** The area of a circle _____

**19.** The volume of a rectangular solid _____

**20.** The volume of a cylinder _____

**21.** The volume of a pyramid _____

**22.** The volume of a cone _____

**23.** The volume of a sphere _____

**24.** In Exercises 20–22, $B$ is the _____ of the base.

**PRACTICE** *In Exercises 25–30, find the value of each expression.*

**25.** $4^2$

**26.** $5^2$

**27.** $6^2$

**28.** $7^3$

**29.** $\left(\dfrac{1}{10}\right)^4$

**30.** $\left(\dfrac{1}{2}\right)^6$

*In Exercises 31–34, use a calculator to find each power.*

**31.** $7.9^3$

**32.** $0.45^4$

**33.** $25.3^2$

**34.** $7.567^3$

*In Exercises 35–42, write each expression as the product of several factors.*

**35.** $x^2$

**36.** $y^3$

**37.** $3z^4$

**38.** $5t^2$

**39.** $(5t)^2$

**40.** $(3z)^4$

**41.** $5(2x)^3$

**42.** $7(3t)^2$

*In Exercises 43–50, find the value of each expression if $x = 3$ and $y = 2$.*

**43.** $4x^2$

**44.** $4y^3$

**45.** $(5y)^3$

**46.** $(2y)^4$

**47.** $2x^y$

**48.** $3y^x$

**49.** $(3y)^x$

**50.** $(2x)^y$

*In Exercises 51–78, simplify each expression by doing the operations.*

**51.** $3 \cdot 5 - 4$

**52.** $4 \cdot 6 + 5$

**53.** $3(5 - 4)$

**54.** $4(6 + 5)$

**55.** $3 + 5^2$

**56.** $4^2 - 2^2$

**57.** $(3 + 5)^2$

**58.** $(5 - 2)^3$

**59.** $2 + 3 \cdot 5 - 4$

**60.** $12 + 2 \cdot 3 + 2$

**61.** $64 \div (3 + 1)$

**62.** $16 \div (5 + 3)$

**63.** $(7 + 9) \div (2 \cdot 4)$

**64.** $(7 + 9) \div 2 \cdot 4$

**65.** $(5 + 7) \div 3 \cdot 4$

**66.** $(5 + 7) \div (3 \cdot 4)$

**67.** $24 \div 4 \cdot 3 + 3$

**68.** $36 \div 9 \cdot 4 - 2$

**69.** $3^2 + 2(1 + 4) - 2$

**70.** $4 \cdot 3 + 2(5 - 2) - 2^3$

**71.** $5^2 - (7 - 3)^2$

**72.** $3^3 + (3 - 1)^3$

**73.** $(2 \cdot 3 - 4)^3$

**74.** $(3 \cdot 5 - 2 \cdot 6)^2$

**75.** $\dfrac{3}{5} \cdot \dfrac{10}{3} + \dfrac{1}{2} \cdot 12$

**76.** $\dfrac{15}{4}\left(1 + \dfrac{3}{5}\right)$

**77.** $\left[\dfrac{1}{3} - \left(\dfrac{1}{2}\right)^2\right]^2$

**78.** $\left[\left(\dfrac{2}{3}\right)^2 - \dfrac{1}{3}\right]^2$

*In Exercises 79–86, use a calculator to simplify each fraction.*

**79.** $\dfrac{(3+5)^2+2}{2(8-5)}$

**80.** $\dfrac{25-(2\cdot 3-1)}{2\cdot 9-8}$

**81.** $\dfrac{(5-3)^2+2}{4^2-(8+2)}$

**82.** $\dfrac{(4^2-2)+7}{5(2+4)-3^2}$

**83.** $\dfrac{2[4+2(3-1)]}{3[3(2\cdot 3-4)]}$

**84.** $\dfrac{3[9-2(7-3)]}{(8-5)(9-7)}$

**85.** $\dfrac{3\cdot 7-5(3\cdot 4-11)}{4(3+2)-3^2+5}$

**86.** $\dfrac{2\cdot 5^2-2^2+3}{2(5-2)^2-11}$

*In Exercises 87–110, evaluate each expression given that $x = 3$, $y = 2$, and $z = 4$.*

**87.** $2x - y$

**88.** $2z + y$

**89.** $10 - 2x$

**90.** $15 - 3z$

**91.** $5z \div 2 + y$

**92.** $5x \div 3 + y$

**93.** $4x - 2z$

**94.** $5y - 3x$

**95.** $x + yz$

**96.** $3z + x - 2y$

**97.** $3(2x + y)$

**98.** $4(x + 3y)$

**99.** $(3 + x)y$

**100.** $(4 + z)y$

**101.** $(z + 1)(x + y)$

**102.** $3(z + 1) \div x$

**103.** $(x + y) \div (z + 1)$

**104.** $(2x + 2y) \div (3z - 2)$

**105.** $xyz + z^2 - 4x$

**106.** $zx + y^2 - 2z$

**107.** $3x^2 + 2y^2$

**108.** $3x^2 + (2y)^2$

**109.** $\dfrac{2x + y^2}{y + 2z}$

**110.** $\dfrac{2z^2 - y}{2x - y^2}$

*In Exercises 111–114, insert parentheses in the expression $3 \cdot 8 + 5 \cdot 3$ to make its value equal to the given number.*

**111.** 39

**112.** 117

**113.** 87

**114.** 69

*In Exercises 115–118, find the perimeter of each figure.*

**115.**

4 in.

4 in.    4 in.

4 in.

**116.**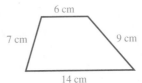

10 cm

3 cm            3 cm

10 cm

**117.**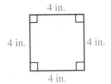

3 m    5 m

7 m

**118.**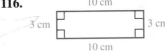

6 cm

7 cm            9 cm

14 cm

*In Exercises 119–122, find the area of each figure.*

**119.**

5 m

5m

**120.**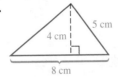

5 cm

4 cm

8 cm

**121.**

6 ft

10 ft

**122.**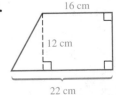

16 cm

12 cm

22 cm

*In Exercises 123–124, find the circumference of each circle. Use $\pi \approx \frac{22}{7}$.*

**123.**

14 m

**124.**

21 cm

*In Exercises 125–126, find the area of each circle. Use $\pi \approx \frac{22}{7}$.*

**125.**

42 ft

**126.**

7 m

*In Exercises 127–132, find the volume of each solid. Use $\pi \approx \frac{22}{7}$.*

**127.**
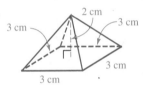
2 cm
3 cm
3 cm
3 cm
3 cm

**128.**
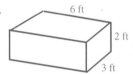
6 ft
2 ft
3 ft

**129.**

6 m

**130.**
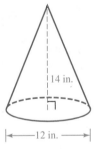
14 in.
12 in.

**131.**
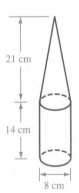
21 cm
14 cm
8 cm

**132.**
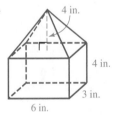
4 in.
4 in.
3 in.
6 in.

*In Exercises 133–138, use a calculator. For $\pi$, use the $\pi$ key. Round to two decimal places.*

**133. Volume of a tank** Find the number of cubic feet of water in a spherical tank with a radius of 21.35 feet.

**134. Storing solvents** A hazardous solvent fills a rectangular tank with dimensions of 12 inches by 9.5 inches by 7.3 inches. For disposal, it must be transferred to a cylindrical canister 7.5 inches in diameter and 18 inches high. How much solvent will be left over?

**135. Volume of a classroom** Thirty students are in a classroom with dimensions of 40 feet by 40 feet by 9 feet. How many cubic feet of air are there for each student?

**136. Wallpapering** One roll of wallpaper covers about 33 square feet. At $27.50 per roll, how much would it cost to paper two walls 8.5 feet high and 17.3 feet long? (*Hint:* Wallpaper can only be purchased in full rolls.)

**137. Focal length** The focal length $f$ of a double-convex thin lens is given by the formula

$$f = \frac{rs}{(r + s)(n - 1)}$$

If $r = 8$, $s = 12$, and $n = 1.6$, find $f$.

**138. Resistance** The total resistance $R$ of two resistors in parallel is given by the formula

$$R = \frac{rs}{r + s}$$

If $r = 170$ and $s = 255$, find $R$.

*WRITING*

**139.** Explain why the symbols $3x$ and $x^3$ have different meanings.

**140.** Students often say that $x^n$ means "$x$ multiplied by itself $n$ times." Explain why this is not correct.

*SOMETHING TO THINK ABOUT*

**141.** If $x$ were greater than 1, would raising $x$ to higher and higher powers produce bigger numbers or smaller numbers?

**142.** What would happen in Exercise 141 if $x$ were a positive number that was less than 1?

# 1.4 Adding and Subtracting Real Numbers

■ ADDING REAL NUMBERS WITH LIKE SIGNS ■ ADDING REAL NUMBERS WITH UNLIKE SIGNS
■ SUBTRACTING REAL NUMBERS ■ USING A CALCULATOR TO ADD AND SUBTRACT REAL NUMBERS

Getting Ready    *Do each operation.*

**1.** $14.32 + 3.2$    **2.** $5.54 - 2.6$    **3.** $4.2 - (3 - 0.8)$

**4.** $(5.42 - 4.22) - 0.2$    **5.** $(437 - 198) - 143$    **6.** $437 - (198 - 143)$

## ■ ADDING REAL NUMBERS WITH LIKE SIGNS

Since the positive direction on the number line is to the right, positive numbers can be represented by arrows pointing to the right. Negative numbers can be represented by arrows pointing to the left.

To add the integers $+2$ and $+3$, we can represent $+2$ with an arrow the length of 2, pointing to the right. We can then represent $+3$ with an arrow of length 3, also pointing to the right. To add the numbers, we place the arrows end to end, as in Figure 1-18. Since the endpoint of the second arrow is the point with coordinate $+5$, we have

$(+2) + (+3) = +5$

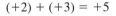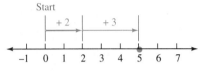

FIGURE 1-18

As a check, we can think of this problem in terms of money. If you had $2 and earned $3 more, you would have a total of $5.

The addition problem

$(-2) + (-3)$

can be represented by the arrows shown in Figure 1-19. Since the endpoint of the final arrow is the point with coordinate $-5$, we have

$(-2) + (-3) = -5$

FIGURE 1-19

As a check, we can think of this problem in terms of money. If you lost $2 and then lost $3 more, you would have lost a total of $5.

Because two real numbers with the same sign can be represented by arrows pointing in the same direction, we have the following rule.

**Adding Real Numbers with Like Signs**
To find the sum of two real numbers with the same sign, add their absolute values and keep their common sign.

EXAMPLE 1     **Adding real numbers**

**a.** $(+4) + (+6) = +(4 + 6)$
$= 10$

**b.** $(-4) + (-6) = -(4 + 6)$
$= -10$

**c.** $+5 + (+10) = +(5 + 10)$
$= 15$

**d.** $-\dfrac{1}{2} + \left(-\dfrac{3}{2}\right) = -\left(\dfrac{1}{2} + \dfrac{3}{2}\right)$
$= -\dfrac{4}{2}$
$= -2$     ■

Self Check    Add   **a.** $(+0.5) + (+1.2)$   and   **b.** $(-3.7) + (-2.3)$.

*Answers*    **a.** 1.7,   **b.** $-6$

**WARNING!**  We do not need to write a $+$ sign in front of a positive number:

$+4 = 4$   and   $+5 = 5$

However, we must always remember to write a $-$ sign in front of a negative number.

## ■ ADDING REAL NUMBERS WITH UNLIKE SIGNS

Real numbers with unlike signs can be represented by arrows on a number line that point in opposite directions. For example, the addition problem

$(-6) + (+2)$

can be represented by the arrows shown in Figure 1-20. Since the endpoint of the final arrow is the point with coordinate $-4$, we have

$(-6) + (+2) = -4$

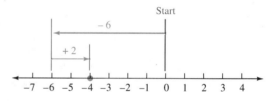

FIGURE 1-20

As a check, we can think of this problem in terms of money. If you lost $6 and then earned $2, you would still have a loss of $4.

The addition problem

$(+7) + (-4)$

can be represented by the arrows shown in Figure 1-21. Since the endpoint of the final arrow is the point with coordinate $+3$, we have

$(+7) + (-4) = +3$

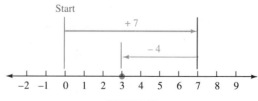

FIGURE 1-21

As a check, you can think of this problem in terms of money. If you had $7 and then lost $4, you would still have a gain of $3.

Because two real numbers with unlike signs can be represented by arrows pointing in opposite directions, we have the following rule.

### Adding Real Numbers with Unlike Signs
To find the sum of two real numbers with unlike signs, subtract their absolute values (the smaller from the larger) and use the sign of the number with the greater absolute value.

**EXAMPLE 2**    **Adding real numbers**

**a.** $(+6) + (-5) = +(6 - 5)$        **b.** $(-2) + (+3) = +(3 - 2)$
$$= 1$$                                    $$= 1$$

**c.** $+6 + (-9) = -(9 - 6)$        **d.** $-\dfrac{2}{3} + \left(+\dfrac{1}{2}\right) = -\left(\dfrac{2}{3} - \dfrac{1}{2}\right)$
$$= -3$$                                    $$= -\left(\dfrac{4}{6} - \dfrac{3}{6}\right)$$
$$= -\dfrac{1}{6}$$            ■

*Self Check*    Add    **a.** $(+3.5) + (-2.6)$    and    **b.** $(-7.2) + (+4.7)$.
*Answers*    **a.** 0.9,    **b.** $-2.5$

**EXAMPLE 3**    **Working with grouping symbols**

**a.** $[(+3) + (-7)] + (-4) = [-4] + (-4)$        Do the work within the brackets first.
$$= -8$$

**b.** $-3 + [(-2) + (-8)] = -3 + [-10]$        Do the work within the brackets first.
$$= -13$$            ■

*Self Check*    Add $-2 + [(+5.2) + (-12.7)]$.
*Answer*    $-9.5$

**EXAMPLE 4**    If $x = -4$, $y = 5$, and $z = -13$, evaluate    **a.** $x + y$    and    **b.** $2y + z$.

*Solution*    We substitute $-4$ for $x$, 5 for $y$, and $-13$ for $z$. Then we simplify.

**a.** $x + y = (-4) + (5)$        **b.** $2y + z = 2 \cdot 5 + (-13)$
$$= 1$$                              $$= 10 + (-13)$$
$$= -3$$            ■

| Self Check | If $x = 5$, $y = -3$, and $z = -4$, evaluate $2y + 3z + x$. |
|---|---|
| Answer | $-13$ |

Sometimes numbers are added vertically, as shown in the next example.

**EXAMPLE 5**    **Adding numbers in a vertical format**

| **a.** $+5$ | **b.** $+5$ | **c.** $-5$ | **d.** $-5$ |
|---|---|---|---|
| $\underline{+2}$ | $\underline{-2}$ | $\underline{+2}$ | $\underline{-2}$ |
| $+7$ | $+3$ | $-3$ | $-7$ |

■

| Self Check | Add    **a.** $+3.2$    and    **b.** $-13.5$ |
|---|---|
| | $\underline{-5.4}$            $\underline{-\ 4.3}$. |
| Answers | **a.** $-2.2$,    **b.** $-17.8$ |

Words and phrases such as *found, gain, credit, up, increase, forward, rises, in the future,* and *to the right* indicate a positive direction. Words and phrases such as *lost, loss, debit, down, backward, falls, in the past,* and *to the left* indicate a negative direction.

**EXAMPLE 6**    **Account balance**    The treasurer of a math club opens a checking account by depositing $350 in the bank. The bank debits the account $9 for check printing, and the treasurer writes a check for $22. Find the balance after these transactions.

Solution    The deposit can be represented by $+350$. The debit of $9 can be represented by $-9$, and the check written for $22 can be represented by $-22$. The balance in the account after these transactions is the sum of 350, $-9$, and $-22$.

$$350 + (-9) + (-22) = 341 + (-22) \qquad \text{Work from left to right.}$$
$$= 319$$

The balance is $319.    ■

| Self Check | Find the balance in Example 6 if another deposit of $17 is made. |
|---|---|
| Answer | $336 |

## ■ SUBTRACTING REAL NUMBERS

In arithmetic, subtraction is a take-away process. For example,

$$7 - 4 = 3$$

can be thought of as taking 4 objects away from 7 objects, leaving 3 objects.
    For algebra, a better approach treats the subtraction problem

$$7 - 4$$

as the equivalent addition problem:

$$7 + (-4)$$

In either case, the answer is 3.

$$7 - 4 = 3 \quad \text{and} \quad 7 + (-4) = 3$$

Thus, to subtract 4 from 7, we can add the negative (or opposite) of 4 to 7. In general, we have the following rule.

---

**Subtracting Real Numbers**
If $a$ and $b$ are two real numbers, then
$$a - b = a + (-b)$$

---

**EXAMPLE 7**    Evaluate    **a.** $12 - 4$,    **b.** $-13 - 5$,    and    **c.** $-14 - (-6)$.

*Solution*    **a.** $12 - 4 = 12 + (-4)$          To subtract 4, add the opposite of 4.
$$= 8$$

**b.** $-13 - 5 = -13 + (-5)$          To subtract 5, add the opposite of 5.
$$= -18$$

**c.** $-14 - (-6) = -14 + [-(-6)]$          To subtract $-6$, add the opposite of $-6$.
$$= -14 + 6$$          The opposite of $-6$ is 6.
$$= -8$$    ∎

---

*Self Check*    Evaluate    **a.** $-12.7 - 8.9$    and    **b.** $15.7 - (-11.3)$.

*Answers*    **a.** $-21.6$,    **b.** 27

---

 b

**EXAMPLE 8**    If $x = -5$ and $y = -3$, evaluate    **a.** $\dfrac{y - x}{7 + x}$    and    **b.** $\dfrac{6 + x}{y - x} - \dfrac{y - 4}{7 + x}$.

*Solution*    We can substitute $-5$ for $x$ and $-3$ for $y$ into each expression and simplify.

**a.** $\dfrac{y - x}{7 + x} = \dfrac{-3 - (-5)}{7 + (-5)}$

$$= \dfrac{-3 + [-(-5)]}{2}$$          To subtract $-5$, add the opposite of $-5$.

$$= \dfrac{-3 + 5}{2}$$          $-(-5) = 5.$

$$= \dfrac{2}{2}$$

$$= 1$$

**b.**
$$\frac{6+x}{y-x} - \frac{y-4}{7+x} = \frac{6+(-5)}{-3-(-5)} - \frac{-3-4}{7+(-5)}$$

$$= \frac{1}{-3+5} - \frac{-3+(-4)}{2} \qquad -(-5) = +5.$$

$$= \frac{1}{2} - \frac{-7}{2}$$

$$= \frac{1-(-7)}{2}$$

$$= \frac{1+[-(-7)]}{2} \qquad \begin{array}{l} \text{To subtract } -7, \text{ add the} \\ \text{opposite of } -7. \end{array}$$

$$= \frac{1+7}{2} \qquad -(-7) = 7.$$

$$= \frac{8}{2}$$

$$= 4 \qquad\qquad\qquad\qquad\qquad \blacksquare$$

**Self Check**    If $a = -3$ and $b = -5$, evaluate $\dfrac{7-a}{b-a+3}$.

**Answer**    10

To use a vertical format for subtracting real numbers, we add the opposite of the number that is to be subtracted by changing the sign of the lower number (called the **subtrahend**) and proceeding as in addition.

**EXAMPLE 9**    Do each subtraction by doing an equivalent addition.

**a.** The subtraction $\;-\;\begin{array}{r} 5 \\ \underline{-4} \end{array}$    becomes the addition $\;+\;\begin{array}{r} 5 \\ \underline{+4} \\ 9 \end{array}$

**b.** The subtraction $\;-\;\begin{array}{r} -8 \\ \underline{+3} \end{array}$    becomes the addition $\;+\;\begin{array}{r} -8 \\ \underline{-3} \\ -11 \end{array}$    $\blacksquare$

**Self Check**    Do the subtraction: $\begin{array}{r} 5.8 \\ \underline{--4.6} \end{array}$.

**Answer**    10.4

EXAMPLE 10    Simplify   **a.** $3 - [4 + (-6)]$   and   **b.** $[-5 + (-3)] - [-2 - (+5)]$.

*Solution*    **a.** $3 - [4 + (-6)] = 3 - (-2)$          Do the addition within the brackets first.

$\qquad\qquad\quad = 3 + [-(-2)]$          To subtract $-2$, add the opposite of $-2$.

$\qquad\qquad\quad = 3 + 2$          $-(-2) = 2$.

$\qquad\qquad\quad = 5$

**b.** $[-5 + (-3)] - [-2 - (+5)]$

$\qquad = [-5 + (-3)] - [-2 + (-5)]$          To subtract $+5$, add the opposite of 5.

$\qquad = -8 - (-7)$          Do the work within the brackets.

$\qquad = -8 + [-(-7)]$          To subtract $-7$, add the opposite of $-7$.

$\qquad = -8 + 7$          $-(-7) = 7$.

$\qquad = -1$          ∎

Self Check    Simplify $[7.2 - (-3)] - [3.2 + (-1.7)]$.

Answer    8.7

EXAMPLE 11    **Temperature change**   At noon, the temperature was 7° above zero. At midnight, the temperature was 4° below zero. Find the difference between these two temperatures.

*Solution*    A temperature of 7° above zero can be represented as $+7$. A temperature of 4° below zero can be represented as $-4$. To find the difference between these temperatures, we can set up a subtraction problem and simplify.

$\qquad\quad 7 - (-4) = 7 + [-(-4)]$          To subtract $-4$, add the opposite of $-4$.

$\qquad\qquad\quad = 7 + 4$          $-(-4) = 4$.

$\qquad\qquad\quad = 11$

The difference between the temperatures is 11°. Figure 1-22 shows this difference.

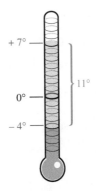

FIGURE 1-22          ∎

Self Check    Find the difference between temperatures of 32° and −10°.

Answer    42°

## ■ USING A CALCULATOR TO ADD AND SUBTRACT REAL NUMBERS

A calculator can add positive and negative numbers.

- You do not have to do anything special to enter positive numbers. When you press 5, for example, a positive 5 is entered.
- To enter −5 into a scientific calculator, you must press the +/− key. This key is called the *plus–minus* or *change-of-sign* key.

To evaluate $-345.678 + (-527.339)$, we enter these numbers and press these keys:

345.678 +/−    +    527.339 +/−    =

The display will read    −873.017 .

Orals    *Find each value.*

**1.** $2 + 3$            **2.** $2 + (-5)$            **3.** $-4 + 7$

**4.** $-5 + (-6)$        **5.** $6 - 2$              **6.** $-8 - 4$

**7.** $-5 - (-7)$                        **8.** $12 - (-4)$

**9.** $-5 + (3 - 4)$                     **10.** $(-5 + 3) - 4$

## EXERCISE 1.4

***REVIEW*** *If $x = 5$, $y = 7$, and $z = 2$, evaluate each expression.*

**1.** $x + 3(y - z)$      **2.** $(x + 3)(y - z)$      **3.** $x + 3y - z$      **4.** $(x + 3)y - z$

***VOCABULARY AND CONCEPTS*** *Fill in each blank to make a true statement.*

**5.** Positive and negative numbers can be represented by _____ on the number line.

**6.** To find the sum of two real numbers with like signs, _____ their absolute values and _____ their common sign.

**7.** To find the sum of two real numbers with unlike signs, _____ their absolute values and use the sign of the number with the _____ absolute value.

**8.** $a - b =$ _____

**9.** To subtract a number, we _____ its _____.

**10.** The subtraction

$$\begin{array}{r} 35 \\ -45 \\ \hline \end{array}$$

is equivalent to the subtraction 35 _____.

*In Exercises 11–26, find each sum.*

**11.** $4 + 8$

**12.** $(-4) + (-2)$

**13.** $(-3) + (-7)$

**14.** $(+4) + 11$

**15.** $6 + (-4)$

**16.** $5 + (-3)$

**17.** $9 + (-11)$

**18.** $10 + (-13)$

**19.** $(-0.4) + 0.9$

**20.** $(-1.2) + (-5.3)$

**21.** $\dfrac{1}{5} + \left(+\dfrac{1}{7}\right)$

**22.** $\dfrac{2}{3} + \left(-\dfrac{1}{4}\right)$

**23.** $\begin{array}{r} 5 \\ +\underline{-4} \end{array}$

**24.** $\begin{array}{r} -20 \\ +\underline{-17} \end{array}$

**25.** $\begin{array}{r} -1.3 \\ +\underline{\ 3.5} \end{array}$

**26.** $\begin{array}{r} 1.3 \\ +\underline{-2.5} \end{array}$

*In Exercises 27–38, evaluate each expression.*

**27.** $5 + [4 + (-2)]$

**28.** $-6 + [(-3) + 8]$

**29.** $-2 + (-4 + 5)$

**30.** $5 + [-4 + (-6)]$

**31.** $[-4 + (-3)] + [2 + (-2)]$

**32.** $[3 + (-1)] + [-2 + (-3)]$

**33.** $-4 + (-3 + 2) + (-3)$

**34.** $5 + [2 + (-5)] + (-2)$

**35.** $-|8 + (-4)| + 7$

**36.** $\left| \dfrac{3}{5} + \left(-\dfrac{4}{5}\right) \right|$

**37.** $-5.2 + |-2.5 + (-4)|$

**38.** $6.8 + |8.6 + (-1.1)|$

*In Exercises 39–52, let $x = 2$, $y = -3$, $z = -4$, and $u = 5$. Evaluate each expression.*

**39.** $x + y$

**40.** $x + z$

**41.** $x + z + u$

**42.** $y + z + u$

**43.** $(x + u) + 3$

**44.** $(y + 5) + x$

**45.** $x + (-1 + z)$

**46.** $-7 + (z + x)$

**47.** $(x + z) + (u + z)$

**48.** $(z + u) + (x + y)$

**49.** $x + [5 + (y + u)]$

**50.** $y + \{[u + (z + (-6)]\} + y$

**51.** $|2x + y|$

**52.** $3|x + y + z|$

*In Exercises 53–68, find each difference.*

**53.** $8 - 4$

**54.** $-8 - 4$

**55.** $8 - (-4)$

**56.** $-9 - (-5)$

**57.** $0 - (-5)$

**58.** $0 - 75$

**59.** $\dfrac{5}{3} - \dfrac{7}{6}$

**60.** $-\dfrac{5}{9} - \dfrac{5}{3}$

**61.** $-3\dfrac{1}{2} - 5\dfrac{1}{4}$

**62.** $2\dfrac{1}{2} - \left(-3\dfrac{1}{2}\right)$

**63.** $-6.7 - (-2.5)$

**64.** $25.3 - 17.5$

**65.** $\begin{array}{r} 8 \\ -\underline{4} \end{array}$

**66.** $\begin{array}{r} 8 \\ -\underline{-3} \end{array}$

**67.** $\begin{array}{r} -10 \\ -\underline{-\ 3} \end{array}$

**68.** $\begin{array}{r} -13 \\ -\underline{\ \ 5} \end{array}$

*In Exercises 69–78, evaluate each quantity.*

**69.** $+3 - [(-4) - 3]$

**70.** $-5 - [4 - (-2)]$

**71.** $(5 - 3) + (3 - 5)$

**72.** $(3 - 5) - [5 - (-3)]$

**73.** $5 - [4 + (-2) - 5]$

**74.** $3 - [-(-2) + 5]$

**75.** $\left(\dfrac{5}{2} - 3\right) - \left(\dfrac{3}{2} - 5\right)$

**76.** $\left(\dfrac{7}{3} - \dfrac{5}{6}\right) - \left[\dfrac{5}{6} - \left(-\dfrac{7}{3}\right)\right]$

**77.** $(5.2 - 2.5) - (5.25 - 5)$

**78.** $\left(3\frac{1}{2} - 2\frac{1}{2}\right) - \left[5\frac{1}{3} - \left(-5\frac{2}{3}\right)\right]$

*In Exercises 79–86, let $x = -4$, $y = 5$, and $z = -6$. Evaluate each quantity.*

**79.** $y - x$

**80.** $y - z$

**81.** $x - y - z$

**82.** $y + z - x$

**83.** $x - (y - z)$

**84.** $y + (z - x)$

**85.** $\dfrac{y - x}{3 - z}$

**86.** $\dfrac{y}{x - z} - \dfrac{x}{8 + z}$

*In Exercises 87–90, let $a = 2$, $b = -3$, and $c = -4$. Evaluate each quantity.*

**87.** $a + b - c$

**88.** $a - b + c$

**89.** $\dfrac{a + b}{b - c}$

**90.** $\dfrac{c - a}{-(a + b)}$

*In Exercises 91–94, use a calculator to evaluate each quantity. Let $x = 2.34$, $y = 3.47$, and $z = 0.72$. Round the answers to one decimal place.*

**91.** $x^3 - y + z^2$

**92.** $y - z^2 - x^2$

**93.** $x^2 - y^2 - z^2$

**94.** $z^3 - x^2 + y^3$

**APPLICATIONS**   *Use signed numbers to solve each problem.*

**95. College tuition**   A student owed $575 in tuition. If she earned a scholarship that would pay $400 of the bill, what did she still owe?

**96. Dieting**   Scott weighed 212 pounds but lost 24 pounds during a diet. What does Scott weigh now?

**97. Temperature**   The temperature rose 13 degrees in 1 hour and then dropped 4 degrees in the next hour. What signed number represents the net change in temperature?

**98. Mountain climbing**   A team of mountaineers climbed 2,347 feet one day but then came down 597 feet to a good spot to make camp. What signed number represents their net change in altitude?

**99. Temperature**   The temperature fell from zero to 14° below one night. By 5:00 P.M. the next day, the temperature had risen 10 degrees. What was the temperature at 5:00 P.M.?

**100. History**   In 1897, Joseph Thompson discovered the electron. Fifty-four years later, the first fission reactor was built. Nineteen years before the reactor was erected, James Chadwick discovered the neutron. In what year was the neutron discovered?

**101. History**   The Greek mathematician Euclid was alive in 300 B.C. The English mathematician Sir Isaac Newton was alive in A.D. 1700. How many years apart did they live?

**102. Banking**   Abdul deposited $212 in a new checking account, wrote a check for $173, and deposited another $312. Find the balance in his account.

**103. Military science**   An army retreated 2,300 meters. After regrouping, it moved forward 1,750 meters. The next day it gained another 1,875 meters. What was the army's net gain?

**104. Football**   A football player gained and lost the following yardage on six consecutive plays: +5, +7, −5, +1, −2, and −6. How many yards were gained or lost?

**105. Aviation**   A pilot flying at 32,000 feet is instructed to descend to 28,000 feet. How many feet must he descend?

**106. Stock market**   Tuesday's high and low prices for Transitronics stock were $37\frac{1}{8}$ and $31\frac{5}{8}$. Find the range of prices for this stock.

**107. Temperature**   Find the difference between a temperature of 32° above zero and a temperature of 27° above zero.

**108. Temperature** Find the difference between a temperature of 3° below zero and a temperature of 21° below zero.

**109. Stock market** At the opening bell on Monday, the Dow Jones Industrial Average was 9,153. At the close, the Dow was down 23 points, but news of a half-point drop in interest rates on Tuesday sent the market up 57 points. What was the Dow average after the market closed on Tuesday?

**110. Stock market** On a Monday morning, the Dow Jones Industrial Average opened at 8,917. For the week, the Dow rose 29 points on Monday and 12 points on Wednesday. However, it fell 53 points on Tuesday and 27 points on both Thursday and Friday. Where did the Dow close on Friday?

**111. Stock splits** A man owned 500 shares of Transitronics Corporation before the company declared a two-for-one stock split. After the split, he sold 300 shares. How many shares does the man now own?

**112. Small business** Maria earned $2,532 in a part-time business. However, $633 of the earnings went for taxes. Find Maria's net earnings.

*In Exercises 113–116, use a calculator.*

**113. Balancing the books** On January 1, Sally had $437.45 in the bank. During the month, she had deposits of $25.17, $37.93, and $45.26, and she had withdrawals of $17.13, $83.44, and $22.58. How much was in her account at the end of the month?

**114. Small business** The owner of a small business has a gross income of $97,345.32. However, he paid $37,675.66 in expenses plus $7,537.45 in taxes, $3,723.41 in health care premiums, and $5,767.99 in pension payments. Find his profit.

**115. Closing a real estate transaction** A woman sold her house for $115,000. Her fees at closing were $78 for preparing a deed, $446 for title work, $216 for revenue stamps, and a sales commission of $7,612.32. In addition, there was a deduction of $23,445.11 to pay off her old mortgage. As part of the deal, the buyer agreed to pay half of the title work. How much money did the woman receive after closing?

**116. Winning the lottery** Mike won $500,000 in a state lottery. He will get $\frac{1}{20}$ of the sum each year for the next 20 years. After he receives his first installment, he plans to pay off a car loan of $7,645.12 and give his son $10,000 for college. By paying off the car loan, he will receive a rebate of 2% of the loan. If he must pay income tax of 28% on his first installment, how much will he have left to spend?

## WRITING

**117.** Explain why the sum of two negative numbers is always negative, and the sum of two positive numbers is always positive.

**118.** Explain why the sum of a negative number and a positive number could be either negative or positive.

## SOMETHING TO THINK ABOUT

**119.** Think of two numbers. First, add the absolute values of the two numbers, and write your answer. Second, add the two numbers, take the absolute value of that sum, and write that answer. Do the two answers agree? Can you find two numbers that produce different answers? When do you get answers that agree, and when don't you?

**120.** "Think of a very small number," requests the teacher. "One one-millionth," answers Charles. "Negative one million," responds Mia. Explain why either answer might be considered correct.

# 1.5 Multiplying and Dividing Real Numbers

■ MULTIPLYING REAL NUMBERS ■ DIVIDING REAL NUMBERS ■ USING A CALCULATOR TO MULTIPLY AND DIVIDE REAL NUMBERS

Getting Ready *Find each product or quotient.*

**1.** $8 \times 7$  **2.** $9 \times 6$  **3.** $8 \times 9$  **4.** $7 \times 9$

**5.** $\dfrac{81}{9}$  **6.** $\dfrac{48}{8}$  **7.** $\dfrac{64}{8}$  **8.** $\dfrac{56}{7}$

## ■ MULTIPLYING REAL NUMBERS

Because the times sign, $\times$, looks like the letter $x$, it is seldom used in algebra. Instead, a dot, parentheses, or no symbol at all is used to denote multiplication. Each of the following expressions indicates the **product** obtained when two real numbers $x$ and $y$ are multiplied.

$$x \cdot y \qquad (x)(y) \qquad x(y) \qquad (x)y \qquad xy$$

To develop rules for multiplying real numbers, we rely on the definition of multiplication. The expression $5 \cdot 4$ indicates that 4 is to be used as a term in a sum five times. That is,

$$5(4) = 4 + 4 + 4 + 4 + 4 = 20$$

Read 5(4) as "5 times 4."

Likewise, the expression $5(-4)$ indicates that $-4$ is to be used as a term in a sum five times. Thus,

$$5(-4) = (-4) + (-4) + (-4) + (-4) + (-4) = -20$$

Read 5(−4) as "5 times negative 4."

If multiplying by a positive number indicates repeated addition, it is reasonable that multiplication by a negative number indicates repeated subtraction. The expression $(-5)4$, for example, means that 4 is to be used as a term in a repeated subtraction five times. That is,

$$
\begin{aligned}
(-5)4 &= -(4) - (4) - (4) - (4) - (4) \\
&= (-4) + (-4) + (-4) + (-4) + (-4) \\
&= -20
\end{aligned}
$$

Likewise, the expression $(-5)(-4)$ indicates that $-4$ is to be used as a term in a repeated subtraction five times. Thus,

$$
\begin{aligned}
(-5)(-4) &= -(-4) - (-4) - (-4) - (-4) - (-4) \\
&= -(-4) + [-(-4)] + [-(-4)] + [-(-4)] + [-(-4)] \\
&= 4 + 4 + 4 + 4 + 4 \\
&= 20
\end{aligned}
$$

The expression $0(-2)$ indicates that $-2$ is to be used zero times as a term in a repeated addition. Thus,

$$0(-2) = 0$$

Finally, the expression $(-3)(1) = -3$ suggests that the product of any number and 1 is the number itself.

The previous results suggest the following rules.

> **Rules for Multiplying Signed Numbers**
> 1. The product of two real numbers with like signs is the product of their absolute values.
> 2. The product of two real numbers with unlike signs is the negative of the product of their absolute values.
> 3. Any number multiplied by 0 is 0: $a \cdot 0 = 0 \cdot a = 0$.
> 4. Any number multiplied by 1 is that number itself: $a \cdot 1 = 1 \cdot a = a$.

**EXAMPLE 1**    Find each product:    **a.** $4(-7)$,    **b.** $(-5)(-4)$,    **c.** $(-7)(6)$,    **d.** $8(6)$, **e.** $(-3)(5)(-4)$,    and    **f.** $(-4)(-2)(-3)$.

*Solution*    **a.** $4(-7) = -(4 \cdot 7)$                    **b.** $(-5)(-4) = +(5 \cdot 4)$
$= -28$                                        $= +20$

**c.** $(-7)(6) = -(7 \cdot 6)$                **d.** $8(6) = +(8 \cdot 6)$
$= -42$                                        $= +48$

**e.** $(-3)(5)(-4) = (-15)(-4)$        **f.** $(-4)(-2)(-3) = 8(-3)$
$= 60$                                        $= -24$    ∎

**Self Check**    Find each product:    **a.** $-7(5)$,    **b.** $-12(-7)$,    and    **c.** $-2(-4)(-9)$.
*Answers*    **a.** $-35$,    **b.** $84$,    **c.** $-72$

**EXAMPLE 2**    If $x = -3$, $y = 2$, and $z = 4$, evaluate    **a.** $y + xz$    and    **b.** $x(y - z)$.

*Solution*    We substitute $-3$ for $x$, 2 for $y$, and 4 for $z$ in each expression and simplify.

**a.** $y + xz = 2 + (-3)(4)$        **b.** $x(y - z) = -3[2 - 4]$
$= 2 + (-12)$                            $= -3[2 + (-4)]$
$= -10$                                    $= -3(-2)$
$= 6$    ∎

**Self Check**    If $x = -4$, $y = -3$, and $z = 5$, evaluate $x - yz$.
*Answer*    $11$

**EXAMPLE 3**   If $x = -2$ and $y = 3$, evaluate   **a.** $x^2 - y^2$   and   **b.** $-x^2$.

*Solution*   **a.** We substitute $-2$ for $x$ and $3$ for $y$ and simplify.

$$x^2 - y^2 = (-2)^2 - 3^2$$
$$= 4 - 9 \qquad \text{Simplify the exponential expressions first.}$$
$$= -5 \qquad \text{Do the subtraction.}$$

**b.** We substitute $-2$ for $x$ and simplify.

$$-x^2 = -(-2)^2$$
$$= -4 \qquad (-2)^2 = 4.$$

*Self Check*   If $a = -3.2$ and $b = -5$, evaluate $a^2 - 2b^3$.
*Answer*   260.24

**EXAMPLE 4**   Find each product:   **a.** $\left(-\dfrac{2}{3}\right)\left(-\dfrac{6}{5}\right)$   and   **b.** $\left(\dfrac{3}{10}\right)\left(-\dfrac{5}{9}\right)$.

*Solution*   **a.** $\left(-\dfrac{2}{3}\right)\left(-\dfrac{6}{5}\right) = +\left(\dfrac{2}{3} \cdot \dfrac{6}{5}\right)$   **b.** $\left(\dfrac{3}{10}\right)\left(-\dfrac{5}{9}\right) = -\dfrac{3}{10} \cdot \dfrac{5}{9}$

$$= +\dfrac{2 \cdot 6}{3 \cdot 5} \qquad\qquad = -\dfrac{3 \cdot 5}{10 \cdot 9}$$

$$= +\dfrac{12}{15} \qquad\qquad = -\dfrac{15}{90}$$

$$= +\dfrac{4}{5} \qquad\qquad = -\dfrac{1}{6}$$

*Self Check*   Evaluate   **a.** $\frac{3}{5}\left(-\frac{10}{9}\right)$   and   **b.** $-\left(\frac{15}{8}\right)\left(-\frac{16}{5}\right)$.
*Answers*   **a.** $-\frac{2}{3}$,   **b.** 6

**EXAMPLE 5**   **Temperature change**   If the temperature is dropping 4° each hour, how much warmer was it 3 hours ago?

*Solution*   A temperature drop of 4° per hour can be represented by $-4$° per hour. "Three hours ago" can be represented by $-3$. The temperature 3 hours ago is the product of $(-3)$ and $(-4)$.

$$(-3)(-4) = +12$$

The temperature was 12° warmer 3 hours ago.

*Self Check*   How much colder will it be after 5 hours?
*Answer*   20° colder

### ■ DIVIDING REAL NUMBERS

We know that 8 divided by 4 is 2 and 18 divided by 6 is 3.

$$\frac{8}{4} = 2, \text{ because } 2 \cdot 4 = 8 \qquad \frac{18}{6} = 3, \text{ because } 3 \cdot 6 = 18$$

These examples suggest that the following rule

$$\frac{a}{b} = c \quad \text{if and only if} \quad c \cdot b = a$$

is true for the division of any real number $a$ by any nonzero real number $b$. For example,

$$\frac{+10}{+2} = +5, \text{ because } (+5)(+2) = +10$$

$$\frac{-10}{-2} = +5, \text{ because } (+5)(-2) = -10$$

$$\frac{+10}{-2} = -5, \text{ because } (-5)(-2) = +10$$

$$\frac{-10}{+2} = -5, \text{ because } (-5)(+2) = -10$$

These examples suggest the rules for dividing real numbers.

---

**Rules for Dividing Signed Numbers**

**1.** The quotient of two real numbers with like signs is the quotient of their absolute values.

**2.** The quotient of two real numbers with unlike signs is the negative of the quotient of their absolute values.

**3.** $\dfrac{a}{0}$ is undefined; $\dfrac{0}{0}$ is indeterminate.

**4.** If $a \neq 0$, then $\dfrac{0}{a} = 0$.

---

**EXAMPLE 6**   Find each quotient:   **a.** $\dfrac{36}{18}$,   **b.** $\dfrac{-44}{11}$,   **c.** $\dfrac{27}{-9}$,   and   **d.** $\dfrac{-64}{-8}$.

*Solution*   **a.** $\dfrac{36}{18} = +\dfrac{36}{18} = 2$

The quotient of two numbers with like signs is the quotient of their absolute values.

**b.** $\dfrac{-44}{11} = -\dfrac{44}{11} = -4$

The quotient of two numbers with unlike signs is the negative of the quotient of their absolute values.

**c.** $\dfrac{27}{-9} = -\dfrac{27}{9} = -3$    The quotient of two numbers with unlike signs is the negative of the quotient of their absolute values.

**d.** $\dfrac{-64}{-8} = +\dfrac{64}{8} = 8$    The quotient of two numbers with like signs is the quotient of their absolute values.    ∎

*Self Check*    Find each quotient:    **a.** $\dfrac{-72.6}{12.1}$    and    **b.** $\dfrac{-24.51}{-4.3}$.

*Answers*    **a.** $-6$,    **b.** $5.7$

**EXAMPLE 7**    If $x = -64$, $y = 16$, and $z = -4$, evaluate

**a.** $\dfrac{yz}{-x}$    and    **b.** $\dfrac{z^3 y}{x}$.

*Solution*    We substitute $-64$ for $x$, $16$ for $y$, and $-4$ for $z$ in each expression and simplify.

**a.** $\dfrac{yz}{-x} = \dfrac{16(-4)}{-(-64)}$        **b.** $\dfrac{z^3 y}{x} = \dfrac{(-4)^3(16)}{-64}$

$\quad\quad = \dfrac{-64}{+64}$            $\quad\quad = \dfrac{(-64)(16)}{(-64)}$

$\quad\quad = -1$                $\quad\quad = 16$    ∎

*Self Check*    Evaluate $\dfrac{x+y}{-z^2}$, given the values in Example 7.

*Answer*    3

**EXAMPLE 8**    If $x = -50$, $y = 10$, and $z = -5$, evaluate

**a.** $\dfrac{xyz}{x - 5z}$    and    **b.** $\dfrac{3xy + 2yz}{2(x + y)}$.

*Solution*    We substitute $-50$ for $x$, $10$ for $y$, and $-5$ for $z$ in each expression and simplify.

**a.** $\dfrac{xyz}{x - 5z} = \dfrac{(-50)(10)(-5)}{-50 - 5(-5)}$        **b.** $\dfrac{3xy + 2yz}{2(x + y)} = \dfrac{3(-50)(10) + 2(10)(-5)}{2(-50 + 10)}$

$\quad\quad = \dfrac{(-500)(-5)}{-50 + 25}$            $\quad\quad = \dfrac{-150(10) + (20)(-5)}{2(-40)}$

$\quad\quad = \dfrac{2,500}{-25}$                $\quad\quad = \dfrac{-1,500 - 100}{-80}$

$\quad\quad = -100$                $\quad\quad = \dfrac{-1,600}{-80}$

$\quad\quad\quad\quad\quad\quad\quad\quad\quad\quad\quad\quad = 20$    ∎

Self Check   Evaluate $\dfrac{2xy - 3z - 5}{3(y - z)}$, given the values in Example 8.

Answer   $-22$

EXAMPLE 9   **Stock reports**   In its annual report, a corporation reports its performance on a per-share basis. When a company with 35 million shares outstanding loses $2.3 million, what will be the per-share loss?

Solution   A loss of $2.3 million can be represented by $-2,300,000$. Because there are 35 million shares, the per-share loss can be represented by the quotient $\frac{-2,300,000}{35,000,000}$.

$$\frac{-2,300,000}{35,000,000} \approx -0.065714285 \qquad \text{Use a calculator.}$$

The company lost about 6.6¢ per share. ∎

Self Check   If the company in Example 9 earns $1.5 million in the following year, find its per-share gain for that year.

Answer   about 4.3¢

Remember these facts about dividing real numbers.

**Division**

1. $\dfrac{a}{0}$ is undefined.   2. If $a \neq 0$, then $\dfrac{0}{a} = 0$.

3. $\dfrac{a}{1} = a$.   4. If $a \neq 0$, then $\dfrac{a}{a} = 1$.

### ■ USING A CALCULATOR TO MULTIPLY AND DIVIDE REAL NUMBERS

A calculator can be used to multiply and divide positive and negative numbers. To evaluate $(-345.678)(-527.339)$, we enter these numbers and press these keys:

345.678 +/− × 527.339 +/− =

The display will read ⌑⌑⌑⌑⌑⌑⌑ 182289.4908 .
To evaluate $\frac{-345.678}{-527.339}$, we enter these numbers and press these keys:

345.678 +/− ÷ 527.339 +/− =

The display will read 0.655513815 .

Orals  *Find each product or quotient.*

**1.** $1(-3)$ **2.** $-2(-5)$ **3.** $-3(-6)$ **4.** $4(-6)$

**5.** $-2(3)(-4)$ **6.** $-2(-3)(-4)$ **7.** $\dfrac{-12}{6}$ **8.** $\dfrac{-10}{-5}$

**9.** $\dfrac{3(6)}{-2}$ **10.** $\dfrac{(-2)(-3)}{-6}$

## EXERCISE 1.5

### REVIEW

**1.** A concrete block weighs $37\frac{1}{2}$ pounds. How much will 30 of these blocks weigh?

**2.** If one brick weighs 1.3 pounds, how much will a skid of 500 bricks weigh?

**3.** If $x = 5$, $y = 8$, and $z = 3$, evaluate $x^3 - yz^2$.

**4.** Put $<$, $=$, or $>$ in the box to make a true statement: $-2(-3 + 4)$ ▢ $-3[3 - (-4)]$

### VOCABULARY AND CONCEPTS  *Fill in each blank to make a true statement.*

**5.** The product of two positive numbers is _____.

**6.** The product of a _____ number and a negative number is negative.

**7.** The product of two negative numbers is _____.

**8.** The quotient of a _____ number and a positive number is negative.

**9.** The quotient of two negative numbers is _____.

**10.** Any number multiplied by ___ is 0.

**11.** $a \cdot 1 =$ ___

**12.** The symbol $\dfrac{a}{0}$ is _____.

**13.** If $a \neq 0$, $\dfrac{0}{a} =$ ___.

**14.** If $a \neq 0$, $\dfrac{a}{a} =$ ___.

### PRACTICE  *In Exercises 15–34, find each product.*

**15.** $(+6)(+8)$ **16.** $(-9)(-7)$ **17.** $(-8)(-7)$ **18.** $(9)(-6)$

**19.** $(+12)(-12)$ **20.** $(-9)(12)$ **21.** $\left(\dfrac{1}{2}\right)(-32)$ **22.** $\left(-\dfrac{3}{4}\right)(12)$

**23.** $\left(-\dfrac{3}{4}\right)\left(-\dfrac{8}{3}\right)$ **24.** $\left(-\dfrac{2}{5}\right)\left(\dfrac{15}{2}\right)$ **25.** $(-3)\left(-\dfrac{1}{3}\right)$ **26.** $(5)\left(-\dfrac{2}{5}\right)$

**27.** $(3)(-4)(-6)$ **28.** $(-1)(-3)(-6)$ **29.** $(-2)(3)(4)$ **30.** $(5)(0)(-3)$

**31.** $(2)(-5)(-6)(-7)$ **32.** $(-3)(-5)(-5)(-2)$

**33.** $(-2)(-2)(-2)(-3)(-4)$ **34.** $(-5)(4)(3)(-2)(-1)$

*In Exercises 35–54, let $x = -1$, $y = 2$, and $z = -3$. Evaluate each expression.*

**35.** $y^2$          **36.** $x^2$          **37.** $-z^2$          **38.** $-xz$

**39.** $xy$          **40.** $yz$          **41.** $y + xz$          **42.** $z - xy$

**43.** $(x + y)z$          **44.** $y(x - z)$          **45.** $(x - z)(x + z)$          **46.** $(y + z)(x - z)$

**47.** $xy + yz$          **48.** $zx - zy$          **49.** $xyz$          **50.** $x^2 y$

**51.** $x^2(y - z)$          **52.** $y^2(x - z)$          **53.** $(-x)(-y) + z^2$          **54.** $(-x)(-z) - y^2$

*In Exercises 55–66, simplify each expression.*

**55.** $\dfrac{80}{-20}$          **56.** $\dfrac{-66}{33}$          **57.** $\dfrac{-110}{-55}$          **58.** $\dfrac{200}{40}$

**59.** $\dfrac{-160}{40}$          **60.** $\dfrac{-250}{-25}$          **61.** $\dfrac{320}{-16}$          **62.** $\dfrac{180}{-36}$

**63.** $\dfrac{8 - 12}{-2}$          **64.** $\dfrac{16 - 2}{2 - 9}$          **65.** $\dfrac{20 - 25}{7 - 12}$          **66.** $\dfrac{2(15)^2 - 2}{-2^3 + 1}$

*In Exercises 67–74, evaluate each expression if $x = -2$, $y = 3$, $z = 4$, $t = 5$, and $w = -18$.*

**67.** $\dfrac{yz}{x}$          **68.** $\dfrac{zt}{x}$          **69.** $\dfrac{tw}{y}$          **70.** $\dfrac{w}{xy}$

**71.** $\dfrac{z + w}{x}$          **72.** $\dfrac{xyz}{y - 1}$          **73.** $\dfrac{xtz}{y + 1}$          **74.** $\dfrac{x + y + z}{t}$

*In Exercises 75–82, evaluate each expression if $x = 4$, $y = -6$, and $z = -3$. Use a calculator.*

**75.** $\dfrac{2x^2 + 2y}{x + y}$          **76.** $\dfrac{y^2 + z^2}{y + z}$          **77.** $\dfrac{2x^2 - 2z^2}{x + z}$          **78.** $\dfrac{8x^3 - 8y^2}{x - z}$

**79.** $\dfrac{y^3 + 4z^3}{(x + y)^2}$          **80.** $\dfrac{x^2 - 2xz + z^2}{x - y + z}$          **81.** $\dfrac{xy^2z + x^2y}{2y - 2z}$          **82.** $\dfrac{(x^2 - 2y)z^2}{-xz}$

*In Exercises 83–90, evaluate each expression if $x = \frac{1}{2}$, $y = -\frac{2}{3}$, and $z = -\frac{3}{4}$.*

**83.** $x + y$          **84.** $y + z$          **85.** $x + y + z$          **86.** $y + x - z$

**87.** $(x + y)(x - y)$          **88.** $(x - z)(x + z)$          **89.** $(x + y + z)(xyz)$          **90.** $xyz(x - y - z)$

**APPLICATIONS**   *Use signed numbers to solve each problem.*

**91. Temperature change** If the temperature is increasing 2 degrees each hour for 3 hours, what product of signed numbers represents the temperature change?

**92. Temperature change** If the temperature is decreasing 2 degrees each hour for 3 hours, what product of signed numbers represents the temperature change?

**93. Gambling** In Las Vegas, Robert lost $30 per hour playing the slot machines for 15 hours. What product of signed numbers represents the change in his financial condition?

**94. Draining a pool** A pool is emptying at the rate of 12 gallons per minute. What product of signed numbers would represent how much more water was in the pool 2 hours ago?

**95. Filling a pool** Water from a pipe is filling a pool at the rate of 23 gallons per minute. What product of signed numbers represents the amount of water in the pool 2 hours ago?

**96. Mowing lawns** Rafael worked all day mowing lawns and was paid $8 per hour. If he had $94 at the end of an 8-hour day, how much did he have before he started working?

**97. Temperature** Suppose that the temperature is dropping at the rate of 3 degrees each hour. If the temperature has dropped 18 degrees, what signed number expresses how many hours the temperature has been falling?

**98. Dieting** A man lost 37.5 pounds. If he lost 2.5 pounds each week, how long has he been dieting?

*Use a calculator and signed numbers to solve each problem.*

**99. Stock market** Over a 7-day period, the Dow Jones Industrial Average had gains of 26, 35, and 17 points. In that period, there were also losses of 25, 31, 12, and 24 points. Find the average daily performance over the 7-day period.

**100. Astronomy** Light travels at the rate of 186,000 miles per second. How long will it take light to travel from the sun to Venus? (*Hint:* The distance from the sun to Venus is 67,000,000 miles.)

**101. Saving for school** A student has saved $15,000 to attend graduate school. If she estimates that her expenses will be $613.50 a month while in school, does she have enough to complete an 18-month master's degree program?

**102. Earnings per share** Over a five-year period, a corporation reported profits of $18 million, $21 million, and $33 million. It also reported losses of $5 million and $71 million. Find the average gain (or loss) each year.

## WRITING

**103.** Explain how you would decide whether the product of several numbers is positive or negative.

**104.** Describe two situations in which negative numbers are useful.

## SOMETHING TO THINK ABOUT

**105.** If the quotient of two numbers is undefined, what would their product be?

**106.** If the product of five numbers is negative, how many of the factors could be negative?

**107.** If $x^5$ is a negative number, can you decide whether $x$ is negative too?

**108.** If $x^6$ is a positive number, can you decide whether $x$ is positive too?

# 1.6 Algebraic Expressions

■ ALGEBRAIC EXPRESSIONS ■ EVALUATING ALGEBRAIC EXPRESSIONS ■ ALGEBRAIC TERMS

Getting Ready *Identify each of the following as a sum, difference, product, or quotient.*

**1.** $x + 3$                     **2.** $57x$

**3.** $\dfrac{x}{9}$                                 **4.** $19 - y$

**5.** $\dfrac{x - 7}{3}$                           **6.** $x - \dfrac{7}{3}$

**7.** $5(x + 2)$                              **8.** $5x + 10$

## ■ ALGEBRAIC EXPRESSIONS

Variables and numbers can be combined with the operations of arithmetic to produce **algebraic expressions.** For example, if $x$ and $y$ are variables, the algebraic expression $x + y$ represents the **sum** of $x$ and $y$, and the algebraic expression $x - y$ represents their **difference.**

    There are many other ways to express addition or subtraction with algebraic expressions, as shown in Tables 1-3 and 1-4.

| The phrase | translates into the algebraic expression |
|:---:|:---:|
| the *sum* of $t$ and 12 | $t + 12$ |
| 5 *plus* $s$ | $5 + s$ |
| 7 *added to* $a$ | $a + 7$ |
| 10 *more than* $q$ | $q + 10$ |
| 12 *greater than* $m$ | $m + 12$ |
| $l$ *increased by* $m$ | $l + m$ |
| *exceeds* $p$ *by* 50 | $p + 50$ |

TABLE 1-3

| The phrase | translates into the algebraic expression |
|:---:|:---:|
| the *difference* of 50 and $r$ | $50 - r$ |
| 1,000 *minus* $q$ | $1,000 - q$ |
| 15 *less than* $w$ | $w - 15$ |
| $t$ *decreased by* $q$ | $t - q$ |
| 12 *reduced by* $m$ | $12 - m$ |
| $l$ *subtracted from* 250 | $250 - l$ |
| 2,000 less $p$ | $2,000 - p$ |

TABLE 1-4

**EXAMPLE 1**      Let $x$ represent a certain number. Write an expression that represents   **a.** the number that is 5 more than $x$   and   **b.** the number 12 decreased by $x$.

*Solution*      **a.** The number "5 more than $x$" is the number found by adding 5 to $x$. It is represented by $x + 5$.

**b.** The number "12 decreased by $x$" is the number found by subtracting $x$ from 12. It is represented by $12 - x$. ∎

Self Check

Let $y$ represent a certain number. Write an expression that represents $y$ increased by 25.

Answer

$y + 25$

EXAMPLE 2

**Income taxes**   Bob worked $x$ hours preparing his income tax return. He worked 3 hours less than that on his son's return. Write an expression that represents   **a.** the number of hours he spent preparing his son's return   and   **b.** the total number of hours he worked.

Solution

**a.** Because he worked $x$ hours on his own return and 3 hours less on his son's return, he worked $(x - 3)$ hours on his son's return.

**b.** Because he worked $x$ hours on his own return and $(x - 3)$ hours on his son's return, the total time he spent on taxes was $[x + (x - 3)]$ hours. ∎

Self Check

Javier deposited $\$d$ in a bank account. Later, he withdrew $\$500$. Write an expression that represents the difference of $d$ and 500.

Answer

$d - 500$

There are several ways to indicate the **product** of two numbers with algebraic expressions, as shown in Table 1-5.

| The phrase | translates into the algebraic expression |
|:---:|:---:|
| the *product* of 100 and $a$ | $100a$ |
| 25 *times* $B$ | $25B$ |
| *twice* $x$ | $2x$ |
| $\dfrac{1}{2}$ *of* $z$ | $\dfrac{1}{2}z$ |
| 12 *multiplied by* $m$ | $12m$ |

TABLE 1-5

EXAMPLE 3

Let $x$ represent a certain number. Denote a number that is   **a.** twice as large as $x$,   **b.** 5 more than 3 times $x$,   and   **c.** 4 less than $\frac{1}{2}$ of $x$.

Solution

**a.** The number "twice as large as $x$" is found by multiplying $x$ by 2. It is represented by $2x$.

**b.** The number "5 more than 3 times $x$" is found by adding 5 to the product of 3 and $x$. It is represented by $3x + 5$.

**c.** The number "4 less than $\frac{1}{2}$ of $x$" is found by subtracting 4 from the product of $\frac{1}{2}$ and $x$. It is represented by $\frac{1}{2}x - 4$.  ∎

| Self Check | Find the product of 40 and $t$. |
|---|---|
| *Answer* | $40t$ |

**EXAMPLE 4**  **Stock valuation**  Jim owns $x$ shares of Transitronic stock, valued at $29 a share; $y$ shares of Positone stock, valued at $32 a share; and 300 shares of Baby Bell, valued at $42 a share.

**a.** How many shares of stock does he own?

**b.** What is the value of his stock?

*Solution*  **a.** Because there are $x$ shares of Transitronic, $y$ shares of Positone, and 300 shares of Baby Bell, his total number of shares is $x + y + 300$.

**b.** The value of $x$ shares of Transitronic is $\$29x$, the value of $y$ shares of Positone is $\$32y$, and the value of 300 shares of Baby Bell is $\$42(300)$. The total value of the stock is $\$(29x + 32y + 12,600)$.  ∎

| Self Check | If water softener salt costs $\$p$ per bag, find the cost of 25 bags. |
|---|---|
| *Answer* | $\$25p$ |

There are also several ways to indicate the **quotient** of two numbers with algebraic expressions, as shown in Table 1-6.

| The phrase | translates into the algebraic expression |
|---|---|
| the *quotient* of 470 and $A$ | $\dfrac{470}{A}$ |
| $B$ *divided by* $C$ | $\dfrac{B}{C}$ |
| the *ratio* of $h$ to 5 | $\dfrac{h}{5}$ |
| $x$ *split into* 5 equals parts | $\dfrac{x}{5}$ |

TABLE 1-6

**EXAMPLE 5**  Let $x$ and $y$ represent two numbers. Write an algebraic expression that represents the sum obtained when 3 times the first number is added to the quotient obtained when the second number is divided by 6.

*Solution*    Three times the first number $x$ is denoted as $3x$. The quotient obtained when the second number $y$ is divided by 6 is the fraction $\frac{y}{6}$. Their sum is expressed as $3x + \frac{y}{6}$. ■

**Self Check**    If the cost $c$ of a meal is split equally among 4 people, what is each person's share?

*Answer*    $\frac{c}{4}$

**EXAMPLE 6**    **Cutting a rope**    A 5-foot section is cut from the end of a rope that is $l$ feet long. If the remaining rope is divided into three equal pieces, find the length of each of the equal pieces.

*Solution*    After a 5-foot section is cut from one end of $l$ feet of rope, the rope that remains is $(l - 5)$ feet long. When that remaining rope is cut into 3 equal pieces, each piece will be $\frac{l-5}{3}$ feet long. See Figure 1-23.

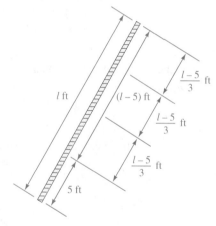

FIGURE 1-23    ■

**Self Check**    If a 7-foot section is cut from a rope that is $l$ feet long and the remaining rope is divided into two equal pieces, how long is each piece?

*Answer*    $\frac{l-7}{2}$ ft

### ■ EVALUATING ALGEBRAIC EXPRESSIONS

Since variables represent numbers, algebraic expressions also represent numbers. We have seen that we can evaluate algebraic expressions when we know the values of the variables.

EXAMPLE 7    If $x = 8$ and $y = 10$, evaluate   **a.** $x + y$,   **b.** $y - x$,   **c.** $3xy$   and   **d.** $\dfrac{5x}{y - 5}$.

*Solution*    We substitute 8 for $x$ and 10 for $y$ in each expression and simplify.

**a.** $x + y = 8 + 10$          **b.** $y - x = 10 - 8$

$\qquad\quad = 18$          $\qquad\qquad = 2$

**c.** $3xy = (3)(8)(10)$

$\qquad = (24)(10)$          Do the multiplications from left to right.

$\qquad = 240$

**d.** $\dfrac{5x}{y - 5} = \dfrac{5 \cdot 8}{10 - 5}$

$\qquad\quad = \dfrac{40}{5}$          Simplify the numerator and the denominator separately.

$\qquad\quad = 8$          Simplify the fraction.

**WARNING!**    After numbers are substituted for the variables in a product, it is often necessary to insert a dot or parentheses to show the multiplication. Otherwise $(3)(8)(10)$, for example, might be mistaken for 3,810, and $5 \cdot 8$ might be mistaken for 58.

■

Self Check    If $a = -2$ and $b = 5$, evaluate $\dfrac{6b + 2}{a + 2b}$.

Answer    4

## ■ ALGEBRAIC TERMS

Numbers without variables, such as 7, 21, and 23, are called **constants.** Expressions such as 37, $xyz$, and $32t$, which are constants, variables, or products of constants and variables, are called **algebraic terms.**

- The expression $3x + 5y$ contains two terms. The first term is $3x$, and the second term is $5y$.
- The expression $xy + (-7)$ contains two terms. The first term is $xy$, and the second term is $-7$.
- The expression $3 + x + 2y$ contains three terms. The first term is 3, the second term is $x$, and the third term is $2y$.

Numbers and variables that are part of a product are called **factors.** For example,

- The product $7x$ has two factors, which are 7 and $x$.
- The product $-3xy$ has three factors, which are $-3$, $x$, and $y$.
- The product $\frac{1}{2}abc$ has four factors, which are $\frac{1}{2}$, $a$, $b$, and $c$.

The number factor of a product is called its **numerical coefficient.** The numerical coefficient (or just the *coefficient*) of $7x$ is 7. The coefficient of $-3xy$ is $-3$, and the coefficient of $\frac{1}{2}abc$ is $\frac{1}{2}$. The coefficient of terms such as $x$, $ab$, and $rst$ is understood to be 1.

$$x = 1x, \qquad ab = 1ab, \qquad \text{and} \qquad rst = 1rst$$

**EXAMPLE 8**
a. The expression $5x + y$ has two terms. The numerical coefficient of its first term is 5. The numerical coefficient of its second term is 1.

b. The expression $-17wxyz$ has one term, which contains the five factors $-17$, $w$, $x$, $y$, and $z$. Its numerical coefficient is $-17$.

c. The expression 37 has one term, the constant 37. Its numerical coefficient is 37. ■

**Self Check**

How many terms does the expression $3x^2 - 2x + 7$ have? Find the sum of the coefficients.

**Answers** 3, 8

**Orals** *If $x = -2$ and $y = 3$, find the value of each expression.*

**1.** $x + y$      **2.** $7x$      **3.** $7x + y$      **4.** $7(x + y)$

**5.** $4x^2$      **6.** $(4x)^2$      **7.** $-3x^2$      **8.** $(-3x)^2$

## EXERCISE 1.6

***REVIEW*** *Evaluate each of the following.*

**1.** 14% of 3,800      **2.** $\frac{3}{5}$ of 4,765      **3.** $\dfrac{-4 + (7 - 9)}{(-9 - 7) + 4}$      **4.** $\dfrac{5}{4}\left(1 - \dfrac{3}{5}\right)$

***VOCABULARY AND CONCEPTS*** *Fill in each blank to make a true statement.*

**5.** The answer to an addition problem is called a ____.

**6.** The answer to a _____ problem is called a difference.

**7.** The answer to a _____ problem is called a product.

**8.** The answer to a division problem is called a _____.

**9.** An _____ expression is a combination of variables, numbers, and the operation symbols for addition, subtraction, multiplication, or division.

**10.** To _____ an algebraic expression, we substitute values for the variables and simplify.

**11.** Letters that stand for numbers are called _____.

**12.** Terms that have no variables are called _____.

**PRACTICE** *In Exercises 13–30, let x, y, and z represent three real numbers. Write an algebraic expression to denote each quantity.*

**13.** The sum of $x$ and $y$

**14.** The product of $x$ and $y$

**15.** The product of $x$ and twice $y$

**16.** The sum of twice $x$ and twice $y$

**17.** The difference obtained when $x$ is subtracted from $y$

**18.** The difference obtained when twice $x$ is subtracted from $y$

**19.** The quotient obtained when $y$ is divided by $x$

**20.** The quotient obtained when the sum of $x$ and $y$ is divided by $z$

**21.** The sum obtained when the quotient of $x$ divided by $y$ is added to $z$

**22.** $y$ decreased by $x$

**23.** $z$ less the product of $x$ and $y$

**24.** $z$ less than the product of $x$ and $y$

**25.** The product of 3, $x$, and $y$

**26.** The quotient obtained when the product of 3 and $z$ is divided by the product of 4 and $x$

**27.** The quotient obtained when the sum of $x$ and $y$ is divided by the sum of $y$ and $z$

**28.** The quotient obtained when the product of $x$ and $y$ is divided by the sum of $x$ and $z$

**29.** The sum of the product $xy$ and the quotient obtained when $y$ is divided by $z$

**30.** The number obtained when $x$ decreased by 4 is divided by the product of 3 and $y$

*In Exercises 31–42, write each algebraic expression as an English phrase.*

**31.** $x + 3$

**32.** $y - 2$

**33.** $\dfrac{x}{y}$

**34.** $xz$

**35.** $2xy$

**36.** $\dfrac{x + y}{2}$

**37.** $\dfrac{5}{x + y}$

**38.** $\dfrac{3x}{y + z}$

**39.** $\dfrac{3 + x}{y}$

**40.** $3 + \dfrac{x}{y}$

**41.** $xy(x + y)$

**42.** $(x + y + z)(xyz)$

*In Exercises 43–50, let x = 8, y = 4, and z = 2. Write each phrase as an algebraic expression, and evaluate it.*

**43.** The sum of $x$ and $z$

**44.** The product of $x$, $y$, and $z$

**45.** $z$ less than $y$

**46.** The quotient obtained when $y$ is divided by $z$

**47.** 3 less than the product of $y$ and $z$

**48.** 7 less than the sum of $x$ and $y$

**49.** The quotient obtained when the product of $x$ and $y$ is divided by $z$

**50.** The quotient obtained when 10 greater than $x$ is divided by $z$

*In Exercises 51–60, give the number of terms in each algebraic expression and also give the numerical coefficient of the first term.*

**51.** $6d$

**52.** $-4c + 3d$

**53.** $-xy - 4t + 35$

**54.** $xy$

**55.** $3ab + bc - cd - ef$

**56.** $-2xyz + cde - 14$

**57.** $-4xyz + 7xy - z$

**58.** $5uvw - 4uv + 8uw$

**59.** $3x + 4y + 2z + 2$

**60.** $7abc - 9ab + 2bc + a - 1$

*In Exercises 61–64, consider the algebraic expression $29xyz + 23xy + 19x$.*

**61.** What are the factors of the third term?

**62.** What are the factors of the second term?

**63.** What are the factors of the first term?

**64.** What factor is common to all three terms?

*In Exercises 65–68, consider the algebraic expression $3xyz + 5xy + 17xz$.*

**65.** What are the factors of the first term?

**66.** What are the factors of the second term?

**67.** What are the factors of the third term?

**68.** What factor is common to all three terms?

*In Exercises 69–72, consider the algebraic expression $5xy + yt + 8xyt$.*

**69.** Find the numerical coefficients of each term.

**70.** What factor is common to all three terms?

**71.** What factors are common to the first and third terms?

**72.** What factors are common to the second and third terms?

*In Exercises 73–76, consider the algebraic expression $3xy + y + 25xyz$.*

**73.** Find the numerical coefficient of each term and find their product.

**74.** Find the numerical coefficient of each term and find their sum.

**75.** What factors are common to the first and third terms?

**76.** What factor is common to all three terms?

## APPLICATIONS

**77. Course load**  A man enrolls in college for $c$ hours of credit, and his sister enrolls for 4 more hours than her brother. Write an expression that represents the number of hours the sister is taking.

**78. Antique cars**  An antique Ford has 25,000 more miles on its odometer than a newer car. If the newer car has traveled $m$ miles, find an expression that represents the mileage on the Ford.

**79. T-bills**  Write an expression that represents the value of $t$ T-bills, each worth \$9,987.

**80. Real estate**  Write an expression that represents the value of $a$ vacant lots if each lot is worth \$35,000.

**81. Cutting rope**  A rope $x$ feet long is cut into 5 equal pieces. Find an expression for the length of each piece.

**82. Plumbing**  A plumber cuts a pipe that is 12 feet long into $x$ equal pieces. Find an expression for the length of each piece.

**83. Comparing assets**  A girl had $d$ dollars, and her brother had $5 more than three times that amount. How much did the brother have?

**84. Comparing investments**  Wendy has $x$ shares of stock. Her sister has 2 fewer shares than twice Wendy's shares. How many shares does her sister have?

*WRITING*

**85.** Distinguish between the meanings of these two phrases: "3 less than $x$" and "3 is less than $x$."

**87.** What is the purpose of using variables? Why aren't ordinary numbers enough?

**86.** Distinguish between *factor* and *term.*

**88.** In words, $xy$ is "the product of $x$ and $y$." However, $\frac{x}{y}$ is "the quotient obtained when $x$ is divided by $y$." Explain why the extra words are needed.

*SOMETHING TO THINK ABOUT*

**89.** If the value of $x$ were doubled, what would happen to the value of $37x$?

**90.** If the values of both $x$ and $y$ were doubled, what would happen to the value of $5xy^2$?

# 1.7  Properties of Real Numbers

■ THE CLOSURE PROPERTIES ■ THE COMMUTATIVE PROPERTIES ■ THE ASSOCIATIVE PROPERTIES
■ THE DISTRIBUTIVE PROPERTY ■ THE IDENTITY ELEMENTS ■ INVERSES FOR ADDITION AND
MULTIPLICATION

Getting Ready  *Do the operations.*

**1.** $3 + (5 + 9)$
**2.** $(3 + 5) + 9$
**3.** $23.7 + 14.9$
**4.** $14.9 + 23.7$
**5.** $7(5 + 3)$
**6.** $7 \cdot 5 + 7 \cdot 3$
**7.** $125.3 + (-125.3)$
**8.** $125.3\left(\dfrac{1}{125.3}\right)$
**9.** $777 + 0$
**10.** $777 \cdot 1$

■ THE CLOSURE PROPERTIES

The **closure properties** guarantee that the sum, difference, product, or quotient (except for division by zero) of any two real numbers is also a real number.

**Closure Properties**

If $a$ and $b$ are real numbers, then

$a + b$ is a real number          $a - b$ is a real number

$ab$ is a real number          $\dfrac{a}{b}$ is a real number    $(b \neq 0)$

**EXAMPLE 1** Assume that $x = 8$ and $y = -4$. Find the real-number answer to show that **a.** $x + y$, **b.** $x - y$, **c.** $xy$, and **d.** $\frac{x}{y}$ all represent real numbers.

*Solution* We substitute 8 for $x$ and $-4$ for $y$ in each expression and simplify.

**a.** $x + y = 8 + (-4)$          **b.** $x - y = 8 - (-4)$
$\qquad\qquad = 4$                    $\qquad\qquad\quad = 8 + 4$
$\qquad\qquad\qquad\qquad\qquad\qquad\qquad\qquad = 12$

**c.** $xy = 8(-4)$          **d.** $\dfrac{x}{y} = \dfrac{8}{-4}$
$\qquad\quad = -32$                $\qquad\qquad\quad = -2$  ■

**Self Check** Assume that $a = -6$ and $b = 3$. Find the real-number answer to show that **a.** $a - b$  and  **b.** $\frac{a}{b}$ are real numbers.

**Answers** **a.** $-9$,  **b.** $-2$

## ■ THE COMMUTATIVE PROPERTIES

The **commutative properties** (from the word *commute,* which means to go back and forth) guarantee that addition or multiplication of two real numbers can be done in either order.

**Commutative Properties**

If $a$ and $b$ are real numbers, then

$a + b = b + a$      commutative property of addition

$ab = ba$      commutative property of multiplication

**EXAMPLE 2** Assume that $x = -3$ and $y = 7$. Show that  **a.** $x + y = y + x$  and  **b.** $xy = yx$.

*Solution* **a.** We can show that the sum $x + y$ is the same as the sum $y + x$ by substituting $-3$ for $x$ and 7 for $y$ in each expression and simplifying.

$x + y = -3 + 7 = 4$      and      $y + x = 7 + (-3) = 4$

**b.** We can show that the product $xy$ is the same as the product $yx$ by substituting $-3$ for $x$ and 7 for $y$ in each expression and simplifying.

$xy = -3(7) = -21$      and      $yx = 7(-3) = -21$  ■

| Self Check | Assume that $a = 6$ and $b = -5$. Show that  **a.** $a + b = b + a$  and  **b.** $ab = ba$. |
|---|---|
| *Answers* | **a.** $a + b = 1$ and $b + a = 1$,  **b.** $ab = -30$ and $ba = -30$ |

## ■ THE ASSOCIATIVE PROPERTIES

The **associative properties** guarantee that three real numbers can be regrouped in an addition or multiplication.

### Associative Properties

If $a$, $b$, and $c$ are real numbers, then

$$(a + b) + c = a + (b + c) \qquad \text{associative property of addition}$$

$$(ab)c = a(bc) \qquad \text{associative property of multiplication}$$

Because of the associative property of addition, we can group (or *associate*) the numbers in a sum in any way that we wish. For example,

$$(3 + 4) + 5 = 7 + 5 \qquad \text{and} \qquad 3 + (4 + 5) = 3 + 9$$
$$= 12 \qquad\qquad\qquad\qquad = 12$$

The answer is 12 regardless of how we group the three numbers.

The associative property of multiplication permits us to group (or *associate*) the numbers in a product in any way that we wish. For example,

$$(3 \cdot 4) \cdot 7 = 12 \cdot 7 \qquad \text{and} \qquad 3 \cdot (4 \cdot 7) = 3 \cdot 28$$
$$= 84 \qquad\qquad\qquad\qquad = 84$$

The answer is 84 regardless of how we group the three numbers.

## ■ THE DISTRIBUTIVE PROPERTY

The **distributive property** shows how to multiply the sum of two numbers by a third number. Because of this property, we can often add first and then multiply, or multiply first and then add.

For example, $2(3 + 7)$ can be calculated in two different ways. We can add and then multiply, or we can multiply each number within the parentheses by 2 and then add.

$$2(3 + 7) = 2(10) \qquad \text{and} \qquad 2(3 + 7) = 2 \cdot 3 + 2 \cdot 7$$
$$= 20 \qquad\qquad\qquad\qquad = 6 + 14$$
$$= 20$$

Either way, the result is 20.

In general, we have the following property.

## Distributive Property

If $a$, $b$, and $c$ are real numbers, then

$$a(b + c) = ab + ac$$

Because multiplication is commutative, the distributive property can also be written in the form

$$(b + c)a = ba + ca$$

FIGURE 1-24

We can interpret the distributive property geometrically. Since the area of the largest rectangle in Figure 1-24 is the product of its width $a$ and its length $b + c$, its area is $a(b + c)$. The areas of the two smaller rectangles are $ab$ and $ac$. Since the area of the largest rectangle is equal to the sum of the areas of the smaller rectangles, we have $a(b + c) = ab + ac$.

**EXAMPLE 3**   Evaluate each expression in two different ways:
**a.** $3(5 + 9)$   and   **b.** $-2(-7 + 3)$.

*Solution*   **a.** $3(5 + 9) = 3(14)$        and        $3(5 + 9) = 3 \cdot 5 + 3 \cdot 9$
$= 42$                                      $= 15 + 27$
$= 42$

**b.** $-2(-7 + 3) = -2(-4)$        and        $-2(-7 + 3) = -2(-7) + (-2)3$
$= 8$                                             $= 14 + (-6)$
$= 8$ ∎

*Self Check*   Evaluate $-5.2(2.7 + 3.5)$ in two different ways.
*Answer*       $-32.24$

The distributive property can be extended to three or more terms. For example, if $a$, $b$, $c$, and $d$ are real numbers, then

$$a(b + c + d) = ab + ac + ad$$

**EXAMPLE 4**   Write $3(x + y + 2)$ without using parentheses.

*Solution*       $3(x + y + 2) = 3x + 3y + 3 \cdot 2$        Distribute the multiplication by 3.
$= 3x + 3y + 6$ ∎

*Self Check*   Write $-6.3(a + 2b + 3.7)$ without using parentheses.
*Answer*       $-6.3a - 12.6b - 23.31$

## ■ THE IDENTITY ELEMENTS

The numbers 0 and 1 play special roles in arithmetic. The number 0 is the only number that can be added to another number (say, $a$) and give an answer of that same number $a$:

$$0 + a = a + 0 = a$$

The number 1 is the only number that can be multiplied by another number (say, $a$) and give an answer of that same number $a$:

$$1 \cdot a = a \cdot 1 = a$$

Because adding 0 to a number or multiplying a number by 1 leaves that number the same (identical), the numbers 0 and 1 are called **identity elements.**

> **Identity Elements**
> 0 is the **identity element for addition.**
>
> 1 is the **identity element for multiplication.**

## ■ INVERSES FOR ADDITION AND MULTIPLICATION

If the sum of two numbers is 0, the numbers are called *negatives,* or **additive inverses,** of each other. Since $3 + (-3) = 0$, the numbers 3 and $-3$ are negatives or additive inverses of each other. In general, because

$$a + (-a) = 0$$

the numbers represented by $a$ and $-a$ are negatives or additive inverses of each other.

If the product of two numbers is 1, the numbers are called **reciprocals,** or **multiplicative inverses,** of each other. Since $7\left(\frac{1}{7}\right) = 1$, the numbers 7 and $\frac{1}{7}$ are reciprocals. Since $(-0.25)(-4) = 1$, the numbers $-0.25$ and $-4$ are reciprocals. In general, because

$$a\left(\frac{1}{a}\right) = 1 \qquad \text{provided } a \neq 0$$

the numbers represented by $a$ and $\frac{1}{a}$ are reciprocals or multiplicative inverses of each other.

> **Additive and Multiplicative Inverses**
> Because $a + (-a) = 0$, the numbers $a$ and $-a$ are called **negatives** or **additive inverses.**
>
> Because $a\left(\frac{1}{a}\right) = 1$  $(a \neq 0)$, the numbers $a$ and $\frac{1}{a}$ are called **reciprocals** or **multiplicative inverses.**

**EXAMPLE 5**   The property in the right column justifies the statement in the left column.

| | |
|---|---|
| $3 + 4$ is a real number | closure property of addition |
| $\dfrac{8}{3}$ is a real number | closure property of division |
| $3 + 4 = 4 + 3$ | commutative property of addition |
| $-3 + (2 + 7) = (-3 + 2) + 7$ | associative property of addition |
| $(5)(-4) = (-4)(5)$ | commutative property of multiplication |
| $(ab)c = a(bc)$ | associative property of multiplication |
| $3(a + 2) = 3a + 3 \cdot 2$ | distributive property |
| $3 + 0 = 3$ | additive identity property |
| $3(1) = 3$ | multiplicative identity property |
| $2 + (-2) = 0$ | additive inverse property |
| $\left(\dfrac{2}{3}\right)\left(\dfrac{3}{2}\right) = 1$ | multiplicative inverse property |

■

**Self Check**   Which property justifies each statement?   **a.** $a + 7 = 7 + a$
**b.** $3(y + 2) = 3y + 3 \cdot 2$   **c.** $3 \cdot (2 \cdot p) = (3 \cdot 2) \cdot p$

**Answers**   **a.** commutative property of addition,   **b.** distributive property,   **c.** associative property of multiplication

The properties of the real numbers are summarized as follows.

**Properties of Real Numbers**
For all real numbers $a$, $b$, and $c$,

| **Closure properties** | $a + b$ is a real number | $a \cdot b$ is a real number |
|---|---|---|
| | $a - b$ is a real number | $a \div b$ is a real number   $(b \neq 0)$ |

| | *Addition* | *Multiplication* |
|---|---|---|
| **Commutative properties** | $a + b = b + a$ | $a \cdot b = b \cdot a$ |
| **Associative properties** | $(a + b) + c = a + (b + c)$ | $(ab)c = a(bc)$ |
| **Identity properties** | $a + 0 = a$ | $a \cdot 1 = a$ |
| **Inverse properties** | $a + (-a) = 0$ | $a \cdot \left(\dfrac{1}{a}\right) = 1$   $(a \neq 0)$ |
| **Distributive property** | $a(b + c) = ab + ac$ | |

Orals   *Give an example of each property.*

**1.** The associative property of multiplication

**2.** The additive identity property

**3.** The distributive property

**4.** The inverse for multiplication

*Provide an example to illustrate each statement.*

**5.** Subtraction is not commutative.

**6.** Division is not associative.

## EXERCISE 1.7

### REVIEW

**1.** Write as a mathematical expression: The sum of $x$ and the square of $y$ is greater than or equal to $z$.

**2.** Write as an English phrase: $3(x + z)$.

*In Exercises 3–4, fill each box with an appropriate symbol.*

**3.** For any number $x$, $|x| \geq$ ___ .

**4.** $x - y = x + ($ ___ $)$

*In Exercises 5–6, fill in each blank to make a true statement.*

**5.** The product of two negative numbers is a _____ number.

**6.** The sum of two negative numbers is a _____ number.

### VOCABULARY AND CONCEPTS   *Fill in each blank to make a true statement.*

**7.** If $a$ and $b$ are real numbers, $a + b$ is a ___ number.

**8.** If $a$ and $b$ are real numbers, $\frac{a}{b}$ is a real number, provided that _____.

**9.** $a + b = b +$ ___

**10.** $a \cdot b =$ ___ $\cdot a$

**11.** $(a + b) + c = a +$ _____

**12.** $(ab)c =$ ___ $\cdot (bc)$

**13.** $a(b + c) = ab +$ ___

**14.** $0 + a =$ ___

**15.** $a \cdot 1 =$ ___

**16.** $0$ is the _____ element for _____.

**17.** $1$ is the identity _____ for _____ .

**18.** If $a + (-a) = 0$, then $a$ and $-a$ are called _____ inverses.

**19.** If $a\left(\dfrac{1}{a}\right) = 1$, then $a$ and ___ are called reciprocals.

**20.** $a(b + c + d) = ab +$ _____

### PRACTICE   *In Exercises 21–28, assume that $x = 12$ and $y = -2$. Show that each expression represents a real number by finding the real-number answer.*

**21.** $x + y$

**22.** $y - x$

**23.** $xy$

**24.** $\dfrac{x}{y}$

**25.** $x^2$

**26.** $y^2$

**27.** $\dfrac{x}{y^2}$

**28.** $\dfrac{2x}{3y}$

*In Exercises 29–34, assume that x = 5 and y = 7. Show that both given expressions have the same value.*

**29.** $x + y; y + x$                   **30.** $xy; yx$                   **31.** $3x + 2y; 2y + 3x$                   **32.** $3xy; 3yx$

**33.** $x(x + y); (x + y)x$                                     **34.** $xy + y^2; y^2 + xy$

*In Exercises 35–40, assume that x = 2, y = −3, and z = 1. Show that the expressions have the same value.*

**35.** $(x + y) + z; x + (y + z)$                           **36.** $(xy)z; x(yz)$
**37.** $(xz)y; x(yz)$                                       **38.** $(x + y) + z; y + (x + z)$
**39.** $x^2(yz^2); (x^2y)z^2$                               **40.** $x(y^2z^3); (xy^2)z^3$

*In Exercises 41–52, use the distributive property to write each expression without parentheses. Simplify each result if possible.*

**41.** $3(x + y)$                   **42.** $4(a + b)$                   **43.** $x(x + 3)$                   **44.** $y(y + z)$
**45.** $-x(a + b)$                  **46.** $a(x + y)$                   **47.** $4(x^2 + x)$                 **48.** $-2(a^2 + 3)$
**49.** $-5(t + 2)$                  **50.** $2x(a - x)$                  **51.** $-2a(x + a)$                **52.** $-p(p - q)$

*In Exercises 53–64, give the additive and the multiplicative inverse of each number when possible.*

**53.** 2                   **54.** 3                   **55.** $\dfrac{1}{3}$                   **56.** $-\dfrac{1}{2}$

**57.** 0                   **58.** −2                  **59.** $-\dfrac{5}{2}$                  **60.** 0.5

**61.** −0.2                **62.** 0.75                **63.** $\dfrac{4}{3}$                   **64.** −1.25

*In Exercises 65–76, state which property of real numbers justifies each statement.*

**65.** $3 + x = x + 3$                                      **66.** $(3 + x) + y = 3 + (x + y)$
**67.** $xy = yx$                                            **68.** $(3)(2) = (2)(3)$
**69.** $-2(x + 3) = -2x + (-2)(3)$                          **70.** $x(y + z) = (y + z)x$
**71.** $(x + y) + z = z + (x + y)$                          **72.** $3(x + y) = 3x + 3y$

**73.** $5 \cdot 1 = 5$                   **74.** $x + 0 = x$                   **75.** $3 + (-3) = 0$                   **76.** $9 \cdot \dfrac{1}{9} = 1$

*In Exercises 77–86, use the given property to rewrite the expression in a different form.*

**77.** $3(x + 2)$; distributive property                          **78.** $x + y$; commutative property of addition
**79.** $y^2x$; commutative property of multiplication             **80.** $x + (y + z)$; associative property of addition

**81.** $(x + y)z$; commutative property of addition              **82.** $x(y + z)$; distributive property

**83.** $(xy)z$; associative property of multiplication

**84.** $1x$; multiplicative identity property

**85.** $0 + x$; additive identity property

**86.** $5 \cdot \dfrac{1}{5}$; multiplicative inverse property

## WRITING

**87.** Explain why division is not commutative.

**88.** Describe two ways of calculating the value of $3(12 + 7)$.

## SOMETHING TO THINK ABOUT

**89.** Suppose there were no other numbers than the odd integers.
  • Would the closure property for addition still be true?
  • Would the closure property for multiplication still be true?
  • Would there still be an identity for addition?
  • Would there still be an identity for multiplication?

**90.** Suppose there were no other numbers than the even integers. Answer the four parts of Exercise 89 again.

■ ■ ■ ■ ■ ■ ■ ■ ■

### MATHEMATICS FOR FUN

In this chapter, we learned that subtraction is not associative. For example, $(10 - 5) - 2 \neq 10 - (5 - 2)$, because

$$(10 - 5) - 2 = 3 \quad \text{but} \quad 10 - (5 - 2) = 7$$

However, in the story on page 2, we expected subtraction to be associative. After the $5 refund on the incorrect $120 room cost, we reasoned incorrectly that

$$
\begin{aligned}
40 + 40 + 40 - 5 &= 120 - 5 \\
&= 120 - (3 + 2) \\
&= (120 - 3) + 2 \qquad \text{This step is false.} \\
&= (40 - 1) + (40 - 1) + (40 - 1) + 2 \\
&= 39 + 39 + 39 + 2 \\
&= 119
\end{aligned}
$$

That $119 is not equal to $120 is irrelevant. The cost of the room was $115.

■ ■ ■ ■ ■ ■ ■ ■ ■   **PROJECTS**

### PROJECT 1

The circumference of any circle (the distance around the circle) and the diameter of the circle (the distance across) are related. When you divide the circumference by the diameter, the quotient is always the same number, **pi,** denoted by the Greek letter $\pi$.

*(continued)*

■ ■ ■ ■ ■ ■ ■ ■ ■ ■  **PROJECTS** *(continued)*

- Carefully measure the circumference of several circles—a quarter, a dinner plate, a bicycle tire—whatever you can find that is round. Then calculate approximations of $\pi$ by dividing (with a calculator) each circle's circumference by its diameter.

- Press the $\boxed{\pi}$ button on the calculator to obtain a more accurate value of $\pi$. How close were your calculations?

PROJECT 2   **a.** The fraction $\frac{22}{7}$ is often used as an approximation of $\pi$. To how many decimal places is this approximation accurate?

**b.** Experiment with your calculator and try to do better. Find another fraction (with no more than three digits in either its numerator or its denominator) that is closer to $\pi$. Who in your class has done best?

# C H A P T E R   S U M M A R Y

## CONCEPTS

### REVIEW EXERCISES

### SECTION 1.1

## *Real Numbers and Their Graphs*

**Natural numbers:**

1, 2, 3, 4, 5, . . .

**Whole numbers:**

0, 1, 2, 3, 4, 5, . . .

**Integers:**

. . . , −3, −2, −1, 0, 1, 2, 3, . . .

**Rational numbers:**
Fractions with integer numerators and nonzero integer denominators

**Real numbers:**
Rational numbers or irrational numbers

**Prime numbers:**

2, 3, 5, 7, 11, 13, 17, . . .

**Composite numbers:**

4, 6, 8, 9, 10, 12, 14, 15, . . .

**Even integers:**

. . . , −6, −4, −2, 0, 2, 4, 6, . . .

**Odd integers:**

. . . , −5, −3, −1, 1, 3, 5, . . .

**1.** Consider the set {0, 1, 2, 3, 4, 5}.
   **a.** Which numbers are natural numbers?
   **b.** Which numbers are prime numbers?
   **c.** Which numbers are odd natural numbers?
   **d.** Which numbers are composite numbers?

**2.** Consider the set $\left\{-6, -\frac{2}{3}, 0, \sqrt{2}, 2.6, \pi, 5\right\}$.
   **a.** Which numbers are integers?
   **b.** Which numbers are rational numbers?
   **c.** Which numbers are prime numbers?
   **d.** Which numbers are real numbers?
   **e.** Which numbers are even integers?
   **f.** Which numbers are odd integers?
   **g.** Which numbers are not rational?

**3.** Place one of the symbols $=$, $<$, or $>$ in each box to make a true statement.

**a.** $-5$ ☐ $12 - 12$

**b.** $\dfrac{24}{6}$ ☐ $5$

**c.** $13 - 13$ ☐ $5 - \dfrac{25}{5}$

**d.** $\dfrac{21}{7}$ ☐ $-33$

**Double negative rule:**
$-(-x) = x$

**4.** Simplify each expression.

**a.** $-(-8)$

**b.** $-(12 - 4)$

Sets of numbers can be graphed on the number line.

**5.** Draw a number line and graph each set of numbers.

**a.** The composite numbers from 14 to 20

14  15  16  17  18  19  20

**b.** The whole numbers between 19 and 25

19  20  21  22  23  24  25

**c.** The real numbers less than or equal to $-3$ or greater than 2

**d.** The real numbers greater than $-4$ and less than 3

The **absolute value** of $x$, denoted as $|x|$, is the distance between $x$ and 0 on the number line.

$|x| \geq 0$

**6.** Find each absolute value.

**a.** $|53 - 42|$

**b.** $|-31|$

---

**SECTION 1.2**     *Fractions*

To simplify a fraction, factor the numerator and the denominator. Then divide out all common factors.

**7.** Simplify each fraction.

**a.** $\dfrac{45}{27}$

**b.** $\dfrac{121}{11}$

To multiply two fractions, multiply their numerators and multiply their denominators.

**8.** Do each operation and simplify the answer, if possible.

**a.** $\dfrac{31}{15} \cdot \dfrac{10}{62}$

**b.** $\dfrac{25}{36} \cdot \dfrac{12}{15} \cdot \dfrac{3}{5}$

To divide two fractions, multiply the first by the reciprocal of the second.

**c.** $\dfrac{18}{21} \div \dfrac{6}{7}$

**d.** $\dfrac{14}{24} \div \dfrac{7}{12} \div \dfrac{2}{5}$

To add (or subtract) two fractions with like denominators, add (or subtract) their numerators and keep their common denominator.

**e.** $\dfrac{7}{12} + \dfrac{9}{12}$

**f.** $\dfrac{13}{24} - \dfrac{5}{24}$

To add (or subtract) two fractions with unlike denominators, rewrite the fractions with the same denominator, add (or subtract) their numerators, and use the common denominator.

Before working with mixed numbers, convert them to improper fractions.

**g.** $\dfrac{1}{3} + \dfrac{1}{7}$

**h.** $\dfrac{5}{7} + \dfrac{4}{9}$

**i.** $\dfrac{2}{3} - \dfrac{1}{7}$

**j.** $\dfrac{4}{5} - \dfrac{2}{3}$

**k.** $3\dfrac{2}{3} + 5\dfrac{1}{4}$

**l.** $7\dfrac{5}{12} - 4\dfrac{1}{2}$

**9.** Do the operations.
   **a.** $32.71 + 15.9$
   **b.** $27.92 - 14.93$
   **c.** $5.3 \cdot 3.5$
   **d.** $21.83 \div 5.9$

**10.** Do each operation and round to two decimal places.
   **a.** $2.7(4.92 - 3.18)$
   **b.** $\dfrac{3.3 + 2.5}{0.22}$
   **c.** $\dfrac{12.5}{14.7 - 11.2}$
   **d.** $(3 - 0.7)(3.63 - 2)$

**11. Average study time**   Four students recorded the time they spent working on a take-home exam: 5.2, 4.7, 9.5, and 8 hours. Find the average time spent. (*Hint:* Add the numbers and divide by 4.)

**12. Absenteeism**   During the height of the flu season, 15% of the 380 university faculty members were sick. How many were ill?

**13. Packaging**   Four steel bands surround the shipping crate in Illustration 1. Find the total length of strapping needed.

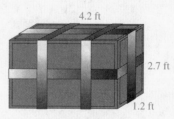

4.2 ft

2.7 ft

1.2 ft

ILLUSTRATION 1

| **SECTION 1.3** | *Exponents and Order of Operations* |

If $n$ is a natural number, then

$$\overbrace{x^n = x \cdot x \cdot x \cdot x \cdots \cdots x}^{n \text{ factors of } x}$$

**14.** Find the value of each expression.

**a.** $3^4$

**b.** $\left(\dfrac{2}{3}\right)^2$

**c.** $(0.5)^2$

**d.** $5^2 + 2^3$

**15.** Let $x = 2$ and $y = 3$ and evaluate each expression.

**a.** $y^4$

**b.** $x^y$

**16.** ▦ **Petroleum storage**   Find the volume of the cylindrical storage tank in Illustration 2. Round to one decimal place.

ILLUSTRATION 2

**Order of operations**

Within each pair of grouping symbols (working from the innermost pair to the outermost pair), do the following operations:

**1.** Evaluate all exponential expressions.

**2.** Do multiplications and divisions, working from left to right.

**3.** Do additions and subtractions, working from left to right.

When the grouping symbols are gone, repeat the above rules to finish the calculation.

In a fraction, simplify the numerator and denominator separately. Then simplify the fraction, if possible.

**17.** Simplify each expression.

**a.** $5 + 3^3$

**b.** $7 \cdot 2 - 7$

**c.** $4 + (8 \div 4)$

**d.** $(4 + 8) \div 4$

**e.** $5^3 - \dfrac{81}{3}$

**f.** $(5 - 2)^2 + 5^2 + 2^2$

**g.** $\dfrac{4 \cdot 3 + 3^4}{31}$

**h.** $\dfrac{4}{3} \cdot \dfrac{9}{2} + \dfrac{1}{2} \cdot 18$

**18.** Let $x = 6$ and $y = 8$ and evaluate each expression.

**a.** $y^2 - x$

**b.** $(y - x)^2$

**c.** $\dfrac{x + y}{x - 4}$

**d.** $\dfrac{xy - 12}{4 + y}$

**19.** Let $x = 2$ and $y = 3$ and evaluate each expression.

**a.** $x^2 + xy^2$

**b.** $\dfrac{x^2 + y}{x^3 - 1}$

| SECTION 1.4 | *Adding and Subtracting Real Numbers* |
|---|---|

To find the sum of two real numbers with the same sign, add their absolute values and keep their common sign.

To add two real numbers with unlike signs, subtract their absolute values (the smaller from the larger) and use the sign of the number with the greater absolute value.

If $x$ and $y$ are two real numbers, then $x - y = x + (-y)$.

**20.** Evaluate each expression.

**a.** $(+7) + (+8)$

**b.** $(-25) + (-32)$

**c.** $(-2.7) + (-3.8)$

**d.** $\dfrac{1}{3} + \dfrac{1}{6}$

**e.** $(+12) + (-24)$

**f.** $(-44) + (+60)$

**g.** $3.7 + (-2.5)$

**h.** $-5.6 + (+2.06)$

**i.** $15 - (-4)$

**j.** $-12 - (-13)$

**k.** $[-5 + (-5)] - (-5)$

**l.** $1 - [5 - (-3)]$

**m.** $\dfrac{5}{6} - \left(-\dfrac{2}{3}\right)$

**n.** $\dfrac{2}{3} - \left(\dfrac{1}{3} - \dfrac{2}{3}\right)$

**o.** $\left|\dfrac{3}{7} - \left(-\dfrac{4}{7}\right)\right|$

**p.** $\dfrac{3}{7} - \left|-\dfrac{4}{7}\right|$

**21.** Let $x = 2$, $y = -3$, and $z = -1$ and evaluate each expression.

**a.** $y + z$

**b.** $x + y$

**c.** $x + (y + z)$

**d.** $x - y$

**e.** $x - (y - z)$

**f.** $(x - y) - z$

| SECTION 1.5 | *Multiplying and Dividing Real Numbers* |
|---|---|

The product of two real numbers with like signs is the positive product of their absolute values.

The product of two real numbers with unlike signs is the negative of the product of their absolute values.

The quotient of two real numbers with like signs is the quotient of their absolute values.

**22.** Evaluate each expression.

**a.** $(+3)(+4)$

**b.** $(-5)(-12)$

**c.** $\left(-\dfrac{3}{14}\right)\left(-\dfrac{7}{6}\right)$

**d.** $(3.75)(0.37)$

**e.** $5(-7)$

**f.** $(-15)(7)$

**g.** $\left(-\dfrac{1}{2}\right)\left(\dfrac{4}{3}\right)$

**h.** $(-12.2)(3.7)$

**i.** $\dfrac{+25}{+5}$

**j.** $\dfrac{-14}{-2}$

**k.** $\dfrac{(-2)(-7)}{4}$

**l.** $\dfrac{-22.5}{-3.75}$

The quotient of two real numbers with unlike signs is the negative of the quotient of their absolute values.

Division by zero is undefined.

**m.** $\dfrac{-25}{5}$

**n.** $\dfrac{(-3)(-4)}{-6}$

**o.** $\left(\dfrac{-10}{2}\right)^2 - (-1)^3$

**p.** $\dfrac{[-3 + (-4)]^2}{10 + (-3)}$

**q.** $\left(\dfrac{-3 + (-3)}{3}\right)\left(\dfrac{-15}{5}\right)$

**r.** $\dfrac{-2 - (-8)}{5 + (-1)}$

**23.** Let $x = 2$, $y = -3$, and $z = -1$ and evaluate each expression.

**a.** $xy$

**b.** $yz$

**c.** $x(x + z)$

**d.** $xyz$

**e.** $y^2z + x$

**f.** $yz^3 + (xy)^2$

**g.** $\dfrac{xy}{z}$

**h.** $\dfrac{|xy|}{3z}$

---

**SECTION 1.6**     *Algebraic Expressions*

**24.** Let $x$, $y$, and $z$ represent three real numbers. Write an algebraic expression that represents each quantity.

**a.** The product of $x$ and $z$

**b.** The sum of $x$ and twice $y$

**c.** Twice the sum of $x$ and $y$

**d.** $x$ decreased by the product of $y$ and $z$

**25.** Write each algebraic expression as an English phrase.

**a.** $3xy$

**b.** $5 - yz$

**c.** $yz - 5$

**d.** $\dfrac{x + y + z}{2xyz}$

**26.** How many terms does the expression $3x + 4y + 9$ have?

**27.** What is the numerical coefficient of the term $7xy$?

**28.** What is the numerical coefficient of the term $xy$?

**29.** Find the sum of the numerical coefficients in $2x^3 + 4x^2 + 3x$.

| **SECTION 1.7** | *Properties of Real Numbers* |
| --- | --- |

**The closure properties:**
$x + y$ is a real number.
$x - y$ is a real number.
$xy$ is a real number.
$\dfrac{x}{y}$ is a real number  $(y \neq 0)$.

**The commutative properties:**
$x + y = y + x$
$xy = yx$

**The associative properties:**
$(x + y) + z = x + (y + z)$
$(xy)z = x(yz)$

**The distributive property:**
$x(y + z) = xy + xz$

**The identity elements:**
0 is the identity for addition.
1 is the identity for multiplication.

**The additive and multiplicative inverse properties:**
$x + (-x) = 0$
$x\left(\dfrac{1}{x}\right) = 1$  $(x \neq 0)$

**30.** Tell which property of real numbers justifies each statement. Assume that all variables represent real numbers.

**a.** $x + y$ is a real number
**b.** $3 \cdot (4 \cdot 5) = (4 \cdot 5) \cdot 3$
**c.** $3 + (4 + 5) = (3 + 4) + 5$
**d.** $5(x + 2) = 5 \cdot x + 5 \cdot 2$
**e.** $a + x = x + a$
**f.** $3 \cdot (4 \cdot 5) = (3 \cdot 4) \cdot 5$
**g.** $3 + (x + 1) = (x + 1) + 3$
**h.** $x \cdot 1 = x$
**i.** $17 + (-17) = 0$
**j.** $x + 0 = x$

## ▪ Chapter Test

**1.** List the prime numbers between 30 and 50.

**2.** What is the only even prime number?

**3.** Graph the composite numbers less than 10 on a number line.

**4.** Graph the real numbers from 5 to 15 on a number line.

**5.** Evaluate $-|23|$.

**6.** Evaluate $-|7| + |-7|$.

*In Problems 7–10, place one of the symbols $=$, $<$, or $>$ in each box to make a true statement.*

**7.** $3(4 - 2)$ ▢ $-2(2 - 5)$

**8.** $1 + 4 \cdot 3$ ▢ $-2(-7)$

**9.** 25% of 136 ▢ $\dfrac{1}{2}$ of 66

**10.** $-13.7$ ▢ $-|-13.7|$

*In Problems 11–16, simplify each expression.*

**11.** $\dfrac{26}{40}$

**12.** $\dfrac{7}{8} \cdot \dfrac{24}{21}$

**13.** $\dfrac{18}{35} \div \dfrac{9}{14}$

**14.** $\dfrac{24}{16} + 3$

**15.** $\dfrac{17 - 5}{36} - \dfrac{2(13 - 5)}{12}$

**16.** $\dfrac{|-7 - (-6)|}{-7 - |-6|}$

**17.** Find 17% of 457 and round the answer to one decimal place.

**18.** Find the area of a rectangle 12.8 feet wide and 23.56 feet long. Round the answer to two decimal places.

**19.** Find the area of the figure in Illustration 1.

**20.** To the nearest cubic inch, find the volume of the solid in Illustration 2.

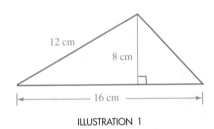

ILLUSTRATION 1

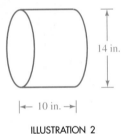

ILLUSTRATION 2

*In Problems 21–26, let $x = -2$, $y = 3$, and $z = 4$. Evaluate each expression.*

**21.** $xy + z$

**22.** $x(y + z)$

**23.** $\dfrac{z + 4y}{2x}$

**24.** $|x^y - z|$

**25.** $x^3 + y^2 + z$

**26.** $|x| - 3|y| - 4|z|$

**27.** Let $x$ and $y$ represent two real numbers. Write an algebraic expression to denote the quotient obtained when the product of the two numbers is divided by their sum.

**28.** Let $x$ and $y$ represent two real numbers. Write an algebraic expression to denote the difference obtained when the sum of $x$ and $y$ is subtracted from the product of 5 and $y$.

**29.** A man lives 12 miles from work and 7 miles from the grocery store. If he made $x$ round trips to work and $y$ round trips to the store, how many miles did he drive?

**30.** A baseball costs $\$a$ and a glove costs $\$b$. How much will it cost a community center to buy 12 baseballs and 8 gloves?

**31.** What is the numerical coefficient of the term $3xy^2$?

**32.** How many terms are in the expression $3x^2y + 5xy^2 + x + 7$?

**33.** What is the identity element for addition?

**34.** What is the multiplicative inverse of $\dfrac{1}{5}$?

*In Problems 35–38, state which property of the real numbers justifies each statement.*

**35.** $(xy)z = z(xy)$

**36.** $3(x + y) = 3x + 3y$

**37.** $2 + x = x + 2$

**38.** $7 \cdot \dfrac{1}{7} = 1$

# 2 Equations and Inequalities

Employees of most corporations participate in retirement plans funded by contributions from both the company and the employee. Persons who are self-employed can also fund a retirement plan. One such plan, called a Simplified Employee Pension (SEP), allows an annual contribution that does not exceed 15% of the income available after deductible expenses. That would seem to be an easy calculation—subtract deductible expenses from gross income and take 15% of what's left. However, the tax code is not so simple. The SEP contribution is considered a deductible expense. The SEP contribution is 15% of what is left after subtracting the deductible expenses and the amount of the SEP contribution. It would seem that to calculate the contribution, you must first know the contribution.

After reading this chapter, you will be able to calculate the maximum annual SEP contribution.

## 2.1 Solving Equations by Addition and Subtraction

■ EQUATIONS ■ SOLVING EQUATIONS ■ MARKDOWN AND MARKUP ■ GEOMETRY

Getting Ready *Fill in each blank to make a true statement.*

**1.** $3 +$ ___ $= 0$     **2.** $(-7) +$ ___ $= 0$     **3.** $(-4) +$ ___ $= 0$

**4.** $7 -$ ___ $= 0$     **5.** $5 + 3(4) =$ ___     **6.** $(-x) +$ ___ $= 0$

### ■ EQUATIONS

An **equation** is a statement indicating that two quantities are equal. Some examples of equations are

$$x + 5 = 21, \qquad 2x - 5 = 11, \qquad \text{and} \qquad 3x^2 - 4x + 5 = 0$$

The statement $3x + 2$ is not an equation, because it does not contain an $=$ sign.

In the equation $x + 5 = 21$, the expression $x + 5$ is called the **left-hand side,** and 21 is called the **right-hand side.** The letter $x$ is called the **variable** (or the **unknown**).

An equation can be true or false. The equation $16 + 5 = 21$ is true, but the equation $10 + 5 = 21$ is false. The equation $2x - 5 = 11$ might be true or false, depending on the value of $x$. If $x = 8$, the equation is true, because when we substitute 8 for $x$, we get 11.

$$2(8) - 5 = 16 - 5$$
$$= 11$$

Any number that makes an equation true when substituted for its variable is said to *satisfy* the equation. All of the numbers that satisfy an equation are called its **solu-**

**tions** or **roots.** Since 8 is the only number that satisfies the equation $2x - 5 = 11$, it is the only solution.

**EXAMPLE 1**   Is 6 a solution of $3x - 5 = 2x$?

*Solution*   We substitute 6 for $x$ and simplify.

$$3x - 5 = 2x$$
$$3 \cdot 6 - 5 \overset{?}{=} 2 \cdot 6 \qquad \text{Substitute 6 for } x.$$
$$18 - 5 \overset{?}{=} 12$$
$$13 = 12$$

Since $13 = 12$ is false, 6 is not a solution. ■

**Self Check**   Is 1 a solution of $2x + 3 = 5$?

*Answer*   yes

## ■ SOLVING EQUATIONS

To **solve an equation** means to find its solutions. To develop an understanding of how to solve equations, we refer to the scales shown in Figure 2-1. We can think of the scale shown in Figure 2-1(a) as representing the equation $x - 5 = 2$. The weight on the left-hand side of the scale is $(x - 5)$ grams, and the weight on the right-hand side is 2 grams. Because these weights are equal, the scale is in balance. To find $x$, we need to isolate it by adding 5 grams to the left-hand side of the scale. To keep the scale in balance, we must also add 5 grams to the right-hand side. After adding 5 grams to both sides of the scale, we can see from Figure 2-1(b) that $x$ grams will

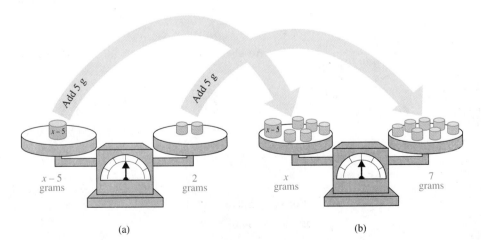

(a)                                        (b)

FIGURE 2-1

be balanced by 7 grams. We say that we have solved the equation and that the solution is 7.

The example suggests the following property of equality: *If the same quantity is added to equal quantities, the results will be equal quantities.* We can express this property in symbols.

---

### Addition Property of Equality

Suppose that $a$, $b$, and $c$ are real numbers. Then

$$\text{If } a = b, \text{ then } a + c = b + c.$$

---

When we use this property, the resulting equation will have the same solutions as the original one. We say that the equations are *equivalent.*

---

### Equivalent Equations

Two equations are **equivalent equations** when they have the same solutions.

---

In the previous example, we found that $x - 5 = 2$ is equivalent to $x = 7$. In the next example, we use the addition property of equality to solve the equation $x - 5 = 2$ algebraically.

**EXAMPLE 2**   Solve $x - 5 = 2$.

*Solution*   To isolate $x$ on one side of the $=$ sign, we undo the subtraction of 5 by adding 5 to both sides of the equation.

$$x - 5 = 2$$
$$x - 5 + 5 = 2 + 5 \qquad \text{Add 5 to both sides of the equation.}$$
$$x = 7 \qquad -5 + 5 = 0 \text{ and } 2 + 5 = 7.$$

We check by substituting 7 for $x$ in the original equation and simplifying.

$$x - 5 = 2$$
$$7 - 5 \overset{?}{=} 2 \qquad \text{Substitute 7 for } x.$$
$$2 = 2$$

Since $2 = 2$, the solution checks.   ■

**Self Check**   Solve $b - 21.8 = 13$.

*Answer*   34.8

■ ■ ■ ■ ■ ■ ■ ■ ■ ■ PERSPECTIVE

To find answers to such questions as How many? How far? How fast? and How heavy?, we often make use of mathematical statements called **equations.** The concept has a long history, and the techniques we will study in this chapter have been developed over many centuries.

The mathematical notation that we use today is the result of thousands of years of development. The ancient Egyptians used a word for variables, best translated as *heap*. Others used the word *res*, which is Latin for *thing*. In the fifteenth century, the letters *p*: and *m*: were used for *plus* and *minus*. What we would now write as $2x + 3 = 5$ might have been written by those early mathematicians as 2 *res p*: 3 *aequalis* 5.

We can think of the scale shown in Figure 2-2(a) as representing the equation $x + 4 = 9$. The weight on the left-hand side of the scale is $(x + 4)$ grams, and the weight on the right-hand side is 9 grams. Because these weights are equal, the scale is in balance. To find $x$, we need to isolate it by removing 4 grams from the left-hand side. To keep the scale in balance, we must also remove 4 grams from the right-hand side. In Figure 2-2(b), we can see that $x$ grams will be balanced by 5 grams. We have found that the solution is 5.

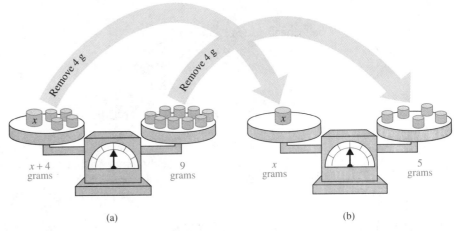

$x + 4$ grams | 9 grams | $x$ grams | 5 grams

(a)                (b)

FIGURE 2-2

The previous example suggests the following property of equality: *If the same quantity is subtracted from equal quantities, the results will be equal quantities.* We can express this property in symbols.

### Subtraction Property of Equality

Suppose that $a$, $b$, and $c$ are real numbers. Then

If $a = b$, then $a - c = b - c$.

When we use this property, the resulting equation will be equivalent to the original one.

In the next example, we use the subtraction property of equality to solve the equation $x + 4 = 9$ algebraically.

**EXAMPLE 3**   Solve $x + 4 = 9$.

*Solution*   To isolate $x$ on one side of the $=$ sign, we undo the addition of 4 by subtracting 4 from both sides of the equation.

$$x + 4 = 9$$
$$x + 4 - 4 = 9 - 4 \qquad \text{Subtract 4 from both sides.}$$
$$x = 5 \qquad 4 - 4 = 0 \text{ and } 9 - 4 = 5.$$

We can check the solution by substituting 5 for $x$ in the original equation and simplifying.

$$x + 4 = 9$$
$$5 + 4 \stackrel{?}{=} 9 \qquad \text{Substitute 5 for } x.$$
$$9 = 9$$

The solution checks. ∎

**Self Check**   Solve $a + 17.5 = 12.2$.
**Answer**   $-5.3$

## ■ MARKDOWN AND MARKUP

When the price of merchandise is reduced, the amount of the reduction is called the **markdown** or the **discount.** To find the sale price of an item, we subtract the markdown from the regular price.

$$\boxed{\text{Sale price}} \quad = \quad \boxed{\text{regular price}} \quad - \quad \boxed{\text{markdown}}$$

**EXAMPLE 4**   **Buying a sofa**   A sofa is on sale for $650. If it has been marked down $325, find its regular price.

*Solution*   We can let $r$ represent the regular price and substitute 650 for the sale price and 325 for the markdown.

| Sale price | = | regular price | − | markdown |
|:---:|:---:|:---:|:---:|:---:|
| 650 | = | $r$ | − | 325 |

We can use the addition property of equality to solve the equation.

$$650 = r - 325$$
$$650 + \mathbf{325} = r - 325 + \mathbf{325} \qquad \text{Add 325 to both sides.}$$
$$975 = r \qquad\qquad 650 + 325 = 975 \text{ and } -325 + 325 = 0.$$

The regular price is $975.  ∎

*Self Check*

Find the regular price of the sofa in Example 4 if the discount were $275.

*Answer*   $925

To make a profit, a merchant must sell an item for more than he or she paid for it. The retail price of the item is the sum of its wholesale cost and the **markup.**

| Retail price | = | wholesale cost | + | markup |
|---|---|---|---|---|

EXAMPLE 5   **Buying a car**   A car with a sticker price of $17,500 has a markup of $3,500. Find the invoice price (the wholesale price) to the dealer.

*Solution*   We can let $w$ represent the wholesale price and substitute 17,500 for the retail price and 3,500 for the markup.

| Retail price | = | wholesale cost | + | markup |
|---|---|---|---|---|
| 17,500 | = | $w$ | + | 3,500 |

We can use the subtraction property of equality to solve the equation.

$$17,500 = w + 3,500$$
$$17,500 - \mathbf{3,500} = w + 3,500 - \mathbf{3,500} \qquad \text{Subtract 3,500 from both sides.}$$
$$14,000 = w \qquad\qquad 17,500 - 3,500 = 14,000 \text{ and } 3,500 - 3,500 = 0.$$

The invoice price is $14,000.  ∎

*Self Check*

Find the invoice price of the car in Example 5 if the markup is $6,700.

*Answer*   $10,800

## ■ GEOMETRY

The geometric figure shown in Figure 2-3(a) is called an **angle.** Angles are measured in **degrees.** The angle shown in Figure 2-3(b) measures 45 degrees (denoted as 45°). If an angle measures 90°, as in Figure 2-3(c), it is called a **right angle.** If an angle measures 180°, it is called a **straight angle.**

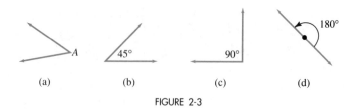

(a)          (b)          (c)          (d)

FIGURE 2-3

**EXAMPLE 6**    Find $x$.

*Solution*   Since the sum of the measures of the angles measuring 37° and $x$ is 75°, we can form and solve the equation

$$x + 37 = 75$$
$$x + 37 - 37 = 75 - 37 \qquad \text{Subtract 37 from both sides.}$$
$$x = 38 \qquad\qquad 37 - 37 = 0 \text{ and } 75 - 37 = 38.$$

Thus, $x = 38°$.                                                                ■

**Self Check**    Find $x$.

*Answer*    47°

**EXAMPLE 7**    Find $x$.

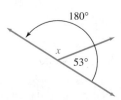

*Solution*   Since the sum of the measures of the angles measuring 53° and $x$ is 180°, we can form and solve the equation

$$x + 53 = 180$$
$$x + 53 - 53 = 180 - 53 \qquad \text{Subtract 53 from both sides.}$$
$$x = 127 \qquad\qquad 53 - 53 = 0 \text{ and } 180 - 53 = 127.$$

Thus, $x = 127°$.                                                              ■

*Self Check*    Find $x$.

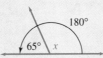

*Answer*    115°

If the sum of two angles is 90°, the angles are called **complementary.** If the sum of two angles is 180°, the angles are called **supplementary.**

**EXAMPLE 8**    Find   **a.** the complement of an angle measuring 30°   and   **b.** the supplement of an angle measuring 50°.

*Solution*    **a.** We can let $x$ represent the complement of 30°. Since the sum of two complementary angles is 90°, we have

$$x + 30 = 90$$
$$x + 30 - 30 = 90 - 30 \qquad \text{Subtract 30 from both sides.}$$
$$x = 60 \qquad\qquad 30 - 30 = 0 \text{ and } 90 - 30 = 60.$$

The complement of a 30° angle is a 60° angle.

**b.** We can let $x$ represent the supplement of 50°. Since the sum of two supplementary angles is 180°, we have

$$x + 50 = 180$$
$$x + 50 - 50 = 180 - 50 \qquad \text{Subtract 50 from both sides.}$$
$$x = 130 \qquad\qquad 50 - 50 = 0 \text{ and } 180 - 50 = 130.$$

The supplement of a 50° angle is a 130° angle.    ■

*Self Check*    Find   **a.** the supplement of 105°   and   **b.** the complement of 15°.
*Answers*       **a.** 75°,   **b.** 75°

*Orals*    *Solve each equation.*

**1.** $x - 9 = 11$          **2.** $x - 3 = 13$          **3.** $w + 5 = 7$

**4.** $x + 32 = 36$          **5.** $x - 2.5 = -2.5$          **6.** $x + 12.4 = 12.4$

**7.** $x + \dfrac{1}{5} = \dfrac{4}{5}$                         **8.** $x - \dfrac{2}{7} = \dfrac{5}{7}$

**9.** Find the complement of a 10°          **10.** Find the supplement of an 80°
     angle.                                     angle.

# EXERCISE 2.1

***REVIEW***   *Do the operations and classify the result as an integer, a prime number, or a composite number.*

**1.** $3[2 - (-3)]$

**2.** $(2 - 4)^4$

**3.** $\dfrac{2^3 - 14}{3^2 - 3}$

**4.** $\dfrac{3 + 5}{3} - \dfrac{5}{7 - 4}$

*Tell which property of real numbers justifies each statement.*

**5.** $3 + 31$ is a real number

**6.** $3(x + y) = 3x + 3y$

**7.** $a + (3 + b) = (3 + b) + a$

**8.** $a + (3 + b) = (a + 3) + b$

*Evaluate each expression.*

**9.** $4^3$

**10.** $(-3)^4$

**11.** $-3(4^2 - 5^2)$

**12.** $-6^2 + 5 \cdot 4$

***VOCABULARY AND CONCEPTS***   *Fill in each blank to make a true statement.*

**13.** An _____ is a statement that two quantities are equal.

**14.** A _____ of an equation is a number that satisfies the equation.

**15.** The answer to an equation is called a solution or a _____ of the equation.

**16.** A letter that represents a number is called a _____.

**17.** If two equations have the same solutions, they are called _____ equations.

**18.** To solve an equation, we isolate the _____ on one side of the equation.

**19.** The equation $3x - 2 = 7$ can be true or false, depending on the value of __.

**20.** If the same quantity is added to _____ quantities, the results will be equal quantities.

**21.** If the same quantity is subtracted from equal quantities, the results will be _____ quantities.

**22.** Sale price = _____ − markdown

**23.** Retail price = wholesale cost + _____

**24.** Another name for markdown is _____.

**25.** If the sum of two angles is 180°, the angles are called _____ angles.

**26.** If the sum of two angles is 90°, the angles are called _____ angles.

*In Exercises 27–34, tell whether each statement is an equation.*

**27.** $x = 2$

**28.** $y - 3$

**29.** $7x < 8$

**30.** $7 + x = 2$

**31.** $x + 7 = 0$

**32.** $3 - 3y > 2$

**33.** $1 + 1 = 3$

**34.** $5 = a + 2$

***PRACTICE***   *In Exercises 35–46, tell whether the given number is a solution of the equation.*

**35.** $x + 2 = 3$; 1

**36.** $x - 2 = 4$; 6

**37.** $a - 7 = 0$; $-7$

**38.** $x + 4 = 4$; 0

**39.** $\dfrac{y}{7} = 4$; 28

**40.** $\dfrac{c}{-5} = -2$; $-10$

**41.** $\dfrac{x}{5} = x$; 0

**42.** $\dfrac{x}{7} = 7x$; 0

**43.** $3k + 5 = 5k - 1;\ 3$

**44.** $2s - 1 = s + 7;\ 6$

**45.** $\dfrac{5 + x}{10} - x = \dfrac{1}{2};\ 0$

**46.** $\dfrac{x - 5}{6} = 12 - x;\ 11$

*In Exercises 47–66, use the addition or the subtraction property of equality to solve each equation.* **Check all solutions.**

**47.** $x + 7 = 13$  **48.** $y + 3 = 7$  **49.** $y - 7 = 12$  **50.** $c - 11 = 22$

**51.** $1 = y - 5$  **52.** $0 = r + 10$  **53.** $p - 404 = 115$  **54.** $41 = 45 + q$

**55.** $-37 + z = 37$  **56.** $-43 + a = -43$  **57.** $-57 = b - 29$  **58.** $-93 = 67 + y$

**59.** $\dfrac{4}{3} = -\dfrac{2}{3} + x$  **60.** $z + \dfrac{5}{7} = -\dfrac{2}{7}$  **61.** $d + \dfrac{2}{3} = \dfrac{3}{2}$  **62.** $s + \dfrac{2}{3} = \dfrac{1}{5}$

**63.** $-\dfrac{3}{5} = x - \dfrac{2}{5}$  **64.** $b + 7 = \dfrac{20}{3}$  **65.** $r - \dfrac{1}{5} = \dfrac{3}{10}$  **66.** $t + \dfrac{4}{7} = \dfrac{11}{14}$

**APPLICATIONS**   *Use an equation to solve each problem.*

**67. Buying a boat**   A boat is on sale for $7,995. Find its regular price if it has been marked down $1,350.

**68. Buying a house**   A house that was priced at $105,000 has been discounted $7,500. Find the new asking price.

**69. Buying clothes**   A sport jacket that sells for $175 has a markup of $85. Find the wholesale price.

**70. Buying a vacuum cleaner**   A vacuum that sells for $97 has a markup of $37. Find the wholesale price.

**71. Banking**   The amount $A$ in an account is given by the formula

$$A = p + i$$

where $p$ is the principal and $i$ is the interest. How much interest has been earned if an original deposit (the principal) of $4,750 has grown to be $5,010?

**72. Depreciation**   The current value $v$ of a car is given by the formula

$$v = c - d$$

where $c$ is the original price and $d$ is the depreciation. Find the original cost of a car that is worth $10,250 after depreciating $7,500.

**73. Appreciation**   The value $v$ of a house is given by the formula

$$v = p + a$$

where $p$ is the original purchase price and $a$ is the appreciation. Find the original purchase price of a house that is worth $110,000 and has appreciated $57,000.

**74. Taxes**   The cost $c$ of an item is given by the formula

$$c = p + t$$

where $p$ is the price and $t$ is the sales tax. Find the tax paid on an item that was priced at $37.10 and cost $39.32.

**75. Buying carpet**   The cost $c$ of carpet is given by the formula

$$c = p + t$$

where $p$ is the price and $t$ is the cost of installation. How much did it cost to install $317 worth of carpet that cost $512?

**76. Selling real estate**    The money $m$ the seller receives from selling a house is given by the formula

$$m = s - c$$

where $s$ is the selling price and $c$ is the agent's commission. Find the selling price of a house if the seller received $217,000 and the agent received $13,020.

**77. Buying real estate**    The cost of a condominium is $57,595 less than the cost of a house. If the house costs $202,744, find the cost of the condominium.

**78. Buying paint**    After reading the ad in Illustration 1, a decorator bought one gallon of primer, one gallon of paint, and a brush. The total cost was $30.44. Find the cost of the brush.

ILLUSTRATION 1

*GEOMETRY*    *In Exercises 79–84, find x.*

**79.**

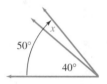

**80.**

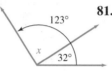

**81.**

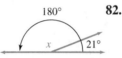

**82.**

**83.**

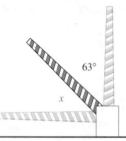

**84.**

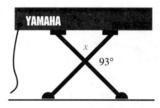

**85.** Find the complement of 37°.

**86.** Find the supplement of 37°.

**87.** Find the supplement of the complement of 40°.

**88.** Find the complement of the supplement of 140°.

### WRITING

**89.** Explain what it means for a number to satisfy an equation.

**90.** How can you tell whether a number is the solution of an equation?

**91.** Explain what Figure 2-1 is trying to show.

**92.** Explain what Figure 2-2 is trying to show.

### SOMETHING TO THINK ABOUT

**93.** If two lines intersect as in Illustration 2, angles 1 and 2 and angles 3 and 4 are called **vertical angles.** Let the measure of angle 1 be various numbers and compute the values of the other three. What do you discover?

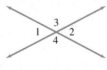

ILLUSTRATION 2

**94.** If two lines meet and form one right angle, the lines are said to be **perpendicular.** See Illustration 3. Find the measures of angles 1, 2, and 3. What do you discover?

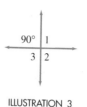

ILLUSTRATION 3

# 2.2  Solving Equations by Multiplication and Division

■ THE DIVISION PROPERTY OF EQUALITY ■ THE MULTIPLICATION PROPERTY OF EQUALITY ■ PERCENT
■ APPLICATIONS OF PERCENT

Getting Ready    *Fill in each blank to make a true statement.*

**1.** $\dfrac{1}{3} \cdot 3 =$ ▮

**2.** $5 \cdot$ ▮ $= 1$

**3.** $\dfrac{-6}{-6} =$ ▮

**4.** $\dfrac{4(2)}{\phantom{x}} = 2$

**5.** $5 \cdot \dfrac{4}{5} =$ ▮

**6.** $\dfrac{-5(3)}{-5} =$ ▮

**7.** $0.07 \cdot 900 =$ ▮

**8.** $0.09 \cdot 800 =$ ▮

## ■ THE DIVISION PROPERTY OF EQUALITY

We will now consider how to solve the equation $2x = 6$. Since $2x$ means $2 \cdot x$, the equation can be written as $2 \cdot x = 6$. We can think of the scale shown in Figure 2-4(a) as representing this equation. The weight on the left-hand side of the scale is $2 \cdot x$ grams, and the weight on the right-hand side is 6 grams. Because these weights

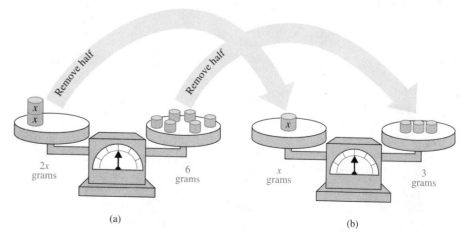

(a)

(b)

FIGURE 2-4

are equal, the scale is in balance. To find $x$, we remove half of the weight from each side. This is equivalent to dividing the weight on both sides by 2. When we do this, the scale will remain in balance. From the scale shown in Figure 2-4(b), we can see that $x$ grams will be balanced by 3 grams. Thus, $x = 3$.

The previous example suggests the following property of equality: *If equal quantities are divided by the same nonzero quantity, the results will be equal quantities.* We can express this property in symbols.

**Division Property of Equality**
Suppose that $a$, $b$, and $c$ are real numbers and that $c \neq 0$. Then

$$\text{If } a = b, \text{ then } \frac{a}{c} = \frac{b}{c}.$$

When we use the division property, the resulting equation will be equivalent to the original one.

To solve the equation $2x = 6$ algebraically, we proceed as in Example 1.

**EXAMPLE 1**   Solve $2x = 6$.

*Solution*   To isolate $x$ on one side of the $=$ sign, we undo the multiplication by 2 by dividing both sides by 2.

$$2x = 6$$
$$\frac{2x}{2} = \frac{6}{2} \qquad \text{Divide both sides by 2.}$$
$$x = 3 \qquad \tfrac{2}{2} = 1 \text{ and } \tfrac{6}{2} = 3.$$

Verify that the solution is 3.    ∎

*Self Check*   Solve $-5x = 15$.
*Answer*   $-3$

## THE MULTIPLICATION PROPERTY OF EQUALITY

We can think of the scale shown in Figure 2-5(a) as representing the equation $\frac{x}{3} = 12$. The weight on the left-hand side of the scale is $\frac{x}{3}$ grams, and the weight on the right-hand side is 12 grams. Because these weights are equal, the scale is in balance. To find $x$, we can triple (or multiply by 3) the weight on each side. When we do this, the scale will remain in balance. From the scale shown in Figure 2-5(b), we can see that $x$ grams will be balanced by 36 grams. Thus, $x = 36$.

The previous example suggests the following property of equality: *If equal quantities are multiplied by the same nonzero quantity, the results will be equal quantities.* We can express this property in symbols.

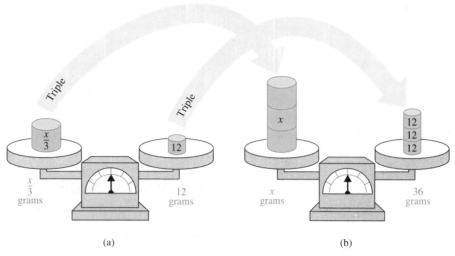

FIGURE 2-5

**Multiplication Property of Equality**

Suppose that $a$, $b$, and $c$ are real numbers, and $c \neq 0$. Then

If $a = b$, then $ca = cb$.

When we use the multiplication property, the resulting equation will be equivalent to the original one.

To solve the equation $\frac{x}{3} = 12$ algebraically, we proceed as in Example 2.

EXAMPLE 2    Solve $\dfrac{x}{3} = 12$.

*Solution*    To find $x$, we undo the division by 3 by multiplying both sides of the equation by 3.

$$\frac{x}{3} = 12$$

$$3 \cdot \frac{x}{3} = 3 \cdot 12 \qquad \text{Multiply both sides by 3.}$$

$$x = 36 \qquad 3 \cdot \tfrac{x}{3} = x \text{ and } 3 \cdot 12 = 36.$$

Verify that the solution checks.

Self Check    Solve $\frac{x}{5} = -7$.

*Answer*    $-35$

## ■ PERCENT

A percent is the numerator of a fraction whose denominator is 100. For example, $6\frac{1}{4}$ percent (written as $6\frac{1}{4}\%$) is the fraction $\frac{6.25}{100}$, or the decimal 0.0625. In problems involving percent, the word *of* usually indicates multiplication. For example, $6\frac{1}{4}\%$ of 8,500 is the product of 0.0625 and 8,500.

$$6\frac{1}{4}\% \text{ of } 8,500 = 0.0625 \cdot 8,500$$
$$= 531.25$$

In the statement $6\frac{1}{4}\%$ of $8,500 = 531.25$, the percent $6\frac{1}{4}\%$ is called a **rate,** 8,500 is called the **base,** and their product, 531.25, is called a **percentage.** Every percent problem is based on the equation **rate · base = percentage.**

> **Percentage Formula**
> The product of a rate $r$ and a base $b$ is called a **percentage.** If $p$ is the percentage, then
>
> $$rb = p$$

Percent problems involve questions such as

- What is 30% of 1,000?
- 45% of what number is 405?
- What percent of 400 is 60?

When we use equations, these problems are easy to solve.

**EXAMPLE 3**    What is 30% of 1,000?

*Solution*    In this problem, the rate $r$ is 30%, and the base is 1,000.

| Rate | · | base | = | percentage |
|------|---|------|---|------------|
| 30% | of | 1,000 | is | the percentage. |

We can substitute these values into the percentage formula and solve for $p$.

$$rb = p$$
$$30\% \cdot 1,000 = p \qquad \text{Substitute 30\% for } r \text{ and 1,000 for } b.$$
$$0.30 \cdot 1,000 = p \qquad \text{Change 30\% to the decimal 0.30.}$$
$$300 = p \qquad \text{Multiply.}$$

Thus, 30% of 1,000 is 300.    ■

*Self Check*    Find 45% of 800.

*Answer*    360

**EXAMPLE 4**

45% of what number is 405?

*Solution*    In this problem, the rate $r$ is 45%, and the percentage $p$ is 405.

| Rate | · | base | = | percentage |
|------|---|------|---|------------|
| 45% | of | what number | is | 405? |

We can substitute these values into the percentage formula and solve for $b$.

$$rb = p$$

$45\% \cdot b = 405$    Substitute 45% for $r$ and 405 for $p$.

$0.45 \cdot b = 405$    Change 45% to a decimal.

$\dfrac{0.45b}{0.45} = \dfrac{405}{0.45}$    To undo the multiplication by 0.45, divide both sides by 0.45.

$b = 900$    $\frac{0.45}{0.45} = 1$ and $\frac{405}{0.45} = 900$.

Thus, 45% of 900 is 405.    ■

**Self Check**    35% of what number is 306.25?
*Answer*    875

**EXAMPLE 5**

What percent of 400 is 60?

*Solution*    In this problem, the base $b$ is 400, and the percentage $p$ is 60.

| Rate | · | base | = | percentage |
|------|---|------|---|------------|
| What percent | of | 400 | is | 60? |

We can substitute these values in the percentage formula and solve for $r$.

$$rb = p$$

$r \cdot 400 = 60$    Substitute 400 for $b$ and 60 for $p$.

$\dfrac{400r}{400} = \dfrac{60}{400}$    To undo the multiplication by 400, divide both sides by 400.

$r = 0.15$    $\frac{400}{400} = 1$ and $\frac{60}{400} = 0.15$.

$r = 15\%$    To change the decimal into a percent, multiply by 100 and insert a % sign.

Thus, 15% of 400 is 60.    ■

## ■ APPLICATIONS OF PERCENT

**EXAMPLE 6**    **Investing**   At a recent stockholders' meeting, 4.5 million shares of stock were voted in favor of a proposal for a mandatory retirement age for members of the board of directors. Since this represented 75% of the number of shares outstanding, the proposal passed. How many shares are outstanding?

*Solution*    Let $b$ represent the number of outstanding shares. Then 75% of $b$ is 4.5 million. We can substitute 75% for $r$ and 4.5 million for $p$ in the formula for percentage and solve for $b$.

$$rb = p$$
$$75\% \cdot b = 4{,}500{,}000 \qquad \text{Substitute 75\% for } r \text{ and 4,500,000 for } p.$$
$$0.75b = 4{,}500{,}000 \qquad \text{Change 75\% to a decimal.}$$
$$\frac{0.75b}{0.75} = \frac{4{,}500{,}000}{0.75} \qquad \begin{array}{l}\text{To undo the multiplication of 0.75, divide both sides} \\ \text{by 0.75.}\end{array}$$
$$b = 6{,}000{,}000 \qquad \frac{0.75}{0.75} = 1 \text{ and } \frac{4{,}500{,}000}{0.75} = 6{,}000{,}000.$$

There are 6 million shares outstanding.    ■

**EXAMPLE 7**    **Quality control**   After examining 240 sweaters, a quality-control inspector found 5 with defective stitching, 8 with mismatched designs, and 2 with incorrect labels. What percent were defective?

*Solution*    Let $r$ represent the percent that are defective. Then the base $b$ is 240, and the percentage $p$ is the number of defective sweaters, which is $5 + 8 + 2 = 15$. We can find $r$ by solving the equation

$$rb = p$$
$$r \cdot 240 = 15 \qquad \text{Substitute 240 for } b \text{ and 15 for } p.$$
$$\frac{240r}{240} = \frac{15}{240} \qquad \text{To undo the multiplication of 240, divide both sides by 240.}$$
$$r = 0.0625 \qquad \tfrac{240}{240} = 1 \text{ and } \tfrac{15}{240} = 0.0625.$$
$$r = 6.25\% \qquad \begin{array}{l}\text{To change 0.0625 to a percent, multiply by 100 and add a} \\ \% \text{ sign.}\end{array}$$

The defect rate is $6\frac{1}{4}\%$.    ■

Self Check    In Example 7, if a second inspector found 3 sweaters with faded colors in addition to the defectives found by inspector 1, what percent were defective?

Answer    $7\frac{1}{2}\%$

Orals    *In Exercises 1–8, solve each equation.*

**1.** $3x = 3$        **2.** $5x = 5$        **3.** $-7x = 14$

**4.** $7.5x = 0$       **5.** $\dfrac{x}{5} = 2$        **6.** $\dfrac{x}{2} = -10$

**7.** $\dfrac{x}{-4} = 3$       **8.** $\dfrac{x}{8} = -3$

**9.** Change 30% to a decimal.       **10.** Change 0.08 to a percent.

# EXERCISE 2.2

***REVIEW***   *Do the operations. Simplify the result when possible.*

**1.** $\dfrac{4}{5} + \dfrac{2}{3}$        **2.** $\dfrac{5}{6} \cdot \dfrac{12}{25}$        **3.** $\dfrac{5}{9} \div \dfrac{3}{5}$        **4.** $\dfrac{15}{7} - \dfrac{10}{3}$

**5.** $2 + 3 \cdot 4$        **6.** $3 \cdot 4^2$        **7.** $3 + 4^3(-5)$        **8.** $\dfrac{5(-4) - 3(-2)}{10 - (-4)}$

*Find the area of each geometric figure.*

**9.** A rectangle with dimensions of 3.5 feet by 7.2 feet.       **10.** A circle with a diameter of 12.45 inches. Give the result to the nearest hundredth.

***VOCABULARY AND CONCEPTS***   *Fill in each blank to make a true statement.*

**11.** If equal quantities are divided by the same nonzero quantity, the results are _____ quantities.

**12.** If $a = b$, then $\dfrac{a}{c} = $ __ , provided that $c \neq$ __.

**13.** If $a = b$, then $ac = $ __.

**14.** If _____ quantities are multiplied by the same nonzero quantity, the results will be equal quantities.

**15.** A percent is the numerator of a fraction whose denominator is _____.

**16.** Rate · _____ = percentage

***PRACTICE***   *In Exercises 17–52, use the division or multiplication property of equality to solve each equation.* **Check all solutions.**

**17.** $6x = 18$        **18.** $25x = 625$        **19.** $-4x = 36$        **20.** $-16y = 64$

**21.** $4t = 108$        **22.** $-66 = -6t$        **23.** $11x = -121$        **24.** $-9y = -9$

**25.** $\dfrac{x}{5} = 5$  **26.** $\dfrac{x}{15} = 3$  **27.** $\dfrac{x}{32} = -2$  **28.** $\dfrac{y}{16} = -5$

**29.** $\dfrac{b}{3} = 5$  **30.** $\dfrac{a}{5} = -3$  **31.** $-3 = \dfrac{s}{11}$  **32.** $\dfrac{s}{-12} = 4$

**33.** $-32z = 64$  **34.** $15 = \dfrac{r}{-5}$  **35.** $18z = -9$  **36.** $-12z = 3$

**37.** $\dfrac{z}{7} = 14$  **38.** $-19x = -57$  **39.** $\dfrac{w}{7} = \dfrac{5}{7}$  **40.** $-17z = -51$

**41.** $\dfrac{s}{-3} = -\dfrac{5}{6}$  **42.** $1{,}228 = \dfrac{x}{0.25}$  **43.** $0.25x = 1{,}228$  **44.** $-255y = 51$

**45.** $\dfrac{b}{3} = \dfrac{1}{3}$  **46.** $\dfrac{a}{13} = \dfrac{1}{26}$  **47.** $-0.2w = -17$  **48.** $1.5a = -14$

**49.** $\dfrac{u}{5} = -\dfrac{3}{10}$  **50.** $\dfrac{t}{-7} = \dfrac{1}{2}$  **51.** $\dfrac{p}{0.2} = 12$  **52.** $\dfrac{t}{0.3} = -36$

*In Exercises 53–68, use the formula rb = p to find each value.*

**53.** What number is 40% of 200?
**54.** What number is 35% of 520?
**55.** What number is 50% of 38?
**56.** What number is 25% of 300?
**57.** 15% of what number is 48?
**58.** 26% of what number is 78?
**59.** 133 is 35% of what number?
**60.** 13.3 is 3.5% of what number?
**61.** 28% of what number is 42?
**62.** 44% of what number is 143?
**63.** What percent of 357.5 is 71.5?
**64.** What percent of 254 is 13.208?
**65.** 0.32 is what percent of 4?
**66.** 3.6 is what percent of 28.8?
**67.** 34 is what percent of 17?
**68.** 39 is what percent of 13?

**APPLICATIONS** *In Exercises 69–78, solve each problem.*

**69. Customer satisfaction** Two-thirds of a movie audience left the theater in disgust. If 78 angry patrons walked out, how many were there originally?

**70. Stock split** After a three-for-two stock split, each shareholder will own 1.5 times as many shares as before. If 555 shares are owned after the split, how many were owned before?

**71. Off-campus housing** Four-sevenths of the senior class is living in off-campus housing. If 868 students live off campus, how large is the senior class?

**72. Union membership** The 2,484 union members represent 90% of a factory's workforce. How many are employed?

**73. Shopper dissatisfaction** Refer to the survey results in Illustration 1. What percent of those surveyed were not pleased?

| Shopper survey results | |
| --- | --- |
| First-time shoppers | 1,731 |
| Major purchase today | 539 |
| Shopped within previous month | 1,823 |
| Satisfied with service | 4,140 |
| Seniors | 2,387 |
| Total surveyed | 9,200 |

ILLUSTRATION 1

**74. Charity overhead** Out of $237,000 donated to a certain charity, $5,925 was used to pay for fundraising expenses. What percent of donations was overhead?

**75. Selling price of a microwave oven** The 5% sales tax on a microwave oven amounts to $13.50. What is the microwave's selling price?

**76. Sales taxes** Sales tax on a $12 compact disc is $0.72. At what rate is sales tax computed?

**77. Hospital occupancy** 18% of hospital patients stay for less than one day. If 1,008 patients in January stayed for less than one day, what total number of patients did the hospital treat in January?

**78. House prices** The average price of houses in one neighborhood decreased 8% since last year, a drop of $7,800. What was the average price of a house last year?

*WRITING*

**79.** Explain how you would decide whether a number is a solution of an equation.

**80.** Distinguish between *percent* and *percentage*.

*SOMETHING TO THINK ABOUT*

**81.** The Ahmes Papyrus mentioned at the beginning of Chapter 1 contains this statement: *A circle nine units in diameter has the same area as a square eight units on a side.* From this statement, determine the ancient Egyptians' approximation of $\pi$.

**82.** Calculate the Egyptians' **percent of error:** What percent of the actual value of $\pi$ is the difference between the values?

---

## 2.3 Solving More Equations

■ SOLVING MORE COMPLICATED EQUATIONS  ■ MARKUP AND MARKDOWN

Getting Ready  *Do the operations.*

**1.** $7 + 3 \cdot 5$  **2.** $3(5 + 7)$  **3.** $\dfrac{3 + 7}{2}$  **4.** $3 + \dfrac{7}{2}$

**5.** $\dfrac{3(5 - 8)}{9}$  **6.** $3 \cdot \dfrac{5 - 8}{9}$  **7.** $\dfrac{3 \cdot 5 - 8}{9}$  **8.** $3 \cdot \dfrac{5}{9} - 8$

■ SOLVING MORE COMPLICATED EQUATIONS

We have solved equations by using the addition, subtraction, multiplication, and division properties of equality. To solve more complicated equations, we need to use several of these properties in succession.

**EXAMPLE 1**   Solve $-12x + 5 = 17$.

*Solution*   The left-hand side of the equation indicates that $x$ is to be multiplied by $-12$ and then 5 is to be added to that product. To isolate $x$, we must undo these operations in the opposite order.

- To undo the addition of 5, we subtract 5 from both sides.
- To undo the multiplication by $-12$, we divide both sides by $-12$.

$$-12x + 5 = 17$$

$$-12x + 5 - 5 = 17 - 5 \qquad \text{To undo the addition of 5, subtract 5 from both sides.}$$

$$-12x = 12 \qquad \text{5} - 5 = 0 \text{ and } 17 - 5 = 12.$$

$$\frac{-12x}{-12} = \frac{12}{-12} \qquad \text{To undo the multiplication by } -12, \text{ divide both sides by } -12.$$

$$x = -1 \qquad \tfrac{-12}{-12} = 1 \text{ and } \tfrac{12}{-12} = -1.$$

*Check:* $-12x + 5 = 17$

$$-12(-1) + 5 \stackrel{?}{=} 17 \qquad \text{Substitute } -1 \text{ for } x.$$

$$12 + 5 \stackrel{?}{=} 17 \qquad \text{Simplify.}$$

$$17 = 17$$

Because $17 = 17$, the solution checks.  ∎

**Self Check**   Solve $2x + 3 = 15$.
**Answer**   6

**EXAMPLE 2**   Solve $\dfrac{x}{3} - 7 = -3$.

*Solution*   The left-hand side of the equation indicates that $x$ is to be divided by 3 and then 7 is to be subtracted from that quotient. To isolate $x$, we must undo these operations in the opposite order.

- To undo the subtraction of 7, we add 7 to both sides.
- To undo the division by 3, we multiply both sides by 3.

$$\frac{x}{3} - 7 = -3$$

$$\frac{x}{3} - 7 + 7 = -3 + 7 \qquad \text{To undo the subtraction of 7, add 7 to both sides.}$$

$$\frac{x}{3} = 4 \qquad -7 + 7 = 0 \text{ and } -3 + 7 = 4.$$

$$3 \cdot \frac{x}{3} = 3 \cdot 4 \qquad \text{To undo the division by 3, multiply both sides by 3.}$$

$$x = 12 \qquad 3 \cdot \tfrac{1}{3} = 1 \text{ and } 3 \cdot 4 = 12.$$

*Check:* $\dfrac{x}{3} - 7 = -3$

$$\frac{12}{3} - 7 \stackrel{?}{=} -3 \qquad \text{Substitute 12 for } x.$$

$$4 - 7 \stackrel{?}{=} -3 \qquad \text{Simplify.}$$

$$-3 = -3$$

Since $-3 = -3$, the solution checks. ■

**Self Check**  Solve $\frac{x}{4} - 3 = 5$.
**Answer**  32

**EXAMPLE 3**  Solve $\dfrac{x-7}{3} = 9$.

*Solution*  The left-hand side of the equation indicates that 7 is to be subtracted from $x$ and that the difference is to be divided by 3. To isolate $x$, we must undo these operations in the opposite order.

• To undo the division by 3, we multiply both sides by 3.
• To undo the subtraction of 7, we add 7 to both sides.

$$\frac{x-7}{3} = 9$$

$$3\left(\frac{x-7}{3}\right) = 3(9) \qquad \text{To undo the division by 3, multiply both sides by 3.}$$

$$x - 7 = 27 \qquad 3 \cdot \tfrac{1}{3} = 1 \text{ and } 3(9) = 27.$$

$$x - 7 + 7 = 27 + 7 \qquad \text{To undo the subtraction of 7, add 7 to both sides.}$$

$$x = 34 \qquad -7 + 7 = 0 \text{ and } 27 + 7 = 34.$$

Verify that the solution checks. ■

**Self Check**  Solve $\frac{a-3}{5} = -2$.
**Answer**  $-7$

**EXAMPLE 4** Solve $\dfrac{3x}{4} + 2 = -7$.

*Solution* The left-hand side of the equation indicates that $x$ is to be multiplied by 3, then $3x$ is to be divided by 4, and then 2 is to be added to that result. To isolate $x$, we must undo these operations in the opposite order.

- To undo the addition of 2, we subtract 2 from both sides.
- To undo the division by 4, we multiply both sides by 4.
- To undo the multiplication by 3, we divide both sides by 3.

$$\frac{3x}{4} + 2 = -7$$

$$\frac{3x}{4} + 2 - 2 = -7 - 2 \qquad \text{To undo the addition of 2, subtract 2 from both sides.}$$

$$\frac{3x}{4} = -9 \qquad 2 - 2 = 0 \text{ and } -7 - 2 = -9.$$

$$4\left(\frac{3x}{4}\right) = 4(-9) \qquad \text{To undo the division by 4, multiply both sides by 4.}$$

$$3x = -36 \qquad 4 \cdot \tfrac{3}{4} = 3 \text{ and } 4(-9) = -36.$$

$$\frac{3x}{3} = \frac{-36}{3} \qquad \text{To undo the multiplication by 3, divide both sides by 3.}$$

$$x = -12 \qquad \tfrac{3}{3} = 1 \text{ and } \tfrac{-36}{3} = -12.$$

Verify that the solution checks. ∎

**Self Check** Solve $\frac{2x}{3} - 4 = 12$.

**Answer** 24

**EXAMPLE 5** **Advertising** A store manager hires a student to distribute advertising circulars door to door. The student will be paid $24 a day plus 12¢ for every ad distributed. How many circulars must she distribute to earn $42 in one day?

*Solution* We can let $a$ represent the number of circulars that the student must distribute. Her earnings can be expressed in two ways: as $24 plus the 12¢-apiece cost of distributing the circulars, and as $42.

| $24 | plus | $a$ ads at $0.12 each | is | $42. | $12¢ = \$0.12.$ |
|------|------|-----------------------|----|----|------------------|
| 24 | + | 0.12$a$ | = | 42 | |

We can solve this equation as follows:

$$24 + 0.12a = 42$$

$$24 - 24 + 0.12a = 42 - 24 \qquad \text{To undo the addition of 24, subtract 24 from both sides.}$$

$$0.12a = 18 \qquad 24 - 24 = 0 \text{ and } 42 - 24 = 18.$$

$$\frac{0.12a}{0.12} = \frac{18}{0.12} \qquad \text{To undo the multiplication by 0.12, divide both sides by 0.12.}$$

$$a = 150 \qquad \frac{0.12}{0.12} = 1 \text{ and } \frac{18}{0.12} = 150.$$

The student must distribute 150 ads. Check the result. ■

| | |
|---|---|
| Self Check | How many circulars must the student in Example 5 deliver in one day to earn $48? |
| *Answer* | 200 |

## ■  MARKUP AND MARKDOWN

We have seen that the retail price of an item is the sum of the cost and the markup.

$$\boxed{\text{Retail price}} \quad = \quad \boxed{\text{cost}} \quad + \quad \boxed{\text{markup}}$$

Often, the markup is expressed as a **percent of cost.**

$$\boxed{\text{Markup}} \quad = \quad \boxed{\text{percent of markup}} \quad \cdot \quad \boxed{\text{cost}}$$

Suppose a store manager buys toasters for $21 and sells them at a 17% markup. To find the retail price, the manager begins with his cost and adds 17% of that cost.

$$\boxed{\text{Retail price}} \quad = \quad \boxed{\text{cost}} \quad + \quad \boxed{\text{markup}}$$

$$= \quad \boxed{\text{cost}} \quad + \quad \boxed{\text{percent of markup}} \quad \cdot \quad \boxed{\text{cost}}$$

$$= \quad 21 \quad + \quad 0.17 \quad \cdot \quad 21$$

$$= 21 + 3.57$$

$$= 24.57$$

The retail price of a toaster is $24.57.

**EXAMPLE 6**   **Antique cars**   In 1956, a Chevrolet BelAir automobile sold for $4,000. Today, it is worth about $28,600. Find the **percent of increase.**

*Solution*   We let $p$ represent the percent of increase, expressed as a decimal.

| Current price | = | original price | + | $p$(original price) |
|---|---|---|---|---|
| 28,600 | = | 4,000 | + | $p(4,000)$ |

$28,600 - 4,000 = 4,000 - 4,000 + 4,000p$  To undo the addition of 4,000, subtract 4,000 from both sides.

$24,600 = 4,000p$  28,600 − 4,000 = 24,600 and 4,000 − 4,000 = 0.

$\dfrac{24,600}{4,000} = \dfrac{4,000p}{4,000}$  To undo the multiplication by 4,000, divide both sides by 4,000.

$6.15 = p$  Simplify.

To convert 6.15 to a percent, we multiply by 100 and insert a % sign. Since the percent of increase is 615%, the car has appreciated 615%.  ∎

**Self Check**  Find the percent of increase in Example 6 if the car sells for $30,000.
**Answer**   650%

We have seen that when the price of merchandise is reduced, the amount of reduction is the **markdown** (also called the **discount**).

| Sale price | = | regular price | − | markdown |
|---|---|---|---|---|

Usually, the markdown is expressed as a percent of the regular price.

| Markdown | = | percent of markdown | · | regular price |
|---|---|---|---|---|

Suppose that a television set that regularly sells for $570 has been marked down 25%. That means the customer will pay 25% less than the regular price. To find the sale price, we use the formula

| Sale price | = | regular price | − | markdown |
|---|---|---|---|---|
| | = | Regular price | − | percent of markdown · regular price |
| | = | $570 | − | 25% of $570 |

$$= \$570 - (0.25)(\$570) \qquad 25\% = 0.25.$$
$$= \$570 - \$142.50$$
$$= \$427.50$$

The television set is selling for $427.50.

**EXAMPLE 7**  **Buying a camera**  A camera that was originally priced at $452 is on sale for $384.20. Find the percent of markdown.

*Solution*  We let $p$ represent the percent of discount, expressed as a decimal, and substitute $384.20 for the sale price and $452 for the regular price.

| Sale price | = | Regular price | − | Percent of markdown | · | Regular price |
|:---:|:---:|:---:|:---:|:---:|:---:|:---:|
| 384.20 | = | 452 | − | $p$ | · | 452 |

$$384.20 - 452 = 452 - 452 - p(452)$$  To undo the addition of 452, subtract 452 from both sides.

$$-67.80 = -p(452)$$  $384.20 - 452 = -67.80$ and $452 - 452 = 0$.

$$\frac{-67.80}{-452} = \frac{-p(452)}{-452}$$  To undo the multiplication by $-452$, divide both sides by $-452$.

$$0.15 = p$$  $\frac{-67.80}{-452} = 0.15$ and $\frac{-452}{-452} = 1$.

The camera is on sale at a 15% markdown.  ∎

**Self Check**  If the camera in Example 7 is reduced another $23, find the percent of discount.
**Answer**  20%

**WARNING!**  When a price increases from $100 to $125, the percent of increase is 25%. When the price *decreases* from $125 to $100, the percent of decrease is 20%. These different results occur because the percent of increase is a percent of the original (smaller) price, $100. The percent of decrease is a percent of the original (larger) price, $125.

**Orals**  *What would you do first when solving each equation?*

**1.** $5x - 7 = -12$

**2.** $15 = \dfrac{x}{5} + 3$

**3.** $\dfrac{x}{7} - 3 = 0$

**4.** $\dfrac{x - 3}{7} = -7$

**5.** $5w - 5 = 5$

**6.** $5w + 5 = 5$

**7.** $\dfrac{x - 7}{3} = 5$

**8.** $\dfrac{3x - 5}{2} + 2 = 0$

*Find the value of the variable in each equation.*

**9.** $7z - 7 = 14$

**10.** $\dfrac{t - 1}{2} = 6$

# EXERCISE 2.3

**REVIEW** *Refer to the formulas given in Section 1.3.*

**1.** Find the perimeter of a rectangle with sides measuring 8.5 cm and 16.5 cm.

**2.** Find the area of a rectangle with sides measuring 2.3 in. and 3.7 in.

**3.** Find the area of a trapezoid with a height of 8.5 in. and bases measuring 6.7 in. and 12.2 in.

**4.** Find the volume of a rectangular solid with dimensions of 8.2 cm by 7.6 cm by 10.2 cm.

**VOCABULARY AND CONCEPTS** *Fill in each blank to make a true statement.*

**5.** Retail price = _____ + markup

**6.** Markup = percent of markup · _____.

**7.** Markdown = _____ of markdown · regular price

**8.** Another word for markdown is _____.

**PRACTICE** *In Exercises 9–60, solve each equation.* **Check all solutions.**

**9.** $5x - 1 = 4$

**10.** $5x + 3 = 8$

**11.** $6x + 2 = -4$

**12.** $4x - 4 = 4$

**13.** $3x - 8 = 1$

**14.** $7x - 19 = 2$

**15.** $11x + 17 = -5$

**16.** $13x - 29 = -3$

**17.** $43t + 72 = 158$

**18.** $96t + 23 = -265$

**19.** $-47 - 21s = 58$

**20.** $-151 + 13s = -229$

**21.** $2y - \dfrac{5}{3} = \dfrac{4}{3}$

**22.** $9y + \dfrac{1}{2} = \dfrac{3}{2}$

**23.** $-4y - 12 = -20$

**24.** $-8y + 64 = -32$

**25.** $\dfrac{x}{3} - 3 = -2$

**26.** $\dfrac{x}{7} + 3 = 5$

**27.** $\dfrac{z}{9} + 5 = -1$

**28.** $\dfrac{y}{5} - 3 = 3$

**29.** $\dfrac{b}{3} + 5 = 2$

**30.** $\dfrac{a}{5} - 3 = -4$

**31.** $\dfrac{s}{11} + 9 = 6$

**32.** $\dfrac{r}{12} + 2 = 4$

**33.** $\dfrac{k}{5} - \dfrac{1}{2} = \dfrac{3}{2}$

**34.** $\dfrac{y}{5} - \dfrac{8}{7} = -\dfrac{1}{7}$

**35.** $\dfrac{w}{16} + \dfrac{5}{4} = 1$

**36.** $\dfrac{m}{7} - \dfrac{1}{14} = \dfrac{1}{14}$

**37.** $\dfrac{b + 5}{3} = 11$

**38.** $\dfrac{2 + a}{13} = 3$

**39.** $\dfrac{r + 7}{3} = 4$

**40.** $\dfrac{t - 2}{7} = -3$

**41.** $\dfrac{u - 2}{5} = 1$

**42.** $\dfrac{v - 7}{3} = -1$

**43.** $\dfrac{x - 4}{4} = -3$

**44.** $\dfrac{3 + y}{5} = -3$

**45.** $\dfrac{3x}{2} - 6 = 9$

**46.** $\dfrac{5x}{7} + 3 = 8$

**47.** $\dfrac{3y}{2} + 5 = 11$

**48.** $\dfrac{5z}{3} + 3 = -2$

**49.** $\dfrac{3x - 12}{2} = 9$

**50.** $\dfrac{5x + 10}{7} = 0$

**51.** $\dfrac{5k - 8}{9} = 1$

**52.** $\dfrac{2x - 1}{3} = -5$

**53.** $\dfrac{3z + 2}{17} = 0$

**54.** $\dfrac{10t - 4}{2} = 1$

**55.** $\dfrac{17k - 28}{21} + \dfrac{4}{3} = 0$

**56.** $\dfrac{5a - 2}{3} = \dfrac{1}{6}$

**57.** $-\dfrac{x}{3} - \dfrac{1}{2} = -\dfrac{5}{2}$

**58.** $\dfrac{17 - 7a}{8} = 2$

**59.** $\dfrac{9 - 5w}{15} = \dfrac{2}{5}$

**60.** $\dfrac{3t - 5}{5} + \dfrac{1}{2} = -\dfrac{19}{2}$

## APPLICATIONS

**61. Integer problem**   Six less than 3 times a certain number is 9. Find the number.

**62. Integer problem**   If a certain number is increased by 7 and that result is divided by 2, the number 5 is obtained. Find the original number.

**63. Apartment rental**   A student moves into a bigger apartment that rents for $400 per month. That rent is $100 less than twice what she had been paying. Find her former rent.

**64. Auto repair**   A mechanic charged $20 an hour to repair the water pump on a car, plus $95 for parts. If the total bill was $155, how many hours did the repair take?

**65. Boarding dogs**   A sportsman boarded his dog at a kennel for $16 plus $12 a day. If the stay cost $100, how many days was the owner gone?

**66. Water billing**   The city's water department charges $7 per month, plus 42¢ for every 100 gallons of water used. Last month, one homeowner used 1,900 gallons and received a bill for $17.98. Was the billing correct?

**67. Telephone charges**   A call to Tucson from a pay phone in Chicago costs 85¢ for the first minute and 27¢ for each additional minute or portion of a minute. If a student has $8.50 in change, how long can she talk?

**68. Monthly sales**   A clerk's sales in February were $2,000 less than three times her sales in January. If her February sales were $7,000, by what amount did her sales increase?

**69. Ticket sales**   A music group charges $1,500 for each performance, plus 20% of the total ticket sales. After a concert, the group received $2,980. How much money did the ticket sales raise?

**70. Getting an A**   To receive a grade of A, the average of four 100-point exams must be 90 or better. If a student received scores of 88, 83, and 92 on the first three exams, what score does he need on the fourth exam to earn an A?

**71. Getting an A**   The grade in history class is based on the average of five 100-point exams. One student received scores of 85, 80, 95, and 78 on the first four exams. With an average of 90 needed, what chance does he have for an A?

**72. Excess inventory**   From the portion of the ad shown in Illustration 1, determine the sale price of a shirt.

ILLUSTRATION 1

**73. Clearance sales**   Sweaters already on sale for 20% off the regular price cost $36 when purchased with a promotional coupon that allows an additional 10% discount. Find the original price. (*Hint:* When you save 20%, you are paying 80%.)

**74. Furniture sale**    A $1,250 sofa is marked down to $900. Find the percent of markdown.

**75. Value of coupons**    The percent discount offered by the coupon in Illustration 2 depends on the amount purchased. Find the range of the percent discount.

**76. Furniture pricing**    A bedroom set selling for $1,900 cost $1,000 wholesale. Find the percent markup.

---

Value coupon
Save $15

on purchases of $100 to $250.

---

ILLUSTRATION 2

## WRITING

**77.** In solving the equation $5x - 3 = 12$, explain why you would add 3 to both sides first, rather than dividing by 5 first.

**78.** To solve the equation $\frac{3x - 4}{7} = 2$, what operations would you perform, and in what order?

## SOMETHING TO THINK ABOUT

**79.** Suppose you must solve the following equation but you can't quite read one number. It reads

$$\frac{7x + \#}{22} = \frac{1}{2}$$

If the solution of the equation is 1, what is the equation?

**80.** A store manager first increases his prices by 30% and then advertises

SALE!! 30% savings!!

What is the real percent discount to customers?

## 2.4  Simplifying Expressions to Solve Equations

■ LIKE TERMS ■ COMBINING LIKE TERMS ■ SOLVING EQUATIONS ■ IDENTITIES AND IMPOSSIBLE EQUATIONS

Getting Ready    *Use the distributive property to remove parentheses.*

**1.** $(3 + 4)x$                     **2.** $(7 + 2)x$
**3.** $(8 - 3)w$                     **4.** $(10 - 4)y$

*Simplify each expression by doing the operations within the parentheses.*

**5.** $(3 + 4)x$                     **6.** $(7 + 2)x$
**7.** $(8 - 3)w$                     **8.** $(10 - 4)y$

## ■  LIKE TERMS

Recall that a *term* is either a number or the product of numbers and variables. Some examples of terms are $7x$, $-3xy$, $y^2$, and 8. The number part of each term is called its **numerical coefficient.**

- The numerical coefficient of $7x$ is $7$.
- The numerical coefficient of $-3xy$ is $-3$.
- The numerical coefficient of $y^2$ is the understood factor of $1$.
- The numerical coefficient of $8$ is $8$.

**Like Terms**
**Like terms,** or **similar terms,** are terms with exactly the same variables and exponents.

The terms $3x$ and $5x$ are **like terms,** as are $9x^2$ and $-3x^2$. The terms $4xy$ and $3x^2$ are **unlike terms,** because they have different variables. The terms $4x$ and $5x^2$ are unlike terms, because the variables have different exponents.

## ■ COMBINING LIKE TERMS

The distributive property can be used to combine terms of algebraic expressions that contain sums or differences of like terms. For example, the terms in $3x + 5x$ and $9xy^2 - 11xy^2$ can be combined as follows:

$$3x + 5x = (3 + 5)x \qquad\qquad 9xy^2 - 11xy^2 = (9 - 11)xy^2$$
$$= 8x \qquad\qquad\qquad\qquad\qquad = -2xy^2$$

These examples suggest the following rule.

**Combining Like Terms**
To combine like terms, add their numerical coefficients and keep the same variables and exponents.

**WARNING!**   If the terms of an expression are unlike terms, they cannot be combined. For example, since the terms in $9xy^2 - 11x^2y$ have variables with different exponents, they are unlike terms and cannot be combined.

**EXAMPLE 1**    Simplify $3(x + 2) + 2(x - 8)$.

*Solution*
$$3(x + 2) + 2(x - 8)$$

| | |
|---|---|
| $= 3x + 3 \cdot 2 + 2x - 2 \cdot 8$ | Use the distributive property to remove parentheses. |
| $= 3x + 6 + 2x - 16$ | $3 \cdot 2 = 6$ and $2 \cdot 8 = 16$. |
| $= 3x + 2x + 6 - 16$ | Use the commutative property of addition: $6 + 2x = 2x + 6$. |
| $= 5x - 10$ | Combine like terms.                     ■ |

Self Check   Simplify $-5(a + 3) + 2(a - 5)$.

Answer   $-3a - 25$

**EXAMPLE 2**   Simplify $3(x - 3) - 5(x + 4)$.

Solution
$$3(x - 3) - 5(x + 4)$$
$$= 3(x - 3) + (-5)(x + 4) \qquad a - b = a + (-b).$$
$$= 3x - 3 \cdot 3 + (-5)x + (-5)4 \qquad \text{Use the distributive property to remove parentheses.}$$
$$= 3x - 9 + (-5x) + (-20) \qquad 3 \cdot 3 = 9 \text{ and } (-5)(4) = -20.$$
$$= -2x - 29 \qquad \text{Combine like terms.} \quad \blacksquare$$

Self Check   Simplify $-3(b - 2) - 4(b - 4)$.

Answer   $-7b + 22$

## ■ SOLVING EQUATIONS

To solve an equation, we must isolate the variable on one side. This is often a multistep process that may require combining like terms. As we solve equations, we will follow these steps.

**Solving Equations**
1. Clear the equation of fractions.
2. Use the distributive property to remove parentheses.
3. Combine like terms if necessary.
4. Undo the operations of addition and subtraction to get the variables on one side and the constants on the other.
5. Combine like terms and undo the operations of multiplication and division to isolate the variable.

**EXAMPLE 3**   Solve $3(x + 2) - 5x = 0$.

Solution
$$3(x + 2) - 5x = 0$$
$$3x + 3 \cdot 2 - 5x = 0 \qquad \text{Use the distributive property to remove parentheses.}$$
$$3x - 5x + 6 = 0 \qquad \text{Rearrange terms and simplify.}$$
$$-2x + 6 = 0 \qquad \text{Combine like terms.}$$

$$-2x + 6 - \mathbf{6} = 0 - \mathbf{6}$$   Subtract 6 from both sides.

$$-2x = -6$$   Combine like terms.

$$\frac{-2x}{\mathbf{-2}} = \frac{-6}{\mathbf{-2}}$$   Divide both sides by $-2$.

$$x = 3$$   Simplify.

*Check:* $3(x + 2) - 5x = 0$

$$3(\mathbf{3} + 2) - 5 \cdot \mathbf{3} \overset{?}{=} 0$$   Substitute 3 for $x$.

$$3 \cdot 5 - 5 \cdot 3 \overset{?}{=} 0$$

$$15 - 15 \overset{?}{=} 0$$

$$0 = 0$$   ■

**Self Check**   Solve $-2(y - 3) - 4y = 0$.
*Answer*   1

**EXAMPLE 4**   Solve $3(x - 5) = 4(x + 9)$.

*Solution*
$$3(x - 5) = 4(x + 9)$$

$$3x - 15 = 4x + 36$$   Remove parentheses.

$$3x - 15 - \mathbf{3x} = 4x + 36 - \mathbf{3x}$$   Subtract $3x$ from both sides.

$$-15 = x + 36$$   Combine like terms.

$$-15 - \mathbf{36} = x + 36 - \mathbf{36}$$   Subtract 36 from both sides.

$$-51 = x$$   Combine like terms.

$$x = -51$$

*Check:* $3(x - 5) = 4(x + 9)$

$$3(\mathbf{-51} - 5) \overset{?}{=} 4(\mathbf{-51} + 9)$$   Substitute $-51$ for $x$.

$$3(-56) \overset{?}{=} 4(-42)$$

$$-168 = -168$$   ■

**Self Check**   Solve $4(z + 3) = -3(z - 4)$.
*Answer*   0

**EXAMPLE 5**   Solve $\dfrac{3x + 11}{5} = x + 3$.

*Solution*   We first multiply both sides by 5 to clear the equation of fractions. When we multiply the right-hand side by 5, we must multiply the *entire* right-hand side by 5.

$$\frac{3x + 11}{5} = x + 3$$

$$5\left(\frac{3x + 11}{5}\right) = 5(x + 3) \qquad \text{Multiply both sides by 5.}$$

$$3x + 11 = 5x + 15 \qquad \text{Remove parentheses.}$$

$$3x + 11 - \mathbf{11} = 5x + 15 - \mathbf{11} \qquad \text{Subtract 11 from both sides.}$$

$$3x = 5x + 4 \qquad \text{Combine like terms.}$$

$$3x - \mathbf{5x} = 5x + 4 - \mathbf{5x} \qquad \text{Subtract } 5x \text{ from both sides.}$$

$$-2x = 4 \qquad \text{Combine like terms.}$$

$$\frac{-2x}{-2} = \frac{4}{-2} \qquad \text{Divide both sides by } -2.$$

$$x = -2 \qquad \text{Simplify.}$$

$$\textit{Check: } \frac{3x + 11}{5} = x + 3$$

$$\frac{3(-2) + 11}{5} \overset{?}{=} (-2) + 3 \qquad \text{Substitute } -2 \text{ for } x.$$

$$\frac{-6 + 11}{5} \overset{?}{=} 1 \qquad \text{Simplify.}$$

$$\frac{5}{5} \overset{?}{=} 1$$

$$1 = 1 \qquad \blacksquare$$

**Self Check**    Solve $\frac{2x - 5}{4} = x - 2$.

**Answer**    $\frac{3}{2}$

 **WARNING!**   Remember that when you multiply one side of an equation by a nonzero number, you must multiply the other side of the equation by the same number.

**EXAMPLE 6**    Solve $0.2x + 0.4(50 - x) = 19$.

*Solution*    Since $0.2 = \frac{2}{10}$ and $0.4 = \frac{4}{10}$, this equation contains fractions. To clear the fractions, we multiply both sides by 10.

$$0.2x + 0.4(50 - x) = 19$$

$$\mathbf{10}[0.2x + 0.4(50 - x)] = \mathbf{10}(19) \qquad \text{Multiply both sides by 10.}$$

$$\mathbf{10}[0.2x] + \mathbf{10}[0.4(50 - x)] = \mathbf{10}(19) \qquad \text{Use the distributive property on the left-hand side.}$$

$$2x + 4(50 - x) = 190 \qquad \text{Do the multiplications.}$$

$$2x + 200 - 4x = 190 \qquad \text{Remove parentheses.}$$

$$-2x + 200 = 190 \qquad \text{Combine like terms.}$$
$$-2x = -10 \qquad \text{Subtract 200 from both sides.}$$
$$x = 5 \qquad \text{Divide both sides by } -2.$$

Verify that the solution checks. ∎

**Self Check** Solve $0.3(20 - x) + 0.5x = 15$.

**Answer** 45

## ■ IDENTITIES AND IMPOSSIBLE EQUATIONS

An equation that is true for all values of its variable is called an **identity.** For example, the equation $x + x = 2x$ is an identity because it is true for all values of $x$.

Because no number can equal a number that is 1 larger than itself, the equation $x = x + 1$ is not true for any number $x$. Such equations are called **impossible equations** or **contradictions.**

The equations in Examples 3–6 are called **conditional equations.** For these equations, some values of $x$ are solutions, but other values of $x$ are not.

**EXAMPLE 7** Solve $3(x + 8) + 5x = 2(12 + 4x)$.

*Solution*
$$3(x + 8) + 5x = 2(12 + 4x)$$
$$3x + 24 + 5x = 24 + 8x \qquad \text{Remove parentheses.}$$
$$8x + 24 = 24 + 8x \qquad \text{Combine like terms.}$$
$$8x + 24 - \mathbf{8x} = 24 + 8x - \mathbf{8x} \qquad \text{Subtract } 8x \text{ from both sides.}$$
$$24 = 24 \qquad \text{Combine like terms.}$$

Since the result $24 = 24$ is true for every number $x$, every number $x$ is a solution of the original equation. This equation is an identity. ∎

**Self Check** Solve $-2(x + 3) - 18x = 5(9 - 4x) - 51$.

**Answer** all values of $x$

**EXAMPLE 8** Solve $3(x + 7) - x = 2(x + 10)$.

*Solution*
$$3(x + 7) - x = 2(x + 10)$$
$$3x + 21 - x = 2x + 20 \qquad \text{Remove parentheses.}$$
$$2x + 21 = 2x + 20 \qquad \text{Combine like terms.}$$
$$2x + 21 - \mathbf{2x} = 2x + 20 - \mathbf{2x} \qquad \text{Subtract } 2x \text{ from both sides.}$$
$$21 = 20 \qquad \text{Combine like terms.}$$

Since the result $21 = 20$ is false, the original equation has no solution. It is an impossible equation. ∎

| Self Check | Solve $5(x - 2) - 2x = 3(x + 7)$. |
|---|---|
| *Answer* | no values of $x$ |

**Orals** *Simplify by combining like terms.*

**1.** $3x + 5x$        **2.** $-2y + 3y$        **3.** $3x + 2x - 5x$

**4.** $3y + 2y - 7y$        **5.** $3(x + 2) - 3x + 6$        **6.** $3(x + 2) + 3x - 6$

*Solve each equation, when possible.*

**7.** $5x = 4x + 3$                       **8.** $2(x - 1) = 2(x + 1)$

**9.** $3x = 2(x + 1)$                  **10.** $x + 2(x + 1) = 3$

# EXERCISE 2.4

***REVIEW*** *Evaluate each expression when $x = -3$, $y = -5$, and $z = 0$.*

**1.** $x^2z(y^3 - z)$        **2.** $z - y^3$        **3.** $\dfrac{x - y^2}{2y - 1 + x}$        **4.** $\dfrac{2y + 1}{x} - x$

*Do the operations.*

**5.** $\dfrac{6}{7} - \dfrac{5}{8}$        **6.** $\dfrac{6}{7} \cdot \dfrac{5}{8}$        **7.** $\dfrac{6}{7} \div \dfrac{5}{8}$        **8.** $\dfrac{6}{7} + \dfrac{5}{8}$

***VOCABULARY AND CONCEPTS*** *Fill in each blank to make a true statement.*

**9.** If terms have the same _____ with the same exponents, they are called ____ terms.

**10.** To combine like terms, ____ their numerical coefficients and _____ the same variables and exponents.

**11.** If an equation is true for all values of its variable, it is called an _____.

**12.** If an equation is true for some values of its variable, but not all, it is called a _____ equation.

***PRACTICE*** *In Exercises 13–34, simplify each expression, when possible.*

**13.** $3x + 17x$        **14.** $12y - 15y$        **15.** $8x^2 - 5x^2$        **16.** $17x^2 + 3x^2$

**17.** $9x + 3y$        **18.** $5x + 5y$        **19.** $3(x + 2) + 4x$        **20.** $9(y - 3) + 2y$

**21.** $5(z - 3) + 2z$                    **22.** $4(y + 9) - 6y$

**23.** $12(x + 11) - 11$              **24.** $-3(3 + z) + 2z$

**25.** $8(y + 7) - 2(y - 3)$         **26.** $9(z + 2) + 5(3 - z)$

**27.** $2x + 4(y - x) + 3y$         **28.** $3y - 6(y + z) + y$

**29.** $(x + 2) - (x - y)$

**30.** $3z + 2(y - z) + y$

**31.** $2\left(4x + \dfrac{9}{2}\right) - 3\left(x + \dfrac{2}{3}\right)$

**32.** $7\left(3x - \dfrac{2}{7}\right) - 5\left(2x - \dfrac{3}{5}\right) + x$

**33.** $8x(x + 3) - 3x^2$

**34.** $2x + x(x + 3)$

*In Exercises 35–72, solve each equation, when possible.* **Check all solutions.**

**35.** $3x + 2 = 2x$

**36.** $5x + 7 = 4x$

**37.** $5x - 3 = 4x$

**38.** $4x + 3 = 5x$

**39.** $9y - 3 = 6y$

**40.** $8y + 4 = 4y$

**41.** $8y - 7 = y$

**42.** $9y - 8 = y$

**43.** $9 - 23w = 4w$

**44.** $y + 4 = -7y$

**45.** $22 - 3r = 8r$

**46.** $14 + 7s = s$

**47.** $3(a + 2) = 4a$

**48.** $4(a - 5) = 3a$

**49.** $5(b + 7) = 6b$

**50.** $8(b + 2) = 9b$

**51.** $2 + 3(x - 5) = 4(x - 1)$

**52.** $2 - (4x + 7) = 3 + 2(x + 2)$

**53.** $10x + 3(2 - x) = 5(x + 2) - 4$

**54.** $11x + 6(3 - x) = 3$

**55.** $3(a + 2) = 2(a - 7)$

**56.** $9(t - 1) = 6(t + 2) - t$

**57.** $9(x + 11) + 5(13 - x) = 0$

**58.** $3(x + 15) + 4(11 - x) = 0$

**59.** $\dfrac{3(t - 7)}{2} = t - 6$

**60.** $\dfrac{2(t + 9)}{3} = t - 8$

**61.** $\dfrac{5(2 - s)}{3} = s + 6$

**62.** $\dfrac{8(5 - s)}{5} = -2s$

**63.** $\dfrac{4(2x - 10)}{3} = 2(x - 4)$

**64.** $\dfrac{11(x - 12)}{2} = 9 - 2x$

**65.** $3.1(x - 2) = 1.3x + 2.8$

**66.** $0.6x - 0.8 = 0.8(2x - 1) - 0.7$

**67.** $2.7(y + 1) = 0.3(3y + 33)$

**68.** $1.5(5 - y) = 3y + 12$

**69.** $19.1x - 4(x + 0.3) = -46.5$

**70.** $18.6x + 7.2 = 1.5(48 - 2x)$

**71.** $14.3(x + 2) + 13.7(x - 3) = 15.5$

**72.** $1.25(x - 1) = 0.5(3x - 1) - 1$

*In Exercises 73–84, solve each equation. If it is an identity or an impossible equation, so indicate.*

**73.** $8x + 3(2 - x) = 5(x + 2) - 4$

**74.** $5(x + 2) = 5x - 2$

**75.** $2(s + 2) = 2(s + 1) + 3$

**76.** $21(b - 1) + 3 = 3(7b - 6)$

**77.** $\dfrac{2(t - 1)}{6} - 2 = \dfrac{t + 2}{6}$

**78.** $\dfrac{2(2r - 1)}{6} + 5 = \dfrac{3(r + 7)}{6}$

**79.** $2(3z + 4) = 2(3z - 2) + 13$

**80.** $x + 7 = \dfrac{2x + 6}{2} + 4$

**81.** $2(y - 3) - \dfrac{y}{2} = \dfrac{3}{2}(y - 4)$

**82.** $\dfrac{20 - a}{2} = \dfrac{3}{2}(a + 4)$

**83.** $\dfrac{3x + 14}{2} = x - 2 + \dfrac{x + 18}{2}$

**84.** $\dfrac{5(x + 3)}{3} - x = \dfrac{2(x + 8)}{3}$

## WRITING

**85.** Explain why $3x^2y$ and $5x^2y$ are like terms.

**86.** Explain why $3x^2y$ and $3xy^2$ are unlike terms.

**87.** Discuss whether $7xxy^3$ and $5x^2yyy$ are like terms.

**88.** Discuss whether $\frac{3}{2}x$ and $\frac{3x}{2}$ are like terms.

## SOMETHING TO THINK ABOUT

**89.** What number is equal to its own double?

**90.** What number is equal to one-half of itself?

# 2.5 Applications of Equations

■ PROBLEM SOLVING ■ NUMBER PROBLEMS ■ GEOMETRIC PROBLEMS ■ INVESTMENT PROBLEMS
■ MOTION PROBLEMS ■ LIQUID MIXTURE PROBLEMS ■ DRY MIXTURE PROBLEMS

**Getting Ready**

**1.** If one part of a pipe is $x$ feet long and the other part is $(x + 2)$ feet long, find an expression that represents the length of the pipe.

**2.** If one part of a board is $x$ feet long and the other part is three times as long, find an expression that represents the length of the board.

**3.** What is the formula for the perimeter of a rectangle?

**4.** Define a triangle.

**5.** Find 7% of $12,000.

**6.** At 55 miles per hour, how far would a car travel in 7 hours?

**7.** If 8 gallons of a mixture of water and alcohol is 70% alcohol, how many gallons of alcohol does the mixture contain?

**8.** At $7.50 per pound, how many pounds of chocolate would be worth $71.25?

## ■ PROBLEM SOLVING

The key to problem solving is to thoroughly understand the problem and then devise a plan to solve it. The following list of steps provides a strategy to follow.

**Problem Solving**

1. **Analyze the problem** by reading it several times to understand the given facts. What information is given? What are you asked to find? Often a sketch, chart, or diagram will help you visualize the facts of the problem.

2. **Form an equation** by picking a variable to represent the quantity to be found. Then express all other unknown quantities in the problem as expressions involving that variable. Finally, write an equation expressing a quantity in two different ways.

3. **Solve the equation.**

4. **State the conclusion.**

5. **Check the result.**

In this section, we will use this five-step strategy to solve many problems.

## ▦ NUMBER PROBLEMS

**EXAMPLE 1**

A plumber wants to cut a 17-foot pipe into three parts. The longest part is to be 3 times as long as the shortest, and the middle-sized part is to be 2 feet longer than the shortest. How long should each part be?

*Analyze the problem*  The information is given in terms of the length of the shortest part. Therefore, we let a variable represent the length of the shortest part and express the other lengths in terms of that variable.

*Form an equation*  Let $x$ represent the length of the shortest part. Then $3x$ represents the length of the longest part, and $x + 2$ represents the length of the middle-sized part. We sketch the pipe as shown in Figure 2-6.

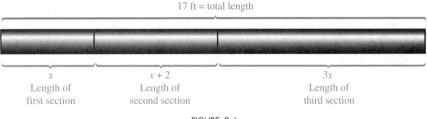

| 17 ft = total length |
|---|

| $x$ | $x + 2$ | $3x$ |
| Length of first section | Length of second section | Length of third section |

FIGURE 2-6

The sum of the lengths of these three parts equals the total length of the pipe.

| The length of part 1 | plus | the length of part 2 | plus | the length of part 3 | equals | the total length. |
|---|---|---|---|---|---|---|
| $x$ | $+$ | $x + 2$ | $+$ | $3x$ | $=$ | 17 |

***Solve the equation***   We can solve this equation as follows:

$$x + x + 2 + 3x = 17 \qquad \text{The equation to solve.}$$
$$5x + 2 = 17 \qquad \text{Combine like terms.}$$
$$5x = 15 \qquad \text{Subtract 2 from both sides.}$$
$$x = 3 \qquad \text{Divide both sides by 5.}$$

***State the conclusion***   The shortest part is 3 feet long. Because the middle-sized part is 2 feet longer than the shortest, it is 5 feet long. Because the longest part is 3 times the shortest, it is 9 feet long.

***Check the result***   Because 3 feet, 5 feet, and 9 feet total 17 feet, the solution checks.   ∎

## ■ GEOMETRIC PROBLEMS

**EXAMPLE 2**   The length of a rectangle is 4 meters more than twice its width. If the perimeter of the rectangle is 26 meters, find its dimensions.

***Analyze the problem***   We can sketch the rectangle as in Figure 2-7. Since the formula for the perimeter of a rectangle is $P = 2l + 2w$, the perimeter of the rectangle in the figure is $2(4 + 2w) + 2w$. We are also told that the perimeter is 26.

$4 + 2w$

$w$

FIGURE 2-7

***Form an equation***   Let $w$ represent the width of the rectangle. Then $4 + 2w$ represents the length of the rectangle. We can form the equation

| 2 · | the length | plus 2 · | the width | equals | the perimeter. |
|---|---|---|---|---|---|
| 2 · | $(4 + 2w)$ | $+$ 2 · | $w$ | $=$ | 26 |

***Solve the equation***   We can solve this equation as follows:

$$2(4 + 2w) + 2w = 26 \qquad \text{The equation to solve.}$$
$$8 + 4w + 2w = 26 \qquad \text{Remove parentheses.}$$
$$6w + 8 = 26 \qquad \text{Combine like terms.}$$
$$6w = 18 \qquad \text{Subtract 8 from both sides.}$$
$$w = 3 \qquad \text{Divide both sides by 6.}$$

***State the conclusion***   The width of the rectangle is 3 meters, and the length, $4 + 2w$, is 10 meters.

***Check the result***   If a rectangle has a width of 3 meters and a length of 10 meters, then the length is 4 meters longer than twice the width ($4 + 2 \cdot 3 = 10$). The perimeter is $2 \cdot 10 + 2 \cdot 3 = 26$ meters. The solution checks.   ■

**EXAMPLE 3**   The vertex angle of an isosceles triangle is 56°. Find the measure of each base angle.

***Analyze the problem***   An **isosceles triangle** has two equal sides, which meet to form the **vertex angle.** See Figure 2-8. The angles opposite those sides, called **base angles,** are also equal. If we let $x$ represent the measure of one base angle, then the measure of the other base angle is also $x$. In any triangle the sum of the three angles is 180°.

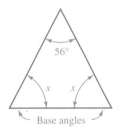

FIGURE 2-8

***Form an equation***   Let $x$ represent the measure of one base angle. Then $x$ also represents the measure of the other base angle. We can form the equation

| One base angle | plus | the other base angle | plus | the vertex angle | equals | 180°. |
|---|---|---|---|---|---|---|
| $x$ | $+$ | $x$ | $+$ | 56 | $=$ | 180 |

***Solve the equation***   We can solve this equation as follows:

$$x + x + 56 = 180 \qquad \text{The equation to solve.}$$
$$2x + 56 = 180 \qquad \text{Combine like terms.}$$
$$2x = 124 \qquad \text{Subtract 56 from both sides.}$$
$$x = 62 \qquad \text{Divide both sides by 2.}$$

*State the conclusion*    The measure of each base angle is 62°.

*Check the result*    The measure of each base angle is 62°, and the vertex angle measures 56°. These three angles total 180°. The solution checks. ■

## ■ INVESTMENT PROBLEMS

**EXAMPLE 4**    A teacher invested part of $12,000 at 6% annual interest, and the rest at 9%. If the annual income from these investments was $945, how much did he invest at each rate?

*Analyze the problem*    The interest $i$ earned by an amount $p$ invested at an annual rate $r$ for $t$ years is given by the formula $i = prt$. In this example, $t = 1$ year. Hence, if $x$ dollars were invested at 6%, the interest earned would be $0.06x$ dollars. If $x$ dollars were invested at 6%, then the rest of the money, $\$(12,000 - x)$, would be invested at 9%. The interest earned on that money would be $0.09(12,000 - x)$ dollars. The total interest earned in dollars can be expressed in two ways: as 945 and as the sum $0.06x + 0.09(12,000 - x)$.

*Form an equation*    Let $x$ represent the amount of money invested at 6%. Then $12,000 - x$ represents the amount of money invested at 9%. We can form an equation as follows:

| The interest earned at 6% | plus | the interest earned at 9% | equals | the total interest. |
|---|---|---|---|---|
| $0.06x$ | $+$ | $0.09(12,000 - x)$ | $=$ | $945$ |

*Solve the equation*    We can solve this equation as follows:

$$0.06x + 0.09(12,000 - x) = 945 \qquad \text{The equation to solve.}$$
$$6x + 9(12,000 - x) = 94,500 \qquad \text{Multiply both sides by 100 to clear the equation of decimals.}$$
$$6x + 108,000 - 9x = 94,500 \qquad \text{Remove parentheses.}$$
$$-3x + 108,000 = 94,500 \qquad \text{Combine like terms.}$$
$$-3x = -13,500 \qquad \text{Subtract 108,000 from both sides.}$$
$$x = 4,500 \qquad \text{Divide both sides by } -3.$$

*State the conclusion*    The teacher invested $4,500 at 6% and $12,000 − $4,500 = $7,500 at 9%.

*Check the result*    The first investment yielded 6% of $4,500, or $270. The second investment yielded 9% of $7,500, or $675. Since the total return was $270 + $675, or $945, the answers check. ■

## ■ MOTION PROBLEMS

**EXAMPLE 5**    Chicago and Green Bay are about 200 miles apart. A car leaves Chicago traveling toward Green Bay at 55 mph at the same time as a truck leaves Green Bay bound for Chicago at 45 mph. How long will it take them to meet?

*Analyze the problem*    Motion problems are based on the formula $d = rt$, where $d$ is the distance traveled, $r$ is the rate, and $t$ is the time. We can organize the information of this problem in chart form, as in Figure 2-9(a).

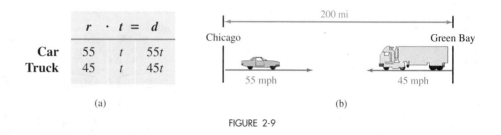

| | $r$ | $\cdot$ $t$ | $=$ $d$ |
|---|---|---|---|
| **Car** | 55 | $t$ | $55t$ |
| **Truck** | 45 | $t$ | $45t$ |

(a)                                                                                   (b)

FIGURE 2-9

We know that the two vehicles travel for the same amount of time—say, $t$ hours. The faster car travels $55t$ miles, and the slower truck travels $45t$ miles. At the time they meet, the total distance can be expressed in two ways: as the sum $55t + 45t$ and as 200 miles.

*Form an equation*    Let $t$ represent the time that each vehicle travels until they meet. Then $55t$ represents the distance traveled by the car, and $45t$ represents the distance traveled by the truck. After referring to Figure 2-9(b), we form the equation

| The distance the car goes | plus | the distance the truck goes | equals | the total distance. |
|---|---|---|---|---|
| $55t$ | $+$ | $45t$ | $=$ | 200 |

*Solve the equation*    We can solve the equation as follows:

$$55t + 45t = 200 \qquad \text{The equation to solve.}$$
$$100t = 200 \qquad \text{Combine like terms.}$$
$$t = 2 \qquad \text{Divide both sides by 100.}$$

*State the conclusion*    The vehicles meet after 2 hours.

*Check the result*    During those 2 hours, the car travels $55 \cdot 2 = 110$ miles, while the truck travels $45 \cdot 2 = 90$ miles. The total distance traveled is $110 + 90 = 200$ miles. Since this is the total distance between Chicago and Green Bay, the answer checks.    ■

### ■ LIQUID MIXTURE PROBLEMS

| | |
|---|---|
| EXAMPLE 6 | A chemist has one solution that is 50% sulfuric acid and another that is 20% sulfuric acid. How much of each should she use to make 12 liters of a solution that is 30% acid? |

*Analyze the problem*   The sulfuric acid present in the final mixture comes from the two solutions to be mixed. If $x$ represents the number of liters of the 50% solution required for the mixture, then the rest of the mixture (($12 - x$) liters) must be the 20% solution. See Figure 2-10. Only 50% of the $x$ liters, and only 20% of the ($12 - x$) liters, is pure sulfuric acid. The total of these amounts is also the amount of acid in the final mixture, which is 30% of 12 liters.

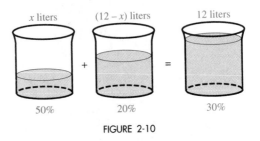

$x$ liters        ($12 - x$) liters        12 liters

50%         20%        30%

FIGURE 2-10

*Form an equation*   Let $x$ represent the required number of liters of the 50% solution. Then $12 - x$ represents the required number of liters of the 20% solution. We can form the equation

| The acid in the 50% solution | plus | the acid in the 20% solution | equals | the acid in the final mixture. |
|---|---|---|---|---|
| 50% of $x$ | + | 20% of ($12 - x$) | = | 30% of 12 |

*Solve the equation*   We can solve this equation as follows:

| | |
|---|---|
| $0.50x + 0.20(12 - x) = 0.30(12)$ | The equation to solve. |
| $5x + 2(12 - x) = 3(12)$ | Multiply both sides by 10 to clear the equation of decimals. |
| $5x + 24 - 2x = 36$ | Remove parentheses. |
| $3x + 24 = 36$ | Combine like terms. |
| $3x = 12$ | Subtract 24 from both sides. |
| $x = 4$ | Divide both sides by 3. |

*State the conclusion*   The chemist must mix 4 liters of the 50% solution and $12 - 4 = 8$ liters of the 20% solution.

Check the result.                                                                                      ■

### ■ DRY MIXTURE PROBLEMS

**EXAMPLE 7**

Fancy cashews are not selling at $9 per pound, because they are too expensive. Filberts are selling at $6 per pound. How many pounds of filberts should be combined with 50 pounds of cashews to obtain a mixture that can be sold at $7 per pound?

*Analyze the problem*

Dry mixture problems are based on the formula $v = pn$, where $v$ is the value of the mixture, $p$ is the price per pound, and $n$ is the number of pounds. Suppose $x$ pounds of filberts are used in the mixture. At $6 per pound, they are worth $6x$. At $9 per pound, the 50 pounds of cashews are worth $9 \cdot 50$, or $450. The mixture will weigh $(50 + x)$ pounds, and at $7 per pound, it will be worth $7(50 + x)$. The value of the ingredients, $(6x + 450)$, is equal to the value of the mixture, $7(50 + x)$. See Figure 2-11.

|          | $v$          | $= p \cdot$ | $n$      |
|----------|--------------|-------------|----------|
| **Filberts** | $6x$     | 6           | $x$      |
| **Cashews**  | $9(50)$  | 9           | 50       |
| **Mixture**  | $7(50 + x)$ | 7        | $50 + x$ |

FIGURE 2-11

*Form an equation*

Let $x$ represent the number of pounds of filberts in the mixture. We can form the equation

| The value of the filberts | plus | the value of the cashews | equals | the value of the mixture. |
|---------------------------|------|--------------------------|--------|---------------------------|
| $6x$ | $+$ | $9 \cdot 50$ | $=$ | $7(50 + x)$ |

*Solve the equation*

We can solve this equation as follows:

$6x + 9 \cdot 50 = 7(50 + x)$    The equation to solve.

$6x + 450 = 350 + 7x$    Remove parentheses and simplify.

$100 = x$    Subtract $6x$ and 350 from both sides.

*State the conclusion*

The storekeeper should use 100 pounds of filberts in the mixture.

*Check the result*

The value of 100 pounds of filberts at $6 per pound is   $ 600
The value of 50 pounds of cashews at $9 per pound is   $ 450
The value of the mixture is   $1,050

The value of 150 pounds of mixture at $7 per pound is also $1,050. ■

*Orals*

**1.** Express the value of 7 pounds of ground coffee worth $d$ per pound.

**2.** Express one year's interest on $18,000, invested at an annual rate $r$.

**3.** Express the length of a rectangle with area of $A$ square feet and width 6 feet.

**4.** Express the length of a rectangle with perimeter of $P$ feet and width of 9 feet.

# EXERCISE 2.5

*REVIEW* *Refer to the formulas in Section 1.3.*

**1.** Find the volume of a pyramid that has a height of 6 centimeters and a square base, 10 centimeters on each side.

**2.** Find the volume of a cone with a height of 6 centimeters and a circular base with radius 6 centimeters. Use $\pi \approx \frac{22}{7}$.

*Simplify each expression.*

**3.** $3(x + 2) + 4(x - 3)$      **4.** $4(x - 2) - 3(x + 1)$      **5.** $\frac{1}{2}(x + 1) - \frac{1}{2}(x + 4)$      **6.** $\frac{3}{2}\left(x + \frac{2}{3}\right) + \frac{1}{2}(x + 8)$

**7.** The amount $A$ on deposit in a bank account bearing simple interest is given by the formula

$$A = P + Prt$$

Find $A$ when $P = \$1{,}200$, $r = 0.08$, and $t = 3$.

**8.** The distance $s$ that a certain object falls in $t$ seconds is given by the formula

$$s = 350 - 16t^2 + vt$$

Determine $s$ when $t = 4$ and $v = -3$.

*VOCABULARY AND CONCEPTS* *Fill in each blank to make a true statement.*

**9.** The perimeter of a rectangle is given by the formula $P = $ _____.

**10.** An _____ triangle is a triangle with two sides of equal length.

**11.** The sides of equal length of an isosceles triangle meet to form the _____ angle.

**12.** The angles opposite the sides of equal length of an isosceles triangle are called _____ angles.

**13.** Motion problems are based on the formula _____.

**14.** The last step in the problem-solving process is to _____ the result.

*APPLICATIONS*

**15. Carpentry** The 12-foot board in Illustration 1 has been cut into two parts, one twice as long as the other. How long is each part?

**16. Plumbing** A 20-foot pipe has been cut into two parts, one 3 times as long as the other. How long is each part?

**17. Triangular bracing** The outside perimeter of the triangular brace shown in Illustration 2 is 57 feet. If all three sides are of equal length, find the length of each side.

ILLUSTRATION 1

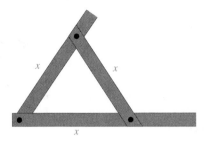

ILLUSTRATION 2

**18. Circuit boards** The perimeter of the circuit board in Illustration 3 is 90 centimeters. Find the dimensions of the board.

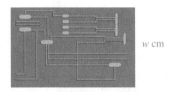

$w$ cm

$(w + 7)$ cm

ILLUSTRATION 3

**19. Swimming pool** The width of a rectangular swimming pool is 11 meters less than the length, and the perimeter is 94 meters. Find its dimensions.

**20. Wooden truss** The truss in Illustration 4 is in the form of an isosceles triangle. Each of the two equal sides is 4 feet less than the third side. If the perimeter is 25 feet, find the length of each side.

ILLUSTRATION 4

**21. Framing pictures** The length of a rectangular picture is 5 inches greater than twice the width. If the perimeter is 112 inches, find the dimensions of the frame.

**22. Guy wires** The two guy wires in Illustration 5 form an isosceles triangle. One of the two equal angles of the triangle is four times the third angle (the vertex angle). Find the measure of the vertex angle.

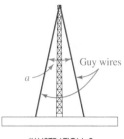

Guy wires

$a$

ILLUSTRATION 5

**23. Equilateral triangles** Find the measure of each angle of an equilateral triangle. (*Hint:* The three angles of an equilateral triangle are equal.)

**24. Land areas** The perimeter of a square piece of land is twice the perimeter of an equilateral (equal-sided) triangular lot. If one side of the square is 60 meters, find the length of a side of the triangle.

**25. Investment problem** A broker invested $24,000 in two mutual funds, one earning 9% annual interest and the other earning 8%. After 1 year, his combined interest is $1,965. How much did he invest at each rate?

**26. Investment problem** A rollover IRA of $18,750 is invested in two mutual funds, one earning 12% interest and the other earning 10%. After 1 year, the combined interest income is $2,117. How much was invested at each rate?

**27. Investment problem** One investment pays 8%, and another pays 11%. If equal amounts are invested in each, the combined interest income for 1 year is $712.50. How much is invested at each rate?

**28. Investment problem** When equal amounts are invested in each of three accounts paying 5%, 6%, and 7%, one year's combined interest income is $882. How much is invested in each account?

**29. Investment problem**  A college professor wants to supplement her retirement income with investment interest. If she invests $15,000 at 6% annual interest, how much more would she have to invest at 7% to achieve a goal of $1,250 in supplemental income?

**30. Investment problem**  A teacher has a choice of two investment plans: an insured fund that has paid an average of 11% interest per year, or a riskier investment that has averaged a 13% return. If the same amount invested at the higher rate would generate an extra $150 per year, how much does the teacher have to invest?

**31. Investment problem**  A financial counselor recommends investing twice as much in CDs as in a bond fund. A client follows his advice and invests $21,000 in CDs paying 1% more interest than the fund. The CDs would generate $840 more interest than the fund. Find the two rates. (*Hint:* 1% = 0.01.)

**32. Investment problem**  The amount of annual interest earned by $8,000 invested at a certain rate is $200 less than $12,000 would earn at a 1% lower rate. At what rate is the $8,000 invested?

**33. Travel time**  Ashford and Bartlett are 315 miles apart. A car leaves Ashford bound for Bartlett at 50 mph. At the same time, another car leaves Bartlett and heads toward Ashford at 55 mph. In how many hours will the two cars meet?

**34. Travel time**  Granville and Preston are 535 miles apart. A car leaves Preston bound for Granville at 47 mph. At the same time, another car leaves Granville and heads toward Preston at 60 mph. How long will it take them to meet?

**35. Travel time**  Two cars leave Peoria at the same time, one heading east at 60 mph and the other west at 50 mph. (See Illustration 6.) How long will it take them to be 715 miles apart?

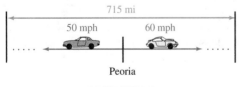

ILLUSTRATION 6

**36. Boating**  Two boats leave port at the same time, one heading north at 35 knots (nautical miles per hour), the other south at 47 knots. How long will it take them to be 738 nautical miles apart?

**37. Travel time**  Two cars start together and head east, one at 42 mph and the other at 53 mph. (See Illustration 7.) In how many hours will the cars be 82.5 miles apart?

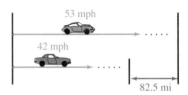

ILLUSTRATION 7

**38. Speed of trains**  Two trains are 330 miles apart, and their speeds differ by 20 mph. They travel toward each other and meet in 3 hours. Find the speed of each train.

**39. Speed of an airplane**  Two planes are 6,000 miles apart, and their speeds differ by 200 mph. They travel toward each other and meet in 5 hours. Find the speed of the slower plane.

**40. Average speed**  An automobile averaged 40 mph for part of a trip and 50 mph for the remainder. If the 5-hour trip covered 210 miles, for how long did the car average 40 mph?

**41. Mixing fuels**  How many gallons of fuel costing $1.15 per gallon must be mixed with 20 gallons of a fuel costing $.85 per gallon to obtain a mixture costing $1 per gallon? (See Illustration 8.)

**42. Mixing paint**  Paint costing $19 per gallon is to be mixed with 5 gallons of a $3-per-gallon thinner to make a paint that can be sold for $14 per gallon. How much paint will be produced?

**43. Brine solution**  How many gallons of a 3% salt solution must be mixed with 50 gallons of a 7% solution to obtain a 5% solution?

**44. Making cottage cheese**  To make low-fat cottage cheese, milk containing 4% butterfat is mixed with 10 gallons of milk containing 1% butterfat to obtain a mixture containing 2% butterfat. How many gallons of the richer milk must be used?

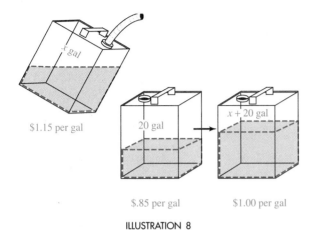

$1.15 per gal

20 gal

$x + 20$ gal

$.85 per gal          $1.00 per gal

ILLUSTRATION 8

**45. Antiseptic solutions**   A nurse wishes to add water to 30 ounces of a 10% solution of benzalkonium chloride to dilute it to an 8% solution. How much water must she add?

**46. Mixing photographic chemicals**   A photographer wishes to mix 2 liters of a 5% acetic acid solution with a 10% solution to get a 7% solution. How many liters of 10% solution must be added?

**47. Mixing candy**   Lemon drops worth $1.90 per pound are to be mixed with jelly beans that cost $1.20 per pound to make 100 pounds of a mixture worth $1.48 per pound. How many pounds of each candy should be used?

**48. Blending gourmet tea**   One grade of tea, worth $3.20 per pound, is to be mixed with another grade worth $2 per pound to make 20 pounds that will sell for $2.72 per pound. How much of each grade of tea must be used?

**49. Mixing nuts**   A bag of peanuts is worth $.30 less than a bag of cashews. Equal amounts of peanuts and cashews are used to make 40 bags of a mixture that sells for $1.05 per bag. How much is a bag of cashews worth?

**50. Mixing candy**   Twenty pounds of lemon drops are to be mixed with cherry chews to make a mixture that will sell for $1.80 per pound. How much of the more expensive candy should be used? See Illustration 9.

| | Price per pound |
|---|---|
| Peppermint patties | $1.35 |
| Lemon drops | $1.70 |
| Licorice lumps | $1.95 |
| Cherry chews | $2.00 |

ILLUSTRATION 9

**51. Coffee blends**   A store sells regular coffee for $4 a pound and gourmet coffee for $7 a pound. To get rid of 40 pounds of the gourmet coffee, the shopkeeper plans to make a gourmet blend that he will put on sale for $5 a pound. How many pounds of regular coffee should be used?

**52. Lawn seed blends**   A garden store sells Kentucky bluegrass seed for $6 per pound and ryegrass seed for $3 per pound. How much rye must be mixed with 100 pounds of bluegrass to obtain a blend that will sell for $5 per pound?

## WRITING

**53.** Describe the steps you would use to analyze and solve a problem.

**55.** Create a mixture problem of your own, and solve it.

**54.** Create a geometry problem that could be solved by solving the equation $2w + 2(w + 5) = 26$.

**56.** In mixture problems, explain why it is important to distinguish between the quantity and the value of the materials being combined.

## SOMETHING TO THINK ABOUT

**57.** Is it possible for the equation of a problem to have a solution, but for the problem to have no solution? For example, is it possible to find two consecutive even integers whose sum is 16?

**58.** Invent a geometric problem that leads to an equation that has a solution, although the problem does not.

**59.** Consider the problem: How many gallons of a 10% and a 20% solution should be mixed to obtain a 30% solution? Without solving it, how do you know that the problem has no solution?

**60.** What happens if you try to solve Exercise 59?

## 2.6 Formulas

■ SOLVING FORMULAS

Getting Ready    *Find the missing number.*

**1.** $\dfrac{3x}{\phantom{0}} = x$     **2.** $\dfrac{-5y}{\phantom{0}} = y$     **3.** $\dfrac{rx}{\phantom{0}} = x$     **4.** $\dfrac{-ay}{\phantom{0}} = y$

**5.** $\phantom{0} \cdot \dfrac{x}{7} = x$     **6.** $\phantom{0} \cdot \dfrac{y}{12} = y$     **7.** $\phantom{0} \cdot \dfrac{x}{d} = x$     **8.** $\phantom{0} \cdot \dfrac{y}{s} = y$

### ■ SOLVING FORMULAS

Equations with several variables are called **literal equations.** Often these equations are **formulas** such as $A = lw$, the formula for finding the area of a rectangle. Suppose that we wish to find the lengths of several rectangles whose areas and widths are known. It would be tedious to substitute values for $A$ and $w$ into the formula and then repeatedly solve the formula for $l$. It would be better to solve the formula $A = lw$ for $l$ first and then substitute values for $A$ and $w$ and compute $l$ directly.

To **solve an equation for a variable** means to isolate that variable on one side of the equation, with all other quantities on the opposite side. We can isolate the variable by using the usual equation-solving techniques.

**EXAMPLE 1**    Solve $A = lw$ for $l$.

*Solution*    To isolate $l$ on the left-hand side, we undo the multiplication by $w$ by dividing both sides of the equation by $w$.

$$A = lw$$

$$\frac{A}{w} = \frac{lw}{w} \qquad \text{To undo the multiplication by } w, \text{ divide both sides by } w.$$

$$\frac{A}{w} = l \qquad \tfrac{w}{w} = 1.$$

$$l = \frac{A}{w}$$

■

Self Check    Solve $A = lw$ for $w$.

Answer    $w = \dfrac{A}{l}$

EXAMPLE 2    Recall that the formula $A = \frac{1}{2}bh$ gives the area of a triangle with base $b$ and height $h$. Solve the formula for $b$.

Solution

$$A = \frac{1}{2}bh$$

$$2A = 2 \cdot \frac{1}{2}bh \qquad \text{To eliminate the fraction, multiply both sides by 2.}$$

$$2A = bh \qquad 2 \cdot \frac{1}{2} = 1.$$

$$\frac{2A}{h} = \frac{bh}{h} \qquad \text{To undo the multiplication by } h \text{, divide both sides by } h.$$

$$\frac{2A}{h} = b \qquad \frac{h}{h} = 1.$$

$$b = \frac{2A}{h}$$

If the area $A$ and the height $h$ of a triangle are known, the base $b$ is given by the formula $b = \frac{2A}{h}$. ■

Self Check    Solve $A = \frac{1}{2}bh$ for $h$.

Answer    $h = \dfrac{2A}{b}$

EXAMPLE 3    The formula $C = \frac{5}{9}(F - 32)$ is used to convert Fahrenheit temperature readings into their Celsius equivalents. Solve the formula for $F$.

Solution

$$C = \frac{5}{9}(F - 32)$$

$$\frac{9}{5}C = \frac{9}{5} \cdot \frac{5}{9}(F - 32) \qquad \text{To eliminate } \tfrac{5}{9} \text{, multiply both sides by } \tfrac{9}{5}.$$

$$\frac{9}{5}C = 1(F - 32) \qquad \frac{9}{5} \cdot \frac{5}{9} = \frac{9 \cdot 5}{5 \cdot 9} = 1.$$

$$\frac{9}{5}C = F - 32 \qquad \text{Remove parentheses.}$$

$$\frac{9}{5}C + 32 = F - 32 + 32 \qquad \text{To undo the subtraction of 32, add 32 to both sides.}$$

$$\frac{9}{5}C + 32 = F \qquad \text{Combine like terms.}$$

$$F = \frac{9}{5}C + 32$$

The formula $F = \frac{9}{5}C + 32$ is used to convert degrees Celsius to degrees Fahrenheit.

■

**Self Check**    Solve $x = \frac{2}{3}(y + 5)$ for $y$.
**Answer**    $y = \frac{3}{2}x - 5$

**EXAMPLE 4**    Recall that the area $A$ of the trapezoid shown in Figure 2-12 is given by the formula

$$A = \frac{1}{2}h(B + b)$$

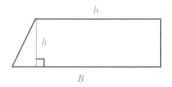

**FIGURE 2-12**

where $B$ and $b$ are its bases and $h$ is its height. Solve the formula for $b$.

*Solution*    *Method 1:*  $A = \frac{1}{2}(B + b)h$

$$2A = 2 \cdot \frac{1}{2}(B + b)h \qquad \text{Multiply both sides by 2.}$$

$$2A = Bh + bh \qquad \text{Simplify and remove parentheses.}$$

$$2A - Bh = Bh + bh - Bh \qquad \text{Subtract } Bh \text{ from both sides.}$$

$$2A - Bh = bh \qquad \text{Combine like terms.}$$

$$\frac{2A - Bh}{h} = \frac{bh}{h} \qquad \text{Divide both sides by } h.$$

$$\frac{2A - Bh}{h} = b \qquad \frac{h}{h} = 1.$$

*Method 2:*  $A = \frac{1}{2}(B + b)h$

$$2 \cdot A = 2 \cdot \frac{1}{2}(B + b)h \qquad \text{Multiply both sides by 2.}$$

$$2A = (B + b)h \qquad \text{Simplify.}$$

$$\frac{2A}{h} = \frac{(B + b)h}{h} \qquad \text{Divide both sides by } h.$$

$$\frac{2A}{h} = B + b \qquad \frac{h}{h} = 1.$$

$$\frac{2A}{h} - B = B + b - B \qquad \text{Subtract } B \text{ from both sides.}$$

$$\frac{2A}{h} - B = b \qquad \text{Combine like terms.}$$

Although they look different, the results of Methods 1 and 2 are equivalent. ∎

**Self Check** Solve $A = \frac{1}{2}h(B + b)$ for $B$.

**Answer** $B = \dfrac{2A - hb}{h} \text{ or } B = \dfrac{2A}{h} - b$

**EXAMPLE 5** Solve the formula $P = 2l + 2w$ for $l$, and then find $l$ when $P = 56$ and $w = 11$.

*Solution* We first solve the formula $P = 2l + 2w$ for $l$.

$$P = 2l + 2w$$
$$P - 2w = 2l + 2w - 2w \qquad \text{Subtract } 2w \text{ from both sides.}$$
$$P - 2w = 2l \qquad \text{Combine like terms.}$$
$$\frac{P - 2w}{2} = \frac{2l}{2} \qquad \text{Divide both sides by 2.}$$
$$\frac{P - 2w}{2} = l \qquad \frac{2}{2} = 1.$$
$$l = \frac{P - 2w}{2}$$

We then substitute 56 for $P$ and 11 for $w$ and simplify.

$$l = \frac{P - 2w}{2}$$
$$l = \frac{56 - 2(11)}{2}$$
$$= \frac{56 - 22}{2}$$
$$= \frac{34}{2}$$
$$= 17$$

Thus, $l = 17$. ∎

**Self Check** Solve $P = 2l + 2w$ for $w$, and then find $w$ when $P = 46$ and $l = 16$.

**Answer** $w = \frac{P - 2l}{2}, 7$

EXAMPLE 6   Recall that the volume $V$ of the right-circular cone shown in Figure 2-13 is given by the formula

$$V = \frac{1}{3}Bh$$

where $B$ is the area of its circular base and $h$ is its height. Solve the formula for $h$, and find the height of a right-circular cone with a volume of 64 cubic centimeters and a base area of 16 square centimeters.

FIGURE 2-13

*Solution*   We first solve the formula for $h$.

$$V = \frac{1}{3}Bh$$

$$3V = 3 \cdot \frac{1}{3}Bh \qquad \text{Multiply both sides by 3.}$$

$$3V = Bh \qquad 3 \cdot \frac{1}{3} = 1.$$

$$\frac{3V}{B} = \frac{Bh}{B} \qquad \text{Divide both sides by } B.$$

$$\frac{3V}{B} = h \qquad \frac{B}{B} = 1.$$

$$h = \frac{3V}{B}$$

We then substitute 64 for $V$ and 16 for $B$ and simplify.

$$h = \frac{3V}{B}$$

$$h = \frac{3(64)}{16}$$

$$= 3(4)$$

$$= 12$$

The height of the cone is 12 centimeters.   ∎

Self Check   Solve $V = \frac{1}{3}Bh$ for $B$, and find the area of the base when the volume is 42 cubic feet and the height is 6 feet.

Answer   $B = \frac{3V}{h}$, 21 square feet

Orals   *Solve the equation $ab + c - d = 0$,*

**1.** for $a$                      **2.** for $b$

**3.** for $c$                      **4.** for $d$

Solve the equation $a + b = \dfrac{c}{d}$,

**5.** for $a$                                     **6.** for $b$

**7.** for $c$                                     **8.** for $d$

## EXERCISE 2.6

***REVIEW*** *Simplify each expression, if possible.*

**1.** $2x - 5y + 3x$

**2.** $2x^2y + 5x^2y^2$

**3.** $\dfrac{3}{5}(x + 5) - \dfrac{8}{5}(10 + x)$

**4.** $\dfrac{2}{11}(22x - y^2) + \dfrac{9}{11}y^2$

***VOCABULARY AND CONCEPTS*** *Fill in each blank to make a true statement.*

**5.** Equations that contain several variables are called _____ equations.

**6.** The equation $A = lw$ is an example of a _____.

**7.** To solve a formula for a variable means to _____ the variable on one side of the formula.

**8.** To solve the formula $d = rt$ for $t$, divide both sides of the formula by __.

**9.** To solve $A = p + i$ for $p$, _____ $i$ from both sides.

**10.** To solve $t = \dfrac{d}{r}$ for $d$, _____ both sides by $r$.

***PRACTICE*** *In Exercises 11–34, solve each formula for the indicated variable.*

**11.** $E = IR$; for $I$

**12.** $i = prt$; for $r$

**13.** $V = lwh$; for $w$

**14.** $K = A + 32$; for $A$

**15.** $P = a + b + c$; for $b$

**16.** $P = 4s$; for $s$

**17.** $P = 2l + 2w$; for $w$

**18.** $d = rt$; for $t$

**19.** $A = P + Prt$; for $t$

**20.** $a = \dfrac{1}{2}(B + b)h$; for $h$

**21.** $C = 2\pi r$; for $r$

**22.** $I = \dfrac{E}{R}$; for $R$

**23.** $K = \dfrac{wv^2}{2g}$; for $w$

**24.** $V = \pi r^2 h$; for $h$

**25.** $P = I^2 R$; for $R$

**26.** $V = \dfrac{1}{3}\pi r^2 h$; for $h$

**27.** $K = \dfrac{wv^2}{2g}$; for $g$

**28.** $P = \dfrac{RT}{mV}$; for $V$

**29.** $F = \dfrac{GMm}{d^2}$; for $M$

**30.** $C = 1 - \dfrac{A}{a}$; for $A$

**31.** $F = \dfrac{GMm}{d^2}$; for $d^2$

**32.** $y = mx + b$; for $x$

**33.** $G = 2(r - 1)b$; for $r$

**34.** $F = f(1 - M)$; for $M$

*In Exercises 35–42, solve each formula for the indicated variable. Then substitute numbers to find the variable's value.*

**35.** $d = rt$     Find $t$ if $d = 135$ and $r = 45$.

**36.** $d = rt$     Find $r$ if $d = 275$ and $t = 5$.

**37.** $i = prt$     Find $t$ if $i = 12$, $p = 100$, and $r = 0.06$.

**38.** $i = prt$     Find $r$ if $i = 120$, $p = 500$, and $t = 6$.

**39.** $P = a + b + c$     Find $c$ if $P = 37$, $a = 15$, and $b = 19$.

**40.** $y = mx + b$     Find $x$ if $y = 30$, $m = 3$, and $b = 0$.

**41.** $K = \dfrac{1}{2}h(a + b)$     Find $h$ if $K = 48$, $a = 7$, and $b = 5$.

**42.** $\dfrac{x}{2} + y = z^2$     Find $x$ if $y = 3$ and $z = 3$.

## APPLICATIONS

**43. Ohm's law**   The formula $E = IR$, called **Ohm's law**, is used in electronics. Solve for $I$, and then calculate the current $I$ if the voltage $E$ is 48 volts and the resistance $R$ is 12 ohms. Current has units of *amperes*.

**44. Volume of a cone**   The volume $V$ of a cone is given by the formula $V = \frac{1}{3}\pi r^2 h$. Solve the formula for $h$, and then calculate the height $h$ if $V$ is $36\pi$ cubic inches and the radius $r$ is 6 inches.

**45. Circumference of a circle**   The circumference $C$ of a circle is given by $C = 2\pi r$, where $r$ is the radius of the circle. Solve the formula for $r$, and then calculate the radius of a circle with a circumference of 14.32 feet. Round to the nearest hundredth of a foot.

**46. Growth of money**   At a simple interest rate $r$, an amount of money $P$ grows to an amount $A$ in $t$ years according to the formula $A = P(1 + rt)$. Solve the formula for $P$. After $t = 3$ years, a girl has an amount $A = \$4{,}357$ on deposit. What amount $P$ did she start with? Assume an interest rate of 6%.

**47. Power loss**   The power $P$ lost when an electric current $I$ passes through a resistance $R$ is given by the formula $P = I^2R$. Solve for $R$. If $P$ is 2,700 watts and $I$ is 14 amperes, calculate $R$ to the nearest hundredth of an ohm.

**48. Geometry**   The perimeter $P$ of a rectangle with length $l$ and width $w$ is given by the formula $P = 2l + 2w$. Solve this formula for $w$. If the perimeter of a certain rectangle is 58.37 meters and its length is 17.23 meters, find its width. Round to two decimal places.

**49. Force of gravity**   The masses of the two objects in Illustration 1 are $m$ and $M$. The force of gravitation $F$ between the masses is given by

$$F = \frac{GmM}{d^2}$$

where $G$ is a constant and $d$ is the distance between them. Solve for $m$.

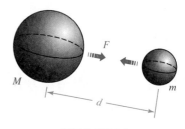

ILLUSTRATION 1

**50. Thermodynamics**   In thermodynamics, the Gibbs free-energy equation is given by

$$G = U - TS + pV$$

Solve this equation for the pressure, $p$.

**51. Pulleys** The approximate length $L$ of a belt joining two pulleys of radii $r$ and $R$ feet with centers $D$ feet apart is given by the formula

$$L = 2D + 3.25(r + R)$$

See Illustration 2. Solve the formula for $D$. If a 25-foot belt joins pulleys with radii of 1 foot and 3 feet, how far apart are the centers of the pulleys?

**52. Geometry** The measure $a$ of an interior angle of a regular polygon with $n$ sides is given by the formula $a = 180°\left(1 - \frac{2}{n}\right)$. See Illustration 3. Solve the formula for $n$. How many sides does a regular polygon have if an interior angle is 108°? (*Hint:* Distribute first.)

ILLUSTRATION 3

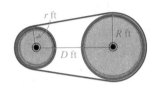

ILLUSTRATION 2

### WRITING

**53.** The formula $P = 2l + 2w$ is also an equation, but an equation such as $2x + 3 = 5$ is not a formula. What equations do you think should be called formulas?

**54.** To solve the equation $s - A(s - 5) = r$ for the variable $s$, one student simply added $A(s - 5)$ to both sides to get $s = r + A(s - 5)$. Explain why this is not correct.

### SOMETHING TO THINK ABOUT

**55.** The energy of an atomic bomb comes from the conversion of matter into energy, according to Einstein's formula $E = mc^2$. The constant $c$ is the speed of light, about 300,000 meters per second. Find the energy in a mass, $m$, of 1 kilogram. Energy has units of **joules.**

**56.** When a car of mass $m$ collides with a wall, the energy of the collision is given by the formula $E = \frac{1}{2}mv^2$. Compare the energy of two collisions: a car striking a wall at 30 mph, and at 60 mph.

## 2.7 Solving Inequalities

■ INEQUALITIES ■ COMPOUND INEQUALITIES ■ APPLICATIONS

Getting Ready    *Graph each set on the number line.*

**1.** All real numbers greater than $-1$.

**2.** All real numbers less than or equal to 5.

**3.** All real numbers between $-2$ and 4.    **4.** All real numbers less than $-2$ or greater than or equal to 4.

## ■ INEQUALITIES

Recall the meaning of the following symbols.

**Inequality Symbols**

| | |
|---|---|
| $<$ means | "is less than" |
| $>$ means | "is greater than" |
| $\leq$ means | "is less than or equal to" |
| $\geq$ means | "is greater than or equal to" |

An **inequality** is a mathematical statement that indicates that two quantities are not necessarily equal. A **solution of an inequality** is any number that makes the inequality true. The number 2 is a solution of the inequality

$$x \leq 3$$

because $2 \leq 3$.

The inequality $x \leq 3$ has many more solutions, because any real number that is less than or equal to 3 will satisfy the inequality. We can use a graph on the number line to represent the solutions of the inequality $x \leq 3$. The colored arrow in Figure 2-14 indicates all those points with coordinates that satisfy the inequality $x \leq 3$.

The bracket at the point with coordinate 3 indicates that the number 3 is a solution of the inequality $x \leq 3$.

The graph of the inequality $x > 1$ appears in Figure 2-15. The colored arrow indicates all those points whose coordinates satisfy the inequality $x > 1$. The parenthesis at the point with coordinate 1 indicates that 1 is not a solution of the inequality $x > 1$.

To solve more complicated inequalities, we need to use the addition, subtraction, multiplication, and division properties of inequalities. When we use any of these properties, the resulting inequality will have the same solutions as the original one.

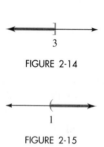

FIGURE 2-14

FIGURE 2-15

**Addition Property of Inequality**

If $a$, $b$, and $c$ are real numbers, and

If $a < b$, then $a + c < b + c$.

Similar statements can be made for the symbols $>$, $\leq$, and $\geq$.

The **addition property of inequality** can be stated this way: *If any quantity is added to both sides of an inequality, the resulting inequality has the same direction as the original inequality.*

**Subtraction Property of Inequality**

If $a$, $b$, and $c$ are real numbers, and

If $a < b$, then $a - c < b - c$.

Similar statements can be made for the symbols $>$, $\leq$, and $\geq$.

The **subtraction property of inequality** can be stated this way: *If any quantity is subtracted from both sides of an inequality, the resulting inequality has the same direction as the original inequality.*

The subtraction property of inequality is included in the addition property: To *subtract* a number $a$ from both sides of an inequality, we could instead *add* the *negative* of $a$ to both sides.

**EXAMPLE 1**    Solve $2x + 5 > x - 4$ and graph the solution on a number line.

*Solution*    To isolate the $x$ on the left-hand side of the $>$ sign, we proceed as if we were solving equations.

$$2x + 5 > x - 4$$
$$2x + 5 - 5 > x - 4 - 5 \qquad \text{Subtract 5 from both sides.}$$
$$2x > x - 9 \qquad \text{Combine like terms.}$$
$$2x - x > x - 9 - x \qquad \text{Subtract } x \text{ from both sides.}$$
$$x > -9 \qquad \text{Combine like terms.}$$

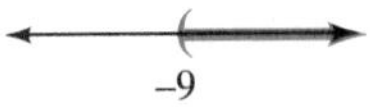

−9

FIGURE 2-16

The graph of the solution (see Figure 2-16) includes all points to the right of $-9$ but does not include $-9$ itself. For this reason, we use a parenthesis at $-9$. ■

**Self Check**
**Answer**

Graph the solution of $3x - 2 < x + 4$.

3

If both sides of the true inequality $2 < 5$ are multiplied by a *positive* number, such as 3, another true inequality results.

$$2 < 5$$
$$3 \cdot 2 < 3 \cdot 5 \qquad \text{Multiply both sides by 3.}$$
$$6 < 15$$

The inequality $6 < 15$ is true. However, if both sides of $2 < 5$ are multiplied by a negative number, such as $-3$, the direction of the inequality symbol must be reversed to produce another true inequality.

$$2 < 5$$
$$-3 \cdot 2 > -3 \cdot 5 \qquad \text{Multiply both sides by the } negative \text{ number } -3 \text{ and} \\ \text{reverse the direction of the inequality.}$$
$$-6 > -15$$

The inequality $-6 > -15$ is true, because $-6$ lies to the right of $-15$ on the number line. These examples suggest the following properties.

### Multiplication Property of Inequality

If $a$, $b$, and $c$ are real numbers, and

     If $a < b$ and $c > 0$, then $ac < bc$.

     If $a < b$ and $c < 0$, then $ac > bc$.

Similar statements can be made for the symbols $>$, $\leq$, and $\geq$.

The multiplication property of inequality can be stated this way:

*If unequal quantities are multiplied by the same positive quantity, the results will be unequal and in the same order.*

*If unequal quantities are multiplied by the same negative quantity, the results will be unequal but in the opposite order.*

There is a similar property for division.

### Division Property of Inequality

If $a$, $b$, and $c$ are real numbers, and

     If $a < b$ and $c > 0$, then $\dfrac{a}{c} < \dfrac{b}{c}$.

     If $a < b$ and $c < 0$, then $\dfrac{a}{c} > \dfrac{b}{c}$.

Similar statements can be made for the symbols $>$, $\leq$, and $\geq$.

The division property of inequality can be stated this way:

*If unequal quantities are divided by the same positive quantity, the results will be unequal and in the same order.*

*If unequal quantities are divided by the same negative quantity, the results will be unequal but in the opposite order.*

To *divide* both sides of an inequality by a nonzero number $c$, we could instead *multiply* both sides by $\frac{1}{c}$.

**WARNING!** If both sides of an inequality are multiplied by a *positive* number, the direction of the resulting inequality remains the same. However, if both sides of an inequality are multiplied by a *negative* number, the direction of the resulting inequality must be reversed.

| | |
|---|---|
| **EXAMPLE 2** | Solve $3x + 7 \leq -5$ and graph the solution. |

*Solution*

$$3x + 7 \leq -5$$
$$3x + 7 - 7 \leq -5 - 7 \qquad \text{Subtract 7 from both sides.}$$
$$3x \leq -12 \qquad \text{Combine like terms.}$$
$$\frac{3x}{3} \leq \frac{-12}{3} \qquad \text{Divide both sides by 3.}$$
$$x \leq -4$$

FIGURE 2-17

The solution consists of all real numbers that are less than or equal to $-4$. The bracket at $-4$ in the graph of Figure 2-17 indicates that $-4$ is one of the solutions. ■

**Self Check**
**Answer**

Graph the solution of $2x - 5 \geq -3$.

**EXAMPLE 3**  Solve $5 - 3x \leq 14$ and graph the solution.

*Solution*

$$5 - 3x \leq 14$$
$$5 - 3x - 5 \leq 14 - 5 \qquad \text{Subtract 5 from both sides.}$$
$$-3x \leq 9 \qquad \text{Combine like terms.}$$
$$\frac{-3x}{-3} \geq \frac{9}{-3} \qquad \text{Divide both sides by } -3 \text{ and reverse the direction of the } \leq \text{ symbol.}$$
$$x \geq -3$$

FIGURE 2-18

Since both sides of the inequality were divided by $-3$, the direction of the inequality was *reversed*. The graph of the solution appears in Figure 2-18. The bracket at $-3$ indicates that $-3$ is one of the solutions. ■

**Self Check**
**Answer**

Graph the solution of $6 - 7x \geq -15$.

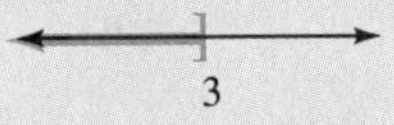

## ■ COMPOUND INEQUALITIES

Two inequalities can often be combined into a **double inequality** or **compound inequality** to indicate that numbers lie *between* two fixed values. For example, the inequality $2 < x < 5$ indicates that $x$ is greater than 2 and that $x$ is also less than 5. The solution of $2 < x < 5$ consists of all numbers that lie *between* 2 and 5. The graph of this set (called an **interval**) appears in Figure 2-19.

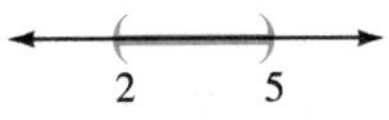

FIGURE 2-19

**EXAMPLE 4**    Solve $-4 < 2(x - 1) \le 4$ and graph the solution.

*Solution*
$$-4 < 2(x - 1) \le 4$$

$-4 < 2x - 2 \le 4$    Remove parentheses.

$-2 < 2x \le 6$    Add 2 to all three parts.

$-1 < x \le 3$    Divide all three parts by 2.

FIGURE 2-20

The graph of the solution appears in Figure 2-20. ■

**Self Check**    Graph the solution of $0 \le 4(x + 5) < 26$.

**Answer**

*(number line graph from -5 to 3/2)*

■ **APPLICATIONS**

**EXAMPLE 5**    A student has scores of 72%, 74%, and 78% on three mathematics examinations. What interval of scores does his last score need to fall in to earn a grade of B (80% or better)?

*Solution*    We can let $x$ represent the score on the fourth (and last) exam. To find the average grade, we add the four scores and divide by 4. To earn a B, this average must be greater than or equal to 80%.

| The average of the four grades | $\ge$ | 80 |
|---|---|---|

$$\frac{72 + 74 + 78 + x}{4} \ge 80$$

We can solve this inequality for $x$.

$\dfrac{224 + x}{4} \ge 80$    $72 + 74 + 78 = 224.$

$224 + x \ge 320$    Multiply both sides by 4.

$x \ge 96$    Subtract 224 from both sides.

A perfect score on the last exam is 100%. To earn a B, the student must score from 96% to 100%. This means that the student's score must be in the interval $96 \le x \le 100$. The graph of this interval appears in Figure 2-21. ■

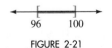

FIGURE 2-21

**EXAMPLE 6**    If the perimeter of an equilateral triangle is less than 15 feet, how long could each side be?

*Solution*    Recall that each side of an equilateral triangle is the same length and that the perimeter of a triangle is the sum of the lengths of its three sides. If we let $x$ represent the length of one of the sides, then $x + x + x$ represents the perimeter. Since the perimeter is to be less than 15 feet, we have the following inequality:

$$x + x + x < 15$$
$$3x < 15 \qquad \text{Combine like terms.}$$
$$x < 5 \qquad \text{Divide both sides by 3.}$$

Each side of the triangle must be less than 5 feet long.    ∎

Orals    *Solve each inequality.*

**1.** $2x < 4$     **2.** $x + 5 \geq 6$
**3.** $-3x \leq -6$     **4.** $-x > 2$
**5.** $2x - 5 < 7$     **6.** $5 - 2x < 7$

# EXERCISE 2.7

***REVIEW***   *Simplify each expression.*

**1.** $3x^2 - 2(y^2 - x^2)$     **2.** $5(xy + 2) - 3xy - 8$
**3.** $\frac{1}{3}(x + 6) - \frac{4}{3}(x - 9)$     **4.** $\frac{4}{5}x(y + 1) - \frac{9}{5}y(x - 1)$

***VOCABULARY AND CONCEPTS***   *Fill in each blank to make a true statement.*

**5.** The symbol $<$ means _____.
**6.** The symbol $>$ means _____.
**7.** The symbol __ means "is greater than or equal to."
**8.** The symbol __ means "is less than or equal to."
**9.** An _____ is a statement indicating that two quantities are not necessarily equal.
**10.** A _____ of an inequality is a number that makes the inequality true.

***PRACTICE***   *In Exercises 11–44, solve each inequality and graph the solution.*

**11.** $x + 2 > 5$     **12.** $x + 5 \geq 2$     **13.** $-x - 3 \leq 7$     **14.** $-x - 9 > 3$

**15.** $3 + x < 2$     **16.** $5 + x \geq 3$     **17.** $2x - 3 \leq 5$     **18.** $-3x - 5 < 4$

**19.** $-3x - 7 > -1$     **20.** $-5x + 7 \leq 12$     **21.** $-4x + 1 > 17$     **22.** $7x - 9 > 5$

**23.** $2x + 9 \leq x + 8$

**24.** $3x + 7 \leq 4x - 2$

**25.** $9x + 13 \geq 8x$

**26.** $7x - 16 < 6x$

**27.** $8x + 4 > 6x - 2$

**28.** $7x + 6 \geq 4x$

**29.** $5x + 7 < 2x + 1$

**30.** $7x + 2 > 4x - 1$

**31.** $7 - x \leq 3x - 1$

**32.** $2 - 3x \geq 6 + x$

**33.** $9 - 2x > 24 - 7x$

**34.** $13 - 17x < 34 - 10x$

**35.** $3(x - 8) < 5x + 6$

**36.** $9(x - 11) > 13 + 7x$

**37.** $8(5 - x) \leq 10(8 - x)$

**38.** $17(3 - x) \geq 3 - 13x$

**39.** $\dfrac{5}{2}(7x - 15) + x \geq \dfrac{13}{2}x - \dfrac{3}{2}$

**40.** $\dfrac{5}{3}(x + 1) \leq -x + \dfrac{2}{3}$

**41.** $\dfrac{3x - 3}{2} < 2x + 2$

**42.** $\dfrac{x + 7}{3} \geq x - 3$

**43.** $\dfrac{2(x + 5)}{3} \leq 3x - 6$

**44.** $\dfrac{3(x - 1)}{4} > x + 1$

*In Exercises 45–62, solve each inequality and graph the solution.*

**45.** $2 < x - 5 < 5$

**46.** $3 < x - 2 < 7$

**47.** $-5 < x + 4 \leq 7$

**48.** $-9 \leq x + 8 < 1$

**49.** $0 \leq x + 10 \leq 10$

**50.** $-8 < x - 8 < 8$

**51.** $4 < -2x < 10$

**52.** $-4 \leq -4x < 12$

**53.** $-3 \leq \dfrac{x}{2} \leq 5$

**54.** $-12 \leq \dfrac{x}{3} < 0$

**55.** $3 \leq 2x - 1 < 5$

**56.** $4 < 3x - 5 \leq 7$

**57.** $0 < 10 - 5x \leq 15$

**58.** $1 \leq -7x + 8 \leq 15$

**59.** $-6 < 3(x + 2) < 9$

**60.** $-18 \leq 9(x - 5) < 27$

**61.** $3 - x < 5 < 7 - x$

**62.** $x + 1 < 2x + 3 < x + 5$

*APPLICATIONS* *In Exercises 63–80, express each solution as an inequality.*

**63. Calculating grades**  A student has test scores of 68%, 75%, and 79%. What must she score on the last exam to earn 80% or better?

**64. Calculating grades**  A student has test scores of 70%, 74%, and 84%. What score does he need on the last exam to maintain 70% or better?

**65. Fleet averages**  An automobile manufacturer produces three sedan models in equal quantities. One model has an economy rating of 17 miles per gallon, and the second model is rated for 19 mpg. If the manufacturer is required to have a fleet average of at least 21 mpg, what economy rating is required for the third model?

**66. Avoiding a service charge**  When the average daily balance of a customer's checking account falls below $500 in any week, the bank assesses a $5 service charge. Bill's account balances for the week were as shown in Illustration 1.

| | |
|---|---|
| Monday | $540.00 |
| Tuesday | $435.50 |
| Wednesday | $345.30 |
| Thursday | $310.00 |

ILLUSTRATION 1

What must Friday's balance be to avoid the service charge?

**67. Geometry**  The perimeter of an equilateral triangle is at most 57 feet. What could be the length of a side? (*Hint:* All three sides of an equilateral triangle are equal.)

**68. Geometry**  The perimeter of a square is no less than 68 centimeters. How long can a side be?

**69. Land elevations**  The land elevations in Nevada range from the 13,143-foot height of Boundary Peak to the Colorado River at 470 feet. To the nearest tenth, what is the range of these elevations in miles? (*Hint:* 1 mile is 5,280 feet.)

**70. Doing homework**  A teacher requires that students do homework at least 2 hours a day. How many minutes should a student work each week?

**71. Plane altitudes**  A pilot plans to fly at an altitude of between 17,500 and 21,700 feet. To the nearest tenth, what will be the range of altitudes in miles? (*Hint:* There are 5,280 feet in 1 mile.)

**72. Getting exercise**  Doctors advise exercising at least 15 minutes but less than 30 minutes per day. Find the range of exercise time for one week.

**73. Comparing temperatures**  To hold the temperature of a room between 19° and 22° Celsius, what Fahrenheit temperatures must be maintained? (*Hint:* Fahrenheit temperature ($F$) and Celsius temperature ($C$) are related by the formula $C = \frac{5}{9}(F - 32)$.)

**74. Melting iron**  To melt iron, the temperature of a furnace must be at least 1,540°C but no more than 1,650°C. What range of Fahrenheit temperatures must be maintained?

**75. Phonograph records**  The radii of phonograph records must lie between 5.9 and 6.1 inches. What variation in circumference can occur? (*Hint:* The circumference of a circle is given by the formula $C = 2\pi r$, where $r$ is the radius. Let $\pi = 3.14$.)

**76. Pythons**  A large snake, the African Rock Python, can grow to a length of 25 feet. To the nearest hundredth, find the snake's range of lengths in meters. (*Hint:* There are about 3.281 feet in 1 meter.)

**77. Comparing weights**  The normal weight of a 6 foot 2 inch man is between 150 and 190 pounds. To the nearest hundredth, what would such a person weigh in kilograms? (*Hint:* There are 2.2 pounds in 1 kilogram.)

**78. Manufacturing**   The time required to assemble a television set at the factory is 2 hours. A stereo receiver requires only 1 hour. The labor force at the factory can supply at least 640 and at most 810 hours of assembly time per week. When the factory is producing 3 times as many television sets as stereos, how many stereos could be manufactured in 1 week?

**79. Geometry**   A rectangle's length is 3 feet less than twice its width, and its perimeter is between 24 and 48 feet. What might be its width?

**80. Geometry**   A rectangle's width is 8 feet less than 3 times its length, and its perimeter is between 8 and 16 feet. What might be its length?

## WRITING

**81.** Explain why multiplying both sides of an inequality by a negative constant reverses the direction of the inequality.

**82.** Explain the use of parentheses and brackets in the graphing of the solution of an inequality.

## SOMETHING TO THINK ABOUT

**83.** To solve the inequality $1 < \frac{1}{x}$, one student multiplies both sides by $x$ to get $x < 1$. Why is this not correct?

**84.** Find the solution of $1 < \frac{1}{x}$. (*Hint:* Will any negative values of $x$ work?)

■ ■ ■ ■ ■ ■ ■ ■ ■ ■

**MATHEMATICS IN RETIREMENT**

To find the maximum allowable annual contribution to an SEP, we first find the net income $N$ by subtracting deductible expenses from gross income. If $C$ is the maximum contribution to the SEP, then

$$C = 0.15(N - C)$$

Solve this equation for $C$.

| | |
|---|---|
| $C = 0.15(N - C)$ | |
| $C = 0.15N - 0.15C$ | Remove parentheses. |
| $C + 0.15C = 0.15N$ | Add $0.15C$ to both sides. |
| $1.15C = 0.15N$ | Combine like terms. |
| $C = \dfrac{0.15}{1.15}N$ | Divide both sides by 1.15. |
| $C = 0.1304N$ | Simplify. |

The maximum allowable annual contribution to an SEP plan is slightly greater than 13% of taxable income.

■ ■ ■ ■ ■ ■ ■ ■ ■ **PROJECTS**

PROJECT 1   Build a scale similar to the one shown in Figure 2-1. Demonstrate to your class how you would use the scale to solve the following equations.

    **a.** $x - 4 = 6$      **b.** $x + 3 = 2$      **c.** $2x = 6$

    **d.** $\dfrac{x}{2} = 3$      **e.** $3x - 2 = 5$      **f.** $\dfrac{x}{3} + 1 = 2$

PROJECT 2   Magicians don't really saw people in half, or pull rabbits out of empty hats. Most magic tricks are just clever illusions, fooling the audience into seeing what isn't really there. The most successful magicians are very believable liars—and it takes a lot of practice to be good.

    Many magic tricks involve cutting ropes in various ways and then restoring them to their original lengths. For example, a magician holds up a long rope for all to see and then cuts it into three separate sections. He displays these to the audience; the sections are of three obviously different lengths, as in Illustration 1.

    The magician then folds the ropes, twists them, coils them around his fist and arm, and utters some magic words—all to distract and confuse the audience. When he holds up the three sections, as in Illustration 2, they are now the same length!

    The secret of the trick lies behind the magician's hand, hidden from the audience. What appears to be two equal lengths of rope in Illustration 3 is only one—the longest of the original three, folded in half. Those "two" sections are equal to the "third," which is just the middle-sized of the original three. What happened to the shortest of the original three? The magician disposed of it when the audience was distracted.

ILLUSTRATION 1

ILLUSTRATION 2

*(continued)*

■ ■ ■ ■ ■ ■ ■ ■ ■ ■ **PROJECTS** *(continued)*

To prepare for this trick, the magician places two marks on an 8-foot rope, so that the two cuts can be made quickly and accurately. A third mark of a different color is the center of the largest section, the point where that rope is to be folded in half.

- If the shortest section is to be 1 foot long, where does the magician make the marks?
- Get some rope, cut an 8-foot piece, mark it as you have determined, and practice the trick. There are several ways to dispose of the shortest segment without being noticed. Try using a stretched rubber band to snap the rope up your sleeve, or fake a distracting sneeze while you slip it into your pocket. It is an easy trick to master, and an effective illusion.

ILLUSTRATION 3

# CHAPTER SUMMARY

CONCEPTS

REVIEW EXERCISES

| SECTION 2.1 | *Solving Equations by Addition and Subtraction* |

An **equation** is a statement indicating that two quantities are equal.

**1.** Tell whether the given number is a solution of the equation.
   **a.** $3x + 7 = 1$; $-2$
   **b.** $5 - 2x = 3$; $-1$
   **c.** $2(x + 3) = x$; $-3$
   **d.** $5(3 - x) = 2 - 4x$; $13$
   **e.** $3(x + 5) = 2(x - 3)$; $-21$
   **f.** $2(x - 7) = x + 14$; $0$

Any real number can be added to (or subtracted from) both sides of an equation to form another equation with the same solutions as the original equation.

**2.** Solve each equation and check all solutions.
   **a.** $x - 7 = -6$
   **b.** $y - 4 = 5$
   **c.** $p + 4 = 20$
   **d.** $x + \dfrac{3}{5} = \dfrac{3}{5}$
   **e.** $y - \dfrac{7}{2} = \dfrac{1}{2}$
   **f.** $z + \dfrac{5}{3} = -\dfrac{1}{3}$

Sale price
= regular price − markdown

**3.** A necklace is on sale for $69.95. If it has been marked down $35.45, what is its regular price?

Retail price
= wholesale cost + markup

**4.** A suit that has been marked up $115.25 sells for $212.95. Find its wholesale price.

If the sum of the measures of two angles is 90°, the angles are complementary.

**5.** Find the complement of an angle that measures 69°.

If the sum of the measures of two angles is 180°, the angles are supplementary.

**6.** Find the supplement of an angle that measures 69°.

| **SECTION 2.2** | *Solving Equations by Multiplication and Division* |

Both sides of an equation can be multiplied (or divided) by any *nonzero* real number to form another equation with the same solutions as the original equation.

**7.** Solve each equation and check all solutions.
  **a.** $3x = 15$
  **b.** $8r = -16$
  **c.** $10z = 5$
  **d.** $14s = 21$
  **e.** $\dfrac{y}{3} = 6$
  **f.** $\dfrac{w}{7} = -5$
  **g.** $\dfrac{a}{-7} = \dfrac{1}{14}$
  **h.** $\dfrac{t}{12} = \dfrac{1}{2}$

Percentage = rate · base

**8.** Solve each problem.
  **a.** What number is 35% of 700?

  **b.** 72% of what number is 936?

  **c.** What percent of 2,300 is 851?

  **d.** 72 is what percent of 576?

**9.** Find the % average of a student with the following scores.

| Test | Number of questions | Number correct |
|------|---------------------|----------------|
| 1 | 89 | 72 |
| 2 | 77 | 53 |
| 3 | 81 | 75 |

| SECTION 2.3 | *Solving More Equations* |

**10.** Solve each equation and check all solutions.

**a.** $5y + 6 = 21$          **b.** $5y - 9 = 1$

**c.** $-12z + 4 = -8$          **d.** $17z + 3 = 20$

**e.** $13 - 13t = 0$          **f.** $10 + 7t = -4$

**g.** $23a - 43 = 3$          **h.** $84 - 21a = -63$

**i.** $3x + 7 = 1$          **j.** $7 - 9x = 16$

**k.** $\dfrac{b + 3}{4} = 2$          **l.** $\dfrac{b - 7}{2} = -2$

**m.** $\dfrac{x - 8}{5} = 1$          **n.** $\dfrac{x + 10}{2} = -1$

**o.** $\dfrac{2y - 2}{4} = 2$          **p.** $\dfrac{3y + 12}{11} = 3$

**q.** $\dfrac{x}{2} + 7 = 11$          **r.** $\dfrac{r}{3} - 3 = 7$

**s.** $\dfrac{a}{2} + \dfrac{9}{4} = 6$          **t.** $\dfrac{x}{8} - 2.3 = 3.2$

**11.** A compact disc player is on sale for $240, a 25% savings from the regular price. Find the regular price.

**12.** A $38 dictionary costs $40.47, with sales tax. Find the tax rate.

**13.** A Turkish rug was purchased for $560. If it is now worth $1,100, find the percent of increase.

**14.** A clock on sale for $215 was regularly priced at $465. Find the percent of discount.

| SECTION 2.4 | *Simplifying Expressions to Solve Equations* |

Like terms can be combined by adding their numerical coefficients and using the same variables and exponents.

**15.** Simplify each expression, if possible.

**a.** $5x + 9x$          **b.** $7a + 12a$

**c.** $18b - 13b$          **d.** $21x - 23x$

**e.** $5y - 7y$          **f.** $19x - 19$

**g.** $7(x + 2) + 2(x - 7)$          **h.** $2(3 - x) + x - 6x$

**i.** $y^2 + 3(y^2 - 2)$          **j.** $2x^2 - 2(x^2 - 2)$

**16.** Solve each equation and check all solutions.

**a.** $2x - 19 = 2 - x$          **b.** $5b - 19 = 2b + 20$

**c.** $3x + 20 = 5 - 2x$          **d.** $0.9x + 10 = 0.7x + 1.8$

**e.** $10(t - 3) = 3(t + 11)$          **f.** $2(5x - 7) = 2(x - 35)$

**g.** $\dfrac{3u - 6}{5} = 3$          **h.** $\dfrac{5v - 35}{3} = -5$

**i.** $\dfrac{7x - 28}{4} = -21$          **j.** $\dfrac{27 + 9y}{5} = -27$

An **identity** is an equation that is true for all values of its variable.

An **impossible equation** or a **contradiction** is an equation that is true for no values of its variable.

**17.** Classify each equation as an identity or a contradiction.
   **a.** $2x - 5 = x - 5 + x$
   **b.** $-3(a + 1) - a = -4a + 3$
   **c.** $2(x - 1) + 4 = 4(1 + x) - (2x + 2)$

| SECTION 2.5 | *Applications of Equations* |
|---|---|

Equations are useful in solving applied problems.

**18.** A carpenter wants to cut an 8-foot board into two pieces so that one piece is 7 feet shorter than twice the longer piece. Where should he make the cut?

**19.** If the length of the rectangular painting in Illustration 1 is 3 inches more than twice the width, how wide is the rectangle?

84 in.

ILLUSTRATION 1

**20.** A woman has $27,000. Part is invested for 1 year in a certificate of deposit paying 7% interest, and the remaining amount in a cash management fund paying 9%. The total interest on the two investments is $2,110. How much does she invest at each rate?

**21.** A bicycle path is 5 miles long. A man walks from one end at the rate of 3 mph. At the same time, a friend bicycles from the other end, traveling at 12 mph. In how many minutes will they meet?

**22.** A container is partly filled with 12 liters of whole milk containing 4% butterfat. How much 1% milk must be added to get a mixture that is 2% butterfat?

**23.** A store manager mixes candy worth 90¢ per pound with gumdrops worth $1.50 per pound to make 20 pounds of a mixture worth $1.20 per pound. How many pounds of each kind of candy must he use?

**24.** The electric company charges $17.50 per month, plus 18¢ for every kilowatt-hour of energy used. One resident's bill was $43.96. How many kilowatt-hours were used that month?

**25.** A contractor charges $35 for the installation of rain gutters, plus $1.50 per foot. If one installation cost $162.50, how many feet of gutter were required?

| SECTION 2.6 | *Formulas* |
| --- | --- |

A literal equation, or formula, can often be solved for any of its variables.

**26.** Solve each equation for the indicated variable.

**a.** $E = IR$; for $R$             **b.** $i = prt$; for $t$

**c.** $P = I^2R$; for $R$           **d.** $d = rt$; for $r$

**e.** $V = lwh$; for $h$            **f.** $y = mx + b$; for $m$

**g.** $V = \pi r^2 h$; for $h$      **h.** $a = 2\pi rh$; for $r$

**i.** $F = \dfrac{GMm}{d^2}$; for $G$    **j.** $P = \dfrac{RT}{mV}$; for $m$

| SECTION 2.7 | *Solving Inequalities* |
| --- | --- |

Inequalities are solved by techniques similar to those used to solve equations, with this exception: *If both sides of an inequality are multiplied or divided by a negative number, the direction of the inequality must be reversed.*

The solution of an inequality can be graphed on the number line.

**27.** Graph the solution to each inequality.

**a.** $3x + 2 < 5$

**b.** $-5x - 8 > 7$

**c.** $5x - 3 \geq 2x + 9$

**d.** $7x + 1 \leq 8x - 5$

**e.** $5(3 - x) \leq 3(x - 3)$

**f.** $3(5 - x) \geq 2x$

**g.** $8 < x + 2 < 13$

**h.** $0 \leq 2 - 2x < 4$

# ■ Chapter Test

*In Problems 1–4, state whether the given number is a solution of the equation.*

**1.** $5x + 3 = -2; -1$

**2.** $3(x + 2) = 2x; -6$

**3.** $-3(2 - x) = 0; -2$

**4.** $3(x + 2) = 2x + 7; 1$

*In Problems 5–14, solve each equation.*

**5.** $x + 17 = -19$

**6.** $a - 15 = 32$

**7.** $12x = -144$

**8.** $\dfrac{x}{7} = -1$

**9.** $8x + 2 = -14$

**10.** $3 = 5 - 2x$

**11.** $\dfrac{2x - 5}{3} = 3$

**12.** $\dfrac{3x - 18}{2} = 6x$

**13.** $23 - 5(x + 10) = -12$

**14.** $\dfrac{7}{8}(x - 4) = 5x - \dfrac{7}{2}$

*In Problems 15–18, simplify each expression.*

**15.** $x + 5(x - 3)$

**16.** $3x - 5(2 - x)$

**17.** $-3x(x + 3) + 3x(x - 3)$

**18.** $-4x(2x - 5) - 7x(4x + 1)$

**19.** A car leaves Rockford at the rate of 65 mph bound for Madison. At the same time, a truck leaves Madison at the rate of 55 mph, bound for Rockford. If the cities are 72 miles apart, how long will it take for the car and the truck to meet?

**20.** How many liters of water must be added to 30 liters of a 10% brine solution to dilute it to an 8% solution?

*In Problems 21–24, solve each equation for the variable indicated.*

**21.** $d = rt$; for $t$

**22.** $P = 2l + 2w$; for $l$

**23.** $A = 2\pi rh$; for $h$

**24.** $A = P + Prt$; for $r$

*In Problems 25–28, graph the solution of each inequality.*

**25.** $8x - 20 \geq 4$

**26.** $x - 2(x + 7) > 14$

**27.** $-4 \leq 2(x + 1) < 10$

**28.** $-2 < 5(x - 1) \leq 10$

# ■ Cumulative Review Exercises

*In Exercises 1–2, classify each number as an integer, a rational number, an irrational number, a real number, a positive number, or a negative number. Each number may be in several classifications.*

**1.** $\dfrac{27}{9}$

**2.** $-0.25$

*In Exercises 3–4, graph each set of numbers on the number line.*

**3.** The natural numbers between 2 and 7

**4.** The real numbers between 2 and 7

*In Exercises 5–8, simplify each expression.*

**5.** $\dfrac{|-3| - |3|}{|-3 - 3|}$

**6.** $\dfrac{5}{7} \cdot \dfrac{14}{3}$

**7.** $2\dfrac{3}{5} + 5\dfrac{1}{2}$

**8.** $35.7 - 0.05$

*In Exercises 9–12, let $x = -5$, $y = 3$, and $z = 0$. Evaluate each expression.*

**9.** $(3x - 2y)z$

**10.** $\dfrac{x - 3y + |z|}{2 - x}$

**11.** $x^2 - y^2 + z^2$

**12.** $\dfrac{x}{y} + \dfrac{y + 2}{3 - z}$

**13.** What is $7\frac{1}{2}\%$ of 330?

**14.** 1,688 is 32% of what number?

*In Exercises 15–16, consider the algebraic expression $3x^3 + 5x^2y + 37y$.*

**15.** Find the coefficient of the second term.

**16.** List the factors of the third term.

*In Exercises 17–20, simplify each expression.*

**17.** $3x - 5x + 2y$

**18.** $3(x - 7) + 2(8 - x)$

**19.** $2x^2y^3 - xy(xy^2)$

**20.** $x^2(3 - y) + x(xy + x)$

*In Exercises 21–24, solve each equation.*

**21.** $3(x - 5) + 2 = 2x$

**22.** $\dfrac{x - 5}{3} - 5 = 7$

**23.** $\dfrac{2x - 1}{5} = \dfrac{1}{2}$

**24.** $2(a - 5) - (3a + 1) = 0$

**25. Auto sales** An auto dealer's promotional ad appears in Illustration 1. One car is selling for $23,499. What was the dealer's invoice?

ILLUSTRATION 1

**26. Furniture pricing** A sofa and a $300 chair are discounted 35%, and are priced at $780 for both. Find the original price of the sofa.

**27. Cost of a car** The total cost of a new car, including an 8.5% sales tax, is $13,725.25. Find the cost before tax.

**28. Manufacturing concrete** Concrete contains 3 times as much gravel as cement. How many pounds of cement are in 500 pounds of dry concrete mix?

**29. Building construction** A 35-foot beam, 1 foot wide and 2 inches thick, is cut into three sections. One section is 14 feet long. Of the remaining two sections, one is twice as long as the other. Will the shortest section span an 8-foot-wide doorway?

**30. Installing solar heating** One solar panel in Illustration 2 is 3.4 feet wider than the other. Find the width of each.

ILLUSTRATION 2

*In Exercises 31–32, solve each formula for the variable indicated.*

**31.** $A = \dfrac{1}{2}h(b + B)$; for $h$

**32.** $y = mx + b$; for $x$

*In Exercises 33–36, evaluate each expression.*

**33.** $4^2 - 5^2$

**34.** $(4 - 5)^2$

**35.** $5(4^3 - 2^3)$

**36.** $-2(5^4 - 7^3)$

*In Exercises 37–38, graph the solutions of each inequality.*

**37.** $8(4 + x) > 10(6 + x)$

**38.** $-9 < 3(x + 2) \le 3$

# 3

# Graphing and Solving Systems of Equations and Inequalities

MATHEMATICS IN ECONOMICS

The number of canoes that boaters will buy depends on price. The higher the price, the fewer canoes people will buy. The equation that relates the retail price of a canoe to the number of canoes bought at that price is called a **demand equation.** Suppose that the demand equation for canoes is

**1.** $p = -\dfrac{1}{2}q + 1{,}300$

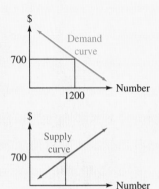

where $p$ is the price and $q$ is the number purchased each month at that price. From the graph of Equation 1, shown in the illustration, we see that boaters will buy 1,200 canoes at a price of $700.

The number of canoes a manufacturer is willing to produce also depends on price. The higher the price, the more canoes manufacturers are willing to produce. The equation that relates the number of canoes produced to the retail price is called a **supply equation.** Suppose that the supply equation for canoes is

**2.** $p = \dfrac{1}{3}q + \dfrac{1{,}400}{3}$

where $p$ is the retail price and $q$ is the number produced for sale at that price. From the graph of this equation, shown in the illustration, we see that manufacturers will produce only 700 canoes at a price of $700.

Since suppliers will produce 700 canoes and people want 1,200, there will be a shortage, and the price will go up. The **equilibrium price** is the price at which supply equals demand. Find the equilibrium price.

After you have read this chapter, you will be able to solve this problem.

It is often said, "A picture is worth a thousand words." In this section, we will show how numerical relationships can be described using mathematical pictures called **graphs.** We will also show how graphs are constructed and how we can obtain important information by reading graphs.

## 3.1 The Rectangular Coordinate System

■ THE RECTANGULAR COORDINATE SYSTEM ■ GRAPHING MATHEMATICAL RELATIONSHIPS
■ READING GRAPHS ■ STEP GRAPHS

Getting Ready   *Graph each set of numbers on the number line.*

**1.** −2, 1, 3                    **2.** All numbers greater than −2

**3.** All numbers less than or equal to 3   **4.** All numbers between $-3$ and 2

## ■ THE RECTANGULAR COORDINATE SYSTEM

When designing the Gateway Arch, shown in Figure 3-1(a), architects created a mathematical model of the arch called a **graph.** This graph, shown in Figure 3-1(b), is drawn on a grid called a **rectangular coordinate system.** This coordinate system is sometimes called a **Cartesian coordinate system** after the 17th-century French mathematician René Descartes.

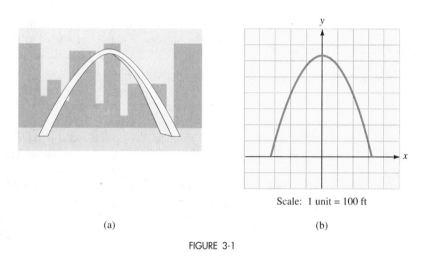

Scale: 1 unit = 100 ft

(a)                                                    (b)

FIGURE 3-1

A rectangular coordinate system (see Figure 3-2) is formed by two perpendicular number lines.

- The horizontal number line is called the ***x*-axis.**
- The vertical number line is called the ***y*-axis.**

The positive direction on the $x$-axis is to the right, and the positive direction on the $y$-axis is upward. The scale on each axis should fit the data. For example, the axes of the graph of the arch shown in Figure 3-1(b) are scaled in units of 100 feet. If no scale is indicated on the axes, we assume that the axes are scaled in units of 1.

The point where the axes cross is called the **origin.** This is the 0 point on each axis. The two axes form a **coordinate plane** and divide it into four regions called **quadrants,** which are numbered as shown in Figure 3-2.

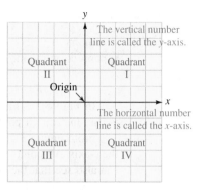

FIGURE 3-2

Each point in a coordinate plane can be identified by a pair of real numbers $x$ and $y$, written as $(x, y)$. The first number in the pair is the **$x$-coordinate,** and the second number is the **$y$-coordinate.** The numbers are called the **coordinates** of the point. Some examples of ordered pairs are $(3, -4)$, $\left(-1, -\frac{3}{2}\right)$, and $(0, 2.5)$.

$$(3, -4)$$

In an ordered pair, the  The $y$-coordinate
$x$-coordinate is listed first.  is listed second.

The process of locating a point in the coordinate plane is called **graphing** or **plotting** the point. In Figure 3-3(a), we show how to graph the point $A$ with coordinates of $(3, -4)$. Since the $x$-coordinate is positive, we start at the origin and move 3 units to the right along the $x$-axis. Since the $y$-coordinate is negative, we then move down 4 units to locate point $A$. Point $A$ is the **graph** of $(3, -4)$ and lies in quadrant IV.

To plot the point $B(-4, 3)$, we start at the origin, move 4 units to the left along the $x$-axis, and then move up 3 units to locate point $B$. Point $B$ lies in quadrant II.

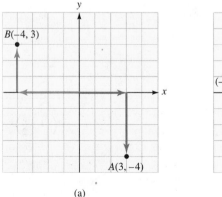

(a)

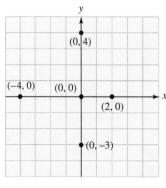

(b)

FIGURE 3-3

**WARNING!**   Note that point $A$ with coordinates of $(3, -4)$ is not the same as point $B$ with coordinates $(-4, 3)$. Since the order of the coordinates of a point is important, we call the pairs **ordered pairs.**

In Figure 3-3(b), we see that the points $(-4, 0)$, $(0, 0)$, and $(2, 0)$ lie on the $x$-axis. In fact, all points with a $y$-coordinate of 0 will lie on the $x$-axis.

From Figure 3-3(b), we also see that the points $(0, -3)$, $(0, 0)$, and $(0, 4)$ lie on the $y$-axis. All points with an $x$-coordinate of 0 lie on the $y$-axis. From the figure, we can also see that the coordinates of the origin are $(0, 0)$.

**EXAMPLE 1**

**Graphing points**   Plot the points   **a.** $A(-2, 3)$,   **b.** $B\left(-1, -\frac{3}{2}\right)$,   **c.** $C(0, 2.5)$, and   **d.** $D(4, 2)$.

*Solution*   **a.** To plot point $A$ with coordinates $(-2, 3)$, we start at the origin, move 2 units to the *left* on the $x$-axis, and move 3 units *up*. Point $A$ lies in quadrant II. (See Figure 3-4.)

**b.** To plot point $B$ with coordinates of $\left(-1, -\frac{3}{2}\right)$, we start at the origin and move 1 unit to the *left* and $\frac{3}{2}$ $\left(\text{or } 1\frac{1}{2}\right)$ units *down*. Point $B$ lies in quadrant III, as shown in Figure 3-4.

**c.** To graph point $C$ with coordinates of $(0, 2.5)$, we start at the origin and move 0 units on the $x$-axis and 2.5 units *up*. Point $C$ lies on the $y$-axis, as shown in Figure 3-4.

**d.** To graph point $D$ with coordinates of $(4, 2)$, we start at the origin and move 4 units to the *right* and 2 units *up*. Point $D$ lies in quadrant I, as shown in Figure 3-4.

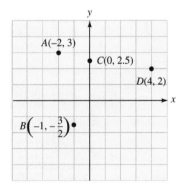

FIGURE 3-4

**Self Check**
**Answer**

Plot the points   **a.** $E(2, -2)$,   **b.** $F(-4, 0)$,   **c.** $G\left(1.5, \frac{5}{2}\right)$,   and   **d.** $H(0, 5)$.

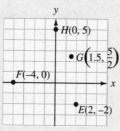

**EXAMPLE 2**

**Orbit of the earth**   The circle shown in Figure 3-5 is an approximate graph of the orbit of the earth. The graph is made up of infinitely many points, each with its own $x$- and $y$-coordinates. Use the graph to find the coordinates of the earth's position during the months of February, May, August, and December.

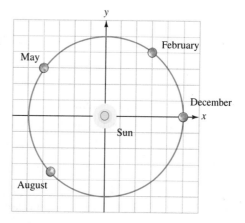

FIGURE 3-5

*Solution*   To find the coordinates of each position, we start at the origin and move left or right along the $x$-axis to find the $x$-coordinate and then up or down to find the $y$-coordinate. See Table 3-1.

| Month | Position of earth on graph | Coordinates |
|---|---|---|
| February | 3 units to the *right*, then 4 units *up* | $(3, 4)$ |
| May | 4 units to the *left*, then 3 units *up* | $(-4, 3)$ |
| August | 3.5 units to the *left*, then 3.5 units *down* | $(-3.5, -3.5)$ |
| December | 5 units *right*, then no units *up* or *down* | $(5, 0)$ |

TABLE 3-1

■ ■ ■ ■ ■ ■ ■ ■ ■ ■ ■ PERSPECTIVE

As a child, René Descartes was frail and often sick. To improve his health, eight-year-old René was sent to a Jesuit school. The headmaster encouraged him to sleep in the morning as long as he wished. As a young man, Descartes spent several years as a soldier and world traveler, but his interests included mathematics and philosophy, as well as science, literature, writing, and taking it easy. The habit of sleeping late continued throughout his life. He claimed that his most productive thinking occurred when he was lying in bed. According to one story, Descartes first thought of analytic geometry as he watched a fly walking on his bedroom ceiling.

Descartes might have lived longer if he had stayed in bed. In 1649, Queen Christina of Sweden decided that she needed a tutor in philosophy, and she requested the services of Descartes. Tutoring would not have been difficult, except that the queen scheduled her lessons before dawn in her library with her windows open. The cold Stockholm mornings were too much for a man who was used to sleeping past noon. Within a few months, Descartes developed a fever and died, probably of pneumonia.

## ■ GRAPHING MATHEMATICAL RELATIONSHIPS

Every day, we deal with quantities that are related.

- The distance that we travel depends on how fast we are going.
- Our weight depends on how much we eat.
- The amount of water in a tub depends on how long the water has been running.

We can often use graphs to visualize relationships between two quantities. For example, suppose that we know the number of gallons of water that are in a tub at several time intervals after the water has been turned on. We can list that information in a **table of values** (see Figure 3-6).

At various times, the amount of water in the tub was measured and recorded in the table of values.

| Time (minutes) | Water in tub (gallons) | |
|---|---|---|
| 0 | 0 | → (0, 0) |
| 1 | 8 | → (1, 8) |
| 3 | 24 | → (3, 24) |
| 4 | 32 | → (4, 32) |

*x*-coordinate    *y*-coordinate    The data in the table can be expressed as ordered pairs (*x*, *y*).

FIGURE 3-6

The information in the table can be used to construct a graph that shows the relationship between the amount of water in the tub and the time the water has been running. Since the amount of water in the tub depends on the time, we will associate *time* with the *x*-axis and the *amount of water* with the *y*-axis.

To construct the graph in Figure 3-7, we plot the four ordered pairs and draw a line through the resulting data points.

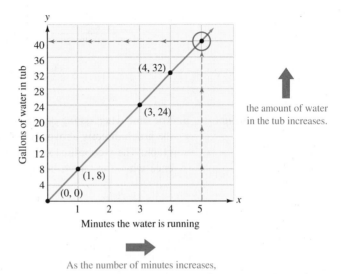

FIGURE 3-7

From the graph, we can see that the amount of water in the tub increases as the water is allowed to run. We can also use the graph to make observations about the amount of water in the tub at other times. For example, the dashed line on the graph shows that in 5 minutes, the tub will contain 40 gallons of water.

## ▪ READING GRAPHS

In the next example, we show that valuable information can be obtained from a graph.

**EXAMPLE 3**

**Reading a graph**   The graph in Figure 3-8 shows the number of people in an audience before, during, and after the taping of a television show. On the *x*-axis, 0 represents the time when taping began. Use the graph to answer the following questions, and record each result in a table of values.

**a.** How many people were in the audience when taping began?

**b.** What was the size of the audience 10 minutes before taping began?

**c.** At what times were there exactly 100 people in the audience?

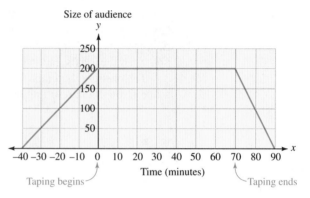

FIGURE 3-8

*Solution*   **a.** The time when taping began is represented by 0 on the *x*-axis. Since the point on the graph directly above 0 has a *y*-coordinate of 200, the point (0, 200) is on the graph. The *y*-coordinate of this point indicates that 200 people were in the audience when the taping began.

| Time | Audience |
|------|----------|
| 0 | 200 |

**b.** Ten minutes before taping began is represented by −10 on the *x*-axis. Since the point on the graph directly above −10 has a *y*-coordinate of 150, the point (−10, 150) is on the graph. The *y*-coordinate of this point indicates that 150 people were in the audience 10 minutes before the taping began.

| Time | Audience |
|------|----------|
| −10 | 150 |

**c.** We can draw a horizontal line passing through 100 on the *y*-axis. Since this line intersects the graph twice, there were two times when 100 people were in the audience. One time was 20 minutes before taping began, and the other was 80 minutes after taping began. So the points (−20, 100) and (80, 100) are on the graph. The *y*-coordinates of these points indicate that there were 100 people in the audience 20 minutes before and 80 minutes after taping began.

| Time | Audience |
|------|----------|
| −20 | 100 |
| 80 | 100 |

**Self Check**   Use the graph in Figure 3-8 to answer the following questions.   **a.** At what times were there exactly 50 people in the audience?   **b.** What was the size of the audience that watched the taping?   **c.** How long did it take for the audience to leave the studio after taping ended?

*Answers*   **a.** 30 min before and 85 min after taping began,   **b.** 200,   **c.** 20 min

## ■ STEP GRAPHS

The graph in Figure 3-9 shows the cost of renting a trailer for different periods of time. For example, the cost of renting the trailer for 4 days is $60, which is the *y*-coordinate of the point with coordinates of $(4, 60)$. For renting the trailer for a period lasting over 4 and up to 5 days, the cost jumps to $70. Since the jumps in cost form steps in the graph, we call the graph a **step graph.**

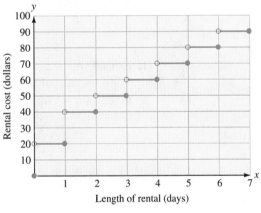

FIGURE 3-9

**EXAMPLE 4**   Use the information in Figure 3-9 to answer the following questions. Write the results in a table of values.

**a.** Find the cost of renting the trailer for 2 days.

**b.** Find the cost of renting the trailer for $5\frac{1}{2}$ days.

**c.** How long can you rent the trailer if you have $50?

**d.** Is the rental cost per day the same?

*Solution*   **a.** We locate 2 days on the *x*-axis and move up to locate the point on the graph directly above the 2. Since the point has coordinates $(2, 40)$, a two-day rental would cost $40. We enter this ordered pair in Table 3-2.

**b.** We locate $5\frac{1}{2}$ days on the *x*-axis and move straight up to locate the point on the graph with coordinates $\left(5\frac{1}{2}, 80\right)$, which indicates that a $5\frac{1}{2}$-day rental would cost $80. We enter this ordered pair in Table 3-2.

**c.** We draw a horizontal line through the point labeled 50 on the y-axis. Since this line intersects one step of the graph, we can look down to the x-axis to find the x-values that correspond to a y-value of 50. From the graph, we see that the trailer can be rented for more than 2 and up to 3 days for $50. We write (3, 50) in Table 3-2.

| Length of rental (days) | Cost (dollars) |
|:---:|:---:|
| 2 | 40 |
| $5\frac{1}{2}$ | 80 |
| 3 | 50 |

TABLE 3-2

**d.** No, the cost per day is not the same. If we look at the y-coordinates, we see that for the first day, the rental fee is $20. For the second day, the cost jumps another $20. For the third day, and all subsequent days, the cost jumps only $10.    ∎

Orals
**1.** Explain why the pair $(-2, 4)$ is called an ordered pair.

**2.** At what point do the coordinate axes intersect?

**3.** In which quadrant does the graph of $(3, -5)$ lie?

**4.** On which axis does the point $(0, 5)$ lie?

# EXERCISE 3.1

## *REVIEW*

**1.** Evaluate $-3 - 3(-5)$.

**2.** Evaluate $(-5)^2 + (-5)$.

**3.** What is the opposite of $-8$?

**4.** Simplify $|-1 - 9|$.

**5.** Solve $-4x + 7 = -21$.

**6.** Solve $P = 2l + 2w$ for $w$.

**7.** Evaluate $(x + 1)(x + y)^2$ for $x = -2$ and $y = -5$.

**8.** Simplify $-6(x - 3) - 2(1 - x)$.

## *VOCABULARY AND CONCEPTS*   *Fill in each blank to make a true statement.*

**9.** The pair of numbers $(-1, -5)$ is called an _____.

**10.** In the ordered pair $\left(-\frac{3}{2}, -5\right)$, $-5$ is called the ___ coordinate.

**11.** The point with coordinates $(0, 0)$ is the _____.

**12.** The x- and y-axes divide the coordinate plane into four regions called _____.

**13.** The point with coordinates $(4, 2)$ can be graphed on a _____ system.

**14.** The process of locating the position of a point on a coordinate plane is called _____ the point.

*In Exercises 15–20, answer each question or fill in each blank to make a true statement.*

**15.** Do (3, 2) and (2, 3) represent the same point?

**16.** In the ordered pair (4, 5), is 4 associated with the horizontal or the vertical axis?

**17.** To plot the point with coordinates (−5, 4.5), we start at the _____, move 5 units to the _____, and then move 4.5 units ____.

**18.** To plot the point with coordinates $\left(6, -\frac{3}{2}\right)$, we start at the _____, move 6 units to the _____, and then move $\frac{3}{2}$ units _____.

**19.** In which quadrant do points with a negative *x*-coordinate and a positive *y*-coordinate lie?

**20.** In which quadrant do points with a positive *x*-coordinate and a negative *y*-coordinate lie?

**21.** Use the graph to complete the table.

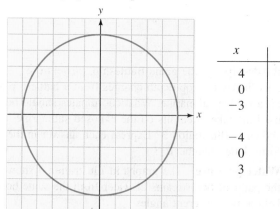

| x | y |
|---|---|
| 4 | |
| 0 | |
| −3 | |
| | 0 |
| −4 | |
| 0 | |
| 3 | |

**22.** Use the graph to complete the table.

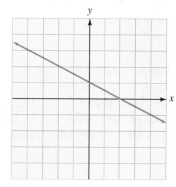

| x | y |
|---|---|
| | 0 |
| | 2 |
| | −1 |
| −4 | |
| | 1 |

*The graph in Illustration 1 gives the heart rate of a woman before, during, and after an aerobic workout. In Exercises 23–30, use the graph to answer the following questions.*

**23.** What information does the point (−10, 60) give us?

**24.** After beginning the workout, how long did it take the woman to reach her training-zone heart rate?

**25.** What was her heart rate one-half hour after beginning the workout?

**26.** For how long did she work out at the training-zone level?

**27.** At what times was her heart rate 100 beats per minute?

**28.** How long was her cooldown period?

**29.** What was the difference in her heart rate before the workout and after the cooldown period?

**30.** What was her approximate heart rate 8 minutes after beginning?

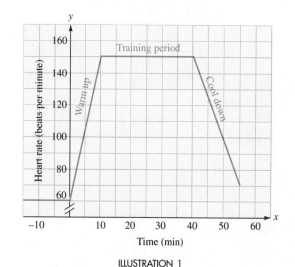

ILLUSTRATION 1

*PRACTICE   In Exercises 31–32, graph each point on the coordinate grid.*

**31.** $A(-3, 4)$, $B(4, 3.5)$, $C\left(-2, -\frac{5}{2}\right)$, $D(0, -4)$, $E\left(\frac{3}{2}, 0\right)$, $F(3, -4)$

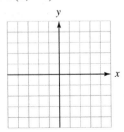

**32.** $G(4, 4)$, $H(0.5, -3)$, $I(-4, -4)$, $J(0, -1)$, $K(0, 0)$, $L(0, 3)$, $M(-2, 0)$

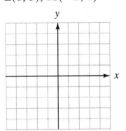

## APPLICATIONS

**33. Road maps**   Road maps usually have a coordinate system to help locate cities. Use the map in Illustration 2 to locate Carbondale, Champaign, Chicago, Peoria, Rockford, Springfield, and St. Louis. Express each answer in the form (number, letter).

ILLUSTRATION 2

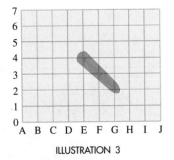

ILLUSTRATION 3

**34. Battleship**   In the game Battleship, players use coordinates to drop depth charges from a battleship to hit a hidden submarine. What coordinates should be used to make three hits on the exposed submarine shown in Illustration 3? Express each answer in the form (letter, number).

**35. Water pressure**   The graphs in Illustration 4 show the paths of two streams of water from the same hose held at two different angles.
   **a.** At which angle does the stream of water shoot higher? How much higher?
   **b.** At which angle does the stream of water shoot out farther? How much farther?

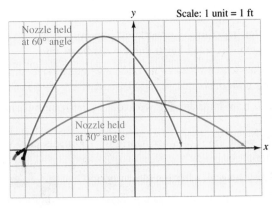

ILLUSTRATION 4

**36. Golf swing** To correct her swing, a golfer was video-taped and then had her image displayed on a computer monitor so that it could be analyzed by a golf pro. See Illustration 5. Give the coordinates of the points that are highlighted on the arc of her swing.

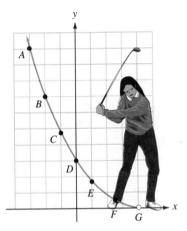

ILLUSTRATION 5

**37. Video rental** The charges for renting a movie are shown in the graph in Illustration 6.

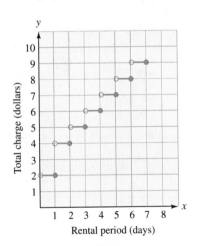

ILLUSTRATION 6

a. Find the charge for a 1-day rental.
b. Find the charge for a 2-day rental.
c. Find the charge if the tape is kept for 5 days.
d. Find the charge if the tape is kept for a week.

**38. Postage rates** The graph shown in Illustration 7 gives the first-class postage rates for mailing parcels weighing up to 5 ounces.

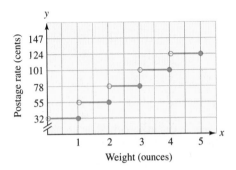

ILLUSTRATION 7

a. Find the cost of postage to mail each of the following letters first class: a 1-ounce letter, a 4-ounce letter, and a $2\frac{1}{2}$-ounce letter.
b. Find the difference in postage for a 3.75-ounce letter and a 4.75-ounce letter.
c. What is the heaviest letter than can be mailed first class for 55¢?

**39. Gas mileage** The table in Illustration 8 gives the number of miles ($y$) that a truck can be driven on $x$ gallons of gasoline. Plot the ordered pairs and draw a line connecting the points.

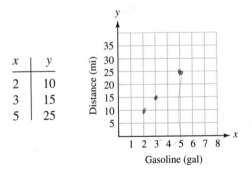

| $x$ | $y$ |
|-----|-----|
| 2 | 10 |
| 3 | 15 |
| 5 | 25 |

ILLUSTRATION 8

a. Estimate how far the truck can go on 7 gallons of gasoline.
b. How many gallons of gas are needed to travel a distance of 20 miles?
c. Estimate how far the truck can go on 6.5 gallons of gasoline.

**40. Wages**   The table in Illustration 9 gives the amount
$y$ (in dollars) that a student can earn by working $x$
hours. Plot the ordered pairs and draw a line
connecting the points.

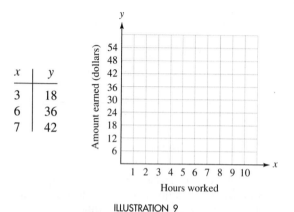

| $x$ | $y$ |
|---|---|
| 3 | 18 |
| 6 | 36 |
| 7 | 42 |

ILLUSTRATION 9

**a.** How much will the student earn in 5 hours?
**b.** How long would the student have to work to earn
$12?
**c.** Estimate how much the student will earn in 3.5
hours.

**41. Value of a car**   The table in Illustration 10 shows
the value $y$ (in thousands of dollars) of a car that is $x$
years old. Plot the ordered pairs and draw a line
connecting the points.
**a.** What does the point $(3, 7)$ on the graph tell you?

**b.** Estimate the value of the car when it is 7 years
old.
**c.** After how many years will the car be worth
$2,500?

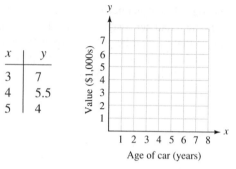

| $x$ | $y$ |
|---|---|
| 3 | 7 |
| 4 | 5.5 |
| 5 | 4 |

ILLUSTRATION 10

**42. Depreciation**   As a piece of farm machinery gets
older, it loses value. The table in Illustration 11
shows the value $y$ of a tractor that is $x$ years old.
Plot the ordered pairs and draw a line connecting
them.

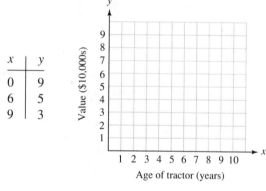

| $x$ | $y$ |
|---|---|
| 0 | 9 |
| 6 | 5 |
| 9 | 3 |

ILLUSTRATION 11

**a.** What does the point $(0, 9)$ on the graph tell you?

**b.** Estimate the value of the tractor in 3 years.

**c.** When will the tractor's value dip below $30,000?

*WRITING*

**43.** Explain why the point with coordinates $(-3, 3)$ is not
the same as the point with coordinates $(3, -3)$.

**44.** Explain what is meant when we say that the
rectangular coordinate graph of the St. Louis Arch is
made up of *infinitely many* points.

**45.** Explain how to plot the point with coordinates
$(-2, 5)$.

**46.** Explain why the coordinates of the origin are $(0, 0)$.

## SOMETHING TO THINK ABOUT

**47.** Could you have a coordinate system where the coordinate axes were not perpendicular? How would it be different?

**48.** René Descartes is famous for saying, "I think. Therefore I am." What do you think he meant by that?

## 3.2 Graphing Linear Equations

■ EQUATIONS WITH TWO VARIABLES ■ CONSTRUCTING TABLES OF VALUES ■ GRAPHING EQUATIONS ■ THE INTERCEPT METHOD OF GRAPHING A LINE ■ GRAPHING HORIZONTAL AND VERTICAL LINES ■ AN APPLICATION OF LINEAR EQUATIONS

*Getting Ready*    *In Problems 1–4, let $y = 2x + 1$.*

**1.** Find $y$ when $x = 0$.

**2.** Find $y$ when $x = 2$.

**3.** Find $y$ when $x = -2$.

**4.** Find $y$ when $x = \dfrac{1}{2}$.

**5.** Find five pairs of numbers with a sum of 8.

**6.** Find five pairs of numbers with a difference of 5.

### ■ EQUATIONS WITH TWO VARIABLES

The equation $x + 2y = 5$ contains the two variables $x$ and $y$. The solutions of such equations are ordered pairs of numbers. For example, the ordered pair $(1, 2)$ is a solution, because the equation is satisfied when $x = 1$ and $y = 2$.

$$x + 2y = 5$$
$$1 + 2(2) = 5 \qquad \text{Substitute 1 for } x \text{ and 2 for } y.$$
$$1 + 4 = 5$$
$$5 = 5$$

**EXAMPLE 1**    Is the pair $(-2, 4)$ a solution of $y = 3x + 9$?

*Solution*    We substitute $-2$ for $x$ and $4$ for $y$ and see whether the resulting equation is true.

$$y = 3x + 9 \qquad \text{The original equation.}$$
$$4 \overset{?}{=} 3(-2) + 9 \qquad \text{Substitute } -2 \text{ for } x \text{ and } 4 \text{ for } y.$$
$$4 \overset{?}{=} -6 + 9 \qquad \text{Do the multiplication: } 3(-2) = -6.$$
$$4 = 3 \qquad \text{Do the addition: } -6 + 9 = 3.$$

Since the equation $4 = 3$ is false, the pair $(-2, 4)$ is not a solution. ■

*Self Check*    Is $(-1, -5)$ a solution of $y = 5x$?

*Answer*    yes

## ■ CONSTRUCTING TABLES OF VALUES

To find solutions of equations in $x$ and $y$, we can pick numbers at random, substitute them for $x$, and find the corresponding values of $y$. For example, to find some ordered pairs that satisfy $y = 5 - x$, we can let $x = 1$ (called the **input value**), substitute 1 for $x$, and solve for $y$ (called the **output value**).

$$y = 5 - x \qquad \text{The original equation.}$$
$$y = 5 - 1 \qquad \text{Substitute the input value of 1 for } x.$$
$$y = 4 \qquad \text{The output is 4.}$$

$y = 5 - x$

| $x$ | $y$ | $(x, y)$ |
|---|---|---|
| 1 | 4 | $(1, 4)$ |

The ordered pair $(1, 4)$ is a solution. As we find solutions, we will list them in a **table of values** like the one shown at the left.

If $x = 2$, we have

$$y = 5 - x \qquad \text{The original equation.}$$
$$y = 5 - 2 \qquad \text{Substitute the input value of 2 for } x.$$
$$y = 3 \qquad \text{The output is 3.}$$

$y = 5 - x$

| $x$ | $y$ | $(x, y)$ |
|---|---|---|
| 1 | 4 | $(1, 4)$ |
| 2 | 3 | $(2, 3)$ |

A second solution is $(2, 3)$. We list it in the table of values at the left.

If $x = 5$, we have

$$y = 5 - x \qquad \text{The original equation.}$$
$$y = 5 - 5 \qquad \text{Substitute the input value of 5 for } x.$$
$$y = 0 \qquad \text{The output is 0.}$$

$y = 5 - x$

| $x$ | $y$ | $(x, y)$ |
|---|---|---|
| 1 | 4 | $(1, 4)$ |
| 2 | 3 | $(2, 3)$ |
| 5 | 0 | $(5, 0)$ |

A third solution is $(5, 0)$. We list it in the table of values at the left.

If $x = -1$, we have

$$y = 5 - x \qquad \text{The original equation.}$$
$$y = 5 - (-1) \qquad \text{Substitute the input value of } -1 \text{ for } x.$$
$$y = 6 \qquad \text{The output is 6.}$$

$y = 5 - x$

| $x$ | $y$ | $(x, y)$ |
|---|---|---|
| 1 | 4 | $(1, 4)$ |
| 2 | 3 | $(2, 3)$ |
| 5 | 0 | $(5, 0)$ |
| -1 | 6 | $(-1, 6)$ |

A fourth solution is $(-1, 6)$. We list it in the table of values at the left.

If $x = 6$, we have

$$y = 5 - x \qquad \text{The original equation.}$$
$$y = 5 - 6 \qquad \text{Substitute the input value 6 for } x.$$
$$y = -1 \qquad \text{The output is } -1.$$

$y = 5 - x$

| $x$ | $y$ | $(x, y)$ |
|---|---|---|
| 1 | 4 | $(1, 4)$ |
| 2 | 3 | $(2, 3)$ |
| 5 | 0 | $(5, 0)$ |
| -1 | 6 | $(-1, 6)$ |
| 6 | -1 | $(6, -1)$ |

A fifth solution is $(6, -1)$. We list it in the table of values at the left.

Since we can choose any real number for $x$, and since any choice of $x$ will give a corresponding value of $y$, it is apparent that the equation $y = 5 - x$ has *infinitely many solutions*.

## ■ GRAPHING EQUATIONS

To graph the equation $y = 5 - x$, we plot the ordered pairs listed in the table on a rectangular coordinate system, as in Figure 3-10. From the figure, we can see that the five points lie on a line.

$y = 5 - x$

| $x$ | $y$ | $(x, y)$ |
|-----|-----|----------|
| 1   | 4   | $(1, 4)$ |
| 2   | 3   | $(2, 3)$ |
| 5   | 0   | $(5, 0)$ |
| $-1$ | 6  | $(-1, 6)$ |
| 6   | $-1$ | $(6, -1)$ |

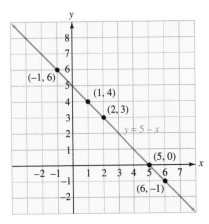

FIGURE 3-10

We draw a line through the points. The arrowheads on the line show that the graph continues forever in both directions. Since the graph of any solution of $y = 5 - x$ will lie on this line, the line is a picture of all of the solutions of the equation $y = 5 - x$. The line is said to be the **graph** of the equation.

Any equation, such as $y = 5 - x$, whose graph is a line is called a **linear equation in two variables.** Any point on the line has coordinates that satisfy the equation, and the graph of any pair $(x, y)$ that satisfies the equation is a point on the line.

Since we will usually choose a number for $x$ first and then find the corresponding value of $y$, the value of $y$ depends on $x$. For this reason, we call $y$ the **dependent variable** and $x$ the **independent variable.** The value of the independent variable is the input value, and the value of the dependent variable is the output value.

Although only two points are needed to graph a linear equation, we often plot a third point as a check. If the three points do not lie on a line, at least one of them is in error.

### Graphing Linear Equations

1. Find two pairs $(x, y)$ that satisfy the equation by picking arbitrary input values for $x$ and solving for the corresponding output values of $y$. A third point provides a check.

2. Plot each resulting pair $(x, y)$ on a rectangular coordinate system. If they do not lie on a line, check your calculations.

3. Draw the line passing through the points.

**EXAMPLE 2**    Graph $y = 3x - 4$.

*Solution*    We find three ordered pairs that satisfy the equation.

| *If x = 1* | *If x = 2* | *If x = 3* |
|------------|------------|------------|
| $y = 3x - 4$ | $y = 3x - 4$ | $y = 3x - 4$ |
| $y = 3(1) - 4$ | $y = 3(2) - 4$ | $y = 3(3) - 4$ |
| $y = -1$ | $y = 2$ | $y = 5$ |

We enter the results in a table of values, plot the points, and draw a line through the points. The graph appears in Figure 3-11.

$y = 3x - 4$

| $x$ | $y$ | $(x, y)$ |
|---|---|---|
| 1 | $-1$ | $(1, -1)$ |
| 2 | 2 | $(2, 2)$ |
| 3 | 5 | $(3, 5)$ |

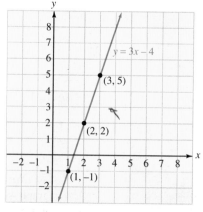

FIGURE 3-11                                               ■

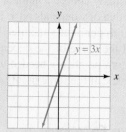

**Self Check**     Graph $y = 3x$.

**Answer**

Note that the graph of $y = 3x$ is 4 units above the graph of $y = 3x - 4$.

**EXAMPLE 3**     Graph $y - 4 = \dfrac{1}{2}(x - 8)$.

*Solution*     We first solve for $y$ and simplify.

$$y - 4 = \frac{1}{2}(x - 8)$$

$$y - 4 = \frac{1}{2}x - 4 \qquad \text{Use the distributive property to remove parentheses.}$$

$$y = \frac{1}{2}x \qquad \text{Add 4 to both sides.}$$

We now find three ordered pairs that satisfy the equation.

**If $x = 0$**          **If $x = 2$**          **If $x = -4$**

$y = \dfrac{1}{2}x$          $y = \dfrac{1}{2}x$          $y = \dfrac{1}{2}x$

$y = \dfrac{1}{2}(0)$          $y = \dfrac{1}{2}(2)$          $y = \dfrac{1}{2}(-4)$

$y = 0$          $y = 1$          $y = -2$

We enter the results in a table of values, plot the points, and draw a line through the points. The graph appears in Figure 3-12.

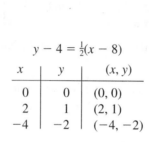

$$y - 4 = \tfrac{1}{2}(x - 8)$$

| $x$ | $y$ | $(x, y)$ |
|---|---|---|
| 0 | 0 | $(0, 0)$ |
| 2 | 1 | $(2, 1)$ |
| $-4$ | $-2$ | $(-4, -2)$ |

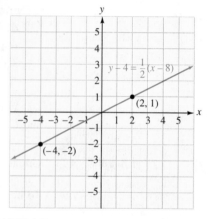

FIGURE 3-12    ■

**Self Check**
**Answer**

Graph $y + 3 = \tfrac{1}{3}(x - 6)$.

## ■ THE INTERCEPT METHOD OF GRAPHING A LINE

The points where a line intersects the $x$- and $y$-axes are called **interce**
line.

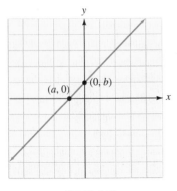

FIGURE 3-13

### x- and y-Intercepts

The **x-intercept** of a line is a point $(a, 0)$ where the line intersects the $x$-axis. (See Figure 3-13.) To find $a$, substitute 0 for $y$ in the equation of the line and solve for $x$.

A **y-intercept** of a line is a point $(0, b)$ where the line intersects the $y$-axis. To find $b$, substitute 0 for $x$ in the equation of the line and solve for $y$.

Plotting the $x$- and $y$-intercepts and drawing a line through them is called the **intercept method of graphing a line.** This method is useful for graphing equations written in **general form.**

### General Form of the Equation of a Line

If $A$, $B$, and $C$ are real numbers and $A$ and $B$ are not both 0, then the equation

$$Ax + By = C$$

is called the **general form** of the equation of a line.

Whenever possible, we will write the general form $Ax + By = C$ so that $A$, $B$, and $C$ are integers and $A \geq 0$.

**EXAMPLE 4**   Graph $3x + 2y = 6$.

*Solution*   To find the $y$-intercept, we let $x = 0$ and solve for $y$.

$$3x + 2y = 6$$
$$3(0) + 2y = 6 \qquad \text{Substitute 0 for } x.$$
$$2y = 6 \qquad \text{Simplify.}$$
$$y = 3 \qquad \text{Divide both sides by 2.}$$

The $y$-intercept is the pair $(0, 3)$. To find the $x$-intercept, we let $y = 0$ and solve for $x$.

$$3x + 2y = 6$$
$$3x + 2(0) = 6 \qquad \text{Substitute 0 for } y.$$
$$3x = 6 \qquad \text{Simplify.}$$
$$x = 2 \qquad \text{Divide both sides by 3.}$$

The $x$-intercept is the pair $(2, 0)$. As a check, we plot one more point. If $x = 4$, then

$$3x + 2y = 6$$
$$3(4) + 2y = 6 \qquad \text{Substitute 4 for } x.$$
$$12 + 2y = 6 \qquad \text{Simplify.}$$
$$2y = -6 \qquad \text{Add } -12 \text{ to both sides.}$$
$$y = -3 \qquad \text{Divide both sides by 2.}$$

The point $(4, -3)$ is on the graph. We plot these three points and join them with a line. The graph of $3x + 2y = 6$ is shown in Figure 3-14.

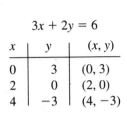

$$3x + 2y = 6$$

| $x$ | $y$ | $(x, y)$ |
|-----|-----|----------|
| 0 | 3 | $(0, 3)$ |
| 2 | 0 | $(2, 0)$ |
| 4 | $-3$ | $(4, -3)$ |

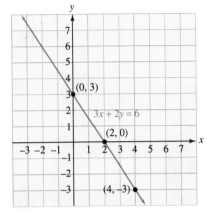

FIGURE 3-14    ■

**Self Check**
**Answer**

Graph $4x + 3y = 6$.

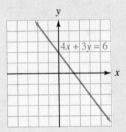

■ GRAPHING HORIZONTAL AND VERTICAL LINES

Equations such as $y = 3$ and $x = -2$ are linear equations, because they can be written in the general form $Ax + By = C$.

$y = 3$      is equivalent to      $0x + 1y = 3$
$x = -2$    is equivalent to      $1x + 0y = -2$

Next, we discuss how to graph these types of linear equations.

**EXAMPLE 5**    Graph    **a.** $y = 3$    and    **b.** $x = -2$.

*Solution*    **a.** We can write the equation $y = 3$ in general form as $0x + y = 3$. Since the coefficient of $x$ is 0, the numbers chosen for $x$ have no effect on $y$. The value of $y$ is always 3. For example, if we substitute $-3$ for $x$, we get

$$0x + y = 3$$
$$0(-3) + y = 3$$
$$0 + y = 3$$
$$y = 3$$

The table in Figure 3-15(a) gives several pairs that satisfy the equation $y = 3$. After plotting these pairs and joining them with a line, we see that the graph of $y = 3$ is a horizontal line that intersects the $y$-axis at 3. The $y$-intercept is $(0, 3)$. There is no $x$-intercept.

**b.** We can write $x = -2$ in general form as $x + 0y = -2$. Since the coefficient of $y$ is 0, the values of $y$ have no effect on $x$. The number $x$ is always $-2$. A table of values and the graph are shown in Figure 3-15(b). The graph of $x = -2$ is a vertical line that intersects the $x$-axis at $-2$. The $x$-intercept is $(-2, 0)$. There is no $y$-intercept.

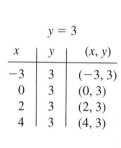

$y = 3$

| $x$ | $y$ | $(x, y)$ |
|---|---|---|
| $-3$ | 3 | $(-3, 3)$ |
| 0 | 3 | $(0, 3)$ |
| 2 | 3 | $(2, 3)$ |
| 4 | 3 | $(4, 3)$ |

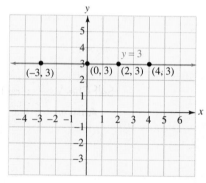

(a)

$x = -2$

| $x$ | $y$ | $(x, y)$ |
|---|---|---|
| $-2$ | $-2$ | $(-2, -2)$ |
| $-2$ | 0 | $(-2, 0)$ |
| $-2$ | 2 | $(-2, 2)$ |
| $-2$ | 3 | $(-2, 3)$ |

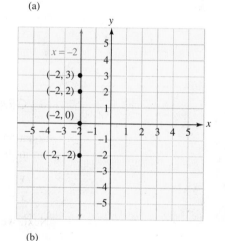

(b)

FIGURE 3-15

*Self Check*  Identify the graph of each equation as a horizontal or a vertical line:  **a.** $x = 5$, **b.** $y = -3$, and **c.** $x = 0$.

*Answers*  **a.** vertical, **b.** horizontal, **c.** vertical

From the results of Example 5, we have the following facts.

**Equations of Horizontal and Vertical Lines**

The equation $y = b$ represents a horizontal line that intersects the $y$-axis at $(0, b)$. If $b = 0$, the line is the $x$-axis.

The equation $x = a$ represents a vertical line that intersects the $x$-axis at $(a, 0)$. If $a = 0$, the line is the $y$-axis.

## ■ AN APPLICATION OF LINEAR EQUATIONS

EXAMPLE 6

**Birthday parties**   A restaurant offers a party package that includes food, drinks, cake, and party favors for a cost of $25 plus $3 per child. Write a linear equation that will give the cost for a party of any size. Then graph the equation.

*Solution*   We can let $c$ represent the cost of the party. Then the cost $c$ will be the sum of the basic charge of $25 and the cost per child times the number of children attending. If the number of children attending is $n$, at $3 per child, the total cost for the children is $3n$.

| The cost | is | the basic $25 charge | plus $3 times | the number of children. |
|----------|-----|----------------------|----------------|--------------------------|
| $c$ | $=$ | 25 | $+ \ 3 \ \cdot$ | $n$ |

For the equation $c = 25 + 3n$, the independent variable (input) is $n$, the number of children. The dependent variable (output) is $c$, the cost of the party. We will find three points on the graph of the equation by choosing $n$-values of 0, 5, and 10 and finding the corresponding $c$-values. The results are recorded in the table.

| If $n = 0$ | If $n = 5$ | If $n = 10$ | $T = 25 + 3n$ | |
|------------|------------|-------------|---------------|---|
| $c = 25 + 3(0)$ | $c = 25 + 3(5)$ | $c = 25 + 3(10)$ | $n$ | $C$ |
| $c = 25$ | $c = 25 + 15$ | $c = 25 + 30$ | 0 | 25 |
| | $c = 40$ | $c = 55$ | 5 | 40 |
| | | | 10 | 55 |

Next, we graph the points in Figure 3-16 and draw a line through them. We don't draw an arrowhead on the left, because it does not make sense to have a *negative* number of children attend a party. We can use the graph to determine the cost of a party of any size. For example, to find the cost of a party with 8 children, we locate 8 on the horizontal axis and then move up to find a point on the graph directly above the 8. Since the coordinates of that point are $(8, 49)$, the cost for 8 children would be $49.

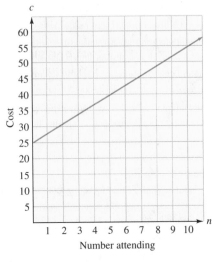

FIGURE 3-16

## Making Tables and Graphs

**GRAPHING CALCULATORS**

TI-83 graphing calculator
(Courtesy of Texas Instruments)

So far, we have graphed equations by making tables of values and plotting points. This method is usually tedious and time-consuming. Fortunately, the task of making tables and graphing equations is much easier when we use a graphing calculator.

Several brands of calculators are available. Although we will use calculators to make tables and graph equations, we will not show complete keystrokes for any specific brand. For these details, please consult your owner's manual.

All graphing calculators have a **viewing window** that is used to display tables and graphs. We will first discuss how to make tables and then discuss how to draw graphs.

**MAKING TABLES** To construct a table of values for the equation $y = x^2$, simply press the Y = key, enter the expression $x^2$, and press the 2nd and TABLE keys to get a screen similar to Figure 3-17(a). You can use the up and down keys to scroll through the table to obtain a screen like Figure 3-17(b).

(a)                    (b)

FIGURE 3-17

**DRAWING GRAPHS** To see the proper picture of a graph, we must often set the minimum and maximum values for the $x$- and $y$-coordinates. The standard window settings of

$$\text{Xmin} = -10 \qquad \text{Xmax} = 10 \qquad \text{Ymin} = -10 \qquad \text{Ymax} = 10$$

indicate that $-10$ is the minimum $x$- and $y$-coordinate to be used in the graph, and that $10$ is the maximum $x$- and $y$-coordinate to be used. We will usually express window values in interval notation. In this notation, the standard settings are

$$X = [-10, 10] \qquad Y = [-10, 10]$$

To graph the equation $2x - 3y = 14$ with a calculator, we must first solve the equation for $y$.

$$2x - 3y = 14$$
$$-3y = -2x + 14 \qquad \text{Subtract } 2x \text{ from both sides.}$$
$$y = \frac{2}{3}x - \frac{14}{3} \qquad \text{Divide both sides by } -3.$$

We now set the standard window values of $X = [-10, 10]$ and $Y = [-10, 10]$, press the $\boxed{Y=}$ key and enter the equation as $(2/3)x - 14/3$, and press $\boxed{\text{GRAPH}}$ to get the line shown in Figure 3-18.

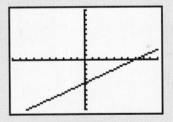

FIGURE 3-18

 **WARNING!** To graph an equation with a graphing calculator, the equation must be solved for $y$.

**USING THE TRACE AND ZOOM FEATURES** With the trace feature, we can find the coordinates of any point on a graph. For example, to find the $x$-intercept of the line shown in Figure 3-18, we press the $\boxed{\text{TRACE}}$ key and move the flashing cursor along the line with the cursor keys until we approach the $x$-intercept, as shown in Figure 3-19(a). The $x$- and $y$-coordinates of the flashing cursor appear at the bottom of the screen.

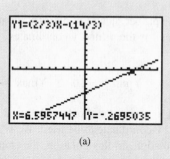 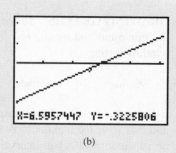 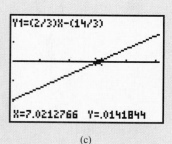

(a)                              (b)                              (c)

FIGURE 3-19

To get better results, we can press the ZOOM key to see a magnified picture of the line, as shown in Figure 3-19(b). We can trace again and move the cursor even closer to the x-intercept, as shown in Figure 3-19(c). Since the y-coordinate shown on the screen is close to 0, the x-coordinate shown on the screen is close to the x-value of the x-intercept. Repeated zooms will show that the x-intercept is (7, 0).

Orals
1. How many points should be plotted to graph a line?
2. Define the intercepts of a line.
3. Find three pairs $(x, y)$ that satisfy $x + y = 8$.
4. Find three pairs $(x, y)$ that satisfy $x - y = 6$.
5. Which lines have no y-intercepts?
6. Which lines have no x-intercepts?

## EXERCISE 3.2

### REVIEW

1. Solve $\dfrac{x}{8} = -12$.
2. Combine like terms: $3t - 4T + 5T - 6t$.
3. Is $\dfrac{x + 5}{6}$ an expression or an equation?
4. Which formula is used to find the perimeter of a rectangle?
5. What number is 0.5% of 250?
6. Solve $-3x + 5 > 17$.
7. Find $-2.5 - (-2.6)$.
8. Evaluate $(-5)^3$.

### VOCABULARY AND CONCEPTS   Fill in each blank to make a true statement.

9. The equation $y = x + 1$ is an equation in _____ variables.
10. An ordered pair is a _____ of an equation if the numbers in the ordered pair satisfy the equation.
11. In equations containing the variables $x$ and $y$, $x$ is called the _____ variable and $y$ is called the _____ variable.
12. When constructing a _____ of values, the values of $x$ are the _____ values and the values of $y$ are the _____ values.

**13.** An equation whose graph is a line and whose variables are to the first power is called a _____ equation.

**14.** The equation $Ax + By = C$ is the _____ form of the equation of a line.

**15.** The _____ of a line is the point $(0, b)$, where the line intersects the $y$-axis.

**16.** The _____ of a line is the point $(a, 0)$, where the line intersects the $x$-axis.

**PRACTICE**  *In Exercises 17–20, tell whether the ordered pair satisfies the equation.*

**17.** $x - 2y = -4$; $(4, 4)$

**18.** $y = 8x - 5$; $(4, 26)$

**19.** $y = \dfrac{2}{3}x + 5$; $(6, 12)$

**20.** $y = -\dfrac{1}{2}x - 2$; $(4, -4)$

*In Exercises 21–24, complete each table of values. Check your work with a graphing calculator.*

**21.** $y = x - 3$

| $x$ | $y$ |
|---|---|
| 0 | |
| 1 | |
| -2 | |

**22.** $y = x + 2$

| $x$ | $x + 2$ |
|---|---|
| 0 | |
| -1 | |
| -2 | |
| 1 | |
| 3 | |

**23.** $y = -2x$

| input | output |
|---|---|
| 0 | |
| 1 | |
| 3 | |
| -1 | |
| -2 | |

**24.** $y = \dfrac{x}{2}$

| $x$ | $\frac{x}{2}$ |
|---|---|
| 0 | |
| 1 | |
| -2 | |
| -4 | |

*In Exercises 25–28, graph each equation. Check your work with a graphing calculator.*

**25.** $y = 2x - 1$

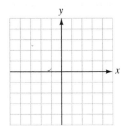

**26.** $y = 3x + 1$

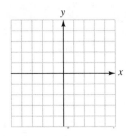

**27.** $y = \dfrac{x}{2} - 2$

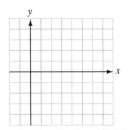

**28.** $y = \dfrac{x}{3} - 3$

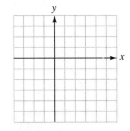

*In Exercises 29–36, write each equation in general form, when necessary. Then graph it using the intercept method.*

**29.** $x + y = 7$

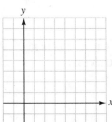

**30.** $x + y = -2$

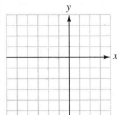

**31.** $x - y = 7$

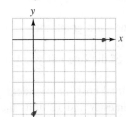

**32.** $x - y = -2$

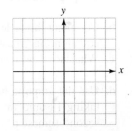

**33.** $y = -2x + 5$

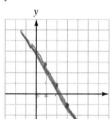

**34.** $y = -3x - 1$

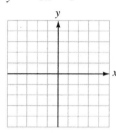

**35.** $2x + 3y = 12$

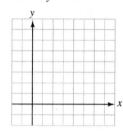

**36.** $3x - 2y = 6$

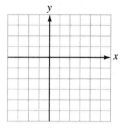

*In Exercises 37–44, graph each equation.*

**37.** $y = -5$

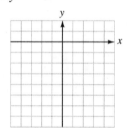

**38.** $x = 4$

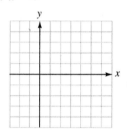

**39.** $x = 5$

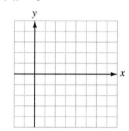

**40.** $y = 4$

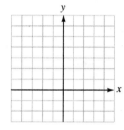

**41.** $y = 0$

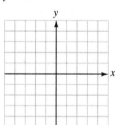

**42.** $x = 0$

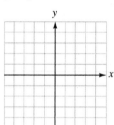

**43.** $2x = 5$

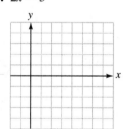

**44.** $3y = 7$

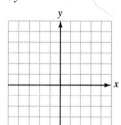

## APPLICATIONS

**45. Educational costs**   Each semester, a college charges a service fee of $50 plus $25 for each unit taken by a student.

   **a.** Write a linear equation that gives the total enrollment cost $c$ for a student taking $u$ units.

   **b.** Complete the table of values and graph the equation. See Illustration 1.

   **c.** What does the $y$-intercept of the line tell you?

   **d.** Use the graph to find the total cost for a student taking 18 units the first semester and 12 units the second semester.

| $u$ | $c$ |
|-----|-----|
| 4 | |
| 8 | |
| 14 | |

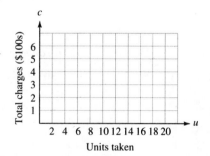

ILLUSTRATION 1

**46. Group rates** To promote the sale of tickets for a cruise to Alaska, a travel agency reduces the regular ticket price of $3,000 by $5 for each individual traveling in the group.

  **a.** Write a linear equation that would find the ticket price $T$ for the cruise if a group of $p$ people travel together.

  **b.** Complete the table of values and then graph the equation. See Illustration 2.

  **c.** As the size of the group increases, what happens to the ticket price?

  **d.** Use the graph to determine the cost of an individual ticket if a group of 25 will be traveling together.

| $p$ | $T$ |
|-----|-----|
| 10  |     |
| 30  |     |
| 60  |     |

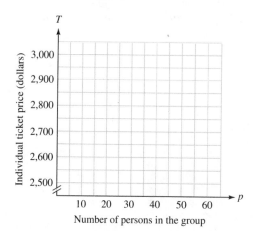

ILLUSTRATION 2

**47. Physiology** Physiologists have found that a woman's height $h$ in inches can be approximated using the linear equation $h = 3.9r + 28.9$, where $r$ represents the length of her radius bone in inches. See Illustration 3.

  **a.** Complete the table of values (round to the nearest tenth), and then graph the equation.

  **b.** Complete this sentence: From the graph, we see that the longer the radius bone, the . . . .

  **c.** From the graph, estimate the height of a girl whose radius bone is 7.5 inches long.

| $r$ | $H$ |
|-----|-----|
| 7   |     |
| 8.5 |     |
| 9   |     |

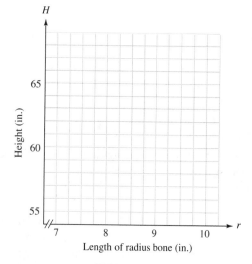

Length of radius bone (in.)

ILLUSTRATION 3

**48. Research** A psychology major found that the time $t$ in seconds that it took a white rat to complete a maze was related to the number of trials $n$ the rat had been given by the equation $t = 25 - 0.25n$. See Illustration 4.

| $n$ | $t$ |
|-----|-----|
| 4   |     |
| 12  |     |
| 16  |     |

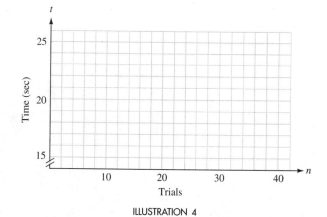

Trials

ILLUSTRATION 4

**a.** Complete the table of values and then graph the equation.

**b.** Complete this sentence: From the graph, we see that the more trials the rat had, the . . .

**c.** From the graph, estimate the time it will take the rat to complete the maze on its 32nd trial.

## WRITING

**49.** From geometry, we know that two points determine a line. Explain why it is good practice when graphing linear equations to find and plot three points instead of just two.

**50.** Explain the process used to find the $x$- and $y$-intercepts of the graph of a line.

**51.** What is a table of values? Why is it often called a table of solutions?

**52.** When graphing an equation in two variables, how many solutions of the equation must be found?

**53.** Give examples of an equation in one variable and an equation in two variables. How do their solutions differ?

**54.** What does it mean when we say that an equation in two variables has infinitely many solutions?

## SOMETHING TO THINK ABOUT
If points $P(a, b)$ and $Q(c, d)$ are two points on a rectangular coordinate system and point $M$ is midway between them, then point $M$ is called the **midpoint** of the line segment joining $P$ and $Q$. (See Illustration 5.) To find the coordinates of the midpoint $M(x_M, y_M)$ of the segment $PQ$, we find the average of the $x$-coordinates and the average of the $y$-coordinates of $P$ and $Q$.

$$x_M = \frac{a + c}{2} \quad \text{and} \quad y_M = \frac{b + d}{2}$$

In Exercises 55–60, find the coordinates of the midpoint of the line segment with the given coordinates.

**55.** $P(5, 3)$ and $Q(7, 9)$

**56.** $P(5, 6)$ and $Q(7, 10)$

**57.** $P(2, -7)$ and $Q(-3, 12)$

**58.** $P(-8, 12)$ and $Q(3, -9)$

**59.** $A(4, 6)$ and $B(10, 6)$

**60.** $A(8, -6)$ and the origin

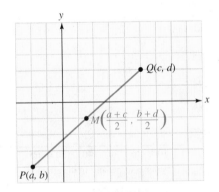

ILLUSTRATION 5

# 3.3  Solving Systems of Equations by Graphing

■ SYSTEMS OF EQUATIONS ■ THE GRAPHING METHOD ■ INCONSISTENT SYSTEMS ■ DEPENDENT EQUATIONS

Getting Ready   If $y = x^2 - 3$, find $y$ when $x =$

**1.** 0

**2.** 1

**3.** $-2$                **4.** 3

## ■ SYSTEMS OF EQUATIONS

The lines graphed in Figure 3-20 approximate the per-person consumption of chicken and beef by Americans for the years 1990 to 1997. We can see that over this period, consumption of chicken increased, while that of beef decreased.

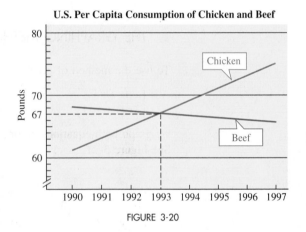

FIGURE 3-20

By graphing this *pair* of lines on the same coordinate system, it is apparent that Americans consumed equal amounts of chicken and beef in 1993—about 67 pounds each. In this section, we will work with pairs of linear equations whose graphs will be intersecting lines.

We have considered equations such as $x + y = 3$ that contain two variables. Because there are infinitely many pairs of numbers whose sum is 3, there are infinitely many pairs $(x, y)$ that will satisfy this equation. Some of these pairs are listed in Table 3-3(a). Likewise, there are infinitely many pairs $(x, y)$ that will satisfy the equation $3x - y = 1$. Some of these pairs are listed in Table 3-3(b).

| $x + y = 3$ | | | $3x - y = 1$ | |
|:---:|:---:|:---:|:---:|:---:|
| $x$ | $y$ | | $x$ | $y$ |
| 0 | 3 | | 0 | $-1$ |
| 1 | 2 | | 1 | 2 |
| 2 | 1 | | 2 | 5 |
| 3 | 0 | | 3 | 8 |
| (a) | | | (b) | |

TABLE 3-3

Although there are infinitely many pairs that satisfy each of these equations, only the pair $(1, 2)$ satisfies both equations. The pair of equations

$$\begin{cases} x + y = 3 \\ 3x - y = 1 \end{cases}$$

is called a **system of equations.** Because the ordered pair $(1, 2)$ satisfies both equations, it is called a **simultaneous solution** or just a **solution of the system of equations.** In this chapter, we will discuss three methods for finding the solution of a system of two equations, each with two variables. In this section, we consider the graphing method.

## ■ THE GRAPHING METHOD

To use the method of graphing to solve the system

$$\begin{cases} x + y = 3 \\ 3x - y = 1 \end{cases}$$

we graph both equations on one set of coordinate axes using the intercept method. See Figure 3-21.

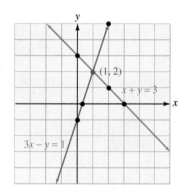

| | $x + y = 3$ | | | $3x - y = 1$ | |
|---|---|---|---|---|---|
| $x$ | $y$ | $(x, y)$ | $x$ | $y$ | $(x, y)$ |
| 0 | 3 | $(0, 3)$ | 0 | $-1$ | $(0, -1)$ |
| 3 | 0 | $(3, 0)$ | $\frac{1}{3}$ | 0 | $\left(\frac{1}{3}, 0\right)$ |
| 2 | 1 | $(2, 1)$ | 2 | 5 | $(2, 5)$ |

FIGURE 3-21

Although there are infinitely many pairs $(x, y)$ that satisfy $x + y = 3$ and infinitely many pairs $(x, y)$ that satisfy $3x - y = 1$, only the coordinates of the point where their graphs intersect satisfy both equations. Thus, the solution of the system is $x = 1$ and $y = 2$, or just $(1, 2)$.

To check the solution, we substitute 1 for $x$ and 2 for $y$ in each equation and verify that the pair $(1, 2)$ satisfies each equation.

| *First equation* | *Second equation* |
|---|---|
| $x + y = 3$ | $3x - y = 1$ |
| $1 + 2 \overset{?}{=} 3$ | $3(1) - 2 \overset{?}{=} 1$ |
| $3 = 3$ | $3 - 2 \overset{?}{=} 1$ |
| | $1 = 1$ |

When the graphs of two equations in a system are different lines, the equations are called **independent equations.** When a system of equations has a solution, the system is called a **consistent system.**

To solve a system of equations in two variables by graphing, we follow these steps.

---

**The Graphing Method**

1. Carefully graph each equation.
2. When possible, find the coordinates of the point where the graphs intersect.
3. Check the solution in the equations of the original system.

---

■ ■ ■ ■ ■ ■ ■ ■ ■ ■ PERSPECTIVE

To schedule a company's workers, managers must consider several factors to match a worker's ability to the demands of various jobs and to match company resources to the requirements of the job. To design bridges or office buildings, engineers must analyze the effects of thousands of forces to ensure that structures won't collapse. A telephone switching network decides which of thousands of possible routes is the most efficient and then rings the correct telephone in seconds. Each of these tasks requires solving systems of equations—not just two equations in two variables, but hundreds of equations in hundreds of variables. These tasks are common in every business, industry, educational institution, and government in the world. All would be much more difficult without a computer.

One of the earliest computers in use was the Mark I, which resulted from a collaboration between IBM and a Harvard mathematician, Howard Aiken. The Mark I was started in 1939 and finished in 1944. It was 8 feet tall, 2 feet thick, and over 50 feet long. It contained over 750,000 parts and performed 3 calculations per second.

Ironically, Aiken could not envision the importance of his invention. He advised the National Bureau of Standards that there was no point in building a better computer, because "there will never be enough work for more than one or two of these machines."

Mark I Relay Computer (1944)
(Courtesy of IBM Corporation)

---

**EXAMPLE 1**    Use graphing to solve $\begin{cases} 2x + 3y = 2 \\ 3x = 2y + 16 \end{cases}$.

*Solution*    Using the intercept method, we graph both equations on one set of coordinate axes, as shown in Figure 3-22.

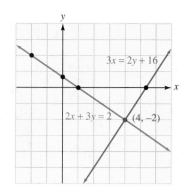

$$2x + 3y = 2$$

| $x$ | $y$ | $(x, y)$ |
|---|---|---|
| 0 | $\frac{2}{3}$ | $\left(0, \frac{2}{3}\right)$ |
| 1 | 0 | $(1, 0)$ |
| $-2$ | 2 | $(-2, 2)$ |

$$3x = 2y + 16$$

| $x$ | $y$ | $(x, y)$ |
|---|---|---|
| 0 | $-8$ | $(0, -8)$ |
| $\frac{16}{3}$ | 0 | $\left(\frac{16}{3}, 0\right)$ |
| 4 | $-2$ | $(4, -2)$ |

FIGURE 3-22

Although there are infinitely many pairs $(x, y)$ that satisfy $2x + 3y = 2$ and infinitely many pairs $(x, y)$ that satisfy $3x = 2y + 16$, only the coordinates of the point where the graphs intersect satisfy both equations. The solution is $x = 4$ and $y = -2$, or just $(4, -2)$.

To check, we substitute 4 for $x$ and $-2$ for $y$ in each equation and verify that the pair $(4, -2)$ satisfies each equation.

$$2x + 3y = 2 \qquad\qquad 3x = 2y + 16$$
$$2(4) + 3(-2) \stackrel{?}{=} 2 \qquad 3(4) \stackrel{?}{=} 2(-2) + 16$$
$$8 - 6 \stackrel{?}{=} 2 \qquad\qquad 12 \stackrel{?}{=} -4 + 16$$
$$2 = 2 \qquad\qquad\qquad 12 = 12$$

The equations in this system are independent equations, and the system is a consistent system of equations. ∎

**Self Check**    Use graphing to solve $\begin{cases} 2x = y - 5 \\ x + y = -1 \end{cases}$.

**Answer**    $(-2, 1)$

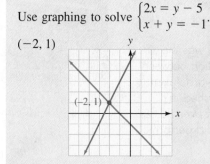

## ■ INCONSISTENT SYSTEMS

Sometimes a system of equations will have no solution. Such systems are called **inconsistent systems.**

EXAMPLE 2    Solve the system $\begin{cases} 2x + y = -6 \\ 4x + 2y = 8 \end{cases}$.

*Solution*    We graph both equations on one set of coordinate axes, as in Figure 3-23.

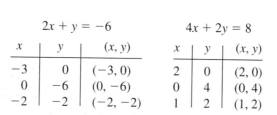

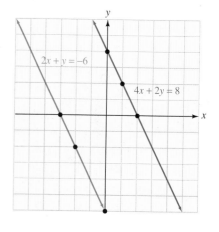

$2x + y = -6$

| $x$ | $y$ | $(x, y)$ |
|-----|-----|----------|
| $-3$ | $0$ | $(-3, 0)$ |
| $0$ | $-6$ | $(0, -6)$ |
| $-2$ | $-2$ | $(-2, -2)$ |

$4x + 2y = 8$

| $x$ | $y$ | $(x, y)$ |
|-----|-----|----------|
| $2$ | $0$ | $(2, 0)$ |
| $0$ | $4$ | $(0, 4)$ |
| $1$ | $2$ | $(1, 2)$ |

FIGURE 3-23

The lines in the figure are parallel. Because parallel lines do not intersect, the system has no solution, and the system is inconsistent. Since the graphs are different lines, the equations of the system are independent. ■

Self Check    Solve $\begin{cases} 2y = 3x \\ 3x - 2y = 6 \end{cases}$.

Answer    Since the lines do not intersect, there is no solution.

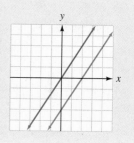

## ■ DEPENDENT EQUATIONS

Sometimes a system will have infinitely many solutions. In this case, we say that the equations of the system are **dependent equations.**

EXAMPLE 3    Solve the system $\begin{cases} y - 2x = 4 \\ 4x + 8 = 2y \end{cases}$.

*Solution*    We graph each equation on one set of axes, as in Figure 3-24.

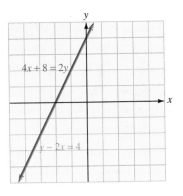

$$y - 2x = 4$$

| $x$ | $y$ | $(x, y)$ |
|---|---|---|
| 0 | 4 | $(0, 4)$ |
| $-2$ | 0 | $(-2, 0)$ |
| 1 | 6 | $(1, 6)$ |

$$4x + 8 = 2y$$

| $x$ | $y$ | $(x, y)$ |
|---|---|---|
| 0 | 4 | $(0, 4)$ |
| $-2$ | 0 | $(-2, 0)$ |
| $-3$ | $-2$ | $(-3, -2)$ |

FIGURE 3-24

The lines in the figure are the same line. Since the lines intersect at infinitely many points, there are infinitely many solutions. Any pair $(x, y)$ that satisfies one of the equations satisfies the other also.

From the graph, we can see that some solutions are $(0, 4)$, $(1, 6)$, and $(-1, 2)$, since each of these points lies on the one line that is the graph of both equations. ∎

**Self Check**    Solve $\begin{cases} 6x - 2y = 4 \\ y + 2 = 3x \end{cases}$.

**Answer**    Since the graphs are the same line, there are infinitely many solutions.

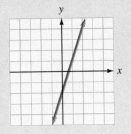

Table 3-4 summarizes the possibilities that can occur when two equations, each with two variables, are graphed.

**EXAMPLE 4**    Solve the system $\begin{cases} \frac{2}{3}x - \frac{1}{2}y = 1 \\ \frac{1}{10}x + \frac{1}{15}y = 1 \end{cases}$.

*Solution*    We can multiply both sides of the first equation by 6 to clear it of fractions.

$$\frac{2}{3}x - \frac{1}{2}y = 1$$

$$6\left(\frac{2}{3}x - \frac{1}{2}y\right) = 6(1)$$

**1.**        $4x - 3y = 6$

| Possible graph | If the | then |
|---|---|---|
| | lines are different and intersect, | the equations are independent and the system is consistent. One solution exists. |
| | lines are different and parallel, | the equations are independent and the system is inconsistent. No solutions exist. |
| | lines coincide (are the same line), | the equations are dependent and the system is consistent. Infinitely many solutions exist. |

TABLE 3-4

We then multiply both sides of the second equation by 30 to clear it of fractions.

$$\frac{1}{10}x + \frac{1}{15}y = 1$$

$$30\left(\frac{1}{10}x + \frac{1}{15}y\right) = 30(1)$$

**2.** $\qquad 3x + 2y = 30$

Equations 1 and 2 form the following equivalent system of equations, which has the same solutions as the original system.

$$\begin{cases} 4x - 3y = 6 \\ 3x + 2y = 30 \end{cases}$$

We can graph each equation of the previous system (see Figure 3-25) and find that their point of intersection has coordinates of $(6, 6)$. The solution of the given system is $x = 6$ and $y = 6$, or just $(6, 6)$.

To verify that $(6, 6)$ satisfies each equation of the original system, we substitute 6 for $x$ and 6 for $y$ in each of the original equations and simplify.

$$\frac{2}{3}x - \frac{1}{2}y = 1 \qquad\qquad \frac{1}{10}x + \frac{1}{15}y = 1$$

$$\frac{2}{3}(6) - \frac{1}{2}(6) \overset{?}{=} 1 \qquad\qquad \frac{1}{10}(6) + \frac{1}{15}(6) \overset{?}{=} 1$$

$$4 - 3 \overset{?}{=} 1 \qquad\qquad \frac{3}{5} + \frac{2}{5} \overset{?}{=} 1$$

$$1 = 1 \qquad\qquad 1 = 1$$

The equations in this system are independent, and the system is consistent.

$$4x - 3y = 6$$

| $x$ | $y$ | $(x, y)$ |
|---|---|---|
| 0 | $-2$ | $(0, -2)$ |
| 3 | 2 | $(3, 2)$ |
| 6 | 6 | $(6, 6)$ |

$$3x + 2y = 30$$

| $x$ | $y$ | $(x, y)$ |
|---|---|---|
| 10 | 0 | $(10, 0)$ |
| 8 | 3 | $(8, 3)$ |
| 6 | 6 | $(6, 6)$ |

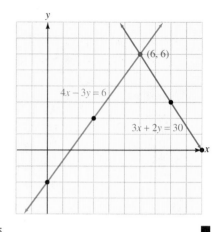

FIGURE 3-25

Self Check

Solve $\begin{cases} -\frac{x}{2} = \frac{y}{4} \\ \frac{1}{4}x - \frac{3}{8}y = -2 \end{cases}$.

Answer

$(-2, 4)$

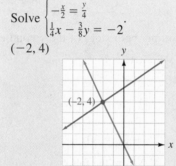

■ ■ ■ ■ ■ ■ ■ ■ ■ ■    **Solving Systems of Equations**

GRAPHING
CALCULATORS

We can use a graphing calculator to solve the system $\begin{cases} 2x + y = 12 \\ 2x - y = -2 \end{cases}$. However, before we can enter the equations into the calculator, we must solve them for $y$.

$$2x + y = 12 \qquad\qquad 2x - y = -2$$
$$y = -2x + 12 \qquad\qquad -y = -2x - 2$$
$$\qquad\qquad\qquad\qquad y = 2x + 2$$

We can now enter the resulting equations into a calculator and graph them. If we use standard window settings of $x = [-10, 10]$ and $y = [-10, 10]$, their graphs will look like Figure 3-26(a). We can trace to see that the coordinates of the intersection point are approximately

$$x = 2.5531915 \qquad \text{and} \qquad y = 6.893617$$

See Figure 3-26(b). For better results, we can zoom in on the intersection point and trace again to find that

$$x = 2.5 \qquad \text{and} \qquad y = 7$$

See Figure 3-26(c). Check the solution.

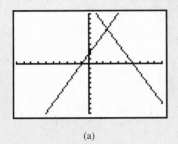

(a)

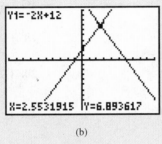

(b)

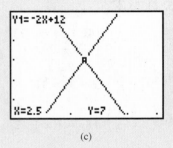

(c)

FIGURE 3-26

Orals    *Tell whether the pair is a solution of the system.*

**1.** $(3, 2)$, $\begin{cases} x + y = 5 \\ x - y = 1 \end{cases}$

**2.** $(1, 2)$, $\begin{cases} x - y = -1 \\ x + y = 3 \end{cases}$

**3.** $(4, 1)$, $\begin{cases} x + y = 5 \\ x - y = 2 \end{cases}$

**4.** $(5, 2)$, $\begin{cases} x - y = 3 \\ x + y = 6 \end{cases}$

# EXERCISE 3.3

**REVIEW**    *Evaluate each expression. Assume that $x = -3$.*

**1.** $(-2)^4$

**2.** $-2^4$

**3.** $3x - x^2$

**4.** $\dfrac{-3 + 2x}{6x}$

**VOCABULARY AND CONCEPTS**    *Fill in each blank to make a true statement.*

**5.** The pair of equations $\begin{cases} x - y = -1 \\ 2x - y = 1 \end{cases}$ is called a _____ of equations.

**6.** Because the ordered pair $(2, 3)$ satisfies both equations in Exercise 5, it is called a _____ of the system.

**7.** When the graphs of two equations in a system are different lines, the equations are called _____ equations.

**8.** When a system of equations has a solution, the system is called a _____.

**9.** Systems of equations that have no solution are called _____ systems.

**10.** When a system has infinitely many solutions, the equations of the system are said to be _____ equations.

*In Exercises 11–22, tell whether the ordered pair is a solution of the given system.*

**11.** $(1, 1)$, $\begin{cases} x + y = 2 \\ 2x - y = 1 \end{cases}$

**12.** $(1, 3)$, $\begin{cases} 2x + y = 5 \\ 3x - y = 0 \end{cases}$

**13.** $(3, -2)$, $\begin{cases} 2x + y = 4 \\ x + y = 1 \end{cases}$

**14.** $(-2, 4)$, $\begin{cases} 2x + 2y = 4 \\ x + 3y = 10 \end{cases}$

**15.** $(4, 5)$, $\begin{cases} 2x - 3y = -7 \\ 4x - 5y = 25 \end{cases}$

**16.** $(2, 3)$, $\begin{cases} 3x - 2y = 0 \\ 5x - 3y = -1 \end{cases}$

**17.** $(-2, -3)$, $\begin{cases} 4x + 5y = -23 \\ -3x + 2y = 0 \end{cases}$

**18.** $(-5, 1)$, $\begin{cases} -2x + 7y = 17 \\ 3x - 4y = -19 \end{cases}$

**19.** $\left(\dfrac{1}{2}, 3\right)$, $\begin{cases} 2x + y = 4 \\ 4x - 3y = 11 \end{cases}$

**20.** $\left(2, \dfrac{1}{3}\right)$, $\begin{cases} x - 3y = 1 \\ -2x + 6y = -6 \end{cases}$

**21.** $\left(-\dfrac{2}{5}, \dfrac{1}{4}\right)$, $\begin{cases} 5x - 4y = -6 \\ 8y = 10x + 12 \end{cases}$

**22.** $\left(-\dfrac{1}{3}, \dfrac{3}{4}\right)$, $\begin{cases} 3x + 4y = 2 \\ 12y = 3(2 - 3x) \end{cases}$

*In Exercises 23–34, solve each system.*

**23.** $\begin{cases} x + y = 2 \\ x - y = 0 \end{cases}$

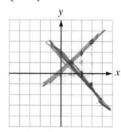

**24.** $\begin{cases} x + y = 4 \\ x - y = 0 \end{cases}$

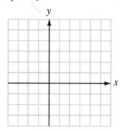

**25.** $\begin{cases} x + y = 2 \\ x - y = 4 \end{cases}$

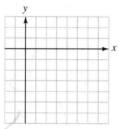

**26.** $\begin{cases} x + y = 1 \\ x - y = -5 \end{cases}$

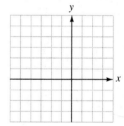

**27.** $\begin{cases} 3x + 2y = -8 \\ 2x - 3y = -1 \end{cases}$

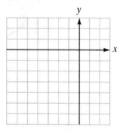

**28.** $\begin{cases} x + 4y = -2 \\ x + y = -5 \end{cases}$

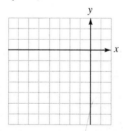

**29.** $\begin{cases} 4x - 2y = 8 \\ y = 2x - 4 \end{cases}$

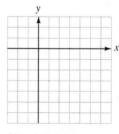

**30.** $\begin{cases} 3x - 6y = 18 \\ x = 2y + 3 \end{cases}$

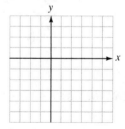

**31.** $\begin{cases} 2x - 3y = -18 \\ 3x + 2y = -1 \end{cases}$

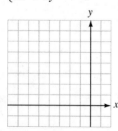

**32.** $\begin{cases} -x + 3y = -11 \\ 3x - y = 17 \end{cases}$

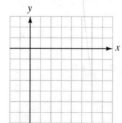

**33.** $\begin{cases} 4x = 3(4 - y) \\ 2y = 4(3 - x) \end{cases}$

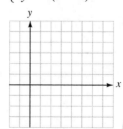

**34.** $\begin{cases} 2x = 3(2 - y) \\ 3y = 2(3 - x) \end{cases}$

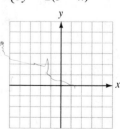

*In Exercises 35–42, solve each system.*

**35.** $\begin{cases} x + 2y = -4 \\ x - \dfrac{1}{2}y = 6 \end{cases}$

**36.** $\begin{cases} \dfrac{2}{3}x - y = -3 \\ 3x + y = 3 \end{cases}$

**37.** $\begin{cases} -\dfrac{3}{4}x + y = 3 \\ \dfrac{1}{4}x + y = -1 \end{cases}$

**38.** $\begin{cases} \dfrac{1}{3}x + y = 7 \\ \dfrac{2}{3}x - y = -4 \end{cases}$

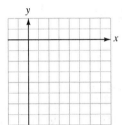
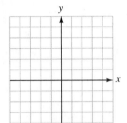
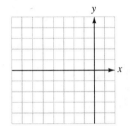
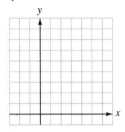

**39.** $\begin{cases} \dfrac{1}{2}x + \dfrac{1}{4}y = 0 \\ \dfrac{1}{4}x - \dfrac{3}{8}y = -2 \end{cases}$

**40.** $\begin{cases} \dfrac{1}{2}x + \dfrac{2}{3}y = -5 \\ \dfrac{3}{2}x - y = 3 \end{cases}$

**41.** $\begin{cases} \dfrac{1}{3}x - \dfrac{1}{2}y = \dfrac{1}{6} \\ \dfrac{2}{5}x + \dfrac{1}{2}y = \dfrac{13}{10} \end{cases}$

**42.** $\begin{cases} \dfrac{3}{4}x + \dfrac{2}{3}y = -\dfrac{19}{6} \\ y - x = -\dfrac{4x}{3} \end{cases}$

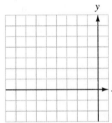
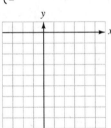
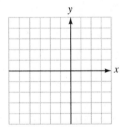
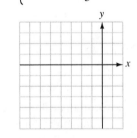

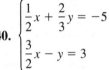

*In Exercises 43–46, use a graphing calculator to solve each system, if possible. If answers are not exact, round to the nearest hundredth.*

**43.** $\begin{cases} y = 4 - x \\ y = 2 + x \end{cases}$

**44.** $\begin{cases} y = x - 2 \\ y = x + 2 \end{cases}$

**45.** $\begin{cases} 3x - 6y = 4 \\ 2x + y = 1 \end{cases}$

**46.** $\begin{cases} 4x + 9y = 4 \\ 6x + 3y = -1 \end{cases}$

## APPLICATIONS

**47. Transplants**   See Illustration 1.
   **a.** What was the relationship between the number of donors and those awaiting a transplant in 1989?

   **b.** In what year was the number of donors and the number waiting for a transplant the same? Estimate the number.

   **c.** Explain the most recent trends.

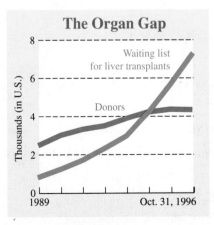

ILLUSTRATION 1

**48. Daily tracking polls**   See Illustration 2.
  **a.** Which candidate was ahead on October 28 and by how much?
  **b.** On what day did the challenger pull even with the incumbent?
  **c.** If the election was held November 4, whom did the poll predict as the winner and by how many percentage points?

ILLUSTRATION 3

**50. Economics**   The graph in Illustration 4 illustrates the law of supply and demand.
  **a.** Complete this sentence: "As the price of an item increases, the *supply* of the item _____."
  **b.** Complete this sentence: "As the price of an item increases, the *demand* for the item _____."
  **c.** For what price will the supply equal the demand? How many items will be supplied for this price?

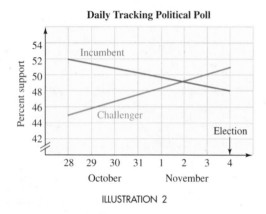

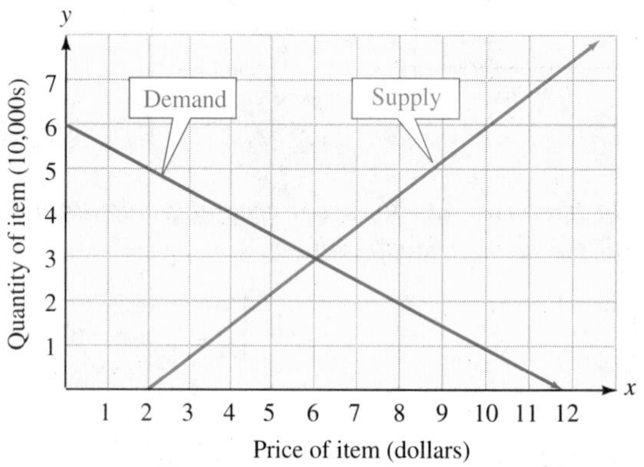

ILLUSTRATION 2

**49. Latitude and longitude**   See Illustration 3.
  **a.** Name three American cities that lie on the latitude line of 30° north.

  **b.** Name three American cities that lie on the longitude line of 90° west.

  **c.** What city lies on both lines?

ILLUSTRATION 4

*WRITING*

**51.** Explain what we mean when we say "inconsistent system."

**52.** Explain what we mean when we say, "The equations of a system are dependent."

*SOMETHING TO THINK ABOUT*

**53.** Use a graphing calculator to solve the system

$$\begin{cases} 11x - 20y = 21 \\ -4x + 7y = 21 \end{cases}$$

What problems did you encounter?

**54.** Can the equations of an inconsistent system with two equations in two variables be dependent?

## 3.4  Solving Systems of Equations by Substitution

■ THE SUBSTITUTION METHOD  ■ INCONSISTENT SYSTEMS  ■ DEPENDENT EQUATIONS

**Getting Ready**    *Remove parentheses.*

**1.** $2(3x + 2)$                    **2.** $5(-5 - 2x)$

*Substitute $x - 2$ for $y$ and remove parentheses.*

**3.** $2y$                       **4.** $3(y - 2)$

### ■ THE SUBSTITUTION METHOD

We now consider the **substitution method** for solving systems of equations.
To solve the system

$$\begin{cases} y = 3x - 2 \\ 2x + y = 8 \end{cases}$$

by the substitution method, we note that $y = 3x - 2$. Because $y = 3x - 2$, we can substitute $3x - 2$ for $y$ in the equation $2x + y = 8$ to get

$$2x + y = 8$$
$$2x + (3x - 2) = 8$$

The resulting equation has only one variable and can be solved for $x$.

| | |
|---|---|
| $2x + (3x - 2) = 8$ | |
| $2x + 3x - 2 = 8$ | Remove parentheses. |
| $5x - 2 = 8$ | Combine like terms. |
| $5x = 10$ | Add 2 to both sides. |
| $x = 2$ | Divide both sides by 5. |

We can find $y$ by substituting 2 for $x$ in either equation of the given system. Because $y = 3x - 2$ is already solved for $y$, it is easier to substitute in this equation.

$$y = 3x - 2$$
$$= 3(2) - 2$$
$$= 6 - 2$$
$$= 4$$

The solution of the given system is $x = 2$ and $y = 4$, or just $(2, 4)$.

*Check:* 
$$y = 3x - 2 \qquad\qquad 2x + y = 8$$
$$4 \overset{?}{=} 3(2) - 2 \qquad\qquad 2(2) + 4 \overset{?}{=} 8$$
$$4 \overset{?}{=} 6 - 2 \qquad\qquad 4 + 4 \overset{?}{=} 8$$
$$4 = 4 \qquad\qquad\qquad 8 = 8$$

Since the pair $x = 2$ and $y = 4$ is a solution, the lines represented by the equations of the given system intersect at the point $(2, 4)$. The equations of this system are independent, and the system is consistent.

To solve a system of equations in $x$ and $y$ by the substitution method, we follow these steps.

> **The Substitution Method**
>
> 1. Solve one of the equations for $x$ or $y$. (This step may not be necessary.)
> 2. Substitute the resulting expression for the variable obtained in Step 1 into the other equation, and solve that equation.
> 3. Find the value of the other variable by substituting the solution found in Step 2 into any equation containing both variables.
> 4. Check the solution in the equations of the original system.

**EXAMPLE 1**   Solve the system $\begin{cases} 2x + y = -5 \\ 3x + 5y = -4 \end{cases}$.

*Solution*   We solve one of the equations for one of its variables. Since the term $y$ in the first equation has a coefficient of 1, we solve the first equation for $y$.

$$2x + y = -5$$
$$y = -5 - 2x \qquad \text{Subtract } 2x \text{ from both sides.}$$

We then substitute $-5 - 2x$ for $y$ in the second equation and solve for $x$.

$$3x + 5y = -4$$
$$3x + 5(-5 - 2x) = -4$$
$$3x - 25 - 10x = -4 \qquad \text{Remove parentheses.}$$
$$-7x - 25 = -4 \qquad \text{Combine like terms.}$$
$$-7x = 21 \qquad \text{Add 25 to both sides.}$$
$$x = -3 \qquad \text{Divide both sides by } -7.$$

We can find $y$ by substituting $-3$ for $x$ in the equation $y = -5 - 2x$.

$$y = -5 - 2x$$
$$= -5 - 2(-3)$$
$$= -5 + 6$$
$$= 1$$

The solution is $x = -3$ and $y = 1$, or just $(-3, 1)$.

*Check:*

| | |
|---|---|
| $2x + y = -5$ | $3x + 5y = -4$ |
| $2(-3) + 1 \overset{?}{=} -5$ | $3(-3) + 5(1) \overset{?}{=} -4$ |
| $-6 + 1 \overset{?}{=} -5$ | $-9 + 5 \overset{?}{=} -4$ |
| $-5 = -5$ | $-4 = -4$ |

■

**Self Check**    Solve $\begin{cases} 2x - 3y = 13 \\ 3x + y = 3 \end{cases}$.

**Answer**    $(2, -3)$

**EXAMPLE 2**    Solve the system $\begin{cases} 2x + 3y = 5 \\ 3x + 2y = 0 \end{cases}$.

*Solution*    We can solve the second equation for $x$:

$$3x + 2y = 0$$

$$3x = -2y \qquad \text{Subtract } 2y \text{ from both sides.}$$

$$x = \frac{-2y}{3} \qquad \text{Divide both sides by 3.}$$

We then substitute $\frac{-2y}{3}$ for $x$ in the other equation and solve for $y$.

$$2x + 3y = 5$$

$$2\left(\frac{-2y}{3}\right) + 3y = 5$$

$$\frac{-4y}{3} + 3y = 5 \qquad \text{Remove parentheses.}$$

$$3\left(\frac{-4y}{3}\right) + 3(3y) = 3(5) \qquad \text{Multiply both sides by 3.}$$

$$-4y + 9y = 15 \qquad \text{Remove parentheses.}$$

$$5y = 15 \qquad \text{Combine like terms.}$$

$$y = 3 \qquad \text{Divide both sides by 5.}$$

We can find $x$ by substituting 3 for $y$ in the equation $x = \frac{-2y}{3}$.

$$x = \frac{-2y}{3}$$

$$= \frac{-2(3)}{3}$$

$$= -2$$

Check the solution $(-2, 3)$ in each equation of the system.    ■

**Self Check**    Solve $\begin{cases} 3x - 2y = -19 \\ 2x + 5y = 0 \end{cases}$.

**Answer**    $(-5, 2)$

**EXAMPLE 3**    Solve the system $\begin{cases} 3(x - y) = 5 \\ x + 3 = -\frac{5}{2}y \end{cases}$.

*Solution*   We begin by writing each equation in general form:

$$3(x - y) = 5 \qquad\qquad x + 3 = -\frac{5}{2}y$$

|   |   |   |   |
|---|---|---|---|
| **1.** | $3x - 3y = 5$ | $2x + 6 = -5y$ | Multiply both sides by 2. |
| **2.** |  | $2x + 5y = -6$ | Add 5y and subtract 6 from both sides. |

To solve the system formed by Equations 1 and 2, we first solve Equation 1 for $x$.

|   |   |   |
|---|---|---|
| **1.** | $3x - 3y = 5$ |  |
|  | $3x = 5 + 3y$ | Add 3y to both sides. |
| **3.** | $x = \dfrac{5 + 3y}{3}$ | Divide both sides by 3. |

We then substitute $\frac{5+3y}{3}$ for $x$ in Equation 2 and proceed as follows:

**2.**
$$2x + 5y = -6$$
$$2\left(\frac{5 + 3y}{3}\right) + 5y = -6$$

|   |   |
|---|---|
| $2(5 + 3y) + 15y = -18$ | Multiply both sides by 3. |
| $10 + 6y + 15y = -18$ | Remove parentheses. |
| $10 + 21y = -18$ | Combine like terms. |
| $21y = -28$ | Subtract 10 from both sides. |
| $y = \dfrac{-28}{21}$ | Divide both sides by 21. |
| $y = -\dfrac{4}{3}$ | Simplify $\frac{-28}{21}$. |

To find $x$, we substitute $-\frac{4}{3}$ for $y$ in Equation 3 and simplify.

$$x = \frac{5 + 3y}{3}$$
$$= \frac{5 + 3\left(-\frac{4}{3}\right)}{3}$$
$$= \frac{5 - 4}{3}$$
$$= \frac{1}{3}$$

Check the solution $\left(\frac{1}{3}, -\frac{4}{3}\right)$ in each equation.  ∎

**Self Check**   Solve $\begin{cases} 2(x + y) = -5 \\ x + 2 = -\frac{3}{5}y \end{cases}$.

**Answer**   $\left(-\frac{5}{4}, -\frac{5}{4}\right)$

### ■ INCONSISTENT SYSTEMS

**EXAMPLE 4**  Solve the system $\begin{cases} x = 4(3 - y) \\ 2x = 4(3 - 2y) \end{cases}$.

*Solution*  Since $x = 4(3 - y)$, we can substitute $4(3 - y)$ for $x$ in the second equation and solve for $y$.

$$2x = 4(3 - 2y)$$
$$2[4(3 - y)] = 4(3 - 2y)$$
$$8(3 - y) = 4(3 - 2y) \qquad 2 \cdot 4 = 8.$$
$$24 - 8y = 12 - 8y \qquad \text{Remove parentheses.}$$
$$24 = 12 \qquad \text{Add } 8y \text{ to both sides.}$$

This impossible result indicates that the equations in this system are independent, but that the system is inconsistent. If each equation in this system were graphed, these graphs would be parallel lines. There are no solutions to this system.  ■

**Self Check**  Solve $\begin{cases} 0.1x - 0.4 = 0.1y \\ -2y = 2(2 - x) \end{cases}$.

*Answer*  no solution

### ■ DEPENDENT EQUATIONS

**EXAMPLE 5**  Solve the system $\begin{cases} 3x = 4(6 - y) \\ 4y + 3x = 24 \end{cases}$.

*Solution*  We can substitute $4(6 - y)$ for $3x$ in the second equation and proceed as follows:

$$4y + 3x = 24$$
$$4y + 4(6 - y) = 24$$
$$4y + 24 - 4y = 24 \qquad \text{Remove parentheses.}$$
$$24 = 24 \qquad \text{Combine like terms.}$$

Although $24 = 24$ is true, we did not find $y$. This result indicates that the equations of this system are dependent. If either equation were graphed, the same line would result.

Because any ordered pair that satisfies one equation satisfies the other also, the system has infinitely many solutions. To find some of them, we substitute 8, 0, and 4 for $x$ in either equation and solve for $y$. The pairs $(8, 0)$, $(0, 6)$, and $(4, 3)$ are solutions.  ■

**Self Check**  Solve $\begin{cases} 3y = -3(x + 4) \\ 3x + 3y = -12 \end{cases}$.

*Answer*  infinitely many solutions

Orals   *Let $y = x + 1$. Find $y$ after each quantity is substituted for $x$.*

**1.** $2z$                                                        **2.** $z + 1$

**3.** $3t + 2$                                                    **4.** $\dfrac{t}{3} + 3$

## EXERCISE 3.4

**REVIEW**   *Let $x = -2$ and $y = 3$ and evaluate each expression.*

**1.** $y^2 - x^2$                 **2.** $-x^2 + y^3$                 **3.** $\dfrac{3x - 2y}{2x + y}$

**4.** $-2x^2y^2$                  **5.** $-x(3y - 4)$                 **6.** $-2y(4x - y)$

**VOCABULARY AND CONCEPTS**   *Fill in each blank to make a true statement.*

**7.** We say the equation $y = 2x + 4$ is solved for ___ or that $y$ is expressed in _____ of $x$.

**8.** To _____ a solution of a system means to see whether the coordinates of the ordered pair satisfy both equations.

**9.** Consider $2(x - 6) = 2x - 12$. The distributive property was applied to _____ parentheses.

**10.** In mathematics, to _____ means to replace an expression with one that is equivalent to it.

**11.** A system with dependent equations has _____ solutions.

**12.** In the term $y$, the _____ is understood to be 1.

**PRACTICE**   *In Exercises 13–54, use the substitution method to solve each system.*

**13.** $\begin{cases} y = 2x \\ x + y = 6 \end{cases}$   **14.** $\begin{cases} y = 3x \\ x + y = 4 \end{cases}$   **15.** $\begin{cases} y = 2x - 6 \\ 2x + y = 6 \end{cases}$   **16.** $\begin{cases} y = 2x - 9 \\ x + 3y = 8 \end{cases}$

**17.** $\begin{cases} y = 2x + 5 \\ x + 2y = -5 \end{cases}$   **18.** $\begin{cases} y = -2x \\ 3x + 2y = -1 \end{cases}$   **19.** $\begin{cases} 2a + 4b = -24 \\ a = 20 - 2b \end{cases}$   **20.** $\begin{cases} 3a + 6b = -15 \\ a = -2b - 5 \end{cases}$

**21.** $\begin{cases} 2a = 3b - 13 \\ b = 2a + 7 \end{cases}$   **22.** $\begin{cases} a = 3b - 1 \\ b = 2a + 2 \end{cases}$   **23.** $\begin{cases} r + 3s = 9 \\ 3r + 2s = 13 \end{cases}$   **24.** $\begin{cases} x - 2y = 2 \\ 2x + 3y = 11 \end{cases}$

**25.** $\begin{cases} 4x + 5y = 2 \\ 3x - y = 11 \end{cases}$   **26.** $\begin{cases} 5u + 3v = 5 \\ 4u - v = 4 \end{cases}$   **27.** $\begin{cases} 2x + y = 0 \\ 3x + 2y = 1 \end{cases}$   **28.** $\begin{cases} 3x - y = 7 \\ 2x + 3y = 1 \end{cases}$

**29.** $\begin{cases} 3x + 4y = -7 \\ 2y - x = -1 \end{cases}$   **30.** $\begin{cases} 4x + 5y = -2 \\ x + 2y = -2 \end{cases}$   **31.** $\begin{cases} 9x = 3y + 12 \\ 4 = 3x - y \end{cases}$   **32.** $\begin{cases} 8y = 15 - 4x \\ x + 2y = 4 \end{cases}$

**33.** $\begin{cases} 2x + 3y = 5 \\ 3x + 2y = 5 \end{cases}$   **34.** $\begin{cases} 3x - 2y = -1 \\ 2x + 3y = -5 \end{cases}$   **35.** $\begin{cases} 2x + 5y = -2 \\ 4x + 3y = 10 \end{cases}$   **36.** $\begin{cases} 3x + 4y = -6 \\ 2x - 3y = -4 \end{cases}$

**37.** $\begin{cases} 2x - 3y = -3 \\ 3x + 5y = -14 \end{cases}$   **38.** $\begin{cases} 4x - 5y = -12 \\ 5x - 2y = 2 \end{cases}$   **39.** $\begin{cases} 7x - 2y = -1 \\ -5x + 2y = -1 \end{cases}$   **40.** $\begin{cases} -8x + 3y = 22 \\ 4x + 3y = -2 \end{cases}$

**41.** $\begin{cases} 2a + 3b = 2 \\ 8a - 3b = 3 \end{cases}$ 

**42.** $\begin{cases} 3a - 2b = 0 \\ 9a + 4b = 5 \end{cases}$ 

**43.** $\begin{cases} y - x = 3x \\ 2(x + y) = 14 - y \end{cases}$ 

**44.** $\begin{cases} y + x = 2x + 2 \\ 2(3x - 2y) = 21 - y \end{cases}$

**45.** $\begin{cases} 3(x - 1) + 3 = 8 + 2y \\ 2(x + 1) = 4 + 3y \end{cases}$ 

**46.** $\begin{cases} 4(x - 2) = 19 - 5y \\ 3(x + 1) - 2y = 2y \end{cases}$

**47.** $\begin{cases} 6a = 5(3 + b + a) - a \\ 3(a - b) + 4b = 5(1 + b) \end{cases}$ 

**48.** $\begin{cases} 5(x + 1) + 7 = 7(y + 1) \\ 5(y + 1) = 6(1 + x) + 5 \end{cases}$

**49.** $\begin{cases} \dfrac{1}{2}x + \dfrac{1}{2}y = -1 \\ \dfrac{1}{3}x - \dfrac{1}{2}y = -4 \end{cases}$ 

**50.** $\begin{cases} \dfrac{2}{3}y + \dfrac{1}{5}z = 1 \\ \dfrac{1}{3}y - \dfrac{2}{5}z = 3 \end{cases}$

**51.** $\begin{cases} 5x = \dfrac{1}{2}y - 1 \\ \dfrac{1}{4}y = 10x - 1 \end{cases}$ 

**52.** $\begin{cases} \dfrac{2}{3}x = 1 - 2y \\ 2(5y - x) + 11 = 0 \end{cases}$

**53.** $\begin{cases} \dfrac{6x - 1}{3} - \dfrac{5}{3} = \dfrac{3y + 1}{2} \\ \dfrac{1 + 5y}{4} + \dfrac{x + 3}{4} = \dfrac{17}{2} \end{cases}$ 

**54.** $\begin{cases} \dfrac{5x - 2}{4} + \dfrac{1}{2} = \dfrac{3y + 2}{2} \\ \dfrac{7y + 3}{3} = \dfrac{x}{2} + \dfrac{7}{3} \end{cases}$

### WRITING

**55.** Explain how to use substitution to solve a system of equations.

**56.** If the equations of a system are written in general form, why is it to your advantage to solve for a variable whose coefficient is 1?

### SOMETHING TO THINK ABOUT

**57.** Could you use substitution to solve the system

$$\begin{cases} y = 2y + 4 \\ x = 3x - 5 \end{cases}$$

How would you solve it?

**58.** What are the advantages and disadvantages of
 **a.** the graphing method?
 **b.** the substitution method?

## 3.5 Solving Systems of Equations by Addition

■ THE ADDITION METHOD ■ INCONSISTENT SYSTEMS ■ DEPENDENT EQUATIONS

Getting Ready     *Add the left-hand sides and the right-hand sides of the equations in each system.*

**1.** $\begin{cases} 2x + 3y = 4 \\ 3x - 3y = 6 \end{cases}$ 

**2.** $\begin{cases} 4x - 2y = 1 \\ -4x + 3y = 5 \end{cases}$

**3.** $\begin{cases} 6x - 5y = 23 \\ -4x + 5y = 10 \end{cases}$         **4.** $\begin{cases} -5x + 6y = 18 \\ 5x + 12y = 10 \end{cases}$

## ■ THE ADDITION METHOD

Another method used to solve systems of equations is the **addition method.** To solve the system

$$\begin{cases} x + y = 8 \\ x - y = -2 \end{cases}$$

by the addition method, we see that the coefficients of $y$ are *opposites* and then add the left-hand sides and the right-hand sides of the equations to eliminate the variable $y$.

$\begin{aligned} x + y &= \phantom{-}8 \\ x - y &= -2 \end{aligned}$    Equal quantities, $x - y$ and $-2$, are added to both sides of the equation $x + y = 8$. By the addition property of equality, the results will be equal.

Now, column by column, we add like terms.

Combine like terms.

$$\begin{aligned} \downarrow\phantom{xx}\downarrow\phantom{xxx}\downarrow\phantom{xx} \\ x + y = \phantom{-}8 \\ \underline{x - y = -2} \\ 2x \phantom{xxx} = \phantom{-}6 \end{aligned}$$  ← Write each result here.

We can then solve the resulting equation for $x$.

$2x = 6$

$\phantom{2}x = 3$         Divide both sides by 2.

To find $y$, we substitute 3 for $x$ in either equation of the system and solve it for $y$.

$x + y = 8$         The first equation of the system.

$3 + y = 8$         Substitute 3 for $x$.

$\phantom{3 + }y = 5$         Subtract 3 from both sides.

We check the solution by verifying that the pair $(3, 5)$ satisfies each equation of the original system.

To solve an equation in $x$ and $y$ by the addition method, we follow these steps.

**The Addition Method**

1. If necessary, write both equations in general form: $Ax + By = C$.

2. If necessary, multiply one or both of the equations by nonzero quantities to make the coefficients of $x$ (or the coefficients of $y$) opposites.

3. Add the equations to eliminate the term involving $x$ (or $y$).

4. Solve the equation resulting from Step 3.

5. Find the value of the other variable by substituting the solution found in Step 4 into any equation containing both variables.

6. Check the solution in the equations of the original system.

EXAMPLE 1    Solve the system $\begin{cases} 3y = 14 + x \\ x + 22 = 5y \end{cases}$.

*Solution*    We can write the equations in the form

$$\begin{cases} -x + 3y = 14 \\ x - 5y = -22 \end{cases}$$

When these equations are added, the terms involving $x$ are eliminated. We solve the resulting equation for $y$.

$$\begin{aligned} -x + 3y &= 14 \\ \underline{x - 5y} &= \underline{-22} \\ -2y &= -8 \\ y &= 4 \qquad \text{Divide both sides by } -2. \end{aligned}$$

To find $x$, we substitute 4 for $y$ in either equation of the system. If we substitute 4 for $y$ in the equation $-x + 3y = 14$, we have

$$\begin{aligned} -x + 3y &= 14 \\ -x + 3(4) &= 14 \\ -x + 12 &= 14 \qquad \text{Simplify.} \\ -x &= 2 \qquad \text{Subtract 12 from both sides.} \\ x &= -2 \qquad \text{Divide both sides by } -1. \end{aligned}$$

Verify that $(-2, 4)$ satisfies each equation.    ■

Self Check    Solve $\begin{cases} 3y = 7 - x \\ 2x - 3y = -22 \end{cases}$.

Answer    $(-5, 4)$

Sometimes we need to multiply both sides of one equation in a system by a number to make the coefficients of one of the variables opposites.

EXAMPLE 2 Solve the system $\begin{cases} 3x + y = 7 \\ x + 2y = 4 \end{cases}$.

Solution If we add the equations as they are, neither variable will be eliminated. We must write the equations so that the coefficients of one of the variables are opposites. To eliminate $x$, we can multiply both sides of the second equation by $-3$ to get

$$\begin{cases} 3x + y = 7 \\ -3(x + 2y) = -3(4) \end{cases} \longrightarrow \begin{cases} 3x + y = 7 \\ -3x - 6y = -12 \end{cases}$$

The coefficients of the terms $3x$ and $-3x$ are opposites. When the equations are added, $x$ is eliminated.

$$
\begin{array}{r}
3x + y = 7 \\
-3x - 6y = -12 \\
\hline
-5y = -5 \\
y = 1
\end{array}
$$
Divide both sides by $-5$.

To find $x$, we substitute 1 for $y$ in the equation $3x + y = 7$.

$$
\begin{array}{ll}
3x + y = 7 & \\
3x + (1) = 7 & \text{Substitute 1 for } y. \\
3x = 6 & \text{Subtract 1 from both sides.} \\
x = 2 & \text{Divide both sides by 3.}
\end{array}
$$

Check the solution $(2, 1)$ in the original system of equations. ∎

Self Check Solve $\begin{cases} 3x + 4y = 25 \\ 2x + y = 10 \end{cases}$.

Answer $(3, 4)$

In some instances, we must multiply both equations by nonzero quantities to make the coefficients of one of the variables opposites.

EXAMPLE 3 Solve the system $\begin{cases} 2a - 5b = 10 \\ 3a - 2b = -7 \end{cases}$.

Solution The equations in the system must be written so that one of the variables will be eliminated when the equations are added.

To eliminate $a$, we can multiply the first equation by 3 and the second equation by $-2$ to get

$$\begin{cases} 3(2a - 5b) = 3(10) \\ -2(3a - 2b) = -2(-7) \end{cases} \longrightarrow \begin{cases} 6a - 15b = 30 \\ -6a + 4b = 14 \end{cases}$$

When these equations are added, the terms $6a$ and $-6a$ are eliminated.

$$
\begin{aligned}
6a - 15b &= 30 \\
\underline{-6a + \phantom{1}4b} &= \underline{14} \\
-11b &= 44 \\
b &= -4 \qquad \text{Divide both sides by } -11.
\end{aligned}
$$

To find $a$, we substitute $-4$ for $b$ in the equation $2a - 5b = 10$.

$$
\begin{aligned}
2a - 5b &= 10 \\
2a - 5(-4) &= 10 \qquad \text{Substitute } -4 \text{ for } b. \\
2a + 20 &= 10 \qquad \text{Simplify.} \\
2a &= -10 \qquad \text{Subtract 20 from both sides.} \\
a &= -5 \qquad \text{Divide both sides by 2.}
\end{aligned}
$$

Check the solution $(-5, -4)$ in the original equations. ∎

**Self Check**   Solve $\begin{cases} 2a + 3b = 7 \\ 5a + 2b = 1 \end{cases}$.

**Answer**   $(-1, 3)$

**EXAMPLE 4**   Solve $\begin{cases} \frac{5}{6}x + \frac{2}{3}y = \frac{7}{6} \\ \frac{10}{7}x - \frac{4}{9}y = \frac{17}{21} \end{cases}$.

*Solution*   To clear the equations of fractions, we multiply both sides of the first equation by 6 and both sides of the second equation by 63. This gives the system

**1.**  $\begin{cases} 5x + 4y = 7 \\ 90x - 28y = 51 \end{cases}$
**2.**

We can solve for $x$ by eliminating the terms involving $y$. To do so, we multiply Equation 1 by 7 and add the result to Equation 2.

$$
\begin{aligned}
35x + 28y &= \phantom{0}49 \\
\underline{90x - 28y} &= \underline{\phantom{0}51} \\
125x \phantom{- 28y} &= 100 \\
x &= \frac{100}{125} \qquad \text{Divide both sides by 125.} \\
x &= \frac{4}{5} \qquad \text{Simplify.}
\end{aligned}
$$

To solve for $y$, we substitute $\frac{4}{5}$ for $x$ in Equation 1 and simplify.

$$5x + 4y = 7$$

$$5\left(\frac{4}{5}\right) + 4y = 7$$

$$4 + 4y = 7 \qquad \text{Simplify.}$$

$$4y = 3 \qquad \text{Subtract 4 from both sides.}$$

$$y = \frac{3}{4} \qquad \text{Divide both sides by 4.}$$

Check the solution of $\left(\frac{4}{5}, \frac{3}{4}\right)$ in the original equations.  ∎

**Self Check**   Solve $\begin{cases} \frac{1}{3}x + \frac{1}{6}y = 1 \\ \frac{1}{2}x - \frac{1}{4}y = 0 \end{cases}$.

**Answer**   $\left(\frac{3}{2}, 3\right)$

## ■ INCONSISTENT SYSTEMS

**EXAMPLE 5**   Solve $\begin{cases} x - \frac{2y}{3} = \frac{8}{3} \\ -\frac{3x}{2} + y = -6 \end{cases}$.

*Solution*   We can multiply both sides of the first equation by 3 and both sides of the second equation by 2 to clear the equations of fractions.

$$\begin{cases} 3\left(x - \dfrac{2y}{3}\right) = 3\left(\dfrac{8}{3}\right) \\ 2\left(-\dfrac{3x}{2} + y\right) = 2(-6) \end{cases} \longrightarrow \begin{cases} 3x - 2y = 8 \\ -3x + 2y = -12 \end{cases}$$

We can add the resulting equations to eliminate the term involving $x$.

$$\begin{array}{r} 3x - 2y = \phantom{-1}8 \\ -3x + 2y = -12 \\ \hline 0 = \phantom{-1}-4 \end{array}$$

Here, the terms involving both $x$ and $y$ drop out, and a false result is obtained. This shows that the equations of the system are independent, but the system itself is inconsistent. This system has no solution.  ∎

**Self Check**   Solve $\begin{cases} x - \frac{y}{3} = \frac{10}{3} \\ 3x - y = \frac{5}{2} \end{cases}$.

**Answer**   no solution

### ■ DEPENDENT EQUATIONS

**EXAMPLE 6**  Solve $\begin{cases} \frac{2x-5y}{2} = \frac{19}{2} \\ -0.2x + 0.5y = -1.9 \end{cases}$.

*Solution*  We can multiply both sides of the first equation by 2 to clear it of fractions and both sides of the second equation by 10 to clear it of decimals.

$$\begin{cases} 2\left(\dfrac{2x-5y}{2}\right) = 2\left(\dfrac{19}{2}\right) \\ 10(-0.2x + 0.5y) = 10(-1.9) \end{cases} \longrightarrow \begin{cases} 2x - 5y = 19 \\ -2x + 5y = -19 \end{cases}$$

We add the resulting equations to get

$$\begin{array}{r} 2x - 5y = \phantom{-}19 \\ -2x + 5y = -19 \\ \hline 0 = \phantom{-}0 \end{array}$$

As in Example 5, both $x$ and $y$ drop out. However, this time a true result is obtained. This shows that the equations are dependent and the system has infinitely many solutions. Any ordered pair that satisfies one equation satisfies the other also. Some solutions are $(2, -3)$, $(12, 1)$, and $\left(0, -\frac{19}{5}\right)$. ■

**Self Check**  Solve $\begin{cases} \frac{3x+y}{6} = \frac{1}{3} \\ -0.3x - 0.1y = -0.2 \end{cases}$.

**Answer**  infinitely many solutions

**Orals**  *Use addition to solve each system for $x$.*

**1.** $\begin{cases} x + y = 1 \\ x - y = 1 \end{cases}$     **2.** $\begin{cases} 2x + y = 4 \\ x - y = 2 \end{cases}$

*Use addition to solve each system for $y$.*

**3.** $\begin{cases} -x + y = 3 \\ x + y = 3 \end{cases}$     **4.** $\begin{cases} x + 2y = 4 \\ -x - y = 1 \end{cases}$

## EXERCISE 3.5

***REVIEW***  *Solve each equation or inequality. For each inequality, give the answer in interval notation and graph the interval.*

**1.** $8(3x - 5) - 12 = 4(2x + 3)$

**2.** $5x - 13 = x - 1$

**3.** $x - 2 = \dfrac{x+2}{3}$

**4.** $\dfrac{3}{2}(y + 4) = \dfrac{20 - y}{2}$

**5.** $7x - 9 \le 5$

**6.** $-2x + 6 > 16$

***VOCABULARY AND CONCEPTS*** *Fill in each blank to make a true statement.*

**7.** The numerical _____ of $-3x$ is $-3$.

**8.** The _____ of 4 is $-4$.

**9.** $Ax + By = C$ is the _____ form of the equation of a line.

**10.** When adding the equations

$$5x - 6y = 10$$
$$\underline{-3x + 6y = 24}$$

the variable $y$ will be _____.

**11.** To clear the equation $\frac{2}{3}x + 4y = -\frac{4}{5}$ of fractions, we must multiply both sides by ___.

**12.** To solve the system

$$\begin{cases} 3x + 12y = 4 \\ 6x - 4y = 8 \end{cases}$$

we would multiply the first equation by ___ and add to eliminate the $x$.

***PRACTICE*** *In Exercises 13–24, use the addition method to solve each system.*

**13.** $\begin{cases} x + y = 5 \\ x - y = -3 \end{cases}$

**14.** $\begin{cases} x - y = 1 \\ x + y = 7 \end{cases}$

**15.** $\begin{cases} x - y = -5 \\ x + y = 1 \end{cases}$

**16.** $\begin{cases} x + y = 1 \\ x - y = 5 \end{cases}$

**17.** $\begin{cases} 2x + y = -1 \\ -2x + y = 3 \end{cases}$

**18.** $\begin{cases} 3x + y = -6 \\ x - y = -2 \end{cases}$

**19.** $\begin{cases} 2x - 3y = -11 \\ 3x + 3y = 21 \end{cases}$

**20.** $\begin{cases} 3x - 2y = 16 \\ -3x + 8y = -10 \end{cases}$

**21.** $\begin{cases} 2x + y = -2 \\ -2x - 3y = -6 \end{cases}$

**22.** $\begin{cases} 3x + 4y = 8 \\ 5x - 4y = 24 \end{cases}$

**23.** $\begin{cases} 4x + 3y = 24 \\ 4x - 3y = -24 \end{cases}$

**24.** $\begin{cases} 5x - 4y = 8 \\ -5x - 4y = 8 \end{cases}$

*In Exercises 25–54, use the addition method to solve each system of equations. If the equations of a system are dependent or if a system is inconsistent, so indicate.*

**25.** $\begin{cases} x + y = 5 \\ x + 2y = 8 \end{cases}$

**26.** $\begin{cases} x + 2y = 0 \\ x - y = -3 \end{cases}$

**27.** $\begin{cases} 2x + y = 4 \\ 2x + 3y = 0 \end{cases}$

**28.** $\begin{cases} 2x + 5y = -13 \\ 2x - 3y = -5 \end{cases}$

**29.** $\begin{cases} 3x + 29 = 5y \\ 4y - 34 = -3x \end{cases}$

**30.** $\begin{cases} 3x - 16 = 5y \\ 33 - 5y = 4x \end{cases}$

**31.** $\begin{cases} 2x = 3(y - 2) \\ 2(x + 4) = 3y \end{cases}$

**32.** $\begin{cases} 3(x - 2) = 4y \\ 2(2y + 3) = 3x \end{cases}$

**33.** $\begin{cases} -2(x + 1) = 3(y - 2) \\ 3(y + 2) = 6 - 2(x - 2) \end{cases}$

**34.** $\begin{cases} 5(x - 1) = 8 - 3(y + 2) \\ 4(x + 2) - 7 = 3(2 - y) \end{cases}$

**35.** $\begin{cases} 4(x + 1) = 17 - 3(y - 1) \\ 2(x + 2) + 3(y - 1) = 9 \end{cases}$

**36.** $\begin{cases} 3(x + 3) + 2(y - 4) = 5 \\ 3(x - 1) = -2(y + 2) \end{cases}$

**37.** $\begin{cases} 2x + y = 10 \\ x + 2y = 10 \end{cases}$

**38.** $\begin{cases} 3x + 2y = 0 \\ 2x - 3y = -13 \end{cases}$

**39.** $\begin{cases} 2x - y = 16 \\ 3x + 2y = 3 \end{cases}$

**40.** $\begin{cases} 3x + 4y = -17 \\ 4x - 3y = -6 \end{cases}$

**41.** $\begin{cases} 4x + 5y = -20 \\ 5x - 4y = -25 \end{cases}$

**42.** $\begin{cases} 3x - 5y = 4 \\ 7x + 3y = 68 \end{cases}$

**43.** $\begin{cases} 6x = -3y \\ 5y = 2x + 12 \end{cases}$

**44.** $\begin{cases} 3y = 4x \\ 5x = 4y - 2 \end{cases}$

**45.** $\begin{cases} 4(2x - y) = 18 \\ 3(x - 3) = 2y - 1 \end{cases}$

**46.** $\begin{cases} 2(2x + 3y) = 5 \\ 8x = 3(1 + 3y) \end{cases}$

**47.** $\begin{cases} \dfrac{3}{5}x + \dfrac{4}{5}y = 1 \\ -\dfrac{1}{4}x + \dfrac{3}{8}y = 1 \end{cases}$

**48.** $\begin{cases} \dfrac{1}{2}x - \dfrac{1}{4}y = 1 \\ \dfrac{1}{3}x + y = 3 \end{cases}$

**49.** $\begin{cases} \dfrac{3}{5}x + y = 1 \\ \dfrac{4}{5}x - y = -1 \end{cases}$

**50.** $\begin{cases} \dfrac{1}{2}x + \dfrac{4}{7}y = -1 \\ 5x - \dfrac{4}{5}y = -10 \end{cases}$

**51.** $\begin{cases} \dfrac{x}{2} - \dfrac{y}{3} = -2 \\ \dfrac{2x - 3}{2} + \dfrac{6y + 1}{3} = \dfrac{17}{6} \end{cases}$

**52.** $\begin{cases} \dfrac{x + 2}{4} + \dfrac{y - 1}{3} = \dfrac{1}{12} \\ \dfrac{x + 4}{5} - \dfrac{y - 2}{2} = \dfrac{5}{2} \end{cases}$

**53.** $\begin{cases} \dfrac{x - 3}{2} + \dfrac{y + 5}{3} = \dfrac{11}{6} \\ \dfrac{x + 3}{3} - \dfrac{5}{12} = \dfrac{y + 3}{4} \end{cases}$

**54.** $\begin{cases} \dfrac{x + 2}{3} = \dfrac{3 - y}{2} \\ \dfrac{x + 3}{2} = \dfrac{2 - y}{3} \end{cases}$

## WRITING

**55.** Why is it usually to your advantage to write the equations of a system in general form before using the addition method to solve it?

**56.** How would you decide whether to use substitution or addition to solve a system of equations?

## SOMETHING TO THINK ABOUT

**57.** If possible, find a solution to the system
$$\begin{cases} x + y = 5 \\ x - y = -3 \\ 2x - y = -2 \end{cases}$$

**58.** If possible, find a solution to the system
$$\begin{cases} x + y = 5 \\ x - y = -3 \\ x - 2y = 0 \end{cases}$$

# 3.6 Applications of Systems of Equations

■ SOLVING PROBLEMS WITH TWO VARIABLES

Getting Ready   *In Problems 1–4, let x and y represent two numbers. Use an algebraic expression to denote each phrase.*

**1.** The sum of $x$ and $y$

**2.** The difference when $y$ is subtracted from $x$

**3.** The product of $x$ and $y$

**4.** The quotient $x$ divided by $y$

**5.** Give the formula for the area of a rectangle.

**6.** Give the formula for the perimeter of a rectangle.

## ■ SOLVING PROBLEMS WITH TWO VARIABLES

We have previously set up equations involving one variable to solve problems. In this section, we consider ways to solve problems by using equations in two variables. The following steps are helpful when solving problems involving two unknown quantities.

> **Problem-Solving Strategy**
>
> **1.** Read the problem several times and *analyze* the facts. Occasionally, a sketch, chart, or diagram will help you visualize the facts of the problem.
>
> **2.** Pick different variables to represent two unknown quantities. *Form two equations* involving each of the two variables. This will give a system of two equations in two variables.
>
> **3.** *Solve the system* using the most convenient method: graphing, substitution, or addition.
>
> **4.** *State the conclusion.*
>
> **5.** *Check the solution* in the words of the problem.

**EXAMPLE 1**

**Farming**    A farmer raises wheat and soybeans on 215 acres. If he wants to plant 31 more acres in wheat than in soybeans, how many acres of each should he plant?

*Analyze the problem*    The farmer plants two fields, one in wheat and one in soybeans. We know that the number of acres of wheat planted plus the number of acres of soybeans planted will equal a total of 215 acres.

*Form two equations*    If $w$ represents the number of acres of wheat and $s$ represents the number of acres of soybeans to be planted, we can form the two equations

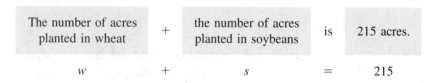

| The number of acres planted in wheat | + | the number of acres planted in soybeans | is | 215 acres. |
|---|---|---|---|---|
| $w$ | + | $s$ | = | 215 |

Since the farmer wants to plant 31 more acres in wheat than in soybeans, we have

| The number of acres planted in wheat | − | the number of acres planted in soybeans | is | 31 acres. |
|---|---|---|---|---|
| $w$ | − | $s$ | = | 31 |

***Solve the system***   We can now solve the system

**1.** $\begin{cases} w + s = 215 \\ w - s = \phantom{0}31 \end{cases}$
**2.**

by the addition method.

$$w + s = 215$$
$$\underline{w - s = \phantom{0}31}$$
$$2w \phantom{+ s} = 246$$
$$w = 123 \qquad \text{Divide both sides by 2.}$$

To find $s$, we substitute 123 for $w$ in Equation 1.

$$w + s = 215$$
$$123 + s = 215 \qquad \text{Substitute 123 for } w.$$
$$s = 92 \qquad \text{Subtract 123 from both sides.}$$

***State the conclusion***   The farmer should plant 123 acres of wheat and 92 acres of soybeans.

***Check the result***   The total acreage planted is $123 + 92$, or 215 acres. The area planted in wheat is 31 acres greater than that planted in soybeans, because $123 - 92 = 31$. The answers check. ∎

EXAMPLE 2

**Lawn care**   An installer of underground irrigation systems wants to cut a 20-foot length of plastic tubing into two pieces. The longer piece is to be 2 feet longer than twice the shorter piece. Find the length of each piece.

***Analyze the problem***   Refer to Figure 3-27, which shows the pipe.

***Form two equations***   We can let $s$ represent the length of the shorter piece and $l$ represent the length of the longer piece. Then we can form the equations

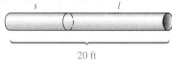

FIGURE 3-27

| The length of the shorter piece | + | the length of the longer piece | is | 20 feet. |
|---|---|---|---|---|
| $s$ | + | $l$ | = | 20 |

Since the longer piece is 2 feet longer than twice the shorter piece, we have

| The length of the longer piece | is | 2 | · | the length of the shorter piece | + | 2 feet. |
|---|---|---|---|---|---|---|
| $l$ | = | 2 | · | $s$ | + | 2 |

*Solve the system*    We can use the substitution method to solve the system

**1.** $\begin{cases} s + l = 20 \\ l = 2s + 2 \end{cases}$
**2.**

$$s + (2s + 2) = 20 \qquad \text{Substitute } 2s + 2 \text{ for } l \text{ in Equation 1.}$$
$$3s + 2 = 20 \qquad \text{Combine like terms.}$$
$$3s = 18 \qquad \text{Subtract 2 from both sides.}$$
$$s = 6 \qquad \text{Divide both sides by 3.}$$

*State the conclusion*    The shorter piece should be 6 feet long. To find the length of the longer piece, we substitute 6 for $s$ in Equation 1 and solve for $l$.

$$s + l = 20$$
$$6 + l = 20 \qquad \text{Substitute 6 for } s.$$
$$l = 14 \qquad \text{Subtract 6 from both sides.}$$

The longer piece should be 14 feet long.

*Check the result*    The sum of 6 and 14 is 20. 14 is 2 more than twice 6. The answers check.    ■

| EXAMPLE 3 |
|---|

**Gardening**    Tom has 150 feet of fencing to enclose a rectangular garden. If the length is to be 5 feet less than 3 times the width, find the area of the garden.

*Analyze the problem*    To find the area of a rectangle, we need to know its length and width. See Figure 3-28.

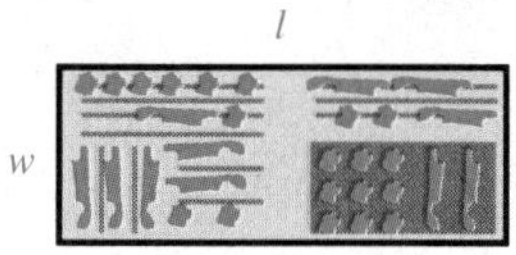

*Form two equations*    We can let $l$ represent the length of the garden and $w$ represent the width. Since the perimeter of a rectangle is two lengths plus two widths, we can form the equations

FIGURE 3-28

| 2 | · | the length of the garden | + | 2 | · | the width of the garden | is | 150 feet. |
|---|---|---|---|---|---|---|---|---|
| 2 | · | $l$ | + | 2 | · | $w$ | = | 150 |

Since the length is 5 feet less than 3 times the width,

| The length of the garden | is | 3 | · | the width of the garden | − | 5 feet. |
|---|---|---|---|---|---|---|
| $l$ | = | 3 | · | $w$ | − | 5 |

*Solve the system*   We can use the substitution method to solve this system.

**1.**  $\begin{cases} 2l + 2w = 150 \\ l = 3w - 5 \end{cases}$
**2.**

$$2(3w - 5) + 2w = 150 \qquad \text{Substitute } 3w - 5 \text{ for } l \text{ in Equation 1.}$$
$$6w - 10 + 2w = 150 \qquad \text{Remove parentheses.}$$
$$8w - 10 = 150 \qquad \text{Combine like terms.}$$
$$8w = 160 \qquad \text{Add 10 to both sides.}$$
$$w = 20 \qquad \text{Divide both sides by 8.}$$

The width of the garden is 20 feet. To find the length, we substitute 20 for $w$ in Equation 2 and simplify.

$$l = 3w - 5$$
$$= 3(20) - 5 \qquad \text{Substitute 20 for } w.$$
$$= 60 - 5$$
$$= 55$$

Since the dimensions of the rectangle are 55 feet by 20 feet, and the area of a rectangle is given by the formula

$$A = l \cdot w \qquad \text{Area = length times width.}$$

we have

$$A = 55 \cdot 20$$
$$= 1,100$$

*State the conclusion*   The garden covers an area of 1,100 square feet.

*Check the result*   Because the dimensions of the garden are 55 feet by 20 feet, the perimeter is

$$P = 2l + 2w$$
$$= 2(55) + 2(20) \qquad \text{Substitute for } l \text{ and } w.$$
$$= 110 + 40$$
$$= 150$$

It is also true that 55 feet is 5 feet less than 3 times 20 feet. The answers check. ∎

---

**EXAMPLE 4**   **Manufacturing**   The setup cost of a machine that mills brass plates is $750. After setup, it costs $0.25 to mill each plate. Management is considering the purchase of a larger machine that can produce the same plate at a cost of $0.20 per plate. If the setup cost of the larger machine is $1,200, how many plates would the company have to produce to make the purchase worthwhile?

*Analyze the problem*   We begin by finding the number of plates (called the **break point**) that will cost equal amounts to produce on either machine.

***Form two equations***   We can let $c$ represent the cost of milling $p$ plates. If we call the machine currently being used machine 1, and the new one machine 2, we can form the two equations

| The cost of making $p$ plates on machine 1 | is | the startup cost of machine 1 | + | the cost per plate on machine 1 | · | the number of plates $p$ to be made. |
|---|---|---|---|---|---|---|
| $c$ | = | 750 | + | 0.25 | · | $p$ |

| The cost of making $p$ plates on machine 2 | is | the startup cost of machine 2 | + | the cost per plate on machine 2 | · | the number of plates $p$ to be made. |
|---|---|---|---|---|---|---|
| $c$ | = | 1,200 | + | 0.20 | · | $p$ |

***Solve the system***   Since the costs at the break point are equal, we can use the substitution method to solve the system

$$\begin{cases} c = 750 + 0.25p \\ c = 1,200 + 0.20p \end{cases}$$

$750 + 0.25p = 1,200 + 0.20p$   Substitute $750 + 0.25p$ for $c$ in the second equation.

$0.25p = 450 + 0.20p$   Subtract 750 from both sides.

$0.05p = 450$   Subtract $0.20p$ from both sides.

$p = 9,000$   Divide both sides by 0.05.

***State the conclusion***   If 9,000 plates are milled, the cost will be the same on either machine. If more than 9,000 plates are milled, the cost will be cheaper on the newer machine, because it mills the plates less expensively than the smaller machine.

***Check the solution***   Figure 3-29 verifies that the break point is 9,000 plates. It also interprets the solution graphically. ∎

**EXAMPLE 5**   **Investing**   Terri and Juan earned $1,150 from a one-year investment of $15,000. If Terri invested some of the money at 8% interest and Juan invested the rest at 7%, how much did each invest?

***Analyze the problem***   We are told that Terri invested an unknown part of the $15,000 at 8% and Juan invested the rest at 7%. Together, these investments earned $1,150.

***Form two equations***   We can let $x$ represent the amount invested by Terri and $y$ represent the amount of money invested by Juan. Because the total investment is $15,000, we have

| The amount invested by Terri | + | the amount invested by Juan | is | $15,000. |
|---|---|---|---|---|
| $x$ | + | $y$ | = | 15,000 |

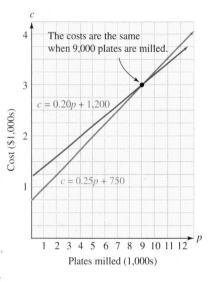

Current machine
$c = 750 + 0.25p$

| $p$ | $c$ |
|---|---|
| 0 | 750 |
| 1,000 | 1,000 |
| 5,000 | 2,000 |

New, larger machine
$c = 1,200 + 0.20p$

| $p$ | $c$ |
|---|---|
| 0 | 1,200 |
| 4,000 | 2,000 |
| 12,000 | 3,600 |

FIGURE 3-29

Since the income on $x$ dollars invested at 8% is $0.08x$, the income on $y$ dollars invested at 7% is $0.07y$, and the combined income is $1,150, we have

| The income on the 8% investment | + | the income on the 7% investment | is | $1,150. |
|---|---|---|---|---|
| $0.08x$ | + | $0.07y$ | = | 1,150 |

Thus, we have the system

**1.**  $\begin{cases} x + y = 15,000 \\ 0.08x + 0.07y = 1,150 \end{cases}$
**2.**

***Solve the system***    To solve the system, we use the addition method.

$$-8x - 8y = -120,000 \qquad \text{Multiply both sides of Equation 1 by } -8.$$
$$\underline{8x + 7y = \phantom{0}115,000} \qquad \text{Multiply both sides of Equation 2 by 100.}$$
$$-y = \phantom{00}-5,000 \qquad \text{Add the equations together.}$$
$$y = 5,000 \qquad \text{Multiply both sides by } -1.$$

To find $x$, we substitute 5,000 for $y$ in Equation 1 and simplify.

$$x + y = 15,000$$
$$x + \mathbf{5,000} = 15,000 \qquad \text{Substitute 5,000 for } y.$$
$$x = 10,000 \qquad \text{Subtract 5,000 from both sides.}$$

***State the conclusion***    Terri invested $10,000, and Juan invested $5,000.

*Check the result*
$10,000 + $5,000 = $15,000     The two investments total $15,000.
0.08($10,000) = $800     Terri earned $800.
0.07($5000) = $350     Juan earned $350.

The combined interest is $800 + $350 = $1,150. The answers check. ∎

EXAMPLE 6

**Boating**  A boat traveled 30 kilometers downstream in 3 hours and made the return trip in 5 hours. Find the speed of the boat in still water.

*Analyze the problem*  Traveling downstream, the speed of the boat will be faster than it would be in still water. Traveling upstream, the speed of the boat will be less than it would be in still water.

*Form two equations*  We can let $s$ represent the speed of the boat in still water and let $c$ represent the speed of the current. Then the rate of speed of the boat while going downstream is $s + c$. The rate of the boat while going upstream is $s - c$. We can organize the information of the problem as in Figure 3-30.

| | Distance = | Rate · | Time |
|---|---|---|---|
| Downstream | 30 | $s + c$ | 3 |
| Upstream | 30 | $s - c$ | 5 |

FIGURE 3-30

Because $d = r \cdot t$, the information in the table gives two equations in two variables.

$$\begin{cases} 30 = 3(s + c) \\ 30 = 5(s - c) \end{cases}$$

After removing parentheses and rearranging terms, we have

1. $\begin{cases} 3s + 3c = 30 \\ 5s - 5c = 30 \end{cases}$
2.

*Solve the system*  To solve this system by addition, we multiply Equation 1 by 5, Equation 2 by 3, add the equations, and solve for $s$.

$$15s + 15c = 150$$
$$\underline{15s - 15c = \phantom{0}90}$$
$$30s \phantom{= 0000} = 240$$
$$s = 8 \qquad \text{Divide both sides by 30.}$$

*State the conclusion*  The speed of the boat in still water is 8 kilometers per hour.

*Check the result*  We leave the check to the reader. ∎

**EXAMPLE 7**    **Medical technology**    A laboratory technician has one batch of antiseptic that is 40% alcohol and a second batch that is 60% alcohol. She would like to make 8 liters of solution that is 55% alcohol. How many liters of each batch should she use?

*Analyze the problem*    Some 60%-alcohol solution must be added to some 40%-alcohol solution to make a 55%-alcohol solution.

*Form two equations*    We can let $x$ represent the number of liters to be used from batch 1, let $y$ represent the number of liters to be used from batch 2, and organize the information of the problem as in Figure 3-31.

| | Fractional part that is alcohol | · | Number of liters of solution | = | Number of liters of alcohol |
|---|---|---|---|---|---|
| **Batch 1** | 0.40 | | $x$ | | $0.40x$ |
| **Batch 2** | 0.60 | | $y$ | | $0.60y$ |
| **Mixture** | 0.55 | | 8 | | $0.55(8)$ |

FIGURE 3-31

The information in Figure 3-31 provides two equations.

**1.**  $x + y = 8$ — The number of liters of batch 1 plus the number of liters of batch 2 equals the total number of liters in the mixture.

**2.**  $0.40x + 0.60y = 0.55(8)$ — The amount of alcohol in batch 1 plus the amount of alcohol in batch 2 equals the amount of alcohol in the mixture.

*Solve the system*    We can use addition to solve this system.

$$
\begin{array}{rl}
-40x - 40y = -320 & \text{Multiply both sides of Equation 1 by } -40. \\
\underline{40x + 60y = \phantom{-}440} & \text{Multiply both sides of Equation 2 by } 100. \\
20y = \phantom{-}120 & \\
y = 6 & \text{Divide both sides by 20.}
\end{array}
$$

To find $x$, we substitute 6 for $y$ in Equation 1 and simplify:

$$
\begin{array}{rl}
x + y = 8 & \\
x + 6 = 8 & \text{Substitute 6 for } y. \\
x = 2 & \text{Subtract 6 from both sides.}
\end{array}
$$

*State the conclusion*    The technician should use 2 liters of the 40% solution and 6 liters of the 60% solution.

*Check the result*    The check is left to the reader.    ■

Orals   *If x and y are integers, express each quantity.*

**1.** Twice $x$                                 **2.** One more than $y$

**3.** The sum of twice $x$ and three times $y$

*If a book costs \$x and a calculator cost \$y, find*

**4.** The cost of 3 books and 2 calculators

**5.** The cost of 4 books and 5 calculators

## EXERCISE 3.6

***REVIEW***   *In Exercises 1–4, graph each inequality.*

**1.** $x < 4$                **2.** $x \geq -3$                **3.** $-1 < x \leq 2$                **4.** $-2 \leq x \leq 0$

*In Exercises 5–8, write each product using exponents.*

**5.** $8 \cdot 8 \cdot 8 \cdot c$        **6.** $4(\pi)(r)(r)$        **7.** $a \cdot a \cdot b \cdot b$        **8.** $(-2)(-2)$

***VOCABULARY AND CONCEPTS***   *Fill in each blank to make a true statement.*

**9.** A _____ is a letter that stands for a number.

**10.** An _____ is a statement indicating that two quantities are equal.

**11.** $\begin{cases} a + b = 20 \\ a = 2b + 4 \end{cases}$ is a _____ of linear equations.

**12.** A _____ of a system of two linear equations satisfies both equations simultaneously.

***PRACTICE***   *In Exercises 13–16, use two equations in two variables to solve each problem.*

**13. Integer problem**   One integer is twice another, and their sum is 96. Find the integers.

**14. Integer problem**   The sum of two integers is 38, and their difference is 12. Find the integers.

**15. Integer problem**   Three times one integer plus another integer is 29. If the first integer plus twice the second is 18, find the integers.

**16. Integer problem**   Twice one integer plus another integer is 21. If the first integer plus 3 times the second is 33, find the integers.

***APPLICATIONS***   *In Exercises 17–54, use two equations in two variables to solve each problem.*

**17. Raising livestock**   A rancher raises five times as many cows as horses. If he has 168 animals, how many cows does he have?

**18. Grass seed mixture**   A landscaper used 100 pounds of grass seed containing twice as much bluegrass as rye. He added 15 more pounds of bluegrass to the mixture before seeding a lawn. How many pounds of bluegrass did he use?

**19. Buying painting supplies**   Two partial receipts for paint supplies appear in Illustration 1. How much did each gallon of paint and each brush cost?

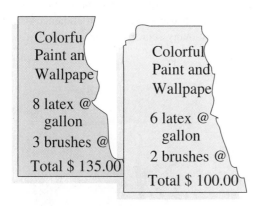

ILLUSTRATION 1

**20. Buying baseball equipment**   One catcher's mitt and ten outfielder's gloves cost $239.50. How much does each cost if one catcher's mitt and five outfielder's gloves cost $134.50?

**21. Buying contact lens cleaner**   Two bottles of contact lens cleaner and three bottles of soaking solution cost $29.40, and three bottles of cleaner and two bottles of soaking solution cost $28.60. Find the cost of each.

**22. Buying clothes**   Two pairs of shoes and four pairs of socks cost $109, and three pairs of shoes and five pairs of socks cost $160. Find the cost of a pair of socks.

**23. Cutting pipe**   A plumber wants to cut the pipe shown in Illustration 2 into two pieces so that one piece is 5 feet longer than the other. How long should each piece be?

25 ft

ILLUSTRATION 2

**24. Cutting lumber**   A carpenter wants to cut a 20-foot board into two pieces so that one piece is 4 times as long as the other. How long should each piece be?

**25. Splitting the lottery**   Maria and Susan pool their resources to buy several lottery tickets. They win $250,000! They agree that Susan should get $50,000 more than Maria, because she gave most of the money. How much will Maria get?

**26. Figuring inheritances**   In his will, a man left his older son $10,000 more than twice as much as he left his younger son. If the estate is worth $497,500, how much did the younger son get?

**27. Television programming**   The producer of a 30-minute documentary about World War I divided it into two parts. Four times as much program time was devoted to the causes of the war as to the outcome. How long was each part of the documentary?

**28. Government**   The salaries of the President and Vice President of the United States total $371,500 a year. If the President makes $28,500 more than the Vice President, find each of their salaries.

**29. Causes of death**   In 1993, the number of Americans dying from cancer was six times the number that died from accidents. If the number of deaths from these two causes totaled 630,000, how many Americans died from each cause?

**30. At the movies**   At an IMAX theater, the giant rectangular movie screen has a width 26 feet less than its length. If its perimeter is 332 feet, find the area of the screen.

**31. Geometry**   The perimeter of the rectangle shown in Illustration 3 is 110 feet. Find its dimensions.

$l = w + 5$

ILLUSTRATION 3

**32. Geometry**   A rectangle is 3 times as long as it is wide, and its perimeter is 80 centimeters. Find its dimensions.

**33. Geometry**   The length of a rectangle is 2 feet more than twice its width. If its perimeter is 34 feet, find its area.

**34. Geometry**   A 50-meter path surrounds the rectangular garden shown in Illustration 4. The width of the garden is two-thirds its length. Find its area.

ILLUSTRATION 4

**35. Choosing a furnace**   A high-efficiency 90+ furnace costs $2,250 and costs an average of $412 per year to operate in Rockford, IL. An 80+ furnace costs only $1,715 but costs $466 per year to operate. Find the break point.

**36. Making tires**   A company has two molds to form tires. One mold has a setup cost of $600 and the other a setup cost of $1,100. The cost to make each tire on the first machine is $15, and the cost per tire on the second machine is $13. Find the break point.

**37. Choosing a furnace**   See Exercise 35. If you intended to live in a house for seven years, which furnace would you choose?

**38. Making tires**   See Exercise 36. If you planned a production run of 500 tires, which mold would you use?

**39. Investing money**   Bill invested some money at 5% annual interest, and Janette invested some at 7%. If their combined interest was $310 on a total investment of $5,000, how much did Bill invest?

**40. Investing money**   Peter invested some money at 6% annual interest, and Martha invested some at 12%. If their combined investment was $6,000 and their combined interest was $540, how much money did Martha invest?

**41. Buying tickets**   Students can buy tickets to a basketball game for $1. The admission for

nonstudents is $2. If 350 tickets are sold and the total receipts are $450, how many student tickets are sold?

**42. Buying tickets**   If receipts for the movie advertised in Illustration 5 were $720 for an audience of 190 people, how many senior citizens attended?

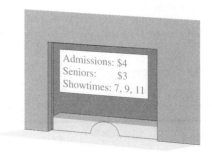

Admissions: $4
Seniors:     $3
Showtimes: 7, 9, 11

ILLUSTRATION 5

**43. Boating**   A boat can travel 24 miles downstream in 2 hours and can make the return trip in 3 hours. Find the speed of the boat in still water.

**44. Aviation**   With the wind, a plane can fly 3,000 miles in 5 hours. Against the same wind, the trip takes 6 hours. Find the airspeed of the plane (the speed in still air).

**45. Aviation**   An airplane can fly downwind a distance of 600 miles in 2 hours. However, the return trip against the same wind takes 3 hours. Find the speed of the wind.

**46. Finding the speed of a current**   It takes a motorboat 4 hours to travel 56 miles down a river, and it takes 3 hours longer to make the return trip. Find the speed of the current.

**47. Mixing chemicals**   A chemist has one solution that is 40% alcohol and another that is 55% alcohol. How much of each must she use to make 15 liters of a solution that is 50% alcohol?

**48. Mixing pharmaceuticals**   A nurse has a solution that is 25% alcohol and another that is 50% alcohol. How much of each must he use to make 20 liters of a solution that is 40% alcohol?

**49. Mixing nuts** A merchant wants to mix the peanuts with the cashews shown in Illustration 6 to get 48 pounds of mixed nuts to sell at $4 per pound. How many pounds of each should the merchant use?

ILLUSTRATION 6

**50. Mixing peanuts and candy** A merchant wants to mix peanuts worth $3 per pound with jelly beans worth $1.50 per pound to make 30 pounds of a mixture worth $2.10 per pound. How many pounds of each should he use?

**51. Selling radios** An electronics store put two types of car radios on sale. One model sold for $87, and the other sold for $119. During the sale, the receipts for the 25 radios sold were $2,495. How many of the less expensive radios were sold?

**52. Selling ice cream** At a store, ice cream cones cost $.90 and sundaes cost $1.65. One day, the receipts for a total of 148 cones and sundaes were $180.45. How many cones were sold?

**53. Investing money** An investment of $950 at one rate of interest and $1,200 at a higher rate together generate an annual income of $205.50. If the investment rates differ by 1%, find the lower rate. (*Hint:* Treat 1% as .01.)

**54. Motion problem** A man drives for a while at 45 mph. Realizing that he is running late, he increases his speed to 60 mph and completes his 405-mile trip in 8 hours. How long does he drive at 45 mph?

## WRITING

**55.** Which problem in the preceding set did you find the hardest? Why?

**56.** Which problem in the preceding set did you find the easiest? Why?

## SOMETHING TO THINK ABOUT

**57.** How many nails will balance one nut in Illustration 7?

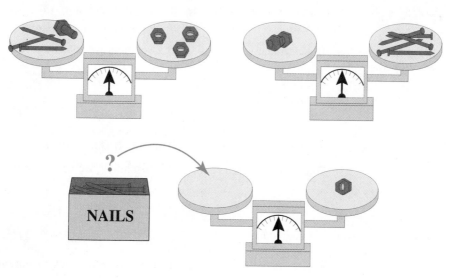

ILLUSTRATION 7

# 3.7 Systems of Linear Inequalities

■ SOLUTIONS OF LINEAR INEQUALITIES ■ GRAPHING LINEAR INEQUALITIES ■ AN APPLICATION OF LINEAR INEQUALITIES ■ SOLVING SYSTEMS OF LINEAR INEQUALITIES ■ AN APPLICATION OF SYSTEMS OF LINEAR INEQUALITIES

**Getting Ready**    *Graph $y = \frac{1}{3}x + 3$ and tell whether the given point lies on the line, above the line, or below the line.*

**1.** $(0, 0)$       **2.** $(0, 4)$       **3.** $(2, 2)$       **4.** $(6, 5)$

**5.** $(-3, 2)$     **6.** $(6, 8)$       **7.** $(-6, 0)$     **8.** $(-9, 5)$

## ■ SOLUTIONS OF LINEAR INEQUALITIES

A **linear inequality** in $x$ and $y$ is an inequality that can be written in one of the following forms:

$$Ax + By > C \qquad Ax + By < C \qquad Ax + By \geq C \qquad Ax + By \leq C$$

where $A$, $B$, and $C$ are real numbers and $A$ and $B$ are not both 0. Some examples of linear inequalities are

$$2x - y > -3 \qquad y < 3 \qquad x + 4y \geq 6 \qquad x \leq -2$$

An ordered pair $(x, y)$ is a solution of an inequality in $x$ and $y$ if a true statement results when the values of $x$ and $y$ are substituted into the inequality.

**EXAMPLE 1**    Determine whether each ordered pair is a solution of $y \geq x - 5$:    **a.** $(4, 2)$    and **b.** $(0, -6)$.

*Solution*    **a.** To determine whether $(4, 2)$ is a solution, we substitute 4 for $x$ and 2 for $y$.

$$y \geq x - 5$$
$$2 \geq 4 - 5$$
$$2 \geq -1$$

Since $2 \geq -1$ is a true inequality, $(4, 2)$ is a solution.

**b.** To determine whether $(0, -6)$ is a solution, we substitute 0 for $x$ and $-6$ for $y$.

$$y \geq x - 5$$
$$-6 \geq 0 - 5$$
$$-6 \geq -5$$

Since $-6 \geq -5$ is a false statement, $(0, -6)$ is not a solution.    ■

Self Check    Use the inequality in Example 1 and determine whether each ordered pair is a solution:    **a.** (8, 2)    and    **b.** (−4, 3).

Answers       **a.** no,    **b.** yes

### ■ GRAPHING LINEAR INEQUALITIES

The graph of $y = x − 5$ is a line consisting of the points whose coordinates satisfy the equation. The graph of the inequality $y \geq x − 5$ is not a line but rather an area bounded by a line, called a **half-plane.** The half-plane consists of the points whose coordinates satisfy the inequality.

EXAMPLE 2      Graph the inequality $y \geq x − 5$.

Solution      Since $y \geq x − 5$ means that $y = x − 5$ or $y > x − 5$, we begin by graphing the equation $y = x − 5$. See Figure 3-32(a).

Because the graph of $y \geq x − 5$ also indicates that $y$ can be greater than $x − 5$, the coordinates of points other than those shown in Figure 3-32(a) satisfy the inequality. For example, the coordinates of the origin satisfy the inequality. We can verify this by letting $x$ and $y$ be 0 in the given inequality:

$$y \geq x − 5$$
$$0 \geq 0 − 5 \qquad \text{Substitute 0 for } x \text{ and 0 for } y.$$
$$0 \geq −5$$

Because $0 \geq −5$ is true, the coordinates of the origin satisfy the original inequality. In fact, the coordinates of every point on the same side of the line as the origin satisfy the inequality. The graph of $y \geq x − 5$ is the half-plane that is shaded in Figure 3-32(b). Since the boundary line $y = x − 5$ is included, we draw it with a solid line.

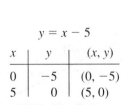

$y = x − 5$

| $x$ | $y$ | $(x, y)$ |
|---|---|---|
| 0 | −5 | $(0, −5)$ |
| 5 | 0 | $(5, 0)$ |

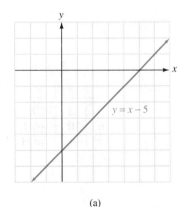

(a)

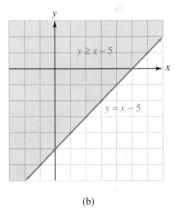

(b)

FIGURE 3-32                                              ■

*Self Check*

Graph $y \geq -x - 2$.

*Answer*

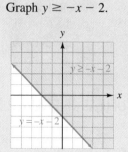

**EXAMPLE 3**

Graph $x + 2y < 6$.

*Solution*    We find the boundary by graphing the equation $x + 2y = 6$. Since the symbol $<$ does not include an $=$ sign, the points on the graph of $x + 2y = 6$ will not be a part of the graph. To show this, we draw the boundary line as a broken line. See Figure 3-33.

To determine which half-plane to shade, we substitute the coordinates of some point that lies on one side of the boundary line into $x + 2y < 6$. The origin is a convenient choice.

$$x + 2y < 6$$
$$0 + 2(0) < 6 \qquad \text{Substitute 0 for } x \text{ and 0 for } y.$$
$$0 < 6$$

Since $0 < 6$ is true, we shade the side of the line that includes the origin. The graph is shown in Figure 3-33.

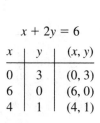

$x + 2y = 6$

| $x$ | $y$ | $(x, y)$ |
|-----|-----|----------|
| 0   | 3   | $(0, 3)$ |
| 6   | 0   | $(6, 0)$ |
| 4   | 1   | $(4, 1)$ |

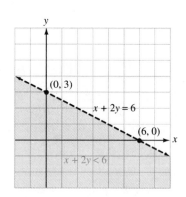

FIGURE 3-33

**Self Check**
**Answer**

Graph $2x - y < 4$.

**EXAMPLE 4**

Graph $y > 2x$.

*Solution*

To find the boundary line, we graph the equation $y = 2x$. Since the symbol $>$ does not include an equal sign, the points on the graph of $y = 2x$ are not a part of the graph of $y > 2x$. To show this, we draw the boundary line as a broken line. See Figure 3-34(a).

To determine which half-plane to shade, we substitute the coordinates of some point that lies on one side of the boundary line into $y > 2x$. Point $T(2, 0)$, for example, is below the boundary line. See Figure 3-34(a). To see if point $T(2, 0)$ satisfies $y > 2x$, we substitute 2 for $x$ and 0 for $y$ in the inequality.

$y > 2x$

$0 > 2(2)$      Substitute 2 for $x$ and 0 for $y$.

$0 > 4$

Since $0 > 4$ is false, the coordinates of point $T$ do not satisfy the inequality, and point $T$ is not on the side of the line we wish to shade. Instead, we shade the other side of the boundary line. The graph of the solution set of $y > 2x$ is shown in Figure 3-34(b).

$y = 2x$

| $x$ | $y$ | $(x, y)$ |
|-----|-----|----------|
| 0 | 0 | $(0, 0)$ |
| $-1$ | $-2$ | $(-1, -2)$ |
| 3 | 6 | $(3, 6)$ |

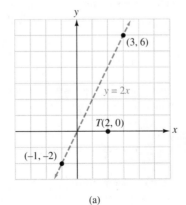

(a)

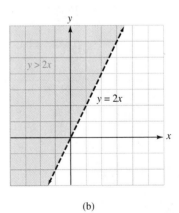

(b)

FIGURE 3-34

*Self Check*
*Answer*

Graph $y < 3x$.

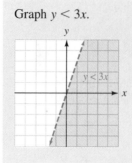

## ■ AN APPLICATION OF LINEAR INEQUALITIES

**EXAMPLE 5**

**Earning money**   Carlos has two part-time jobs, one paying $5 per hour and the other paying $6 per hour. He must earn at least $120 per week to pay his expenses while attending college. Write an inequality that shows the various ways he can schedule his time to achieve his goal.

*Solution*   If we let $x$ represent the number of hours he works on the first job and $y$ the number of hours he works on the second job, we have

| The hourly rate on the first job | · | the hours worked on the first job | + | the hourly rate on the second job | · | the hours worked on the second job | is at least | $120. |
|---|---|---|---|---|---|---|---|---|
| $5 | · | $x$ | + | $6 | · | $y$ | ≥ | $120 |

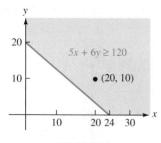

FIGURE 3-35

The graph of the inequality $5x + 6y \geq 120$ is shown in Figure 3-35. Any point in the shaded region indicates a possible way he can schedule his time and earn $120 or more per week. For example, if he works 20 hours on the first job and 10 hours on the second job, he will earn

$$\$5(20) + \$6(10) = \$100 + \$60$$
$$= \$160$$

Since Carlos cannot work a negative number of hours, the graph in the figure has no meaning when either $x$ or $y$ is negative. ■

## ■ SOLVING SYSTEMS OF LINEAR INEQUALITIES

We have seen that the graph of a linear inequality in two variables is a half-plane. Therefore, we would expect the graph of a system of two linear inequalities to be two overlapping half-planes. For example, to solve the system

$$\begin{cases} x + y \geq 1 \\ x - y \geq 1 \end{cases}$$

we graph each inequality and then superimpose the graphs on one set of coordinate axes.

The graph of $x + y \geq 1$ includes the graph of the equation $x + y = 1$ and all points above it. Because the boundary line is included, we draw it with a solid line. See Figure 3-36(a).

The graph of $x - y \geq 1$ includes the graph of the equation $x - y = 1$ and all points below it. Because the boundary line is included, we draw it with a solid line. See Figure 3-36(b).

$x + y = 1$

| $x$ | $y$ | $(x, y)$ |
|---|---|---|
| 0 | 1 | $(0, 1)$ |
| 1 | 0 | $(1, 0)$ |
| 2 | $-1$ | $(2, -1)$ |

$x - y = 1$

| $x$ | $y$ | $(x, y)$ |
|---|---|---|
| 0 | $-1$ | $(0, -1)$ |
| 1 | 0 | $(1, 0)$ |
| 2 | 1 | $(2, 1)$ |

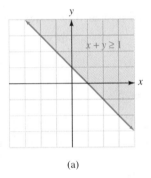

(a)

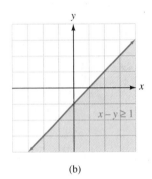

(b)

FIGURE 3-36

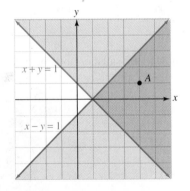

FIGURE 3-37

In Figure 3-37, we show the result when the graphs are superimposed on one coordinate system. The area that is shaded twice represents the set of solutions of the given system. Any point in the doubly shaded region has coordinates that satisfy both of the inequalities.

To see that this is true, we can pick a point, such as point $A$, that lies in the doubly shaded region and show that its coordinates satisfy both inequalities. Because point $A$ has coordinates $(4, 1)$, we have

$$x + y \geq 1 \qquad \text{and} \qquad x - y \geq 1$$
$$4 + 1 \geq 1 \qquad \qquad \qquad 4 - 1 \geq 1$$
$$5 \geq 1 \qquad \qquad \qquad \quad 3 \geq 1$$

Since the coordinates of point $A$ satisfy each equation, point $A$ is a solution. If we pick a point that is not in the doubly shaded region, its coordinates will not satisfy both of the inequalities.

In general, to solve systems of linear inequalities, we will take the following steps.

## Solving Systems of Inequalities

**1.** Graph each inequality in the system on the same coordinate axes.

**2.** Find the region where the graphs overlap.

**3.** Pick a test point from the region to verify the solution.

**EXAMPLE 6**    Graph the solution set of $\begin{cases} 2x + y < 4 \\ -2x + y > 2 \end{cases}$.

*Solution*    We graph each inequality on one set of coordinate axes, as in Figure 3-38.

- The graph of $2x + y < 4$ includes all points below the line $2x + y = 4$. Since the boundary is not included, we draw it as a broken line.
- The graph of $-2x + y > 2$ includes all points above the line $-2x + y = 2$. Since the boundary is not included, we draw it as a broken line.

The area that is shaded twice represents the set of solutions of the given system.

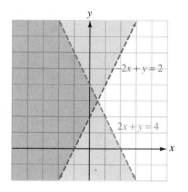

| $2x + y = 4$ | | |
|---|---|---|
| $x$ | $y$ | $(x, y)$ |
| 0 | 4 | $(0, 4)$ |
| 1 | 2 | $(1, 2)$ |
| 2 | 0 | $(2, 0)$ |

| $-2x + y = 2$ | | |
|---|---|---|
| $x$ | $y$ | $(x, y)$ |
| $-1$ | 0 | $(-1, 0)$ |
| 0 | 2 | $(0, 2)$ |
| 2 | 6 | $(2, 6)$ |

FIGURE 3-38

Pick a point in the doubly shaded region and show that it satisfies both inequalities. ■

*Self Check*    Graph the solution of $\begin{cases} x + 3y < 6 \\ -x + 3y < 6 \end{cases}$.

*Answer*

**EXAMPLE 7**  Graph the solution set of $\begin{cases} x \le 2 \\ y > 3 \end{cases}$.

*Solution*  We graph each inequality on one set of coordinate axes, as in Figure 3-39.

- The graph of $x \le 2$ includes all points on the line $x = 2$ and all points to the left of the line. Since the boundary line is included, we draw it as a solid line.
- The graph $y > 3$ includes all points above the line $y = 3$. Since the boundary is not included, we draw it as a broken line.

The area that is shaded twice represents the set of solutions of the given system.

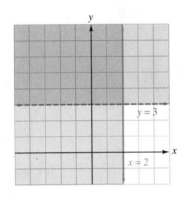

| $x = 2$ | | |
|---|---|---|
| $x$ | $y$ | $(x, y)$ |
| 2 | 0 | (2, 0) |
| 2 | 2 | (2, 2) |
| 2 | 4 | (2, 4) |

| $y = 3$ | | |
|---|---|---|
| $x$ | $y$ | $(x, y)$ |
| 0 | 3 | (0, 3) |
| 1 | 3 | (1, 3) |
| 4 | 3 | (4, 3) |

FIGURE 3-39

Pick a point in the doubly shaded region and show that this is true.  ■

**Self Check**  Solve $\begin{cases} y \ge 1 \\ x > 2 \end{cases}$.

*Answer*

**EXAMPLE 8**  Graph the solution set of the system $\begin{cases} y < 3x - 1 \\ y \ge 3x + 1 \end{cases}$.

*Solution*  We graph each inequality, as in Figure 3-40.

- The graph of $y < 3x - 1$ includes all of the points below the broken line $y = 3x - 1$.

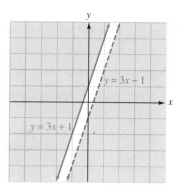

FIGURE 3-40

- The graph of $y \geq 3x + 1$ includes all of the points on and above the solid line $y = 3x + 1$.

Since the graphs of these inequalities do not intersect, there are no solutions. ∎

**Self Check**  Solve $\begin{cases} y \geq -\frac{1}{2}x + 1 \\ y \leq -\frac{1}{2}x - 1 \end{cases}$.

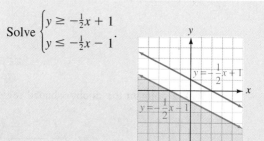

*Answer*  no solutions

## ■ AN APPLICATION OF SYSTEMS OF LINEAR INEQUALITIES

**EXAMPLE 9**

**Landscaping**  A homeowner budgets from \$300 to \$600 for trees and bushes to landscape his yard. After shopping around, he finds that good trees cost \$150 and mature bushes cost \$75. What combinations of trees and bushes can he afford to buy?

*Analyze the problem*  The homeowner wants to spend *at least* \$300 but *not more than* \$600 for trees and bushes.

*Form two inequalities*  We can let $x$ represent the number of trees purchased and $y$ the number of bushes purchased. We can then form the following system of inequalities.

| The cost of a tree | · | the number of trees purchased | + | the cost of a bush | · | the number of bushes purchased | should be at least | $300. |
|---|---|---|---|---|---|---|---|---|
| $150 | · | $x$ | + | $75 | · | $y$ | $\geq$ | $300 |

| The cost of a tree | · | the number of trees purchased | + | the cost of a bush | · | the number of bushes purchased | should not be more than | $600. |
|---|---|---|---|---|---|---|---|---|
| $150 | · | $x$ | + | $75 | · | $y$ | $\leq$ | $600 |

***Solve the system***   We graph the system

$$\begin{cases} 150x + 75y \geq 300 \\ 150x + 75y \leq 600 \end{cases}$$

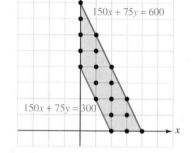

FIGURE 3-41

as in Figure 3-41. The coordinates of each point shown in the graph give a possible combination of the number of trees ($x$) and the number of bushes ($y$) that can be purchased. These possibilities are

$(0, 4)$, $(0, 5)$, $(0, 6)$, $(0, 7)$, $(0, 8)$

$(1, 2)$, $(1, 3)$, $(1, 4)$, $(1, 5)$, $(1, 6)$

$(2, 0)$, $(2, 1)$, $(2, 2)$, $(2, 3)$, $(2, 4)$

$(3, 0)$, $(3, 1)$, $(3, 2)$, $(4, 0)$

Only these points can be used, because the homeowner cannot buy part of a tree or part of a bush.   ∎

Orals   *Tell whether the following coordinates satisfy $y > 3x + 2$.*

**1.** $(0, 0)$        **2.** $(5, 5)$        **3.** $(-2, 4)$        **4.** $(-3, -6)$

*Tell whether the following coordinates satisfy the inequality $y \leq \frac{1}{2}x - 1$.*

**5.** $(0, 0)$        **6.** $(2, 0)$        **7.** $(4, 3)$        **8.** $(-4, -3)$

# EXERCISE 3.7

**REVIEW**

**1.** Solve $3x + 5 = 14$.

**2.** Solve $2(x - 4) \leq -12$.

**3.** Solve $A = P + Prt$ for $t$.

**4.** Does the graph of the line $y = -x$ pass through the origin?

*Simplify each expression.*

**5.** $2a + 5(a - 3)$

**6.** $2t - 3(3 + t)$

**7.** $4(b - a) + 3b + 2a$

**8.** $3p + 2(q - p) + q$

***VOCABULARY AND CONCEPTS*** *Fill in each blank to make a true statement.*

**9.** $2x - y \leq 4$ is a linear _____ in $x$ and $y$.

**10.** The symbol $\leq$ means _____ or _____.

**11.** In the accompanying graph, the line $2x - y = 4$ is the _____ of the graph $2x - y \leq 4$.

**12.** In the accompanying graph, the line $2x - y = 4$ divides the rectangular coordinate system into two _____.

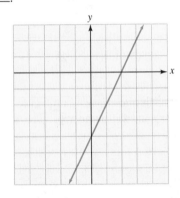

**13.** $\begin{cases} x + y > 2 \\ x + y < 4 \end{cases}$ This is a system of linear _____.

**14.** The _____ of a system of linear inequalities are all the ordered pairs that make all of the inequalities of the system true at the same time.

**15.** Any point in the _____ region of the graph of the solution of a system of two linear inequalities has coordinates that satisfy both of the inequalities of the system.

**16.** To graph a linear inequality such as $x + y > 2$, first graph the boundary. Then pick a test _____ to determine which half-plane to shade.

*Answer each question.*

**17.** Tell whether each ordered pair is a solution of $5x - 3y \geq 0$.
  **a.** $(1, 1)$
  **b.** $(-2, -3)$
  **c.** $(0, 0)$
  **d.** $\left(\dfrac{1}{5}, \dfrac{4}{3}\right)$

**18.** Tell whether each ordered pair is a solution of $x + 4y < -1$.
  **a.** $(3, 1)$
  **b.** $(-2, 0)$
  **c.** $(0.5, 0.2)$
  **d.** $\left(-2, \dfrac{1}{4}\right)$

**19.** Tell whether the graph of each linear inequality includes the boundary line.
  **a.** $y > -x$
  **b.** $5x - 3y \leq -2$

**20.** If a false statement results when the coordinates of a test point are substituted into a linear inequality, which half-plane should be shaded to represent the solution of the inequality?

***PRACTICE*** *In Exercises 21–28, complete the graph by shading the correct half-plane.*

**21.** $y \leq x + 2$

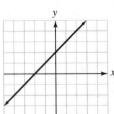

**22.** $y > x - 3$

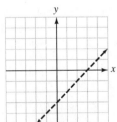

**23.** $y > 2x - 4$

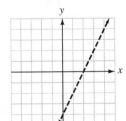

**24.** $y \leq -x + 1$

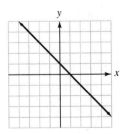

**25.** $x - 2y \le 4$

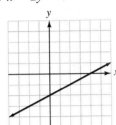

**26.** $3x + 2y \ge 12$

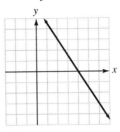

**27.** $y \le 4x$

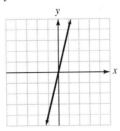

**28.** $y + 2x < 0$

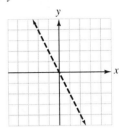

*In Exercises 29–44, graph each inequality.*

**29.** $y \ge 3 - x$

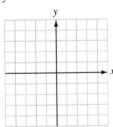

**30.** $y < 2 - x$

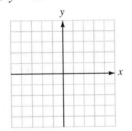

**31.** $y < 2 - 3x$

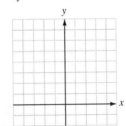

**32.** $y \ge 5 - 2x$

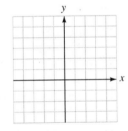

**33.** $y \ge 2x$

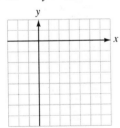

**34.** $y < 3x$

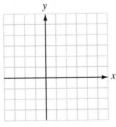

**35.** $2y - x < 8$

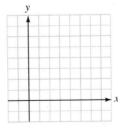

**36.** $y + 9x \ge 3$

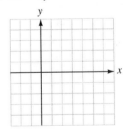

**37.** $3x - 4y > 12$

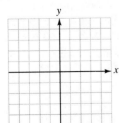

**38.** $4x + 3y \le 12$

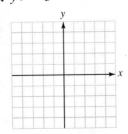

**39.** $5x + 4y \ge 20$

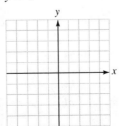

**40.** $7x - 2y < 21$

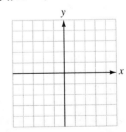

**41.** $x < 2$

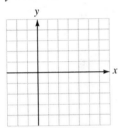

**42.** $y > -3$

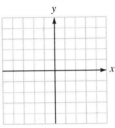

**43.** $y \le 1$

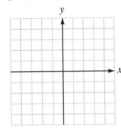

**44.** $x \ge -4$

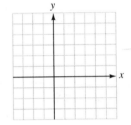

*APPLICATIONS* *In Exercises 45–50, graph each inequality for nonnegative values of x and y. Then give some ordered pairs that satisfy the inequality.*

**45. Production planning** It costs a bakery $3 to make a cake and $4 to make a pie. Production costs cannot exceed $120 per day. Find an inequality that shows the possible combinations of cakes (*x*) and pies (*y*) that can be made, and graph it in Illustration 1.

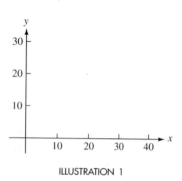

ILLUSTRATION 1

**46. Hiring baby sitters** Mary has a choice of two babysitters. Sitter 1 charges $6 per hour, and sitter 2 charges $7 per hour. Mary can afford no more than $42 per week for sitters. Find an inequality that shows the possible ways that she can hire sitter 1 (*x*) and sitter 2 (*y*), and graph it in Illustration 2.

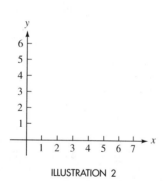

ILLUSTRATION 2

**47. Inventory** A clothing store advertises that it maintains an inventory of at least $4,400 worth of men's jackets. A leather jacket costs $100, and a nylon jacket costs $88. Find an inequality that shows

the possible ways that leather jackets (*x*) and nylon jackets (*y*) can be stocked, and graph it in Illustration 3.

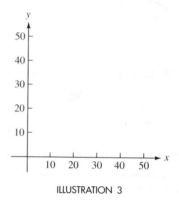

ILLUSTRATION 3

**48. Making sporting goods** To keep up with demand, a sporting goods manufacturer allocates at least 2,400 units of time per day to make baseballs and footballs. It takes 20 units of time to make a baseball and 30 units of time to make a football. Find an inequality that shows the possible ways to schedule the time to make baseballs (*x*) and footballs (*y*), and graph it in Illustration 4.

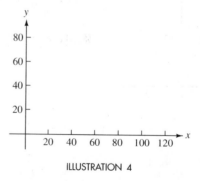

ILLUSTRATION 4

**49. Investing** Robert has up to $8,000 to invest in two companies. Stock in Robotronics sells for $40 per share, and stock in Macrocorp sells for $50 per share. Find an inequality that shows the possible ways that he can buy shares of Robotronics (*x*) and Macrocorp (*y*), and graph it in Illustration 5.

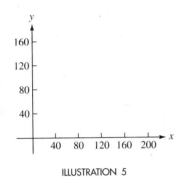

ILLUSTRATION 5

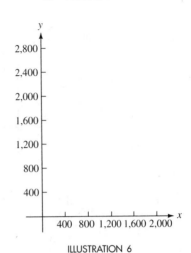

ILLUSTRATION 6

**50. Buying tickets**   Tickets to the Rockford Rox baseball games cost $6 for reserved seats and $4 for general admission. Nightly receipts must average at least $10,200 to meet expenses. Find an inequality that shows the possible ways that the Rox can sell reserved seats (*x*) and general admission tickets (*y*), and graph it in Illustration 6.

**PRACTICE**   *In Exercises 51–68, find the solution set of each system of inequalities, when possible.*

**51.** $\begin{cases} x + 2y \le 3 \\ 2x - y \ge 1 \end{cases}$

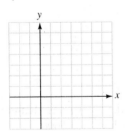

**52.** $\begin{cases} 2x + y \ge 3 \\ x - 2y \le -1 \end{cases}$

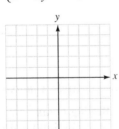

**53.** $\begin{cases} x + y < -1 \\ x - y > -1 \end{cases}$

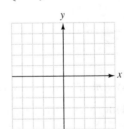

**54.** $\begin{cases} x + y > 2 \\ x - y < -2 \end{cases}$

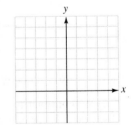

**55.** $\begin{cases} 2x - y < 4 \\ x + y \ge -1 \end{cases}$

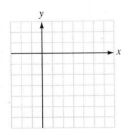

**56.** $\begin{cases} x - y \ge 5 \\ x + 2y < -4 \end{cases}$

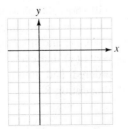

**57.** $\begin{cases} x > 2 \\ y \le 3 \end{cases}$

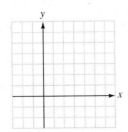

**58.** $\begin{cases} x \ge -1 \\ y > -2 \end{cases}$

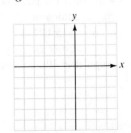

**59.** $\begin{cases} x + y < 1 \\ x + y > 3 \end{cases}$

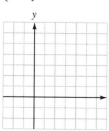

**60.** $\begin{cases} x \le 0 \\ y < 0 \end{cases}$

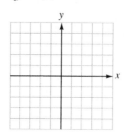

**61.** $\begin{cases} 3x + 4y > -7 \\ 2x - 3y \ge 1 \end{cases}$

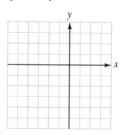

**62.** $\begin{cases} 3x + y \le 1 \\ 4x - y > -8 \end{cases}$

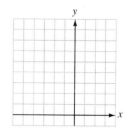

**63.** $\begin{cases} 2x - 4y > -6 \\ 3x + y \ge 5 \end{cases}$

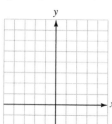

**64.** $\begin{cases} 2x - 3y < 0 \\ 2x + 3y \ge 12 \end{cases}$

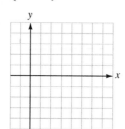

**65.** $\begin{cases} 3x - y \le -4 \\ 3y > -2(x + 5) \end{cases}$

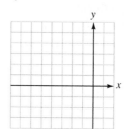

**66.** $\begin{cases} 3x + y < -2 \\ y > 3(1 - x) \end{cases}$

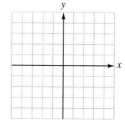

**67.** $\begin{cases} \dfrac{x}{2} + \dfrac{y}{3} \ge 2 \\ \dfrac{x}{2} - \dfrac{y}{2} < -1 \end{cases}$

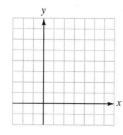

**68.** $\begin{cases} \dfrac{x}{3} - \dfrac{y}{2} < -3 \\ \dfrac{x}{3} + \dfrac{y}{2} > -1 \end{cases}$

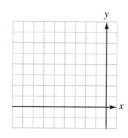

*In Exercises 69–72, graph each system of inequalities and give two possible solutions to each problem.*

**69. Buying compact discs**   Melodic Music has compact discs on sale for either $10 or $15. A customer wants to spend at least $30 but no more than $60 on CDs. Find a system of inequalities whose graph will show the possible combinations of $10 CDs (*x*) and $15 CDs (*y*) that the customer can buy, and graph it in Illustration 7.

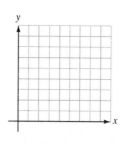

ILLUSTRATION 7

**70. Buying boats**   Dry Boatworks wholesales aluminum boats for $800 and fiberglass boats for $600. Northland Marina wants to order at least $2,400 but no more than $4,800 worth of boats. Find a system of inequalities whose graph will show the possible combinations of aluminum boats (*x*) and fiberglass boats (*y*) that can be ordered, and graph it in Illustration 8.

ILLUSTRATION 8

**71. Buying furniture**   A distributor wholesales desk chairs for $150 and side chairs for $100. Best Furniture wants to order no more than $900 worth of chairs and wants to order more side chairs than desk chairs. Find a system of inequalities whose graph will show the possible combinations of desk chairs ($x$) and side chairs ($y$) that can be ordered, and graph it in Illustration 9.

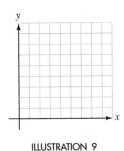

ILLUSTRATION 9

**72. Ordering furnace equipment**   J. Bolden Heating Company wants to order no more than $2,000 worth of electronic air cleaners and humidifiers from a wholesaler that charges $500 for aircleaners and $200 for humidifiers. Bolden wants more humidifiers than air cleaners. Find a system of inequalities whose graph will show the possible combinations of air cleaners ($x$) and humidifiers ($y$) that can be ordered, and graph it in Illustration 10.

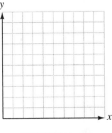

ILLUSTRATION 10

## WRITING

**73.** Explain how to find the boundary for the graph of an inequality.

**75.** Explain how to use graphing to solve a system of inequalities.

**74.** Explain how to decide which side of the boundary line to shade.

**76.** Explain when a system of inequalities will have no solutions.

## SOMETHING TO THINK ABOUT

**77.** What are some limitations of the graphing method for solving inequalities?

**79.** Can a system of inequalities have
   **a.** no solutions?
   **b.** exactly one solution?
   **c.** infinitely many solutions?

**78.** Graph $y = 3x + 1$, $y < 3x + 1$, and $y > 3x + 1$. What do you discover?

**80.** Find a system of two inequalities that has a solution of $(2, 0)$ but no solutions of the form $(x, y)$ where $y < 0$.

■ ■ ■ ■ ■ ■ ■ ■ ■ ■

**MATHEMATICS IN ECONOMICS**   Demand will equal supply at the point where the graphs of the demand equation and the supply equation intersect. We can find this point by solving the following system.

$$\begin{cases} p = -\dfrac{1}{2}q + 1{,}300 & \text{The demand equation.} \\[2mm] p = \dfrac{1}{3}q + \dfrac{1{,}400}{3} & \text{The supply equation.} \end{cases}$$

If we substitute the expression for $p$ in the first equation for $p$ in the second equation, we have

$$-\frac{1}{2}q + 1{,}300 = \frac{1}{3}q + \frac{1{,}400}{3}$$

$$-3q + 7{,}800 = 2q + 2{,}800 \qquad \text{Multiply both sides by 6.}$$

$$5{,}000 = 5q \qquad \text{Subtract 2,800 from both sides and add } 3q \text{ to both sides.}$$

$$1{,}000 = q \qquad \text{Divide both sides by 5.}$$

The quantity $q$ that will determine the equilibrium price is 1,000 canoes. If we substitute 1,000 for $q$ in either of the original equations, we get

$$p = \frac{1}{3}q + \frac{1{,}400}{3}$$

$$p = \frac{1}{3}(1{,}000) + \frac{1{,}400}{3}$$

$$p = \frac{2{,}400}{3}$$

$$p = 800$$

The equilibrium price is \$800. When 1,000 canoes are manufactured each month and priced at \$800, supply will equal demand.

# ■ ■ ■ ■ ■ ■ ■ ■ ■ ■ **PROJECT**

The graphing method of solving a system of equations is not as accurate as algebraic methods, and some systems are more difficult than others to solve accurately. For example, the two lines in Illustration 1(a) could be drawn carelessly, and the point of intersection would not be far from the correct location. If the lines in Illustration 1(b) were drawn carelessly, the point of intersection could move substantially from its correct location.

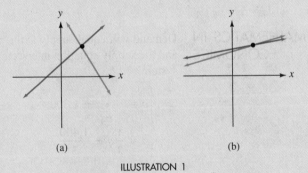

(a)                              (b)

ILLUSTRATION 1

- Carefully solve each of these systems of equations graphically (by hand, not with a graphing calculator). Indicate your best estimate of the solution of each system.

$$\begin{cases} 2x - 4y = -7 \\ 4x + 2y = 11 \end{cases} \qquad \begin{cases} 5x - 4y = -1 \\ 12x - 10y = -3 \end{cases}$$

- Solve each system algebraically. How close were your graphical solutions to the actual solutions? Write a paragraph explaining any differences.
- Create a system of equations with the solutions $x = 3$, $y = 2$ for which an accurate solution could be obtained graphically.
- Create a system of equations with the solutions $x = 3$, $y = 2$ that is more difficult to solve accurately than the previous system, and write a paragraph explaining why.

# CHAPTER SUMMARY

## CONCEPTS

## REVIEW EXERCISES

### SECTION 3.1
### The Rectangular Coordinate System

Any ordered pair of real numbers represents a point on the rectangular coordinate system.

**1.** Plot each point on the rectangular coordinate system in Illustration 1.
 **a.** $A(1, 3)$          **b.** $B(1, -3)$
 **c.** $C(-3, 1)$        **d.** $D(-3, -1)$
 **e.** $E(0, 5)$          **f.** $F(-5, 0)$

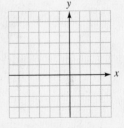

ILLUSTRATION 1

**2.** Find the coordinates of each point in Illustration 2.
 **a.** $A$          **b.** $B$
 **c.** $C$          **d.** $D$
 **e.** $E$          **f.** $F$
 **g.** $G$          **h.** $H$

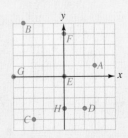

ILLUSTRATION 2

| SECTION 3.2 | *Graphing Linear Equations* |

An ordered pair of real numbers is a **solution** if it satisfies the equation.

To graph a linear equation,
1. Find three pairs $(x, y)$ that satisfy the equation.
2. Plot each pair on the rectangular coordinate system.
3. Draw a line passing through the three points.

**3.** Tell whether each pair satisfies the equation $3x - 4y = 12$.

**a.** $(2, 1)$

**b.** $\left(3, -\dfrac{3}{4}\right)$

**4.** Graph each equation on a rectangular coordinate system.

**a.** $y = x - 5$

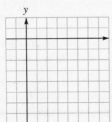

**b.** $y = 2x + 1$

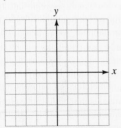

**c.** $y = \dfrac{x}{2} + 2$

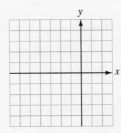

**d.** $y = 3$

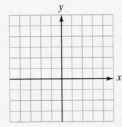

**e.** $x + y = 4$

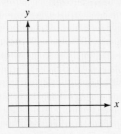

**f.** $x - y = -3$

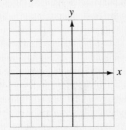

**g.** $3x + 5y = 15$

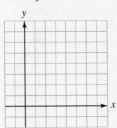

**h.** $7x - 4y = 28$

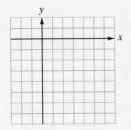

| **SECTION 3.3** | *Solving Systems of Equations by Graphing* |
|---|---|

To solve a system of equations graphically, carefully graph each equation of the system. If the lines intersect, the coordinates of the point of intersection give the solution of the system.

**5.** Tell whether the ordered pair is a solution of the system.

**a.** $(1, 5)$, $\begin{cases} 3x - y = -2 \\ 2x + 3y = 17 \end{cases}$      **b.** $(-2, 4)$, $\begin{cases} 5x + 3y = 2 \\ -3x + 2y = 16 \end{cases}$

**c.** $\left(14, \dfrac{1}{2}\right)$, $\begin{cases} 2x + 4y = 30 \\ \dfrac{x}{4} - y = 3 \end{cases}$      **d.** $\left(\dfrac{7}{2}, -\dfrac{2}{3}\right)$, $\begin{cases} 4x - 6y = 18 \\ \dfrac{x}{3} + \dfrac{y}{2} = \dfrac{5}{6} \end{cases}$

**6.** Use the graphing method to solve each system.

**a.** $\begin{cases} x + y = 7 \\ 2x - y = 5 \end{cases}$      **b.** $\begin{cases} \dfrac{x}{3} + \dfrac{y}{5} = -1 \\ x - 3y = -3 \end{cases}$

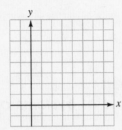
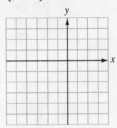

**c.** $\begin{cases} 3x + 6y = 6 \\ x + 2y = 2 \end{cases}$      **d.** $\begin{cases} 6x + 3y = 12 \\ 2x + y = 2 \end{cases}$

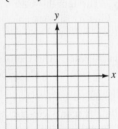
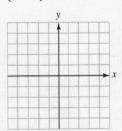

| **SECTION 3.4** | *Solving Systems of Equations by Substitution* |
|---|---|

To solve a system of equations by substitution, solve one of the equations of the system for one of its variables, substitute the resulting expression into the other equation, and solve for the other variable.

**7.** Use the substitution method to solve each system.

**a.** $\begin{cases} x = 3y + 5 \\ 5x - 4y = 3 \end{cases}$      **b.** $\begin{cases} 3x - \dfrac{2y}{5} = 2(x - 2) \\ 2x - 3 = 3 - 2y \end{cases}$

**c.** $\begin{cases} 8x + 5y = 3 \\ 5x - 8y = 13 \end{cases}$      **d.** $\begin{cases} 6(x + 2) = y - 1 \\ 5(y - 1) = x + 2 \end{cases}$

## SECTION 3.5 — Solving Systems of Equations by Addition

To solve a system of equations by addition, first multiply one or both of the equations by suitable constants, if necessary, to eliminate one of the variables when the equations are added. The equation that results can be solved for its single variable. Then substitute the value obtained back into one of the original equations and solve for the other variable.

**8.** Use the addition method to solve each system.

a. $\begin{cases} 2x + y = 1 \\ 5x - y = 20 \end{cases}$

b. $\begin{cases} x + 8y = 7 \\ x - 4y = 1 \end{cases}$

c. $\begin{cases} 5x + y = 2 \\ 3x + 2y = 11 \end{cases}$

d. $\begin{cases} x + y = 3 \\ 3x = 2 - y \end{cases}$

e. $\begin{cases} 11x + 3y = 27 \\ 8x + 4y = 36 \end{cases}$

f. $\begin{cases} 9x + 3y = 5 \\ 3x = 4 - y \end{cases}$

g. $\begin{cases} 9x + 3y = 5 \\ 3x + y = \dfrac{5}{3} \end{cases}$

h. $\begin{cases} \dfrac{x}{3} + \dfrac{y + 2}{2} = 1 \\ \dfrac{x + 8}{8} + \dfrac{y - 3}{3} = 0 \end{cases}$

## SECTION 3.6 — Applications of Systems of Equations

Systems of equations are useful in solving many different types of problems.

**9. Integer problem**  One number is 5 times another, and their sum is 18. Find the numbers.

**10. Geometry**  The length of a rectangle is 3 times its width, and its perimeter is 24 feet. Find its dimensions.

**11. Buying grapefruit**  A grapefruit costs 15 cents more than an orange. Together, they cost 85 cents. Find the cost of a grapefruit.

**12. Utility bills**  A man's electric bill for January was $23 less than his gas bill. The two utilities cost him a total of $109. Find the amount of his gas bill.

**13. Buying groceries**  Two gallons of milk and 3 dozen eggs cost $6.80. Three gallons of milk and 2 dozen eggs cost $7.35. How much does each gallon of milk cost?

**14. Investing money**  Carlos invested part of $3,000 in a 10% certificate of deposit account and the rest in a 6% passbook account. If the total annual interest from both accounts is $270, how much did he invest at 6%?

| SECTION 3.7 | *Systems of Linear Inequalities* |

To graph a system of inequalities, first graph the individual inequalities of the system. The final solution, if one exists, is that region where all the individual graphs intersect.

**15.** Graph each inequality.

**a.** $y \geq x + 2$

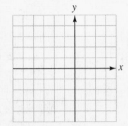

**b.** $x < 3$

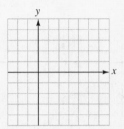

**16.** Solve each system of inequalities.

**a.** $\begin{cases} 5x + 3y < 15 \\ 3x - y > 3 \end{cases}$

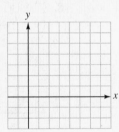

**b.** $\begin{cases} 5x - 3y \geq 5 \\ 3x + 2y \geq 3 \end{cases}$

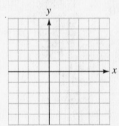

**c.** $\begin{cases} x \geq 3y \\ y < 3x \end{cases}$

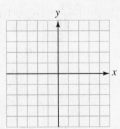

**d.** $\begin{cases} x > 0 \\ x \leq 3 \end{cases}$

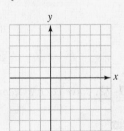

# ■ Chapter Test

*In Problems 1–4, graph each equation.*

**1.** $y = \dfrac{x}{2} + 1$   **2.** $2(x + 1) - y = 4$   **3.** $x = 1$   **4.** $2y = 8$

 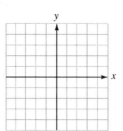 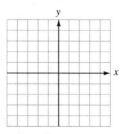 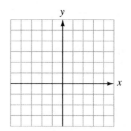

*In Problems 5–6, tell whether the given ordered pair is a solution of the given system.*

**5.** $(2, -3),\ \begin{cases} 3x - 2y = 12 \\ 2x + 3y = -5 \end{cases}$

**6.** $(-2, -1),\ \begin{cases} 4x + y = -9 \\ 2x - 3y = -7 \end{cases}$

*In Problems 7–8, solve each system by graphing.*

**7.** $\begin{cases} 3x + y = 7 \\ x - 2y = 0 \end{cases}$

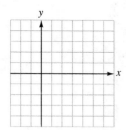

**8.** $\begin{cases} x + \dfrac{y}{2} = 1 \\ y = 1 - 3x \end{cases}$

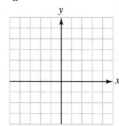

*In Problems 9–10, solve each system by substitution.*

**9.** $\begin{cases} y = x - 1 \\ x + y = -7 \end{cases}$

**10.** $\begin{cases} \dfrac{x}{6} + \dfrac{y}{10} = 3 \\ \dfrac{5x}{16} - \dfrac{3y}{16} = \dfrac{15}{8} \end{cases}$

*In Problems 11–12, solve each system by addition.*

**11.** $\begin{cases} 3x - y = 2 \\ 2x + y = 8 \end{cases}$

**12.** $\begin{cases} 4x + 3 = -3y \\ \dfrac{-x}{7} + \dfrac{4y}{21} = 1 \end{cases}$

*In Problems 13–14, classify each system as consistent or inconsistent.*

13. $\begin{cases} 2x + 3(y - 2) = 0 \\ -3y = 2(x - 4) \end{cases}$

14. $\begin{cases} \dfrac{x}{3} + y - 4 = 0 \\ -3y = x - 12 \end{cases}$

*In Problems 15–16, use a system of equations in two variables to solve each problem.*

15. The sum of two numbers is $-18$. One number is 2 greater than 3 times the other. Find the product of the numbers.

16. A woman invested some money at 8% and some at 9%. The interest on the combined investment of $10,000 was $840. How much was invested at 9%?

*In Problems 17–18, solve each system of inequalities by graphing.*

17. $\begin{cases} x + y < 3 \\ x - y < 1 \end{cases}$

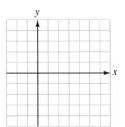

18. $\begin{cases} 2x + 3y \le 6 \\ x \ge 2 \end{cases}$

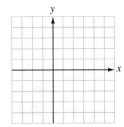

# 4  Polynomials

MATHEMATICS
IN MEDICINE

The red cells of our blood pick up oxygen in the lungs and carry it to all parts of the body. Each red cell is a tiny disc with an approximate radius of 0.00015 inch. Because the amount of oxygen carried depends on the surface area of the cells, and the cells are so tiny, a very great number is needed—25 trillion in an average adult.

What is the total surface area of all the red blood cells in the body?

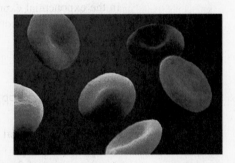

After reading this chapter, you will be able to answer this question.

## 4.1 Natural-Number Exponents

■ EXPONENTS ■ POWERS OF EXPRESSIONS ■ THE PRODUCT RULE FOR EXPONENTS ■ THE POWER RULES FOR EXPONENTS ■ THE QUOTIENT RULE FOR EXPONENTS

Getting Ready    *Evaluate each expression.*

**1.** $2^3$ **2.** $3^2$ **3.** $3(2)$ **4.** $2(3)$

**5.** $2^3 + 2^2$ **6.** $2^3 \cdot 2^2$ **7.** $3^3 - 3^2$ **8.** $\dfrac{3^3}{3^2}$

### ■ EXPONENTS

We have used natural-number exponents to indicate repeated multiplication. For example,

$$2^5 = 2 \cdot 2 \cdot 2 \cdot 2 \cdot 2 = 32 \qquad\qquad (-7)^3 = (-7)(-7)(-7) = -343$$
$$x^4 = x \cdot x \cdot x \cdot x \qquad\qquad\qquad -y^5 = -y \cdot y \cdot y \cdot y \cdot y$$

These examples suggest a definition for $x^n$, where $n$ is a natural number.

**Natural-Number Exponents**

If $n$ is a natural number, then

$$x^n = \overbrace{x \cdot x \cdot x \cdot \cdots \cdot x}^{n \text{ factors of } x}$$

In the exponential expression $x^n$, $x$ is called the **base** and $n$ is called the **exponent**. The entire expression is called a **power of $x$.**

$$\text{Base} \longrightarrow x^n \longleftarrow \text{Exponent}$$

## ■ POWERS OF EXPRESSIONS

If an exponent is a natural number, it tells how many times its base is to be used as a factor. An exponent of 1 indicates that its base is to be used one time as a factor, an exponent of 2 indicates that its base is to be used two times as a factor, and so on.

$$3^1 = 3 \qquad (-y)^1 = -y \qquad (-4z)^2 = (-4z)(-4z) \qquad \text{and} \qquad (t^2)^3 = t^2 \cdot t^2 \cdot t^2$$

**EXAMPLE 1**   Show that   **a.**  $-2^4$   and   **b.**  $(-2)^4$ have different values.

*Solution*   We find each power and show that the results are different.

$$\begin{aligned}
-2^4 &= -(2^4) & (-2)^4 &= (-2)(-2)(-2)(-2) \\
&= -(2 \cdot 2 \cdot 2 \cdot 2) & &= 16 \\
&= -16
\end{aligned}$$

Since $-16 \neq 16$, if follows that $-2^4 \neq (-2)^4$.   ■

*Self Check*   Show that   **a.**  $(-4)^3$   and   **b.**  $-4^3$ have the same value.

*Answers*   **a.**  $-64$,   **b.**  $-64$

**EXAMPLE 2**   Write each expression without using exponents:   **a.**  $r^3$,   **b.**  $(-2s)^4$,   and
**c.**  $\left(\frac{1}{3}ab\right)^5$.

*Solution*   **a.**  $r^3 = r \cdot r \cdot r$

**b.**  $(-2s)^4 = (-2s)(-2s)(-2s)(-2s)$

**c.**  $\left(\frac{1}{3}ab\right)^5 = \left(\frac{1}{3}ab\right)\left(\frac{1}{3}ab\right)\left(\frac{1}{3}ab\right)\left(\frac{1}{3}ab\right)\left(\frac{1}{3}ab\right)$   ■

*Self Check*   Write each expression without using exponents:   **a.**  $x^4$ and   **b.**  $\left(-\frac{1}{2}xy\right)^3$.

*Answers*   **a.**  $x \cdot x \cdot x \cdot x$,   **b.**  $\left(-\frac{1}{2}\right)\left(-\frac{1}{2}\right)\left(-\frac{1}{2}\right)x \cdot x \cdot x \cdot y \cdot y \cdot y$

■ THE PRODUCT RULE FOR EXPONENTS

To develop a rule for multiplying exponential expressions with the same base, we consider the product $x^2 \cdot x^3$. Since the expression $x^2$ means that $x$ is to be used as a factor two times and the expression $x^3$ means that $x$ is to be used as a factor three times, we have

$$x^2 x^3 = \overbrace{x \cdot x}^{2 \text{ factors of } x} \cdot \overbrace{x \cdot x \cdot x}^{3 \text{ factors of } x}$$

$$= \overbrace{x \cdot x \cdot x \cdot x \cdot x}^{5 \text{ factors of } x}$$

$$= x^5$$

In general,

$$x^m \cdot x^n = \overbrace{x \cdot x \cdot x \cdots x}^{m \text{ factors of } x} \cdot \overbrace{x \cdot x \cdot x \cdot x \cdots x}^{n \text{ factors of } x}$$

$$= \overbrace{x \cdot x \cdot x \cdot x \cdot x \cdot x \cdots x \cdot x \cdot x}^{m + n \text{ factors of } x}$$

$$= x^{m+n}$$

This discussion suggests the following rule: *To multiply two exponential expressions with the same base, keep the base and add the exponents.*

> **Product Rule for Exponents**
> If $m$ and $n$ are natural numbers, then
> $$x^m x^n = x^{m+n}$$

**EXAMPLE 3**  Simplify each expression.

**a.** $x^3 x^4 = x^{3+4}$       Keep the base and add the exponents.

     $= x^7$       $3 + 4 = 7$.

**b.** $y^2 y^4 y = (y^2 y^4) y$       Use the associative property to group $y^2$ and $y^4$ together.

     $= (y^{2+4}) y$       Keep the base and add the exponents.

     $= y^6 y$       $2 + 4 = 6$.

     $= y^{6+1}$       Keep the base and add the exponents; $y = y^1$.

     $= y^7$       $6 + 1 = 7$.       ■

**Self Check**

**Answers**

Simplify each expression:   **a.** $zz^3$   and   **b.** $x^2 x^3 x^6$.

**a.** $z^4$,   **b.** $x^{11}$

**EXAMPLE 4**  Simplify $(2y^3)(3y^2)$.

*Solution*  $(2y^3)(3y^2) = 2(3)y^3y^2$     Use the commutative and associative properties to group the numbers together and the variables together.

$= 6y^{3+2}$     Multiply the coefficients. Keep the base and add the exponents.

$= 6y^5$     $3 + 2 = 5$.     ■

Self Check  Simplify $(4x)(-3x^2)$.
Answer  $-12x^3$

**WARNING!**  The product rule for exponents applies only to exponential expressions with the same base. An expression such as $x^2y^3$ cannot be simplified, because $x^2$ and $y^3$ have different bases.

## ■ THE POWER RULES FOR EXPONENTS

To find another rule of exponents, we consider the expression $(x^3)^4$, which can be written as $x^3 \cdot x^3 \cdot x^3 \cdot x^3$. Because each of the four factors of $x^3$ contains three factors of $x$, there are $4 \cdot 3$ (or 12) factors of $x$. Thus, the product can be written as $x^{12}$.

$$(x^3)^4 = x^3 \cdot x^3 \cdot x^3 \cdot x^3$$

$$= \overbrace{x \cdot x \cdot x \cdot x \cdot x \cdot x \cdot x \cdot x \cdot x \cdot x \cdot x \cdot x}^{\text{12 factors of } x}$$
$$\underbrace{\phantom{x \cdot x \cdot x}}_{x^3} \underbrace{\phantom{x \cdot x \cdot x}}_{x^3} \underbrace{\phantom{x \cdot x \cdot x}}_{x^3} \underbrace{\phantom{x \cdot x \cdot x}}_{x^3}$$

$$= x^{12}$$

In general,

$$(x^m)^n = \overbrace{x^m \cdot x^m \cdot x^m \cdot \cdots \cdot x^m}^{n \text{ factors of } x^m}$$

$$= \overbrace{x \cdot x \cdot x \cdot x \cdot x \cdot x \cdot \cdots \cdot x}^{m \cdot n \text{ factors of } x}$$

$$= x^{m \cdot n}$$

This discussion suggests the following rule: *To raise an exponential expression to a power, keep the base and multiply the exponents.*

### First Power Rule for Exponents
If $m$ and $n$ are natural numbers, then
$$(x^m)^n = x^{m \cdot n}$$

EXAMPLE 5    Write each expression using one exponent.

**a.** $(2^3)^7 = 2^{3 \cdot 7}$    Keep the base and multiply the exponents.

$\quad\quad = 2^{21}$    $3 \cdot 7 = 21.$

**b.** $(z^7)^7 = z^{7 \cdot 7}$    Keep the base and multiply the exponents.

$\quad\quad = z^{49}$    $7 \cdot 7 = 49.$    ■

Self Check    Write each expression using one exponent:    **a.** $(y^5)^2$    and    **b.** $(u^x)^y$.

Answers    **a.** $y^{10}$,    **b.** $u^{xy}$

In the next example, the product and power rules of exponents are both used.

EXAMPLE 6    Write each expression using one exponent.

**a.** $(x^2x^5)^2 = (x^7)^2$          **b.** $(y^6y^2)^3 = (y^8)^3$

$\quad\quad\quad = x^{14}$          $\quad\quad\quad\quad = y^{24}$

**c.** $(z^2)^4(z^3)^3 = z^8z^9$          **d.** $(x^3)^2(x^5x^2)^3 = x^6(x^7)^3$

$\quad\quad\quad = z^{17}$          $\quad\quad\quad\quad\quad = x^6x^{21}$

$\quad\quad\quad\quad\quad\quad\quad\quad\quad\quad\quad\quad = x^{27}$    ■

Self Check    Write each expression using one exponent:    **a.** $(a^4a^3)^3$    and    **b.** $(a^3)^3(a^4)^2$.

Answers    **a.** $a^{21}$,    **b.** $a^{17}$

To find two more rules for exponents, we consider the expressions $(2x)^3$ and $\left(\frac{2}{x}\right)^3$.

$$(2x)^3 = (2x)(2x)(2x)$$          $$\left(\frac{2}{x}\right)^3 = \left(\frac{2}{x}\right)\left(\frac{2}{x}\right)\left(\frac{2}{x}\right) \quad (x \neq 0)$$

$$= (2 \cdot 2 \cdot 2)(x \cdot x \cdot x)$$          $$= \frac{2 \cdot 2 \cdot 2}{x \cdot x \cdot x}$$

$$= 2^3x^3$$          $$= \frac{2^3}{x^3}$$

$$= 8x^3$$          $$= \frac{8}{x^3}$$

These examples suggest the following rules: *To raise a product to a power, we raise each factor of the product to that power*, and *to raise a fraction to a power, we raise both the numerator and denominator to that power.*

**More Power Rules for Exponents**

If $n$ is a natural number, then

$$(xy)^n = x^ny^n \quad\quad \text{and if } y \neq 0, \text{ then} \quad \left(\frac{x}{y}\right)^n = \frac{x^n}{y^n}$$

EXAMPLE 7

Write each expression without using parentheses. Assume there are no divisions by zero.

**a.** $(ab)^4 = a^4b^4$

**b.** $(3c)^3 = 3^3c^3$
$$= 27c^3$$

**c.** $(x^2y^3)^5 = (x^2)^5(y^3)^5$
$$= x^{10}y^{15}$$

**d.** $(-2x^3y)^2 = (-2)^2(x^3)^2y^2$
$$= 4x^6y^2$$

**e.** $\left(\dfrac{4}{k}\right)^3 = \dfrac{4^3}{k^3}$
$$= \dfrac{64}{k^3}$$

**f.** $\left(\dfrac{3x^2}{2y^3}\right)^5 = \dfrac{3^5(x^2)^5}{2^5(y^3)^5}$
$$= \dfrac{243x^{10}}{32y^{15}}$$

∎

Self Check

Write each expression without using parentheses: **a.** $(3x^2y)^2$ and

**b.** $\left(\dfrac{2x^3}{3y^2}\right)^4.$

Answers

**a.** $9x^4y^2$, **b.** $\dfrac{16x^{12}}{81y^8}$

## ■ THE QUOTIENT RULE FOR EXPONENTS

To find a rule for dividing exponential expressions, we consider the fraction $\dfrac{4^5}{4^2}$, where the exponent in the numerator is greater than the exponent in the denominator. We can simplify the fraction as follows:

$$\dfrac{4^5}{4^2} = \dfrac{4 \cdot 4 \cdot 4 \cdot 4 \cdot 4}{4 \cdot 4}$$

$$= \dfrac{\overset{1}{\cancel{4}} \cdot \overset{1}{\cancel{4}} \cdot 4 \cdot 4 \cdot 4}{\underset{1}{\cancel{4}} \cdot \underset{1}{\cancel{4}}}$$

$$= 4^3$$

The result of $4^3$ has a base of 4 and an exponent of $5 - 2$ (or 3). This suggests that *to divide exponential expressions with the same base, we keep the base and subtract the exponents.*

**Quotient Rule for Exponents**
If $m$ and $n$ are natural numbers, $m > n$, and $x \neq 0$, then

$$\dfrac{x^m}{x^n} = x^{m-n}$$

b,d

**EXAMPLE 8** Simplify each expression. Assume that there are no divisions by 0.

**a.** $\dfrac{x^4}{x^3} = x^{4-3}$

$= x^1$

$= x$

**b.** $\dfrac{8y^2y^6}{4y^3} = \dfrac{8y^8}{4y^3}$

$= \dfrac{8}{4}y^{8-3}$

$= 2y^5$

**c.** $\dfrac{a^3a^5a^7}{a^4a} = \dfrac{a^{15}}{a^5}$

$= a^{15-5}$

$= a^{10}$

**d.** $\dfrac{(a^3b^4)^2}{ab^5} = \dfrac{a^6b^8}{ab^5}$

$= a^{6-1}b^{8-5}$

$= a^5b^3$  ∎

Self Check  Simplify  **a.** $\dfrac{a^5}{a^3}$,  **b.** $\dfrac{6b^2b^3}{2b^4}$,  and  **c.** $\dfrac{(x^2y^3)^2}{x^3y^4}$.

Answers  **a.** $a^2$,  **b.** $3b$,  **c.** $xy^2$

We summarize the rules for positive exponents as follows.

**Properties of Exponents**

If $n$ is a natural number, then

$$x^n = \overbrace{x \cdot x \cdot x \cdot \cdots \cdot x}^{n \text{ factors of } x}$$

If $m$ and $n$ are natural numbers and there are no divisions by 0, then

$$x^m x^n = x^{m+n} \qquad (x^m)^n = x^{m \cdot n} \qquad (xy)^n = x^n y^n \qquad \left(\dfrac{x}{y}\right)^n = \dfrac{x^n}{y^n}$$

$$\dfrac{x^m}{x^n} = x^{m-n} \quad \text{provided } m > n.$$

Orals  *Find the base and the exponent in each expression.*

**1.** $x^3$       **2.** $3^x$       **3.** $ab^c$       **4.** $(ab)^c$

*Evaluate each expression.*

**5.** $6^2$       **6.** $(-6)^2$       **7.** $2^3 + 1^3$       **8.** $(2 + 1)^3$

# EXERCISE 4.1

## *REVIEW*

**1.** Graph the real numbers $-3$, $0$, $2$, and $-\frac{3}{2}$ on a number line.

$$\overset{\hspace{1.2em}\text{'}\hspace{1.2em}\text{'}\hspace{1.7em}\text{'}\hspace{0.8em}\text{'}}{\underset{-4\ \ -3\ \ -2\ \ -1\ \ \ 0\ \ \ 1\ \ \ 2\ \ \ 3}{\longleftrightarrow}}$$

**2.** Graph the interval $(-2, 3]$ on a number line.

$$\underset{-3\ \ -2\ \ -1\ \ \ 0\ \ \ 1\ \ \ 2\ \ \ 3}{\longleftrightarrow}$$

*Write each algebraic expression as an English phrase.*

**3.** $3(x + y)$

**4.** $3x + y$

*Write each English phrase as an algebraic expression.*

**5.** Three greater than the absolute value of twice $x$

**6.** The sum of the numbers $y$ and $z$ decreased by the sum of their squares

## *VOCABULARY AND CONCEPTS*   *Fill in each blank to make a true statement.*

**7.** The base of the exponential expression $(-5)^3$ is ____. The exponent is __.

**8.** The base of the exponential expression $-5^3$ is __. The exponent is __.

**9.** $(3x)^4$ means _____.

**10.** Write $(-3y)(-3y)(-3y)$ as a power. _____

**11.** $y^5 =$ _____

**12.** $x^m x^n =$ _____

**13.** $(xy)^n =$ _____

**14.** $\left(\dfrac{a}{b}\right)^n =$ ___

**15.** $(a^b)^c =$ _____

**16.** $\dfrac{x^m}{x^n} =$ _____

**17.** The area of the square shown in Illustration 1 is $s \cdot s$. Why do you think the symbol $s^2$ is called "$s$ squared"?

**18.** The volume of the cube shown in Illustration 2 is $s \cdot s \cdot s$. Why do you think the symbol $s^3$ is called "$s$ cubed"?

$s$

$s$

ILLUSTRATION 1

$s$

$s$

$s$

ILLUSTRATION 2

*In Exercises 19–30, identify the base and the exponent in each expression.*

**19.** $4^3$

**20.** $(-5)^2$

**21.** $x^5$

**22.** $y^8$

**23.** $(2y)^3$

**24.** $(-3x)^2$

**25.** $-x^4$

**26.** $(-x)^4$

**27.** $x$

**28.** $(xy)^3$

**29.** $2x^3$

**30.** $-3y^6$

In Exercises 31–38, write each expression without using exponents.

**31.** $5^3$          **32.** $-4^5$          **33.** $x^7$          **34.** $3x^3$

**35.** $-4x^5$          **36.** $(-2y)^4$          **37.** $(3t)^5$          **38.** $a^3b^2$

In Exercises 39–46, write each expression using exponents.

**39.** $2 \cdot 2 \cdot 2$          **40.** $5 \cdot 5$          **41.** $x \cdot x \cdot x \cdot x$          **42.** $y \cdot y \cdot y \cdot y \cdot y \cdot y$

**43.** $(2x)(2x)(2x)$          **44.** $(-4y)(-4y)$          **45.** $-4t \cdot t \cdot t \cdot t$          **46.** $5 \cdot u \cdot u$

**PRACTICE**  In Exercises 47–54, evaluate each expression.

**47.** $5^4$          **48.** $(-3)^3$          **49.** $2^2 + 3^2$          **50.** $2^3 - 2^2$

**51.** $5^4 - 4^3$          **52.** $2(4^3 + 3^2)$          **53.** $-5(3^4 + 4^3)$          **54.** $-5^2(4^3 - 2^6)$

In Exercises 55–70, write each expression as an expression involving only one exponent.

**55.** $x^4x^3$          **56.** $y^5y^2$          **57.** $x^5x^5$          **58.** $yy^3$

**59.** $tt^2$          **60.** $w^3w^5$          **61.** $a^3a^4a^5$          **62.** $b^2b^3b^5$

**63.** $y^3(y^2y^4)$          **64.** $(y^4y)y^6$          **65.** $4x^2(3x^5)$          **66.** $-2y(y^3)$

**67.** $(-y^2)(4y^3)$          **68.** $(-4x^3)(-5x)$          **69.** $6x^3(-x^2)(-x^4)$          **70.** $-2x(-x^2)(-3x)$

In Exercises 71–86, write each expression as an expression involving only one exponent.

**71.** $(3^2)^4$          **72.** $(4^3)^3$          **73.** $(y^5)^3$          **74.** $(b^3)^6$

**75.** $(a^3)^7$          **76.** $(b^2)^3$          **77.** $(x^2x^3)^5$          **78.** $(y^3y^4)^4$

**79.** $(3zz^2z^3)^5$          **80.** $(4t^3t^6t^2)^2$          **81.** $(x^5)^2(x^7)^3$          **82.** $(y^3y)^2(y^2)^2$

**83.** $(r^3r^2)^4(r^3r^5)^2$          **84.** $(s^2)^3(s^3)^2(s^4)^4$          **85.** $(s^3)^3(s^2)^2(s^5)^4$          **86.** $(yy^3)^3(y^2y^3)^4(y^3y^3)^2$

In Exercises 87–102, write each expression without using parentheses.

**87.** $(xy)^3$          **88.** $(uv^2)^4$          **89.** $(r^3s^2)^2$          **90.** $(a^3b^2)^3$

**91.** $(4ab^2)^2$          **92.** $(3x^2y)^3$          **93.** $(-2r^2s^3t)^3$          **94.** $(-3x^2y^4z)^2$

**95.** $\left(\dfrac{a}{b}\right)^3$          **96.** $\left(\dfrac{r^2}{s}\right)^4$          **97.** $\left(\dfrac{x^2}{y^3}\right)^5$          **98.** $\left(\dfrac{u^4}{v^2}\right)^6$

**99.** $\left(\dfrac{-2a}{b}\right)^5$          **100.** $\left(\dfrac{2t}{3}\right)^4$          **101.** $\left(\dfrac{b^2}{3a}\right)^3$          **102.** $\left(\dfrac{a^3b}{c^4}\right)^5$

In Exercises 103–118, simplify each expression.

**103.** $\dfrac{x^5}{x^3}$          **104.** $\dfrac{a^6}{a^3}$          **105.** $\dfrac{y^3y^4}{yy^2}$          **106.** $\dfrac{b^4b^5}{b^2b^3}$

**107.** $\dfrac{12a^2a^3a^4}{4(a^4)^2}$

**108.** $\dfrac{16(aa^2)^3}{2a^2a^3}$

**109.** $\dfrac{(ab^2)^3}{(ab)^2}$

**110.** $\dfrac{(m^3n^4)^3}{(mn^2)^3}$

**111.** $\dfrac{20(r^4s^3)^4}{6(rs^3)^3}$

**112.** $\dfrac{15(x^2y^5)^5}{21(x^3y)^2}$

**113.** $\dfrac{17(x^4y^3)^8}{34(x^5y^2)^4}$

**114.** $\dfrac{35(r^3s^2)^2}{49r^2s^4}$

**115.** $\left(\dfrac{y^3y}{2yy^2}\right)^3$

**116.** $\left(\dfrac{3t^3t^4t^5}{4t^2t^6}\right)^3$

**117.** $\left(\dfrac{-2r^3r^3}{3r^4r}\right)^3$

**118.** $\left(\dfrac{-6y^4y^5}{5y^3y^5}\right)^2$

## APPLICATIONS

**119. Bouncing ball**   When a certain ball is dropped, it always rebounds to one-half of its previous height. If the ball is dropped from a height of 32 feet, explain why the expression $32\left(\frac{1}{2}\right)^4$ represents the height of the ball on the fourth bounce. Find the height of the fourth bounce.

**120. Having babies**   The probability that a couple will have $n$ baby boys in a row is given by the formula $\left(\frac{1}{2}\right)^n$. Find the probability that a couple will have four baby boys in a row.

**121. Investing**   If an investment of $1,000 doubles every seven years, find the value of the investment after 28 years.

**122.**   **Investing**   Guess the answer to the following problem. Then use a calculator to find the correct answer. Were you close?
*If the value of 1¢ is to double every day, what will the penny be worth after 31 days?*

## WRITING

**123.** Describe how you would multiply two exponential expressions with like bases.

**124.** Describe how you would divide two exponential expressions with like bases.

## SOMETHING TO THINK ABOUT

**125.** Is the operation of raising to a power commutative? That is, is $a^b = b^a$? Explain.

**126.** Is the operation of raising to a power associative? That is, is $(a^b)^c = a^{(b^c)}$? Explain.

# 4.2  Zero and Negative-Integer Exponents

■ ZERO EXPONENTS ■ NEGATIVE-INTEGER EXPONENTS ■ EXPONENTS WITH VARIABLES
■ FINDING PRESENT VALUE

Getting Ready   *Simplify by dividing out common factors.*

**1.** $\dfrac{3 \cdot 3 \cdot 3}{3 \cdot 3 \cdot 3 \cdot 3}$

**2.** $\dfrac{2yy}{2yyy}$

**3.** $\dfrac{3xx}{3xx}$

**4.** $\dfrac{xxy}{xxxyy}$

### ■ ZERO EXPONENTS

When we discussed the quotient rule for exponents in the previous section, the exponent in the numerator was always greater than the exponent in the denominator. We now consider what happens when the exponents are equal.

If we apply the quotient rule to the fraction $\frac{5^3}{5^3}$, where the exponents in the numerator and denominator are equal, we obtain $5^0$. However, because any nonzero number divided by itself equals 1, we also obtain 1.

$$\frac{5^3}{5^3} = 5^{3-3} = 5^0 \qquad \frac{5^3}{5^3} = \frac{\overset{1}{5}\cdot\overset{1}{5}\cdot\overset{1}{5}}{\underset{1}{5}\cdot\underset{1}{5}\cdot\underset{1}{5}} = 1$$

These are equal.

For this reason, we will define $5^0$ to be equal to 1. In general, the following is true.

> **Zero Exponents**
> If $x$ is any nonzero real number, then
> $$x^0 = 1$$
> Since $x \neq 0$, $0^0$ is undefined.

**EXAMPLE 1**  Write each expression without using exponents.

**a.** $\left(\frac{1}{13}\right)^0 = 1$ 

**b.** $\dfrac{x^5}{x^5} = x^{5-5} \quad (x \neq 0)$
$$= x^0$$
$$= 1$$

**c.** $3x^0 = 3(1)$
$$= 3$$

**d.** $(3x)^0 = 1$

**e.** $\dfrac{6^n}{6^n} = 6^{n-n}$
$$= 6^0$$
$$= 1$$

**f.** $\dfrac{y^m}{y^m} = y^{m-m} \quad (y \neq 0)$
$$= y^0$$
$$= 1$$

Parts **c** and **d** point out that $3x^0 \neq (3x)^0$. ■

**Self Check**  Write each expression without using exponents:  **a.** $(-0.115)^0$,
**b.** $\dfrac{4^2}{4^2}$,  and  **c.** $\dfrac{x^m}{x^m}$, $(x \neq 0)$.

**Answers**  **a.** 1,  **b.** 1,  **c.** 1

### ■ NEGATIVE-INTEGER EXPONENTS

If we apply the quotient rule to $\frac{6^2}{6^5}$, where the exponent in the numerator is less than the exponent in the denominator, we obtain $6^{-3}$. However, by dividing out two factors of 6, we also obtain $\frac{1}{6^3}$.

$$\frac{6^2}{6^5} = 6^{2-5} = 6^{-3}$$

$$\frac{6^2}{6^5} = \frac{\overset{1}{\cancel{6}} \cdot \overset{1}{\cancel{6}}}{\underset{1}{\cancel{6}} \cdot \underset{1}{\cancel{6}} \cdot 6 \cdot 6 \cdot 6} = \frac{1}{6^3}$$

These are equal.

For these reasons, we define $6^{-3}$ to be equal to $\frac{1}{6^3}$. In general, the following is true.

> **Negative Exponents**
> If $x$ is any nonzero number and $n$ is a natural number, then
> $$x^{-n} = \frac{1}{x^n}$$

b,c,e

**EXAMPLE 2** Express each quantity without using negative exponents or parentheses. Assume that no denominators are zero.

**a.** $3^{-5} = \dfrac{1}{3^5}$
$= \dfrac{1}{243}$

**b.** $x^{-4} = \dfrac{1}{x^4}$

**c.** $(2x)^{-2} = \dfrac{1}{(2x)^2}$
$= \dfrac{1}{4x^2}$

**d.** $2x^{-2} = 2\left(\dfrac{1}{x^2}\right)$
$= \dfrac{2}{x^2}$

**e.** $(-3a)^{-4} = \dfrac{1}{(-3a)^4}$
$= \dfrac{1}{81a^4}$

**f.** $(x^3x^2)^{-3} = (x^5)^{-3}$
$= \dfrac{1}{(x^5)^3}$
$= \dfrac{1}{x^{15}}$

■

**Self Check** Write each expression without using negative exponents or parentheses:
**a.** $a^{-5}$, **b.** $(3y)^{-3}$, and **c.** $(a^4a^3)^{-2}$.

**Answers** **a.** $\dfrac{1}{a^5}$, **b.** $\dfrac{1}{27y^3}$, **c.** $\dfrac{1}{a^{14}}$

Because of the definitions of negative and zero exponents, the product, power, and quotient rules are true for all integer exponents.

**Properties of Exponents**

If $m$ and $n$ are integers and there are no divisions by 0, then

$$x^m x^n = x^{m+n} \qquad (x^m)^n = x^{m \cdot n} \qquad (xy)^n = x^n y^n \qquad \left(\frac{x}{y}\right)^n = \frac{x^n}{y^n}$$

$$x^0 = 1 \quad (x \neq 0) \qquad x^{-n} = \frac{1}{x^n} \qquad \frac{x^m}{x^n} = x^{m-n}$$

 c,e

**EXAMPLE 3**    Simplify and write the result without using negative exponents. Assume that no denominators are zero.

**a.** $(x^{-3})^2 = x^{-6}$

$= \dfrac{1}{x^6}$

**b.** $\dfrac{x^3}{x^7} = x^{3-7}$

$= x^{-4}$

$= \dfrac{1}{x^4}$

**c.** $\dfrac{y^{-4}y^{-3}}{y^{-20}} = \dfrac{y^{-7}}{y^{-20}}$

$= y^{-7-(-20)}$

$= y^{-7+20}$

$= y^{13}$

**d.** $\dfrac{12a^3 b^4}{4a^5 b^2} = 3a^{3-5}b^{4-2}$

$= 3a^{-2}b^2$

$= \dfrac{3b^2}{a^2}$

**e.** $\left(-\dfrac{x^3 y^2}{xy^{-3}}\right)^{-2} = (-x^{3-1}y^{2-(-3)})^{-2}$

$= (-x^2 y^5)^{-2}$

$= \dfrac{1}{(-x^2 y^5)^2}$

$= \dfrac{1}{x^4 y^{10}}$

■

**Self Check**    Simplify and write the result without using negative exponents:

**a.** $(x^4)^{-3}$,   **b.** $\dfrac{a^4}{a^8}$,   **c.** $\dfrac{a^{-4}a^{-5}}{a^{-3}}$,   and   **d.** $\dfrac{20x^5 y^3}{5x^3 y^6}$.

*Answers*    **a.** $\dfrac{1}{x^{12}}$,   **b.** $\dfrac{1}{a^4}$,   **c.** $\dfrac{1}{a^6}$,   **d.** $\dfrac{4x^2}{y^3}$

## ■ EXPONENTS WITH VARIABLES

These properties of exponents are also true when the exponents are algebraic expressions.

**EXAMPLE 4**     Simplify each expression.

**a.** $x^{2m}x^{3m} = x^{2m+3m}$
$= x^{5m}$

**b.** $\dfrac{y^{2m}}{y^{4m}} = y^{2m-4m}$   $(y \neq 0)$
$= y^{-2m}$
$= \dfrac{1}{y^{2m}}$

**c.** $a^{2m-1}a^{2m} = a^{2m-1+2m}$
$= a^{4m-1}$

**d.** $(b^{m+1})^{2m} = b^{(m+1)2m}$
$= b^{2m^2+2m}$     ■

**Self Check**     Simplify each expression:   **a.** $z^{3n}z^{2n}$,   **b.** $\dfrac{z^{3n}}{z^{5n}}$,   and   **c.** $(x^{m+2})^{3m}$.

**Answers**     **a.** $z^{5n}$,   **b.** $\dfrac{1}{z^{2n}}$,   **c.** $x^{3m^2+6m}$

## ■ FINDING PRESENT VALUE

■ ■ ■ ■ ■ ■ ■ ■ ■     **Finding Present Value**

CALCULATORS     To find out how much money $P$ must be invested at an annual rate $i$ (expressed as a decimal) to have $\$A$ in $n$ years, we use the formula $P = A(1 + i)^{-n}$. To find out how much we must invest at 6% to have $50,000 in 10 years, we substitute 50,000 for $A$, 0.06 (6%) for $i$, and 10 for $n$ to get

$$P = A(1 + i)^{-n}$$
$$P = 50,000(1 + 0.06)^{-10}$$

To evaluate $P$ with a scientific calculator, we enter these numbers and press these keys:

**Keystrokes**

( 1 + .06 )   $y^x$   10   +/−   ×   50000          $27919.73885$

We must invest $27,919.74 to have $50,000 in 10 years.

Orals *Simplify each quantity.*

**1.** $2^{-1}$

**2.** $2^{-2}$

**3.** $\left(\dfrac{1}{2}\right)^{-1}$

**4.** $\left(\dfrac{7}{9}\right)^{0}$

**5.** $x^{-1}x^{2}$

**6.** $y^{-2}y^{-5}$

**7.** $\dfrac{x^{5}x^{2}}{x^{7}}$

**8.** $\left(\dfrac{x}{y}\right)^{-1}$

## EXERCISE 4.2

### REVIEW

**1.** If $a = -2$ and $b = 3$, evaluate $\dfrac{3a^{2} + 4b + 8}{a + 2b^{2}}$.

**2.** Evaluate $|-3 + 5 \cdot 2|$.

*Solve each equation.*

**3.** $5\left(x - \dfrac{1}{2}\right) = \dfrac{7}{2}$

**4.** $\dfrac{5(2 - x)}{6} = \dfrac{x + 6}{2}$

**5.** Solve $P = L + \dfrac{s}{f}i$ for $s$.

**6.** Solve $P = L + \dfrac{s}{f}i$ for $i$.

### VOCABULARY AND CONCEPTS  *Fill in each blank to make a true statement.*

**7.** If $x$ is any nonzero real number, then $x^{0} =$ ___.

**8.** If $x$ is any nonzero real number, then $x^{-n} =$ ___.

**9.** Since $\dfrac{6^{4}}{6^{4}} = 6^{4-4} = 6^{0}$ and $\dfrac{6^{4}}{6^{4}} = 1$, we define $6^{0}$ to be ___.

**10.** Since $\dfrac{8^{3}}{8^{5}} = 8^{3-5} = 8^{-2}$ and $\dfrac{8^{3}}{8^{5}} = \dfrac{8 \cdot 8 \cdot 8}{8 \cdot 8 \cdot 8 \cdot 8 \cdot 8} = \dfrac{1}{8^{2}}$, we define $8^{-2}$ to be ___.

### PRACTICE  *In Exercises 11–74, simplify each expression. Write each answer without using parentheses or negative exponents.*

**11.** $2^{5} \cdot 2^{-2}$

**12.** $10^{2} \cdot 10^{-4} \cdot 10^{5}$

**13.** $4^{-3} \cdot 4^{-2} \cdot 4^{5}$

**14.** $3^{-4} \cdot 3^{5} \cdot 3^{-3}$

**15.** $\dfrac{3^{5} \cdot 3^{-2}}{3^{3}}$

**16.** $\dfrac{6^{2} \cdot 6^{-3}}{6^{-2}}$

**17.** $\dfrac{2^{5} \cdot 2^{7}}{2^{6} \cdot 2^{-3}}$

**18.** $\dfrac{5^{-2} \cdot 5^{-4}}{5^{-6}}$

**19.** $2x^{0}$

**20.** $(2x)^{0}$

**21.** $(-x)^{0}$

**22.** $-x^{0}$

**23.** $\left(\dfrac{a^{2}b^{3}}{ab^{4}}\right)^{0}$

**24.** $\dfrac{2}{3}\left(\dfrac{xyz}{x^{2}y}\right)^{0}$

**25.** $\dfrac{x^{0} - 5x^{0}}{2x^{0}}$

**26.** $\dfrac{4a^{0} + 2a^{0}}{3a^{0}}$

**27.** $x^{-2}$

**28.** $y^{-3}$

**29.** $b^{-5}$

**30.** $c^{-4}$

**31.** $(2y)^{-4}$

**32.** $(-3x)^{-1}$

**33.** $(ab^2)^{-3}$

**34.** $(m^2n^3)^{-2}$

**35.** $\dfrac{y^4}{y^5}$

**36.** $\dfrac{t^7}{t^{10}}$

**37.** $\dfrac{(r^2)^3}{(r^3)^4}$

**38.** $\dfrac{(b^3)^4}{(b^5)^4}$

**39.** $\dfrac{y^4y^3}{y^4y^{-2}}$

**40.** $\dfrac{x^{12}x^{-7}}{x^3x^4}$

**41.** $\dfrac{a^4a^{-2}}{a^2a^0}$

**42.** $\dfrac{b^0b^3}{b^{-3}b^4}$

**43.** $(ab^2)^{-2}$

**44.** $(c^2d^3)^{-2}$

**45.** $(x^2y)^{-3}$

**46.** $(-xy^2)^{-4}$

**47.** $(x^{-4}x^3)^3$

**48.** $(y^{-2}y)^3$

**49.** $(y^3y^{-2})^{-2}$

**50.** $(x^{-3}x^{-2})^2$

**51.** $(a^{-2}b^{-3})^{-4}$

**52.** $(y^{-3}z^5)^{-6}$

**53.** $(-2x^3y^{-2})^{-5}$

**54.** $(-3u^{-2}v^3)^{-3}$

**55.** $\left(\dfrac{a^3}{a^{-4}}\right)^2$

**56.** $\left(\dfrac{a^4}{a^{-3}}\right)^3$

**57.** $\left(\dfrac{b^5}{b^{-2}}\right)^{-2}$

**58.** $\left(\dfrac{b^{-2}}{b^3}\right)^{-3}$

**59.** $\left(\dfrac{4x^2}{3x^{-5}}\right)^4$

**60.** $\left(\dfrac{-3r^4r^{-3}}{r^{-3}r^7}\right)^3$

**61.** $\left(\dfrac{12y^3z^{-2}}{3y^{-4}z^3}\right)^2$

**62.** $\left(\dfrac{6xy^3}{3x^{-1}y}\right)^3$

**63.** $\left(\dfrac{2x^3y^{-2}}{4xy^2}\right)^7$

**64.** $\left(\dfrac{9u^2v^3}{18u^{-3}v}\right)^4$

**65.** $\left(\dfrac{14u^{-2}v^3}{21u^{-3}v}\right)^4$

**66.** $\left(\dfrac{-27u^{-5}v^{-3}w}{18u^3v^{-2}}\right)^4$

**67.** $\left(\dfrac{6a^2b^3}{2ab^2}\right)^{-2}$

**68.** $\left(\dfrac{15r^2s^{-2}t}{3r^{-3}s^3}\right)^{-3}$

**69.** $\left(\dfrac{18a^2b^3c^{-4}}{3a^{-1}b^2c}\right)^{-3}$

**70.** $\left(\dfrac{21x^{-2}y^2z^{-2}}{7x^3y^{-1}}\right)^{-2}$

**71.** $\dfrac{(2x^{-2}y)^{-3}}{(4x^2y^{-1})^3}$

**72.** $\dfrac{(ab^{-2}c)^2}{(a^{-2}b)^{-3}}$

**73.** $\dfrac{(17x^5y^{-5}z)^{-3}}{(17x^{-5}y^3z^2)^{-4}}$

**74.** $\dfrac{16(x^{-2}yz)^{-2}}{(2x^{-3}z^0)^4}$

*In Exercises 75–90, write each expression with a single exponent.*

**75.** $x^{2m}x^m$

**76.** $y^{3m}y^{2m}$

**77.** $u^{2m}v^{3n}u^{3m}v^{-3n}$

**78.** $r^{2m}s^{-3}r^{3m}s^3$

**79.** $y^{3m+2}y^{-m}$

**80.** $x^{m+1}x^m$

**81.** $\dfrac{y^{3m}}{y^{2m}}$

**82.** $\dfrac{z^{4m}}{z^{2m}}$

**83.** $\dfrac{x^{3n}}{x^{6n}}$

**84.** $\dfrac{x^m}{x^{5m}}$

**85.** $(x^{m+1})^2$

**86.** $(y^2)^{m+1}$

**87.** $(x^{3-2n})^{-4}$

**88.** $(y^{1-n})^{-3}$

**89.** $(y^{2-n})^{-4}$

**90.** $(x^{3-4n})^{-2}$

## APPLICATIONS

**91.** ▦ **Present value**   How much money must be invested at 7% to have $100,000 in 40 years?

**92.** ▦ **Present value**   How much money must be invested at 8% to have $100,000 in 40 years?

**93.** ⬛ **Present value**  How much money must be invested at 9% to have \$100,000 in 40 years?

**94. Biology**  During bacterial reproduction, the time required for a population to double is called the **generation time.** If $b$ bacteria are introduced into a medium, then after the generation time has elapsed, there will be $2b$ bacteria. After $n$ generations, there will be $b \cdot 2^n$ bacteria. Give the meaning of this expression when $n = 0$.

## WRITING

**95.** Tell how you would help a friend understand that $2^{-3}$ is not equal to $-8$.

**96.** Describe how you would verify on a calculator that

$$2^{-3} = \frac{1}{2^3}$$

## SOMETHING TO THINK ABOUT

**97.** If a positive number $x$ is raised to a negative power, is the result greater than, equal to, or less than $x$? Explore the possibilities.

**98.** We know that $x^{-n} = \dfrac{1}{x^n}$. Is it also true that $x^n = \dfrac{1}{x^{-n}}$? Explain.

# 4.3  Scientific Notation

⬛ SCIENTIFIC NOTATION ⬛ WRITING NUMBERS IN SCIENTIFIC NOTATION ⬛ CHANGING FROM SCIENTIFIC NOTATION TO STANDARD NOTATION ⬛ USING SCIENTIFIC NOTATION TO SIMPLIFY COMPUTATIONS

**Getting Ready**  *Evaluate each expression.*

**1.** $10^2$     **2.** $10^3$     **3.** $10^1$     **4.** $10^{-2}$

**5.** $5(10^2)$     **6.** $8(10^3)$     **7.** $3(10^1)$     **8.** $7(10^{-2})$

## ⬛ SCIENTIFIC NOTATION

Scientists often deal with extremely large and extremely small numbers. For example,

- The distance from the earth to the sun is approximately 150,000,000 kilometers.

- Ultraviolet light emitted from a mercury arc has a wavelength of approximately 0.000025 centimeter.

The large number of zeros in these numbers makes them difficult to read and hard to remember. In this section, we will discuss a notation that will make these numbers easier to work with.

Scientific notation provides a compact way of writing large and small numbers.

> **Scientific Notation**
> A number is written in **scientific notation** if it is written as the product of a number between 1 (including 1) and 10 and an integer power of 10.

Each of the following numbers is written in scientific notation.

$$3.67 \times 10^6 \qquad 2.24 \times 10^{-4} \qquad \text{and} \qquad 9.875 \times 10^{22}$$

Every number that is written in scientific notation has the following form:

An integer exponent

$$\underline{\phantom{0}}.\underline{\phantom{0}} \times 10^{-}$$

A decimal between 1 and 10

### ■ WRITING NUMBERS IN SCIENTIFIC NOTATION

**EXAMPLE 1**     Change 150,000,000 to scientific notation.

*Solution*     We note that 1.5 lies between 1 and 10. To obtain 150,000,000, the decimal point in 1.5 must be moved eight places to the right. Because multiplying a number by 10 moves the decimal point one place to the right, we can accomplish this by multiplying 1.5 by 10 eight times.

$$1.50000000$$
8 places to the right

150,000,000 written in scientific notation is $1.5 \times 10^8$. ■

*Self Check*     Change 93,000,000 to scientific notation.
*Answer*     $9.3 \times 10^7$

**EXAMPLE 2**     Change 0.000025 to scientific notation.

*Solution*     We note that 2.5 is between 1 and 10. To obtain 0.000025, the decimal point in 2.5 must be moved five places to the left. We can accomplish this by dividing 2.5 by $10^5$, which is equivalent to multiplying 2.5 by $\frac{1}{10^5}$ (or by $10^{-5}$).

$$0.0002.5$$
5 places to the left

0.000025 written in scientific notation is $2.5 \times 10^{-5}$. ■

Self Check | Write 0.00125 in scientific notation.
*Answer* | $1.25 \times 10^{-3}$

---

EXAMPLE 3 | Write **a.** 235,000 and **b.** 0.00000235 in scientific notation.

*Solution* | **a.** $235,000 = 2.35 \times 10^5$, because $2.35 \times 10^5 = 235,000$ and 2.35 is between 1 and 10.

**b.** $0.00000235 = 2.35 \times 10^{-6}$, because $2.35 \times 10^{-6} = 0.00000235$ and 2.35 is between 1 and 10. ∎

Self Check | Write **a.** 17,500 and **b.** 0.657 in scientific notation.
*Answers* | **a.** $1.75 \times 10^4$, **b.** $6.57 \times 10^{-1}$

---

■ ■ ■ ■ ■ ■ ■ ■ ■ PERSPECTIVE

**The Metric System**

A common metric unit of length is the kilometer, which is 1,000 meters. Because 1,000 is $10^3$, we can write $1 \text{ km} = 10^3 \text{ m}$. Similarly, 1 centimeter is one-hundredth of a meter: $1 \text{ cm} = 10^{-2} \text{ m}$. In the metric system, prefixes such as *kilo* and *centi* refer to powers of 10. Other prefixes are used in the metric system, as shown in the table.

| Prefix | Symbol | Meaning | |
|--------|--------|---------|---|
| peta   | P      | $10^{15}$ | = 1,000,000,000,000,000. |
| tera   | T      | $10^{12}$ | = 1,000,000,000,000. |
| giga   | G      | $10^{9}$  | = 1,000,000,000. |
| mega   | M      | $10^{6}$  | = 1,000,000. |
| kilo   | k      | $10^{3}$  | = 1,000. |
| deci   | d      | $10^{-1}$ | = 0.1 |
| centi  | c      | $10^{-2}$ | = 0.01 |
| milli  | m      | $10^{-3}$ | = 0.001 |
| micro  | $\mu$  | $10^{-6}$ | = 0.000 001 |
| nano   | n      | $10^{-9}$ | = 0.000 000 001 |
| pico   | p      | $10^{-12}$ | = 0.000 000 000 001 |
| femto  | f      | $10^{-15}$ | = 0.000 000 000 000 001 |
| atto   | a      | $10^{-18}$ | = 0.000 000 000 000 000 001 |

To appreciate the magnitudes involved, consider these facts: Light, which travels 186,000 miles every second, will travel about one foot in one nanosecond. The distance to the nearest star is 43 petameters, and the diameter of an atom is about 10 nanometers. To measure some quantities, however, even these units are inadequate. The sun, for example, radiates $5 \times 10^{26}$ watts. That's a lot of light bulbs!

EXAMPLE 4   Write $432.0 \times 10^5$ in scientific notation.

*Solution*   The number $432.0 \times 10^5$ is not written in scientific notation, because 432.0 is not a number between 1 and 10. To write the number in scientific notation, we proceed as follows:

$$432.0 \times 10^5 = 4.32 \times 10^2 \times 10^5 \qquad \text{Write 432.0 in scientific notation.}$$
$$= 4.32 \times 10^7 \qquad\qquad 10^2 \times 10^5 = 10^7. \qquad ■$$

*Self Check*   Write $85 \times 10^{-3}$ in scientific notation.
*Answer*   $8.5 \times 10^{-2}$

## ■ CHANGING FROM SCIENTIFIC NOTATION TO STANDARD NOTATION

We can change a number written in scientific notation to **standard notation.** For example, to write $9.3 \times 10^7$ in standard notation, we multiply 9.3 by $10^7$.

$$9.3 \times 10^7 = 9.3 \times 10,000,000$$
$$= 93,000,000$$

EXAMPLE 5   Write   **a.** $3.4 \times 10^5$   and   **b.** $2.1 \times 10^{-4}$   in standard notation.

*Solution*   **a.** $3.4 \times 10^5 = 3.4 \times 100,000$          **b.** $2.1 \times 10^{-4} = 2.1 \times \dfrac{1}{10^4}$
$\qquad\qquad\quad = 340,000$

$$\qquad\qquad\qquad\qquad\qquad\qquad = 2.1 \times \frac{1}{10,000}$$
$$\qquad\qquad\qquad\qquad\qquad\qquad = 0.00021 \qquad ■$$

*Self Check*   Write   **a.** $4.76 \times 10^5$   and   **b.** $9.8 \times 10^{-3}$   in standard notation.
*Answers*   **a.** 476,000,   **b.** 0.0098

Each of the following numbers is written in both scientific and standard notation. In each case, the exponent gives the number of places that the decimal point moves, and the sign of the exponent indicates the direction that it moves.

$5.32 \times 10^5 = 5\,3\,2\,0\,0\,0.$          5 places to the right.

$2.37 \times 10^6 = 2\,3\,7\,0\,0\,0\,0.$          6 places to the right.

$8.95 \times 10^{-4} = 0.0\,0\,0\,8\,9\,5$          4 places to the left.

$8.375 \times 10^{-3} = 0.0\,0\,8\,3\,7\,5$          3 places to the left.

$9.77 \times 10^0 = 9.77$          No movement of the decimal point.

## ■ USING SCIENTIFIC NOTATION TO SIMPLIFY COMPUTATIONS

Another advantage of scientific notation becomes apparent when we simplify fractions such as

$$\frac{(0.0032)(25,000)}{0.00040}$$

that contain very large or very small numbers. Although we can simplify this fraction by using arithmetic, scientific notation provides an easier way. First, we write each number in scientific notation; then we do the arithmetic on the numbers and the exponential expressions separately. Finally, we write the result in standard form, if desired.

$$\frac{(0.0032)(25,000)}{0.00040} = \frac{(3.2 \times 10^{-3})(2.5 \times 10^4)}{4.0 \times 10^{-4}}$$

$$= \frac{(3.2)(2.5)}{4.0} \times \frac{10^{-3}10^4}{10^{-4}}$$

$$= \frac{8.0}{4.0} \times 10^{-3+4-(-4)}$$

$$= 2.0 \times 10^5$$

$$= 200,000$$

---

■ ■ ■ ■ ■ ■ ■ ■ ■ ■ **Finding Powers of Decimals**

CALCULATORS   To find the value of $(453.46)^5$, we can use a scientific calculator and enter these numbers and press these keys:

**Keystrokes**

453.46  $y^x$  5  =                    $1.917321395 \quad ^{13}$

So we have $(453.46)^5 = 1.917321395 \times 10^{13}$. Since this number is too large to show on the calculator display, the calculator gives the result in scientific notation.

---

As the previous example shows, scientific calculators can do operations using scientific notation. Consult your owner's manual to see how to enter numbers written in scientific notation into the calculator.

EXAMPLE 6    **Speed of light**  In a vacuum, light travels 1 meter in approximately 0.000000003 second. How long does it take for light to travel 500 kilometers?

*Solution*    Because 1 kilometer = 1,000 meters, the length of time for light to travel 500 kilometers (500 · 1,000 meters) is given by

$$(0.000000003)(500)(1,000) = (3 \times 10^{-9})(5 \times 10^2)(1 \times 10^3)$$
$$= 3(5) \times 10^{-9+2+3}$$
$$= 15 \times 10^{-4}$$
$$= 1.5 \times 10^1 \times 10^{-4}$$
$$= 1.5 \times 10^{-3}$$
$$= 0.0015$$

Light travels 500 kilometers in approximately 0.0015 second.    ■

*Orals*    *Tell which number of each pair is the larger.*

**1.** 37.2 or $3.72 \times 10^2$

**2.** 37.2 or $3.72 \times 10^{-1}$

**3.** $3.72 \times 10^3$ or $4.72 \times 10^3$

**4.** $3.72 \times 10^3$ or $4.72 \times 10^2$

**5.** $3.72 \times 10^{-1}$ or $4.72 \times 10^{-2}$

**6.** $3.72 \times 10^{-3}$ or $2.72 \times 10^{-2}$

## EXERCISE 4.3

### REVIEW

**1.** If $y = -1$, find the value of $-5y^{55}$.

**2.** Evaluate $\dfrac{3a^2 - 2b}{2a + 2b}$ if $a = 4$ and $b = 3$.

*Tell which property of real numbers justifies each statement.*

**3.** $5 + z = z + 5$

**4.** $7(u + 3) = 7u + 7 \cdot 3$

*Solve each equation.*

**5.** $3(x - 4) - 6 = 0$

**6.** $8(3x - 5) - 4(2x + 3) = 12$

### VOCABULARY AND CONCEPTS    *Fill in each blank to make a true statement.*

**7.** A number is written in _____ when it is written as the product of a number between 1 (including 1) and 10 and an integer power of 10.

**8.** The number 125,000 is written in _____ notation.

### PRACTICE    *In Exercises 9–20, write each number in scientific notation.*

**9.** 23,000

**10.** 4,750

**11.** 1,700,000

**12.** 290,000

**13.** 0.062

**14.** 0.00073

**15.** 0.0000051

**16.** 0.04

**17.** $42.5 \times 10^2$    **18.** $0.3 \times 10^3$    **19.** $0.25 \times 10^{-2}$    **20.** $25.2 \times 10^{-3}$

*In Exercises 21–32, write each number in standard notation.*

**21.** $2.3 \times 10^2$    **22.** $3.75 \times 10^4$    **23.** $8.12 \times 10^5$    **24.** $1.2 \times 10^3$
**25.** $1.15 \times 10^{-3}$    **26.** $4.9 \times 10^{-2}$    **27.** $9.76 \times 10^{-4}$    **28.** $7.63 \times 10^{-5}$

**29.** $25 \times 10^6$    **30.** $0.07 \times 10^3$    **31.** $0.51 \times 10^{-3}$    **32.** $617 \times 10^{-2}$

*In Exercises 33–38, use scientific notation to simplify each expression. Give all answers in standard notation.*

**33.** $(3.4 \times 10^2)(2.1 \times 10^3)$    **34.** $(4.1 \times 10^{-3})(3.4 \times 10^4)$

**35.** $\dfrac{9.3 \times 10^2}{3.1 \times 10^{-2}}$    **36.** $\dfrac{7.2 \times 10^6}{1.2 \times 10^8}$

**37.** $\dfrac{96,000}{(12,000)(0.00004)}$    **38.** $\dfrac{(0.48)(14,400,000)}{96,000,000}$

## APPLICATIONS

**39. Distance to Alpha Centauri**  The distance from the earth to the nearest star outside our solar system is approximately 25,700,000,000,000 miles. Write this number in scientific notation.

**40. Speed of sound**  The speed of sound in air is 33,100 centimeters per second. Write this number in scientific notation.

**41. Distance to Mars**  The distance from Mars to the sun is approximately $1.14 \times 10^8$ miles. Write this number in standard notation.

**42. Distance to Venus**  The distance from Venus to the sun is approximately $6.7 \times 10^7$ miles. Write this number in standard notation.

**43. Length of one meter**  One meter is approximately 0.00622 mile. Use scientific notation to express this number.

**44. Angstrom**  One angstrom is $1 \times 10^{-7}$ millimeter. Write this number in standard notation.

**45. Distance between Mercury and the sun**  The distance from Mercury to the sun is approximately $3.6 \times 10^7$ miles. Use scientific notation to express this distance in feet. (*Hint:* 5,280 feet = 1 mile.)

**46. Mass of a proton**  The mass of one proton is approximately $1.7 \times 10^{-24}$ gram. Use scientific notation to express the mass of 1 million protons.

**47. Speed of sound**  The speed of sound in air is approximately $3.3 \times 10^4$ centimeters per second. Use scientific notation to express this speed in kilometers per second. (*Hint:* 100 centimeters = 1 meter and 1,000 meters = 1 kilometer.)

**48. Light year**  One light year is approximately $5.87 \times 10^{12}$ miles. Use scientific notation to express this distance in feet. (*Hint:* 5,280 feet = 1 mile.)

## WRITING

**49.** In what situations would scientific notation be more convenient than standard notation?

**50.** To multiply a number by a power of 10, we move the decimal point. Which way, and how far? Explain.

### SOMETHING TO THINK ABOUT

**51.** Two positive numbers are written in scientific notation. How could you decide which is larger, without converting either to standard notation?

**52.** The product $1 \cdot 2 \cdot 3 \cdot 4 \cdot 5$, or 120, is called **5 factorial,** written 5!. Similarly, the number $6! = 6 \cdot 5 \cdot 4 \cdot 3 \cdot 2 \cdot 1 = 620$. Factorials get large very quickly. Calculate 30!, and write the number in standard notation. (*Hint:* Experiment with the $x!$ key on a calculator.) How large a factorial can you compute with a calculator?

## 4.4  Polynomials

■ POLYNOMIALS ■ MONOMIALS, BINOMIALS, AND TRINOMIALS ■ DEGREE OF A POLYNOMIAL ■ EVALUATING POLYNOMIALS ■ FUNCTIONS ■ FUNCTION NOTATION ■ GRAPHING POLYNOMIAL FUNCTIONS

**Getting Ready**   *Write each expression using exponents.*

**1.** $2xxyyy$

**2.** $3xyyy$

**3.** $2xx + 3yy$

**4.** $xxx + yyy$

**5.** $(3xxy)(2xyy)$

**6.** $(5xyzzz)(xyz)$

**7.** $3(5xy)\left(\dfrac{1}{3}xy\right)$

**8.** $(xy)(xz)(yz)(xyz)$

### ■ POLYNOMIALS

Recall that expressions such as

$$3x \qquad 4y^2 \qquad -8x^2y^3 \qquad \text{and} \qquad 25$$

with constant and/or variable factors are called **algebraic terms.** The numerical coefficients of the first three of these terms are 3, 4, and $-8$, respectively. Because $25 = 25x^0$, 25 is considered to be the numerical coefficient of the term 25.

> **Polynomials**
> A **polynomial** is an algebraic expression that is the sum of one or more terms containing whole-number exponents on the variables.

Here are some examples of polynomials:

$$8xy^2t \qquad 3x + 2 \qquad 4y^2 - 2y + 3 \qquad \text{and} \qquad 3a - 4b - 4c + 8d$$

**WARNING!**   The expression $2x^3 - 3y^{-2}$ is not a polynomial, because the second term contains a negative exponent on a variable base.

EXAMPLE 1     Tell whether each expression is a polynomial.

     **a.** $x^2 + 2x + 1$        Yes.

     **b.** $3x^{-1} - 2x - 3$      No. The first term has a negative exponent on a variable base.

     **c.** $\dfrac{1}{2}x^3 - 2.3x + 5$      Yes.                             ■

Self Check     Tell whether each expression is a polynomial:   **a.** $3x^{-4} + 2x^2 - 3$   and
                 **b.** $7.5x^3 - 4x^2 - 3x$

Answers     **a.** no,   **b.** yes

## ■ MONOMIALS, BINOMIALS, AND TRINOMIALS

A polynomial with one term is called a **monomial.** A polynomial with two terms is called a **binomial.** A polynomial with three terms is called a **trinomial.** Here are some examples.

| Monomials | Binomials | Trinomials |
|:---:|:---:|:---:|
| $5x^2y$ | $3u^3 - 4u^2$ | $-5t^2 + 4t + 3$ |
| $-6x$ | $18a^2b + 4ab$ | $27x^3 - 6x - 2$ |
| $29$ | $-29z^{17} - 1$ | $-32r^6 + 7y^3 - z$ |

EXAMPLE 2     Classify each polynomial as a monomial, a binomial, or a trinomial.

     **a.** $5x^4 + 3x$           Since the polynomial has two terms, it is a binomial.

     **b.** $7x^4 - 5x^3 - 2$      Since the polynomial has three terms, it is a trinomial.

     **c.** $-5x^2y^3$            Since the polynomial has one term, it is a monomial.       ■

Self Check     Classify each polynomial as a monomial, a binomial, or a trinomial:   **a.** $5x$,
                 **b.** $-5x^2 + 2x - 5$,   and   **c.** $16x^2 - 9y^2$.

Answers     **a.** monomial,   **b.** trinomial,   **c.** binomial

## ■ DEGREE OF A POLYNOMIAL

The monomial $7x^6$ is called a **monomial of sixth degree** or a **monomial of degree 6,** because the variable $x$ occurs as a factor six times. The monomial $3x^3y^4$ is a monomial of the seventh degree, because the variables $x$ and $y$ occur as factors a total of seven times. Other examples are

$-2x^3$ is a monomial of degree 3.

$47x^2y^3$ is a monomial of degree 5.

$18x^4y^2z^8$ is a monomial of degree 14.

8 is a monomial of degree 0, because $8 = 8x^0$.

These examples illustrate the following definition.

### Degree of a Monomial
If $a$ is a nonzero constant, the **degree of the monomial** $ax^n$ is $n$.

The **degree of a monomial with several variables** is the sum of the exponents on those variables.

**WARNING!**    Note that the degree of $ax^n$ is not defined when $a = 0$. Since $ax^n = 0$ when $a = 0$, the constant 0 has no defined degree.

Because each term of a polynomial is a monomial, we define the degree of a polynomial by considering the degree of each of its terms.

### Degree of a Polynomial
The **degree of a polynomial** is the same as the degree of its term with largest degree.

For example,

- $x^2 + 2x$ is a binomial of degree 2, because the degree of its first term is 2 and the degree of its other term is less than 2.
- $3x^3y^2 + 4x^4y^4 - 3x^3$ is a trinomial of degree 8, because the degree of its second term is 8 and the degree of each of its other terms is less than 8.
- $25x^4y^3z^7 - 15xy^8z^{10} - 32x^8y^8z^3 + 4$ is a polynomial of degree 19, because its second and third terms are of degree 19. Its other terms have degrees less than 19.

**EXAMPLE 3**    Find the degree of each polynomial.

**a.** $-4x^3 - 5x^2 + 3x$        3

**b.** $5x^4y^2 + 7xy^2 - 16x^3y^5$        8

**c.** $-17a^2b^3c^4 + 12a^3b^4c$        9        ■

*Self Check* Find the degree of each polynomial: **a.** $15p^3q^4 - 25p^4q^2$ and **b.** $-14rs^3t^4 + 12r^3s^3t^3$.

*Answers* **a.** 7, **b.** 9

## ■ EVALUATING POLYNOMIALS

When a number is substituted for the variable in a polynomial, the polynomial takes on a numerical value. Finding that value is called **evaluating the polynomial.**

 a,d

**EXAMPLE 4** Evaluate the polynomial $3x^2 + 2$ when **a.** $x = 0$, **b.** $x = 2$, **c.** $x = -3$, and **d.** $x = -\frac{1}{5}$.

*Solution*

**a.** $3x^2 + 2 = 3(0)^2 + 2$
$= 3(0) + 2$
$= 0 + 2$
$= 2$

**b.** $3x^2 + 2 = 3(2)^2 + 2$
$= 3(4) + 2$
$= 12 + 2$
$= 14$

**c.** $3x^2 + 2 = 3(-3)^2 + 2$
$= 3(9) + 2$
$= 27 + 2$
$= 29$

**d.** $3x^2 + 2 = 3\left(-\frac{1}{5}\right)^2 + 2$
$= 3\left(\frac{1}{25}\right) + 2$
$= \frac{3}{25} + \frac{50}{25}$
$= \frac{53}{25}$ ■

*Self Check* Evaluate $3x^2 + x - 2$ when **a.** $x = 2$ and **b.** $x = -1$.

*Answers* **a.** 12, **b.** 0

When we evaluate a polynomial function for several values of its variable, we often write the results in a table.

**EXAMPLE 5** Evaluate the polynomial $x^3 + 1$ when **a.** $x = -2$, **b.** $x = -1$, **c.** $x = 0$, **d.** $x = 1$, and **e.** $x = 2$. Write the results in a table.

*Solution*

| $x$ | $x^3 + 1$ | |
|---|---|---|
| **a.** $-2$ | $-7$ | $x^3 + 1 = (-2)^3 + 1 = -7.$ |
| **b.** $-1$ | $0$ | $x^3 + 1 = (-1)^3 + 1 = 0.$ |
| **c.** $0$ | $1$ | $x^3 + 1 = (0)^3 + 1 = 1.$ |
| **d.** $1$ | $2$ | $x^3 + 1 = (1)^3 + 1 = 2.$ |
| **e.** $2$ | $9$ | $x^3 + 1 = (2)^3 + 1 = 9.$ |

■

Self Check

Consider the polynomial $-x^3 + 1$. Complete the following table.

| $x$ | $-x^3 + 1$ |
|-----|------------|
| $-2$ | |
| $-1$ | |
| $0$ | |
| $1$ | |
| $2$ | |

*Answers*   9, 2, 1, 0, −7

## ■ FUNCTIONS

The results of Examples 4 and 5 illustrate that for every input value $x$ that we substitute into a polynomial with the variable $x$, there is exactly one output value. Whenever we consider a polynomial equation such as $y = 3x^2 + 2$, where each input value $x$ determines a single output value $y$, we say that $y$ is a *function* of $x$.

> **Function**
> Any equation in $x$ and $y$ where each value of $x$ (the input) determines one value of $y$ (the output) is called a **function.** In this case, we say that $y$ is a function of $x$.
>
> The set of all input values $x$ is called the **domain** of the function, and the set of all output values $y$ is called the **range.**

Since each output value $y$ depends on some input value of $x$, we call $y$ the **dependent variable** and $x$ the **independent variable.** Here are some equations that define $y$ to be a function of $x$.

**1.** $y = 2x - 3$    Note that each input value of $x$ determines a single output value of $y$. For example, if $x = 4$, then $y = 5$. Since any real number can be substituted for $x$, the domain is the set of real numbers. We will soon show that the range is also the set of real numbers.

**2.** $y = x^2$    Note that each input value of $x$ determines a single output value of $y$. For example, if $x = 3$, then $y = 9$. Since any real number can be substituted for $x$, the domain is the set of real numbers. Since the square of any real number is positive or 0, the range is the set of all numbers $y$ such that $y \geq 0$.

**3.** $y = x^3$    Note that each input value of $x$ determines a single output value of $y$. For example, if $x = -2$, then $y = -8$. Since any number can be substituted for $x$, the domain is the set of real numbers. We will soon show that the range is also the set of real numbers.

### ■ FUNCTION NOTATION

There is a special notation for functions that uses the symbol $f(x)$, read as "$f$ of $x$."

**Function Notation**
The notation $y = f(x)$ denotes that the variable $y$ is a function of $x$.

 **WARNING!**   The notation $f(x)$ does not mean "$f$ times $x$."

The notation $y = f(x)$ provides a way to denote the values of $y$ in a function that correspond to individual values of $x$. For example, if $y = f(x)$, the value of $y$ that is determined by $x = 3$ is denoted as $f(3)$. Similarly, $f(-1)$ represents the value of $y$ that corresponds to $x = -1$.

**EXAMPLE 6**    Let $y = f(x) = 2x - 3$ and find   **a.** $f(3)$,   **b.** $f(-1)$,   **c.** $f(0)$,   and   **d.** $f(0.2)$.

*Solution*   **a.** We replace $x$ with 3.

$$f(x) = 2x - 3$$
$$f(3) = 2(3) - 3$$
$$= 6 - 3$$
$$= 3$$

**b.** We replace $x$ with $-1$.

$$f(x) = 2x - 3$$
$$f(-1) = 2(-1) - 3$$
$$= -2 - 3$$
$$= -5$$

**c.** We replace $x$ with 0.

$$f(x) = 2x - 3$$
$$f(0) = 2(0) - 3$$
$$= 0 - 3$$
$$= -3$$

**d.** We replace $x$ with 0.2.

$$f(x) = 2x - 3$$
$$f(0.2) = 2(0.2) - 3$$
$$= 0.4 - 3$$
$$= -2.6$$    ■

**Self Check**    Using the function of Example 6, find   **a.** $f(-2)$   and   **b.** $f\left(\frac{3}{2}\right)$.
**Answers**    **a.** $-7$,   **b.** 0

■ ■ ■ ■ ■ ■ ■ ■ ■    **Height of a Rocket**

CALCULATORS    The height $h$ (in feet) of a toy rocket launched straight up into the air with an initial velocity of 64 feet per second is given by the polynomial function

$$h = f(t) = -16t^2 + 64t$$

In this case, the height $h$ is the dependent variable, and the time $t$ is the independent variable. To find the height of the rocket 3.5 seconds after launch, we substitute 3.5 for $t$ and evaluate $h$.

$$h = -16t^2 + 64t$$
$$h = -16(3.5)^2 + 64(3.5)$$

To evaluate $h$ using a scientific calculator, we enter these numbers and press these keys:

**Keystrokes**

16   +/−   ×   3.5   $x^2$   +   (   64   ×   3.5   )   =

The display will read                    28.

After 3.5 seconds, the rocket will be 28 feet above the ground.

## ■  GRAPHING POLYNOMIAL FUNCTIONS

Since the right-hand sides of the functions $y = f(x) = 2x - 3$, $y = f(x) = x^2$, and $y = f(x) = x^3$ are polynomials, they are called **polynomial functions.** We can graph these functions as we graphed equations in Section 3.2. We make a table of values, plot points, and draw the line or curve that passes through those points.

In the next example, we graph the function $y = f(x) = 2x - 3$. Since its graph is a line, we call this function a **linear function.**

**EXAMPLE 7**      Graph $y = f(x) = 2x - 3$.

*Solution*      We substitute numbers for $x$, compute the corresponding values of $f(x)$, and list the results in a table, as in Figure 4-1. We then plot the pairs $(x, y)$ and draw a line

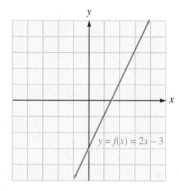

$$y = f(x) = 2x - 3$$

| $x$ | $y$ | $(x, y)$ |
|-----|-----|----------|
| $-3$ | $-9$ | $(-3, -9)$ |
| $-2$ | $-7$ | $(-2, -7)$ |
| $-1$ | $-5$ | $(-1, -5)$ |
| $0$ | $-3$ | $(0, -3)$ |
| $1$ | $-1$ | $(1, -1)$ |
| $2$ | $1$ | $(2, 1)$ |
| $3$ | $3$ | $(3, 3)$ |

FIGURE 4-1

through the points, as shown in the figure. From the graph, we can see that $x$ can be any value. This confirms that the domain is the set of all real numbers. We can also see that $y$ can be any value. This confirms that the range is also the set of all real numbers. ∎

*Self Check* Graph $y = f(x) = \frac{1}{2}x + 3$ and tell whether it is a linear function.

*Answer* It is a linear function.

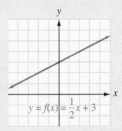

In the next example, we graph the function $y = f(x) = x^2$, called the **squaring function**. Since the polynomial on the right-hand side is of second degree, we call this function a **quadratic function.**

**EXAMPLE 8** Graph $y = f(x) = x^2$.

*Solution* We substitute numbers for $x$, compute the corresponding values of $f(x)$, and list the results in a table, as in Figure 4-2. We then plot the pairs $(x, y)$ and draw a smooth curve through the points, as shown in the figure. This curve is called a **parabola.** From the graph, we can see that $x$ can be any value. This confirms that the domain is the set of all real numbers. We can also see that $y$ is always a positive number or 0. This confirms that the range is the set of all real numbers such that $y \geq 0$.

$$y = f(x) = x^2$$

| $x$ | $y$ | $(x, y)$ |
|-----|-----|----------|
| $-3$ | 9 | $(-3, 9)$ |
| $-2$ | 4 | $(-2, 4)$ |
| $-1$ | 1 | $(-1, 1)$ |
| 0 | 0 | $(0, 0)$ |
| 1 | 1 | $(1, 1)$ |
| 2 | 4 | $(2, 4)$ |
| 3 | 9 | $(3, 9)$ |

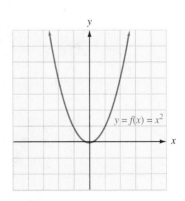

FIGURE 4-2 ∎

Self Check    Graph $y = f(x) = x^2 - 3$ and compare the graph to the graph of $y = f(x) = x^2$ shown in Figure 4-2.

Answer    The graph has the same shape but is 3 units lower.

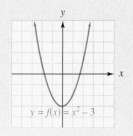

In the next example, we graph the function $y = f(x) = x^3$, called the **cubing function**.

EXAMPLE 9    Graph $y = f(x) = x^3$.

Solution    We substitute numbers for $x$, compute the corresponding values of $f(x)$, and list the results in a table, as in Figure 4-3. We then plot the pairs $(x, y)$ and draw a smooth curve through the points, as shown in the figure.

$y = f(x) = x^3$

| $x$ | $y$ | $(x, y)$ |
|-----|-----|----------|
| $-2$ | $-8$ | $(-2, -8)$ |
| $-1$ | $-1$ | $(-1, -1)$ |
| $0$ | $0$ | $(0, 0)$ |
| $1$ | $1$ | $(1, 1)$ |
| $2$ | $8$ | $(2, 8)$ |

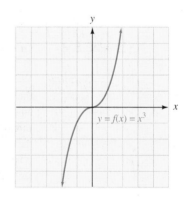

FIGURE 4-3

Self Check    Graph $y = f(x) = x^3 + 3$ and compare the graph to the graph of $y = f(x) = x^3$ shown in Figure 4-3.

Answer    The graph has the same shape but is 3 units higher.

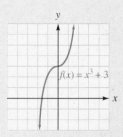

■ ■ ■ ■ ■ ■ ■ ■ ■ ■ **Graphing Polynomial Functions**

GRAPHING
CALCULATORS

It is possible to use a graphing calculator to generate tables and graphs for polynomial functions. For example, Figure 4-4 shows calculator tables and graphs of $y = f(x) = 2x - 3$, $y = f(x) = x^2$, and $y = f(x) = x^3$.

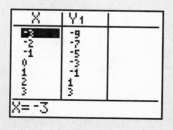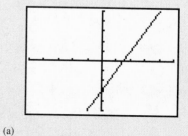

(a)

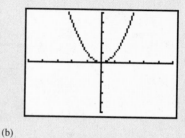

(b)

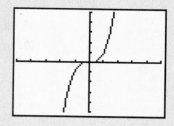

(c)

FIGURE 4-4

**EXAMPLE 10**    Graph $y = f(x) = x^2 - 2x$.

*Solution*    We substitute numbers for $x$, compute the corresponding values of $f(x)$, and list the results in a table, as in Figure 4-5. We then plot the pairs $(x, y)$ and draw a smooth curve through the points, as shown in the figure.

$$y = f(x) = x^2 - 2x$$

| $x$ | $y$ | $(x, y)$ |
|-----|-----|----------|
| $-2$ | 8 | $(-2, 8)$ |
| $-1$ | 3 | $(-1, 3)$ |
| 0 | 0 | $(0, 0)$ |
| 1 | $-1$ | $(1, -1)$ |
| 2 | 0 | $(2, 0)$ |
| 3 | 3 | $(3, 3)$ |
| 4 | 8 | $(4, 8)$ |

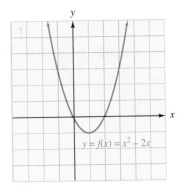

FIGURE 4-5

**Self Check Answer**   Use a graphing calculator to graph $y = f(x) = x^2 - 2x$.

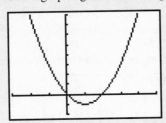

**Orals**   *Give an example of a polynomial that is . . .*

**1.** a binomial

**2.** a monomial

**3.** a trinomial

**4.** not a monomial, a binomial, or a trinomial

**5.** of degree 3

**6.** of degree 1

**7.** of degree 0

**8.** has no defined degree

# EXERCISE 4.4

***REVIEW***   *Solve each equation.*

**1.** $5(u - 5) + 9 = 2(u + 4)$

**2.** $8(3a - 5) - 12 = 4(2a + 3)$

*Solve each inequality and graph the solution set.*

**3.** $-4(3y + 2) \le 28$

**4.** $-5 < 3t + 4 \le 13$

*Write each expression without using parentheses or negative exponents.*

**5.** $(x^2x^4)^3$

**6.** $(a^2)^3(a^3)^2$

**7.** $\left(\dfrac{y^2y^5}{y^4}\right)^3$

**8.** $\left(\dfrac{2t^3}{t}\right)^{-4}$

***VOCABULARY AND CONCEPTS*** *Fill in each blank to make a true statement.*

**9.** An expression with a constant and/or a variable is called an _____ term.

**10.** The numerical coefficient of the term $-25x^2y^3$ is _____.

**11.** A _____ is an algebraic expression that is the sum of one or more terms containing whole-number exponents.

**12.** A _____ is a polynomial with two terms.

**13.** A _____ is a polynomial with three terms.

**14.** A _____ is a polynomial with one term.

**15.** If $a \neq 0$, the _____ of $ax^n$ is $n$.

**16.** The degree of a monomial with several variables is the _____ of the exponents on those variables.

**17.** Any equation in $x$ and $y$ where each input value $x$ determines exactly one output value $y$ is called a _____.

**18.** $f(x)$ is read as _____.

**19.** In a function, the set of all input values is called the _____.

**20.** In a function, the set of all output values is called the _____.

*Tell whether each expression is a polynomial.*

**21.** $x^3 - 5x^2 - 2$

**22.** $x^{-4} - 5x$

**23.** $\frac{1}{2}x^3 + 3$

**24.** $x^3 - 1$

*In Exercises 25–36, classify each polynomial as a monomial, a binomial, a trinomial, or none of these.*

**25.** $3x + 7$

**26.** $3y - 5$

**27.** $3y^2 + 4y + 3$

**28.** $3xy$

**29.** $3z^2$

**30.** $3x^4 - 2x^3 + 3x - 1$

**31.** $5t - 32$

**32.** $9x^2y^3z^4$

**33.** $s^2 - 23s + 31$

**34.** $12x^3 - 12x^2 + 36x - 3$

**35.** $3x^5 - 2x^4 - 3x^3 + 17$

**36.** $x^3$

*In Exercises 37–48, give the degree of each polynomial.*

**37.** $3x^4$

**38.** $3x^5 - 4x^2$

**39.** $-2x^2 + 3x^3$

**40.** $-5x^5 + 3x^2 - 3x$

**41.** $3x^2y^3 + 5x^3y^5$

**42.** $-2x^2y^3 + 4x^3y^2z$

**43.** $-5r^2s^2t - 3r^3st^2 + 3$

**44.** $4r^2s^3t^3 - 5r^2s^8$

**45.** $x^{12} + 3x^2y^3z^4$

**46.** $17^2x$

**47.** $38$

**48.** $-25$

***PRACTICE*** *In Exercises 49–52, evaluate $5x - 3$ for each value.*

**49.** $x = 2$

**50.** $x = 0$

**51.** $x = -1$

**52.** $x = -2$

*In Exercises 53–56, evaluate $-x^2 - 4$ for each value.*

**53.** $x = 0$

**54.** $x = 1$

**55.** $x = -1$

**56.** $x = -2$

*In Exercises 57–60, evaluate $x^2 - 2x + 3$ for each value.*

**57.** $x = 0$

**58.** $x = 3$

**59.** $x = -2$

**60.** $x = -1$

*In Exercises 61–64, complete each table.*

**61.**

| $x$ | $x^2 - 3$ |
|-----|-----------|
| $-2$ | |
| $-1$ | |
| $0$ | |
| $1$ | |
| $2$ | |

**62.**

| $x$ | $-x^2 + 3$ |
|-----|-----------|
| $-2$ | |
| $-1$ | |
| $0$ | |
| $1$ | |
| $2$ | |

**63.**

| $x$ | $x^3 + 2$ |
|-----|-----------|
| $-2$ | |
| $-1$ | |
| $0$ | |
| $1$ | |
| $2$ | |

**64.**

| $x$ | $-x^3 + 2$ |
|-----|-----------|
| $-2$ | |
| $-1$ | |
| $0$ | |
| $1$ | |
| $2$ | |

*In Exercises 65–68, graph each polynomial function. Check your work with a graphing calculator.*

**65.** $f(x) = x^2 - 1$

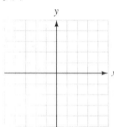

**66.** $f(x) = x^2 + 2$

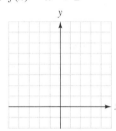

**67.** $f(x) = x^3 + 2$

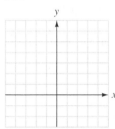

**68.** $f(x) = x^3 - 2$

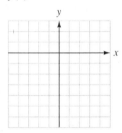

**APPLICATIONS**    *Use a calculator to help solve each problem.*

**69. Height of a rocket**   See the Calculators section on page 289. Find the height of the rocket 2 seconds after launch.

**70. Height of a rocket**   Again referring to page 289, make a table of values to find the rocket's height at various times. For what values of $t$ will the height of the rocket be 0?

**71. Stopping distance**   The number of feet that a car travels before stopping depends on the driver's reaction time and the braking distance. (See Illustration 1.) For one driver, the stopping distance $d$ is given by the function $d = f(v) = 0.04v^2 + 0.9v$, where $v$ is the velocity of the car. Find the stopping distance when the driver is traveling at 30 mph.

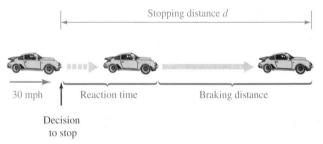

ILLUSTRATION 1

**72. Stopping distance**   Find the stopping distance of the car discussed in Exercise 71 when the driver is going 70 mph.

**WRITING**

**73.** Describe how to determine the degree of a polynomial.

**74.** Describe how to classify a polynomial as a monomial, a binomial, a trinomial, or none of these.

**SOMETHING TO THINK ABOUT**

**75.** Find a polynomial whose value will be 1 if you substitute $\frac{3}{2}$ for $x$.

**76.** Graph the function $y = f(x) = -x^2$. What do you discover?

## 4.5 Adding and Subtracting Polynomials

■ ADDING MONOMIALS ■ SUBTRACTING MONOMIALS ■ ADDING POLYNOMIALS ■ SUBTRACTING POLYNOMIALS ■ ADDING AND SUBTRACTING MULTIPLES OF POLYNOMIALS ■ AN APPLICATION OF ADDING POLYNOMIALS

**Getting Ready** *Combine like terms and simplify, if possible.*

**1.** $3x + 2x$      **2.** $5y - 3y$      **3.** $19x + 6x$      **4.** $8z - 3z$

**5.** $9r + 3r$      **6.** $4r - 3s$      **7.** $7r - 7r$      **8.** $17r - 17r^2$

### ■ ADDING MONOMIALS

Recall that like terms have the same variables with the same exponents. For example,

$3xyz^2$ and $-2xyz^2$ are like terms.

$\frac{1}{2}ab^2c$ and $\frac{1}{3}a^2bd^2$ are unlike terms.

Also recall that to combine like terms, we add (or subtract) their coefficients and keep the same variables with the same exponents. For example,

$$2y + 5y = (2 + 5)y \qquad \text{and} \qquad -3x^2 + 7x^2 = (-3 + 7)x^2$$
$$= 7y \qquad\qquad\qquad\qquad = 4x^2$$

Likewise,

$$4x^3y^2 + 9x^3y^2 = 13x^3y^2 \qquad \text{and} \qquad 4r^2s^3t^4 + 7r^2s^3t^4 = 11r^2s^3t^4$$

These examples suggest that to add like monomials, we simply combine like terms.

 a,c

**EXAMPLE 1**

**a.** $5xy^3 + 7xy^3 = 12xy^3$

**b.** $-7x^2y^2 + 6x^2y^2 + 3x^2y^2 = -x^2y^2 + 3x^2y^2$
$$= 2x^2y^2$$

**c.** $(2x^2)^2 + 81x^4 = 4x^4 + 81x^4$     $(2x^2)^2 = (2x^2)(2x^2) = 4x^4$.
$$= 85x^4$$ ■

**Self Check** Do the following additions:   **a.** $6a^3b^2 + 5a^3b^2$,   **b.** $-2pq^2 + 5pq^2 + 8pq^2$, and   **c.** $27x^6 + (2x^2)^3$.

**Answers**   **a.** $11a^3b^2$,   **b.** $11pq^2$,   **c.** $35x^6$

## ■ SUBTRACTING MONOMIALS

To subtract one monomial from another, we add the negative of the monomial that is to be subtracted. In symbols, $x - y = x + (-y)$.

**EXAMPLE 2**

**a.** $8x^2 - 3x^2 = 8x^2 + (-3x^2)$
$$= 5x^2$$

**b.** $6x^3y^2 - 9x^3y^2 = 6x^3y^2 + (-9x^3y^2)$
$$= -3x^3y^2$$

**c.** $-3r^2st^3 - 5r^2st^3 = -3r^2st^3 + (-5r^2st^3)$
$$= -8r^2st^3$$ ■

**Self Check**
**Answers**

Find each difference:   **a.** $12m^3 - 7m^3$   and   **b.** $-4p^3q^2 - 8p^3q^2$.
**a.** $5m^3$,   **b.** $-12p^3q^2$

## ■ ADDING POLYNOMIALS

Because of the distributive property, we can remove parentheses enclosing several terms when the sign preceding the parentheses is a + sign. We simply drop the parentheses.

$$+(3x^2 + 3x - 2) = +1(3x^2 + 3x - 2)$$
$$= 1(3x^2) + 1(3x) + 1(-2)$$
$$= 3x^2 + 3x + (-2)$$
$$= 3x^2 + 3x - 2$$

We can add polynomials by removing parentheses, if necessary, and then combining any like terms that are contained within the polynomials.

**EXAMPLE 3**

Add $(3x^2 - 3x + 2) + (2x^2 + 7x - 4)$.

**Solution**

$$(3x^2 - 3x + 2) + (2x^2 + 7x - 4)$$
$$= 3x^2 - 3x + 2 + 2x^2 + 7x - 4$$
$$= 3x^2 + 2x^2 - 3x + 7x + 2 + (-4)$$
$$= 5x^2 + 4x - 2$$ ■

**Self Check**
**Answer**

Add $(2a^2 - a + 4) + (5a^2 + 6a - 5)$.
$7a^2 + 5a - 1$

Problems such as Example 3 are often written with like terms aligned vertically. We can then add column by column.

$$3x^2 - 3x + 2$$
$$\underline{2x^2 + 7x - 4}$$
$$5x^2 + 4x - 2$$

**EXAMPLE 4**   Add:   $4x^2y + 8x^2y^2 - 3x^2y^3$
$$\underline{3x^2y - 8x^2y^2 + 8x^2y^3}$$
$$7x^2y \qquad\quad + 5x^2y^3$$ ■

**Self Check**   Add:   $4pq^2 + 6pq^3 - 7pq^4$
$$\underline{2pq^2 - 8pq^3 + 9pq^4}$$

**Answer**   $6pq^2 - 2pq^3 + 2pq^4$

## ■ SUBTRACTING POLYNOMIALS

Because of the distributive property, we can remove parentheses enclosing several terms when the sign preceding the parentheses is a − sign. We simply drop the minus sign and the parentheses, and *change the sign of every term within the parentheses.*

$$-(3x^2 + 3x - 2) = -1(3x^2 + 3x - 2)$$
$$= -1(3x^2) + (-1)(3x) + (-1)(-2)$$
$$= -3x^2 + (-3x) + 2$$
$$= -3x^2 - 3x + 2$$

This suggests that the way to subtract polynomials is to remove parentheses and combine like terms.

**EXAMPLE 5**   **a.** $(3x - 4) - (5x + 7) = 3x - 4 - 5x - 7$
$$= -2x - 11$$

**b.** $(3x^2 - 4x - 6) - (2x^2 - 6x + 12) = 3x^2 - 4x - 6 - 2x^2 + 6x - 12$
$$= x^2 + 2x - 18$$

**c.** $(-4rt^3 + 2r^2t^2) - (-3rt^3 + 2r^2t^2) = -4rt^3 + 2r^2t^2 + 3rt^3 - 2r^2t^2$
$$= -rt^3$$ ■

**Self Check**   Find the difference: $(-2a^2b + 5ab^2) - (-5a^2b - 7ab^2)$.
**Answer**   $3a^2b + 12ab^2$

To subtract polynomials in vertical form, we add the negative of the **subtrahend** (the bottom polynomial) to the **minuend** (the top polynomial).

**EXAMPLE 6**    Subtract $3x^2y - 2xy^2$ from $2x^2y + 4xy^2$.

*Solution*    We write the subtraction in vertical form, change the signs of the terms of the subtrahend, and add:

$$
\begin{array}{r}
2x^2y + 4xy^2 \\
- \underline{3x^2y - 2xy^2}
\end{array}
\qquad
\begin{array}{r}
2x^2y + 4xy^2 \\
+ \underline{-3x^2y + 2xy^2} \\
- x^2y + 6xy^2
\end{array}
$$

In horizontal form, the solution is

$$2x^2y + 4xy^2 - (3x^2y - 2xy^2) = 2x^2y + 4xy^2 - 3x^2y + 2xy^2$$
$$= -x^2y + 6xy^2$$

■

**Self Check**    Find the difference:     $\begin{array}{r} 5p^2q - 6pq + 7q \\ - \underline{2p^2q + 2pq - 8q} \end{array}$

**Answer**    $3p^2q - 8pq + 15q$

**EXAMPLE 7**    Subtract $6xy^2 + 4x^2y^2 - x^3y^2$ from $-2xy^2 - 3x^3y^2$.

*Solution*    
$$
\begin{array}{r}
-2xy^2 \qquad\quad - 3x^3y^2 \\
- \underline{6xy^2 + 4x^2y^2 - x^3y^2}
\end{array}
\qquad
\begin{array}{r}
-2xy^2 \qquad\quad - 3x^3y^2 \\
+ \underline{-6xy^2 - 4x^2y^2 + x^3y^2} \\
-8xy^2 - 4x^2y^2 - 2x^3y^2
\end{array}
$$

In horizontal form, the solution is

$$-2xy^2 - 3x^3y^2 - (6xy^2 + 4x^2y^2 - x^3y^2)$$
$$= -2xy^2 - 3x^3y^2 - 6xy^2 - 4x^2y^2 + x^3y^2$$
$$= -8xy^2 - 4x^2y^2 - 2x^3y^2$$

■

**Self Check**    Subtract $-2pq^2 - 2p^2q^2 + 3p^3q^2$ from $5pq^2 + 3p^2q^2 - p^3q^2$.
**Answer**    $7pq^2 + 5p^2q^2 - 4p^3q^2$

## ■ ADDING AND SUBTRACTING MULTIPLES OF POLYNOMIALS

Because of the distributive property, we can remove parentheses enclosing several terms when a monomial precedes the parentheses. We simply multiply every term within the parentheses by that monomial. For example, to add $3(2x + 5)$ and $2(4x - 3)$, we proceed as follows:

$$3(2x + 5) + 2(4x - 3) = 6x + 15 + 8x - 6$$
$$= 6x + 8x + 15 - 6 \qquad 15 + 8x = 8x + 15.$$
$$= 14x + 9 \qquad \text{Combine like terms.}$$

**EXAMPLE 8**

**a.** $3(x^2 + 4x) + 2(x^2 - 4) = 3x^2 + 12x + 2x^2 - 8$
$$= 5x^2 + 12x - 8$$

**b.** $8(y^2 - 2y + 3) - 4(2y^2 + y - 3) = 8y^2 - 16y + 24 - 8y^2 - 4y + 12$
$$= -20y + 36$$

**c.** $-4x(xy^2 - xy + 3) - x(xy^2 - 2) + 3(x^2y^2 + 2x^2y)$
$$= -4x^2y^2 + 4x^2y - 12x - x^2y^2 + 2x + 3x^2y^2 + 6x^2y$$
$$= -2x^2y^2 + 10x^2y - 10x \qquad ∎$$

*Self Check*    Remove parentheses and simplify:    **a.** $2(a^3 - 3a) + 5(a^3 + 2a)$    and
**b.** $5x(xy + 2x) - x^2(y - 3)$.

*Answers*    **a.** $7a^3 + 4a$,    **b.** $4x^2y + 13x^2$

## ∎ AN APPLICATION OF ADDING POLYNOMIALS

**EXAMPLE 9**

**Property values**    A house purchased for \$95,000 is expected to appreciate according to the formula $y = 2{,}500x + 95{,}000$, where $y$ is the value of the house after $x$ years. A second house purchased for \$125,000 is expected to appreciate according to the formula $y = 4{,}500x + 125{,}000$. Find one formula that will give the value of both properties after $x$ years.

*Solution*    The value of the first house after $x$ years is given by the polynomial $2{,}500x + 95{,}000$. The value of the second house after $x$ years is given by the polynomial $4{,}500x + 125{,}000$. The value of both houses will be the sum of these two polynomials.

$$2{,}500x + 95{,}000 + 4{,}500x + 125{,}000 = 7{,}000x + 220{,}000$$

The total value of the properties is given by the formula $7{,}000x + 220{,}000$.    ∎

*Orals*    *Simplify.*

**1.** $x^3 + 3x^3$

**2.** $3xy + xy$

**3.** $(x + 3y) - (x + y)$

**4.** $5(1 - x) + 3(x - 1)$

**5.** $(2x - y^2) - (2x + y^2)$

**6.** $5(x^2 + y) + (x^2 - y)$

**7.** $3x^2 + 2y + x^2 - y$

**8.** $2x^2y + y - (2x^2y - y)$

# EXERCISE 4.5

**REVIEW**    *Let a = 3, b = −2, c = −1, and d = 2. Evaluate each expression.*

**1.** $ab + cd$

**2.** $ad + bc$

**3.** $a(b + c)$

**4.** $d(b + a)$

**5.** Solve the inequality $−4(2x − 9) ≥ 12$ and graph the solution set.

**6.** The **kinetic energy** of a moving object is given by the formula

$$K = \frac{mv^2}{2}$$

Solve the formula for $m$.

**VOCABULARY AND CONCEPTS**    *Fill in each blank to make a true statement.*

**7.** A _____ is a polynomial with one term.

**8.** If two polynomials are subtracted in vertical form, the bottom polynomial is called the _____, and the top polynomial is called the _____.

**9.** To add like monomials, add the numerical _____ and keep the _____.

**10.** $a − b = a +$ _____

**11.** To add two polynomials, combine any _____ contained in the polynomials.

**12.** To subtract polynomials, remove parentheses and combine _____.

*In Exercises 13–24, tell whether the terms are like or unlike terms. If they are like terms, add them.*

**13.** $3y, 4y$

**14.** $3x^2, 5x^2$

**15.** $3x, 3y$

**16.** $3x^2, 6x$

**17.** $3x^3, 4x^3, 6x^3$

**18.** $−2y^4, −6y^4, 10y^4$

**19.** $−5x^3y^2, 13x^3y^2$

**20.** $23, 12x$

**21.** $−23t^6, 32t^6, 56t^6$

**22.** $32x^5y^3, −21x^5y^3, −11x^5y^3$

**23.** $−x^2y, xy, 3xy^2$

**24.** $4x^3y^2z, −6x^3y^2z, 2x^3y^2z$

**PRACTICE**    *In Exercises 25–42, simplify each expression if possible.*

**25.** $4y + 5y$

**26.** $−2x + 3x$

**27.** $−8t^2 − 4t^2$

**28.** $15x^2 + 10x^2$

**29.** $32u^3 − 16u^3$

**30.** $25xy^2 − 7xy^2$

**31.** $18x^5y^2 − 11x^5y^2$

**32.** $17x^6y − 22x^6y$

**33.** $3rst + 4rst + 7rst$

**34.** $−2ab + 7ab − 3ab$

**35.** $−4a^2bc + 5a^2bc − 7a^2bc$

**36.** $(xy)^2 + 4x^2y^2 − 2x^2y^2$

**37.** $(3x)^2 − 4x^2 + 10x^2$

**38.** $(2x)^4 − (3x^2)^2$

**39.** $5x^2y^2 + 2(xy)^2 − (3x^2)y^2$

**40.** $−3x^3y^6 + 2(xy^2)^3 − (3x)^3y^6$

**41.** $(−3x^2y)^4 + (4x^4y^2)^2 − 2x^8y^4$

**42.** $5x^5y^{10} − (2xy^2)^5 + (3x)^5y^{10}$

*In Exercises 43–74, do the operations and simplify.*

**43.** $(3x + 7) + (4x - 3)$

**44.** $(2y - 3) + (4y + 7)$

**45.** $(4a + 3) - (2a - 4)$

**46.** $(5b - 7) - (3b + 5)$

**47.** $(2x + 3y) + (5x - 10y)$

**48.** $(5x - 8y) - (2x + 5y)$

**49.** $(-8x - 3y) - (11x + y)$

**50.** $(-4a + b) + (5a - b)$

**51.** $(3x^2 - 3x - 2) + (3x^2 + 4x - 3)$

**52.** $(3a^2 - 2a + 4) - (a^2 - 3a + 7)$

**53.** $(2b^2 + 3b - 5) - (2b^2 - 4b - 9)$

**54.** $(4c^2 + 3c - 2) + (3c^2 + 4c + 2)$

**55.** $(2x^2 - 3x + 1) - (4x^2 - 3x + 2) + (2x^2 + 3x + 2)$

**56.** $(-3z^2 - 4z + 7) + (2z^2 + 2z - 1) - (2z^2 - 3z + 7)$

**57.** $2(x + 3) + 3(x + 3)$

**58.** $5(x + y) + 7(x + y)$

**59.** $-8(x - y) + 11(x - y)$

**60.** $-4(a - b) - 5(a - b)$

**61.** $2(x^2 - 5x - 4) - 3(x^2 - 5x - 4) + 6(x^2 - 5x - 4)$

**62.** $7(x^2 + 3x + 1) + 9(x^2 + 3x + 1) - 5(x^2 + 3x + 1)$

**63.** Add: $3x^2 + 4x + 5$ $2x^2 - 3x + 6$

**64.** Add: $2x^3 + 2x^2 - 3x + 5$ $3x^3 - 4x^2 - x - 7$

**65.** Add: $2x^3 - 3x^2 + 4x - 7$ $-9x^3 - 4x^2 - 5x + 6$

**66.** Add: $-3x^3 + 4x^2 - 4x + 9$ $2x^3 + 9x - 3$

**67.** Add: $-3x^2y + 4xy + 25y^2$ $5x^2y - 3xy - 12y^2$

**68.** Add: $-6x^3z - 4x^2z^2 + 7z^3$ $-7x^3z + 9x^2z^2 - 21z^3$

**69.** Subtract: $3x^2 + 4x - 5$ $-2x^2 - 2x + 3$

**70.** Subtract: $3y^2 - 4y + 7$ $6y^2 - 6y - 13$

**71.** Subtract: $4x^3 + 4x^2 - 3x + 10$ $5x^3 - 2x^2 - 4x - 4$

**72.** Subtract: $3x^3 + 4x^2 + 7x + 12$ $-4x^3 + 6x^2 + 9x - 3$

**73.** Subtract: $-2x^2y^2 - 4xy + 12y^2$ $10x^2y^2 + 9xy - 24y^2$

**74.** Subtract: $25x^3 - 45x^2z + 31xz^2$ $12x^3 + 27x^2z - 17xz^2$

**75.** Find the sum when $x^2 + x - 3$ is added to the sum of $2x^2 - 3x + 4$ and $3x^2 - 2$.

**76.** Find the sum when $3y^2 - 5y + 7$ is added to the sum of $-3y^2 - 7y + 4$ and $5y^2 + 5y - 7$.

**77.** Find the difference when $t^3 - 2t^2 + 2$ is subtracted from the sum of $3t^3 + t^2$ and $-t^3 + 6t - 3$.

**78.** Find the difference when $-3z^3 - 4z + 7$ is subtracted from the sum of $2z^2 + 3z - 7$ and $-4z^3 - 2z - 3$.

**79.** Find the sum when $3x^2 + 4x - 7$ is added to the sum of $-2x^2 - 7x + 1$ and $-4x^2 + 8x - 1$.

**80.** Find the difference when $32x^2 - 17x + 45$ is subtracted from the sum of $23x^2 - 12x - 7$ and $-11x^2 + 12x + 7$.

*In Exercises 81–90, simplify each expression.*

**81.** $2(x + 3) + 4(x - 2)$

**82.** $3(y - 4) - 5(y + 3)$

**83.** $-2(x^2 + 7x - 1) - 3(x^2 - 2x + 7)$

**84.** $-5(y^2 - 2y - 6) + 6(2y^2 + 2y - 5)$

**85.** $2(2y^2 - 2y + 2) - 4(3y^2 - 4y - 1) + 4y(y^2 - y - 1)$

**86.** $-4(z^2 - 5z) - 5(4z^2 - 1) + 6(2z - 3)$

**87.** $2a(ab^2 - b) - 3b(a + 2ab) + b(b - a + a^2b)$

**88.** $3y(xy + y) - 2y^2(x - 4 + y) + 2(y^3 + y^2)$

**89.** $-4xy^2(x + y + z) - 2x(xy^2 - 4y^2z) - 2y(8xy^2 - 1)$

**90.** $-3uv(u - v^2 + w) + 4w(uv + w) - 3w(w + uv)$

**APPLICATIONS** *In Exercises 91–96, consider the following information: If a house was purchased for $105,000 and is expected to appreciate $900 per year, its value y after x years is given by the formula $y = 900x + 105,000$.*

**91. Value of a house** Find the expected value of the house in 10 years.

**92. Value of a house** A second house was purchased for $120,000 and was expected to appreciate $1,000 per year. Find a polynomial equation that will give the value $y$ of the house in $x$ years.

**93. Value of a house** Find the value of the house discussed in Exercise 92 after 12 years.

**94. Value of a house** Find one polynomial equation that will give the combined value $y$ of both houses after $x$ years.

**95. Value of two houses** Find the value of the two houses after 20 years by

  **a.** substituting 20 into the polynomial equations $y = 900x + 105,000$ and $y = 1,000x + 120,000$ and adding the results.

  **b.** substituting into the result of Exercise 94.

**96. Value of two houses** Find the value of the two houses after 25 years by

  **a.** substituting 25 into the polynomial equations $y = 900x + 105,000$ and $y = 1,000x + 120,000$ and adding the results.

  **b.** substituting into the result of Exercise 94.

*In Exercises 97–100, consider the following information: A business bought two computers, one for $6,600 and the other for $9,200. The first computer is expected to depreciate $1,100 per year and the second $1,700 per year.*

**97. Value of a computer** Write a polynomial equation that will give the value of the first computer after $x$ years.

**98. Value of a computer** Write a polynomial equation that will give the value of the second computer after $x$ years.

**99. Value of two computers** Find one polynomial equation that will give the value of both computers after $x$ years.

**100. Value of two computers** In two ways, find the value of the computers after 3 years.

**WRITING**

**101.** How do you recognize like terms?

**102.** How do you add like terms?

**SOMETHING TO THINK ABOUT** *In Exercises 103–104, let $P(x) = 3x - 5$. Find each value.*

**103.** $P(x + h) + P(x)$

**104.** $P(x + h) - P(x)$

**105.** If $P(x) = x^{23} + 5x^2 + 73$ and $Q(x) = x^{23} + 4x^2 + 73$, find $P(7) - Q(7)$.

**106.** If two numbers written in scientific notation have the same power of 10, they can be added as similar terms:

$$2 \times 10^3 + 3 \times 10^3 = 5 \times 10^3$$

Without converting to standard form, how could you add

$$2 \times 10^3 + 3 \times 10^4$$

## 4.6 Multiplying Polynomials

■ MULTIPLYING MONOMIALS ■ MULTIPLYING A POLYNOMIAL BY A MONOMIAL ■ MULTIPLYING A BINOMIAL BY A BINOMIAL ■ THE FOIL METHOD ■ MULTIPLYING A POLYNOMIAL BY A BINOMIAL ■ MULTIPLYING BINOMIALS TO SOLVE EQUATIONS ■ AN APPLICATION OF MULTIPLYING POLYNOMIALS

**Getting Ready** *Simplify.*

**1.** $(2x)(3)$

**2.** $(3xxx)(x)$

**3.** $5x^2 \cdot x$

**4.** $8x^2x^3$

*Use the distributive property to remove parentheses.*

**5.** $3(x + 5)$

**6.** $x(x + 5)$

**7.** $4(y - 3)$

**8.** $2y(y - 3)$

### ■ MULTIPLYING MONOMIALS

We have previously multiplied monomials by other monomials. For example, to multiply $4x^2$ by $-2x^3$, we use the commutative and associative properties of multiplication to group the numerical factors together and the variable factors together. Then we multiply the numerical factors and multiply the variable factors.

$$4x^2(-2x^3) = 4(-2)x^2x^3$$
$$= -8x^5$$

This example suggests the following rule.

**Multiplying Monomials**
To multiply two monomials, multiply the numerical factors and then multiply the variable factors.

a,c

**EXAMPLE 1**   Multiply   **a.** $3x^5(2x^5)$,   **b.** $-2a^2b^3(5ab^2)$,   and   **c.** $-4y^5z^2(2y^3z^3)(3yz)$.

*Solution*   **a.** $3x^5(2x^5) = 3(2)x^5x^5$
$$= 6x^{10}$$

**b.** $-2a^2b^3(5ab^2) = -2(5)a^2ab^3b^2$
$$= -10a^3b^5$$

**c.** $-4y^5z^2(2y^3z^3)(3yz) = -4(2)(3)y^5y^3yz^2z^3z$
$$= -24y^9z^6$$  ∎

*Self Check*   Multiply   **a.** $(5a^2b^3)(6a^3b^4)$   and   **b.** $(-15p^3q^2)(5p^3q^2)$.
*Answers*   **a.** $30a^5b^7$,   **b.** $-75p^6q^4$

## ■ MULTIPLYING A POLYNOMIAL BY A MONOMIAL

To find the product of a monomial and a polynomial with more than one term, we use the distributive property. To multiply $2x + 4$ by $5x$, for example, we proceed as follows:

$$5x(2x + 4) = 5x \cdot 2x + 5x \cdot 4 \qquad \text{Use the distributive property.}$$

$$= 10x^2 + 20x \qquad \text{Multiply the monomials: } 5x \cdot 2x = 10x^2 \text{ and } 5x \cdot 4 = 20x.$$

This example suggests the following rule.

> **Multiplying Polynomials by Monomials**
> To multiply a polynomial with more than one term by a monomial, use the distributive property to remove parentheses and simplify.

**EXAMPLE 2**   Multiply   **a.** $3a^2(3a^2 - 5a)$   and   **b.** $-2xz^2(2x - 3z + 2z^2)$.

*Solution*   **a.** $3a^2(3a^2 - 5a) = 3a^2 \cdot 3a^2 - 3a^2 \cdot 5a$     Use the distributive property.

$$= 9a^4 - 15a^3 \qquad \text{Multiply: } 3a^2 \cdot 3a^2 = 9a^4 \text{ and } 3a^2 \cdot 5a = 15a^3.$$

**b.**  $-2xz^2(2x - 3z + 2z^2)$

$$= -2xz^2 \cdot 2x - (-2xz^2) \cdot 3z + (-2xz^2) \cdot 2z^2 \qquad \text{Use the distributive}$$
$$\text{property.}$$

$$= -4x^2z^2 - (-6xz^3) + (-4xz^4) \qquad \begin{array}{l} \text{Multiply:} \\ -2xz^2 \cdot 2x = -4x^2z^2, \\ -2xz^2 \cdot 3z = -6xz^3, \\ \text{and } -2xz^2 \cdot 2z^2 = \\ -4xz^4. \end{array}$$

$$= -4x^2z^2 + 6xz^3 - 4xz^4 \qquad\qquad\qquad \blacksquare$$

**Self Check**  Multiply  **a.** $2p^3(3p^2 - 5p)$  and  **b.** $-5a^2b(3a + 2b - 4ab)$.

**Answers**  **a.** $6p^5 - 10p^4$,  **b.** $-15a^3b - 10a^2b^2 + 20a^3b^2$

■ MULTIPLYING A BINOMIAL BY A BINOMIAL

To multiply two binomials, we must use the distributive property more than once. For example, to multiply $2a - 4$ by $3a + 5$, we proceed as follows.

$$(2a - 4)(3a + 5) = (2a - 4) \cdot 3a + (2a - 4) \cdot 5 \qquad \text{Use the distributive} \\ \text{property.}$$

$$= 3a(2a - 4) + 5(2a - 4) \qquad \begin{array}{l} \text{Use the commutative} \\ \text{property of} \\ \text{multiplication.} \end{array}$$

$$= 3a \cdot 2a - 3a \cdot 4 + 5 \cdot 2a - 5 \cdot 4 \qquad \begin{array}{l} \text{Use the distributive} \\ \text{property.} \end{array}$$

$$= 6a^2 - 12a + 10a - 20 \qquad \begin{array}{l} \text{Do the} \\ \text{multiplications.} \end{array}$$

$$= 6a^2 - 2a - 20 \qquad\qquad \text{Combine like terms.}$$

This example suggests the following rule.

**Multiplying Two Binomials**
To multiply two binomials, multiply each term of one binomial by each term of the other binomial and combine like terms.

■ THE FOIL METHOD

We can use a shortcut method, called the **FOIL** method, to multiply binomials. FOIL is an acronym for **F**irst terms, **O**uter terms, **I**nner terms, and **L**ast terms. To use the FOIL method to multiply $2a - 4$ by $3a + 5$, we

**1.** multiply the **First** terms $2a$ and $3a$ to obtain $6a^2$,

**2.** multiply the **Outer** terms $2a$ and $5$ to obtain $10a$,

**3.** multiply the **Inner** terms $-4$ and $3a$ to obtain $-12a$, and

**4.** multiply the **Last** terms $-4$ and $5$ to obtain $-20$.

Then we simplify the resulting polynomial, if possible.

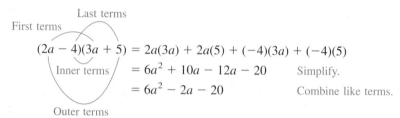

$$(2a - 4)(3a + 5) = 2a(3a) + 2a(5) + (-4)(3a) + (-4)(5)$$
$$= 6a^2 + 10a - 12a - 20 \qquad \text{Simplify.}$$
$$= 6a^2 - 2a - 20 \qquad \text{Combine like terms.}$$

 b,c

**EXAMPLE 3**    Find each product.

**a.** $(3x + 4)(2x - 3) = 3x(2x) + 3x(-3) + 4(2x) + 4(-3)$
$$= 6x^2 - 9x + 8x - 12$$
$$= 6x^2 - x - 12$$

**b.** $(2y - 7)(5y - 4) = 2y(5y) + 2y(-4) + (-7)(5y) + (-7)(-4)$
$$= 10y^2 - 8y - 35y + 28$$
$$= 10y^2 - 43y + 28$$

**c.** $(2r - 3s)(2r + t) = 2r(2r) + 2r(t) - 3s(2r) - 3s(t)$
$$= 4r^2 + 2rt - 6rs - 3st$$

■

**Self Check**    Find each product:   **a.** $(2a - 1)(3a + 2)$   and   **b.** $(5y - 2z)(2y + 3z)$.
**Answers**     **a.** $6a^2 + a - 2$,   **b.** $10y^2 + 11yz - 6z^2$

**EXAMPLE 4**    Simplify each expression.

**a.** $3(2x - 3)(x + 1)$

$$= 3(2x^2 + 2x - 3x - 3) \qquad \text{Use FOIL to multiply the binomials.}$$

$$= 3(2x^2 - x - 3) \qquad \text{Combine like terms.}$$
$$= 6x^2 - 3x - 9 \qquad \text{Use the distributive property to remove parentheses.}$$

**b.** $(x + 1)(x - 2) - 3x(x + 3)$

$$= x^2 - 2x + x - 2 - 3x^2 - 9x$$
$$= -2x^2 - 10x - 2 \qquad \text{Combine like terms.} \quad \blacksquare$$

**Self Check**    Simplify $(x + 3)(2x - 1) + 2x(x - 1)$.

**Answer**    $4x^2 + 3x - 3$

The products discussed in Example 5 are called **special products.**

**EXAMPLE 5**    Find each product.

**a.** $(x + y)^2 = (x + y)(x + y)$

$$= x^2 + xy + xy + y^2$$
$$= x^2 + 2xy + y^2$$

The square of the sum of two quantities has three terms: the square of the first quantity, plus twice the product of the quantities, plus the square of the second quantity.

**b.** $(x - y)^2 = (x - y)(x - y)$

$$= x^2 - xy - xy + y^2$$
$$= x^2 - 2xy + y^2$$

The square of the difference of two quantities has three terms: the square of the first quantity, minus twice the product of the quantities, plus the square of the second quantity.

**c.** $(x + y)(x - y) = x^2 - xy + xy - y^2$
$$= x^2 - y^2$$

The product of a sum and a difference of two quantities is a binomial. It is the product of the first quantities minus the product of the second quantities. Binomials that have the same terms, but different signs, are often called **conjugate binomials.**    $\blacksquare$

**Self Check**    Find each product:   **a.** $(p + 2)^2$,   **b.** $(p - 2)^2$,   and   **c.** $(p + 2q)(p - 2q)$.

**Answers**    **a.** $p^2 + 4p + 4$,   **b.** $p^2 - 4p + 4$,   **c.** $p^2 - 4q^2$

Because the products discussed in Example 5 occur so often, it is wise to learn their forms.

### Special Products

$$(x + y)^2 = x^2 + 2xy + y^2$$
$$(x - y)^2 = x^2 - 2xy + y^2$$
$$(x + y)(x - y) = x^2 - y^2$$

**WARNING!**   Note that $(x + y)^2 \neq x^2 + y^2$ and $(x - y)^2 \neq x^2 - y^2$.

### ■ MULTIPLYING A POLYNOMIAL BY A BINOMIAL

We must use the distributive property more than once to multiply a polynomial by a binomial. For example, to multiply $3x^2 + 3x - 5$ by $2x + 3$, we proceed as follows:

$$\begin{aligned}(2x + 3)(3x^2 + 3x - 5) &= (2x + 3)3x^2 + (2x + 3)3x - (2x + 3)5 \\ &= 3x^2(2x + 3) + 3x(2x + 3) - 5(2x + 3) \\ &= 6x^3 + 9x^2 + 6x^2 + 9x - 10x - 15 \\ &= 6x^3 + 15x^2 - x - 15\end{aligned}$$

This example suggests the following rule.

### Multiplying Polynomials

To multiply one polynomial by another, multiply each term of one polynomial by each term of the other polynomial and combine like terms.

It is often convenient to organize the work vertically.

**EXAMPLE 6**

**a.** Multiply:

$$\begin{array}{r} 3a^2 - 4a + 7 \\ 2a + 5 \\ \hline \end{array}$$

$2a(3a^2 - 4a + 7) \longrightarrow 6a^3 - 8a^2 + 14a$

$5(3a^2 - 4a + 7) \longrightarrow \underline{\phantom{6a^3} + 15a^2 - 20a + 35}$

$$6a^3 + 7a^2 - 6a + 35$$

**b.** Multiply:

$$\begin{array}{r} 3y^2 - 5y + 4 \\ - 4y^2 - 3 \\ \hline \end{array}$$

$-4y^2(3y^2 - 5y + 4) \longrightarrow -12y^4 + 20y^3 - 16y^2$

$-3(3y^2 - 5y + 4) \longrightarrow \underline{\phantom{-12y^4 + 20y^3} - 9y^2 + 15y - 12}$

$$-12y^4 + 20y^3 - 25y^2 + 15y - 12$$

■

Self Check   Multiply   **a.** $(3x + 2)(2x^2 - 4x + 5)$   and   **b.** $(-2x^2 + 3)(2x^2 - 4x - 1)$.
Answers     **a.** $6x^3 - 8x^2 + 7x + 10$,   **b.** $-4x^4 + 8x^3 + 8x^2 - 12x - 3$

### ■ MULTIPLYING BINOMIALS TO SOLVE EQUATIONS

To solve an equation such as $(x + 2)(x + 3) = x(x + 7)$, we can first use the FOIL method to remove the parentheses on the left-hand side, use the distributive property to remove parentheses on the right-hand side, and proceed as follows:

$$(x + 2)(x + 3) = x(x + 7)$$
$$x^2 + 3x + 2x + 6 = x^2 + 7x$$

$$3x + 2x + 6 = 7x \qquad \text{Subtract } x^2 \text{ from both sides.}$$
$$5x + 6 = 7x \qquad \text{Combine like terms.}$$
$$6 = 2x \qquad \text{Subtract } 5x \text{ from both sides.}$$
$$3 = x \qquad \text{Divide both sides by 2.}$$

*Check:*  $(x + 2)(x + 3) = x(x + 7)$
$$(3 + 2)(3 + 3) \stackrel{?}{=} 3(3 + 7) \qquad \text{Replace } x \text{ with 3.}$$
$$5(6) \stackrel{?}{=} 3(10) \qquad \text{Do the additions within parentheses.}$$

$$30 = 30$$

**EXAMPLE 7**   Solve $(x + 5)(x + 4) = (x + 9)(x + 10)$.

*Solution*   We use the FOIL method to remove parentheses on both sides of the equation. Then we proceed as follows:

$$(x + 5)(x + 4) = (x + 9)(x + 10)$$
$$x^2 + 4x + 5x + 20 = x^2 + 10x + 9x + 90$$
$$9x + 20 = 19x + 90 \qquad \text{Subtract } x^2 \text{ from both sides and combine like terms.}$$
$$20 = 10x + 90 \qquad \text{Subtract } 9x \text{ from both sides.}$$
$$-70 = 10x \qquad \text{Subtract 90 from both sides.}$$
$$-7 = x \qquad \text{Divide both sides by 10.}$$

*Check:*  $(x + 5)(x + 4) = (x + 9)(x + 10)$
$$(-7 + 5)(-7 + 4) \stackrel{?}{=} (-7 + 9)(-7 + 10) \qquad \text{Replace } x \text{ with } -7.$$
$$(-2)(-3) \stackrel{?}{=} (2)(3) \qquad \text{Do the additions within parentheses.}$$

$$6 = 6$$

■

Self Check  Solve $(x + 2)(x - 4) = (x + 6)(x - 3)$.

Answer  2

## ■ AN APPLICATION OF MULTIPLYING POLYNOMIALS

EXAMPLE 8  **Dimensions of a painting**  A square painting is surrounded by a border 2 inches wide. If the area of the border is 96 square inches, find the dimensions of the painting.

*Analyze the problem*  Refer to Figure 4-6, which shows a square painting surrounded by a border 2 inches wide. We know that the area of this border is 96 square inches, and we are to find the dimensions of the painting.

$x$          $x + 4$

FIGURE 4-6

*Form an equation*  Let $x$ represent the length of each side of the square painting. The outer rectangle is also a square, and its dimensions are $(x + 4)$ by $(x + 4)$ inches. Since the area of a square is the product of its length and width, the area of the larger square is $(x + 4)(x + 4)$, and the area of the painting is $x \cdot x$. If we subtract the area of the painting from the area of the larger square, the difference is 96 (the area of the border).

| The area of the large square | minus | the area of the square painting | = | the area of the border. |
|---|---|---|---|---|
| $(x + 4)(x + 4)$ | $-$ | $x \cdot x$ | $=$ | 96 |

*Solve the equation*

$(x + 4)(x + 4) - x^2 = 96$    $x \cdot x = x^2$.

$x^2 + 8x + 16 - x^2 = 96$    $(x + 4)(x + 4) = x^2 + 8x + 16$.

$8x + 16 = 96$    Combine like terms.

$8x = 80$    Subtract 16 from both sides.

$x = 10$    Divide both sides by 8.

*State the conclusion*  The dimensions of the painting are 10 inches by 10 inches.

*Check the result*  Check the result.

■

Orals *Find each product.*

**1.** $2x^2(3x - 1)$     **2.** $5y(2y^2 - 3)$     **3.** $7xy(x + y)$     **4.** $-2y(2x - 3y)$

**5.** $(x + 3)(x + 2)$                 **6.** $(x - 3)(x + 2)$

**7.** $(2x + 3)(x + 2)$             **8.** $(3x - 1)(3x + 1)$

**9.** $(x + 3)^2$                     **10.** $(x - 5)^2$

## EXERCISE 4.6

***REVIEW*** *In Exercises 1–4, tell which property of real numbers justifies each statement.*

**1.** $3(x + 5) = 3x + 3 \cdot 5$            **2.** $(x + 3) + y = x + (3 + y)$

**3.** $3(ab) = (ab)3$                  **4.** $a + 0 = a$

**5.** Solve $\frac{5}{3}(5y + 6) - 10 = 0$.         **6.** Solve $F = \dfrac{GMm}{d^2}$ for $m$.

***VOCABULARY AND CONCEPTS*** *Fill in each blank to make a true statement.*

**7.** A polynomial with one term is called a _____.

**8.** A _____ is a polynomial with two terms.

**9.** A polynomial with three terms is called a _____.

**10.** In the acronym FOIL, F stands for _____, O stands for _____, I stands for _____, and L stands for _____.

*In Exercises 11–14, consider the product $(2x + 5)(3x - 4)$.*

**11.** The product of the first terms is ____.

**12.** The product of the outer terms is ____.

**13.** The product of the inner terms is ____.

**14.** The product of the last terms is ____.

***PRACTICE*** *In Exercises 15–26, find each product.*

**15.** $(3x^2)(4x^3)$       **16.** $(-2a^3)(3a^2)$       **17.** $(3b^2)(-2b)(4b^3)$       **18.** $(3y)(2y^2)(-y^4)$

**19.** $(2x^2y^3)(3x^3y^2)$       **20.** $(-x^3y^6z)(x^2y^2z^7)$       **21.** $(x^2y^5)(x^2z^5)(-3y^2z^3)$       **22.** $(-r^4st^2)(2r^2st)(rst)$

**23.** $(x^2y^3)^5$       **24.** $(a^3b^2c)^4$       **25.** $(a^3b^2c)(abc^3)^2$       **26.** $(xyz^3)(xy^2z^2)^3$

*In Exercises 27–44, find each product.*

**27.** $3(x + 4)$       **28.** $-3(a - 2)$       **29.** $-4(t + 7)$       **30.** $6(s^2 - 3)$

**31.** $3x(x - 2)$       **32.** $4y(y + 5)$       **33.** $-2x^2(3x^2 - x)$       **34.** $4b^3(2b^2 - 2b)$

**35.** $3xy(x + y)$      **36.** $-4x^2(3x^2 - x)$      **37.** $2x^2(3x^2 + 4x - 7)$      **38.** $3y^3(2y^2 - 7y - 8)$

**39.** $\frac{1}{4}x^2(8x^5 - 4)$      **40.** $\frac{4}{3}a^2b(6a - 5b)$      **41.** $-\frac{2}{3}r^2t^2(9r - 3t)$      **42.** $-\frac{4}{5}p^2q(10p + 15q)$

**43.** $(3xy)(-2x^2y^3)(x + y)$      **44.** $(-2a^2b)(-3a^3b^2)(3a - 2b)$

*In Exercises 45–62, use the FOIL method to find each product.*

**45.** $(a + 4)(a + 5)$      **46.** $(y - 3)(y + 5)$      **47.** $(3x - 2)(x + 4)$      **48.** $(t + 4)(2t - 3)$

**49.** $(2a + 4)(3a - 5)$      **50.** $(2b - 1)(3b + 4)$      **51.** $(3x - 5)(2x + 1)$      **52.** $(2y - 5)(3y + 7)$

**53.** $(x + 3)(2x - 3)$      **54.** $(2x + 3)(2x - 5)$      **55.** $(2s + 3t)(3s - t)$      **56.** $(3a - 2b)(4a + b)$

**57.** $(x + y)(x + z)$      **58.** $(a - b)(x + y)$      **59.** $(u + v)(u + 2t)$      **60.** $(x - 5y)(a + 2y)$

**61.** $(-2r - 3s)(2r + 7s)$      **62.** $(-4a + 3)(-2a - 3)$

*In Exercises 63–70, find each product.*

**63.** $4x + 3$
$\underline{\quad x + 2}$

**64.** $5r + 6$
$\underline{\quad 2r - 1}$

**65.** $4x - 2y$
$\underline{3x + 5y}$

**66.** $5r + 6s$
$\underline{2r - \ s}$

**67.** $x^2 + x + 1$
$\underline{\qquad x - 1}$

**68.** $4x^2 - 2x + 1$
$\underline{\qquad 2x + 1}$

**69.** $(2x + 1)(x^2 + 3x - 1)$      **70.** $(3x - 2)(2x^2 - x + 2)$

*In Exercises 71–88, find each special product.*

**71.** $(x + 4)(x + 4)$      **72.** $(a + 3)(a + 3)$      **73.** $(t - 3)(t - 3)$      **74.** $(z - 5)(z - 5)$

**75.** $(r + 4)(r - 4)$      **76.** $(b + 2)(b - 2)$      **77.** $(x + 5)^2$      **78.** $(y - 6)^2$

**79.** $(2s + 1)(2s + 1)$      **80.** $(3t - 2)(3t - 2)$      **81.** $(4x + 5)(4x - 5)$      **82.** $(5z + 1)(5z - 1)$

**83.** $(x - 2y)^2$      **84.** $(3a + 2b)^2$      **85.** $(2a - 3b)^2$      **86.** $(2x + 5y)^2$

**87.** $(4x + 5y)(4x - 5y)$      **88.** $(6p + 5q)(6p - 5q)$

*In Exercises 89–98, find each product.*

**89.** $2(x - 4)(x + 1)$

**90.** $-3(2x + 3y)(3x - 4y)$

**91.** $3a(a + b)(a - b)$

**92.** $-2r(r + s)(r + s)$

**93.** $(4t + 3)(t^2 + 2t + 3)$

**94.** $(3x + y)(2x^2 - 3xy + y^2)$

**95.** $(-3x + y)(x^2 - 8xy + 16y^2)$

**96.** $(3x - y)(x^2 + 3xy - y^2)$

**97.** $(x - 2y)(x^2 + 2xy + 4y^2)$

**98.** $(2m + n)(4m^2 - 2mn + n^2)$

*In Exercises 99–108, simplify each expression.*

**99.** $2t(t + 2) + 3t(t - 5)$

**100.** $3y(y + 2) + (y + 1)(y - 1)$

**101.** $3xy(x + y) - 2x(xy - x)$

**102.** $(a + b)(a - b) - (a + b)(a + b)$

**103.** $(x + y)(x - y) + x(x + y)$

**104.** $(2x - 1)(2x + 1) + x(2x + 1)$

**105.** $(x + 2)^2 - (x - 2)^2$

**106.** $(x - 3)^2 - (x + 3)^2$

**107.** $(2s - 3)(s + 2) + (3s + 1)(s - 3)$

**108.** $(3x + 4)(2x - 2) - (2x + 1)(x + 3)$

*In Exercises 109–118, solve each equation.*

**109.** $(s - 4)(s + 1) = s^2 + 5$

**110.** $(y - 5)(y - 2) = y^2 - 4$

**111.** $z(z + 2) = (z + 4)(z - 4)$

**112.** $(z + 3)(z - 3) = z(z - 3)$

**113.** $(x + 4)(x - 4) = (x - 2)(x + 6)$

**114.** $(y - 1)(y + 6) = (y - 3)(y - 2) + 8$

**115.** $(a - 3)^2 = (a + 3)^2$

**116.** $(b + 2)^2 = (b - 1)^2$

**117.** $4 + (2y - 3)^2 = (2y - 1)(2y + 3)$

**118.** $7s^2 + (s - 3)(2s + 1) = (3s - 1)^2$

## APPLICATIONS

**119. Millstones** The radius of one millstone in Illustration 1 is 3 meters greater than the radius of the other, and their areas differ by $15\pi$ square meters. Find the radius of the larger millstone.

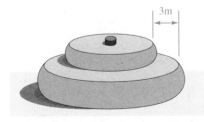

ILLUSTRATION 1

**120. Bookbinding** Two square sheets of cardboard used for making book covers differ in area by 44 square inches. An edge of the larger square is 2 inches greater than an edge of the smaller square. Find the length of an edge of the smaller square.

**121. Baseball** In major league baseball, the distance between bases is 30 feet greater than it is in softball. The bases in major league baseball mark the corners of a square that has an area 4,500 square feet greater than for softball. Find the distance between the bases in baseball.

**122. Pulley design** The radius of one pulley in Illustration 2 is 1 inch greater than the radius of the second pulley, and their areas differ by $4\pi$ square inches. Find the radius of the smaller pulley.

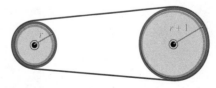

ILLUSTRATION 2

## WRITING

**123.** Describe the steps involved in finding the product of a binomial and its conjugate.

**124.** Writing the expression $(x + y)^2$ as $x^2 + y^2$ illustrates a common error. Explain.

## SOMETHING TO THINK ABOUT

**125.** The area of the square in Illustration 3 is the total of the areas of the four smaller regions. The picture illustrates the product $(x + y)^2$. Explain.

**126.** Illustration 4 represents the product of two binomials. Explain.

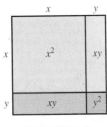

ILLUSTRATION 3

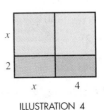

ILLUSTRATION 4

## 4.7 Dividing Polynomials by Monomials

■ DIVIDING A MONOMIAL BY A MONOMIAL ■ DIVIDING A POLYNOMIAL BY A MONOMIAL ■ AN APPLICATION OF DIVIDING A POLYNOMIAL BY A MONOMIAL

**Getting Ready** *Simplify each fraction.*

**1.** $\dfrac{4x^2y^3}{2xy}$

**2.** $\dfrac{9xyz}{9xz}$

**3.** $\dfrac{15x^2y}{10x}$

**4.** $\dfrac{6x^2y}{6xy^2}$

**5.** $\dfrac{(2x^2)(5y^2)}{10xy}$

**6.** $\dfrac{(5x^3y)(6xy^3)}{10x^4y^4}$

## ■ DIVIDING A MONOMIAL BY A MONOMIAL

We have seen that dividing by a number is equivalent to multiplying by its reciprocal. For example, dividing the number 8 by 2 gives the same answer as multiplying 8 by $\frac{1}{2}$.

$$\frac{8}{2} = 4 \qquad \text{and} \qquad \frac{1}{2} \cdot 8 = 4$$

In general, the following is true.

**Division**

$$\frac{a}{b} = \frac{1}{b} \cdot a \quad (b \neq 0)$$

Recall that to simplify a fraction, we write both its numerator and denominator as the product of several factors and then divide out all common factors. For example,

$$\frac{4}{6} = \frac{2 \cdot 2}{2 \cdot 3} \qquad \text{Factor: } 4 = 2 \cdot 2 \text{ and } 6 = 2 \cdot 3. \qquad \frac{20}{25} = \frac{4 \cdot 5}{5 \cdot 5} \qquad \text{Factor: } 20 = 4 \cdot 5 \text{ and } 25 = 5 \cdot 5.$$

$$= \frac{\overset{1}{\cancel{2}} \cdot 2}{\underset{1}{\cancel{2}} \cdot 3} \qquad \begin{array}{l}\text{Divide out the} \\ \text{common factor of 2.}\end{array} \qquad = \frac{4 \cdot \overset{1}{\cancel{5}}}{\underset{1}{\cancel{5}} \cdot 5} \qquad \begin{array}{l}\text{Divide out the} \\ \text{common factor of 5.}\end{array}$$

$$= \frac{2}{3} \qquad \tfrac{2}{2} = 1. \qquad = \frac{4}{5} \qquad \tfrac{5}{5} = 1.$$

We can use the same method to simplify algebraic fractions that contain variables.

$$\frac{3p^2q}{6pq^3} = \frac{3 \cdot p \cdot p \cdot q}{2 \cdot 3 \cdot p \cdot q \cdot q \cdot q} \qquad \text{Factor: } p^2 = p \cdot p, \ 6 = 2 \cdot 3, \text{ and } q^3 = q \cdot q \cdot q.$$

$$= \frac{\overset{1}{\cancel{3}} \cdot \overset{1}{\cancel{p}} \cdot p \cdot \overset{1}{\cancel{q}}}{2 \cdot \underset{1}{\cancel{3}} \cdot \underset{1}{\cancel{p}} \cdot \underset{1}{\cancel{q}} \cdot q \cdot q} \qquad \text{Divide out the common factors of 3, } p, \text{ and } q.$$

$$= \frac{p}{2q^2} \qquad \tfrac{3}{3} = 1, \ \tfrac{p}{p} = 1, \text{ and } \tfrac{q}{q} = 1.$$

To divide monomials, we can either use the previous method used for simplifying arithmetic fractions or use the rules of exponents.

 b    **EXAMPLE 1**    Simplify    **a.** $\dfrac{x^2y}{xy^2}$    and    **b.** $\dfrac{-8a^3b^2}{4ab^3}$.

*Solution*    ***Using Fractions***              ***Using the Rules of Exponents***

**a.** $\dfrac{x^2y}{xy^2} = \dfrac{x \cdot x \cdot y}{x \cdot y \cdot y}$              $\dfrac{x^2y}{xy^2} = x^{2-1}y^{1-2}$

$\quad\quad = \dfrac{\overset{1}{\cancel{x}} \cdot x \cdot \overset{1}{\cancel{y}}}{\underset{1}{\cancel{x}} \cdot y \cdot \underset{1}{\cancel{y}}}$              $= x^1y^{-1}$

$\quad\quad = \dfrac{x}{y}$              $= \dfrac{x}{y}$

***Using Fractions***             ***Using the Rules of Exponents***

**b.** $\dfrac{-8a^3b^2}{4ab^3} = \dfrac{-2 \cdot 4 \cdot a \cdot a \cdot a \cdot b \cdot b}{4 \cdot a \cdot b \cdot b \cdot b}$      $\dfrac{-8a^3b^2}{4ab^3} = \dfrac{(-1)2^3a^3b^2}{2^2ab^3}$

$$= \dfrac{-2 \cdot \overset{1}{\cancel{4}} \cdot \overset{1}{\cancel{a}} \cdot a \cdot a \cdot \overset{1}{\cancel{b}} \cdot \overset{1}{\cancel{b}}}{\underset{1}{\cancel{4}} \cdot \underset{1}{\cancel{a}} \cdot \underset{1}{\cancel{b}} \cdot b \cdot \underset{1}{\cancel{b}}}$$

$$= (-1)2^{3-2}a^{3-1}b^{2-3}$$
$$= (-1)2^1a^2b^{-1}$$

$$= \dfrac{-2a^2}{b} \qquad\qquad\qquad = \dfrac{-2a^2}{b} \qquad ■$$

**Self Check**    Simplify $\dfrac{-5p^2q^3}{10pq^4}$.

**Answer**    $\dfrac{-p}{2q}$

## ■ DIVIDING A POLYNOMIAL BY A MONOMIAL

To divide a polynomial with more than one term by a monomial, we write the division as a product, use the distributive property to remove parentheses, and simplify each resulting fraction.

**EXAMPLE 2**    Simplify $\dfrac{9x + 6y}{3xy}$.

**Solution**    $\dfrac{9x + 6y}{3xy} = \dfrac{1}{3xy}(9x + 6y)$

$$= \dfrac{9x}{3xy} + \dfrac{6y}{3xy} \qquad \text{Remove parentheses.}$$

$$= \dfrac{3}{y} + \dfrac{2}{x} \qquad \text{Simplify each fraction.} \qquad ■$$

**Self Check**    Simplify $\dfrac{4a - 8b}{4ab}$.

**Answer**    $\dfrac{1}{b} - \dfrac{2}{a}$

**EXAMPLE 3**    Simplify $\dfrac{6x^2y^2 + 4x^2y - 2xy}{2xy}$.

**Solution**    $\dfrac{6x^2y^2 + 4x^2y - 2xy}{2xy}$

$$= \dfrac{1}{2xy}(6x^2y^2 + 4x^2y - 2xy)$$

$$= \frac{6x^2y^2}{2xy} + \frac{4x^2y}{2xy} - \frac{2xy}{2xy} \qquad \text{Remove parentheses.}$$

$$= 3xy + 2x - 1 \qquad \text{Simplify each fraction.} \quad \blacksquare$$

**Self Check**  Simplify $\dfrac{9a^2b - 6ab^2 + 3ab}{3ab}$.

**Answer**  $3a - 2b + 1$

**EXAMPLE 4**  Simplify $\dfrac{12a^3b^2 - 4a^2b + a}{6a^2b^2}$.

**Solution**  $\dfrac{12a^3b^2 - 4a^2b + a}{6a^2b^2}$

$$= \frac{1}{6a^2b^2}(12a^3b^2 - 4a^2b + a)$$

$$= \frac{12a^3b^2}{6a^2b^2} - \frac{4a^2b}{6a^2b^2} + \frac{a}{6a^2b^2} \qquad \text{Remove parentheses.}$$

$$= 2a - \frac{2}{3b} + \frac{1}{6ab^2} \qquad \text{Simplify each fraction.} \quad \blacksquare$$

**Self Check**  Simplify $\dfrac{14p^3q + pq^2 - p}{7p^2q}$.

**Answer**  $2p + \frac{q}{7p} - \frac{1}{7pq}$

**EXAMPLE 5**  Simplify $\dfrac{(x - y)^2 - (x + y)^2}{xy}$.

**Solution**  $\dfrac{(x - y)^2 - (x + y)^2}{xy}$

$$= \frac{x^2 - 2xy + y^2 - (x^2 + 2xy + y^2)}{xy} \qquad \begin{array}{l}\text{Multiply the binomials in the}\\ \text{numerator.}\end{array}$$

$$= \frac{x^2 - 2xy + y^2 - x^2 - 2xy - y^2}{xy} \qquad \text{Remove parentheses.}$$

$$= \frac{-4xy}{xy} \qquad \text{Combine like terms.}$$

$$= -4 \qquad \text{Divide out } xy. \quad \blacksquare$$

**Self Check**  Simplify $\dfrac{(x + y)^2 - (x - y)^2}{xy}$.

**Answer**  4

## ■ AN APPLICATION OF DIVIDING A POLYNOMIAL BY A MONOMIAL

The area of the trapezoidal drainage ditch shown in Figure 4-7 is given by the formula $A = \frac{1}{2}h(B + b)$, where $B$ and $b$ are its bases and $h$ is its height. To solve the formula for $b$, we proceed as follows.

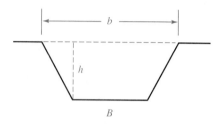

$$A = \frac{1}{2}h(B + b)$$

FIGURE 4-7

$$2A = 2 \cdot \frac{1}{2}h(B + b) \qquad \text{Multiply both sides by 2.}$$

$$2A = h(B + b) \qquad \text{Simplify: } 2 \cdot \frac{1}{2} = \frac{2}{2} = 1.$$

$$2A = hB + hb \qquad \text{Use the distributive property to remove parentheses.}$$

$$2A - hB = hB - hB + hb \qquad \text{Subtract } hB \text{ from both sides.}$$

$$2A - hB = hb \qquad \text{Combine like terms: } hB - hB = 0.$$

$$\frac{2A - hB}{h} = \frac{hb}{h} \qquad \text{Divide both sides by } h.$$

$$\frac{2A - hB}{h} = b \qquad \frac{hb}{h} = b.$$

**EXAMPLE 6**   Another student worked the previous problem in a different way and got a result of $b = \frac{2A}{h} - B$. Is this result correct?

*Solution*   To show that this result is correct, we must show that $\frac{2A - hB}{h} = \frac{2A}{h} - B$. We can do this by dividing $2A - hB$ by $h$.

$$\frac{2A - hB}{h} = \frac{1}{h}(2A - hB)$$

$$= \frac{2A}{h} - \frac{hB}{h} \qquad \text{Use the distributive property to remove parentheses.}$$

$$= \frac{2A}{h} - B \qquad \text{Simplify } \frac{hB}{h} = B.$$

The results are the same.                                                ■

*Self Check*   In Example 6, suppose another student got $2A - B$. Is this result correct?

*Answer*   no

Orals *Simplify each fraction.*

**1.** $\dfrac{4x^3y}{2xy}$　　　**2.** $\dfrac{6x^3y^2}{3x^3y}$　　　**3.** $\dfrac{35ab^2c^3}{7abc}$　　　**4.** $\dfrac{-14p^2q^5}{7pq^4}$

**5.** $\dfrac{(x+y)+(x-y)}{2x}$　　　　　**6.** $\dfrac{(2x^2-z)+(x^2+z)}{x}$

## EXERCISE 4.7

**REVIEW** *In Exercises 1–4, identify each polynomial as a monomial, a binomial, a trinomial, or none of these.*

**1.** $5a^2b + 2ab^2$　　　　　　　　　　**2.** $-3x^3y$

**3.** $-2x^3 + 3x^2 - 4x + 12$　　　　　**4.** $17t^2 - 15t + 27$

**5.** What is the degree of the trinomial $3x^2 - 2x + 4$?　　**6.** What is the numerical coefficient of the second term of the trinomial $-7t^2 - 5t + 17$?

**VOCABULARY AND CONCEPTS** *Fill in each blank to make a true statement.*

**7.** A _____ is an algebraic expression in which the exponents on the variables are whole numbers.

**8.** A _____ is a polynomial with one algebraic term.

**9.** A binomial is a polynomial with ____ terms.

**10.** A trinomial is a polynomial with _____ terms.

**11.** $\dfrac{1}{b} \cdot a = $ ___

**12.** $\dfrac{15x - 6y}{6xy} = $ ___ $\cdot (15x - 6y)$

**PRACTICE** *In Exercises 13–24, simplify each fraction.*

**13.** $\dfrac{5}{15}$　　　**14.** $\dfrac{64}{128}$　　　**15.** $\dfrac{-125}{75}$　　　**16.** $\dfrac{-98}{21}$

**17.** $\dfrac{120}{160}$　　　**18.** $\dfrac{70}{420}$　　　**19.** $\dfrac{-3,612}{-3,612}$　　　**20.** $\dfrac{-288}{-112}$

**21.** $\dfrac{-90}{360}$　　　**22.** $\dfrac{8,423}{-8,423}$　　　**23.** $\dfrac{5,880}{2,660}$　　　**24.** $\dfrac{-762}{366}$

*In Exercises 25–52, do each division by simplifying each fraction. Write all answers without using negative or zero exponents.*

**25.** $\dfrac{xy}{yz}$　　　**26.** $\dfrac{a^2b}{ab^2}$　　　**27.** $\dfrac{r^3s^2}{rs^3}$　　　**28.** $\dfrac{y^4z^3}{y^2z^2}$

**29.** $\dfrac{8x^3y^2}{4xy^3}$　　　**30.** $\dfrac{-3y^3z}{6yz^2}$　　　**31.** $\dfrac{12u^5v}{-4u^2v^3}$　　　**32.** $\dfrac{16rst^2}{-8rst^3}$

**33.** $\dfrac{-16r^3y^2}{-4r^2y^4}$　　　**34.** $\dfrac{35xyz^2}{-7x^2yz}$　　　**35.** $\dfrac{-65rs^2t}{15r^2s^3t}$　　　**36.** $\dfrac{112u^3z^6}{-42u^3z^6}$

**37.** $\dfrac{x^2x^3}{xy^6}$

**38.** $\dfrac{(xy)^2}{x^2y^3}$

**39.** $\dfrac{(a^3b^4)^3}{ab^4}$

**40.** $\dfrac{(a^2b^3)^3}{a^6b^6}$

**41.** $\dfrac{15(r^2s^3)^2}{-5(rs^5)^3}$

**42.** $\dfrac{-5(a^2b)^3}{10(ab^2)^3}$

**43.** $\dfrac{-32(x^3y)^3}{128(x^2y^2)^3}$

**44.** $\dfrac{68(a^6b^7)^2}{-96(abc^2)^3}$

**45.** $\dfrac{(5a^2b)^3}{(2a^2b^2)^3}$

**46.** $\dfrac{-(4x^3y^3)^2}{(x^2y^4)^8}$

**47.** $\dfrac{-(3x^3y^4)^3}{-(9x^4y^5)^2}$

**48.** $\dfrac{(2r^3s^2t)^2}{-(4r^2s^2t^2)^2}$

**49.** $\dfrac{(a^2a^3)^4}{(a^4)^3}$

**50.** $\dfrac{(b^3b^4)^5}{(bb^2)^2}$

**51.** $\dfrac{(z^3z^{-4})^3}{(z^{-3})^2}$

**52.** $\dfrac{(t^{-3}t^5)}{(t^2)^{-3}}$

*In Exercises 53–66, do each division.*

**53.** $\dfrac{6x+9y}{3xy}$

**54.** $\dfrac{8x+12y}{4xy}$

**55.** $\dfrac{5x-10y}{25xy}$

**56.** $\dfrac{2x-32}{16x}$

**57.** $\dfrac{3x^2+6y^3}{3x^2y^2}$

**58.** $\dfrac{4a^2-9b^2}{12ab}$

**59.** $\dfrac{15a^3b^2-10a^2b^3}{5a^2b^2}$

**60.** $\dfrac{9a^4b^3-16a^3b^4}{12a^2b}$

**61.** $\dfrac{4x-2y+8z}{4xy}$

**62.** $\dfrac{5a^2+10b^2-15ab}{5ab}$

**63.** $\dfrac{12x^3y^2-8x^2y-4x}{4xy}$

**64.** $\dfrac{12a^2b^2-8a^2b-4ab}{4ab}$

**65.** $\dfrac{-25x^2y+30xy^2-5xy}{-5xy}$

**66.** $\dfrac{-30a^2b^2-15a^2b-10ab^2}{-10ab}$

*In Exercises 67–76, simplify each numerator and do the division.*

**67.** $\dfrac{5x(4x-2y)}{2y}$

**68.** $\dfrac{9y^2(x^2-3xy)}{3x^2}$

**69.** $\dfrac{(-2x)^3+(3x^2)^2}{6x^2}$

**70.** $\dfrac{(-3x^2y)^3+(3xy^2)^3}{27x^3y^4}$

**71.** $\dfrac{4x^2y^2-2(x^2y^2+xy)}{2xy}$

**72.** $\dfrac{-5a^3b-5a(ab^2-a^2b)}{10a^2b^2}$

**73.** $\dfrac{(3x-y)(2x-3y)}{6xy}$

**74.** $\dfrac{(2m-n)(3m-2n)}{-3m^2n^2}$

**75.** $\dfrac{(a+b)^2-(a-b)^2}{2ab}$

**76.** $\dfrac{(x-y)^2+(x+y)^2}{2x^2y^2}$

**APPLICATIONS**

**77. Reconciling formulas** Are the formulas
$$l=\dfrac{P-2w}{2} \quad \text{and} \quad l=\dfrac{P}{2}-w$$
the same?

**78. Reconciling formulas** Are the formulas
$$r=\dfrac{G+2b}{2b} \quad \text{and} \quad r=\dfrac{G}{2b}+b$$
the same?

**79. Phone bills** On a phone bill, the following formulas are given to compute the average cost per minute of $x$ minutes of phone usage. Are they equivalent?

$$C = \frac{0.15x + 12}{x} \quad \text{and} \quad C = 0.15 + \frac{12}{x}$$

**80. Electric bills** On an electric bill, the following formulas are given to compute the average cost of $x$ kwh of electricity. Are they equivalent?

$$C = \frac{0.08x + 5}{x} \quad \text{and} \quad C = 0.08x + \frac{5}{x}$$

*WRITING*

**81.** Describe how you would simplify the fraction

$$\frac{4x^2y + 8xy^2}{4xy}$$

**82.** A fellow student attempts to simplify the fraction $\frac{3x + 5}{x + 5}$ by dividing out the $x + 5$:

$$\frac{3x + 5}{x + 5} = \frac{3\cancel{x + 5}}{\cancel{x + 5}} = 3$$

What would you say to him?

*SOMETHING TO THINK ABOUT*

**83.** If $x = 501$, evaluate $\dfrac{x^{500} - x^{499}}{x^{499}}$.

**84.** An exercise reads as follows:

$$\text{Simplify } \frac{3x^3y + 6xy^2}{3xy^3}$$

It contains a misprint: one mistyped letter or digit. The correct answer is $\frac{x^2}{y} + 2$. Fix the exercise.

# 4.8 Dividing Polynomials by Polynomials

■ DIVIDING POLYNOMIALS BY POLYNOMIALS ■ WRITING POWERS IN DESCENDING ORDER ■ THE CASE OF THE MISSING TERMS

Getting Ready    *Divide.*

**1.** $12\overline{)156}$    **2.** $17\overline{)357}$    **3.** $13\overline{)247}$    **4.** $19\overline{)247}$

## ■ DIVIDING POLYNOMIALS BY POLYNOMIALS

To divide one polynomial by another, we use a method similar to long division in arithmetic. We will illustrate the method with several examples.

**EXAMPLE 1**    Divide $x^2 + 5x + 6$ by $x + 2$.

*Solution*    Here the divisor is $x + 2$, and the dividend is $x^2 + 5x + 6$.

*Step 1:*
$$x + 2\overline{)x^2 + 5x + 6}$$
with $x$ above the division symbol

How many times does $x$ divide $x^2$? $x^2/x = x$. Place the $x$ above the division symbol.

*Step 2:*
$$
\begin{array}{r}
x \phantom{+ 5x + 6} \\
x + 2 \overline{\smash{)}x^2 + 5x + 6} \\
x^2 + 2x \phantom{+ 6}
\end{array}
$$

Multiply each term in the divisor by $x$. Place the product under $x^2 + 5x$ and draw a line.

*Step 3:*
$$
\begin{array}{r}
x \phantom{+ 5x + 6} \\
x + 2 \overline{\smash{)}x^2 + 5x + 6} \\
\underline{x^2 + 2x} \phantom{+ 6} \\
3x + 6
\end{array}
$$

Subtract $x^2 + 2x$ from $x^2 + 5x$ by adding the negative of $x^2 + 2x$ to $x^2 + 5x$.

Bring down the 6.

*Step 4:*
$$
\begin{array}{r}
x \phantom{+} + 3 \phantom{6} \\
x + 2 \overline{\smash{)}x^2 + 5x + 6} \\
x^2 + 2x \phantom{+ 6} \\
3x + 6
\end{array}
$$

How many times does $x$ divide $3x$? $3x/x = +3$. Place the $+3$ above the division symbol.

*Step 5:*
$$
\begin{array}{r}
x \phantom{+} + 3 \phantom{6} \\
x + 2 \overline{\smash{)}x^2 + 5x + 6} \\
\underline{x^2 + 2x} \phantom{+ 6} \\
3x + 6 \\
3x + 6
\end{array}
$$

Multiply each term in the divisor by 3. Place the product under the $3x + 6$ and draw a line.

*Step 6:*
$$
\begin{array}{r}
x \phantom{+} + 3 \phantom{6} \\
x + 2 \overline{\smash{)}x^2 + 5x + 6} \\
\underline{x^2 + 2x} \phantom{+ 6} \\
3x + 6 \\
\underline{3x + 6} \\
0
\end{array}
$$

Subtract $3x + 6$ from $3x + 6$ by adding the negative of $3x + 6$.

The quotient is $x + 3$, and the remainder is 0.

*Step 7* Check the work by verifying that $x + 2$ times $x + 3$ is $x^2 + 5x + 6$.

$$(x + 2)(x + 3) = x^2 + 3x + 2x + 6$$
$$= x^2 + 5x + 6$$

The answer checks. ∎

**Self Check**

**Answer**

Divide $x^2 + 7x + 12$ by $x + 3$.

$x + 4$

**EXAMPLE 2**

Divide $\dfrac{6x^2 - 7x - 2}{2x - 1}$.

*Solution*

Here the divisor is $2x - 1$, and the dividend is $6x^2 - 7x - 2$.

*Step 1:*
$$
\begin{array}{r}
3x \phantom{6x^2 - 7x - 2} \\
2x - 1 \overline{\smash{)}6x^2 - 7x - 2}
\end{array}
$$

How many times does $2x$ divide $6x^2$? $6x^2/2x = 3x$. Place the $3x$ above the division symbol.

*Step 2:*

$$\begin{array}{r} 3x \phantom{0000000} \\ 2x - 1{\overline{\smash{\big)}\,6x^2 - 7x - 2}} \\ \underline{6x^2 - 3x} \phantom{000} \end{array}$$

Multiply each term in the divisor by $3x$. Place the product under $6x^2 - 7x$ and draw a line.

*Step 3:*

$$\begin{array}{r} 3x \phantom{0000000} \\ 2x - 1{\overline{\smash{\big)}\,6x^2 - 7x - 2}} \\ \underline{6x^2 - 3x} \phantom{000} \\ - 4x - 2 \end{array}$$

Subtract $6x^2 - 3x$ from $6x^2 - 7x$ by adding the negative of $6x^2 - 3x$ to $6x^2 - 7x$.

Bring down the $-2$.

*Step 4:*

$$\begin{array}{r} 3x \phantom{00} - 2 \phantom{00} \\ 2x - 1{\overline{\smash{\big)}\,6x^2 - 7x - 2}} \\ \underline{6x^2 - 3x} \phantom{000} \\ - 4x - 2 \end{array}$$

How many times does $2x$ divide $-4x$? $-4x/2x = -2$. Place the $-2$ above the division symbol.

*Step 5:*

$$\begin{array}{r} 3x \phantom{00} - 2 \phantom{00} \\ 2x - 1{\overline{\smash{\big)}\,6x^2 - 7x - 2}} \\ \underline{6x^2 - 3x} \phantom{000} \\ - 4x - 2 \\ \underline{- 4x + 2} \end{array}$$

Multiply each term in the divisor by $-2$. Place the product under the $-4x - 2$ and draw a line.

*Step 6:*

$$\begin{array}{r} 3x \phantom{00} - 2 \phantom{00} \\ 2x - 1{\overline{\smash{\big)}\,6x^2 - 7x - 2}} \\ \underline{6x^2 - 3x} \phantom{000} \\ - 4x - 2 \\ \underline{- 4x + 2} \\ - 4 \end{array}$$

Subtract $-4x + 2$ from $-4x - 2$ by adding the negative of $-4x + 2$.

Here the quotient is $3x - 2$, and the remainder is $-4$. It is common to write the answer in quotient $+ \frac{\text{remainder}}{\text{divisor}}$ form:

$$3x - 2 + \frac{-4}{2x - 1}$$

where the fraction $\dfrac{-4}{2x - 1}$ is formed by dividing the remainder by the divisor.

*Step 7:* To check the answer, we multiply $3x - 2 + \frac{-4}{2x-1}$ by $2x - 1$. The product should be the dividend.

$$\begin{aligned} (2x - 1)\left( 3x - 2 + \frac{-4}{2x - 1} \right) &= (2x - 1)(3x - 2) + (2x - 1)\left( \frac{-4}{2x - 1} \right) \\ &= (2x - 1)(3x - 2) - 4 \\ &= 6x^2 - 4x - 3x + 2 - 4 \\ &= 6x^2 - 7x - 2 \end{aligned}$$

Because the result is the dividend, the answer checks.    ■

Self Check    Divide $\dfrac{8x^2 + 6x - 3}{2x + 3}$.

Answer    $4x - 3 + \dfrac{6}{2x+3}$

### ■ WRITING POWERS IN DESCENDING ORDER

The division method works best when exponents of the terms in the divisor and the dividend are written in descending order. This means that the term involving the highest power of $x$ appears first, the term involving the second-highest power of $x$ appears second, and so on. For example, the terms in

$$3x^3 + 2x^2 - 7x + 5$$

have their exponents written in descending order.

If the powers in the dividend or divisor are not in descending order, we can use the commutative property of addition to write them that way.

EXAMPLE 3    Divide $4x^2 + 2x^3 + 12 - 2x$ by $x + 3$.

Solution    We write the dividend so that the exponents are in descending order and divide.

$$
\begin{array}{r}
2x^2 - 2x\ + 4 \\
x + 3{\overline{\smash{\big)}\,2x^3 + 4x^2 - 2x + 12}} \\
\underline{2x^3 + 6x^2} \\
-2x^2 - 2x \\
\underline{-2x^2 - 6x} \\
+4x + 12 \\
\underline{+4x + 12}
\end{array}
$$

*Check:* $(x + 3)(2x^2 - 2x + 4) = 2x^3 - 2x^2 + 4x + 6x^2 - 6x + 12$
$= 2x^3 + 4x^2 - 2x + 12$    ■

Self Check    Divide $x^2 - 10x + 6x^3 + 4$ by $2x - 1$.
Answer    $3x^2 + 2x - 4$

### ■ THE CASE OF THE MISSING TERMS

When we write the terms of a dividend in descending powers of $x$, we may notice that some powers of $x$ are missing. For example, in the dividend of

$$x + 1{\overline{\smash{\big)}\,3x^4 - 7x^2 - 3x + 15}}$$

the term involving $x^3$ is missing. When this happens, we should either write the term with a coefficient of 0 or leave a blank space for it. In this case, we would write the dividend as

$$3x^4 + 0x^3 - 7x^2 - 3x + 15 \qquad \text{or} \qquad 3x^4 \qquad - 7x^2 - 3x + 15$$

**EXAMPLE 4**  Divide $\dfrac{x^2 - 4}{x + 2}$.

*Solution*  Since $x^2 - 4$ does not have a term involving $x$, we must either include the term $0x$ or leave a space for it.

$$
\begin{array}{r}
x \phantom{.} - 2 \\
x + 2 \overline{) x^2 + 0x - 4} \\
\underline{x^2 + 2x} \phantom{aaa} \\
-2x - 4 \\
\underline{-2x - 4}
\end{array}
$$

*Check:* $(x + 2)(x - 2) = x^2 - 2x + 2x - 4$
$$= x^2 - 4 \qquad \blacksquare$$

**Self Check**  Divide $\dfrac{x^2 - 9}{x - 3}$.

*Answer*  $x + 3$

**EXAMPLE 5**  Divide $x^3 + y^3$ by $x + y$.

*Solution*  We write $x^3 + y^3$ leaving spaces for the missing terms and proceed as follows.

$$
\begin{array}{r}
x^2 - xy \phantom{a} + y^2 \\
x + y \overline{) x^3 \phantom{aaaaaaaaaa} + y^3} \\
\underline{x^3 + x^2y} \phantom{aaaaaaaa} \\
-x^2y \phantom{aaaaaaa} \\
\underline{-x^2y - xy^2} \phantom{aa} \\
+xy^2 + y^3 \\
\underline{xy^2 + y^3}
\end{array}
$$

*Check:* $(x + y)(x^2 - xy + y^2) = x^3 - x^2y + xy^2 + x^2y - xy^2 + y^3$
$$= x^3 + y^3 \qquad \blacksquare$$

**Self Check**  Divide $x^3 - y^3$ by $x - y$.
*Answer*  $x^2 + xy + y^2$

Orals *Divide, and give the answer in* quotient $+ \frac{\text{remainder}}{\text{divisor}}$ *form.*

**1.** $x\overline{)2x + 3}$  **2.** $x\overline{)3x - 5}$  **3.** $x + 1\overline{)2x + 3}$

**4.** $x + 1\overline{)3x + 5}$  **5.** $x + 1\overline{)x^2 + x}$  **6.** $x + 2\overline{)x^2 + 2x}$

# EXERCISE 4.8

## REVIEW

**1.** List the composite numbers between 20 and 30.

**2.** Graph the set of prime numbers between 10 and 20 on a number line.

$$\xleftarrow{\quad} \overset{10 \quad 11 \quad 12 \quad 13 \quad 14 \quad 15 \quad 16 \quad 17 \quad 18 \quad 19 \quad 20}{\mid \quad \mid \quad \mid \quad \mid \quad \mid \quad \mid \quad \mid \quad \mid \quad \mid \quad \mid \quad \mid} \xrightarrow{\quad}$$

*Let $a = -2$ and $b = 3$. Evaluate each expression.*

**3.** $|a - b|$  **4.** $|a + b|$  **5.** $-|a^2 - b^2|$  **6.** $a - |-b|$

*Simplify each expression.*

**7.** $3(2x^2 - 4x + 5) + 2(x^2 + 3x - 7)$

**8.** $-2(y^3 + 2y^2 - y) - 3(3y^3 + y)$

## VOCABULARY AND CONCEPTS *Fill in each blank to make a true statement.*

**9.** In the division $x + 1\overline{)x^2 + 2x + 1}$, $x + 1$ is called the _____, and $x^2 + 2x + 1$ is called the _____.

**10.** The answer to a division problem is called the _____.

**11.** If a division does not come out even, the leftover part is called a _____.

**12.** The exponents in $2x^4 + 3x^3 + 4x^2 - 7x - 2$ are said to be written in _____ order.

*Write each polynomial with the powers in descending order.*

**13.** $4x^3 + 7x - 2x^2 + 6$

**14.** $5x^2 + 7x^3 - 3x - 9$

**15.** $9x + 2x^2 - x^3 + 6x^4$

**16.** $7x^5 + x^3 - x^2 + 2x^4$

*Identify the missing terms in each polynomial.*

**17.** $5x^4 + 2x^2 - 1$

**18.** $-3x^5 - 2x^3 + 4x - 6$

## PRACTICE *In Exercises 19–24, do each division.*

**19.** Divide $x^2 + 4x + 4$ by $x + 2$.

**20.** Divide $x^2 - 5x + 6$ by $x - 2$.

**21.** Divide $y^2 + 13y + 12$ by $y + 1$.

**22.** Divide $z^2 - 7z + 12$ by $z - 3$.

**23.** Divide $a^2 + 2ab + b^2$ by $a + b$.

**24.** Divide $a^2 - 2ab + b^2$ by $a - b$.

*In Exercises 25–30, do each division.*

**25.** $\dfrac{6a^2 + 5a - 6}{2a + 3}$

**26.** $\dfrac{8a^2 + 2a - 3}{2a - 1}$

**27.** $\dfrac{3b^2 + 11b + 6}{3b + 2}$

**28.** $\dfrac{3b^2 - 5b + 2}{3b - 2}$

**29.** $\dfrac{2x^2 - 7xy + 3y^2}{2x - y}$

**30.** $\dfrac{3x^2 + 5xy - 2y^2}{x + 2y}$

*In Exercises 31–42, write the powers of x in descending order and do each division.*

**31.** $5x + 3\overline{)11x + 10x^2 + 3}$

**32.** $2x - 7\overline{)-x - 21 + 2x^2}$

**33.** $4 + 2x\overline{)-10x - 28 + 2x^2}$

**34.** $1 + 3x\overline{)9x^2 + 1 + 6x}$

**35.** $2x - y\overline{)xy - 2y^2 + 6x^2}$

**36.** $2y + x\overline{)3xy + 2x^2 - 2y^2}$

**37.** $x + 3y\overline{)2x^2 - 3y^2 + 5xy}$

**38.** $2x - 3y\overline{)2x^2 - 3y^2 - xy}$

**39.** $3x - 2y\overline{)-10y^2 + 13xy + 3x^2}$

**40.** $2x + 3y\overline{)-12y^2 + 10x^2 + 7xy}$

**41.** $4x + y\overline{)-19xy + 4x^2 - 5y^2}$

**42.** $x - 4y\overline{)5x^2 - 4y^2 - 19xy}$

*In Exercises 43–48, do each division.*

**43.** $2x + 3\overline{)2x^3 + 7x^2 + 4x - 3}$

**44.** $2x - 1\overline{)2x^3 - 3x^2 + 5x - 2}$

**45.** $3x + 2\overline{)6x^3 + 10x^2 + 7x + 2}$

**46.** $4x + 3\overline{)4x^3 - 5x^2 - 2x + 3}$

**47.** $2x + y\overline{)2x^3 + 3x^2y + 3xy^2 + y^3}$

**48.** $3x - 2y\overline{)6x^3 - x^2y + 4xy^2 - 4y^3}$

*In Exercises 49–58, do each division. If there is a remainder, leave the answer in* quotient $+ \frac{\text{remainder}}{\text{divisor}}$ *form.*

**49.** $\dfrac{2x^2 + 5x + 2}{2x + 3}$

**50.** $\dfrac{3x^2 - 8x + 3}{3x - 2}$

**51.** $\dfrac{4x^2 + 6x - 1}{2x + 1}$

**52.** $\dfrac{6x^2 - 11x + 2}{3x - 1}$

**53.** $\dfrac{x^3 + 3x^2 + 3x + 1}{x + 1}$

**54.** $\dfrac{x^3 + 6x^2 + 12x + 8}{x + 2}$

**55.** $\dfrac{2x^3 + 7x^2 + 4x + 3}{2x + 3}$

**56.** $\dfrac{6x^3 + x^2 + 2x + 1}{3x - 1}$

**57.** $\dfrac{2x^3 + 4x^2 - 2x + 3}{x - 2}$

**58.** $\dfrac{3y^3 - 4y^2 + 2y + 3}{y + 3}$

*In Exercises 59–68, do each division.*

**59.** $\dfrac{x^2 - 1}{x - 1}$

**60.** $\dfrac{x^2 - 9}{x + 3}$

**61.** $\dfrac{4x^2 - 9}{2x + 3}$

**62.** $\dfrac{25x^2 - 16}{5x - 4}$

**63.** $\dfrac{x^3 + 1}{x + 1}$

**64.** $\dfrac{x^3 - 8}{x - 2}$

**65.** $\dfrac{a^3 + a}{a + 3}$

**66.** $\dfrac{y^3 - 50}{y - 5}$

**67.** $3x - 4\overline{)15x^3 - 23x^2 + 16x}$

**68.** $2y + 3\overline{)21y^2 + 6y^3 - 20}$

## WRITING

**69.** Distinguish among *dividend, divisor, quotient,* and *remainder.*

**70.** How would you check the results of a division?

## SOMETHING TO THINK ABOUT

**71.** What's wrong here?

$$
\begin{array}{r}
x + 1 \phantom{xxx} \\
x - 2 \overline{)\ x^2 + 3x - 2} \\
\underline{x^2 - 2x} \phantom{xx} \\
x - 2 \\
\underline{x - 2} \\
0
\end{array}
$$

**72.** What's wrong here?

$$
\begin{array}{r}
3x \phantom{xxxx} \\
x + 2 \overline{)\ 3x^2 + 10x + 7} \\
\underline{3x^2 + \ 9x} \phantom{x} \\
x + 7
\end{array}
$$

The quotient is $3x$ and the remainder is $x + 7$.

■ ■ ■ ■ ■ ■ ■ ■ ■

MATHEMATICS
IN MEDICINE

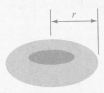

ILLUSTRATION 1

The area of a circle is given by the formula $A = \pi r^2$. The bulk of the surface area of the red blood cell in Illustration 1 is contained on its top and bottom. That area is $2\pi r^2$, twice the area of one circle. If there are $N$ discs, their total surface area $T$ will be $N$ times the surface area of a single disc: $T = 2N\pi r^2$.

To find the total surface area of the oxygen-carrying red cells, we first express the given quantities in scientific notation.

Radius $= r = 0.00015$ in. $= 1.5 \times 10^{-4}$ in.

Quantity $= N = 25$ trillion $= 2.5 \times 10^{13}$

Then we substitute these values into the formula for total surface area.

$T = 2N\pi r^2$

$T = 2(2.5 \times 10^{13})(3.14)(1.5 \times 10^{-4})^2.$     Use 3.14 for $\pi$.

$= 2(2.5)(3.14)(1.5)^2 \times 10^{13} \times 10^{-8}$

$\approx 35.3 \times 10^5$

$\approx 3.53 \times 10^6$

$\approx 3,530,000$

The total surface area of the red blood cells is over $3\frac{1}{2}$ million square inches, or approximately 24,500 square feet—almost one-half the area of a football field!

■ ■ ■ ■ ■ ■ ■ ■ ■ ■ **PROJECT**

There is a pattern in the behavior of polynomials. To discover it, consider the polynomial $2x^2 - 3x - 5$. First, evaluate the polynomial at $x = 1$ and $x = 3$. Then divide the polynomial by $x - 1$ and again by $x - 3$.

1. What do you notice about the remainders of these divisions?

2. Try others. For example, evaluate the polynomial at $x = 2$ and then divide by $x - 2$.

3. Can you make the pattern hold when you evaluate the polynomial at $x = -2$?

4. Does the pattern hold for other polynomials? Try some polynomials of your own, experiment, and report your conclusions.

# C H A P T E R   S U M M A R Y

## CONCEPTS

## REVIEW EXERCISES

### SECTION 4.1

## Natural-Number Exponents

If $n$ is a natural number, then

$$\overbrace{x^n = x \cdot x \cdot x \cdot \cdots \cdot x}^{n \text{ factors of } x}$$

**1.** Write each expression without using exponents.
   **a.** $(-3x)^4$
   **b.** $\left(\dfrac{1}{2}pq\right)^3$

**2.** Evaluate each expression.
   **a.** $5^3$    **b.** $3^5$
   **c.** $(-8)^2$    **d.** $-8^2$
   **e.** $3^2 + 2^2$    **f.** $(3 + 2)^2$

If $m$ and $n$ are integers, then

$$x^m x^n = x^{m+n}$$
$$(x^m)^n = x^{m \cdot n}$$
$$(xy)^n = x^n y^n$$
$$\left(\frac{x}{y}\right)^n = \frac{x^n}{y^n} \quad (y \neq 0)$$
$$\frac{x^m}{x^n} = x^{m-n} \quad (x \neq 0)$$

**3.** Do the operations and simplify.
   **a.** $x^3 x^2$    **b.** $x^2 x^7$
   **c.** $(y^7)^3$    **d.** $(x^{21})^2$
   **e.** $(ab)^3$    **f.** $(3x)^4$
   **g.** $b^3 b^4 b^5$    **h.** $-z^2(z^3 y^2)$
   **i.** $(16s)^2 s$    **j.** $-3y(y^5)$
   **k.** $(x^2 x^3)^3$    **l.** $(2x^2 y)^2$
   **m.** $\dfrac{x^7}{x^3}$    **n.** $\left(\dfrac{x^2 y}{xy^2}\right)^2$
   **o.** $\dfrac{8(y^2 x)^2}{4(yx^2)^2}$    **p.** $\dfrac{(5y^2 z^3)^3}{25(yz)^5}$

### SECTION 4.2

## Zero and Negative-Integer Exponents

$$x^0 = 1 \quad (x \neq 0)$$

$$x^{-n} = \frac{1}{x^n} \quad (x \neq 0)$$

**4.** Write each expression without using negative exponents or parentheses.
   **a.** $x^0$    **b.** $(3x^2 y^2)^0$
   **c.** $(3x^0)^2$    **d.** $(3x^2 y^0)^2$
   **e.** $x^{-3}$    **f.** $x^{-2} x^3$

**g.** $y^4y^{-3}$

**h.** $\dfrac{x^3}{x^{-7}}$

**i.** $(x^{-3}x^4)^{-2}$

**j.** $(a^{-2}b)^{-3}$

**k.** $\left(\dfrac{x^2}{x}\right)^{-5}$

**l.** $\left(\dfrac{15z^4}{5z^3}\right)^{-2}$

---

| SECTION 4.3 | *Scientific Notation* |
|---|---|

A number is written in scientific notation if it is written as the product of a number between 1 (including 1) and 10 and an integer power of 10.

**5.** Write each number in scientific notation.
  **a.** 728
  **b.** 9,370
  **c.** 0.0136
  **d.** 0.00942
  **e.** 7.73
  **f.** $753 \times 10^3$
  **g.** $0.018 \times 10^{-2}$
  **h.** $600 \times 10^2$

**6.** Write each number in standard notation.
  **a.** $7.26 \times 10^5$
  **b.** $3.91 \times 10^{-4}$
  **c.** $2.68 \times 10^0$
  **d.** $5.76 \times 10^1$
  **e.** $739 \times 10^{-2}$
  **f.** $0.437 \times 10^{-3}$
  **g.** $\dfrac{(0.00012)(0.00004)}{0.00000016}$
  **h.** $\dfrac{(4,800)(20,000)}{600,000}$

---

| SECTION 4.4 | *Polynomials* |
|---|---|

**7.** Find the degree of each polynomial and classify it as a monomial, a binomial, or a trinomial.
  **a.** $13x^7$
  **b.** $5^3x + x^2$
  **c.** $-3x^5 + x - 1$
  **d.** $9xy + 21x^3y^2$

When a number is substituted for the variable in a polynomial, the polynomial takes on a numerical value.

**8.** Evaluate $3x + 2$ for each value of $x$.
  **a.** $x = 3$
  **b.** $x = 0$
  **c.** $x = -2$
  **d.** $x = \dfrac{2}{3}$

**9.** Evaluate $5x^4 - x$ for each value of $x$.
  **a.** $x = 3$
  **b.** $x = 0$
  **c.** $x = -2$
  **d.** $x = (-0.3)$

Any equation in $x$ and $y$ where each value of $x$ determines a single value of $y$ is a **function**. We say that $y$ is a function of $x$.

**10.** If $y = f(x) = x^2 - 4$, find each value.
  **a.** $f(0)$
  **b.** $f(5)$
  **c.** $f(-2)$
  **d.** $f\left(\dfrac{1}{2}\right)$

**11.** Graph each polynomial function.

**a.** $y = f(x) = x^2 - 5$

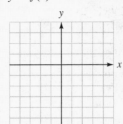

**b.** $y = f(x) = x^3 - 2$

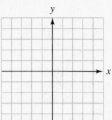

## SECTION 4.5 — Adding and Subtracting Polynomials

When adding or subtracting polynomials, combine like terms by adding or subtracting the numerical coefficients and using the same variables and the same exponents.

**12.** Simplify each expression.

**a.** $3x + 5x - x$

**b.** $3x + 2y$

**c.** $(xy)^2 + 3x^2y^2$

**d.** $-2x^2yz + 3yx^2z$

**e.** $(3x^2 + 2x) + (5x^2 - 8x)$

**f.** $(7a^2 + 2a - 5) - (3a^2 - 2a + 1)$

**g.** $3(9x^2 + 3x + 7) - 2(11x^2 - 5x + 9)$

**h.** $4(4x^3 + 2x^2 - 3x - 8) - 5(2x^3 - 3x + 8)$

## SECTION 4.6 — Multiplying Polynomials

To multiply two monomials, first multiply the numerical factors and then multiply the variable factors using the properties of exponents.

**13.** Find each product.

**a.** $(2x^2y^3)(5xy^2)$

**b.** $(xyz^3)(x^3z)^2$

To multiply a polynomial with more than one term by a monomial, multiply each term of the polynomial by the monomial and simplify.

**14.** Find each product.

**a.** $5(x + 3)$

**b.** $3(2x + 4)$

**c.** $x^2(3x^2 - 5)$

**d.** $2y^2(y^2 + 5y)$

**e.** $-x^2y(y^2 - xy)$

**f.** $-3xy(xy - x)$

To multiply two binomials, use the **FOIL method.**

**15.** Find each product.

**a.** $(x + 3)(x + 2)$

**b.** $(2x + 1)(x - 1)$

**c.** $(3a - 3)(2a + 2)$

**d.** $6(a - 1)(a + 1)$

**e.** $(a - b)(2a + b)$

**f.** $(3x - y)(2x + y)$

**Special products:**

$$(x + y)^2 = x^2 + 2xy + y^2$$
$$(x - y)^2 = x^2 - 2xy + y^2$$
$$(x + y)(x - y) = x^2 - y^2$$

To multiply one polynomial by another, multiply each term of one polynomial by each term of the other polynomial, and simplify.

**16.** Find each product.
   **a.** $(x + 3)(x + 3)$
   **b.** $(x + 5)(x - 5)$
   **c.** $(y - 2)(y + 2)$
   **d.** $(x + 4)^2$
   **e.** $(x - 3)^2$
   **f.** $(y - 1)^2$
   **g.** $(2y + 1)^2$
   **h.** $(y^2 + 1)(y^2 - 1)$

**17.** Find each product.
   **a.** $(3x + 1)(x^2 + 2x + 1)$
   **b.** $(2a - 3)(4a^2 + 6a + 9)$

**18.** Solve each equation.
   **a.** $x^2 + 3 = x(x + 3)$
   **b.** $x^2 + x = (x + 1)(x + 2)$
   **c.** $(x + 2)(x - 5) = (x - 4)(x - 1)$
   **d.** $(x - 1)(x - 2) = (x - 3)(x + 1)$
   **e.** $x^2 + x(x + 2) = x(2x + 1) + 1$
   **f.** $(x + 5)(3x + 1) = x^2 + (2x - 1)(x - 5)$

## SECTION 4.7 — Dividing Polynomials by Monomials

To divide a polynomial by a monomial, write the division as a product, use the distributive property to remove parentheses, and simplify each resulting fraction.

**19.** Do each division.
   **a.** $\dfrac{3x + 6y}{2xy}$
   **b.** $\dfrac{14xy - 21x}{7xy}$
   **c.** $\dfrac{15a^2bc + 20ab^2c - 25abc^2}{-5abc}$
   **d.** $\dfrac{(x + y)^2 + (x - y)^2}{-2xy}$

## SECTION 4.8 — Dividing Polynomials by Polynomials

Use long division to divide one polynomial by another.

**20.** Do each division.
   **a.** $x + 2 \overline{)x^2 + 3x + 5}$
   **b.** $x - 1 \overline{)x^2 - 6x + 5}$
   **c.** $x + 3 \overline{)2x^2 + 7x + 3}$
   **d.** $3x - 1 \overline{)3x^2 + 14x - 2}$
   **e.** $2x - 1 \overline{)6x^3 + x^2 + 1}$
   **f.** $3x + 1 \overline{)-13x - 4 + 9x^3}$

## Chapter Test

**1.** Use exponents to rewrite $2xxxyyy$.

**2.** Evaluate $3^2 + 5^3$.

*In Problems 3–6, write each expression as an expression containing only one exponent.*

**3.** $y^2(yy^3)$

**4.** $(-3b^2)(2b^3)(-b^2)$

**5.** $(2x^3)^5(x^2)^3$

**6.** $(2rr^2r^3)^3$

*In Problems 7–10, simplify each expression. Write answers without using parentheses or negative exponents.*

**7.** $3x^0$

**8.** $2y^{-5}y^2$

**9.** $\dfrac{y^2}{yy^{-2}}$

**10.** $\left(\dfrac{a^2b^{-1}}{4a^3b^{-2}}\right)^{-3}$

**11.** Write 28,000 in scientific notation.

**12.** Write 0.0025 in scientific notation.

**13.** Write $7.4 \times 10^3$ in standard notation.

**14.** Write $9.3 \times 10^{-5}$ in standard notation.

**15.** Classify $3x^2 + 2$ as a monomial, a binomial, or a trinomial.

**16.** Find the degree of the polynomial $3x^2y^3z^4 + 2x^3y^2z - 5x^2y^3z^5$

**17.** Evaluate $x^2 + x - 2$ when $x = -2$.

**18.** Graph the polynomial function $y = f(x) = x^2 + 2$.

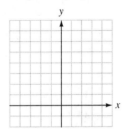

**19.** Simplify $-6(x - y) + 2(x + y) - 3(x + 2y)$

**20.** Simplify $-2(x^2 + 3x - 1) - 3(x^2 - x + 2) + 5(x^2 + 2)$

**21.** Add $\quad 3x^3 + 4x^2 - \phantom{3}x - 7$
$\phantom{\textbf{21.} \text{Add} \quad} 2x^3 - 2x^2 + 3x + 2$

**22.** Subtract $\quad 2x^2 - 7x + 3$
$\phantom{\textbf{22.} \text{Subtract} \quad} 3x^2 - 2x - 1$

*In Problems 23–26, find each product.*

**23.** $(-2x^3)(2x^2y)$

**24.** $3y^2(y^2 - 2y + 3)$

**25.** $(2x - 5)(3x + 4)$

**26.** $(2x - 3)(x^2 - 2x + 4)$

**27.** Solve the equation $(a + 2)^2 = (a - 3)^2$.

**28.** Simplify $\dfrac{8x^2y^3z^4}{16x^3y^2z^4}$.

**29.** Simplify $\dfrac{6a^2 - 12b^2}{24ab}$.

**30.** Divide $2x + 3\overline{)2x^2 - x - 6}$.

## ■ Cumulative Review Exercises

*In Exercises 1–4, evaluate each expression. Assume that $x = 2$ and $y = -5$.*

**1.** $5 + 3 \cdot 2$

**2.** $3 \cdot 5^2 - 4$

**3.** $\dfrac{3x - y}{xy}$

**4.** $\dfrac{x^2 - y^2}{x + y}$

*In Exercises 5–8, solve each equation.*

**5.** $\dfrac{4}{5}x + 6 = 18$

**6.** $x - 2 = \dfrac{x + 2}{3}$

**7.** $2(5x + 2) = 3(3x - 2)$

**8.** $4(y + 1) = -2(4 - y)$

*In Exercises 9–12, graph the solution of each inequality.*

**9.** $5x - 3 > 7$

**10.** $7x - 9 < 5$

**11.** $-2 < -x + 3 < 5$

**12.** $0 \le \dfrac{4 - x}{3} \le 2$

*In Exercises 13–14, solve each formula for the indicated variable.*

**13.** $A = p + prt$, for $r$

**14.** $A = \dfrac{1}{2}bh$, for $h$

*In Exercises 15–16, graph each equation.*

**15.** $3x - 4y = 12$

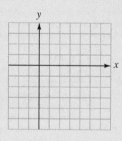

**16.** $y - 2 = \dfrac{1}{2}(x - 4)$

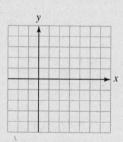

*In Exercises 17–18, solve each system by graphing.*

**17.** $\begin{cases} x - y = 4 \\ 2x + y = 5 \end{cases}$

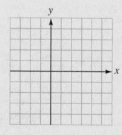

**18.** $\begin{cases} 3x + 2y \ge 6 \\ x + 3y \le 6 \end{cases}$

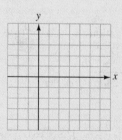

*In Exercises 19–20, solve each system of equations by an algebraic method.*

**19.** $\begin{cases} x + y = 1 \\ x - y = 7 \end{cases}$

**20.** $\begin{cases} 4x + 9y = 8 \\ 2x - 6y = -3 \end{cases}$

*In Exercises 21–24, write each expression as an expression using only one exponent.*

**21.** $(y^3 y^5)y^6$

**22.** $\dfrac{x^3 y^4}{x^2 y^3}$

**23.** $\dfrac{a^4 b^{-3}}{a^{-3} b^3}$

**24.** $\left(\dfrac{-x^{-2} y^3}{x^{-3} y^2}\right)^2$

*In Exercises 25–28, do each operation.*

**25.** $(3x^2 + 2x - 7) - (2x^2 - 2x + 7)$

**26.** $(3x - 7)(2x + 8)$

**27.** $(x - 2)(x^2 + 2x + 4)$

**28.** $x - 3 \overline{)2x^2 - 5x - 3}$

**29. Astronomy** The **parsec,** a unit of distance used in astronomy, is $3 \times 10^{16}$ meters. The distance to Betelgeuse, a star in the constellation Orion, is $1.6 \times 10^2$ parsecs. Use scientific notation to express this distance in meters.

**30. Surface area** The total surface area $A$ of a box with dimensions $l$, $w$, and $d$ (see Illustration 1) is given by the formula

$$A = 2lw + 2wd + 2ld$$

If $A = 202$ square inches, $l = 9$ inches, and $w = 5$ inches, find $d$.

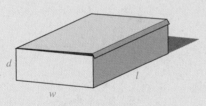

ILLUSTRATION 1

**31. Concentric circles** The area of the ring between the two concentric circles of radius $r$ and $R$ (see Illustration 2) is given by the formula

$$A = \pi(R + r)(R - r)$$

If $r = 3$ inches and $R = 17$ inches, find $A$ to the nearest tenth.

ILLUSTRATION 2

**32. Employee discounts** Employees at an appliance store can purchase merchandise at 25% less than the regular price. An employee buys a color TV set for $414.72, including 8% sales tax. Find the regular price of the TV.

# 5 *Factoring Polynomials*

MATHEMATICS IN
ECOLOGY

Many types of bacteria cannot survive in air. In one of the steps in waste treatment, sewage is exposed to the air by placing it in large, shallow, circular aeration pools. One sewage processing plant has two such pools, with diameters of 40 and 42 meters. To meet new clean-water standards, the plant must double its capacity, which includes building another aeration pool. How large a pool should the design engineers specify to double the capacity of this phase of sewage treatment?

After reading this chapter, you will be able to answer this question.

## 5.1 Factoring Out the Greatest Common Factor

■ FACTORING NATURAL NUMBERS ■ FACTORING MONOMIALS ■ FACTORING OUT A COMMON MONOMIAL ■ FACTORING OUT A NEGATIVE FACTOR ■ QUADRATIC EQUATIONS

Getting Ready    *Simplify each expression by removing parentheses.*

**1.** $5(x + 3)$       **2.** $7(y - 8)$       **3.** $x(3x - 2)$       **4.** $y(5y + 9)$

**5.** $a(b + 9)$       **6.** $x(3 + x + y)$       **7.** $xy(x - 4)$       **8.** $xy^2(2x - 5y)$

### ■ FACTORING NATURAL NUMBERS

In this chapter, we shall reverse the operation of multiplication and show how to find the factors of a known product. The process of finding the individual factors of a product is called **factoring.**

Because 4 divides 12 exactly, 4 is called a **factor** of 12. The numbers 1, 2, 3, 4, 6, and 12 are the natural-number factors of 12, because each one divides 12 exactly. Recall that a natural number greater than 1 whose only factors are 1 and the number itself is called a **prime number.** For example, 19 is a prime number, because

**1.** 19 is a natural number greater than 1, and

**2.** The only two natural number factors of 19 are 1 and 19.

The prime numbers less than 50 are

   2, 3, 5, 7, 11, 13, 17, 19, 23, 29, 31, 37, 41, 43, and 47

A natural number is said to be in **prime-factored form** if it is written as the product of factors that are prime numbers.

To find the prime-factored form of a natural number, we can use a **factoring tree.** For example, to find the prime-factored form of 60, we proceed as follows:

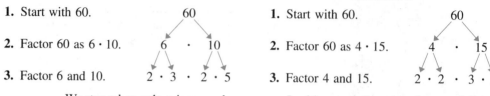

|  *Solution 1* |  *Solution 2* |
| --- | --- |
| **1.** Start with 60. | **1.** Start with 60. |
| **2.** Factor 60 as $6 \cdot 10$. | **2.** Factor 60 as $4 \cdot 15$. |
| **3.** Factor 6 and 10. | **3.** Factor 4 and 15. |

We stop when only prime numbers appear. In either case, the prime factors of 60 are $2 \cdot 2 \cdot 3 \cdot 5$. Thus, the prime-factored form of 60 is $2^2 \cdot 3 \cdot 5$. This illustrates the **fundamental theorem of arithmetic,** which states that there is only one prime factorization for any natural number greater than 1.

The right-hand sides of the equations

$$42 = 2 \cdot 3 \cdot 7$$
$$60 = 2^2 \cdot 3 \cdot 5$$
$$90 = 2 \cdot 3^2 \cdot 5$$

show the prime-factored forms (or **prime factorizations**) of 42, 60, and 90. The largest natural number that divides each of these numbers is called their **greatest common factor (GCF).** The GCF of 42, 60, and 90 is 6, because 6 is the largest natural number that divides each of these numbers:

$$\frac{42}{6} = 7 \qquad \frac{60}{6} = 10 \qquad \text{and} \qquad \frac{90}{6} = 15$$

## ■ FACTORING MONOMIALS

Algebraic monomials can also have a greatest common factor. The right-hand sides of the equations

$$6a^2b^3 = 2 \cdot 3 \cdot a \cdot a \cdot b \cdot b \cdot b$$
$$4a^3b^2 = 2 \cdot 2 \cdot a \cdot a \cdot a \cdot b \cdot b$$
$$18a^2b = 2 \cdot 3 \cdot 3 \cdot a \cdot a \cdot b$$

show the prime factorizations of $6a^2b^3$, $4a^3b^2$, and $18a^2b$. Since all three of these monomials have one factor of 2, two factors of $a$, and one factor of $b$, the GCF is

$$2 \cdot a \cdot a \cdot b \qquad \text{or} \qquad 2a^2b$$

To find the GCF of several monomials, we follow these steps.

### Strategy for Finding the Greatest Common Factor (GCF)

**1.** Find the prime factorization of each monomial.

**2.** List each common factor the least number of times it appears in any one monomial.

**3.** Find the product of the factors found in the list to obtain the GCF.

EXAMPLE 1    Find the GCF of $10x^3y^2$, $60x^2y$, and $30xy^2$.

*Solution*    **1.** Find the prime factorization of each monomial.

$$10x^3y^2 = 2 \cdot 5 \cdot x \cdot x \cdot x \cdot y \cdot y$$
$$60x^2y = 2 \cdot 2 \cdot 3 \cdot 5 \cdot x \cdot x \cdot y$$
$$30xy^2 = 2 \cdot 3 \cdot 5 \cdot x \cdot y \cdot y$$

**2.** List each common factor the least number of times it appears in any one monomial: 2, 5, $x$, and $y$.

**3.** Find the product of the factors in the list:

$$2 \cdot 5 \cdot x \cdot y = 10xy$$    ∎

Self Check    Find the GCF of $20a^2b^3$, $12ab^4$, and $8a^3b^2$.
Answer        $4ab^2$

■ ■ ■ ■ ■ ■ ■ ■ ■ ■  PERSPECTIVE

Much of the mathematics that we have inherited from earlier times is the result of teamwork. In a battle early in the 12th century, control of the Spanish city of Toledo was taken from the Mohammedans, who had ruled there for four centuries. Libraries in this great city contained many books written in Arabic, full of knowledge that was unknown in Europe.

The Archbishop of Toledo wanted to share this knowledge with the rest of the world. He knew that these books should be translated into Latin, the universal language of scholarship. But what European scholar could read Arabic? The citizens of Toledo knew both Arabic and Spanish, and most scholars of Europe could read Spanish.

Teamwork saved the day. A citizen of Toledo read the Arabic text aloud, in Spanish. The scholars listened to the Spanish version and wrote it down in Latin. One of these scholars was an Englishman, Robert of Chester. It was he who translated al-Khowarazmi's book, *Ihm al-jabr wa'l muqabalah*, the beginning of the subject we now know as algebra.

■ FACTORING OUT A COMMON MONOMIAL

Recall that the distributive property provides a way to multiply a polynomial by a monomial. For example,

$$3x^2(2x - 3y) = 3x^2 \cdot 2x - 3x^2 \cdot 3y$$
$$= 6x^3 - 9x^2y$$

To reverse this process and factor the product $6x^3 - 9x^2y$, we can find the GCF of each monomial (which is $3x^2$) and then use the distributive property in reverse.

$$6x^3 - 9x^2y = 3x^2 \cdot 2x - 3x^2 \cdot 3y$$
$$= 3x^2(2x - 3y)$$

This process is called **factoring out the greatest common factor.**

**EXAMPLE 2**     Factor $12y^2 + 20y$.

*Solution*     To find the GCF, we find the prime factorization of $12y^2$ and $20y$.

$$\left.\begin{array}{l} 12y^2 = 2 \cdot 2 \cdot 3 \cdot y \cdot y \\ 20y = 2 \cdot 2 \cdot 5 \cdot y \end{array}\right\} \text{GCF} = 4y$$

We can use the distributive property to factor out the GCF of $4y$.

$$\begin{aligned} 12y^2 + 20y &= 4y \cdot 3y + 4y \cdot 5 \\ &= 4y(3y + 5) \end{aligned}$$

Check by verifying that $4y(3y + 5) = 12y^2 + 20y$.  ∎

**Self Check**     Factor $15x^3 - 20x^2$.
**Answer**          $5x^2(3x - 4)$

**EXAMPLE 3**     Factor $35a^3b^2 - 14a^2b^3$.

*Solution*     To find the GCF, we find the prime factorization of $35a^3b^2$ and $-14a^2b^3$.

$$\left.\begin{array}{l} 35a^3b^2 = 5 \cdot 7 \cdot a \cdot a \cdot a \cdot b \cdot b \\ -14a^2b^3 = -2 \cdot 7 \cdot a \cdot a \cdot b \cdot b \cdot b \end{array}\right\} \text{GCF} = 7a^2b^2$$

We factor out the GCF of $7a^2b^2$.

$$\begin{aligned} 35a^3b^2 - 14a^2b^3 &= 7a^2b^2 \cdot 5a - 7a^2b^2 \cdot 2b \\ &= 7a^2b^2(5a - 2b) \end{aligned}$$

Check by verifying that $7a^2b^2(5a - 2b) = 35a^3b^2 - 14a^2b^3$.  ∎

**Self Check**     Factor $40x^2y^3 + 15x^3y^2$.
**Answer**          $5x^2y^2(8y + 3x)$

**EXAMPLE 4**     Factor $a^2b^2 - ab$.

*Solution*     We factor out the GCF, which is $ab$.

$$\begin{aligned} a^2b^2 - ab &= ab \cdot ab - ab \cdot 1 \\ &= ab(ab - 1) \end{aligned}$$

**WARNING!**  The last term of $a^2b^2 - ab$ has an implied coefficient of 1. When $ab$ is factored out, we must write the coefficient of 1.

We check by verifying that $ab(ab - 1) = a^2b^2 - ab$.  ∎

**Self Check**   Factor $x^3y^5 + x^2y^3$.
**Answer**   $x^2y^3(xy^2 + 1)$

**EXAMPLE 5**   Factor $12x^3y^2z + 6x^2yz - 3xz$.

*Solution*   We factor out the GCF, which is $3xz$.

$$12x^3y^2z + 6x^2yz - 3xz = 3xz \cdot 4x^2y^2 + 3xz \cdot 2xy - 3xz \cdot 1$$
$$= 3xz(4x^2y^2 + 2xy - 1)$$

Check by verifying that

$$3xz(4x^2y^2 + 2xy - 1) = 12x^3y^2z + 6x^2yz - 3xz$$   ■

**Self Check**   Factor $6ab^2c - 12a^2bc + 3ab$.
**Answer**   $3ab(2bc - 4ac + 1)$

## ■ FACTORING OUT A NEGATIVE FACTOR

It is often useful to factor out the negative of a monomial.

**EXAMPLE 6**   Factor $-1$ out of $-a^3 + 2a^2 - 4$.

*Solution*   $-a^3 + 2a^2 - 4$
$= (-1)a^3 + (-1)(-2a^2) + (-1)4$     $(-1)(-2a^2) = +2a^2.$
$= -1(a^3 - 2a^2 + 4)$     Factor out $-1$.
$= -(a^3 - 2a^2 + 4)$     The coefficient of 1 need not be written.

Check by verifying that

$$-(a^3 - 2a^2 + 4) = -a^3 + 2a^2 - 4$$   ■

**Self Check**   Factor $-1$ out of $-b^4 - 3b^2 + 2$.
**Answer**   $-(b^4 + 3b^2 - 2)$

**EXAMPLE 7**   Factor out the negative of the GCF: $-18a^2b + 6ab^2 - 12a^2b^2$.

*Solution*   The GCF is $6ab$. To factor out its negative, we factor out $-6ab$.

$$-18a^2b + 6ab^2 - 12a^2b^2 = (-6ab)3a - (-6ab)b + (-6ab)2ab$$
$$= -6ab(3a - b + 2ab)$$

Check by verifying that

$$-6ab(3a - b + 2ab) = -18a^2b + 6ab^2 - 12a^2b^2$$   ■

Self Check  Factor out the negative of the GCF: $-25xy^2 - 15x^2y + 30x^2y^2$.

Answer  $-5xy(5y + 3x - 6xy)$

## ■ QUADRATIC EQUATIONS

Equations such as $9x - 6 = 0$ that involve first-degree polynomials are called **linear equations.** Equations such as $9x^2 - 6x = 0$ that involve second-degree polynomials are called **quadratic equations.**

> **Quadratic Equations**
> A **quadratic equation** is an equation of the form
> $$ax^2 + bx + c = 0$$
> where $a$, $b$, and $c$ are real numbers, and $a \neq 0$.

The techniques that we have used to solve linear equations cannot be used to solve quadratic equations. For example, these techniques cannot be used to isolate $x$ on one side of the equation $9x^2 - 6x = 0$. However, we can often solve quadratic equations by factoring and using the following property of real numbers.

> **Zero-Factor Property of Real Numbers**
> Suppose $a$ and $b$ represent two real numbers. Then
> > If $ab = 0$, then $a = 0$ or $b = 0$.

We already know that if either of two numbers is 0, their product is 0. The zero-factor property says that if the product of two numbers is 0, then at least one of them must be 0.

For example, the equation $(x - 4)(x + 5) = 0$ indicates that a product is equal to 0. By the zero-factor property, one of the factors must be 0:

$$x - 4 = 0 \quad \text{or} \quad x + 5 = 0$$

We can solve each of these linear equations to get

$$x = 4 \quad \text{or} \quad x = -5$$

The equation $(x - 4)(x + 5) = 0$ has two solutions: 4 and $-5$.

EXAMPLE 8  Solve $9x^2 - 6x = 0$.

Solution  We begin by factoring the left-hand side of the equation.

$$9x^2 - 6x = 0$$
$$3x(3x - 2) = 0$$

By the zero-factor theorem, we have

$$3x = 0 \quad \text{or} \quad 3x - 2 = 0$$

We can solve each of these equations to get

$$x = 0 \quad \text{or} \quad x = \frac{2}{3}$$

*Check:* To check, we substitute these results for $x$ in the original equation, and simplify.

**For $x = 0$**
$$9x^2 - 6x = 0$$
$$9(0)^2 - 6(0) \stackrel{?}{=} 0$$
$$0 - 0 \stackrel{?}{=} 0$$
$$0 = 0$$

**For $x = \frac{2}{3}$**
$$9x^2 - 6x = 0$$
$$9\left(\frac{2}{3}\right)^2 - 6\left(\frac{2}{3}\right) \stackrel{?}{=} 0$$
$$9\left(\frac{4}{9}\right) - 6\left(\frac{2}{3}\right) \stackrel{?}{=} 0$$
$$4 - 4 \stackrel{?}{=} 0$$
$$0 = 0$$

Both solutions check. ■

**Self Check**
**Answer**
Solve $5y^2 + 10y = 0$.
$0, -2$

**Orals** *Find the prime factorization of each number.*

**1.** 36　　**2.** 27　　**3.** 81　　**4.** 45

*Find the greatest common factor:*

**5.** 3, 6, and 9　　**6.** $3a^2b$, $6ab$, and $9ab^2$

*Factor out the greatest common factor:*

**7.** $15xy + 10$　　**8.** $15xy + 10xy^2$

## EXERCISE 5.1

**REVIEW** *Solve each equation and check all solutions.*

**1.** $3x - 2(x + 1) = 5$　　**2.** $5(y - 1) + 1 = y$

**3.** $\dfrac{2x - 7}{5} = 3$　　**4.** $2x - \dfrac{x}{2} = 5x$

***VOCABULARY AND CONCEPTS*** *Fill in each blank to make a true statement.*

5. A natural number greater than 1 whose only factors are 1 and itself is called a _____ number.

6. If a natural number is written as the product of prime numbers, it is written in _____ form.

7. The GCF of several natural numbers is the _____ number that divides each of the numbers.

8. An equation of the form $ax^2 + bx + c = 0$, where $a \neq 0$, is called a _____ equation.

9. If $ab = 0$, then $a = \_\_$ or $b = \_\_$.

10. A quadratic equation contains a _____-degree polynomial.

***PRACTICE*** *In Exercises 11–22, find the prime factorization of each number.*

11. 12

12. 24

13. 15

14. 20

15. 40

16. 62

17. 98

18. 112

19. 225

20. 144

21. 288

22. 968

*In Exercises 23–28, complete each factorization.*

23. $4a + 12 = \boxed{\phantom{x}}(a + 3)$

24. $3t - 27 = 3\left(t - \boxed{\phantom{x}}\right)$

25. $r^4 + r^2 = r^2\left(\boxed{\phantom{x}} + 1\right)$

26. $a^3 - a^2 = \boxed{\phantom{x}}(a - 1)$

27. $4y^2 + 8y - 2xy = 2y\left(2y + \boxed{\phantom{x}} - \boxed{\phantom{x}}\right)$

28. $3x^2 - 6xy + 9xy^2 = \boxed{\phantom{x}}\left(\boxed{\phantom{x}} - 2y + 3y^2\right)$

*In Exercises 29–56, factor out the greatest common factor.*

29. $3x + 6$

30. $2y - 10$

31. $xy - xz$

32. $uv + ut$

33. $t^3 + 2t^2$

34. $b^3 - 3b^2$

35. $r^4 - r^2$

36. $a^3 + a^2$

37. $a^3b^3z^3 - a^2b^3z^2$

38. $r^3s^6t^9 + r^2s^2t^2$

39. $24x^2y^3z^4 + 8xy^2z^3$

40. $3x^2y^3 - 9x^4y^3z$

41. $12uvw^3 - 18uv^2w^2$

42. $14xyz - 16x^2y^2z$

43. $3x + 3y - 6z$

44. $2x - 4y + 8z$

45. $ab + ac - ad$

46. $rs - rt + ru$

47. $4y^2 + 8y - 2xy$

48. $3x^2 - 6xy + 9xy^2$

49. $12r^2 - 3rs + 9r^2s^2$

50. $6a^2 - 12a^3b + 36ab$

51. $abx - ab^2x + abx^2$

52. $a^2b^2x^2 + a^3b^2x^2 - a^3b^3x^3$

53. $4x^2y^2z^2 - 6xy^2z^2 + 12xyz^2$

54. $32xyz + 48x^2yz + 36xy^2z$

55. $70a^3b^2c^2 + 49a^2b^3c^3 - 21a^2b^2c^2$

56. $8a^2b^2 - 24ab^2c + 9b^2c^2$

*In Exercises 57–68, factor out $-1$ from each polynomial.*

57. $-a - b$

58. $-x - 2y$

59. $-2x + 5y$

60. $-3x + 8z$

61. $-2a + 3b$

62. $-2x + 5y$

63. $-3m - 4n + 1$

64. $-3r + 2s - 3$

65. $-3xy + 2z + 5w$

66. $-4ab + 3c - 5d$

67. $-3ab - 5ac + 9bc$

68. $-6yz + 12xz - 5xy$

*In Exercises 69–78, factor out the greatest common factor, including* $-1$.

**69.** $-3x^2y - 6xy^2$        **70.** $-4a^2b^2 + 6ab^2$

**71.** $-4a^2b^3 + 12a^3b^2$     **72.** $-25x^4y^3z^2 + 30x^2y^3z^4$

**73.** $-4a^2b^2c^2 + 14a^2b^2c - 10ab^2c^2$     **74.** $-10x^4y^3z^2 + 8x^3y^2z - 20x^2y$

**75.** $-14a^6b^6 + 49a^2b^3 - 21ab$     **76.** $-35r^9s^9t^9 + 25r^6s^6t^6 + 75r^3s^3t^3$

**77.** $-5a^2b^3c + 15a^3b^4c^2 - 25a^4b^3c$     **78.** $-7x^5y^4z^3 + 49x^5y^5z^4 - 21x^6y^4z^3$

*In Exercises 79–86, solve each equation.*

**79.** $(x - 2)(x + 3) = 0$     **80.** $(x - 3)(x - 2) = 0$

**81.** $(x - 4)(x + 1) = 0$     **82.** $(x + 5)(x + 2) = 0$

**83.** $(2x - 5)(3x + 6) = 0$     **84.** $(3x - 4)(x + 1) = 0$

**85.** $(x - 1)(x + 2)(x - 3) = 0$     **86.** $(x + 2)(x + 3)(x - 4) = 0$

*In Exercises 87–98, solve each equation.*

**87.** $x^2 - 3x = 0$     **88.** $x^2 + 5x = 0$

**89.** $2x^2 - 5x = 0$     **90.** $5x^2 + 7x = 0$

**91.** $x^2 - 7x = 0$     **92.** $x^2 - 8x = 0$

**93.** $3x^2 + 8x = 0$     **94.** $5x^2 - x = 0$

**95.** $8x^2 - 16x = 0$     **96.** $15x^2 - 20x = 0$

**97.** $10x^2 + 2x = 0$     **98.** $5x^2 + x = 0$

## WRITING

**99.** When we add $5x$ and $7x$, we combine like terms: $5x + 7x = 12x$. Explain how this is related to factoring out a common factor.

**100.** One student summarized the zero-factor property of real numbers by saying, "Anything times zero is zero." This answer is true, but it does not describe the zero-factor property. Explain.

## SOMETHING TO THINK ABOUT

**101.** Think of two positive integers. Divide their product by their greatest common factor. Why do you think the result is called the **lowest common multiple** of the two integers? (*Hint:* The **multiples** of an integer such as 5 are 5, 10, 15, 20, 25, 30, and so on.)

**102.** Two integers are **relatively prime** if their greatest common factor is 1. For example, 6 and 25 are relatively prime, but 6 and 15 are not. If the greatest common factor of three integers is 1, must any two of them be relatively prime? Explain.

## 5.2  Factoring by Grouping

■ FACTORING OUT A POLYNOMIAL  ■ FACTORING BY GROUPING

Getting Ready    *Remove parentheses and simplify.*

**1.** $3(x + y) + a(x + y)$                    **2.** $x(y + 1) + 5(y + 1)$

**3.** $5(x + 1) - y(x + 1)$                    **4.** $x(x + 2) - y(x + 2)$

**5.** $(3x - y)x + (3x - y)y$                  **6.** $5(y - 7) - y(y - 7)$

### ■  FACTORING OUT A POLYNOMIAL

If the GCF of several terms is a polynomial, we can factor out the common polynomial factor. For example, since $a + b$ is a common factor of $(a + b)x$ and $(a + b)y$, we can factor out the $a + b$.

$$(a + b)x + (a + b)y = (a + b)(x + y)$$

We can check by verifying that $(a + b)(x + y) = (a + b)x + (a + b)y$.

EXAMPLE 1    Factor $a + 3$ out of $(a + 3) + (a + 3)^2$.

Solution    Recall that $a + 3$ is equal to $(a + 3)1$ and that $(a + 3)^2$ is equal to $(a + 3)(a + 3)$. We can factor out $a + 3$ and simplify.

$$(a + 3) + (a + 3)^2 = (a + 3)1 + (a + 3)(a + 3)$$
$$= (a + 3)[1 + (a + 3)]$$
$$= (a + 3)(a + 4)$$ ■

Self Check    Factor out $y + 2$: $(y + 2)^2 - 3(y + 2)$.
Answer    $(y + 2)(y - 1)$

EXAMPLE 2    Factor $6a^2b^2(x + 2y) - 9ab(x + 2y)$.

Solution    The GCF of $6a^2b^2$ and $9ab$ is $3ab$. We can factor out this GCF as well as $(x + 2y)$.

$$6a^2b^2(x + 2y) - 9ab(x + 2y)$$
$$= 3ab \cdot 2ab(x + 2y) - 3ab \cdot 3(x + 2y)$$
$$= 3ab(x + 2y)(2ab - 3)$$    Factor out $3ab(x + 2y)$.    ■

Self Check    Factor $4p^3q^2(2a + b) + 8p^2q^3(2a + b)$.

Answer       $4p^2q^2(2a + b)(p + 2q)$

## ■ FACTORING BY GROUPING

Suppose we wish to factor

$$ax + ay + cx + cy$$

Although no factor is common to all four terms, there is a common factor of $a$ in $ax + ay$ and a common factor of $c$ in $cx + cy$. We can factor out the $a$ and the $c$ to obtain

$$ax + ay + cx + cy = a(x + y) + c(x + y) \qquad \text{Factor out } x + y.$$
$$= (x + y)(a + c)$$

We can check the result by multiplication.

$$(x + y)(a + c) = ax + cx + ay + cy$$
$$= ax + ay + cx + cy$$

Thus, $ax + ay + cx + cy$ factors as $(x + y)(a + c)$. This type of factoring is called **factoring by grouping.**

EXAMPLE 3    Factor $2c + 2d - cd - d^2$.

Solution     $2c + 2d - cd - d^2 = 2(c + d) - d(c + d)$    Factor out 2 from $2c + 2d$ and $-d$ from $-cd - d^2$.

$$= (c + d)(2 - d) \qquad \text{Factor out } c + d.$$

Check: $(c + d)(2 - d) = 2c - cd + 2d - d^2$
$$= 2c + 2d - cd - d^2 \qquad ■$$

Self Check    Factor $3a + 3b - ac - bc$.

Answer       $(a + b)(3 - c)$

EXAMPLE 4    Factor $x^2y - ax - xy + a$.

Solution     $x^2y - ax - xy + a = x(xy - a) - 1(xy - a)$    Factor out $x$ from $x^2y - ax$ and $-1$ from $-xy + a$.

$$= (xy - a)(x - 1) \qquad \text{Factor out } xy - a.$$

Check by multiplication.                                    ■

<table>
<tr><td>Self Check</td><td>Factor $pq^2 + tq + 2pq + 2t$.</td></tr>
<tr><td>Answer</td><td>$(pq + t)(q + 2)$</td></tr>
</table>

**WARNING!**   When factoring expressions such as those in the previous two examples, don't think that $2(c + d) - d(c + d)$ or $x(xy - a) - 1(xy - a)$ are in factored form. To be in factored form, the final result must be a product.

Factoring by grouping often works on polynomials with more than four terms.

**EXAMPLE 5**   Factor $6am - 6bm + 6cm + 5an - 5bn + 5cn$.

*Solution*   Factor $6m$ from the first three terms and $5n$ from the last three terms to obtain

$$6am - 6bm + 6cm + 5an - 5bn + 5cn = 6m(a - b + c) + 5n(a - b + c)$$

Then factor out the common factor of $(a - b + c)$.

$$6am - 6bm + 6cm + 5an - 5bn + 5cn = (a - b + c)(6m + 5n)$$

Check by multiplication.                                                      ■

<table>
<tr><td>Self Check</td><td>Factor $2ap + 2aq - 2at - bp - bq + bt$.</td></tr>
<tr><td>Answer</td><td>$(p + q - t)(2a - b)$</td></tr>
</table>

 b

**EXAMPLE 6**   Factor   **a.** $a(c - d) + b(d - c)$   and   **b.** $ac + bd - ad - bc$.

*Solution*   **a.**  $a(c - d) + b(d - c) = a(c - d) - b(-d + c)$      Factor $-1$ from $d - c$.
$\qquad\qquad\qquad\qquad = a(c - d) - b(c - d)$      $-d + c = c - d$.
$\qquad\qquad\qquad\qquad = (c - d)(a - b)$      Factor out $(c - d)$.

**b.** In this example, we cannot factor anything from the first two terms or the last two terms. However, if we rearrange the terms, the factoring is routine:

$$ac + bd - ad - bc = ac - ad + bd - bc \qquad bd - ad = -ad + bd.$$
$$= a(c - d) + b(d - c) \qquad \text{Factor } a \text{ from } ac - ad \text{ and } b \text{ from } bd - bc.$$
$$= (c - d)(a - b) \qquad \text{See part a.} \qquad ■$$

<table>
<tr><td>Self Check</td><td>Factor $ax - by - ay + bx$.</td></tr>
<tr><td>Answer</td><td>$(a + b)(x - y)$</td></tr>
</table>

Orals  *Find the common factor of the given terms.*

1. $a(x + 3)$ and $3(x + 3)$
2. $5(a - 1)$ and $xy(a - 1)$
3. $b(x - 2)$ and $(x - 2)^2$
4. $(y + 5)$ and $(y + 5)^2$
5. $a(x - 7)$, $9(x - 7)$, and $x(x - 7)$
6. $5(2y + 9)$, $y(2y + 9)$, and $y^2(2y + 9)$

## EXERCISE 5.2

**REVIEW**  *Simplify each expression and write all results without using negative exponents.*

1. $u^3 u^2 u^4$
2. $\dfrac{y^6}{y^8}$
3. $\dfrac{a^3 b^4}{a^2 b^5}$
4. $(3x^5)^0$

**VOCABULARY AND CONCEPTS**  *Fill in each blank to make a true statement.*

5. The GCF of $x(a + b) - y(a + b)$ is _____.

6. Check the results of a factoring problem by _____.

*In Exercises 7–10, complete each factorization.*

7. $a(x + y) + b(x + y) = (x + y)$_____
8. $p(m - n) - q(m - n) = $_____$(p - q)$
9. $(r - s)p - (r - s)q = (r - s)$_____
10. $ax + bx + ap + bp = x$_____$+ p$_____
    $= $_____$(x + p)$

**PRACTICE**  *In Exercises 11–30, factor each expression.*

11. $(x + y)2 + (x + y)b$
12. $(a - b)c + (a - b)d$
13. $3(x + y) - a(x + y)$
14. $x(y + 1) - 5(y + 1)$
15. $3(r - 2s) - x(r - 2s)$
16. $x(a + 2b) + y(a + 2b)$
17. $(x - 3)^2 + (x - 3)$
18. $(3t + 5)^2 - (3t + 5)$
19. $2x(a^2 + b) + 2y(a^2 + b)$
20. $3x(c - 3d) + 6y(c - 3d)$
21. $3x^2(r + 3s) - 6y^2(r + 3s)$
22. $9a^2 b^2(3x - 2y) - 6ab(3x - 2y)$

23. $3x(a + b + c) - 2y(a + b + c)$
24. $2m(a - 2b + 3c) - 21xy(a - 2b + 3c)$

25. $14x^2 y(r + 2s - t) - 21xy(r + 2s - t)$
26. $15xy^3(2x - y + 3z) + 25xy^2(2x - y + 3z)$

27. $(x + 3)(x + 1) - y(x + 1)$
28. $x(x^2 + 2) - y(x^2 + 2)$
29. $(3x - y)(x^2 - 2) + (x^2 - 2)$
30. $(x - 5y)(a + 2) - (x - 5y)$

*In Exercises 31–50, factor each expression.*

31. $2x + 2y + ax + ay$
32. $bx + bz + 5x + 5z$
33. $7r + 7s - kr - ks$
34. $9p - 9q + mp - mq$

**35.** $xr + xs + yr + ys$

**36.** $pm - pn + qm - qn$

**37.** $2ax + 2bx + 3a + 3b$

**38.** $3xy + 3xz - 5y - 5z$

**39.** $2ab + 2ac + 3b + 3c$

**40.** $3ac + a + 3bc + b$

**41.** $2x^2 + 2xy - 3x - 3y$

**42.** $3ab + 9a - 2b - 6$

**43.** $3tv - 9tw + uv - 3uw$

**44.** $ce - 2cf + 3de - 6df$

**45.** $9mp + 3mq - 3np - nq$

**46.** $ax + bx - a - b$

**47.** $mp - np - m + n$

**48.** $6x^2u - 3x^2v + 2yu - yv$

**49.** $x(a - b) + y(b - a)$

**50.** $p(m - n) - q(n - m)$

*In Exercises 51–58, factor each expression. Factor out all common factors first, if they exist.*

**51.** $ax^3 + bx^3 + 2ax^2y + 2bx^2y$

**52.** $x^3y^2 - 2x^2y^2 + 3xy^2 - 6y^2$

**53.** $4a^2b + 12a^2 - 8ab - 24a$

**54.** $-4abc - 4ac^2 + 2bc + 2c^2$

**55.** $x^3 + 2x^2 + x + 2$

**56.** $y^3 - 3y^2 - 5y + 15$

**57.** $x^3y - x^2y - xy^2 + y^2$

**58.** $2x^3z - 4x^2z + 32xz - 64z$

*In Exercises 59–66, factor each expression completely.*

**59.** $x^2 + xy + x + 2x + 2y + 2$

**60.** $ax + ay + az + bx + by + bz$

**61.** $am + bm + cm - an - bn - cn$

**62.** $x^2 + xz - x - xy - yz + y$

**63.** $ad - bd - cd + 3a - 3b - 3c$

**64.** $ab + ac - ad - b - c + d$

**65.** $ax^2 - ay + bx^2 - by + cx^2 - cy$

**66.** $a^2x - bx - a^2y + by + a^2z - bz$

*In Exercises 67–78, factor each expression completely. You may have to rearrange some terms first.*

**67.** $2r - bs - 2s + br$

**68.** $5x + ry + rx + 5y$

**69.** $ax + by + bx + ay$

**70.** $mr + ns + ms + nr$

**71.** $ac + bd - ad - bc$

**72.** $sx - ry + rx - sy$

**73.** $ar^2 - brs + ars - br^2$

**74.** $a^2bc + a^2c + abc + ac$

**75.** $ba + 3 + a + 3b$

**76.** $xy + 7 + y + 7x$

**77.** $pr + qs - ps - qr$

**78.** $ac - bd - ad + bc$

## WRITING

**79.** Explain why $a - b$ and $b - a$ are negatives of each other.

**80.** Explain how you would factor $x(a - b) + y(b - a)$.

## SOMETHING TO THINK ABOUT

**81.** Factor $ax + ay + bx + by$ by grouping the first two terms and the last two terms. Then rearrange the terms as $ax + bx + ay + by$, and factor again by grouping the first two and the last two. Do the results agree?

**82.** Factor $2xy + 2xz - 3y - 3z$ by grouping in two different ways.

## 5.3 Factoring the Difference of Two Squares

■ FACTORING THE DIFFERENCE OF TWO SQUARES ■ MULTISTEP FACTORING ■ SOLVING EQUATIONS

**Getting Ready** *Multiply the binomials.*

**1.** $(a + b)(a - b)$
**2.** $(2r + s)(2r - s)$
**3.** $(3x + 2y)(3x - 2y)$
**4.** $(4x^2 + 3)(4x^2 - 3)$

### ■ FACTORING THE DIFFERENCE OF TWO SQUARES

Whenever we multiply a binomial of the form $x + y$ by a binomial of the form $x - y$, we obtain a binomial of the form $x^2 - y^2$.

$$(x + y)(x - y) = x^2 - xy + xy - y^2$$
$$= x^2 - y^2$$

The binomial $x^2 - y^2$ is called the **difference of two squares,** because $x^2$ is the square of $x$ and $y^2$ is the square of $y$. The difference of the squares of two quantities always factors into the sum of those two quantities multiplied by the difference of those two quantities.

**Factoring the Difference of Two Squares**
$$x^2 - y^2 = (x + y)(x - y)$$

If we think of the difference of two squares as the square of a **First** quantity minus the square of a **Last** quantity, we have the formula

$$F^2 - L^2 = (F + L)(F - L)$$

and we say, *To factor the square of a First quantity minus the square of a Last quantity, we multiply the First plus the Last by the First minus the Last.*

To factor $x^2 - 9$, we note that it can be written in the form $x^2 - 3^2$ and use the formula for factoring the difference of two squares:

$$F^2 - L^2 = (F + L)(F - L)$$
$$x^2 - 3^2 = (x + 3)(x - 3)$$

We can check by verifying that $(x + 3)(x - 3) = x^2 - 9$.

To factor the difference of two squares, it is helpful to know the integers that are perfect squares. The number 400, for example, is a perfect square, because $20^2 = 400$. The perfect integer squares less than 400 are

**1, 4, 9, 16, 25, 36, 49, 64, 81, 100, 121, 144, 169, 196, 225, 256, 289, 324, 361**

Expressions containing variables such as $x^4y^2$ are also perfect squares, because they can be written as the square of a quantity:

$$x^4y^2 = (x^2y)^2$$

**EXAMPLE 1**

Factor $25x^2 - 49$.

*Solution*

We can write $25x^2 - 49$ in the form $(5x)^2 - 7^2$ and use the formula for factoring the difference of two squares:

$$F^2 - L^2 = (F + L)(F - L)$$
$$\downarrow \quad \downarrow \quad \downarrow \quad \downarrow \quad \downarrow \quad \downarrow$$
$$(5x)^2 - 7^2 = (5x + 7)(5x - 7) \qquad \text{Substitute } 5x \text{ for F and 7 for L.}$$

We can check by multiplying $5x + 7$ and $5x - 7$.

$$(5x + 7)(5x - 7) = 25x^2 - 35x + 35x - 49$$
$$= 25x^2 - 49 \qquad \blacksquare$$

*Self Check*

Factor $16a^2 - 81$.

*Answer*

$(4a + 9)(4a - 9)$

**EXAMPLE 2**

Factor $4y^4 - 25z^2$.

*Solution*

We can write $4y^4 - 25z^2$ in the form $(2y^2)^2 - (5z)^2$ and use the formula for factoring the difference of two squares:

$$F^2 - L^2 = (F + L)(F - L)$$
$$\downarrow \quad \downarrow \quad \downarrow \quad \downarrow \quad \downarrow \quad \downarrow$$
$$(2y^2)^2 - (5z)^2 = (2y^2 + 5z)(2y^2 - 5z)$$

Check by multiplication. $\qquad \blacksquare$

*Self Check*

Factor $9m^2 - 64n^4$.

*Answer*

$(3m + 8n^2)(3m - 8n^2)$

■ **MULTISTEP FACTORING**

We can often factor out a greatest common factor before factoring the difference of two squares. To factor $8x^2 - 32$, for example, we factor out the GCF of 8 and then factor the resulting difference of two squares.

$$8x^2 - 32 = 8(x^2 - 4) \qquad \text{Factor out 8.}$$
$$= 8(x^2 - 2^2) \qquad \text{Write 4 as } 2^2.$$
$$= 8(x + 2)(x - 2) \qquad \text{Factor the difference of two squares.}$$

We can check by multiplication:

$$8(x + 2)(x - 2) = 8(x^2 - 4)$$
$$= 8x^2 - 32$$

**EXAMPLE 3**   Factor $2a^2x^3y - 8b^2xy$.

*Solution*   We factor out the GCF of $2xy$ and then factor the resulting difference of two squares.

$$2a^2x^3y - 8b^2xy$$
$$= 2xy \cdot a^2x^2 - 2xy \cdot 4b^2 \qquad \text{The GCF is } 2xy.$$
$$= 2xy(a^2x^2 - 4b^2) \qquad \text{Factor out } 2xy.$$
$$= 2xy[(ax)^2 - (2b)^2] \qquad \text{Write } a^2x^2 \text{ as } (ax)^2 \text{ and } 4b^2 \text{ as } (2b)^2.$$
$$= 2xy(ax + 2b)(ax - 2b) \qquad \text{Factor the difference of two squares.}$$

We check by multiplication. ∎

**Self Check**   Factor $2p^2q^2s - 18r^2s$.
**Answer**   $2s(pq + 3r)(pq - 3r)$

Sometimes we must factor a difference of two squares more than once to factor a polynomial. For example, the binomial $625a^4 - 81b^4$ can be written in the form $(25a^2)^2 - (9b^2)^2$, which factors as

$$625a^4 - 81b^4 = (25a^2)^2 - (9b^2)^2$$
$$= (25a^2 + 9b^2)(25a^2 - 9b^2)$$

Since the factor $25a^2 - 9b^2$ can be written in the form $(5a)^2 - (3b)^2$, it is the difference of two squares and can be factored as $(5a + 3b)(5a - 3b)$. Thus,

$$625a^4 - 81b^4 = (25a^2 + 9b^2)(5a + 3b)(5a - 3b)$$

**WARNING!**   The binomial $25a^2 + 9b^2$ is the **sum of two squares,** because it can be written in the form $(5a)^2 + (3b)^2$. If we are limited to integer coefficients, binomials that are the sum of two squares cannot be factored.
   Polynomials that do not factor over the integers are called **prime polynomials.**

**EXAMPLE 4**   Factor $2x^4y - 32y$.

*Solution*

$$2x^4y - 32y = 2y \cdot x^4 - 2y \cdot 16$$
$$= 2y(x^4 - 16) \qquad \text{Factor out the GCF of } 2y.$$
$$= 2y(x^2 + 4)(x^2 - 4) \qquad \text{Factor } x^4 - 16.$$
$$= 2y(x^2 + 4)(x + 2)(x - 2) \qquad \text{Factor } x^2 - 4. \text{ Note that } x^2 + 4$$
$$\text{does not factor.} \quad ∎$$

**Self Check** Factor $48a^5 - 3ab^4$.
**Answer** $3a(4a^2 + b^2)(2a + b)(2a - b)$

Example 5 requires the techniques of factoring out a common factor, factoring by grouping, and factoring the difference of two squares.

**EXAMPLE 5** Factor $2x^3 - 8x + 2yx^2 - 8y$.

*Solution*

$$2x^3 - 8x + 2yx^2 - 8y = 2(x^3 - 4x + yx^2 - 4y)$$ Factor out 2.
$$= 2[x(x^2 - 4) + y(x^2 - 4)]$$ Factor out $x$ from $x^3 - 4x$ and $y$ from $yx^2 - 4y$.
$$= 2[(x^2 - 4)(x + y)]$$ Factor out $x^2 - 4$.
$$= 2(x + 2)(x - 2)(x + y)$$ Factor $x^2 - 4$.

Check by multiplication. ∎

**Self Check** Factor $3a^3 - 12a + 3a^2b - 12b$.
**Answer** $3(a + 2)(a - 2)(a + b)$

 **WARNING!** To *factor* an expression means to factor the expression *completely.*

## ■ SOLVING EQUATIONS

We can use factoring the difference of two squares to solve many quadratic equations.

 **EXAMPLE 6** Solve $4x^2 = 36$.

*Solution* Before we can use the zero-factor theorem, we must subtract 36 from both sides to make the right-hand side 0.

$$4x^2 = 36$$
$$4x^2 - 36 = 0$$ Subtract 36 from both sides.
$$x^2 - 9 = 0$$ Divide both sides by 4.
$$(x + 3)(x - 3) = 0$$ Factor $x^2 - 9$.
$$x + 3 = 0 \quad \text{or} \quad x - 3 = 0$$ Set each factor equal to 0.
$$x = -3 \qquad x = 3$$ Solve each linear equation.

Check each solution.

| *For x = −3* | *For x = 3* |
|---|---|
| $4x^2 = 36$ | $4x^2 = 36$ |
| $4(-3)^2 \stackrel{?}{=} 36$ | $4(3)^2 \stackrel{?}{=} 36$ |
| $4(9) \stackrel{?}{=} 36$ | $4(9) \stackrel{?}{=} 36$ |
| $36 = 36$ | $36 = 36$ |

Both solutions check.                                           ■

**Self Check**    Solve $9p^2 = 64$.

**Answer**    $\frac{8}{3}, -\frac{8}{3}$

Orals    *Factor each binomial.*

**1.** $x^2 - 9$              **2.** $y^2 - 36$

**3.** $z^2 - 4$              **4.** $p^2 - q^2$

**5.** $25 - t^2$            **6.** $36 - r^2$

**7.** $100 - y^2$          **8.** $100 - y^4$

# EXERCISE 5.3

## REVIEW

**1.** In the study of the flow of fluids, Bernoulli's law is given by the equation

$$\frac{p}{w} + \frac{v^2}{2g} + h = k$$

Solve the equation for $p$.

**2.** Solve Bernoulli's law for $h$. (See Exercise 1.)

$$h = k - \frac{p}{w} - \frac{v^2}{2g}$$

## VOCABULARY AND CONCEPTS    *Fill in each blank to make a true statement.*

**3.** A binomial of the form $a^2 - b^2$ is called the

_____.

**4.** A binomial of the form $a^2 + b^2$ is called the

_____.

**5.** $p^2 - q^2 = (p + q)$_____

**6.** The _____ of two squares cannot be factored by using only integer coefficients.

*Complete each factorization.*

**7.** $x^2 - 9 = (x + 3)$_____

**8.** $p^2 - q^2 =$ _____$(p - q)$

**9.** $4m^2 - 9n^2 = (2m + 3n)$_____

**10.** $16p^2 - 25q^2 =$ _____$(4p - 5q)$

**PRACTICE**   *In Exercises 11–30, factor each expression, if possible.*

**11.** $x^2 - 16$

**12.** $x^2 - 25$

**13.** $y^2 - 49$

**14.** $y^2 - 81$

**15.** $4y^2 - 49$

**16.** $9z^2 - 4$

**17.** $9x^2 - y^2$

**18.** $4x^2 - z^2$

**19.** $25t^2 - 36u^2$

**20.** $49u^2 - 64v^2$

**21.** $16a^2 - 25b^2$

**22.** $36a^2 - 121b^2$

**23.** $a^2 + b^2$

**24.** $121a^2 - 144b^2$

**25.** $a^4 - 4b^2$

**26.** $9y^2 + 16z^2$

**27.** $49y^2 - 225z^4$

**28.** $25x^2 + 36y^2$

**29.** $196x^4 - 169y^2$

**30.** $144a^4 + 169b^4$

*In Exercises 31–46, factor each expression.*

**31.** $8x^2 - 32y^2$

**32.** $2a^2 - 200b^2$

**33.** $2a^2 - 8y^2$

**34.** $32x^2 - 8y^2$

**35.** $3r^2 - 12s^2$

**36.** $45u^2 - 20v^2$

**37.** $x^3 - xy^2$

**38.** $a^2b - b^3$

**39.** $4a^2x - 9b^2x$

**40.** $4b^2y - 16c^2y$

**41.** $3m^3 - 3mn^2$

**42.** $2p^2q - 2q^3$

**43.** $4x^4 - x^2y^2$

**44.** $9xy^2 - 4xy^4$

**45.** $2a^3b - 242ab^3$

**46.** $50c^4d^2 - 8c^2d^4$

*In Exercises 47–58, factor each expression.*

**47.** $x^4 - 81$

**48.** $y^4 - 625$

**49.** $a^4 - 16$

**50.** $b^4 - 256$

**51.** $a^4 - b^4$

**52.** $m^4 - 16n^4$

**53.** $81r^4 - 256s^4$

**54.** $x^8 - y^4$

**55.** $a^4 - b^8$

**56.** $16y^8 - 81z^4$

**57.** $x^8 - y^8$

**58.** $x^8y^8 - 1$

*In Exercises 59–78, factor each expression.*

**59.** $2x^4 - 2y^4$

**60.** $a^5 - ab^4$

**61.** $a^4b - b^5$

**62.** $m^5 - 16mn^4$

**63.** $48m^4n - 243n^5$

**64.** $2x^4y - 512y^5$

**65.** $3a^5y + 6ay^5$

**66.** $2p^{10}q - 32p^2q^5$

**67.** $3a^{10} - 3a^2b^4$

**68.** $2x^9y + 2xy^9$

**69.** $2x^8y^2 - 32y^6$

**70.** $3a^8 - 243a^4b^8$

**71.** $a^6b^2 - a^2b^6c^4$

**72.** $a^2b^3c^4 - a^2b^3d^4$

**73.** $a^2b^7 - 625a^2b^3$

**74.** $16x^3y^4z - 81x^3y^4z^5$

**75.** $243r^5s - 48rs^5$    **76.** $1,024m^5n - 324mn^5$

**77.** $16(x - y)^2 - 9$    **78.** $9(x + 1)^2 - y^2$

*In Exercises 79–88, factor each expression.*

**79.** $a^3 - 9a + 3a^2 - 27$    **80.** $b^3 - 25b - 2b^2 + 50$
**81.** $y^3 - 16y - 3y^2 + 48$    **82.** $a^3 - 49a + 2a^2 - 98$
**83.** $3x^3 - 12x + 3x^2 - 12$    **84.** $2x^3 - 18x - 6x^2 + 54$
**85.** $3m^3 - 3mn^2 + 3am^2 - 3an^2$    **86.** $ax^3 - axy^2 - bx^3 + bxy^2$
**87.** $2m^3n^2 - 32mn^2 + 8m^2 - 128$    **88.** $2x^3y + 4x^2y - 98xy - 196y$

*In Exercises 89–100, solve each equation.*

**89.** $x^2 - 25 = 0$    **90.** $x^2 - 36 = 0$    **91.** $y^2 - 49 = 0$    **92.** $z^2 - 121 = 0$

**93.** $4x^2 - 1 = 0$    **94.** $9y^2 - 1 = 0$    **95.** $9y^2 - 4 = 0$    **96.** $16z^2 - 25 = 0$

**97.** $x^2 = 49$    **98.** $z^2 = 25$    **99.** $4x^2 = 81$    **100.** $9y^2 = 64$

## WRITING

**101.** Explain how to factor the difference of two squares.

**102.** Explain why $x^4 - y^4$ is not completely factored as $(x^2 + y^2)(x^2 - y^2)$.

## SOMETHING TO THINK ABOUT

**103.** It is easy to multiply 399 by 401 without a calculator: The product is $400^2 - 1$, or 159,999. Explain.

**104.** Use the method in the previous exercise to find $498 \cdot 502$ without a calculator.

## 5.4  Factoring Trinomials with Lead Coefficients of 1

■ FACTORING TRINOMIALS OF THE FORM $x^2 + bx + c$  ■ FACTORING OUT $-1$  ■ PRIME TRINOMIALS
■ MULTISTEP FACTORING  ■ FACTORING PERFECT-SQUARE TRINOMIALS  ■ SOLVING EQUATIONS

Getting Ready    *Multiply the binomials.*

**1.** $(x + 6)(x + 6)$    **2.** $(y - 7)(y - 7)$    **3.** $(a - 3)(a - 3)$

**4.** $(x + 4)(x + 5)$      **5.** $(r - 2)(r - 5)$      **6.** $(m + 3)(m - 7)$

**7.** $(a - 3b)(a + 4b)$      **8.** $(u - 3v)(u - 5v)$      **9.** $(x + 4y)(x - 6y)$

## ■ FACTORING TRINOMIALS OF THE FORM $x^2 + bx + c$

The product of two binomials is often a trinomial. For example,

$$(x + 3)(x + 3) = x^2 + 6x + 9 \quad \text{and} \quad (x - 4y)(x - 4y) = x^2 - 8xy + 16y^2$$

For this reason, we should not be surprised that many trinomials factor into the product of two binomials. To develop a method for factoring trinomials, we multiply $(x + a)$ and $(x + b)$.

$$(x + a)(x + b) = x^2 + bx + ax + ab \qquad \text{Use the FOIL method.}$$
$$= x^2 + ax + bx + ab \qquad \text{Write } bx + ax \text{ as } ax + bx.$$
$$= x^2 + (a + b)x + ab \qquad \text{Factor } x \text{ out of } ax + bx.$$

From the result, we can see that

- the coefficient of the middle term is the sum of $a$ and $b$, and
- the last term is the product of $a$ and $b$

We can use these facts to factor trinomials with lead coefficients of 1.

**EXAMPLE 1**    Factor $x^2 + 5x + 6$.

*Solution*    To factor this trinomial, we will write it as the product of two binomials. Since the first term of the trinomial is $x^2$, the first term of each binomial factor must be $x$. To fill in the following blanks, we must find two integers whose product is $+6$ and whose sum is $+5$.

$$x^2 + 5x + 6 = (x \quad\quad)(x \quad\quad)$$

The positive factorizations of 6 and the sums of the factors are shown in the following table.

| Product of the factors | Sum of the factors |
|---|---|
| $1(6) = 6$ | $1 + 6 = 7$ |
| $2(3) = 6$ | $2 + 3 = 5$ |

The last row contains the integers $+2$ and $+3$, whose product is $+6$ and whose sum is $+5$. So we can fill in the blanks with $+2$ and $+3$.

$$x^2 + 5x + 6 = (x + 2)(x + 3)$$

To check the result, we verify that $(x + 2)$ times $(x + 3)$ is $x^2 + 5x + 6$.

$$(x + 2)(x + 3) = x^2 + 3x + 2x + 2 \cdot 3$$
$$= x^2 + 5x + 6 \qquad \blacksquare$$

**Self Check**  Factor $y^2 + 5y + 4$.

*Answer*  $(y + 1)(y + 4)$

In Example 1, the factors can be written in either order. An equivalent factorization is $x^2 + 5x + 6 = (x + 3)(x + 2)$.

**EXAMPLE 2**  Factor $y^2 - 7y + 12$.

*Solution*  Since the first term of the trinomial is $y^2$, the first term of each binomial factor must be $y$. To fill in the following blanks, we must find two integers whose product is $+12$ and whose sum is $-7$.

$$y^2 - 7y + 12 = (y \qquad )(y \qquad )$$

The two-integer factorizations of 12 and the sums of the factors are shown in the following table.

| Product of the factors | Sum of the factors |
|:---:|:---:|
| $1(12) = 12$ | $1 + 12 = 13$ |
| $2(6) = 12$ | $2 + 6 = 8$ |
| $3(4) = 12$ | $3 + 4 = 7$ |
| $-1(-12) = 12$ | $-1 + (-12) = -13$ |
| $-2(-6) = 12$ | $-2 + (-6) = -8$ |
| $-3(-4) = 12$ | $-3 + (-4) = -7$ |

The last row contains the integers $-3$ and $-4$, whose product is $+12$ and whose sum is $-7$. So we can fill in the blanks with $-3$ and $-4$.

$$y^2 - 7y + 12 = (y - 3)(y - 4)$$

To check the result, we verify that $(y - 3)$ times $(y - 4)$ is $y^2 - 7y + 12$.

$$(y - 4)(y - 3) = y^2 - 3y - 4y + 12$$
$$= y^2 - 7y + 12 \qquad \blacksquare$$

**Self Check**  Factor $p^2 - 5p + 6$.

*Answer*  $(p - 3)(p - 2)$

**EXAMPLE 3**  Factor $a^2 + 2a - 15$.

*Solution*  Since the first term is $a^2$, the first term of each binomial factor must be $a$. To fill in the following blanks, we must find two integers whose product is $-15$ and whose sum is $+2$.

$$a^2 + 2a - 15 = (a \qquad)(a \qquad)$$

The possible factorizations of $-15$ and the sums of the factors are shown in the following table.

| Product of the factors | Sum of the factors |
|:---:|:---:|
| $1(-15) = -15$ | $1 + (-15) = -14$ |
| $3(-5) = -15$ | $3 + (-5) = -2$ |
| $5(-3) = -15$ | $5 + (-3) = 2$ |
| $15(-1) = -15$ | $15 + (-1) = 14$ |

The third row contains the integers $+5$ and $-3$, whose product is $-15$ and whose sum is $+2$. So we can fill in the blanks with $+5$ and $-3$.

$$a^2 + 2a - 15 = (a + 5)(a - 3)$$

We can check by multiplying $a + 5$ and $a - 3$.

$$(a + 5)(a - 3) = a^2 - 3a + 5a - 15$$
$$= a^2 + 2a - 15 \qquad \blacksquare$$

**Self Check**  Factor $p^2 + 3p - 18$.
**Answer**  $(p + 6)(p - 3)$

**EXAMPLE 4**  Factor $z^2 - 4z - 21$.

*Solution*  Since the first term is $z^2$, the first term of each binomial factor must be $z$. To fill in the following blanks, we must find two integers whose product is $-21$ and whose sum is $-4$.

$$z^2 - 4y - 21 = (z \qquad)(z \qquad)$$

The factorizations of $-21$ and the sums of the factors are shown in the following table.

| Product of the factors | Sum of the factors |
|:---:|:---:|
| $1(-21) = -21$ | $1 + (-21) = -20$ |
| $3(-7) = -21$ | $3 + (-7) = -4$ |
| $7(-3) = -21$ | $7 + (-3) = 4$ |
| $21(-1) = -21$ | $21 + (-1) = 20$ |

The second row contains the integers $+3$ and $-7$, whose product is $-21$ and whose sum is $-4$. So we can fill in the blanks with $+3$ and $-7$.

$$z^2 - 4z - 21 = (z + 3)(z - 7)$$

To check, we multiply $z + 3$ and $z - 7$.

$$(z + 3)(z - 7) = z^2 - 7z + 3z - 21$$
$$= z^2 - 4z - 21 \qquad \blacksquare$$

**Self Check**

*Answer*

Factor $q^2 - 2q - 24$.

$(q + 4)(q - 6)$

The next example has two variables.

**EXAMPLE 5**    Factor $x^2 + xy - 6y^2$.

*Solution*    Since the first term is $x^2$, the first term of each binomial factor must be $x$. Since the last term is $-6y^2$, the second term of each binomial factor has a factor of $y$. To fill in the following blanks, we must find coefficients whose product is $-6$ that will give a middle term of $xy$.

$$x^2 + xy - 6y^2 = (x \quad\quad y\ )(x \quad\quad y\ )$$

The possible factorizations of $-6$ and the sums of the factors are shown in the following table.

| Product of the factors | Sum of the factors |
|:---:|:---:|
| $1(-6) = -6$ | $1 + (-6) = -5$ |
| $2(-3) = -6$ | $2 + (-3) = -1$ |
| $3(-2) = -6$ | $3 + (-2) = 1$ |
| $6(-1) = -6$ | $6 + (-1) = 5$ |

The third row contains the integers $3$ and $-2$. These are the only integers whose product is $-6$ and will give the correct middle term of $xy$. So we can fill in the blanks with $3$ and $-2$.

$$x^2 + xy - 6y^2 = (x + 3y)(x - 2y)$$

We can check by multiplying $x + 3y$ and $x - 2y$.

$$(x + 3y)(x - 2y) = x^2 - 2xy + 3xy - 6y^2$$
$$= x^2 + xy - 6y^2 \qquad \blacksquare$$

**Self Check**

*Answer*

Factor $a^2 + ab - 12b^2$.

$(a - 3b)(a + 4b)$

### ■ FACTORING OUT −1

When the coefficient of the first term is −1, we begin by factoring out −1.

**EXAMPLE 6**   Factor $-x^2 + 2x + 15$.

*Solution*   We factor out −1 and then factor the trinomial.

$$-x^2 + 2x + 15 = -(x^2 - 2x - 15) \qquad \text{Factor out } -1.$$
$$= -(x - 5)(x + 3) \qquad \text{Factor } x^2 - 2x - 15.$$

We check by multiplying −1, $x - 5$, and $x + 3$.

$$-(x - 5)(x + 3) = -(x^2 + 3x - 5x - 15)$$
$$= -(x^2 - 2x - 15)$$
$$= -x^2 + 2x + 15$$

*Self Check*   Factor $-x^2 + 11x - 18$.
*Answer*   $-(x - 9)(x - 2)$

### ■ PRIME TRINOMIALS

If a trinomial cannot be factored using only integers, it is called a **prime polynomial.**

**EXAMPLE 7**   Factor $x^2 + 2x + 3$, if possible.

*Solution*   To factor the trinomial, we must find two integers whose product is +3 and whose sum is +2. The possible factorizations of 3 and the sums of the factors are shown in the following table.

| Product of the factors | Sum of the factors |
| --- | --- |
| $1(3) = 3$ | $1 + 3 = 4$ |
| $-1(-3) = 3$ | $-1 + (-3) = -4$ |

Since two integers whose product is +3 and whose sum is +2 do not exist, $x^2 + 2x + 3$ cannot be factored. It is a prime trinomial.

*Self Check*   Factor $x^2 - 4x + 6$, if possible.
*Answer*   It is prime.

### ■ MULTISTEP FACTORING

The following examples require more than one step.

**EXAMPLE 8**    Factor $-3ax^2 + 9a - 6ax$.

*Solution*    We write the trinomial in descending powers of $x$ and factor out the common factor of $-3a$.

$$-3ax^2 + 9a - 6ax = -3ax^2 - 6ax + 9a$$
$$= -3a(x^2 + 2x - 3)$$

Finally, we factor the trinomial $x^2 + 2x - 3$.

$$-3ax^2 + 9a - 6ax = -3a(x + 3)(x - 1)$$

We can check by multiplying.

$$-3a(x + 3)(x - 1) = -3a(x^2 + 2x - 3)$$
$$= -3ax^2 - 6ax + 9a$$
$$= -3ax^2 + 9a - 6ax \qquad ■$$

**Self Check**    Factor $-2pq^2 + 6p - 4pq$.
**Answer**    $-2p(q + 3)(q - 1)$

**EXAMPLE 9**    Factor $m^2 - 2mn + n^2 - 64a^2$.

*Solution*    We group the first three terms together and factor the resulting trinomial.

$$m^2 - 2mn + n^2 - 64a^2 = (m - n)(m - n) - 64a^2$$
$$= (m - n)^2 - (8a)^2$$

Then we factor the resulting difference of two squares:

$$m^2 - 2mn + n^2 - 64a^2 = (m - n)^2 - (8a)^2$$
$$= (m - n + 8a)(m - n - 8a) \qquad ■$$

**Self Check**    Factor $p^2 + 4pq + 4q^2 - 25y^2$.
**Answer**    $(p + 2q + 5y)(p + 2q - 5y)$

### ■ FACTORING PERFECT-SQUARE TRINOMIALS

We have discussed the following special product formulas used to square binomials.

**Special Product Formulas**
$$(x + y)^2 = x^2 + 2xy + y^2$$
$$(x - y)^2 = x^2 - 2xy + y^2$$

These formulas can be used in reverse order to factor perfect-square trinomials.

1. $x^2 + 2xy + y^2 = (x + y)^2$
2. $x^2 - 2xy + y^2 = (x - y)^2$

In words, Formula 1 states that *if a trinomial is the square of one quantity, plus twice the product of two quantities, plus the square of the second quantity, it factors into the square of the sum of the quantities.*

Formula 2 states that *if a trinomial is the square of one quantity, minus twice the product of two quantities, plus the square of the second quantity, it factors into the square of the difference of the quantities.*

The trinomials on the left-hand sides of the previous equations are called **perfect-square trinomials,** because they are the results of squaring a binomial. Although we can factor perfect-square trinomials by using the techniques discussed earlier in this section, we can usually factor them by inspecting their terms. For example, $x^2 + 8x + 16$ is a perfect-square trinomial, because

- The first term $x^2$ is the square of $x$.
- The last term 16 is the square of 4.
- The middle term $8x$ is twice the product of $x$ and 4.

Thus,

$$x^2 + 8x + 16 = x^2 + 2(x)(4) + 4^2$$
$$= (x + 4)^2$$

**EXAMPLE 10**  Factor $x^2 - 10x + 25$.

*Solution*  $x^2 - 10x + 25$ is a perfect-square trinomial, because

- The first term $x^2$ is the square of $x$.
- The last term 25 is the square of 5.
- The middle term $-10x$ is the negative of twice the product of $x$ and 5.

Thus,

$$x^2 - 10x + 25 = x^2 - 2(x)(5) + 5^2$$
$$= (x - 5)^2$$

∎

**Self Check**  Factor $x^2 + 10x + 25$.

*Answer*  $(x + 5)^2$

### ■ SOLVING EQUATIONS

We can use the factoring of trinomials and the zero-factor property to solve many equations.

**EXAMPLE 11**  Solve $x^3 - 2x^2 - 63x = 0$.

*Solution*

$$x^3 - 2x^2 - 63x = 0$$
$$x(x^2 - 2x - 63) = 0 \qquad \text{Factor out } x.$$
$$x(x + 7)(x - 9) = 0 \qquad \text{Factor the trinomial.}$$
$$x = 0 \quad \text{or} \quad x + 7 = 0 \quad \text{or} \quad x - 9 = 0 \qquad \text{Set each factor equal to 0.}$$
$$x = -7 \qquad\qquad x = 9 \qquad \text{Solve each linear equation.}$$

The solutions are 0, $-7$, and 9. Check each one. ■

*Self Check*   Solve $x^3 - x^2 - 2x = 0$.

*Answers*   0, $-1$, 2

*Orals*   *Finish each factoring problem.*

**1.** $x^2 + 5x + 4 = (x + 1)(x + \phantom{)})$
**2.** $x^2 - 5x + 6 = (x \phantom{space} 2)(x \phantom{space} 3)$
**3.** $x^2 + x - 6 = (x \phantom{space} 2)(x + \phantom{)})$
**4.** $x^2 - x - 6 = (x \phantom{space} 3)(x + \phantom{)})$
**5.** $x^2 + 5x - 6 = (x + \phantom{)})(x - \phantom{)})$
**6.** $x^2 - 7x + 6 = (x - \phantom{)})(x - \phantom{)})$

## EXERCISE 5.4

***REVIEW***   *Graph the solution of each inequality on a number line.*

**1.** $x - 3 > 5$
**2.** $x + 4 \leq 3$
**3.** $-3x - 5 \geq 4$
**4.** $2x - 3 < 7$

**5.** $\dfrac{3(x - 1)}{4} < 12$
**6.** $\dfrac{-2(x + 3)}{3} \geq 9$
**7.** $-2 < x \leq 4$
**8.** $-5 \leq x + 1 < 5$

***VOCABULARY AND CONCEPTS***   *Complete each formula.*

**9.** $x^2 + 2xy + y^2 = $ _____
**10.** $x^2 - 2xy + y^2 = $ _____

*Complete each factorization.*

**11.** $y^2 + 6y + 8 = (y + \phantom{)})(y + \phantom{)})$
**12.** $z^2 - 3z - 10 = (z + \phantom{)})(z - \phantom{)})$
**13.** $x^2 - xy - 2y^2 = (x + \phantom{)})(x - \phantom{)})$
**14.** $a^2 + ab - 6b^2 = (a + \phantom{)})(a - \phantom{)})$

*In Exercises 15–42, factor each trinomial, if possible. Use the FOIL method to check each result.*

**15.** $x^2 + 3x + 2$     **16.** $y^2 + 4y + 3$     **17.** $z^2 + 12z + 11$     **18.** $x^2 + 7x + 10$

**19.** $a^2 - 4a - 5$     **20.** $b^2 + 6b - 7$     **21.** $t^2 - 9t + 14$     **22.** $c^2 - 9c + 8$

**23.** $u^2 + 10u + 15$     **24.** $v^2 + 9v + 15$     **25.** $y^2 - y - 30$     **26.** $x^2 - 3x - 40$

**27.** $a^2 + 6a - 16$     **28.** $x^2 + 5x - 24$     **29.** $t^2 - 5t - 50$     **30.** $a^2 - 10a - 39$

**31.** $r^2 - 9r - 12$     **32.** $s^2 + 11s - 26$     **33.** $y^2 + 2yz + z^2$     **34.** $r^2 - 2rs + 4s^2$

**35.** $x^2 + 4xy + 4y^2$     **36.** $a^2 + 10ab + 9b^2$     **37.** $m^2 + 3mn - 10n^2$     **38.** $m^2 - mn - 12n^2$

**39.** $a^2 - 4ab - 12b^2$     **40.** $p^2 + pq - 6q^2$     **41.** $u^2 + 2uv - 15v^2$     **42.** $m^2 + 3mn - 10n^2$

*In Exercises 43–54, factor each trinomial. Factor out $-1$ first.*

**43.** $-x^2 - 7x - 10$     **44.** $-x^2 + 9x - 20$     **45.** $-y^2 - 2y + 15$     **46.** $-y^2 - 3y + 18$

**47.** $-t^2 - 15t + 34$     **48.** $-t^2 - t + 30$     **49.** $-r^2 + 14r - 40$     **50.** $-r^2 + 14r - 45$

**51.** $-a^2 - 4ab - 3b^2$     **52.** $-a^2 - 6ab - 5b^2$     **53.** $-x^2 + 6xy + 7y^2$     **54.** $-x^2 - 10xy + 11y^2$

*In Exercises 55–66, write each trinomial in descending powers of one variable, and then factor.*

**55.** $4 - 5x + x^2$     **56.** $y^2 + 5 + 6y$     **57.** $10y + 9 + y^2$     **58.** $x^2 - 13 - 12x$

**59.** $c^2 - 5 + 4c$     **60.** $b^2 - 6 - 5b$     **61.** $-r^2 + 2s^2 + rs$     **62.** $u^2 - 3v^2 + 2uv$

**63.** $4rx + r^2 + 3x^2$     **64.** $-a^2 + 5b^2 + 4ab$     **65.** $-3ab + a^2 + 2b^2$     **66.** $-13yz + y^2 - 14z^2$

*In Exercises 67–78, completely factor each trinomial. Factor out any common monomials first (including $-1$, if necessary).*

**67.** $2x^2 + 10x + 12$     **68.** $3y^2 - 21y + 18$     **69.** $3y^3 + 6y^2 + 3y$     **70.** $4x^4 + 16x^3 + 16x^2$

**71.** $-5a^2 + 25a - 30$     **72.** $-2b^2 + 20b - 18$     **73.** $3z^2 - 15tz + 12t^2$     **74.** $5m^2 + 45mn - 50n^2$

**75.** $12xy + 4x^2y - 72y$     **76.** $48xy + 6xy^2 + 96x$     **77.** $-4x^2y - 4x^3 + 24xy^2$     **78.** $3x^2y^3 + 3x^3y^2 - 6xy^4$

*In Exercises 79–86, completely factor each expression.*

**79.** $ax^2 + 4ax + 4a + bx + 2b$

**80.** $mx^2 + mx - 6m + nx - 2n$

**81.** $a^2 + 8a + 15 + ab + 5b$

**82.** $x^2 + 2xy + y^2 + 2x + 2y$

**83.** $a^2 + 2ab + b^2 - 4$

**84.** $a^2 + 6a + 9 - b^2$

**85.** $b^2 - y^2 - 4y - 4$

**86.** $c^2 - a^2 + 8a - 16$

*In Exercises 87–98, factor each perfect square trinomial.*

**87.** $x^2 + 6x + 9$

**88.** $x^2 + 10x + 25$

**89.** $y^2 - 8y + 16$

**90.** $z^2 - 2z + 1$

**91.** $t^2 + 20t + 100$

**92.** $r^2 + 24r + 144$

**93.** $u^2 - 18u + 81$

**94.** $v^2 - 14v + 49$

**95.** $x^2 + 4xy + 4y^2$

**96.** $a^2 + 6ab + 9b^2$

**97.** $r^2 - 10rs + 25s^2$

**98.** $m^2 - 12mn + 36n^2$

*In Exercises 99–116, solve each equation.*

**99.** $x^2 - 13x + 12 = 0$

**100.** $x^2 + 7x + 6 = 0$

**101.** $x^2 - 2x - 15 = 0$

**102.** $x^2 - x - 20 = 0$

**103.** $-4x - 21 + x^2 = 0$

**104.** $2x + x^2 - 15 = 0$

**105.** $x^2 + 8 - 9x = 0$

**106.** $45 + x^2 - 14x = 0$

**107.** $a^2 + 8a = -15$

**108.** $a^2 - a = 56$

**109.** $2y - 8 = -y^2$

**110.** $-3y + 18 = y^2$

**111.** $x^3 + 3x^2 + 2x = 0$

**112.** $x^3 - 7x^2 + 10x = 0$

**113.** $x^3 - 27x - 6x^2 = 0$

**114.** $x^3 - 22x - 9x^2 = 0$

**115.** $(x - 1)(x^2 + 5x + 6) = 0$

**116.** $(x - 2)(x^2 - 8x + 7) = 0$

## WRITING

**117.** Explain how you would write a trinomial in descending order.

**118.** Explain how to use the FOIL method to check the factoring of a trinomial.

## SOMETHING TO THINK ABOUT

**119.** Two students factor $2x^2 + 20x + 42$ and get two different answers: $(2x + 6)(x + 7)$, and $(x + 3)(2x + 14)$. Do both answers check? Why don't they agree? Is either completely correct?

**120.** Find the error:

| | |
|---|---|
| $x = y$ | |
| $x^2 = xy$ | Multiply both sides by $x$. |
| $x^2 - y^2 = xy - y^2$ | Subtract $y^2$ from both sides. |
| $(x + y)(x - y) = y(x - y)$ | Factor. |
| $x + y = y$ | Divide both sides by $(x - y)$. |
| $y + y = y$ | Substitute $y$ for its equal, $x$. |
| $2y = y$ | Combine like terms. |
| $2 = 1$ | Divide both sides by $y$. |

## 5.5 Factoring General Trinomials

■ FACTORING TRINOMIALS OF THE FORM $ax^2 + bx + c$ ■ FACTORING PERFECT-SQUARE
TRINOMIALS ■ SOLVING EQUATIONS

Getting Ready    *Multiply and combine like terms.*

**1.** $(2x + 1)(3x + 2)$        **2.** $(3y - 2)(2y - 5)$        **3.** $(4t - 3)(2t + 3)$

**4.** $(2r + 5)(2r - 3)$        **5.** $(2m - 3)(3m - 2)$        **6.** $(4a + 3)(4a + 1)$

### ■ FACTORING TRINOMIALS OF THE FORM $ax^2 + bx + c$

We must consider more combinations of factors when we factor trinomials with lead
coefficients other than 1.

EXAMPLE 1    Factor $2x^2 + 5x + 3$.

Solution    Since the first term is $2x^2$, the first terms of the binomial factors must be $2x$ and $x$.
To fill in the following blanks, we must find two factors of $+3$ that will give a
middle term of $+5x$.

$$(2x \qquad )(x \qquad )$$

Since the sign of each term of the trinomial is $+$, we need to consider only positive
factors of the last term (3). Since the positive factors of 3 are 1 and 3, there are two
possible factorizations.

$$(2x + 1)(x + 3) \qquad \text{or} \qquad (2x + 3)(x + 1)$$

The first possibility is incorrect, because it gives a middle term of $7x$. The second
possibility is correct, because it gives a middle term of $5x$. Thus,

$$2x^2 + 5x + 3 = (2x + 3)(x + 1)$$

Check by multiplication.    ■

Self Check    Factor $3x^2 + 7x + 2$.
Answer       $(3x + 1)(x + 2)$

EXAMPLE 2    Factor $6x^2 - 17x + 5$.

Solution    Since the first term is $6x^2$, the first terms of the binomial factors must be $6x$ and $x$
or $3x$ and $2x$. To fill in the following blanks, we must find two factors of $+5$ that will
give a middle term of $-17x$.

$$(6x \qquad )(x \qquad ) \qquad \text{or} \qquad (3x \qquad )(2x \qquad )$$

Since the sign of the third term is + and the sign of the middle term is −, we need to consider only negative factors of the last term (5). Since the negative factors of 5 are −1 and −5, there are four possible factorizations.

$$(6x - 1)(x - 5) \qquad (6x - 5)(x - 1)$$
$$(3x - 1)(2x - 5) \qquad (3x - 5)(2x - 1)$$

Only the possibility printed in color gives the correct middle term of −17x. Thus,

$$6x^2 - 17x + 5 = (3x - 1)(2x - 5)$$

Check by multiplication. ∎

**Self Check** Factor $6x^2 - 7x + 2$.

**Answer** $(3x - 2)(2x - 1)$

**EXAMPLE 3** Factor $3y^2 - 4y - 4$.

*Solution* Since the first term is $3y^2$, the first terms of the binomial factors must be $3y$ and $y$. To fill in the following blanks, we must find two factors of −4 that will give a middle term of −4y.

$$(3y \quad\quad )(y \quad\quad )$$

Since the sign of the third term is −, the signs inside the binomial factors will be different. Because the factors of the last term (4) are 1, 2, and 4, there are six possibilities to consider.

$$(3y + 1)(y - 4) \qquad (3y + 4)(y - 1)$$
$$(3y - 1)(y + 4) \qquad (3y - 4)(y + 1)$$
$$(3y - 2)(y + 2) \qquad (3y + 2)(y - 2)$$

Only the possibility printed in color gives the correct middle term of −4y. Thus,

$$3y^2 - 4y - 4 = (3y + 2)(y - 2)$$

Check by multiplication. ∎

**Self Check** Factor $5a^2 - 7a - 6$.

**Answer** $(5a + 3)(a - 2)$

**EXAMPLE 4** Factor $6b^2 + 7b - 20$.

*Solution* Since the first term is $6b^2$, the first terms of the binomial factors must be $6b$ and $b$ or $3b$ and $2b$. To fill in the following blanks, we must find two factors of −20 that will give a middle term of +7b.

$$(6b \quad\quad )(b \quad\quad ) \qquad \text{or} \qquad (3b \quad\quad )(2b \quad\quad )$$

Since the sign of the third term is $-$, the signs inside the binomial factors will be different. Because the factors of the last term (20) are 1, 2, 5, 10, and 20, there are many possible combinations for the last terms. We must try to find one that will give a last term of $-20$ and a sum of the products of the outer terms and inner terms of $+7b$.

If we pick factors of $6b$ and $b$ for the first terms and $-5$ and $+4$ for the last terms, we have

$$(6b - 5)(b + 4)$$

$$\begin{array}{r} -5b \\ \underline{24b} \\ 19b \end{array}$$

which gives a wrong middle term of $19b$.

If we pick factors of $3b$ and $2b$ for the first terms and $-4$ and $+5$ for the last terms, we have

$$(3b - 4)(2b + 5)$$

$$\begin{array}{r} -8b \\ \underline{15b} \\ 7b \end{array}$$

which gives the correct middle term of $+7b$ and the correct last term of $-20$. Thus,

$$6b^2 + 7b - 20 = (3b - 4)(2b + 5)$$

Check by multiplication. ∎

**Self Check** Factor $4x^2 + 4x - 3$.

**Answer** $(2x + 3)(2x - 1)$

The next example has two variables.

**EXAMPLE 5** Factor $2x^2 + 7xy + 6y^2$.

**Solution** Since the first term is $2x^2$, the first terms of the binomial factors must be $2x$ and $x$. To fill in the following blanks, we must find two factors of $6y^2$ that will give a middle term of $+7xy$.

$$(2x \qquad )(x \qquad )$$

Since the sign of each term is $+$, the signs inside the binomial factors will be $+$. The possible factors of the last term $(6y^2)$ are $y$, $2y$, $3y$, and $6y$. We must try to find one that will give a last term of $+6y^2$ and a sum of the products of the outer terms and inner terms of $+7xy$.

If we pick factors of $6y$ and $y$, we have

$$(2x + y)(x + 6y)$$

$$\begin{array}{c} xy \\ \underline{12xy} \\ 13xy \end{array}$$

which gives a wrong middle term of $13xy$.

If we pick factors of $3y$ and $2y$, we have

$$(2x + 3y)(x + 2y)$$

$$\begin{array}{c} 3xy \\ \underline{4xy} \\ 7xy \end{array}$$

which gives a correct middle term of $7xy$. Thus,

$$2x^2 + 7xy + 6y^2 = (2x + 3y)(x + 2y)$$

Check by multiplication.  ■

**Self Check**   Factor $4x^2 + 8xy + 3y^2$.
**Answer**   $(2x + 3y)(2x + y)$

Because some guesswork is often necessary, it is difficult to give specific rules for factoring trinomials. However, the following hints are often helpful.

### Factoring General Trinomials

1. Write the trinomial in descending powers of one variable.
2. Factor out any GCF (including $-1$ if that is necessary to make the coefficient of the first term positive).
3. If the sign of the third term is $+$, the signs between the terms of the binomial factors are the same as the sign of the middle term. If the sign of the third term is $-$, the signs between the terms of the binomial factors are opposite.
4. Try combinations of first terms and last terms until you find one that works, or until you exhaust all the possibilities. If no combination works, the trinomial is prime.
5. Check the factorization by multiplication.

**EXAMPLE 6**   Factor $2x^2y - 8x^3 + 3xy^2$.

*Solution*   *Step 1:* Write the trinomial in descending powers of $x$.

$$-8x^3 + 2x^2y + 3xy^2$$

*Step 2:* Factor out the negative of the GCF, which is $-x$.

$$-8x^3 + 2x^2y + 3xy^2 = -x(8x^2 - 2xy - 3y^2)$$

*Step 3:* Because the sign of the third term of the trinomial factor is $-$, the signs within its binomial factors will be opposites.

*Step 4:* Find the binomial factors of the trinomial.

$$\begin{aligned} -8x^3 + 2x^2y + 3xy^2 &= -x(8x^2 - 2xy - 3y^2) \\ &= -x(2x + y)(4x - 3y) \end{aligned}$$

*Step 5:* Check by multiplication.

$$\begin{aligned} -x(2x + y)(4x - 3y) &= -x(8x^2 - 6xy + 4xy - 3y^2) \\ &= -x(8x^2 - 2xy - 3y^2) \\ &= -8x^3 + 2x^2y + 3xy^2 \\ &= 2x^2y - 8x^3 + 3xy^2 \end{aligned}$$    ∎

**Self Check**   Factor $12y - 2y^3 - 2y^2$.
**Answer**   $-2y(y + 3)(y - 2)$

## ■ FACTORING PERFECT-SQUARE TRINOMIALS

As before, we can factor perfect-square trinomials by inspection.

**EXAMPLE 7**   Factor $4x^2 - 20x + 25$.

*Solution*   $4x^2 - 20x + 25$ is a perfect-square trinomial, because

- The first term $4x^2$ is the square of $2x$: $(2x)^2 = 4x^2$.
- The last term 25 is the square of 5: $5^2 = 25$.
- The middle term $-20x$ is the negative of twice the product of $2x$ and 5.

Thus,

$$4x^2 - 20x + 25 = (2x)^2 - 2(2x)(5) + 5^2$$
$$= (2x - 5)^2$$ ∎

**Self Check**   Factor $9x^2 - 12x + 4$.
**Answer**   $(3x - 2)^2$

The next example combines the techniques of factoring by grouping, factoring a perfect-square trinomial, and factoring the difference of two squares.

**EXAMPLE 8**   Factor $4x^2 - 4xy + y^2 - 9$.

**Solution**
$$4x^2 - 4xy + y^2 - 9$$

| | |
|---|---|
| $= (4x^2 - 4xy + y^2) - 9$ | Group the first three terms. |
| $= (2x - y)^2 - 9$ | Factor the perfect-square trinomial. |
| $= [(2x - y) + 3][(2x - y) - 3]$ | Factor the difference of two squares. |
| $= (2x - y + 3)(2x - y - 3)$ | Remove parentheses. |

Check by multiplication. ∎

**Self Check**   Factor $x^2 + 4x + 4 - y^2$.
**Answer**   $(x + 2 + y)(x + 2 - y)$

**EXAMPLE 9**   Factor $9 - 4x^2 - 4xy - y^2$.

**Solution**

| | |
|---|---|
| $9 - 4x^2 - 4xy - y^2 = 9 - (4x^2 + 4xy + y^2)$ | Factor $-1$ from the trinomial. |
| $= 9 - (2x + y)(2x + y)$ | Factor the perfect-square trinomial. |
| $= 9 - (2x + y)^2$ | $(2x + y)(2x + y) = (2x + y)^2$. |
| $= [3 + (2x + y)][3 - (2x + y)]$ | Factor the difference of two squares. |
| $= (3 + 2x + y)(3 - 2x - y)$ | Remove parentheses. |

Check by multiplication. ∎

**Self Check**   Factor $16 - a^2 - 2a - 1$.
**Answer**   $(a + 5)(3 - a)$

### ■ SOLVING EQUATIONS

**EXAMPLE 10**   Solve $2x^2 + 3x = 2$.

*Solution*   We write the equation in the form $ax^2 + bx + c = 0$ and solve for $x$.

$$2x^2 + 3x = 2$$
$$2x^2 + 3x - 2 = 0 \qquad \text{Add } -2 \text{ to both sides.}$$
$$(2x - 1)(x + 2) = 0 \qquad \text{Factor } 2x^2 + 3x - 2.$$
$$2x - 1 = 0 \quad \text{or} \quad x + 2 = 0 \qquad \text{Set each factor equal to 0.}$$
$$2x = 1 \qquad\qquad x = -2 \qquad \text{Solve each linear equation.}$$
$$x = \frac{1}{2}$$

Check each solution.   ■

**Self Check**   Solve $3x^2 - 5x - 2 = 0$.
**Answers**   $2, -\frac{1}{3}$

**EXAMPLE 11**   Solve $6x^3 + 12x = 17x^2$.

*Solution*
$$6x^3 + 12x = 17x^2$$
$$6x^3 - 17x^2 + 12x = 0 \qquad\qquad \text{Subtract } 17x^2 \text{ from both sides.}$$
$$x(6x^2 - 17x + 12) = 0 \qquad\qquad \text{Factor out } x.$$
$$x(2x - 3)(3x - 4) = 0 \qquad\qquad \text{Factor } 6x^2 - 17x + 12.$$
$$x = 0 \quad \text{or} \quad 2x - 3 = 0 \quad \text{or} \quad 3x - 4 = 0 \qquad \text{Set each factor equal to 0.}$$
$$x = 0 \qquad\qquad 2x = 3 \qquad\qquad 3x = 4 \qquad \text{Solve the linear equations.}$$
$$x = \frac{3}{2} \qquad\qquad x = \frac{4}{3}$$

Check each solution.   ■

**Self Check**   Solve $6x^3 + 7x^2 = 5x$.
**Answers**   $0, \frac{1}{2}, -\frac{5}{3}$

**Orals**   *Finish factoring each problem.*

**1.** $2x^2 + 5x + 3 = (\phantom{x}x + \phantom{)})(x + 1)$   **2.** $6x^2 + 5x + 1 = (\phantom{x}x + 1)(3x + 1)$

**3.** $6x^2 + 5x - 1 = (x \phantom{1)(6x} 1)(6x \phantom{1)} 1)$   **4.** $6x^2 + x - 1 = (2x \phantom{1)(3x} 1)(3x \phantom{1)} 1)$

**5.** $4x^2 + 4x - 3 = (2x + \phantom{)})(2x - \phantom{)})$   **6.** $4x^2 - x - 3 = (4x + \phantom{)})(x - \phantom{)})$

## EXERCISE 5.5

### REVIEW

**1.** The $n$th term $l$ of an arithmetic sequence is

$$l = f + (n - 1)d$$

where $f$ is the first term and $d$ is the common difference. Remove the parentheses and solve the equation for $n$.

**2.** The sum $S$ of $n$ consecutive terms of an arithmetic sequence is

$$S = \frac{n}{2}(f + l)$$

where $f$ is the first term and $l$ is the $n$th term. Solve for $f$.

### VOCABULARY AND CONCEPTS   *Fill in each blank to make a true statement.*

**3.** To factor a general trinomial, first write the trinomial in _____ powers of one variable.

**4.** If the sign of the first and third terms of a trinomial are $+$, the signs within the binomial factors are _____ as the sign of the middle term.

**5.** If the sign of the first term of a trinomial is $+$ and the sign of the third term is $-$, the signs within the binomial factors are _____.

**6.** Always check factorizations by _____.

*Complete each factorization.*

**7.** $6x^2 + x - 2 = (3x + \quad)(2x - \quad)$

**8.** $15x^2 - 7x - 4 = (5x - \quad)(3x + \quad)$

**9.** $12x^2 - 7xy + y^2 = (3x - \quad)(4x - \quad)$

**10.** $6x^2 + 5xy - 6y^2 = (2x + \quad)(3x - \quad)$

### PRACTICE   *In Exercises 11–34, factor each trinomial.*

**11.** $2x^2 - 3x + 1$

**12.** $2y^2 - 7y + 3$

**13.** $3a^2 + 13a + 4$

**14.** $2b^2 + 7b + 6$

**15.** $4z^2 + 13z + 3$

**16.** $4t^2 - 4t + 1$

**17.** $6y^2 + 7y + 2$

**18.** $4x^2 + 8x + 3$

**19.** $6x^2 - 7x + 2$

**20.** $4z^2 - 9z + 2$

**21.** $3a^2 - 4a - 4$

**22.** $8u^2 - 2u - 15$

**23.** $2x^2 - 3x - 2$

**24.** $12y^2 - y - 1$

**25.** $2m^2 + 5m - 12$

**26.** $10u^2 - 13u - 3$

**27.** $10y^2 - 3y - 1$

**28.** $6m^2 + 19m + 3$

**29.** $12y^2 - 5y - 2$

**30.** $10x^2 + 21x - 10$

**31.** $5t^2 + 13t + 6$

**32.** $16y^2 + 10y + 1$

**33.** $16m^2 - 14m + 3$

**34.** $16x^2 + 16x + 3$

*In Exercises 35–46, factor each trinomial.*

**35.** $3x^2 - 4xy + y^2$

**36.** $2x^2 + 3xy + y^2$

**37.** $2u^2 + uv - 3v^2$

**38.** $2u^2 + 3uv - 2v^2$

**39.** $4a^2 - 4ab + b^2$

**40.** $2b^2 - 5bc + 2c^2$

**41.** $6r^2 + rs - 2s^2$

**42.** $3m^2 + 5mn + 2n^2$

**43.** $4x^2 + 8xy + 3y^2$

**44.** $4b^2 + 15bc - 4c^2$

**45.** $4a^2 - 15ab + 9b^2$

**46.** $12x^2 + 5xy - 3y^2$

*In Exercises 47–62, write the terms of each trinomial in descending powers of one variable. Then factor the trinomial, if possible.*

**47.** $-13x + 3x^2 - 10$

**48.** $-14 + 3a^2 - a$

**49.** $15 + 8a^2 - 26a$

**50.** $16 - 40a + 25a^2$

**51.** $12y^2 + 12 - 25y$

**52.** $12t^2 - 1 - 4t$

**53.** $3x^2 + 6 + x$

**54.** $25 + 2u^2 + 3u$

**55.** $2a^2 + 3b^2 + 5ab$

**56.** $11uv + 3u^2 + 6v^2$

**57.** $pq + 6p^2 - q^2$

**58.** $-11mn + 12m^2 + 2n^2$

**59.** $b^2 + 4a^2 + 16ab$

**60.** $3b^2 + 3a^2 - ab$

**61.** $12x^2 + 10y^2 - 23xy$

**62.** $5ab + 25a^2 - 2b^2$

*In Exercises 63–78, factor each polynomial.*

**63.** $4x^2 + 10x - 6$

**64.** $9x^2 + 21x - 18$

**65.** $y^3 + 13y^2 + 12y$

**66.** $2xy^2 + 8xy - 24x$

**67.** $6x^3 - 15x^2 - 9x$

**68.** $9y^3 + 3y^2 - 6y$

**69.** $30r^5 + 63r^4 - 30r^3$

**70.** $6s^5 - 26s^4 - 20s^3$

**71.** $4a^2 - 4ab - 8b^2$

**72.** $6x^2 + 3xy - 18y^2$

**73.** $8x^2 - 12xy - 8y^2$

**74.** $24a^2 + 14ab + 2b^2$

**75.** $-16m^3n - 20m^2n^2 - 6mn^3$

**76.** $-84x^4 - 100x^3y - 24x^2y^2$

**77.** $-28u^3v^3 + 26u^2v^4 - 6uv^5$

**78.** $-16x^4y^3 + 30x^3y^4 + 4x^2y^5$

*In Exercises 79–84, factor each perfect-square trinomial.*

**79.** $4x^2 + 12x + 9$

**80.** $4x^2 - 4x + 1$

**81.** $9x^2 + 12x + 4$

**82.** $4x^2 - 20x + 25$

**83.** $16x^2 - 8xy + y^2$

**84.** $25x^2 + 20xy + 4y^2$

*In Exercises 85–90, factor each polynomial.*

**85.** $4x^2 + 4xy + y^2 - 16$

**86.** $9x^2 - 6x + 1 - d^2$

**87.** $9 - a^2 - 4ab - 4b^2$

**88.** $25 - 9a^2 + 6ac - c^2$

**89.** $4x^2 + 4xy + y^2 - a^2 - 2ab - b^2$

**90.** $a^2 - 2ab + b^2 - x^2 + 2x - 1$

*In Exercises 91–106, solve each equation.*

**91.** $2x^2 - 5x + 2 = 0$

**92.** $2x^2 + x - 3 = 0$

**93.** $5x^2 - 6x + 1 = 0$

**94.** $6x^2 - 5x + 1 = 0$

**95.** $3x^2 - 8x = 3$

**96.** $2x^2 - 11x = 21$

**97.** $15x^2 - 2 = 7x$

**98.** $8x^2 + 10x = 3$

**99.** $x(6x + 5) = 6$      **100.** $x(2x - 3) = 14$      **101.** $(x + 1)(8x + 1) = 18x$    **102.** $4x(3x + 2) = x + 12$

**103.** $2x(3x^2 + 10x) = -6x$    **104.** $2x^3 = 2x(x + 2)$      **105.** $x^3 + 7x^2 = x^2 - 9x$     **106.** $x^2(x + 10) = 2x(x - 8)$

## WRITING

**107.** Describe an organized approach to finding all of the possibilities when you attempt to factor $12x^2 - 4x + 9$.

**108.** Explain how to determine whether a trinomial is prime.

## SOMETHING TO THINK ABOUT

**109.** For what values of $b$ will the trinomial $6x^2 + bx + 6$ be factorable?

**110.** Create a quadratic equation with the two solutions $x = 3$ and $x = \frac{3}{2}$.

## 5.6 Factoring the Sum and Difference of Two Cubes

■ FACTORING THE SUM OF TWO CUBES ■ FACTORING THE DIFFERENCE OF TWO CUBES
■ MULTISTEP FACTORING

*Getting Ready*    *Find each product.*

**1.** $(x - 3)(x^2 + 3x + 9)$          **2.** $(x + 2)(x^2 - 2x + 4)$
**3.** $(y + 4)(y^2 - 4y + 16)$       **4.** $(r - 5)(r^2 + 5r + 25)$
**5.** $(a - b)(a^2 + ab + b^2)$      **6.** $(a + b)(a^2 - ab + b^2)$

Recall that the difference of the squares of two quantities factors into the product of two binomials. One binomial is the sum of the quantities, and the other is the difference of the quantities.

$$x^2 - y^2 = (x + y)(x - y) \qquad \text{or} \qquad \text{F}^2 - \text{L}^2 = (\text{F} + \text{L})(\text{F} - \text{L})$$

There are similar formulas for factoring the sum of two cubes and the difference of two cubes.

### ■ FACTORING THE SUM OF TWO CUBES

To find the formula for factoring the sum of two cubes, we need to find the following product:

$$(x + y)(x^2 - xy + y^2) = (x + y)x^2 - (x + y)xy + (x + y)y^2 \qquad \text{Use the distributive property.}$$

$$= x^3 + x^2y - x^2y - xy^2 + xy^2 + y^3$$
$$= x^3 + y^3$$

This result justifies the formula for factoring the **sum of two cubes.**

**Factoring the Sum of Two Cubes**

$$x^3 + y^3 = (x + y)(x^2 - xy + y^2)$$

If we think of the sum of two cubes as the cube of a **First** quantity plus the cube of a **Last** quantity, we have the formula

$$F^3 + L^3 = (F + L)(F^2 - FL + L^2)$$

In words, we say, *To factor the cube of a **First** quantity plus the cube of a **Last** quantity, we multiply the **First** plus the **Last** by*

- *the **First** squared*
- *minus the **First** times the **Last***
- *plus the **Last** squared.*

To factor the sum of two cubes, it is helpful to know the cubes of the numbers from 1 to 10:

**1, 8, 27, 64, 125, 216, 343, 512, 729, 1,000**

Expressions containing variables such as $x^6y^3$ are also perfect cubes, because they can be written as the cube of a quantity:

$$x^6y^3 = (x^2y)^3$$

**EXAMPLE 1**   Factor $x^3 + 8$.

*Solution*   The binomial $x^3 + 8$ is the sum of two cubes, because

$$x^3 + 8 = x^3 + 2^3$$

Thus, $x^3 + 8$ factors as $(x + 2)$ times the trinomial $x^2 - 2x + 2^2$.

$$F^3 + L^3 = (F + L)(F^2 - F\ L\ + L^2)$$
$$\downarrow\quad\downarrow\qquad\downarrow\quad\downarrow\ \downarrow\quad\downarrow\downarrow\qquad\downarrow$$
$$x^3 + 2^3 = (x + 2)(x^2 - x \cdot 2 + 2^2)$$
$$= (x + 2)(x^2 - 2x + 4)$$

Check by multiplication.

$$(x + 2)(x^2 - 2x + 4) = (x + 2)x^2 - (x + 2)2x + (x + 2)4$$
$$= x^3 + 2x^2 - 2x^2 - 4x + 4x + 8$$
$$= x^3 + 8$$ ∎

*Self Check*   Factor $p^3 + 64$.

*Answer*   $(p + 4)(p^2 - 4p + 16)$

**EXAMPLE 2**  Factor $8b^3 + 27c^3$.

*Solution*  The binomial $8b^3 + 27c^3$ is the sum of two cubes, because

$$8b^3 + 27c^3 = (2b)^3 + (3c)^3$$

Thus, $8b^3 + 27c^3$ factors as $(2b + 3c)$ times the trinomial $(2b)^2 - (2b)(3c) + (3c)^2$.

$$\text{F}^3 \; + \; \text{L}^3 \; = ( \text{F} + \text{L} ) \, (\text{F}^2 \; - \; \text{F} \;\; \text{L} \; + \; \text{L}^2)$$

$$(2b)^3 + (3c)^3 = (2b + 3c)[(2b)^2 - (2b)(3c) + (3c)^2]$$
$$= (2b + 3c)(4b^2 - 6bc + 9c^2)$$

Check by multiplication.

$$(2b + 3c)(4b^2 - 6bc + 9c^2)$$
$$= (2b + 3c)4b^2 - (2b + 3c)6bc + (2b + 3c)9c^2$$
$$= 8b^3 + 12b^2c - 12b^2c - 18bc^2 + 18bc^2 + 27c^3$$
$$= 8b^3 + 27c^3 \qquad\blacksquare$$

**Self Check**   Factor $1{,}000p^3 + q^3$.

**Answer**   $(10p + q)(100p^2 - 10pq + q^2)$

## ■ FACTORING THE DIFFERENCE OF TWO CUBES

To find the formula for factoring the difference of two cubes, we need to find the following product:

$$(x - y)(x^2 + xy + y^2) = (x - y)x^2 + (x - y)xy + (x - y)y^2 \qquad \text{Use the distributive property.}$$

$$= x^3 - x^2y + x^2y - xy^2 + xy^2 - y^3$$
$$= x^3 - y^3$$

This result justifies the formula for factoring the **difference of two cubes.**

**Factoring the Difference of Two Cubes**
$$x^3 - y^3 = (x - y)(x^2 + xy + y^2)$$

If we think of the difference of two cubes as the cube of a **First** quantity minus the cube of a **Last** quantity, we have the formula

$$\text{F}^3 - \text{L}^3 = (\text{F} - \text{L})(\text{F}^2 + \text{FL} + \text{L}^2)$$

In words, we say, *To factor the cube of a **First** quantity minus the cube of a **Last** quantity, we multiply the **First** minus the **Last** by*

- *the **First** squared*
- *plus the **First** times the **Last***
- *plus the **Last** squared.*

**EXAMPLE 3**     Factor $a^3 - 64b^3$.

*Solution*     The binomial $a^3 - 64b^3$ is the difference of two cubes.

$$a^3 - 64b^3 = a^3 - (4b)^3$$

Thus, its factors are the difference $a - 4b$ and the trinomial $a^2 + a(4b) + (4b)^2$.

$$\mathbf{F}^3 - \mathbf{L}^3 \ = (\mathbf{F} - \mathbf{L})(\mathbf{F}^2 + \mathbf{F} \ \mathbf{L} \ + \ \mathbf{L}^2)$$
$$\downarrow \quad \downarrow \quad \quad \downarrow \quad \downarrow \ \downarrow \quad \downarrow \ \downarrow \quad \quad \downarrow$$
$$a^3 - (4b)^3 = (a - 4b)[a^2 + a(4b) + (4b)^2]$$
$$= (a - 4b)(a^2 + 4ab + 16b^2)$$

Check by multiplication.

$$(a - 4b)(a^2 + 4ab + 16b^2)$$
$$= (a - 4b)a^2 + (a - 4b)4ab + (a - 4b)16b^2$$
$$= a^3 - 4a^2b + 4a^2b - 16ab^2 + 16ab^2 - 64b^3$$
$$= a^3 - 64b^3 \qquad \blacksquare$$

*Self Check*     Factor $27p^3 - 8$.
*Answer*     $(3p - 2)(9p^2 + 6p + 4)$

## ■ MULTISTEP FACTORING

Sometimes we must factor out a greatest common factor before factoring a sum or difference of two cubes.

**EXAMPLE 4**     Factor $-2t^5 + 128t^2$.

*Solution*     $-2t^5 + 128t^2 = -2t^2(t^3 - 64)$     Factor out $-2t^2$.
$\qquad\qquad\qquad = -2t^2(t - 4)(t^2 + 4t + 16)$     Factor $t^3 - 64$.

Verify this factorization by multiplication.     $\blacksquare$

*Self Check*     Factor $-3p^4 + 81p$.
*Answer*     $-3p(p - 3)(p^2 + 3p + 9)$

**EXAMPLE 5**    Factor $x^6 - 64$.

*Solution*    The binomial $x^6 - 64$ is both the difference of two squares and the difference of two cubes. Since it is easier to factor the difference of two squares first, the expression factors into the product of a sum and a difference.

$$x^6 - 64 = (x^3)^2 - 8^2$$
$$= (x^3 + 8)(x^3 - 8)$$

Because $x^3 + 8$ is the sum of two cubes and $x^3 - 8$ is the difference of two cubes, each of these binomials can be factored.

$$x^6 - 64 = (x^3 + 8)(x^3 - 8)$$
$$= (x + 2)(x^2 - 2x + 4)(x - 2)(x^2 + 2x + 4)$$

Verify this factorization by multiplication.    ∎

*Self Check*    Factor $a^6 - 1$.
*Answer*    $(a + 1)(a^2 - a + 1)(a - 1)(a^2 + a + 1)$

*Orals*    *Factor each sum or difference of two cubes.*

**1.** $x^3 - y^3$                    **2.** $x^3 + y^3$
**3.** $a^3 + 8$                    **4.** $b^3 - 27$
**5.** $1 + 8x^3$                    **6.** $8 - r^3$
**7.** $x^3y^3 + 1$                    **8.** $125 - 8t^3$

# EXERCISE 5.6

## REVIEW

**1.** The length of one Fermi is $1 \times 10^{-13}$ centimeter, approximately the radius of a proton. Express this number in standard notation.

**2.** In the 14th century, the Black Plague killed about 25,000,000 people, which was 25% of the population of Europe. Find the population at that time, expressed in scientific notation.

## VOCABULARY AND CONCEPTS    *Complete each formula.*

**3.** $x^3 + y^3 = (x + y)$_____

**4.** $x^3 - y^3 = (x - y)$_____

## PRACTICE    *In Exercises 5–24, factor each expression.*

**5.** $y^3 + 1$          **6.** $x^3 - 8$          **7.** $a^3 - 27$          **8.** $b^3 + 125$

**9.** $8 + x^3$      **10.** $27 - y^3$      **11.** $s^3 - t^3$      **12.** $8u^3 + w^3$

**13.** $27x^3 + y^3$      **14.** $x^3 - 27y^3$      **15.** $a^3 + 8b^3$      **16.** $27a^3 - b^3$

**17.** $64x^3 - 27$      **18.** $27x^3 + 125$

**19.** $27x^3 - 125y^3$      **20.** $64x^3 + 27y^3$

**21.** $a^6 - b^3$      **22.** $a^3 + b^6$

**23.** $x^9 + y^6$      **24.** $x^3 - y^9$

*In Exercises 25–40, factor each expression. Factor out any greatest common factors first.*

**25.** $2x^3 + 54$      **26.** $2x^3 - 2$      **27.** $-x^3 + 216$      **28.** $-x^3 - 125$

**29.** $64m^3x - 8n^3x$      **30.** $16r^4 + 128rs^3$

**31.** $x^4y + 216xy^4$      **32.** $16a^5 - 54a^2b^3$

**33.** $81r^4s^2 - 24rs^5$      **34.** $4m^5n + 500m^2n^4$

**35.** $125a^6b^2 + 64a^3b^5$      **36.** $216a^4b^4 - 1{,}000ab^7$

**37.** $y^7z - yz^4$      **38.** $x^{10}y^2 - xy^5$

**39.** $2mp^4 + 16mpq^3$      **40.** $24m^5n - 3m^2n^4$

*In Exercises 41–44, factor each expression completely. Factor a difference of two squares first.*

**41.** $x^6 - 1$      **42.** $x^6 - y^6$

**43.** $x^{12} - y^6$      **44.** $a^{12} - 64$

*In Exercises 45–52, factor each expression completely.*

**45.** $3(x^3 + y^3) - z(x^3 + y^3)$      **46.** $x(8a^3 - b^3) + 4(8a^3 - b^3)$

**47.** $(m^3 + 8n^3) + (m^3x + 8n^3x)$      **48.** $(a^3x + b^3x) - (a^3y + b^3y)$

**49.** $(a^4 + 27a) - (a^3b + 27b)$      **50.** $(x^4 + xy^3) - (x^3y + y^4)$

**51.** $y^3(y^2 - 1) - 27(y^2 - 1)$      **52.** $z^3(y^2 - 4) + 8(y^2 - 4)$

## *WRITING*

**53.** Explain how to factor $a^3 + b^3$.      **54.** Explain the difference between $x^3 - y^3$ and $(x - y)^3$.

***SOMETHING TO THINK ABOUT***

**55.** Use a calculator to verify that

$$a^3 - b^3 = (a - b)(a^2 + ab + b^2)$$

when $a = 11$ and $b = 7$.

**56.** What difficulty do you encounter when you solve $x^3 - 8 = 0$ by factoring?

## 5.7  Summary of Factoring Techniques

■ IDENTIFYING FACTORING TYPES

Getting Ready    *Factor each polynomial.*

**1.** $3ax^2 + 3a^2x$

**2.** $x^2 - 9y^2$

**3.** $x^3 - 8$

**4.** $2x^2 - 8$

**5.** $x^2 - 3x - 10$

**6.** $6x^2 - 13x + 6$

**7.** $6x^2 - 14x + 4$

**8.** $ax^2 + bx^2 - ay^2 - by^2$

### ■ IDENTIFYING FACTORING TYPES

In this brief section, we will discuss ways to approach a randomly chosen factoring problem. For example, suppose we wish to factor the trinomial

$$x^4y + 7x^3y - 18x^2y$$

We begin by attempting to identify the problem type. The first type we look for is **factoring out a common factor.** Because the trinomial has a common factor of $x^2y$, we factor it out:

$$x^4y + 7x^3y - 18x^2y = x^2y(x^2 + 7x - 18)$$

We can factor the remaining trinomial $x^2 + 7x - 18$ as $(x + 9)(x - 2)$. Thus,

$$x^4y + 7x^3y - 18x^2y = x^2y(x^2 + 7x - 18)$$
$$= x^2y(x + 9)(x - 2)$$

To identify the type of factoring problem, we follow these steps.

**Factoring a Polynomial**

**1.** Factor out all common factors.

**2.** If an expression has two terms, check to see if the problem type is

    **a.** the **difference of two squares:** $x^2 - y^2 = (x + y)(x - y)$.

    **b.** the **sum of two cubes:** $x^3 + y^3 = (x + y)(x^2 - xy + y^2)$.

    **c.** the **difference of two cubes:** $x^3 - y^3 = (x - y)(x^2 + xy + y^2)$.

*continued*

**Factoring a Polynomial** *continued*

**3.** If an expression has three terms, check to see if it is a **perfect trinomial square:**

$$x^2 + 2xy + y^2 = (x + y)(x + y)$$
$$x^2 - 2xy + y^2 = (x - y)(x - y).$$

If the trinomial is not a trinomial square, attempt to factor the trinomial as a **general trinomial.**

**4.** If an expression has four or more terms, try to factor the expression by **grouping.**

**5.** Continue factoring until each individual factor is prime.

**6.** Check the results by multiplying.

EXAMPLE 1    Factor $x^5y^2 - xy^6$.

*Solution*    We begin by factoring out the common factor of $xy^2$:

$$x^5y^2 - xy^6 = xy^2(x^4 - y^4)$$

Since the expression $x^4 - y^4$ has two terms, we check to see if it is the difference of two squares, which it is. As the difference of two squares, it factors as $(x^2 + y^2)(x^2 - y^2)$.

$$x^5y^2 - xy^6 = xy^2(x^4 - y^4)$$
$$= xy^2(x^2 + y^2)(x^2 - y^2)$$

The binomial $x^2 + y^2$ is the sum of two squares and cannot be factored. However, $x^2 - y^2$ is the difference of two squares and factors as $(x + y)(x - y)$.

$$x^5y^2 - xy^6 = xy^2(x^4 - y^4)$$
$$= xy^2(x^2 + y^2)(x^2 - y^2)$$
$$= xy^2(x^2 + y^2)(x + y)(x - y)$$

Since each individual factor is prime, the given expression is in completely factored form.   ∎

Self Check    Factor $-a^5b + ab^5$.
Answer        $-ab(a^2 + b^2)(a + b)(a - b)$

EXAMPLE 2    Factor $x^6 - x^4y^2 - x^3y^3 + xy^5$.

*Solution*    We begin by factoring out the common factor of $x$.

$$x^6 - x^4y^2 - x^3y^3 + xy^5 = x(x^5 - x^3y^2 - x^2y^3 + y^5)$$

Since $x^5 - x^3y^2 - x^2y^3 + y^5$ has four terms, we try factoring it by grouping:

$$x^6 - x^4y^2 - x^3y^3 + xy^5 = x(x^5 - x^3y^2 - x^2y^3 + y^5)$$
$$= x[x^3(x^2 - y^2) - y^3(x^2 - y^2)]$$
$$= x(x^2 - y^2)(x^3 - y^3) \qquad \text{Factor out } x^2 - y^2.$$

Finally, we factor the difference of two squares and the difference of two cubes:

$$x^6 - x^4y^2 - x^3y^3 + xy^5 = x(x + y)(x - y)(x - y)(x^2 + xy + y^2)$$

Since each factor is prime, the given expression is in prime-factored form. ∎

**Self Check**   Factor $2a^5 - 2a^2b^3 - 8a^3 + 8b^3$.
**Answer**   $2(a + 2)(a - 2)(a - b)(a^2 + ab + b^2)$

**Orals**   *Indicate which factoring technique you would use first, if any.*

**1.** $2x^2 - 4x$    **2.** $16 - 25y^2$    **3.** $125 + r^3s^3$    **4.** $ax + ay - x - y$

**5.** $x^2 + 4$    **6.** $8x^2 - 50$    **7.** $25r^2 - s^4$    **8.** $8a^3 - 27b^3$

## EXERCISE 5.7

***REVIEW***   *Solve each equation, if possible.*

**1.** $2(t - 5) + t = 3(2 - t)$    **2.** $5 + 3(2x - 1) = 2(4 + 3x) - 24$
**3.** $5x^2 - 35x = 0$    **4.** $6x^2 - x = 35$

***VOCABULARY AND CONCEPTS***   *Fill in each blank to make a true statement.*

**5.** The first step in any factoring problem is to factor out all common _____, if possible.

**6.** If a polynomial has two terms, check to see if it is the _____, the sum of two cubes, or the _____ of two cubes.

**7.** If a polynomial has three terms, try to factor it as the product of two _____.

**8.** If a polynomial has four or more terms, try factoring by _____.

***PRACTICE***   *In Exercises 9–58, factor each expression.*

**9.** $6x + 3$    **10.** $x^2 - 9$    **11.** $x^2 - 6x - 7$    **12.** $a^3 + b^3$

**13.** $6t^2 + 7t - 3$    **14.** $3rs^2 - 6r^2st$    **15.** $4x^2 - 25$    **16.** $ac + ad + bc + bd$

**17.** $t^2 - 2t + 1$        **18.** $6p^2 - 3p - 2$        **19.** $a^3 - 8$        **20.** $2x^2 - 32$

**21.** $x^2y^2 - 2x^2 - y^2 + 2$                      **22.** $a^2c + a^2d^2 + bc + bd^2$
**23.** $70p^4q^3 - 35p^4q^2 + 49p^5q^2$           **24.** $a^2 + 2ab + b^2 - x^2 - 2xy - y^2$

**25.** $2ab^2 + 8ab - 24a$       **26.** $t^4 - 16$             **27.** $-8p^3q^7 - 4p^2q^3$        **28.** $8m^2n^3 - 24mn^4$

**29.** $4a^2 - 4ab + b^2 - 9$     **30.** $3rs + 6r^2 - 18s^2$        **31.** $x^2 + 7x + 1$          **32.** $3a^3 + 24b^3$

**33.** $-2x^5 + 128x^2$         **34.** $16 - 40z + 25z^2$       **35.** $14t^3 - 40t^2 + 6t^4$      **36.** $6x^2 + 7x - 20$

**37.** $a^2(x - a) - b^2(x - a)$                **38.** $5x^3y^3z^4 + 25x^2y^3z^2 - 35x^3y^2z^5$

**39.** $8p^6 - 27q^6$                            **40.** $2c^2 - 5cd - 3d^2$
**41.** $125p^3 - 64y^3$                         **42.** $8a^2x^3y - 2b^2xy$
**43.** $-16x^4y^2z + 24x^5y^3z^4 - 15x^2y^3z^7$      **44.** $2ac + 4ad + bc + 2bd$

**45.** $81p^4 - 16q^4$                          **46.** $6x^2 - x - 16$
**47.** $4x^2 + 9y^2$                              **48.** $30a^4 + 5a^3 - 200a^2$
**49.** $54x^3 + 250y^6$                         **50.** $6a^3 + 35a^2 - 6a$
**51.** $10r^2 - 13r - 4$                       **52.** $4x^2 + 4x + 1 - y^2$
**53.** $21t^3 - 10t^2 + t$                      **54.** $16x^2 - 40x^3 + 25x^4$
**55.** $x^5 - x^3y^2 + x^2y^3 - y^5$           **56.** $a^3x^3 - a^3y^3 + b^3x^3 - b^3y^3$

**57.** $2a^2c - 2b^2c + 4a^2d - 4b^2d$      **58.** $3a^2x^2 + 6a^2x + 3a^2 - 6b^2x^2 - 12b^2x - 6b^2$

## WRITING

**59.** Explain how to identify the type of factoring required to factor a polynomial.

**60.** Which factoring technique do you find most difficult? Why?

## SOMETHING TO THINK ABOUT

**61.** Write $x^6 - y^6$ as $(x^3)^2 - (y^3)^2$, factor it as the difference of two squares, and show that you get

$$(x + y)(x^2 - xy + y^2)(x - y)(x^2 + xy + y^2)$$

Write $x^6 - y^6$ as $(x^2)^3 - (y^2)^3$, factor it as the difference of two cubes, and show that you get

$$(x + y)(x - y)(x^4 + x^2y^2 + y^4)$$

**62.** Verify that the results of Exercise 61 agree by showing that the colored parts agree. Which do you think is completely factored?

# 5.8  Problem Solving

■ INTEGER PROBLEMS ■ BALLISTICS ■ GEOMETRIC PROBLEMS

*Getting Ready*

**1.** One side of a square is $s$ inches long. Find an expression that represents its area.

**2.** The length of a rectangle is 4 centimeters more than twice the width. If $w$ represents the width, find an expression that represents the length.

**3.** If $x$ represents the smaller of two consecutive integers, find an expression that represents their product.

**4.** The length of a rectangle is 3 inches greater than the width. If $w$ represents the width of the rectangle, find an expression that represents the area.

■  INTEGER PROBLEMS

**EXAMPLE 1**    One negative integer is 5 less than another, and their product is 84. Find the integers.

*Analyze the problem*    Let $x$ represent the larger number. Then $x - 5$ represents the smaller number. We know that the product of the negative integers is 84.

*Form an equation*    Since their product is 84, we can form the equation $x(x - 5) = 84$.

*Solve the equation*    To solve the equation, we proceed as follows.

$$
\begin{aligned}
x(x - 5) &= 84 & &\text{Remove parentheses.}\\
x^2 - 5x &= 84 & &\text{Remove parentheses.}\\
x^2 - 5x - 84 &= 0 & &\text{Subtract 84 from both sides.}\\
(x - 12)(x + 7) &= 0 & &\text{Factor.}\\
x - 12 = 0 \quad &\text{or} \quad x + 7 = 0 & &\text{Set each factor equal to 0.}\\
x = 12 \qquad & \qquad x = -7 & &\text{Solve each linear equation.}
\end{aligned}
$$

*State the conclusion*    Since we need two negative numbers, we discard the result $x = 12$. The two negative integers are

$$x = -7 \qquad \text{and} \qquad x - 5 = -7 - 5$$
$$= -12$$

*Check the result*    The number $-12$ is five less than $-7$, and $(-12)(-7) = 84$.    ■

### ■ BALLISTICS

**EXAMPLE 2**   If an object is thrown straight up into the air with an initial velocity of 112 feet per second, its height after $t$ seconds is given by the formula

$$h = 112t - 16t^2$$

where $h$ represents the height of the object in feet. After this object has been thrown, in how many seconds will it hit the ground?

**Analyze the problem**   Before the object is thrown, its height above the ground is 0. When it is thrown, it will go up and then come down. When it hits the ground, its height will again be 0.

**Form an equation**   Thus, we set $h$ equal to 0 in the formula $h = 112t - 16t^2$ to form the equation $0 = 112t - 16t^2$.

$$h = 112t - 16t^2$$
$$0 = 112t - 16t^2$$

**Solve the equation**   We then solve the equation as follows.

$$0 = 112t - 16t^2$$
$$0 = 16t(7 - t) \qquad \text{Factor out } 16t.$$
$$16t = 0 \quad \text{or} \quad 7 - t = 0 \qquad \text{Set each factor equal to } 0.$$
$$t = 0 \qquad\qquad t = 7 \qquad \text{Solve each linear equation.}$$

**State the conclusion**   When $t = 0$, the object's height above the ground is 0 feet, because it has just been released. When $t = 7$, the height is again 0 feet. The object has hit the ground. The solution is 7 seconds.

Check the result.   ■

### ■ GEOMETRIC PROBLEMS

Recall that the area of a rectangle is given by the formula

$$A = lw$$

where $A$ represents the area, $l$ the length, and $w$ the width of the rectangle. The perimeter of a rectangle is given by the formula

$$P = 2l + 2w$$

where $P$ represents the perimeter of the rectangle, $l$ the length, and $w$ the width of the rectangle.

**EXAMPLE 3**   Assume that the rectangle in Figure 5-1 has an area of 52 square centimeters and that its length is 1 centimeter more than 3 times its width. Find the perimeter of the rectangle.

$3w + 1$

$w$ | $A = 52 \text{ cm}^2$

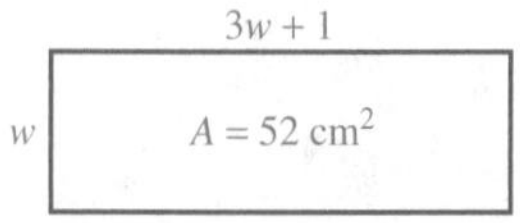

FIGURE 5-1

*Analyze the problem*   Let $w$ represent the width of the rectangle. Then $3w + 1$ represents its length. Its area is 52 square centimeters. We can use this fact to find the values of its width and length. Then we can find the perimeter.

*Form and solve an equation*   Because the area is 52 square centimeters, we substitute 52 for $A$ and $3w + 1$ for $l$ in the formula $A = lw$ and solve for $w$.

$$A = lw$$
$$52 = (3w + 1)w$$
$$52 = 3w^2 + w \qquad \text{Remove parentheses.}$$
$$0 = 3w^2 + w - 52 \qquad \text{Subtract 52 from both sides.}$$
$$0 = (3w + 13)(w - 4) \qquad \text{Factor.}$$

$$3w + 13 = 0 \quad \text{or} \quad w - 4 = 0 \qquad \text{Set each factor equal to 0.}$$
$$3w = -13 \qquad\qquad w = 4 \qquad \text{Solve each linear equation.}$$
$$w = -\frac{13}{3}$$

Because the width of a rectangle cannot be negative, we discard the result $w = -\frac{13}{3}$. Thus, the width of the rectangle is 4, and the length is given by

$$3w + 1 = 3(4) + 1$$
$$= 12 + 1$$
$$= 13$$

The dimensions of the rectangle are 4 centimeters by 13 centimeters. We find the perimeter by substituting 13 for $l$ and 4 for $w$ in the formula for the perimeter.

$$P = 2l + 2w$$
$$= 2(13) + 2(4)$$
$$= 26 + 8$$
$$= 34$$

*State the conclusion*   The perimeter of the rectangle is 34 centimeters.

*Check the result*   A rectangle with dimensions of 13 centimeters by 4 centimeters does have an area of 52 square centimeters, and the length is 1 centimeter more than 3 times the width. A rectangle with these dimensions has a perimeter of 34 centimeters.   ■

**EXAMPLE 4**   The triangle in Figure 5-2 has an area of 10 square centimeters and a height that is 3 centimeters less than twice the length of its base. Find the length of the base and the height of the triangle.

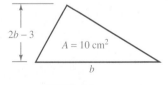

FIGURE 5-2

*Analyze the problem*  Let $b$ represent the length of the base of the triangle. Then $2b - 3$ represents the height. Because the area is 10 square centimeters, we can substitute 10 for $A$ and $2b - 3$ for $h$ in the formula $A = \frac{1}{2}bh$ and solve for $b$.

*Form and solve an equation*

$$A = \frac{1}{2}bh$$

$$10 = \frac{1}{2}b(2b - 3)$$

| | |
|---|---|
| $20 = b(2b - 3)$ | Multiply both sides by 2. |
| $20 = 2b^2 - 3b$ | Remove parentheses. |
| $0 = 2b^2 - 3b - 20$ | Subtract 20 from both sides. |
| $0 = (2b + 5)(b - 4)$ | Factor. |

$$2b + 5 = 0 \quad \text{or} \quad b - 4 = 0 \qquad \text{Set both factors equal to 0.}$$

$$2b = -5 \qquad\qquad b = 4 \qquad \text{Solve each linear equation.}$$

$$b = -\frac{5}{2}$$

*State the conclusion*  Because a triangle cannot have a negative number for the length of its base, we discard the result $b = -\frac{5}{2}$. The length of the base of the triangle is 4 centimeters. Its height is $2(4) - 3$, or 5 centimeters.

*Check the result*  If the base of the triangle has a length of 4 centimeters and the height of the triangle is 5 centimeters, its height is 3 centimeters less than twice the length of its base. Its area is 10 centimeters.

$$A = \frac{1}{2}bh$$

$$= \frac{1}{2}(4)(5)$$

$$= 2(5)$$

$$= 10$$

■

Orals  *Give the formula for . . .*

**1.** The area of a rectangle

**2.** The area of a triangle

**3.** The area of a square

**4.** The area of a rectangular solid

**5.** The perimeter of a rectangle

**6.** The perimeter of a square

EXERCISE 5.8

***REVIEW*** *Solve each equation.*

**1.** $-2(5z + 2) = 3(2 - 3z)$

**2.** $3(2a - 1) - 9 = 2a$

**3.** A rectangle is 3 times as long as it is wide, and its perimeter is 120 centimeters. Find its area.

**4.** A woman invested $15,000, part at 7% annual interest and part at 8% annual interest. If she receives $1,100 interest per year, how much did she invest at 7%?

***VOCABULARY AND CONCEPTS*** *Fill in each blank to make a true statement.*

**5.** The first step in the problem-solving process is to _____ the problem.

**6.** The last step in the problem-solving process is to _____.

***PRACTICE*** *Solve each problem.*

**7. Integer problem** One positive integer is 2 more than another. Their product is 35. Find the integers.

**8. Integer problem** One positive integer is 5 less than 4 times another. Their product is 21. Find the integers.

**9. Integer problem** If 4 is added to the square of a composite integer, the result is 5 less than 10 times that integer. Find the integer.

**10. Integer problem** If 3 times the square of a certain natural number is added to the number itself, the result is 14. Find the number.

*In Exercises 11–14, an object has been thrown straight up into the air. The formula $h = vt - 16t^2$ gives the height h of the object above the ground after t seconds when it is thrown upward with an initial velocity v.*

**11. Time of flight** After how many seconds will an object hit the ground if it was thrown with a velocity of 144 feet per second?

**12. Time of flight** After how many seconds will an object hit the ground if it was thrown with a velocity of 160 feet per second?

**13. Ballistics** If a cannonball is fired with an upward velocity of 220 feet per second, at what times will it be at a height of 600 feet?

**14. Ballistics** A cannonball's initial upward velocity is 128 feet per second. At what times will it be 192 feet above the ground?

***APPLICATIONS*** *Solve each problem.*

**15. Exhibition diving** At a resort, tourists watch swimmers dive from a cliff to the water 64 feet below. A diver's height $h$ above the water $t$ seconds after diving is given by $h = -16t^2 + 64$. How long does a dive last?

**16. Forensic medicine** The kinetic energy $E$ of a moving object is given by $E = \frac{1}{2}mv^2$, where $m$ is the

mass of the object (in kilograms) and $v$ is the object's velocity (in meters per second). Kinetic energy is measured in Joules. By the damage done to a victim, a police pathologist determines that the energy of a 3-kilogram mass at impact was 54 Joules. Find the velocity at impact.

17. **Insulation**   The area of the rectangular slab of foam insulation in Illustration 1 is 36 square meters. Find the dimensions of the slab.

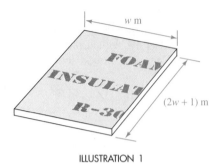

ILLUSTRATION 1

18. **Shipping pallets**   The length of a rectangular shipping pallet is 2 feet less than 3 times its width. Its area is 21 square feet. Find the dimensions of the pallet.

19. **Carpentry**   A room containing 143 square feet is 2 feet longer than it is wide. How long a crown molding is needed to trim the perimeter of the ceiling?

20. **Designing a tent**   The length of the base of the triangular sheet of canvas above the door of the tent in Illustration 2 is 2 feet more than twice its height. The area is 30 square feet. Find the height and the length of the base of the triangle.

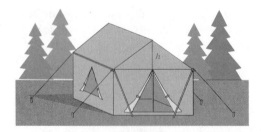

ILLUSTRATION 2

21. **Dimensions of a triangle**   The height of a triangle is 2 inches less than 5 times the length of its base. The area is 36 square inches. Find the length of the base and the height of the triangle.

22. **Area of a triangle**   The base of a triangle is numerically 3 less than its area, and the height is numerically 6 less than its area. Find the area of the triangle.

23. **Area of a triangle**   The length of the base and the height of a triangle are numerically equal. Their sum is 6 less than the number of units in the area of the triangle. Find the area of the triangle.

24. **Dimensions of a parallelogram**   The formula for the area of a parallelogram is $A = bh$. The area of the parallelogram in Illustration 3 is 200 square centimeters. If its base is twice its height, how long is the base?

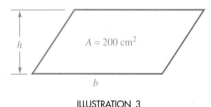

ILLUSTRATION 3

25. **Swimming pool border**   The owners of the rectangular swimming pool in Illustration 4 want to surround the pool with a crushed-stone border of uniform width. They have enough stone to cover 74 square meters. How wide should they make the border? (*Hint:* The area of the larger rectangle minus the area of the smaller is the area of the border.)

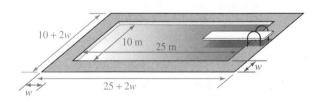

ILLUSTRATION 4

26. **House construction**   The formula for the area of a trapezoid is $A = \frac{h(B + b)}{2}$. The area of the trapezoidal truss in Illustration 5 is 24 square meters. Find the height of the trapezoid if one base is 8 meters and the other base is the same as the height.

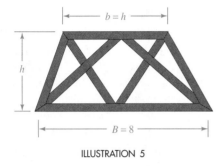

ILLUSTRATION 5

**27. Volume of a solid** The volume of a rectangular solid is given by the formula $V = lwh$, where $l$ is the length, $w$ is the width, and $h$ is the height. The volume of the rectangular solid in Illustration 6 is 210 cubic centimeters. Find the width of the rectangular solid if its length is 10 centimeters and its height is 1 centimeter longer than twice its width.

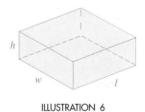

ILLUSTRATION 6

**28. Volume of a pyramid** The volume of a pyramid is given by the formula $V = \frac{Bh}{3}$, where $B$ is the area of its base and $h$ is its height. The volume of the pyramid in

Illustration 7 is 192 cubic centimeters. Find the dimensions of its rectangular base if one edge of the base is 2 centimeters longer than the other, and the height of the pyramid is 12 centimeters.

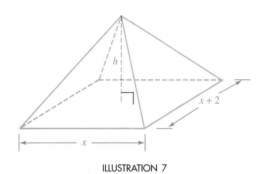

ILLUSTRATION 7

**29. Volume of a pyramid** The volume of a pyramid is 84 cubic centimeters. Its height is 9 centimeters, and one side of its rectangular base is 3 centimeters shorter than the other. Find the dimensions of its base. (See Exercise 28.)

**30. Volume of a solid** The volume of a rectangular solid is 72 cubic centimeters. Its height is 4 centimeters, and its width is 3 centimeters shorter than its length. Find the sum of its length and width. (See Exercise 27.)

*WRITING*

**31.** Explain the steps you would use to set up and solve an application problem.

**32.** Explain how you should check the solution to an application problem.

*SOMETHING TO THINK ABOUT*

**33.** Here is an easy-sounding problem:
The length of a rectangle is 2 feet greater than the width, and the area is 18 square feet. Find the width of the rectangle.
Set up the equation. Can you solve it? Why not?

**34.** Does the equation in Exercise 33 have a solution, even if you can't find it? If it does, find an estimate of the solution.

■ ■ ■ ■ ■ ■ ■ ■ ■ ■

MATHEMATICS IN
ECOLOGY

To double the existing capacity of the sewage plant discussed at the beginning of the chapter, the surface area of the new pool must equal the total of the surface areas of the two existing pools. The area of a circle is given by $A = \pi r^2$. Because the radius of the smaller pool is 20 meters (one-half of the diameter), its area is $\pi 20^2$, or $400\pi$ square meters. Similarly, the area of the larger pool is $\pi 21^2$, or $441\pi$ square meters. The total of these areas is $841\pi$ square meters. The engineers must design a third pool with an area of $841\pi$ square meters.

Let $r$ represent the radius of the new pool.

| The area of the new pool | equals | the total existing area. |
|:---:|:---:|:---:|

$$\pi r^2 = 841\pi$$  The area of a circle is $\pi r^2$.

$$r^2 = 841$$  Divide both sides by $\pi$.

$$r^2 - 841 = 0$$  Subtract 841 from both sides.

$$(r + 29)(r - 29) = 0$$  Factor the difference of two squares. $(841 = 29^2.)$

$$r + 29 = 0 \quad \text{or} \quad r - 29 = 0$$  Set each factor equal to 0.

$$r = -29 \qquad\qquad r = 29$$

Because the radius of a pool cannot be negative, we discard the negative solution. The design engineers should specify a pool with a radius of 29 meters.

■ ■ ■ ■ ■ ■ ■ ■ ■ ■  **PROJECT**

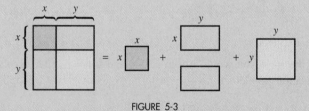

FIGURE 5-3

Because the length of each side of the largest square in Figure 5-3 is $x + y$, its area is $(x + y)^2$. This area is also the sum of four smaller areas, which illustrates the factorization

$$x^2 + 2xy + y^2 = (x + y)^2$$

What factorization is illustrated by each of the following figures?

**1.**

**3.**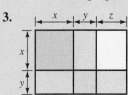

**2.**

**4.**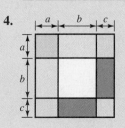

**5.** Factor the expression

$$a^2 + ac + 2a + ab + bc + 2b$$

and draw a figure that illustrates the factorization.

**6.** Verify the factorization

$$x^3 + 3x^2y + 3xy^2 + y^3 = (x + y)^3$$

*Hint:* Expand the right-hand side: $(x + y)^3 = (x + y)(x + y)(x + y)$

Then draw a figure that illustrates the factorization.

# C H A P T E R   S U M M A R Y

## CONCEPTS

## REVIEW EXERCISES

### SECTION 5.1

### *Factoring Out the Greatest Common Factor*

A natural number is in **prime-factored form** if it is written as the product of prime-number factors.

The **greatest common factor (GCF)** of several monomials is found by taking each common prime factor and variable factor the fewest number of times it appears in any one monomial.

**1.** Find the prime factorization of each number.
   **a.** 35
   **b.** 45
   **c.** 96
   **d.** 102
   **e.** 87
   **f.** 99
   **g.** 2,050
   **h.** 4,096

**2.** Factor each expression completely.
   **a.** $3x + 9y$
   **b.** $5ax^2 + 15a$
   **c.** $7x^2 + 14x$
   **d.** $3x^2 - 3x$
   **e.** $2x^3 + 4x^2 - 8x$
   **f.** $ax + ay - az$
   **g.** $ax + ay - a$
   **h.** $x^2yz + xy^2z$

**Zero-factor property:**
If $a$ and $b$ represent two real numbers and if $ab = 0$, then $a = 0$ or $b = 0$.

**3.** Solve each equation.
  **a.** $x^2 + 2x = 0$          **b.** $2x^2 - 6x = 0$

---

*Factoring by Grouping*

If a polynomial has four or more terms, consider factoring it by grouping.

**4.** Factor each polynomial.
  **a.** $(x + y)a + (x + y)b$

  **b.** $(x + y)^2 + (x + y)$

  **c.** $2x^2(x + 2) + 6x(x + 2)$

  **d.** $3x(y + z) - 9x(y + z)^2$

  **e.** $3p + 9q + ap + 3aq$

  **f.** $ar - 2as + 7r - 14s$

  **g.** $x^2 + ax + bx + ab$

  **h.** $xy + 2x - 2y - 4$

  **i.** $xa + yb + ya + xb$

---

*Factoring the Difference of Two Squares*

To factor the difference of two squares, use the pattern
$$x^2 - y^2 = (x + y)(x - y)$$

**5.** Factor each expression.
  **a.** $x^2 - 9$          **b.** $x^2 y^2 - 16$

  **c.** $(x + 2)^2 - y^2$

  **d.** $z^2 - (x + y)^2$

  **e.** $6x^2 y - 24y^3$

  **f.** $(x + y)^2 - z^2$

**6.** Solve each equation.
  **a.** $x^2 - 9 = 0$          **b.** $x^2 - 25 = 0$

| **SECTIONS 5.4–5.5** | *Factoring Trinomials* |

Factor trinomials by trying these steps:

1. Write the trinomial with the exponents of one variable in descending order.

2. Factor out any greatest common factor (including −1 if that is necessary to make the coefficient of the first term positive).

3. If the sign of the third term of the trinomial is plus (+), the signs between the terms of each binomial factor are the same as the sign of the trinomial's second term. If the sign of the third term is minus (−), the signs between the terms of the binomials are opposite.

4. Mentally try various combinations of first terms and last terms until you find the one that works or you exhaust all the possibilities. In that case, the trinomial is prime.

5. Check by multiplication.

**7.** Factor each polynomial.
  **a.** $x^2 + 10x + 21$
  **b.** $x^2 + 4x - 21$
  **c.** $x^2 + 2x - 24$
  **d.** $x^2 - 4x - 12$

**8.** Factor each polynomial.
  **a.** $2x^2 - 5x - 3$
  **b.** $3x^2 - 14x - 5$
  **c.** $6x^2 + 7x - 3$
  **d.** $6x^2 + 3x - 3$
  **e.** $6x^3 + 17x^2 - 3x$
  **f.** $4x^3 - 5x^2 - 6x$

**9.** Solve each equation.
  **a.** $a^2 - 7a + 12 = 0$
  **b.** $x^2 - 2x - 15 = 0$
  **c.** $2x - x^2 + 24 = 0$
  **d.** $16 + x^2 - 10x = 0$
  **e.** $2x^2 - 5x - 3 = 0$
  **f.** $2x^2 + x - 3 = 0$
  **g.** $4x^2 = 1$
  **h.** $9x^2 = 4$
  **i.** $x^3 - 7x^2 + 12x = 0$
  **j.** $x^3 + 5x^2 + 6x = 0$
  **k.** $2x^3 + 5x^2 = 3x$
  **l.** $3x^3 - 2x = x^2$

| **SECTION 5.6** | *Factoring the Sum and Difference of Two Cubes* |

The sum and difference of two cubes factor according to the patterns
$$x^3 + y^3 = (x + y)(x^2 - xy + y^2)$$
$$x^3 - y^3 = (x - y)(x^2 + xy + y^2)$$

**10.** Factor each polynomial.
  **a.** $c^3 - 27$
  **b.** $d^3 + 8$
  **c.** $2x^3 + 54$
  **d.** $2ab^4 - 2ab$

## Summary of Factoring Techniques

**Steps for factoring polynomials:**
1. Factor out all common factors.
2. If an expression has two terms, check to see if the problem type is
   a. the **difference of two squares:**
   $a^2 - b^2 = (a + b)(a - b)$
   b. the **sum of two cubes:**
   $a^3 + b^3 = (a + b)(a^2 - ab + b^2)$
   c. the **difference of two cubes:**
   $a^3 - b^3 = (a - b)(a^2 + ab + b^2)$
3. If an expression has three terms, check to see if the problem type is a **perfect trinomial square:**
   $a^2 + 2ab + b^2 = (a + b)(a + b)$
   $a^2 - 2ab + b^2 = (a - b)(a - b)$
   If the trinomial is not a trinomial square, attempt to factor the trinomial as a **general trinomial.**
4. If an expression has four or more terms, try to factor it by **grouping.**

**11.** Factor each polynomial.

  **a.** $3x^2y - xy^2 - 6xy + 2y^2$

  **b.** $5x^2 + 10x - 15xy - 30y$

  **c.** $2a^2x + 2abx + a^3 + a^2b$

  **d.** $x^2 + 2ax + a^2 - y^2$

  **e.** $ax^2 + 4ax + 3a - bx - b$

  **f.** $ax^6 - ay^6$

## Problem Solving

**12. Number problem** The sum of two numbers is 12, and their product is 35. Find the numbers.

**13. Number problem** If 3 times the square of a positive number is added to 5 times the number, the result is 2. Find the number.

**14. Dimensions of a rectangle** A rectangle is 2 feet longer than it is wide, and its area is 48 square feet. Find its dimensions.

**15. Gardening** A rectangular flower bed is 3 feet longer than twice its width, and its area is 27 square feet. Find its dimensions.

**16. Geometry** A rectangle is 3 feet longer than it is wide. Its area is numerically equal to its perimeter. Find its dimensions.

# Chapter Test

**1.** Find the prime factorization of 196.

**2.** Find the prime factorization of 111.

*In Problems 3–4, factor out the greatest common factor.*

**3.** $60ab^2c^3 + 30a^3b^2c - 25a$

**4.** $3x^2(a + b) - 6xy(a + b)$

*In Problems 5–20, factor each expression.*

**5.** $ax + ay + bx + by$

**6.** $x^2 - 25$

**7.** $3a^2 - 27b^2$

**8.** $16x^4 - 81y^4$

**9.** $x^2 + 4x + 3$

**10.** $x^2 - 9x - 22$

**11.** $x^2 + 10xy + 9y^2$

**12.** $6x^2 - 30xy + 24y^2$

**13.** $3x^2 + 13x + 4$

**14.** $2a^2 + 5a - 12$

**15.** $2x^2 + 3xy - 2y^2$

**16.** $12 - 25x + 12x^2$

**17.** $12a^2 + 6ab - 36b^2$

**18.** $x^3 - 64$

**19.** $216 + 8a^3$

**20.** $x^9z^3 - y^3z^6$

*In Problems 21–28, solve each equation.*

**21.** $x^2 + 3x = 0$

**22.** $2x^2 + 5x + 3 = 0$

**23.** $9y^2 - 81 = 0$

**24.** $-3(y - 6) + 2 = y^2 + 2$

**25.** $10x^2 - 13x = 9$

**26.** $10x^2 - x = 9$

**27.** $10x^2 + 43x = 9$

**28.** $10x^2 - 89x = 9$

**29. Cannon fire**  A cannonball is fired straight up into the air with a velocity of 192 feet per second. In how many seconds will it hit the ground? (Its height above the ground is given by the formula $h = vt - 16t^2$, where $v$ is the velocity and $t$ is the time in seconds.)

**30. Base of a triangle**  The base of a triangle with an area of 40 square meters is 2 meters longer than it is high. Find the base of the triangle.

# 6 Proportion and Rational Expressions

MATHEMATICS IN ARCHITECTURE

An architect is designing a home with the combined living room–dining room shown in the sketch. The total length available is 32 feet, and the shortest wall in the living room must be 4 feet to accommodate a custom bookshelf. The area of the living room is to be 288 square feet, and the area of the dining room is 168 square feet.

What must be the design width $w$ of the dining room?

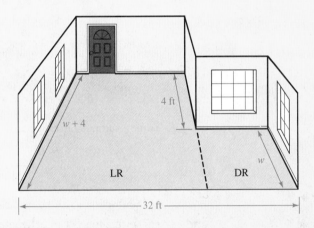

After you have read this chapter, you will be able to answer this question.

## 6.1 Ratios

■ RATIOS ■ UNIT COSTS ■ RATES

Getting Ready  *Simplify each fraction.*

**1.** $\dfrac{2}{4}$     **2.** $\dfrac{8}{12}$     **3.** $-\dfrac{20}{25}$     **4.** $\dfrac{-45}{81}$

### ■ RATIOS

Ratios appear often in real-life situations. For example,

- To prepare fuel for a Lawnboy lawnmower, gasoline must be mixed with oil in the ratio of 50 to 1.
- To make 14-karat jewelry, gold is mixed with other metals in the ratio of 14 to 10.
- In the stock market, winning stocks might outnumber losing stocks in the ratio of 7 to 4.
- At Rock Valley College, the ratio of students to faculty is 16 to 1.

Ratios give us a way to compare numerical quantities.

> **Ratios**
> A **ratio** is the comparison of two numbers by their indicated quotient. In symbols,
>
> If $a$ and $b$ are two numbers, the ratio of $a$ to $b$ is $\dfrac{a}{b}$.

> **WARNING!** The denominator $b$ cannot be 0 in the fraction $\frac{a}{b}$, but $b$ can be 0 in the ratio $\frac{a}{b}$. For example, the ratio of women to men on a women's softball team could be 25 to 0. However, these applications are rare.

Some examples of ratios are

$$\frac{7}{9}, \quad \frac{21}{27}, \quad \text{and} \quad \frac{2{,}290}{1{,}317}$$

- The fraction $\frac{7}{9}$ can be read as "the ratio of 7 to 9."
- The fraction $\frac{21}{27}$ can be read as "the ratio of 21 to 27."
- The ratio $\frac{2{,}290}{1{,}317}$ can be read as "the ratio of 2,290 to 1,317."

Because $\frac{7}{9}$ and $\frac{21}{27}$ represent equal numbers, they are **equal ratios.**

**EXAMPLE 1**    Express each phrase as a ratio in lowest terms:   **a.** the ratio of 15 to 12   and **b.** the ratio of 0.3 to 1.2.

*Solution*   **a.** The ratio of 15 to 12 can be written as the fraction $\frac{15}{12}$. After simplifying, the ratio is $\frac{5}{4}$.

**b.** The ratio of 0.3 to 1.2 can be written as the fraction $\frac{0.3}{1.2}$. We can simplify this fraction as follows:

$$\frac{0.3}{1.2} = \frac{0.3 \cdot \mathbf{10}}{1.2 \cdot \mathbf{10}} \qquad \text{To clear the decimal, multiply both numerator and denominator by 10.}$$

$$= \frac{3}{12} \qquad \text{Multiply: } 0.3 \cdot 10 = 3 \text{ and } 1.2 \cdot 10 = 12.$$

$$= \frac{1}{4} \qquad \text{Simplify the fraction: } \frac{3}{12} = \frac{\overset{1}{\cancel{3}} \cdot 1}{\underset{1}{\cancel{3}} \cdot 4} = \frac{1}{4}.$$

∎

*Self Check*   Express each ratio in lowest terms:   **a.** the ratio of 8 to 12   and   **b.** the ratio of 3.2 to 16.

*Answers*   **a.** $\frac{2}{3}$,   **b.** $\frac{1}{5}$

EXAMPLE 2    Express each phrase as a ratio in lowest terms:    **a.** the ratio of 3 meters to 8 meters and    **b.** the ratio of 4 ounces to 1 pound.

*Solution*    **a.** The ratio of 3 meters to 8 meters can be written as the fraction $\frac{3 \text{ meters}}{8 \text{ meters}}$, or just $\frac{3}{8}$.

**b.** When possible, we should express ratios in the same units. Since there are 16 ounces in 1 pound, the proper ratio is $\frac{4 \text{ ounces}}{16 \text{ ounces}}$, which simplifies to $\frac{1}{4}$. ∎

Self Check    Express each ratio in lowest terms:    **a.** the ratio of 8 ounces to 2 pounds    and
**b.** the ratio of 1 foot to 2 yards. (*Hint:* 3 feet = 1 yard.)

Answers    **a.** $\frac{1}{4}$,    **b.** $\frac{1}{6}$

EXAMPLE 3    At a college, there are 2,772 students and 154 faculty members. Write a fraction in simplified form that expresses the ratio of students per faculty member.

*Solution*    The ratio of students to faculty is 2,772 to 154. We can write this ratio as the fraction $\frac{2,772}{154}$ and simplify it.

$$\frac{2{,}772}{154} = \frac{18 \cdot \overset{1}{\mathbf{154}}}{1 \cdot \underset{1}{\mathbf{154}}}$$

$$= \frac{18}{1} \qquad \tfrac{154}{154} = 1.$$

The ratio of students to faculty is 18 to 1. ∎

Self Check    In a college graduating class, 224 students out of 632 went on to graduate school. Write a fraction in simplified form that expresses the ratio of the number of students going on to the number in the graduating class.

Answer    $\frac{28}{79}$

## ■ UNIT COSTS

The *unit cost* of an item is the ratio of its cost to its quantity. For example, the unit cost (the cost per pound) of 5 pounds of coffee priced at $20.75 is given by the ratio

$$\frac{\$20.75}{5 \text{ pounds}} = \frac{\$2{,}075}{500 \text{ pounds}} \qquad \begin{array}{l}\text{To eliminate the decimal, multiply numerator} \\ \text{and denominator by 100.}\end{array}$$

$$= \$4.15 \text{ per pound} \qquad \$2{,}070 \div 500 = \$4.15.$$

The unit cost is $4.15 per pound.

**EXAMPLE 4**    Olives come packaged in a 12-ounce jar, which sells for $3.09, or in a 6-ounce jar, which sells for $1.53. Which is the better buy?

*Solution*    To find the better buy, we must find each unit cost. The unit cost of the 12-ounce jar is

$$\frac{\$3.09}{12 \text{ ounces}} = \frac{309\text{¢}}{12 \text{ ounces}} \qquad \text{Change \$3.09 to 309 cents.}$$

$$= 25.75\text{¢ per ounce}$$

The unit cost of the 6-ounce jar is

$$\frac{\$1.53}{6 \text{ ounces}} = \frac{153\text{¢}}{6 \text{ ounces}} \qquad \text{Change \$1.53 to 153 cents.}$$

$$= 25.5\text{¢ per ounce}$$

Since the unit cost is less when olives are packaged in 6-ounce jars, that is the better buy.    ∎

**Self Check**    A fast-food restaurant sells a 12-ounce soft drink for 79¢ and a 16-ounce soft drink for 99¢. Which is the better buy?

*Answer*    the 16-oz drink

■ **RATES**

When ratios are used to compare quantities with different units, they are called *rates*. For example, if we drive 413 miles in 7 hours, the average rate of speed is the ratio of the miles driven to the length of time of the trip.

$$\text{Average rate of speed} = \frac{413 \text{ miles}}{7 \text{ hours}} = \frac{59 \text{ miles}}{1 \text{ hour}} \qquad \frac{413}{7} = \frac{\cancel{7} \cdot 59}{\cancel{7} \cdot 1} = \frac{59}{1}.$$

The ratio $\frac{59 \text{ miles}}{1 \text{ hour}}$ can be expressed in any of the following forms:

$$59 \, \frac{\text{miles}}{\text{hour}}, \qquad 59 \text{ miles per hour}, \qquad 59 \text{ miles/hour}, \qquad \text{or} \qquad 59 \text{ mph}$$

**EXAMPLE 5**    Find the hourly rate of pay for a student who earns $370 for working 40 hours.

*Solution*    We can write the rate of pay as the ratio

$$\text{Rate of pay} = \frac{\$370}{40 \text{ hours}}$$

and simplify by dividing 370 by 40.

$$\text{Rate of pay} = 9.25 \, \frac{\text{dollars}}{\text{hour}}$$

The rate is $9.25 per hour.    ∎

Self Check

Joan earns $316 per 40-hour week managing a dress shop. Set up a ratio and find her hourly rate of pay.

*Answer*

$7.90 per hour

**EXAMPLE 6**

One household used 813.75 kilowatt hours of electricity during a 31-day period. Find the rate of energy consumption in kilowatt hours per day.

*Solution*

We can write the rate of energy consumption as the ratio

$$\text{Rate of energy consumption} = \frac{813.75 \text{ kilowatt hours}}{31 \text{ days}}$$

and simplify by dividing 813.75 by 31.

$$\text{Rate of energy consumption} = 26.25 \, \frac{\text{kilowatt hours}}{\text{day}}$$

The rate of consumption is 26.25 kilowatt hours per day.  ■

Self Check

To heat a house for 30 days, a furnace burned 72 therms of natural gas. Find the rate of gas consumption in therms per day.

*Answer*

2.4 therms per day

**EXAMPLE 7**

A textbook costs $49.22, including sales tax. If the tax was $3.22, find the sales tax rate.

*Solution*

Since the tax was $3.22, the cost of the book alone was

$49.22 − $3.22 = $46.00

We can write the sales tax rate as the ratio

$$\text{Sales tax rate} = \frac{\text{amount of sales tax}}{\text{cost of the book, without tax}}$$

$$= \frac{\$3.22}{\$46}$$

and simplify by dividing 3.22 by 46.

Sales tax rate = 0.07

The tax rate is 0.07, or 7%.  ■

Self Check

A sport coat costs $160.50, including sales tax. If the cost of the coat without tax is $150, find the sales tax rate.

*Answer*

7%

## Computing Gas Mileage

CALCULATORS A man drove a total of 775 miles. Along the way, he stopped for gas three times, pumping 10.5, 11.3, and 8.75 gallons of gas. He started with the tank half-full and ended with the tank half-full. To find how many miles he got per gallon, we need to divide the total distance by the total number of gallons of gas consumed.

$$\frac{775}{10.5 + 11.3 + 8.75} \begin{matrix}\leftarrow \text{Total distance}\\ \leftarrow \text{Total number of gallons consumed}\end{matrix}$$

We can make this calculation by pressing these keys on a scientific calculator.

**Keystrokes**

775 ÷ ( 10.5 + 11.3 + 8.75 ) =

The display will read $25.36824877$. To the nearest one-hundredth, he got 25.37 mpg.

Orals *Express as a ratio in lowest terms.*

**1.** 5 to 7    **2.** 50 to 1    **3.** 3 to 9    **4.** 7 to 10

## EXERCISE 6.1

*REVIEW*   *Solve each equation.*

**1.** $2x + 4 = 38$    **2.** $\frac{x}{2} - 4 = 38$    **3.** $3(x + 2) = 24$    **4.** $\frac{x - 6}{3} = 20$

*Factor each expression.*

**5.** $2x + 6$    **6.** $x^2 - 49$    **7.** $2x^2 - x - 6$    **8.** $x^3 + 27$

*VOCABULARY AND CONCEPTS*   *Fill in each blank to make a true statement.*

**9.** A ratio is a _____ of two numbers by their indicated _____.

**10.** The _____ of an item is the ratio of its cost to its quantity.

**11.** The ratios $\frac{2}{3}$ and $\frac{4}{6}$ are _____ ratios.

**12.** The ratio $\frac{500 \text{ miles}}{15 \text{ hours}}$ is called a _____.

**13.** Give three examples of ratios that you have encountered this past week.

**14.** Suppose that a basketball player made 8 free throws out of 12 tries. The ratio of $\frac{8}{12}$ can be simplified as $\frac{2}{3}$. Interpret this result.

*PRACTICE*   *In Exercises 15–30, express each phrase as a ratio in lowest terms.*

**15.** 5 to 7    **16.** 3 to 5    **17.** 17 to 34    **18.** 19 to 38
**19.** 22 to 33    **20.** 14 to 21    **21.** 7 to 24.5    **22.** 0.65 to 0.15

**23.** 4 ounces to 12 ounces

**24.** 3 inches to 15 inches

**25.** 12 minutes to 1 hour

**26.** 8 ounces to 1 pound

**27.** 3 days to 1 week

**28.** 4 inches to 2 yards

**29.** 18 months to 2 years

**30.** 8 feet to 4 yards

*In Exercises 31–34, refer to the monthly family budget shown in Illustration 1. Give each ratio in lowest terms.*

**31.** Find the total amount of the budget.

**32.** Find the ratio of the amount budgeted for rent to the total budget.

**33.** Find the ratio of the amount budgeted for entertainment to the total budget.

**34.** Find the ratio of the amount budgeted for phone to the amount budgeted for entertainment.

| Item | Amount |
|------|--------|
| Rent | $750 |
| Food | $652 |
| Gas and electric | $188 |
| Phone | $125 |
| Entertainment | $110 |

ILLUSTRATION 1

*In Exercises 35–38, refer to the tax deductions listed in Illustration 2. Give each ratio in lowest terms.*

**35.** Find the total amount of deductions.

**36.** Find the ratio of real estate tax deductions to the total deductions.

**37.** Find the ratio of the contributions to the total deductions.

**38.** Find the ratio of the mortgage interest deduction to the union dues deduction.

| Item | Amount |
|------|--------|
| Medical | $ 995 |
| Real estate tax | $1,245 |
| Contributions | $1,680 |
| Mortgage interest | $4,580 |
| Union dues | $ 225 |

ILLUSTRATION 2

**APPLICATIONS**    *In Exercises 39–56, find each ratio and express it in lowest terms. You may use a calculator when it is helpful.*

**39. Faculty-to-student ratio**    At a college, there are 125 faculty members and 2,000 students. Find the faculty-to-student ratio.

**40. Ratio of men to women**    In a state senate, there are 94 men and 24 women. Find the ratio of men to women.

**41. Unit cost of gasoline**    A driver pumped 17 gallons of gasoline into his tank at a cost of $21.59. Write a ratio of dollars to gallons, and give the unit cost of gasoline.

**42. Unit cost of grass seed**    A 50-pound bag of grass seed costs $222.50. Write a ratio of dollars to pounds, and give the unit cost of grass seed.

**43. Unit cost of cranberry juice**    A 12-ounce can of cranberry juice sells for 84¢. Give the unit cost in cents per ounce.

**44. Unit cost of beans**    A 24-ounce package of green beans sells for $1.29. Give the unit cost in cents per ounce.

**45. Comparative shopping**    A 6-ounce can of orange juice sells for 89¢, and an 8-ounce can sells for $1.19. Which is the better buy?

**46. Comparing speeds**    A car travels 345 miles in 6 hours, and a truck travels 376 miles in 6.2 hours. Which vehicle travels faster?

**47. Comparing reading speeds** One seventh-grader read a 54-page book in 40 minutes, and another read an 80-page book in 62 minutes. If the books were equally difficult, which student read faster?

**48. Comparative shopping** A 30-pound bag of fertilizer costs $12.25, and an 80-pound bag costs $30.25. Which is the better buy?

**49. Emptying a tank** An 11,880-gallon tank can be emptied in 27 minutes. Write a ratio of gallons to minutes, and give the rate of flow in gallons per minute.

**50. Rate of pay** Ricardo worked for 27 hours to help insulate a hockey arena. For his work, he received $337.50. Write a ratio of dollars to hours, and find his hourly rate of pay.

**51. Sales tax** A sweater cost $36.75 after sales tax had been added. Find the tax rate as a percent if the sweater retailed for $35.

**52. Real estate taxes** The real estate taxes on a summer home assessed at $75,000 were $1,500. Find the tax rate as a percent.

**53. Rate of speed** A car travels 325 miles in 5 hours. Find its rate of speed in miles per hour.

**54. Rate of speed** An airplane travels from Chicago to San Francisco, a distance of 1,883 miles, in 3.5 hours. Find the average rate of speed of the plane.

**55. Comparing gas mileage** One car went 1,235 miles on 51.3 gallons of gasoline, and another went 1,456 miles on 55.78 gallons. Which car had the better mpg rating?

**56. Comparing electric rates** In one community, a bill for 575 kilowatt hours (kwh) of electricity was $38.81. In a second community, a bill for 831 kwh was $58.10. In which community is electricity cheaper?

## WRITING

**57.** Some people think that the word *ratio* comes from the words *rational number.* Explain why this may be true.

**58.** In the fraction $\frac{a}{b}$, $b$ cannot be 0. Explain why. In the ratio $\frac{a}{b}$, $b$ can be 0. Explain why.

## SOMETHING TO THINK ABOUT

**59.** Which ratio is the larger? How can you tell?

$$\frac{17}{19} \quad \text{or} \quad \frac{19}{21}$$

**60.** Which ratio is the smaller? How can you tell?

$$-\frac{13}{29} \quad \text{or} \quad -\frac{17}{31}$$

# 6.2 Proportions and Similar Triangles

■ PROPORTIONS ■ MEANS AND EXTREMES OF A PROPORTION ■ SOLVING PROPORTIONS
■ PROBLEM SOLVING ■ SIMILAR TRIANGLES

Getting Ready    *Solve each equation.*

**1.** $\dfrac{5}{2} = \dfrac{x}{4}$    **2.** $\dfrac{7}{9} = \dfrac{y}{3}$    **3.** $\dfrac{y}{10} = \dfrac{2}{7}$    **4.** $\dfrac{1}{x} = \dfrac{8}{40}$

**5.** $\dfrac{w}{14} = \dfrac{7}{21}$    **6.** $\dfrac{c}{12} = \dfrac{5}{12}$    **7.** $\dfrac{3}{q} = \dfrac{1}{7}$    **8.** $\dfrac{16}{3} = \dfrac{8}{z}$

### ■ PROPORTIONS

Consider Table 6-1, in which we are given the costs of various numbers of gallons of gasoline.

| Number of gallons | Cost |
|:---:|:---:|
| 2 | $ 2.72 |
| 5 | $ 6.80 |
| 8 | $10.88 |
| 12 | $16.32 |
| 20 | $27.20 |

TABLE 6-1

If we find the ratios of the costs to the numbers of gallons purchased, we will see that they are equal. In this example, each ratio represents the cost of 1 gallon of gasoline, which is $1.36 per gallon.

$$\frac{\$2.72}{2} = \$1.36, \qquad \frac{\$6.80}{5} = \$1.36, \qquad \frac{\$10.88}{8} = \$1.36, \qquad \frac{\$16.32}{12} = \$1.36,$$

$$\text{and} \qquad \frac{\$27.20}{20} = \$1.36$$

When two ratios such as $\frac{\$2.72}{2}$ and $\frac{\$6.80}{5}$ are equal, they form a *proportion*. In this section, we will discuss proportions and use them to solve problems.

> **Proportions**
> A **proportion** is a statement that two ratios are equal.

Some examples of proportions are

$$\frac{1}{2} = \frac{3}{6}, \qquad \frac{7}{3} = \frac{21}{9}, \qquad \frac{8x}{1} = \frac{40x}{5}, \qquad \text{and} \qquad \frac{a}{b} = \frac{c}{d}$$

- The proportion $\dfrac{1}{2} = \dfrac{3}{6}$ can be read as "1 is to 2 as 3 is to 6."

- The proportion $\dfrac{7}{3} = \dfrac{21}{9}$ can be read as "7 is to 3 as 21 is to 9."

- The proportion $\dfrac{8x}{1} = \dfrac{40x}{5}$ can be read as "8x is to 1 as 40x is to 5."

- The proportion $\dfrac{a}{b} = \dfrac{c}{d}$ can be read as "a is to b as c is to d."

The terms of the proportion $\frac{a}{b} = \frac{c}{d}$ are numbered as follows:

First term ⟶    $\dfrac{a}{b} = \dfrac{c}{d}$    ⟵ Third term
Second term ⟶           ⟵ Fourth term

## ■ MEANS AND EXTREMES OF A PROPORTION

In the proportion $\frac{1}{2} = \frac{3}{6}$, the numbers 1 and 6 are called the **extremes,** and the numbers 2 and 3 are called the **means.**

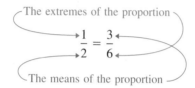

The extremes of the proportion

$$\frac{1}{2} = \frac{3}{6}$$

The means of the proportion

In this proportion, the product of the extremes is equal to the product of the means.

$$1 \cdot 6 = 6 \qquad \text{and} \qquad 2 \cdot 3 = 6$$

This illustrates a fundamental property of proportions.

> **Fundamental Property of Proportions**
> In any proportion, the product of the extremes is equal to the product of the means.

In the proportion $\frac{a}{b} = \frac{c}{d}$, $a$ and $d$ are the extremes, and $b$ and $c$ are the means. We can show that the product of the extremes ($ad$) is equal to the product of the means ($bc$) by multiplying both sides of the proportion by $bd$ and observing that $ad = bc$.

$$\frac{a}{b} = \frac{c}{d}$$

$$\frac{bd}{1} \cdot \frac{a}{b} = \frac{bd}{1} \cdot \frac{c}{d} \qquad \text{To eliminate the fractions, multiply both sides by } \tfrac{bd}{1}.$$

$$\frac{abd}{b} = \frac{bcd}{d} \qquad \text{Multiply the numerators and multiply the denominators.}$$

$$ad = bc \qquad \text{Divide out the common factors: } \tfrac{b}{b} = 1 \text{ and } \tfrac{d}{d} = 1.$$

Since $ad = bc$, the product of the extremes equals the product of the means.

To determine whether an equation is a proportion, we can check to see whether the product of the extremes is equal to the product of the means.

**EXAMPLE 1**   Determine whether each equation is a proportion:   **a.** $\dfrac{3}{7} = \dfrac{9}{21}$   and
**b.** $\dfrac{8}{3} = \dfrac{13}{5}$.

*Solution*   In each case, we check to see whether the product of the extremes is equal to the product of the means.

**a.** The product of the extremes is $3 \cdot 21 = 63$. The product of the means is $7 \cdot 9 = 63$. Since the products are equal, the equation is a proportion: $\frac{3}{7} = \frac{9}{21}$.

**b.** The product of the extremes is $8 \cdot 5 = 40$. The product of the means is $3 \cdot 13 = 39$. Since the products are not equal, the equation is not a proportion: $\frac{8}{3} \neq \frac{13}{5}$. ∎

**Self Check**   Determine whether the equation is a proportion: $\frac{6}{13} = \frac{24}{53}$.

*Answer*   no

When two pairs of numbers such as 2, 3 and 8, 12 form a proportion, we say that they are **proportional.** To show that 2, 3, 8, and 12 are proportional, we check to see whether the equation

$$\frac{2}{3} = \frac{8}{12}$$

is a proportion. To do so, we find the product of the extremes and the product of the means:

$$2 \cdot 12 = 24 \qquad\qquad 3 \cdot 8 = 24$$

Since the products are equal, the equation is a proportion, and the numbers are proportional.

**EXAMPLE 2**   Determine whether 3, 7, 36, and 91 are proportional.

*Solution*   We check to see whether $\frac{3}{7} = \frac{36}{91}$ is a proportion by finding two products:

$$3 \cdot 91 = 273 \qquad \text{The product of the extremes.}$$
$$7 \cdot 36 = 252 \qquad \text{The product of the means.}$$

Since the products are not equal, the numbers are not proportional. ∎

**Self Check**   Determine whether 6, 11, 54, and 99 are proportional.

*Answer*   yes

## ■ SOLVING PROPORTIONS

Suppose that we know three terms in the proportion

$$\frac{x}{5} = \frac{24}{20}$$

To find the unknown term, we multiply the extremes and multiply the means, set them equal, and solve for $x$:

$$\frac{x}{5} = \frac{24}{20}$$

$20x = 5 \cdot 24$     In a proportion, the product of the extremes is equal to the product of the means.

$20x = 120$     Multiply: $5 \cdot 24 = 120$.

$$\frac{20x}{20} = \frac{120}{20}$$     To undo the multiplication by 20, divide both sides by 20.

$x = 6$     Simplify: $\frac{20}{20} = 1$ and $\frac{120}{20} = 6$.

The first term is 6.

| EXAMPLE 3 | Solve $\dfrac{12}{18} = \dfrac{3}{x}$. |

*Solution*

$$\frac{12}{18} = \frac{3}{x}$$

$12 \cdot x = 18 \cdot 3$     In a proportion, the product of the extremes equals the product of the means.

$12x = 54$     Multiply: $18 \cdot 3 = 54$.

$$\frac{12x}{12} = \frac{54}{12}$$     To undo the multiplication by 12, divide both sides by 12.

$x = \dfrac{9}{2}$     Simplify: $\frac{12}{12} = 1$ and $\frac{54}{12} = \frac{9}{2}$.

Thus, $x = \dfrac{9}{2}$.     ■

*Self Check*   Solve $\frac{15}{x} = \frac{25}{40}$.

*Answer*       24

**EXAMPLE 4**    Find the third term of the proportion $\dfrac{3.5}{7.2} = \dfrac{x}{15.84}$.

*Solution*

$$\frac{3.5}{7.2} = \frac{x}{15.84}$$

$3.5(15.84) = 7.2x$    In a proportion, the product of the extremes equals the product of the means.

$55.44 = 7.2x$    Multiply: $3.5 \cdot 15.84 = 55.44$.

$\dfrac{55.44}{7.2} = \dfrac{7.2x}{7.2}$    To undo the multiplication by 7.2, divide both sides by 7.2.

$7.7 = x$    Simplify: $\frac{55.44}{7.2} = 7.7$ and $\frac{7.2}{7.2} = 1$.

The third term is 7.7.    ∎

**Self Check**    Find the second term of the proportion $\dfrac{6.7}{x} = \dfrac{33.5}{38}$.

**Answer**    7.6

■ ■ ■ ■ ■ ■ ■ ■ ■    **Solving Equations with a Calculator**

CALCULATORS    To solve the equation in Example 4 with a calculator, we can proceed as follows.

$$\frac{3.5}{7.2} = \frac{x}{15.84}$$

$\dfrac{3.5(15.84)}{7.2} = x$    Multiply both sides by 15.84.

We can find $x$ by pressing these keys on a scientific calculator.

**Keystrokes**

3.5 × 15.84 ÷ 7.2 =

The display will read    $7.7$ . Thus, $x = 7.7$.

**EXAMPLE 5**    Solve $\dfrac{2x + 1}{4} = \dfrac{10}{8}$.

*Solution*

$$\frac{2x + 1}{4} = \frac{10}{8}$$

$$8(2x + 1) = 40$$ — In a proportion, the product of the extremes equals the product of the means.

$$16x + 8 = 40$$ — Use the distributive property to remove parentheses.

$$16x + 8 - 8 = 40 - 8$$ — To undo the addition of 8, subtract 8 from both sides.

$$16x = 32$$ — Simplify: $8 - 8 = 0$ and $40 - 8 = 32$.

$$\frac{16x}{16} = \frac{32}{16}$$ — To undo the multiplication by 16, divide both sides by 16.

$$x = 2$$ — Simplify: $\frac{16}{16} = 1$ and $\frac{32}{16} = 2$.

Thus, $x = 2$. ∎

**Self Check**    Solve $\dfrac{3x - 1}{2} = \dfrac{12.5}{5}$.

**Answer**    2

### ■ PROBLEM SOLVING

We can use proportions to solve problems.

**EXAMPLE 6**    If 6 apples cost \$1.38, how much will 16 apples cost?

*Solution*    Let $c$ represent the cost of 16 apples. The ratios of the numbers of apples to their costs are equal.

6 apples is to \$1.38 as 16 apples is to \$$c$.

$$\text{6 apples} \to \frac{6}{1.38} = \frac{16}{c} \gets \text{16 apples}$$
Cost of 6 apples → ; ← Cost of 16 apples

$$6 \cdot c = 1.38(16)$$ — In a proportion, the product of the extremes is equal to the product of the means.

$$6c = 22.08$$ — Do the multiplication: $1.38 \cdot 16 = 22.08$.

$$\frac{6c}{6} = \frac{22.08}{6}$$ — To undo the multiplication by 6, divide both sides by 6.

$$c = 3.68$$ — Simplify: $\frac{6}{6} = 1$ and $\frac{22.08}{6} = 3.68$.

Sixteen apples will cost \$3.68. ∎

**Self Check**    If 9 tickets to a concert cost \$112.50, how much will 15 tickets cost?

**Answer**    \$187.50

**EXAMPLE 7**    A solution contains 2 quarts of antifreeze and 5 quarts of water. How many quarts of antifreeze must be mixed with 18 quarts of water to have the same concentration?

*Solution*    Let $q$ represent the number of quarts of antifreeze to be mixed with the water. The ratios of the quarts of antifreeze to the quarts of water are equal.

2 quarts antifreeze is to 5 quarts water as $q$ quarts antifreeze is to 18 quarts water.

$$\text{2 quarts antifreeze} \rightarrow \frac{2}{5} = \frac{q}{18} \leftarrow q \text{ quarts of antifreeze} \\ \text{5 quarts water} \rightarrow \qquad\qquad \leftarrow 18 \text{ quarts of water}$$

$$2 \cdot 18 = 5q \qquad \text{In a proportion, the product of the extremes is equal to the product of the means.}$$

$$36 = 5q \qquad \text{Do the multiplication: } 2 \cdot 18 = 36.$$

$$\frac{36}{5} = \frac{5q}{5} \qquad \text{To undo the multiplication by 5, divide both sides by 5.}$$

$$\frac{36}{5} = q \qquad \text{Simplify: } \tfrac{5}{5} = 1.$$

The mixture should contain $\frac{36}{5}$ or 7.2 quarts of antifreeze.    ■

**Self Check**    A solution should contain 2 ounces of alcohol for every 7 ounces of water. How much alcohol should be added to 20 ounces of water to get the proper concentration?

*Answer*    $\frac{40}{7}$ oz

**EXAMPLE 8**    A recipe for rhubarb cake calls for $1\frac{1}{4}$ cups of sugar for every $2\frac{1}{2}$ cups of flour. How many cups of flour are needed if the baker intends to use 3 cups of sugar?

*Solution*    Let $f$ represent the number of cups of flour to be mixed with the sugar. The ratios of the cups of sugar to the cups of flour are equal.

$1\frac{1}{4}$ cups sugar is to $2\frac{1}{2}$ cups flour as 3 cups sugar is to $f$ cups flour.

$$\text{$1\frac{1}{4}$ cups sugar} \rightarrow \frac{1\frac{1}{4}}{2\frac{1}{2}} = \frac{3}{f} \leftarrow \text{3 cups sugar} \\ \text{$2\frac{1}{2}$ cups flour} \rightarrow \qquad\qquad \leftarrow f \text{ cups flour}$$

$$\frac{1.25}{2.5} = \frac{3}{f} \qquad \text{Change the fractions to decimals.}$$

$$1.25f = 2.5 \cdot 3 \qquad \text{In a proportion, the product of the extremes is equal to the product of the means.}$$

$$1.25f = 7.5 \qquad \text{Do the multiplication: } 2.5 \cdot 3 = 7.5.$$

$$\frac{1.25f}{1.25} = \frac{7.5}{1.25} \qquad \text{To undo the multiplication by 1.25, divide both sides by 1.25.}$$

$$f = 6 \qquad \text{Divide: } \tfrac{1.25}{1.25} = 1 \text{ and } \tfrac{7.5}{1.25} = 6.$$

The baker should use 6 cups of flour.    ■

**Self Check**    In Example 8, how many cups of sugar will be needed to make several cakes that will require a total of 25 cups of flour?

**Answer**    12.5 cups

**EXAMPLE 9**    In a manufacturing process, 15 parts out of 90 were found to be defective. How many defective parts will be expected in a run of 120 parts?

*Solution*    Let $d$ represent the expected number of defective parts. In each run, the ratio of the defective parts to the total number of parts should be the same.

15 defective parts is to 90 as $d$ defective parts is to 120.

$$\text{15 defective parts} \rightarrow \frac{15}{90} = \frac{d}{120} \leftarrow \text{120 parts}$$
$$\text{90 parts} \rightarrow$$

$15 \cdot 120 = 90d$    In a proportion, the product of the extremes is equal to the product of the means.

$1,800 = 90d$    Do the multiplication: $15 \cdot 120 = 1,800$.

$\dfrac{1,800}{90} = \dfrac{90d}{90}$    To undo the multiplication by 90, divide both sides by 90.

$20 = d$    Divide: $\frac{1,800}{90} = 20$ and $\frac{90}{90} = 1$.

The expected number of defective parts is 20.    ■

**Self Check**    In Example 9, how many defective parts will be expected in a run of 3,000 parts?

**Answer**    500

### ■ SIMILAR TRIANGLES

If two angles of one triangle have the same measure as two angles of a second triangle, the triangles will have the same shape. Triangles with the same shape are called **similar triangles.** In Figure 6-1, $\triangle ABC \sim \triangle DEF$ (read the symbol $\sim$ as "is similar to.")

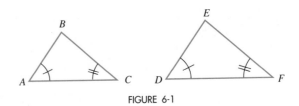

FIGURE 6-1

**Property of Similar Triangles**
If two triangles are similar, all pairs of corresponding sides are in proportion.

In the similar triangles shown in Figure 6-1, the following proportions are true.

$$\frac{\overline{AB}}{\overline{DE}} = \frac{\overline{BC}}{\overline{EF}}, \qquad \frac{\overline{BC}}{\overline{EF}} = \frac{\overline{CA}}{\overline{FD}}, \qquad \text{and} \qquad \frac{\overline{CA}}{\overline{FD}} = \frac{\overline{AB}}{\overline{DE}}$$

**EXAMPLE 10**    A tree casts a shadow 18 feet long at the same time as a woman 5 feet tall casts a shadow that is 1.5 feet long. Find the height of the tree.

*Solution*    Figure 6-2 shows the triangles determined by the tree and its shadow and the woman and her shadow.

FIGURE 6-2

Since the triangles have the same shape, they are similar, and the lengths of their corresponding sides are in proportion. If we let *h* represent the height of the tree, we can find *h* by solving the following proportion.

$$\frac{h}{5} = \frac{18}{1.5} \qquad \frac{\text{Height of the tree}}{\text{Height of the woman}} = \frac{\text{Shadow of the tree}}{\text{Shadow of the woman}}.$$

$$1.5h = 5(18) \qquad \text{In a proportion, the product of the extremes is equal to the product of the means.}$$

$$h = 60 \qquad \text{To undo the multiplication by 1.5, divide both sides by 1.5 and simplify.}$$

The tree is 60 feet tall.                                                                    ■

**Self Check**    Find the height of the tree in Example 10 if the woman is 5 feet 6 inches tall and her shadow is still 1.5 feet long.

*Answer*    66 ft

Orals    *Indicate which are proportions.*

**1.** $\dfrac{3}{5} = \dfrac{6}{10}$      **2.** $\dfrac{1}{2} = \dfrac{1}{3}$      **3.** $\dfrac{1}{2} + \dfrac{2}{4}$      **4.** $\dfrac{1}{x} = \dfrac{2}{2x}$

# EXERCISE 6.2

## REVIEW

**1.** Change $\dfrac{9}{10}$ to a percent.

**2.** Change $\dfrac{7}{8}$ to a percent.

**3.** Change $33\frac{1}{3}\%$ to a fraction.

**4.** Change $75\%$ to a fraction.

**5.** Find $30\%$ of $1{,}600$.

**6.** Find $\dfrac{1}{2}\%$ of $520$.

**7. Shopping**   If Maria bought a dress for $25\%$ off the original price of $98$, how much did the dress cost?

**8. Shopping**   Bill purchased a shirt on sale for $17.50. Find the original cost of the shirt if it was marked down $30\%$.

## VOCABULARY AND CONCEPTS   *Fill in each blank to make a true statement.*

**9.** A _____ is a statement that two _____ are equal.

**10.** The first and fourth terms of a proportion are called the _____ of the proportion.

**11.** The second and third terms of a proportion are called the _____ of the proportion.

**12.** When two pairs of numbers form a proportion, we say that the numbers are _____.

**13.** If two triangles have the same _____, they are said to be similar.

**14.** If two triangles are similar, the lengths of their corresponding sides are in _____.

**15.** The equation $\frac{a}{b} = \frac{c}{d}$ is a proportion if the product ___ is equal to the product ___.

**16.** If $3 \cdot 10 = 17 \cdot x$, then _____ is a proportion. (Note that answers may differ.)

**17.** Read $\triangle ABC$ as _____ $ABC$.

**18.** The symbol $\sim$ is read as _____.

## PRACTICE   *In Exercises 19–26, tell whether each statement is a proportion.*

**19.** $\dfrac{9}{7} = \dfrac{81}{70}$

**20.** $\dfrac{5}{2} = \dfrac{20}{8}$

**21.** $\dfrac{-7}{3} = \dfrac{14}{-6}$

**22.** $\dfrac{13}{-19} = \dfrac{-65}{95}$

**23.** $\dfrac{9}{19} = \dfrac{38}{80}$

**24.** $\dfrac{40}{29} = \dfrac{29}{22}$

**25.** $\dfrac{10.4}{3.6} = \dfrac{41.6}{14.4}$

**26.** $\dfrac{13.23}{3.45} = \dfrac{39.96}{11.35}$

*In Exercises 27–42, solve for the variable in each proportion.*

**27.** $\dfrac{2}{3} = \dfrac{x}{6}$

**28.** $\dfrac{3}{6} = \dfrac{x}{8}$

**29.** $\dfrac{5}{10} = \dfrac{3}{c}$

**30.** $\dfrac{7}{14} = \dfrac{2}{b}$

**31.** $\dfrac{-6}{x} = \dfrac{8}{4}$

**32.** $\dfrac{4}{x} = \dfrac{2}{8}$

**33.** $\dfrac{x}{3} = \dfrac{9}{3}$

**34.** $\dfrac{x}{2} = \dfrac{-18}{6}$

**35.** $\dfrac{x+1}{5} = \dfrac{3}{15}$

**36.** $\dfrac{x-1}{7} = \dfrac{2}{21}$

**37.** $\dfrac{x+3}{12} = \dfrac{-7}{6}$

**38.** $\dfrac{x+7}{-4} = \dfrac{3}{12}$

**39.** $\dfrac{4-x}{13} = \dfrac{11}{26}$

**40.** $\dfrac{5-x}{17} = \dfrac{13}{34}$

**41.** $\dfrac{2x+1}{18} = \dfrac{14}{3}$

**42.** $\dfrac{2x-1}{18} = \dfrac{9}{54}$

*APPLICATIONS* 📇 *In Exercises 43–62, set up and solve a proportion. Use a calculator when it is helpful.*

**43. Grocery shopping**   If 3 pints of yogurt cost $1, how much will 51 pints cost?

**44. Shopping for clothes**   If shirts are on sale at two for $25, how much will 5 shirts cost?

**45. Gardening**   Garden seed is on sale at 3 packets for 50¢. How much will 39 packets cost?

**46. Cooking**   A recipe for spaghetti sauce requires four 16-ounce bottles of catsup to make two gallons of sauce. How many bottles of catsup are needed to make 10 gallons of sauce?

**47. Mixing perfume**   A perfume is to be mixed in the ratio of 3 drops of pure essence to 7 drops of alcohol. How many drops of pure essence should be mixed with 56 drops of alcohol?

**48. Making cologne**   A cologne can be made by mixing 2 drops of pure essence with 5 drops of distilled water. How many drops of water should be used with 15 drops of pure essence?

**49. Making cookies**   A recipe for chocolate chip cookies calls for $1\frac{1}{4}$ cups of flour and 1 cup of sugar. The recipe will make $3\frac{1}{2}$ dozen cookies. How many cups of flour will be needed to make 12 dozen cookies?

**50. Making brownies**   A recipe for brownies calls for 4 eggs and $1\frac{1}{2}$ cups of flour. If the recipe make 15 brownies, how many cups of flour will be needed to make 130 brownies?

**51. Quality control**   In a manufacturing process, 95% of the parts made are to be within specifications. How many defective parts would be expected in a run of 940 pieces?

**52. Quality control**   Out of a sample of 500 men's shirts, 17 were rejected because of crooked collars. How many crooked collars would you expect to find in a run of 15,000 shirts?

**53. Gas consumption**   If a car can travel 42 miles on 1 gallon of gas, how much gas will it need to travel 315 miles?

**54. Gas consumption**   If a truck gets 12 miles per gallon of gas, how far can it go on 17 gallons?

**55. Computing paychecks**   Bill earns $412 for a 40-hour week. If he missed 10 hours of work last week, how much did he get paid?

**56. Model railroading**   An HO-scale model railroad engine is 9 inches long. If HO scale is 87 feet to 1 foot, how long is a real engine?

**57. Model railroading**   An N-scale model railroad caboose is 3.5 inches long. If N scale is 169 feet to 1 foot, how long is a real caboose?

**58. Model houses**   A model house is built to a scale of 1 inch to 8 inches. If a model house is 36 inches wide, how wide is the real house?

**59. Staffing**   A school board determined that there should be 3 teachers for every 50 students. How many teachers are needed for an enrollment of 2,700 students?

**60. Drafting**   In a scale drawing, a 280-foot antenna tower is drawn 7 inches high. The building next to it is drawn 2 inches high. How tall is the actual building?

**61. Mixing fuel**   The instructions on a can of oil intended to be added to lawnmower gasoline read:

| Recommended | Gasoline | Oil |
|:---:|:---:|:---:|
| 50 to 1 | 6 gal | 16 oz |

Are these instructions correct? (*Hint:* There are 128 ounces in 1 gallon.)

**62. Mixing fuel**   See Exercise 61. How much oil should be mixed with 28 gallons of gas?

*In Exercises 63–70, use similar triangles to solve each problem.*

**63. Height of a tree**   A tree casts a shadow of 26 feet at the same time as a 6-foot man casts a shadow of 4 feet. (See Illustration 1.) Find the height of the tree.

**64. Height of a flagpole**   A man places a mirror on the ground and sees the reflection of the top of a flagpole, as in Illustration 2. The two triangles in the illustration are similar. Find the height $h$ of the flagpole.

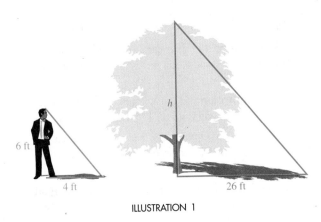

ILLUSTRATION 1

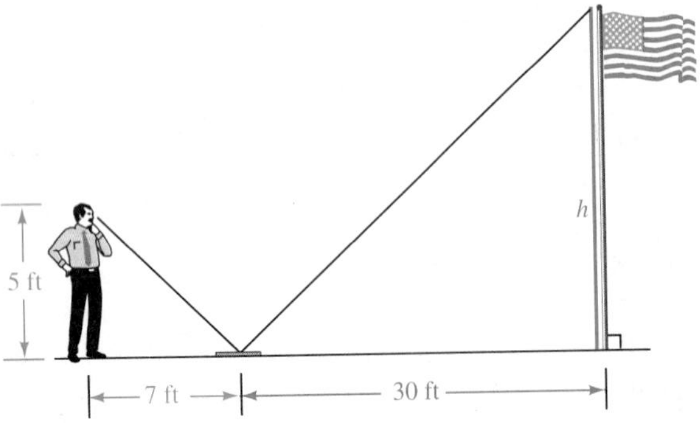

**66. Flight path**   An airplane ascends 100 feet as it flies a horizontal distance of 1,000 feet. How much altitude will it gain as it flies a horizontal distance of 1 mile? See Illustration 4. (*Hint:* 5,280 feet = 1 mile.)

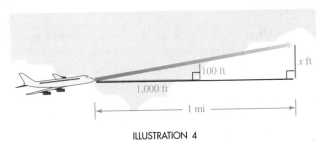

ILLUSTRATION 4

**67. Flight path**   An airplane descends 1,350 feet as it flies a horizontal distance of 1 mile. How much altitude is lost as it flies a horizontal distance of 5 miles?

**68. Ski runs**   A ski course falls 100 feet in every 300 feet of horizontal run. If the total horizontal run is $\frac{1}{2}$ mile, find the height of the hill.

**69. Mountain travel**   A road ascends 750 feet in every 2,500 feet of travel. By how much will the road rise in a trip of 10 miles?

**70. Photo enlargements**   The 3-by-5 photo in Illustration 5 is to be blown up to the larger size. Find $x$.

ILLUSTRATION 2

**65. Width of a river**   Use the dimensions in Illustration 3 to find $w$, the width of the river. The two triangles in the illustration are similar.

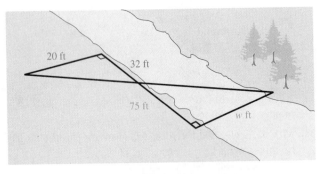

ILLUSTRATION 3

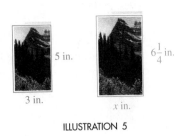

ILLUSTRATION 5

**WRITING**

**71.** Explain the difference between a ratio and a proportion.

**72.** Explain how to tell whether the equation $\frac{3.2}{3.7} = \frac{5.44}{6.29}$ is a proportion.

## SOMETHING TO THINK ABOUT

**73.** Verify that $\frac{3}{5} = \frac{12}{20} = \frac{3+12}{5+20}$. Is the following rule always true?

$$\frac{a}{b} = \frac{c}{d} = \frac{a+c}{b+d}$$

**74.** Verify that since $\frac{3}{5} = \frac{9}{15}$, then $\frac{3+5}{5} = \frac{9+15}{15}$. Is the following rule always true?

$$\text{If } \frac{a}{b} = \frac{c}{d}, \text{ then } \frac{a+b}{b} = \frac{c+d}{d}.$$

## 6.3 Simplifying Fractions

■ SIMPLIFYING FRACTIONS   ■ DIVISION BY 1   ■ DIVIDING POLYNOMIALS THAT ARE NEGATIVES

Getting Ready  *Simplify.*

**1.** $\dfrac{12}{16}$    **2.** $\dfrac{16}{8}$    **3.** $\dfrac{25}{55}$    **4.** $\dfrac{36}{72}$

### ■ SIMPLIFYING FRACTIONS

Ratios such as $\frac{1}{2}$ and $\frac{3}{4}$ that are the quotient of two integers are *rational numbers.* Expressions such as

$$\frac{x}{x+2} \quad \text{and} \quad \frac{5a^2 + b^2}{3a - b}$$

where the numerators and denominators are polynomials, are called **rational expressions.**

We have seen that a fraction can be simplified by dividing out common factors shared by its numerator and denominator. For example,

$$\frac{18}{30} = \frac{3 \cdot 6}{5 \cdot 6} = \frac{3 \cdot \overset{1}{\cancel{6}}}{5 \cdot \cancel{6}} = \frac{3}{5} \quad \text{and} \quad -\frac{6}{15} = -\frac{3 \cdot 2}{3 \cdot 5} = -\frac{\overset{1}{\cancel{3}} \cdot 2}{\cancel{3} \cdot 5} = -\frac{2}{5}$$

To simplify the fraction $\frac{ac}{bc}$, we can divide out the common factor of $c$ to obtain

$$\frac{ac}{bc} = \frac{a\overset{1}{\cancel{c}}}{b\cancel{c}} = \frac{a}{b}$$

This fact establishes the fundamental property of fractions.

**Fundamental Property of Fractions**

If $a$ is a real number and $b$ and $c$ are nonzero real numbers, then

$$\frac{ac}{bc} = \frac{a}{b}$$

The fundamental property of fractions implies that factors common to both the numerator and denominator of a fraction can be divided out. When all common factors have been divided out, we say that the fraction has been **expressed in lowest terms.** To **simplify a fraction** means to write it in lowest terms.

**EXAMPLE 1**   Simplify $\dfrac{21x^2y}{14xy^2}$.

*Solution*   To simplify a fraction means to write it in lowest terms.

$$\frac{21x^2y}{14xy^2} = \frac{3 \cdot 7 \cdot x \cdot x \cdot y}{2 \cdot 7 \cdot x \cdot y \cdot y} \qquad \text{Factor the numerator and denominator.}$$

$$= \frac{3 \cdot \overset{1}{\cancel{7}} \cdot \overset{1}{\cancel{x}} \cdot x \cdot \overset{1}{\cancel{y}}}{2 \cdot \underset{1}{\cancel{7}} \cdot \underset{1}{\cancel{x}} \cdot y \cdot \underset{1}{\cancel{y}}} \qquad \text{Divide out the common factors of 7, } x, \text{ and } y.$$

$$= \frac{3x}{2y}$$

This fraction can also be simplified by using the rules of exponents:

$$\frac{21x^2y}{14xy^2} = \frac{3 \cdot 7}{2 \cdot 7}x^{2-1}y^{1-2} \qquad \frac{x^2}{x} = x^{2-1}; \frac{y}{y^2} = y^{1-2}.$$

$$= \frac{3}{2}xy^{-1} \qquad 2 - 1 = 1; 1 - 2 = -1.$$

$$= \frac{3}{2} \cdot \frac{x}{y} \qquad y^{-1} = \frac{1}{y}.$$

$$= \frac{3x}{2y} \qquad \text{Multiply.} \qquad \blacksquare$$

*Self Check*   Simplify $\dfrac{32a^3b^2}{24ab^4}$.

*Answer*   $\dfrac{4a^2}{3b^2}$

The fraction $\frac{8}{4}$ is equal to 2, because $4 \cdot 2 = 8$. The expression $\frac{8}{0}$ is undefined, because there is no number $x$ for which $0 \cdot x = 8$. The expression $\frac{0}{0}$ presents a different problem, however, because $\frac{0}{0}$ seems to equal any number. For example, $\frac{0}{0} = 17$, because $0 \cdot 17 = 0$. Similarly, $\frac{0}{0} = \pi$, because $0 \cdot \pi = 0$. Since "no answer" and "any answer" are both unacceptable, division by 0 is not allowed.

Although $\frac{0}{0}$ represents many numbers, there is often one best answer. In the 17th century, mathema-

ticians such as Sir Isaac Newton (1642–1727) and Gottfried Wilhelm von Leibniz (1646–1716) began to look more closely at expressions related to the fraction $\frac{0}{0}$. They discovered that under certain conditions, there was one best answer. Expressions related to $\frac{0}{0}$ are called **indeterminate forms.** One of these expressions, called a **derivative,** is the foundation of **calculus,** an important area of mathematics discovered independently by both Newton and Leibniz.

| EXAMPLE 2 | Write $\dfrac{x^2 + 3x}{3x + 9}$ in lowest terms. |

*Solution*

$$\frac{x^2 + 3x}{3x + 9} = \frac{x(x + 3)}{3(x + 3)} \qquad \text{Factor the numerator and the denominator.}$$

$$= \frac{x(\cancel{x + 3})}{3(\cancel{x + 3})} \qquad \text{Divide out the common factor of } x + 3.$$

$$= \frac{x}{3}$$

*Self Check*  Simplify $\dfrac{x^2 - 5x}{5x - 25}$.

*Answer*  $\frac{x}{5}$

## ■ DIVISION BY 1

Any number divided by the number 1 remains unchanged. For example,

$$\frac{37}{1} = 37, \qquad \frac{5x}{1} = 5x, \qquad \text{and} \qquad \frac{3x + y}{1} = 3x + y$$

In general, for any real number $a$, the following is true.

**Division by 1**

$$\frac{a}{1} = a$$

EXAMPLE 3    Simplify $\dfrac{x^3 + x^2}{x + 1}$.

*Solution*    $\dfrac{x^3 + x^2}{x + 1} = \dfrac{x^2(x + 1)}{x + 1}$    Factor the numerator.

$= \dfrac{x^2\cancel{(x + 1)}}{\cancel{x + 1}}$    Divide out the common factor of $x + 1$.

$= \dfrac{x^2}{1}$

$= x^2$    Denominators of 1 need not be written.    ∎

Self Check    Simplify $\dfrac{x^2 - x}{x - 1}$.

Answer    $x$

## ■ DIVIDING POLYNOMIALS THAT ARE NEGATIVES

If the terms of two polynomials are the same, except for sign, the polynomials are called **negatives** of each other. For example,

$x - y$ and $-x + y$ are negatives,
$2a - 1$ and $-2a + 1$ are negatives, and
$3x^2 - 2x + 5$ and $-3x^2 + 2x - 5$ are negatives.

Example 4 shows why the quotient of two binomials that are negatives is always $-1$.

EXAMPLE 4    Simplify   **a.** $\dfrac{x - y}{y - x}$   and   **b.** $\dfrac{2a - 1}{1 - 2a}$.

*Solution*    We can rearrange terms in each numerator, factor out $-1$, and proceed as follows:

**a.** $\dfrac{x - y}{y - x} = \dfrac{-y + x}{y - x}$          **b.** $\dfrac{2a - 1}{1 - 2a} = \dfrac{-1 + 2a}{1 - 2a}$

$= \dfrac{-(y - x)}{y - x}$                   $= \dfrac{-(1 - 2a)}{1 - 2a}$

$= \dfrac{-\cancel{(y - x)}}{\cancel{y - x}}$                   $= \dfrac{-\cancel{(1 - 2a)}}{\cancel{1 - 2a}}$

$= -1$                          $= -1$    ∎

Self Check    Simplify $\dfrac{3p - 2q}{2q - 3p}$.

Answer    $-1$

In general, we have this important result.

**Division of Negatives**
The quotient of any nonzero expression and its negative is $-1$.

EXAMPLE 5    Simplify $\dfrac{x^2 + 13x + 12}{x^2 - 144}$.

Solution

$$\dfrac{x^2 + 13x + 12}{x^2 - 144} = \dfrac{(x + 1)(x + 12)}{(x + 12)(x - 12)}$$    Factor the numerator and denominator.

$$= \dfrac{(x + 1)\cancel{(x + 12)}^{1}}{\cancel{(x + 12)}_{1}(x - 12)}$$    Divide out the common factor of $x + 12$.

$$= \dfrac{x + 1}{x - 12}$$    ∎

Self Check    Simplify $\dfrac{x^2 - 9}{x^3 - 3x^2}$.

Answer    $\dfrac{x + 3}{x^2}$

**WARNING!**    Remember that only *factors* that are common to the *entire numerator* and the *entire denominator* can be divided out. *Terms* that are common to both the numerator and denominator *cannot* be divided out. For example, consider the correct simplification

$$\dfrac{5 + 8}{5} = \dfrac{13}{5}$$

It would be incorrect to divide out the common *term* of 5 in the above simplification. Doing so gives an incorrect answer.

$$\dfrac{5 + 8}{5} \neq \dfrac{\cancel{5} + 8}{\cancel{5}} = \dfrac{1 + 8}{1} = 9$$

**EXAMPLE 6**   Write $\dfrac{5(x + 3) - 5}{7(x + 3) - 7}$ in lowest terms.

*Solution*   We cannot divide out $x + 3$, because it is not a factor of the entire numerator, nor is it a factor of the entire denominator. Instead, we simplify the numerator and denominator, factor them, and then divide out any common factors.

$$\frac{5(x + 3) - 5}{7(x + 3) - 7} = \frac{5x + 15 - 5}{7x + 21 - 7} \qquad \text{Remove parentheses.}$$

$$= \frac{5x + 10}{7x + 14} \qquad \text{Combine like terms.}$$

$$= \frac{5(x + 2)}{7(x + 2)} \qquad \text{Factor the numerator and denominator.}$$

$$= \frac{5(\overset{1}{\cancel{x + 2}})}{7(\underset{1}{\cancel{x + 2}})} \qquad \text{Divide out the common factor of } x + 2.$$

$$= \frac{5}{7} \qquad\qquad\qquad \blacksquare$$

*Self Check*   Simplify $\dfrac{4(x - 2) + 4}{3(x - 2) + 3}$.

*Answer*   $\frac{4}{3}$

**EXAMPLE 7**   Simplify $\dfrac{x(x + 3) - 3(x - 1)}{x^2 + 3}$.

*Solution*   $\dfrac{x(x + 3) - 3(x - 1)}{x^2 + 3} = \dfrac{x^2 + 3x - 3x + 3}{x^2 + 3}$   Remove parentheses in the numerator.

$$= \frac{x^2 + 3}{x^2 + 3} \qquad \text{Combine like terms in the numerator.}$$

$$= \frac{\overset{1}{\cancel{x^2 + 3}}}{\underset{1}{\cancel{x^2 + 3}}} \qquad \text{Divide out the common factor of } x^2 + 3.$$

$$= 1 \qquad\qquad\qquad \blacksquare$$

*Self Check*   Simplify $\dfrac{a(a + 2) - 2(a - 1)}{a^2 + 2}$.

*Answer*   1

Sometimes a fraction does not simplify. Such a fraction is already in lowest terms. For example, to attempt to simplify

$$\frac{x^2 + x - 2}{x^2 + x}$$

we factor the numerator and denominator.

$$\frac{x^2 + x - 2}{x^2 + x} = \frac{(x + 2)(x - 1)}{x(x + 1)}$$

Because there are no factors common to the numerator and denominator, this fraction is already in lowest terms.

Orals  *Simplify each fraction.*

**1.** $\dfrac{14}{21}$    **2.** $\dfrac{34}{17}$    **3.** $\dfrac{xyz}{wxy}$    **4.** $\dfrac{8x^2}{4x}$

**5.** $\dfrac{6x^2y}{6xy^2}$    **6.** $\dfrac{x^2y^3}{x^2y^4}$    **7.** $\dfrac{x + y}{y + x}$    **8.** $\dfrac{x - y}{y - x}$

## EXERCISE 6.3

*REVIEW*

**1.** State the associative property of addition.

**2.** State the distributive property.

**3.** What is the additive identity?

**4.** What is the multiplicative identity?

**5.** Find the additive inverse of $-\dfrac{5}{3}$.

**6.** Find the multiplicative inverse of $-\dfrac{5}{3}$.

*VOCABULARY AND CONCEPTS*  *Fill in each blank to make a true statement.*

**7.** In a fraction, the part above the fraction bar is called the _____.

**8.** In a fraction, the part below the fraction bar is called the _____.

**9.** The denominator of a fraction cannot be __.

**10.** A fraction that has polynomials in its numerator and denominator is called a _____ expression.

**11.** $x - 2$ and $2 - x$ are called _____ of each other.

**12.** To *simplify* a fraction means to write it in _____ terms.

**13.** The fundamental property of fractions states that

$$\frac{ac}{bc} = \frac{}{}.$$

**14.** Any number $x$ divided by 1 is __.

**15.** To simplify a fraction, we _____ the numerator and denominator and divide out _____ factors.

**16.** A fraction cannot be simplified when it is written in _____.

**PRACTICE** *In Exercises 17–84, write each fraction in lowest terms. If a fraction is already in lowest terms, so indicate. Assume that no denominators are 0.*

17. $\dfrac{8}{10}$    18. $\dfrac{16}{28}$    19. $\dfrac{28}{35}$    20. $\dfrac{14}{20}$

21. $\dfrac{8}{52}$    22. $\dfrac{15}{21}$    23. $\dfrac{10}{45}$    24. $\dfrac{21}{35}$

25. $\dfrac{-18}{54}$    26. $\dfrac{16}{40}$    27. $\dfrac{4x}{2}$    28. $\dfrac{2x}{4}$

29. $\dfrac{-6x}{18}$    30. $\dfrac{-25y}{5}$    31. $\dfrac{45}{9a}$    32. $\dfrac{48}{16y}$

33. $\dfrac{7+3}{5z}$    34. $\dfrac{(3-18)k}{25}$    35. $\dfrac{(3+4)a}{24-3}$    36. $\dfrac{x+x}{2}$

37. $\dfrac{2x}{3x}$    38. $\dfrac{5y}{7y}$    39. $\dfrac{6x^2}{4x^2}$    40. $\dfrac{9xy}{6xy}$

41. $\dfrac{2x^2}{3y}$    42. $\dfrac{5y^2}{2y^2}$    43. $\dfrac{15x^2y}{5xy^2}$    44. $\dfrac{12xz}{4xz^2}$

45. $\dfrac{28x}{32y}$    46. $\dfrac{14xz^2}{7x^2z^2}$    47. $\dfrac{x+3}{3(x+3)}$    48. $\dfrac{2(x+7)}{x+7}$

49. $\dfrac{5x+35}{x+7}$    50. $\dfrac{x-9}{3x-27}$    51. $\dfrac{x^2+3x}{2x+6}$    52. $\dfrac{xz-2x}{yz-2y}$

53. $\dfrac{15x-3x^2}{25y-5xy}$    54. $\dfrac{3y+xy}{3x+xy}$    55. $\dfrac{6a-6b+6c}{9a-9b+9c}$    56. $\dfrac{3a-3b-6}{2a-2b-4}$

57. $\dfrac{x-7}{7-x}$    58. $\dfrac{d-c}{c-d}$    59. $\dfrac{6x-3y}{3y-6x}$    60. $\dfrac{3c-4d}{4c-3d}$

61. $\dfrac{a+b-c}{c-a-b}$    62. $\dfrac{x-y-z}{z+y-x}$    63. $\dfrac{x^2+3x+2}{x^2+x-2}$    64. $\dfrac{x^2+x-6}{x^2-x-2}$

65. $\dfrac{x^2-8x+15}{x^2-x-6}$    66. $\dfrac{x^2-6x-7}{x^2+8x+7}$    67. $\dfrac{2x^2-8x}{x^2-6x+8}$    68. $\dfrac{3y^2-15y}{y^2-3y-10}$

69. $\dfrac{xy+2x^2}{2xy+y^2}$    70. $\dfrac{3x+3y}{x^2+xy}$    71. $\dfrac{x^2+3x+2}{x^3+x^2}$    72. $\dfrac{6x^2-13x+6}{3x^2+x-2}$

73. $\dfrac{x^2-8x+16}{x^2-16}$    74. $\dfrac{3x+15}{x^2-25}$    75. $\dfrac{2x^2-8}{x^2-3x+2}$    76. $\dfrac{3x^2-27}{x^2+3x-18}$

77. $\dfrac{x^2-2x-15}{x^2+2x-15}$    78. $\dfrac{x^2+4x-77}{x^2-4x-21}$    79. $\dfrac{x^2-3(2x-3)}{9-x^2}$    80. $\dfrac{x(x-8)+16}{16-x^2}$

81. $\dfrac{4(x+3)+4}{3(x+2)+6}$    82. $\dfrac{4+2(x-5)}{3x-5(x-2)}$

**83.** $\dfrac{x^2 - 9}{(2x + 3) - (x + 6)}$

**84.** $\dfrac{x^2 + 5x + 4}{2(x + 3) - (x + 2)}$

### *WRITING*

**85.** Explain why $\dfrac{x - 7}{7 - x} = -1$.

**86.** Explain why $\dfrac{x + 7}{7 + x} = 1$.

### *SOMETHING TO THINK ABOUT*

**87.** Exercise 79,

$$\dfrac{x^2 - 3(2x - 3)}{9 - x^2}$$

has two possible answers: $\dfrac{3 - x}{3 + x}$ and $-\dfrac{x - 3}{x + 3}$.
Why is either answer correct?

**88.** Find two different-looking but correct answers for the following problem.

Simplify $\dfrac{y^2 + 5(2y + 5)}{25 - y^2}$.

<br>

## 6.4  Multiplying and Dividing Fractions

■ MULTIPLYING FRACTIONS ■ MULTIPLYING A FRACTION BY A POLYNOMIAL ■ DIVIDING FRACTIONS ■ DIVIDING A FRACTION BY A POLYNOMIAL ■ COMBINED OPERATIONS

**Getting Ready**  *Multiply the fractions and simplify.*

**1.** $\dfrac{3}{7} \cdot \dfrac{14}{9}$    **2.** $\dfrac{21}{15} \cdot \dfrac{10}{3}$    **3.** $\dfrac{19}{38} \cdot 6$    **4.** $42 \cdot \dfrac{3}{21}$

**5.** $\dfrac{4}{9} \cdot \dfrac{45}{8}$    **6.** $\dfrac{11}{7} \cdot \dfrac{14}{22}$    **7.** $\dfrac{75}{12} \cdot \dfrac{6}{50}$    **8.** $\dfrac{13}{5} \cdot \dfrac{20}{26}$

### ■ MULTIPLYING FRACTIONS

Recall that to multiply fractions, we multiply their numerators and multiply their denominators. For example, to find the product of $\frac{4}{7}$ and $\frac{3}{5}$, we proceed as follows.

$$\dfrac{4}{7} \cdot \dfrac{3}{5} = \dfrac{4 \cdot 3}{7 \cdot 5} \qquad \text{Multiply the numerators and multiply the denominators.}$$

$$= \dfrac{12}{35} \qquad 4 \cdot 3 = 12 \text{ and } 7 \cdot 5 = 35.$$

In general, the following is true.

## Rule for Multiplying Fractions

If $a$, $b$, $c$, and $d$ are real numbers and $b \neq 0$ and $d \neq 0$, then

$$\frac{a}{b} \cdot \frac{c}{d} = \frac{ac}{bd}$$

**EXAMPLE 1** Multiply **a.** $\dfrac{1}{3} \cdot \dfrac{2}{5}$, **b.** $\dfrac{7}{9} \cdot \dfrac{-5}{3x}$, **c.** $\dfrac{x^2}{2} \cdot \dfrac{3}{y^2}$, and **d.** $\dfrac{t+1}{t} \cdot \dfrac{t-1}{t-2}$.

*Solution* **a.** $\dfrac{1}{3} \cdot \dfrac{2}{5} = \dfrac{1 \cdot 2}{3 \cdot 5}$ **b.** $\dfrac{7}{9} \cdot \dfrac{-5}{3x} = \dfrac{7(-5)}{9 \cdot 3x}$

$= \dfrac{2}{15}$ $= \dfrac{-35}{27x}$

**c.** $\dfrac{x^2}{2} \cdot \dfrac{3}{y^2} = \dfrac{x^2 \cdot 3}{2 \cdot y^2}$ **d.** $\dfrac{t+1}{t} \cdot \dfrac{t-1}{t-2} = \dfrac{(t+1)(t-1)}{t(t-2)}$

$= \dfrac{3x^2}{2y^2}$ ■

*Self Check* Multiply $\dfrac{3x}{4} \cdot \dfrac{p-3}{y}$.

*Answer* $\dfrac{3x(p-3)}{4y}$

**EXAMPLE 2** Multiply $\dfrac{35x^2y}{7y^2z} \cdot \dfrac{z}{5xy}$.

*Solution* $\dfrac{35x^2y}{7y^2z} \cdot \dfrac{z}{5xy} = \dfrac{35x^2y \cdot z}{7y^2z \cdot 5xy}$    Multiply the numerators and multiply the denominators.

$= \dfrac{5 \cdot 7 \cdot x \cdot x \cdot y \cdot z}{7 \cdot y \cdot y \cdot z \cdot 5 \cdot x \cdot y}$    Factor.

$= \dfrac{\overset{1}{\cancel{5}} \cdot \overset{1}{\cancel{7}} \cdot \overset{1}{\cancel{x}} \cdot x \cdot \overset{1}{\cancel{y}} \cdot \overset{1}{\cancel{z}}}{\underset{1}{\cancel{7}} \cdot \underset{1}{\cancel{y}} \cdot y \cdot \underset{1}{\cancel{z}} \cdot \underset{1}{\cancel{5}} \cdot \underset{1}{\cancel{x}} \cdot y}$    Divide out common factors.

$= \dfrac{x}{y^2}$ ■

*Self Check* Multiply $\dfrac{a^2b^2}{2a} \cdot \dfrac{9a^3}{3b^3}$.

*Answer* $\dfrac{3a^4}{2b}$

**EXAMPLE 3**    Multiply $\dfrac{x^2 - x}{2x + 4} \cdot \dfrac{x + 2}{x}$.

*Solution*    $\dfrac{x^2 - x}{2x + 4} \cdot \dfrac{x + 2}{x} = \dfrac{(x^2 - x)(x + 2)}{(2x + 4)(x)}$    Multiply the numerators and multiply the denominators.

$$= \dfrac{x(x - 1)(x + 2)}{2(x + 2)x}$$    Factor.

$$= \dfrac{\overset{1}{\cancel{x}}(x - 1)\overset{1}{\cancel{(x + 2)}}}{2\cancel{(x + 2)}\cancel{x}}$$    Divide out common factors.

$$= \dfrac{x - 1}{2}$$    ∎

**Self Check**    Multiply $\dfrac{x^2 + x}{3x + 6} \cdot \dfrac{x + 2}{x + 1}$.

*Answer*    $\frac{x}{3}$

**EXAMPLE 4**    Multiply $\dfrac{x^2 - 3x}{x^2 - x - 6}$ and $\dfrac{x^2 + x - 2}{x^2 - x}$.

*Solution*    $\dfrac{x^2 - 3x}{x^2 - x - 6} \cdot \dfrac{x^2 + x - 2}{x^2 - x}$

$$= \dfrac{(x^2 - 3x)(x^2 + x - 2)}{(x^2 - x - 6)(x^2 - x)}$$    Multiply the numerators and multiply the denominators.

$$= \dfrac{x(x - 3)(x + 2)(x - 1)}{(x + 2)(x - 3)x(x - 1)}$$    Factor.

$$= \dfrac{\overset{1}{\cancel{x}}\overset{1}{\cancel{(x - 3)}}\overset{1}{\cancel{(x + 2)}}\overset{1}{\cancel{(x - 1)}}}{\underset{1}{\cancel{(x + 2)}}\underset{1}{\cancel{(x - 3)}}\underset{1}{\cancel{x}}\underset{1}{\cancel{(x - 1)}}}$$    Divide out common factors.

$$= 1$$    ∎

**Self Check**    Multiply $\dfrac{a^2 + a}{a^2 - 4} \cdot \dfrac{a^2 - a - 2}{a^2 + 2a + 1}$.

*Answer*    $\frac{a}{a + 2}$

## ■ MULTIPLYING A FRACTION BY A POLYNOMIAL

Since any number divided by 1 remains unchanged, we can write any polynomial as a fraction by inserting a denominator of 1.

**EXAMPLE 5** Multiply $\dfrac{x^2 + x}{x^2 + 8x + 7} \cdot x + 7$.

*Solution*

$$\frac{x^2 + x}{x^2 + 8x + 7} \cdot (x + 7) = \frac{x^2 + x}{x^2 + 8x + 7} \cdot \frac{x + 7}{1}$$

Write $x + 7$ as a fraction with a denominator of 1.

$$= \frac{x(x + 1)(x + 7)}{(x + 1)(x + 7)1}$$

Multiply the fractions and factor where possible.

$$= \frac{x \overset{1}{(x + 1)} \overset{1}{(x + 7)}}{1 \underset{1}{(x + 1)} \underset{1}{(x + 7)}}$$

Divide out all common factors.

$$= x \qquad \blacksquare$$

*Self Check* Multiply $a - 7 \cdot \dfrac{a^2 - a}{a^2 - 8a + 7}$.

*Answer* $a$

## ■ DIVIDING FRACTIONS

Recall that division by a nonzero number is equivalent to multiplying by the reciprocal of that number. Thus, to divide two fractions, we can invert the **divisor** (the fraction following the $\div$ sign) and multiply. For example, to divide $\frac{4}{7}$ by $\frac{3}{5}$, we proceed as follows:

$$\frac{4}{7} \div \frac{3}{5} = \frac{4}{7} \cdot \frac{5}{3}$$

Invert $\frac{3}{5}$ and change the division to a multiplication.

$$= \frac{20}{21}$$

Multiply the numerators and multiply the denominators.

In general, the following is true.

> **Division of Fractions**
> If $a$ is a real number and $b$, $c$, and $d$ are nonzero real numbers, then
> $$\frac{a}{b} \div \frac{c}{d} = \frac{a}{b} \cdot \frac{d}{c} = \frac{ad}{bc}$$

**EXAMPLE 6** Do the divisions: **a.** $\dfrac{7}{13} \div \dfrac{21}{26}$ and **b.** $\dfrac{-9x}{35y} \div \dfrac{15x^2}{14}$.

*Solution*  **a.** $\dfrac{7}{13} \div \dfrac{21}{26} = \dfrac{7}{13} \cdot \dfrac{26}{21}$    Invert the divisor and multiply.

$= \dfrac{7 \cdot 2 \cdot 13}{13 \cdot 3 \cdot 7}$    Multiply the fractions and factor where possible.

$= \dfrac{\overset{1}{\cancel{7}} \cdot 2 \cdot \overset{1}{\cancel{13}}}{\cancel{13} \cdot 3 \cdot \cancel{7}}$    Divide out common factors.

$= \dfrac{2}{3}$

**b.** $\dfrac{-9x}{35y} \div \dfrac{15x^2}{14} = \dfrac{-9x}{35y} \cdot \dfrac{14}{15x^2}$    Invert the divisor and multiply.

$= \dfrac{-3 \cdot 3 \cdot x \cdot 2 \cdot 7}{5 \cdot 7 \cdot y \cdot 3 \cdot 5 \cdot x \cdot x}$    Multiply the fractions and factor where possible.

$= \dfrac{-3 \cdot \overset{1}{\cancel{3}} \cdot \overset{1}{\cancel{x}} \cdot 2 \cdot \overset{1}{\cancel{7}}}{5 \cdot \cancel{7} \cdot y \cdot \cancel{3} \cdot 5 \cdot \cancel{x} \cdot x}$    Divide out common factors.

$= -\dfrac{6}{25xy}$    Multiply the remaining factors.    ∎

---

**Self Check**    Divide $\dfrac{-8a}{3b} \div \dfrac{16a^2}{9b^2}$.

**Answer**    $-\dfrac{3b}{2a}$

---

**EXAMPLE 7**    Divide $\dfrac{x^2 + x}{3x - 15} \div \dfrac{x^2 + 2x + 1}{6x - 30}$.

*Solution*    $\dfrac{x^2 + x}{3x - 15} \div \dfrac{x^2 + 2x + 1}{6x - 30}$

$= \dfrac{x^2 + x}{3x - 15} \cdot \dfrac{6x - 30}{x^2 + 2x + 1}$    Invert the divisor and multiply.

$= \dfrac{x(x + 1) \cdot 2 \cdot 3(x - 5)}{3(x - 5)(x + 1)(x + 1)}$    Multiply the fractions and factor.

$= \dfrac{x(\overset{1}{\cancel{x + 1}}) \cdot 2 \cdot \overset{1}{\cancel{3}}(\overset{1}{\cancel{x - 5}})}{\underset{1}{\cancel{3}}(\underset{1}{\cancel{x - 5}})(\underset{1}{\cancel{x + 1}})(x + 1)}$    Divide out all common factors.

$= \dfrac{2x}{x + 1}$    ∎

Self Check    Divide $\dfrac{a^2 - 1}{a^2 + 4a + 3} \div \dfrac{a - 1}{a^2 + 2a - 3}$.

Answer    $a - 1$

## ■ DIVIDING A FRACTION BY A POLYNOMIAL

To divide a fraction by a polynomial, we write the polynomial as a fraction by inserting a denominator of 1 and then divide the fractions.

**EXAMPLE 8**    Divide $\dfrac{2x^2 - 3x - 2}{2x + 1} \div (4 - x^2)$.

Solution
$$\dfrac{2x^2 - 3x - 2}{2x + 1} \div (4 - x^2)$$

$$= \dfrac{2x^2 - 3x - 2}{2x + 1} \div \dfrac{4 - x^2}{1}$$     Write $4 - x^2$ as a fraction with a denominator of 1.

$$= \dfrac{2x^2 - 3x - 2}{2x + 1} \cdot \dfrac{1}{4 - x^2}$$     Invert the divisor and multiply.

$$= \dfrac{(2x + 1)(x - 2) \cdot 1}{(2x + 1)(2 + x)(2 - x)}$$     Multiply the fractions and factor where possible.

$$= \dfrac{\overset{1}{(2x + 1)}\overset{-1}{(x - 2)} \cdot 1}{\underset{1}{(2x + 1)}(2 + x)\underset{1}{(2 - x)}}$$     Divide out common factors: $\frac{x - 2}{2 - x} = -1$.

$$= \dfrac{-1}{2 + x}$$

$$= -\dfrac{1}{2 + x}$$     ■

Self Check    Divide $(b - a) \div \dfrac{a^2 - b^2}{a^2 + ab}$.

Answer    $-a$

## ■ COMBINED OPERATIONS

Unless parentheses indicate otherwise, we do multiplications and divisions in order from left to right.

**EXAMPLE 9** Simplify $\dfrac{x^2 - x - 6}{x - 2} \div \dfrac{x^2 - 4x}{x^2 - x - 2} \cdot \dfrac{x - 4}{x^2 + x}$.

*Solution* Since there are no parentheses to indicate otherwise, we do the division first.

$$\dfrac{x^2 - x - 6}{x - 2} \div \dfrac{x^2 - 4x}{x^2 - x - 2} \cdot \dfrac{x - 4}{x^2 + x}$$

$$= \dfrac{x^2 - x - 6}{x - 2} \cdot \dfrac{x^2 - x - 2}{x^2 - 4x} \cdot \dfrac{x - 4}{x^2 + x}$$    Invert the divisor and multiply.

$$= \dfrac{(x + 2)(x - 3)(x + 1)(x - 2)(x - 4)}{(x - 2)x(x - 4)x(x + 1)}$$    Multiply the fractions and factor.

$$= \dfrac{(x + 2)(x - 3)(x + 1)(x - 2)(x - 4)}{(x - 2)x(x - 4)x(x + 1)}$$    Divide out all common factors.

$$= \dfrac{(x + 2)(x - 3)}{x^2}$$    ∎

**Self Check** Simplify $\dfrac{a^2 + ab}{ab - b^2} \cdot \dfrac{a^2 - b^2}{a^2 + ab} \div \dfrac{a + b}{b}$.

*Answer* 1

**EXAMPLE 10** Simplify $\dfrac{x^2 + 6x + 9}{x^2 - 2x}\left(\dfrac{x^2 - 4}{x^2 + 3x} \div \dfrac{x + 2}{x}\right)$.

*Solution* We do the division within the parentheses first.

$$\dfrac{x^2 + 6x + 9}{x^2 - 2x}\left(\dfrac{x^2 - 4}{x^2 + 3x} \div \dfrac{x + 2}{x}\right)$$

$$= \dfrac{x^2 + 6x + 9}{x^2 - 2x}\left(\dfrac{x^2 - 4}{x^2 + 3x} \cdot \dfrac{x}{x + 2}\right)$$    Invert the divisor and multiply.

$$= \dfrac{(x + 3)(x + 3)(x + 2)(x - 2)x}{x(x - 2)x(x + 3)(x + 2)}$$    Multiply the fractions and factor where possible.

$$= \dfrac{(x + 3)(x + 3)(x + 2)(x - 2)x}{x(x - 2)x(x + 3)(x + 2)}$$    Divide out all common factors.

$$= \dfrac{x + 3}{x}$$    ∎

Self Check   Simplify $\dfrac{x^2 - 2x}{x^2 + 6x + 9} \div \left( \dfrac{x^2 - 4}{x^2 + 3x} \cdot \dfrac{x}{x + 2} \right)$.

Answer   $\dfrac{x}{x+3}$

Orals   *Do the operations and simplify.*

**1.** $\dfrac{x}{2} \cdot \dfrac{3}{x}$

**2.** $\dfrac{x+1}{5} \cdot \dfrac{7}{x+1}$

**3.** $\dfrac{5}{x+7} \cdot (x+7)$

**4.** $\dfrac{3}{7} \div \dfrac{3}{7}$

**5.** $\dfrac{3}{4} \div 3$

**6.** $(x+1) \div \dfrac{x+1}{x}$

## EXERCISE 6.4

**REVIEW**   *Simplify each expression. Write all answers without using negative exponents.*

**1.** $2x^3y^2(-3x^2y^4z)$

**2.** $\dfrac{8x^4y^5}{-2x^3y^2}$

**3.** $(3y)^{-4}$

**4.** $(a^{-2}a)^{-3}$

**5.** $\dfrac{x^{3m}}{x^{4m}}$

**6.** $(3x^2y^3)^0$

*Do the operations and simplify.*

**7.** $-4(y^3 - 4y^2 + 3y - 2) + 6(-2y^2 + 4) - 4(-2y^3 - y)$

**8.** $y - 5\overline{)5y^3 - 3y^2 + 4y - 1}$

**VOCABULARY AND CONCEPTS**   *Fill in each blank to make a true statement.*

**9.** In a fraction, the part above the fraction bar is called the _____.

**10.** In a fraction, the part below the fraction bar is called the _____.

**11.** To multiply fractions, we multiply their _____ and multiply their _____.

**12.** $\dfrac{a}{b} \cdot \dfrac{c}{d} = $ ___

**13.** To write a polynomial in fractional form, we insert a denominator of __.

**14.** $\dfrac{a}{b} \div \dfrac{c}{d} = \dfrac{a}{b} \cdot $ ___

**15.** To divide two fractions, invert the _____ and _____.

**16.** Unless parentheses indicate otherwise, do multiplications and divisions in order from ___ to ___.

**PRACTICE**   *In Exercises 17–62, do the multiplications. Simplify answers if possible.*

**17.** $\dfrac{5}{7} \cdot \dfrac{9}{13}$

**18.** $\dfrac{2}{7} \cdot \dfrac{5}{11}$

**19.** $\dfrac{25}{35} \cdot \dfrac{-21}{55}$

**20.** $-\dfrac{27}{24} \cdot \left( -\dfrac{56}{35} \right)$

**21.** $\dfrac{2}{3} \cdot \dfrac{15}{2} \cdot \dfrac{1}{7}$

**22.** $\dfrac{2}{5} \cdot \dfrac{10}{9} \cdot \dfrac{3}{2}$

**23.** $\dfrac{3x}{y} \cdot \dfrac{y}{2}$

**24.** $\dfrac{2y}{z} \cdot \dfrac{z}{3}$

**25.** $\dfrac{5y}{7} \cdot \dfrac{7x}{5z}$

**26.** $\dfrac{4x}{3y} \cdot \dfrac{3y}{7x}$

**27.** $\dfrac{7z}{9z} \cdot \dfrac{4z}{2z}$

**28.** $\dfrac{8z}{2x} \cdot \dfrac{16x}{3x}$

**29.** $\dfrac{2x^2y}{3xy} \cdot \dfrac{3xy^2}{2}$

**30.** $\dfrac{2x^2z}{z} \cdot \dfrac{5x}{z}$

**31.** $\dfrac{8x^2y^2}{4x^2} \cdot \dfrac{2xy}{2y}$

**32.** $\dfrac{9x^2y}{3x} \cdot \dfrac{3xy}{3y}$

**33.** $\dfrac{-2xy}{x^2} \cdot \dfrac{3xy}{2}$

**34.** $\dfrac{-3x}{x^2} \cdot \dfrac{2xz}{3}$

**35.** $\dfrac{ab^2}{a^2b} \cdot \dfrac{b^2c^2}{abc} \cdot \dfrac{abc^2}{a^3c^2}$

**36.** $\dfrac{x^3y}{z} \cdot \dfrac{xz^3}{x^2y^2} \cdot \dfrac{yz}{xyz}$

**37.** $\dfrac{10r^2st^3}{6rs^2} \cdot \dfrac{3r^3t}{2rst} \cdot \dfrac{2s^3t^4}{5s^2t^3}$

**38.** $\dfrac{3a^3b}{25cd^3} \cdot \dfrac{-5cd^2}{6ab} \cdot \dfrac{10abc^2}{2bc^2d}$

**39.** $\dfrac{z+7}{7} \cdot \dfrac{z+2}{z}$

**40.** $\dfrac{a-3}{a} \cdot \dfrac{a+3}{5}$

**41.** $\dfrac{x-2}{2} \cdot \dfrac{2x}{x-2}$

**42.** $\dfrac{y+3}{y} \cdot \dfrac{3y}{y+3}$

**43.** $\dfrac{x+5}{5} \cdot \dfrac{x}{x+5}$

**44.** $\dfrac{y-9}{y+9} \cdot \dfrac{y}{9}$

**45.** $\dfrac{(x+1)^2}{x+1} \cdot \dfrac{x+2}{x+1}$

**46.** $\dfrac{(y-3)^2}{y-3} \cdot \dfrac{y-3}{y-3}$

**47.** $\dfrac{2x+6}{x+3} \cdot \dfrac{3}{4x}$

**48.** $\dfrac{3y-9}{y-3} \cdot \dfrac{y}{3y^2}$

**49.** $\dfrac{x^2-x}{x} \cdot \dfrac{3x-6}{3x-3}$

**50.** $\dfrac{5z-10}{z+2} \cdot \dfrac{3}{3z-6}$

**51.** $\dfrac{7y-14}{y-2} \cdot \dfrac{x^2}{7x}$

**52.** $\dfrac{y^2+3y}{9} \cdot \dfrac{3x}{y+3}$

**53.** $\dfrac{x^2+x-6}{5x} \cdot \dfrac{5x-10}{x+3}$

**54.** $\dfrac{z^2+4z-5}{5z-5} \cdot \dfrac{5z}{z+5}$

**55.** $\dfrac{m^2-2m-3}{2m+4} \cdot \dfrac{m^2-4}{m^2+3m+2}$

**56.** $\dfrac{p^2-p-6}{3p-9} \cdot \dfrac{p^2-9}{p^2+6p+9}$

**57.** $\dfrac{x^2+7xy+12y^2}{x^2+2xy-8y^2} \cdot \dfrac{x^2-xy-2y^2}{x^2+4xy+3y^2}$

**58.** $\dfrac{m^2+9mn+20n^2}{m^2-25n^2} \cdot \dfrac{m^2-9mn+20n^2}{m^2-16n^2}$

**59.** $\dfrac{abc^2}{a+1} \cdot \dfrac{c}{a^2b^2} \cdot \dfrac{a^2+a}{ac}$

**60.** $\dfrac{x^3yz^2}{4x+8} \cdot \dfrac{x^2-4}{2x^2y^2z^2} \cdot \dfrac{8yz}{x-2}$

**61.** $\dfrac{3x^2+5x+2}{x^2-9} \cdot \dfrac{x-3}{x^2-4} \cdot \dfrac{x^2+5x+6}{6x+4}$

**62.** $\dfrac{x^2-25}{3x+6} \cdot \dfrac{x^2+x-2}{2x+10} \cdot \dfrac{6x}{3x^2-18x+15}$

*In Exercises 63–92, do each division. Simplify answers when possible.*

**63.** $\dfrac{1}{3} \div \dfrac{1}{2}$

**64.** $\dfrac{3}{4} \div \dfrac{1}{3}$

**65.** $\dfrac{21}{14} \div \dfrac{5}{2}$

**66.** $\dfrac{14}{3} \div \dfrac{10}{3}$

**67.** $\dfrac{2}{y} \div \dfrac{4}{3}$

**68.** $\dfrac{3}{a} \div \dfrac{a}{9}$

**69.** $\dfrac{3x}{2} \div \dfrac{x}{2}$

**70.** $\dfrac{y}{6} \div \dfrac{2}{3y}$

**71.** $\dfrac{3x}{y} \div \dfrac{2x}{4}$

**72.** $\dfrac{3y}{8} \div \dfrac{2y}{4y}$

**73.** $\dfrac{4x}{3x} \div \dfrac{2y}{9y}$

**74.** $\dfrac{14}{7y} \div \dfrac{10}{5z}$

**75.** $\dfrac{x^2}{3} \div \dfrac{2x}{4}$

**76.** $\dfrac{z^2}{z} \div \dfrac{z}{3z}$

**77.** $\dfrac{x^2y}{3xy} \div \dfrac{xy^2}{6y}$

**78.** $\dfrac{2xz}{z} \div \dfrac{4x^2}{z^2}$

**79.** $\dfrac{x+2}{3x} \div \dfrac{x+2}{2}$

**80.** $\dfrac{z-3}{3z} \div \dfrac{z+3}{z}$

**81.** $\dfrac{(z-2)^2}{3z^2} \div \dfrac{z-2}{6z}$

**82.** $\dfrac{(x+7)^2}{x+7} \div \dfrac{(x-3)^2}{x+7}$

**83.** $\dfrac{(z-7)^2}{z+2} \div \dfrac{z(z-7)}{5z^2}$

**84.** $\dfrac{y(y+2)}{y^2(y-3)} \div \dfrac{y^2(y+2)}{(y-3)^2}$

**85.** $\dfrac{x^2-4}{3x+6} \div \dfrac{x-2}{x+2}$

**86.** $\dfrac{x^2-9}{5x+15} \div \dfrac{x-3}{x+3}$

**87.** $\dfrac{x^2-1}{3x-3} \div \dfrac{x+1}{3}$

**88.** $\dfrac{x^2-16}{x-4} \div \dfrac{3x+12}{x}$

**89.** $\dfrac{5x^2+13x-6}{x+3} \div \dfrac{5x^2-17x+6}{x-2}$

**90.** $\dfrac{x^2-x-6}{2x^2+9x+10} \div \dfrac{x^2-25}{2x^2+15x+25}$

**91.** $\dfrac{2x^2+8x-42}{x-3} \div \dfrac{2x^2+14x}{x^2+5x}$

**92.** $\dfrac{x^2-2x-35}{3x^2+27x} \div \dfrac{x^2+7x+10}{6x^2+12x}$

*In Exercises 93–106, do the operations.*

**93.** $\dfrac{x}{3} \cdot \dfrac{9}{4} \div \dfrac{x^2}{6}$

**94.** $\dfrac{y^2}{2} \div \dfrac{4}{y} \cdot \dfrac{y^2}{8}$

**95.** $\dfrac{x^2}{18} \div \dfrac{x^3}{6} \div \dfrac{12}{x^2}$

**96.** $\dfrac{y^3}{3y} \cdot \dfrac{3y^2}{4} \div \dfrac{15}{20}$

**97.** $\dfrac{x^2-1}{x^2-9} \cdot \dfrac{x+3}{x+2} \div \dfrac{5}{x+2}$

**98.** $\dfrac{2}{3x-3} \div \dfrac{2x+2}{x-1} \cdot \dfrac{5}{x+1}$

**99.** $\dfrac{x^2-4}{2x+6} \div \dfrac{x+2}{4} \cdot \dfrac{x+3}{x-2}$

**100.** $\dfrac{x^2-5x}{x+1} \cdot \dfrac{x+1}{x^2+3x} \div \dfrac{x-5}{x-3}$

**101.** $\dfrac{x-x^2}{x^2-4} \left( \dfrac{2x+4}{x+2} \div \dfrac{5}{x+2} \right)$

**102.** $\dfrac{2}{3x-3} \div \left( \dfrac{2x+2}{x-1} \cdot \dfrac{5}{x+1} \right)$

**103.** $\dfrac{y^2}{x+1} \cdot \dfrac{x^2+2x+1}{x^2-1} \div \dfrac{3y}{xy-y}$

**104.** $\dfrac{x^2-y^2}{x^4-x^3} \div \dfrac{x-y}{x^2} \div \dfrac{x^2+2xy+y^2}{x+y}$

**105.** $\dfrac{x^2+x-6}{x^2-4} \cdot \dfrac{x^2+2x}{x-2} \div \dfrac{x^2+3x}{x+2}$

**106.** $\dfrac{x^2-x-6}{x^2+6x-7} \cdot \dfrac{x^2+x-2}{x^2+2x} \div \dfrac{x^2+7x}{x^2-3x}$

## WRITING

**107.** Explain how to multiply two fractions and how to simplify the result.

**108.** Explain why any mathematical expression can be written as a fraction.

**109.** To divide fractions, you must first know how to multiply fractions. Explain.

**110.** Explain how to do the division $\dfrac{a}{b} \div \dfrac{c}{d} \div \dfrac{e}{f}$.

## SOMETHING TO THINK ABOUT

**111.** Let $x$ equal a number of your choosing. Without simplifying first, use a calculator to evaluate

$$\frac{x^2 + x - 6}{x^2 + 3x} \cdot \frac{x^2}{x - 2}$$

Try again, with a different value of $x$. If you were to simplify the expression, what do you think you would get?

**112.** Simplify the expression in Exercise 111 to determine whether your guess was correct.

## 6.5   Adding and Subtracting Fractions

■ ADDING FRACTIONS WITH LIKE DENOMINATORS ■ SUBTRACTING FRACTIONS WITH LIKE DENOMINATORS ■ COMBINED OPERATIONS ■ THE LCD ■ ADDING FRACTIONS WITH UNLIKE DENOMINATORS ■ SUBTRACTING FRACTIONS WITH UNLIKE DENOMINATORS ■ COMBINED OPERATIONS

Getting Ready     *Add the fractions and simplify.*

**1.** $\dfrac{1}{5} + \dfrac{3}{5}$     **2.** $\dfrac{3}{7} + \dfrac{4}{7}$     **3.** $\dfrac{3}{8} + \dfrac{4}{8}$     **4.** $\dfrac{18}{19} + \dfrac{20}{19}$

*Subtract the fractions and simplify.*

**5.** $\dfrac{5}{9} - \dfrac{4}{9}$     **6.** $\dfrac{7}{12} - \dfrac{1}{12}$     **7.** $\dfrac{7}{13} - \dfrac{9}{13}$     **8.** $\dfrac{20}{10} - \dfrac{7}{10}$

## ■ ADDING FRACTIONS WITH LIKE DENOMINATORS

To add fractions with a common denominator, we add their numerators and keep the common denominator. For example,

$$\frac{2x}{7} + \frac{3x}{7} = \frac{2x + 3x}{7} \qquad \text{Add the numerators and keep the common denominator.}$$

$$= \frac{5x}{7} \qquad 2x + 3x = 5x.$$

In general, we have the following result.

## Adding Fractions with Like Denominators

If $a$, $b$, and $d$ represent real numbers, then

$$\frac{a}{d} + \frac{b}{d} = \frac{a+b}{d} \quad (d \neq 0)$$

**EXAMPLE 1**  Do each addition.

**a.** $\dfrac{xy}{8z} + \dfrac{3xy}{8z} = \dfrac{xy + 3xy}{8z}$    Add the numerators and keep the common denominator.

$\qquad\qquad = \dfrac{4xy}{8z}$    Combine like terms.

$\qquad\qquad = \dfrac{xy}{2z}$    $\frac{4xy}{8z} = \frac{4xy}{4 \cdot 2z} = \frac{xy}{2z}$, because $\frac{4}{4} = 1$.

**b.** $\dfrac{3x+y}{5x} + \dfrac{x+y}{5x} = \dfrac{3x+y+x+y}{5x}$    Add the numerators and keep the common denominator.

$\qquad\qquad = \dfrac{4x+2y}{5x}$    Combine like terms. ∎

**Self Check**   Add   **a.** $\frac{x}{7} + \frac{y}{7}$ and **b.** $\frac{3x}{7y} + \frac{4x}{7y}$.

**Answers**   **a.** $\frac{x+y}{7}$, **b.** $\frac{x}{y}$

**EXAMPLE 2**  Add $\dfrac{3x+21}{5x+10} + \dfrac{8x+1}{5x+10}$.

*Solution*   Because the fractions have the same denominator, we add their numerators and keep the common denominator.

$$\frac{3x+21}{5x+10} + \frac{8x+1}{5x+10} = \frac{3x+21+8x+1}{5x+10} \quad \text{Add the fractions.}$$

$$= \frac{11x+22}{5x+10} \quad \text{Combine like terms.}$$

$$= \frac{11(x+2)}{5(x+2)} \quad \text{Factor and divide out the common factor of } x+2.$$

$$= \frac{11}{5} \quad ∎$$

**Self Check**   Add $\frac{x+4}{6x-12} + \frac{x-8}{6x-12}$.

**Answer**   $\frac{1}{3}$

## ■ SUBTRACTING FRACTIONS WITH LIKE DENOMINATORS

To subtract fractions with a common denominator, we subtract their numerators and keep the common denominator.

> **Subtracting Fractions with Like Denominators**
> If $a$, $b$, and $d$ represent real numbers, then
> $$\frac{a}{d} - \frac{b}{d} = \frac{a-b}{d} \quad (d \neq 0)$$

**EXAMPLE 3**    Subtract   **a.** $\dfrac{5x}{3} - \dfrac{2x}{3}$  and  **b.** $\dfrac{5x+1}{x-3} - \dfrac{4x-2}{x-3}$.

*Solution*    In each part, the fractions have the same denominator. To subtract them, we subtract their numerators and keep the common denominator.

**a.** $\dfrac{5x}{3} - \dfrac{2x}{3} = \dfrac{5x-2x}{3}$    Subtract the fractions.

$\qquad\qquad = \dfrac{3x}{3}$    Combine like terms.

$\qquad\qquad = \dfrac{x}{1}$    $\frac{3}{3} = 1$.

$\qquad\qquad = x$    Denominators of 1 need not be written.

**b.** $\dfrac{5x+1}{x-3} - \dfrac{4x-2}{x-3} = \dfrac{(5x+1)-(4x-2)}{x-3}$    Subtract the fractions.

$\qquad\qquad\qquad = \dfrac{5x+1-4x+2}{x-3}$    Remove parentheses.

$\qquad\qquad\qquad = \dfrac{x+3}{x-3}$    Combine like terms.    ■

**Self Check**    Subtract $\dfrac{2y+1}{y+5} - \dfrac{y-4}{y+5}$.

**Answer**    1

## ■ COMBINED OPERATIONS

To add and/or subtract three or more fractions, we follow the rules for order of operations.

**EXAMPLE 4**    Simplify $\dfrac{3x+1}{x-7} - \dfrac{5x+2}{x-7} + \dfrac{2x+1}{x-7}$.

*Solution*     This example involves both addition and subtraction of fractions. Unless parentheses indicate otherwise, we do additions and subtractions from left to right.

$$\frac{3x + 1}{x - 7} - \frac{5x + 2}{x - 7} + \frac{2x + 1}{x - 7}$$

$$= \frac{(3x + 1) - (5x + 2) + (2x + 1)}{x - 7} \qquad \text{Combine the numerators and keep the common denominator.}$$

$$= \frac{3x + 1 - 5x - 2 + 2x + 1}{x - 7} \qquad \text{Remove parentheses.}$$

$$= \frac{0}{x - 7} \qquad \text{Combine like terms.}$$

$$= 0 \qquad \text{Simplify.} \qquad \blacksquare$$

*Self Check*     Simplify $\frac{2a - 3}{a - 5} + \frac{3a + 2}{a - 5} - \frac{24}{a - 5}$.

*Answer*     5

Example 4 illustrates that if the numerator of a fraction is 0, its value is 0.

## ■ THE LCD

Since the denominators of the fractions in the addition $\frac{4}{7} + \frac{3}{5}$ are different, we cannot add the fractions in their present form.

four-sevenths     +     three-fifths

└── Different denominators ──┘

To add these fractions, we need to find a common denominator. The smallest common denominator (called the **least** or **lowest common denominator**) is the easiest one to work with.

> **Least Common Denominator**
> The **least common denominator** (**LCD**) for a set of fractions is the smallest number that each denominator will divide exactly.

In the addition $\frac{4}{7} + \frac{3}{5}$, the denominators are 7 and 5. The smallest number that 7 and 5 will divide evenly is 35. This is the LCD. We now build each fraction into a fraction with a denominator of 35.

$$\frac{4}{7} + \frac{3}{5} = \frac{4 \cdot 5}{7 \cdot 5} + \frac{3 \cdot 7}{5 \cdot 7} \qquad \text{Multiply numerator and denominator of } \frac{4}{7} \text{ by 5, and multiply numerator and denominator of } \frac{3}{5} \text{ by 7.}$$

$$= \frac{20}{35} + \frac{21}{35} \qquad \text{Do the multiplications.}$$

Now that the fractions have a common denominator, we can add them.

$$\frac{20}{35} + \frac{21}{35} = \frac{20 + 21}{35} = \frac{41}{35}$$

**EXAMPLE 5** Change **a.** $\dfrac{1}{2y}$, **b.** $\dfrac{3y}{5}$, and **c.** $\dfrac{7x}{10y}$ into fractions with a common denominator of $30y$.

*Solution* To build each fraction, we multiply the numerator and denominator by what it takes to make the denominator $30y$.

**a.** $\dfrac{1}{2y} = \dfrac{1 \cdot 15}{2y \cdot 15} = \dfrac{15}{30y}$

**b.** $\dfrac{3y}{5} = \dfrac{3y \cdot 6y}{5 \cdot 6y} = \dfrac{18y^2}{30y}$

**c.** $\dfrac{7x}{10y} = \dfrac{7x \cdot 3}{10y \cdot 3} = \dfrac{21x}{30y}$ ∎

*Self Check* Change $\frac{5a}{6b}$ into a fraction with a denominator of $30ab$.

*Answer* $\dfrac{25a^2}{30ab}$

There is a process that we can use to find the least common denominator of several fractions.

> **Finding the Least Common Denominator (LCD)**
>
> **1.** List the different denominators that appear in the fractions.
>
> **2.** Completely factor each denominator.
>
> **3.** Form a product using each different factor obtained in Step 2. Use each different factor the *greatest* number of times it appears in any one factorization. The product formed by multiplying these factors is the LCD.

**EXAMPLE 6** Find the LCD of $\dfrac{5a}{24b}$, $\dfrac{11a}{18b}$, and $\dfrac{35a}{36b}$.

*Solution* We list and factor each denominator into the product of prime numbers.

$$24b = 2 \cdot 2 \cdot 2 \cdot 3 \cdot b = 2^3 \cdot 3 \cdot b$$
$$18b = 2 \cdot 3 \cdot 3 \cdot b = 2 \cdot 3^2 \cdot b$$
$$36b = 2 \cdot 2 \cdot 3 \cdot 3 \cdot b = 2^2 \cdot 3^2 \cdot b$$

We then form a product with factors of 2, 3, and $b$. To find the LCD, we use each of these factors the greatest number of times it appears in any one factorization. We use 2 three times, because it appears three times as a factor of 24. We use 3 twice, because it occurs twice as a factor of 18 and 36. We use $b$ once.

$$\text{LCD} = 2 \cdot 2 \cdot 2 \cdot 3 \cdot 3 \cdot b$$
$$= 8 \cdot 9 \cdot b$$
$$= 72b$$ ∎

**Self Check**

Find the LCD of $\frac{3y}{28z}$ and $\frac{5x}{21z}$.

**Answer**    $84z$

## ■ ADDING FRACTIONS WITH UNLIKE DENOMINATORS

The following list of steps summarizes how to add fractions that have unlike denominators.

> **Adding Fractions with Unlike Denominators**
> To add fractions with different denominators,
> **1.** Find the LCD.
> **2.** Write each fraction as a fraction with a denominator that is the LCD.
> **3.** Add the resulting fractions and simplify the result, if possible.

To add $\frac{4x}{7}$ and $\frac{3x}{5}$, we first find the LCD, which is 35. We then build the fractions so that each one has a denominator of 35. Finally, we add the resulting fractions.

$$\frac{4x}{7} + \frac{3x}{5} = \frac{4x \cdot 5}{7 \cdot 5} + \frac{3x \cdot 7}{5 \cdot 7}$$    Multiply numerator and denominator of $\frac{4x}{7}$ by 5 and numerator and denominator of $\frac{3x}{5}$ by 7.

$$= \frac{20x}{35} + \frac{21x}{35}$$    Do the multiplications.

$$= \frac{41x}{35}$$    Add the numerators and keep the common denominator.

**EXAMPLE 7**    Add $\dfrac{5a}{24b}, \dfrac{11a}{18b},$ and $\dfrac{35a}{36b}.$

*Solution*  In Example 6, we saw that the LCD of these fractions is $2 \cdot 2 \cdot 2 \cdot 3 \cdot 3 \cdot b = 72b$. To add the fractions, we first factor each denominator:

$$\frac{5a}{24b} + \frac{11a}{18b} + \frac{35a}{36b} = \frac{5a}{2 \cdot 2 \cdot 2 \cdot 3 \cdot b} + \frac{11a}{2 \cdot 3 \cdot 3 \cdot b} + \frac{35a}{2 \cdot 2 \cdot 3 \cdot 3 \cdot b}$$

In each resulting fraction, we multiply the numerator and the denominator by whatever it takes to build the denominator to the lowest common denominator of $2 \cdot 2 \cdot 2 \cdot 3 \cdot 3 \cdot b$.

$$= \frac{5a \cdot 3}{2 \cdot 2 \cdot 2 \cdot 3 \cdot b \cdot 3} + \frac{11a \cdot 2 \cdot 2}{2 \cdot 3 \cdot 3 \cdot b \cdot 2 \cdot 2} + \frac{35a \cdot 2}{2 \cdot 2 \cdot 3 \cdot 3 \cdot b \cdot 2}$$

$$= \frac{15a + 44a + 70a}{72b} \qquad \text{Do the multiplications and add the fractions.}$$

$$= \frac{129a}{72b} \qquad \text{Simplify.} \qquad \blacksquare$$

**Self Check**

**Answer**

Add $\frac{3y}{28z} + \frac{5x}{21z}$.

$\frac{9y + 20x}{84z}$

**EXAMPLE 8**  Add $\dfrac{5y}{14x} + \dfrac{2y}{21x}$.

*Solution*  We first find the LCD.

$$\left.\begin{array}{l} 14x = 2 \cdot 7 \cdot x \\ 21x = 3 \cdot 7 \cdot x \end{array}\right\} \text{LCD} = 2 \cdot 3 \cdot 7 \cdot x = 42x$$

We then build the fractions so that each one has a denominator of $42x$.

$$\frac{5y}{14x} + \frac{2y}{21x} = \frac{5y \cdot 3}{14x \cdot 3} + \frac{2y \cdot 2}{21x \cdot 2} \qquad \begin{array}{l} \text{Multiply the numerator and denominator} \\ \text{of } \frac{5y}{14x} \text{ by 3 and those of } \frac{2y}{21x} \text{ by 2.} \end{array}$$

$$= \frac{15y}{42x} + \frac{4y}{42x} \qquad \text{Do the multiplications.}$$

$$= \frac{19y}{42x} \qquad \text{Add the fractions.} \qquad \blacksquare$$

**Self Check**

**Answer**

Add $\frac{3y}{4x} + \frac{2y}{3x}$.

$\frac{17y}{12x}$

**EXAMPLE 9**   Add $\dfrac{1}{x} + \dfrac{x}{y}$.

*Solution*   By inspection, the LCD is $xy$.

$$\dfrac{1}{x} + \dfrac{x}{y} = \dfrac{1(y)}{x(y)} + \dfrac{(x)x}{(x)y}$$   Build the fractions to get the common denominator of $xy$.

$$= \dfrac{y}{xy} + \dfrac{x^2}{xy}$$   Do the multiplications.

$$= \dfrac{y + x^2}{xy}$$   Add the fractions. ∎

*Self Check*   Add $\dfrac{a}{b} + \dfrac{3}{a}$.

*Answer*   $\dfrac{a^2 + 3b}{ab}$

## ■ SUBTRACTING FRACTIONS WITH UNLIKE DENOMINATORS

To subtract fractions with unlike denominators, we first change them into fractions with the same denominator.

**EXAMPLE 10**   Subtract $\dfrac{x}{x + 1} - \dfrac{3}{x}$.

*Solution*   By inspection, the least common denominator is $(x + 1)x$.

$$\dfrac{x}{x + 1} - \dfrac{3}{x} = \dfrac{x(x)}{(x + 1)x} - \dfrac{3(x + 1)}{x(x + 1)}$$   Build the fractions to get the common denominator.

$$= \dfrac{x(x) - 3(x + 1)}{(x + 1)x}$$   Subtract the numerators and keep the common denominator.

$$= \dfrac{x^2 - 3x - 3}{(x + 1)x}$$   Do the multiplications in the numerator. ∎

*Self Check*   Subtract $\dfrac{a}{a - 1} - \dfrac{5}{b}$.

*Answer*   $\dfrac{ab - 5a + 5}{(a - 1)b}$

**EXAMPLE 11**   Subtract $\dfrac{a}{a - 1} - \dfrac{2}{a^2 - 1}$.

*Solution* We factor $a^2 - 1$ and discover that the LCD is $(a + 1)(a - 1)$.

$$\frac{a}{a - 1} - \frac{2}{a^2 - 1}$$

$$= \frac{a(a + 1)}{(a - 1)(a + 1)} - \frac{2}{(a + 1)(a - 1)}$$ 
Build the first fraction and factor the denominator of the second fraction.

$$= \frac{a(a + 1) - 2}{(a - 1)(a + 1)}$$ 
Subtract the numerators and keep the common denominator.

$$= \frac{a^2 + a - 2}{(a - 1)(a + 1)}$$ 
Remove parentheses.

$$= \frac{(a + 2)\overset{1}{\cancel{(a - 1)}}}{\underset{1}{\cancel{(a - 1)}}(a + 1)}$$ 
Factor.

$$= \frac{a + 2}{a + 1}$$ 
Divide out the common factor of $a - 1$. ∎

**Self Check** Subtract $\dfrac{b}{b + 1} - \dfrac{3}{b^2 - 1}$.

**Answer** $\dfrac{b^2 - b - 3}{(b + 1)(b - 1)}$

**EXAMPLE 12** Subtract $\dfrac{3}{x - y} - \dfrac{x}{y - x}$.

*Solution* We note that the second denominator is the negative of the first. So we can multiply the numerator and denominator of the second fraction by $-1$ to get

$$\frac{3}{x - y} - \frac{x}{y - x} = \frac{3}{x - y} - \frac{-1x}{-1(y - x)}$$ 
Multiply numerator and denominator by $-1$.

$$= \frac{3}{x - y} - \frac{-x}{-y + x}$$ 
Remove parentheses.

$$= \frac{3}{x - y} - \frac{-x}{x - y}$$ 
$-y + x = x - y$.

$$= \frac{3 - (-x)}{x - y}$$ 
Subtract the numerators and keep the common denominator.

$$= \frac{3 + x}{x - y}$$ 
$-(-x) = x$. ∎

**Self Check** Subtract $\dfrac{5}{a - b} - \dfrac{2}{b - a}$.

**Answer** $\dfrac{7}{a - b}$

### ■ COMBINED OPERATIONS

To add and/or subtract three or more fractions, we follow the rules for order of operations.

**EXAMPLE 13**   Do the operations: $\dfrac{3}{x^2y} + \dfrac{2}{xy} - \dfrac{1}{xy^2}$.

*Solution*   Find the least common denominator.

$$\left.\begin{array}{l} x^2y = x \cdot x \cdot y \\ xy = x \cdot y \\ xy^2 = x \cdot y \cdot y \end{array}\right\} \qquad \text{Factor each denominator.}$$

In any one of these denominators, the factor $x$ occurs at most twice, and the factor $y$ occurs at most twice. Thus,

$$\begin{aligned} \text{LCD} &= x \cdot x \cdot y \cdot y \\ &= x^2y^2 \end{aligned}$$

We build each fraction into a fraction with a denominator of $x^2y^2$.

$$\begin{aligned} &\frac{3}{x^2y} + \frac{2}{xy} - \frac{1}{xy^2} \\ &= \frac{3 \cdot y}{x \cdot x \cdot y \cdot y} + \frac{2 \cdot x \cdot y}{x \cdot y \cdot x \cdot y} - \frac{1 \cdot x}{x \cdot y \cdot y \cdot x} \qquad \begin{array}{l}\text{Factor each denominator}\\\text{and build each fraction.}\end{array} \\ &= \frac{3y + 2xy - x}{x^2y^2} \qquad \begin{array}{l}\text{Do the multiplications and}\\\text{combine the numerators.}\end{array} \quad ■ \end{aligned}$$

*Self Check*   Combine $\dfrac{5}{ab^2} - \dfrac{b}{a} + \dfrac{a}{b}$.

*Answer*   $\dfrac{5 - b^3 + a^2b}{ab^2}$

**EXAMPLE 14**   Do the operations: $\dfrac{3}{x^2 - y^2} + \dfrac{2}{x - y} - \dfrac{1}{x + y}$.

*Solution*   Find the least common denominator.

$$\left.\begin{array}{l} x^2 - y^2 = (x - y)(x + y) \\ x - y = x - y \\ x + y = x + y \end{array}\right\} \qquad \text{Factor each denominator, where possible.}$$

Since the least common denominator is $(x - y)(x + y)$, we build each fraction into a new fraction with that common denominator.

$$\frac{3}{x^2 - y^2} + \frac{2}{x - y} - \frac{1}{x + y}$$

$$= \frac{3}{(x - y)(x + y)} + \frac{2}{x - y} - \frac{1}{x + y} \qquad \text{Factor.}$$

$$= \frac{3}{(x - y)(x + y)} + \frac{2(x + y)}{(x - y)(x + y)} - \frac{1(x - y)}{(x + y)(x - y)} \qquad \text{Build each fraction to get a common denominator.}$$

$$= \frac{3 + 2(x + y) - 1(x - y)}{(x - y)(x + y)} \qquad \text{Combine the numerators and keep the common denominator.}$$

$$= \frac{3 + 2x + 2y - x + y}{(x - y)(x + y)} \qquad \text{Remove parentheses.}$$

$$= \frac{3 + x + 3y}{(x - y)(x + y)} \qquad \text{Combine like terms.} \quad \blacksquare$$

**Self Check**   Combine $\dfrac{5}{a^2 - b^2} - \dfrac{3}{a + b} + \dfrac{4}{a - b}$.

**Answer**   $\frac{a + 7b + 5}{(a + b)(a - b)}$

Orals   *Indicate whether the fractions are equal.*

**1.** $\dfrac{1}{2}, \dfrac{6}{12}$     **2.** $\dfrac{3}{8}, \dfrac{15}{40}$     **3.** $\dfrac{7}{9}, \dfrac{14}{27}$     **4.** $\dfrac{5}{10}, \dfrac{15}{30}$

**5.** $\dfrac{x}{3}, \dfrac{3x}{9}$     **6.** $\dfrac{5}{3}, \dfrac{5x}{3y}$     **7.** $\dfrac{5}{3}, \dfrac{5x}{3x}$     **8.** $\dfrac{5y}{10}, \dfrac{y}{2}$

## EXERCISE 6.5

***REVIEW***   *Write each number in prime-factored form.*

**1.** 49       **2.** 64       **3.** 136       **4.** 242

**5.** 102      **6.** 315      **7.** 144      **8.** 145

***VOCABULARY AND CONCEPTS***   *Fill in each blank to make a true statement.*

**9.** The _____ for a set of fractions is the smallest number that each denominator divides exactly.

**10.** When we multiply the numerator and denominator of a fraction by some number to get a common denominator, we say that we are _____ the fraction.

**11.** To add two fractions with like denominators, we add their _____ and keep the _____.

**12.** To subtract two fractions with _____ denominators, we need to find a common denominator.

***PRACTICE***   *In Exercises 13–24, do each addition. Simplify answers, if possible.*

**13.** $\dfrac{1}{3} + \dfrac{1}{3}$

**14.** $\dfrac{3}{4} + \dfrac{3}{4}$

**15.** $\dfrac{2}{9} + \dfrac{1}{9}$

**16.** $\dfrac{5}{7} + \dfrac{9}{7}$

**17.** $\dfrac{2x}{y} + \dfrac{2x}{y}$

**18.** $\dfrac{4y}{3x} + \dfrac{2y}{3x}$

**19.** $\dfrac{4}{7y} + \dfrac{10}{7y}$

**20.** $\dfrac{x^2}{4y} + \dfrac{x^2}{4y}$

**21.** $\dfrac{y+2}{5z} + \dfrac{y+4}{5z}$

**22.** $\dfrac{x+3}{x^2} + \dfrac{x+5}{x^2}$

**23.** $\dfrac{3x-5}{x-2} + \dfrac{6x-13}{x-2}$

**24.** $\dfrac{8x-7}{x+3} + \dfrac{2x+37}{x+3}$

*In Exercises 25–36, do each subtraction. Simplify answers, if possible.*

**25.** $\dfrac{5}{7} - \dfrac{4}{7}$

**26.** $\dfrac{5}{9} - \dfrac{3}{9}$

**27.** $\dfrac{35}{72} - \dfrac{44}{72}$

**28.** $\dfrac{35}{99} - \dfrac{13}{99}$

**29.** $\dfrac{2x}{y} - \dfrac{x}{y}$

**30.** $\dfrac{7y}{5} - \dfrac{4y}{5}$

**31.** $\dfrac{9y}{3x} - \dfrac{6y}{3x}$

**32.** $\dfrac{5r^2}{2r} - \dfrac{r^2}{2r}$

**33.** $\dfrac{6x-5}{3xy} - \dfrac{3x-5}{3xy}$

**34.** $\dfrac{7x+7}{5y} - \dfrac{2x+7}{5y}$

**35.** $\dfrac{3y-2}{y+3} - \dfrac{2y-5}{y+3}$

**36.** $\dfrac{5x+8}{x+5} - \dfrac{3x-2}{x+5}$

*In Exercises 37–44, do the operations. Simplify answers, if possible.*

**37.** $\dfrac{13x}{15} + \dfrac{12x}{15} - \dfrac{5x}{15}$

**38.** $\dfrac{13y}{32} + \dfrac{13y}{32} - \dfrac{10y}{32}$

**39.** $\dfrac{x}{3y} + \dfrac{2x}{3y} - \dfrac{x}{3y}$

**40.** $\dfrac{5y}{8x} + \dfrac{4y}{8x} - \dfrac{y}{8x}$

**41.** $\dfrac{3x}{y+2} - \dfrac{3y}{y+2} + \dfrac{x+y}{y+2}$

**42.** $\dfrac{3y}{x-5} + \dfrac{x}{x-5} - \dfrac{y-x}{x-5}$

**43.** $\dfrac{x+1}{x-2} - \dfrac{2(x-3)}{x-2} + \dfrac{3(x+1)}{x-2}$

**44.** $\dfrac{3xy}{x-y} - \dfrac{x(3y-x)}{x-y} - \dfrac{x(x-y)}{x-y}$

*In Exercises 45–56, build each fraction into an equivalent fraction with the indicated denominator.*

**45.** $\dfrac{25}{4}$; 20

**46.** $\dfrac{5}{y}$; $xy$

**47.** $\dfrac{8}{x}$; $x^2y$

**48.** $\dfrac{7}{y}$; $xy^2$

**49.** $\dfrac{3x}{x+1}$; $(x+1)^2$

**50.** $\dfrac{5y}{y-2}$; $(y-2)^2$

**51.** $\dfrac{2y}{x}$; $x^2+x$

**52.** $\dfrac{3x}{y}$; $y^2-y$

**53.** $\dfrac{z}{z-1}; z^2-1$  **54.** $\dfrac{y}{y+2}; y^2-4$  **55.** $\dfrac{2}{x+1}; x^2+3x+2$  **56.** $\dfrac{3}{x-1}; x^2+x-2$

*In Exercises 57–66, several denominators are given. Find the LCD.*

**57.** $2x, 6x$  
**58.** $3y, 9y$  
**59.** $3x, 6y, 9xy$  
**60.** $2x^2, 6y, 3xy$  
**61.** $x^2-1, x+1$  
**62.** $y^2-9, y-3$  
**63.** $x^2+6x, x+6, x$  
**64.** $xy^2-xy, xy, y-1$  
**65.** $x^2-4x-5, x^2-25$  
**66.** $x^2-x-6, x^2-9$

*In Exercises 67–96, do the operations. Simplify answers, if possible.*

**67.** $\dfrac{1}{2}+\dfrac{2}{3}$  **68.** $\dfrac{2}{3}-\dfrac{5}{6}$  **69.** $\dfrac{2y}{9}+\dfrac{y}{3}$  **70.** $\dfrac{8a}{15}-\dfrac{5a}{12}$

**71.** $\dfrac{21x}{14}-\dfrac{5x}{21}$  **72.** $\dfrac{7y}{6}+\dfrac{10y}{9}$  **73.** $\dfrac{4x}{3}+\dfrac{2x}{y}$  **74.** $\dfrac{2y}{5x}-\dfrac{y}{2}$

**75.** $\dfrac{2}{x}-3x$  **76.** $14+\dfrac{10}{y^2}$  **77.** $\dfrac{y+2}{5y}+\dfrac{y+4}{15y}$  **78.** $\dfrac{x+3}{x^2}+\dfrac{x+5}{2x}$

**79.** $\dfrac{x+5}{xy}-\dfrac{x-1}{x^2y}$  **80.** $\dfrac{y-7}{y^2}-\dfrac{y+7}{2y}$  **81.** $\dfrac{x}{x+1}+\dfrac{x-1}{x}$  **82.** $\dfrac{3x}{xy}+\dfrac{x+1}{y-1}$

**83.** $\dfrac{x-1}{x}+\dfrac{y+1}{y}$  **84.** $\dfrac{a+2}{b}+\dfrac{b-2}{a}$  **85.** $\dfrac{x}{x-2}+\dfrac{4+2x}{x^2-4}$  **86.** $\dfrac{y}{y+3}-\dfrac{2y-6}{y^2-9}$

**87.** $\dfrac{x+1}{x-1}+\dfrac{x-1}{x+1}$  **88.** $\dfrac{2x}{x+2}+\dfrac{x+1}{x-3}$  **89.** $\dfrac{2x+2}{x-2}-\dfrac{2x}{2-x}$  **90.** $\dfrac{y+3}{y-1}-\dfrac{y+4}{1-y}$

**91.** $\dfrac{2x}{x^2-3x+2}+\dfrac{2x}{x-1}-\dfrac{x}{x-2}$  **92.** $\dfrac{4a}{a-2}-\dfrac{3a}{a-3}+\dfrac{4a}{a^2-5a+6}$

**93.** $\dfrac{2x}{x-1}+\dfrac{3x}{x+1}-\dfrac{x+3}{x^2-1}$  **94.** $\dfrac{a}{a-1}-\dfrac{2}{a+2}+\dfrac{3(a-2)}{a^2+a-2}$

**95.** $\dfrac{x+1}{2x+4}-\dfrac{x^2}{2x^2-8}$  **96.** $\dfrac{x+1}{x+2}-\dfrac{x^2+1}{x^2-x-6}$

## WRITING

**97.** Explain how to add fractions with the same denominator.

**98.** Explain how to subtract fractions with the same denominator.

**99.** Explain how to find a lowest common denominator.

**100.** Explain how to add two fractions with different denominators.

***SOMETHING TO THINK ABOUT***

**101.** Find the mistake:

$$\frac{2x+3}{x+5}-\frac{x+2}{x+5}=\frac{2x+3-x+2}{x+5}$$
$$=\frac{x+5}{x+5}$$
$$=1$$

**102.** Find the mistake:

$$\frac{5x-4}{y}+\frac{x}{y}=\frac{5x-4+x}{y+y}$$
$$=\frac{6x-4}{2y}$$
$$=\frac{3x-2}{y}$$

*In Exercises 103–104, show that each formula is true.*

**103.** $\dfrac{a}{b}+\dfrac{c}{d}=\dfrac{ad+bc}{bd}$

**104.** $\dfrac{a}{b}-\dfrac{c}{d}=\dfrac{ad-bc}{bd}$

## 6.6 Complex Fractions

■ SIMPLIFYING COMPLEX FRACTIONS ■ SIMPLIFYING FRACTIONS WITH TERMS CONTAINING NEGATIVE EXPONENTS

Getting Ready    *Use the distributive property to remove parentheses, and simplify.*

**1.** $3\left(1+\dfrac{1}{3}\right)$    **2.** $10\left(\dfrac{1}{5}-2\right)$    **3.** $4\left(\dfrac{3}{2}+\dfrac{1}{4}\right)$    **4.** $14\left(\dfrac{3}{7}-1\right)$

**5.** $x\left(\dfrac{3}{x}+3\right)$    **6.** $y\left(\dfrac{2}{y}-1\right)$    **7.** $4x\left(3-\dfrac{1}{2x}\right)$    **8.** $6xy\left(\dfrac{1}{2x}+\dfrac{1}{3y}\right)$

■ **SIMPLIFYING COMPLEX FRACTIONS**

Fractions such as

$$\frac{\frac{1}{3}}{4},\quad \frac{\frac{5}{3}}{\frac{2}{9}},\quad \frac{x+\frac{1}{2}}{3-x},\quad\text{and}\quad \frac{\frac{x+1}{2}}{x+\frac{1}{x}}$$

that contain fractions in their numerators or denominators are called **complex fractions.** Complex fractions can often be simplified. For example, we can simplify the complex fraction

$$\frac{\frac{5x}{3}}{\frac{2y}{9}}$$

by doing the division:

$$\frac{\dfrac{5x}{3}}{\dfrac{2y}{9}} = \frac{5x}{3} \div \frac{2y}{9} = \frac{5x}{3} \cdot \frac{9}{2y} = \frac{5x \cdot \overset{1}{\cancel{3}} \cdot 3}{\underset{1}{\cancel{3}} \cdot 2y} = \frac{15x}{2y}$$

There are two methods that we can use to simplify complex fractions.

**Simplifying Complex Fractions**

*Method 1*

Write the numerator and the denominator of the complex fraction as single fractions. Then divide the fractions and simplify.

*Method 2*

Multiply the numerator and denominator of the complex fraction by the LCD of the fractions in its numerator and denominator. Then simplify the results, if possible.

To simplify the complex fraction $\dfrac{\dfrac{3x}{5} + 1}{2 - \dfrac{x}{5}}$ by using Method 1, we proceed as

follows:

$$\frac{\dfrac{3x}{5} + 1}{2 - \dfrac{x}{5}} = \frac{\dfrac{3x}{5} + \dfrac{5}{5}}{\dfrac{10}{5} - \dfrac{x}{5}} \qquad \text{Change 1 to } \tfrac{5}{5} \text{ and 2 to } \tfrac{10}{5}.$$

$$= \frac{\dfrac{3x + 5}{5}}{\dfrac{10 - x}{5}} \qquad \begin{array}{l}\text{Add the fractions in the numerator and} \\ \text{subtract the fractions in the denominator.}\end{array}$$

$$= \frac{3x + 5}{5} \div \frac{10 - x}{5} \qquad \begin{array}{l}\text{Write the complex fraction as an equivalent} \\ \text{division problem.}\end{array}$$

$$= \frac{3x + 5}{5} \cdot \frac{5}{10 - x} \qquad \text{Invert the divisor and multiply.}$$

$$= \frac{(3x + 5)5}{5(10 - x)} \qquad \text{Multiply the fractions.}$$

$$= \frac{3x + 5}{10 - x} \qquad \text{Divide out the common factor of 5; } \tfrac{5}{5} = 1.$$

To use Method 2, we proceed as follows:

$$\frac{\dfrac{3x}{5} + 1}{2 - \dfrac{x}{5}} = \frac{5\left(\dfrac{3x}{5} + 1\right)}{5\left(2 - \dfrac{x}{5}\right)}$$

Multiply both numerator and denominator by 5, the LCD of $\frac{3x}{5}$ and $\frac{x}{5}$.

$$= \frac{5 \cdot \dfrac{3x}{5} + 5 \cdot 1}{5 \cdot 2 - 5 \cdot \dfrac{x}{5}}$$

Remove parentheses.

$$= \frac{3x + 5}{10 - x}$$

Do the multiplications.

In this example, Method 2 is easier than Method 1. Any complex fraction can be simplified by using either method. With practice, you will be able to see which method is best to use in any given situation.

**EXAMPLE 1**    Simplify $\dfrac{\dfrac{x}{3}}{\dfrac{y}{3}}$.

*Solution*        **Method 1**                              **Method 2**

$$\frac{\dfrac{x}{3}}{\dfrac{y}{3}} = \frac{x}{3} \div \frac{y}{3} \qquad\qquad \frac{\dfrac{x}{3}}{\dfrac{y}{3}} = \frac{3\left(\dfrac{x}{3}\right)}{3\left(\dfrac{y}{3}\right)}$$

$$= \frac{x}{3} \cdot \frac{3}{y} \qquad\qquad\qquad = \frac{\dfrac{x}{1}}{\dfrac{y}{1}}$$

$$= \frac{3x}{3y} \qquad\qquad\qquad\qquad = \frac{x}{y}$$

$$= \frac{x}{y}$$

■

Self Check    Simplify $\dfrac{\dfrac{a}{4}}{\dfrac{5}{b}}$.

Answer      $\frac{ab}{20}$

**EXAMPLE 2**    Simplify $\dfrac{\dfrac{x}{x+1}}{\dfrac{y}{x}}$.

Solution

**Method 1**

$$\frac{\dfrac{x}{x+1}}{\dfrac{y}{x}} = \frac{x}{x+1} \div \frac{y}{x}$$

$$= \frac{x}{x+1} \cdot \frac{x}{y}$$

$$= \frac{x^2}{y(x+1)}$$

**Method 2**

$$\frac{\dfrac{x}{x+1}}{\dfrac{y}{x}} = \frac{x(x+1)\left(\dfrac{x}{x+1}\right)}{x(x+1)\left(\dfrac{y}{x}\right)}$$

$$= \frac{\dfrac{x^2}{1}}{\dfrac{y(x+1)}{1}}$$

$$= \frac{x^2}{y(x+1)}$$ ∎

Self Check

Simplify $\dfrac{\frac{x}{y}}{\frac{x}{y+1}}$.

Answer

$\frac{y+1}{y}$

EXAMPLE 3

Simplify $\dfrac{1+\dfrac{1}{x}}{1-\dfrac{1}{x}}$.

Solution

**Method 1**

$$\frac{1+\dfrac{1}{x}}{1-\dfrac{1}{x}} = \frac{\dfrac{x}{x}+\dfrac{1}{x}}{\dfrac{x}{x}-\dfrac{1}{x}}$$

$$= \frac{\dfrac{x+1}{x}}{\dfrac{x-1}{x}}$$

$$= \frac{x+1}{x} \div \frac{x-1}{x}$$

$$= \frac{x+1}{x} \cdot \frac{x}{x-1}$$

$$= \frac{(x+1)x}{x(x-1)}$$

$$= \frac{x+1}{x-1}$$

**Method 2**

$$\frac{1+\dfrac{1}{x}}{1-\dfrac{1}{x}} = \frac{x\left(1+\dfrac{1}{x}\right)}{x\left(1-\dfrac{1}{x}\right)}$$

$$= \frac{x+1}{x-1}$$ ∎

Self Check Simplify $\dfrac{\frac{1}{x} + 1}{\frac{1}{x} - 1}$.

Answer $\dfrac{1 + x}{1 - x}$

EXAMPLE 4 Simplify $\dfrac{1}{1 + \dfrac{1}{x + 1}}$.

*Solution* Use Method 2.

$$\dfrac{1}{1 + \dfrac{1}{x + 1}} = \dfrac{(x + 1) \cdot 1}{(x + 1)\left(1 + \dfrac{1}{x + 1}\right)} \qquad \text{Multiply numerator and denominator by } x + 1.$$

$$= \dfrac{x + 1}{(x + 1)1 + 1} \qquad \text{Simplify.}$$

$$= \dfrac{x + 1}{x + 2} \qquad \text{Simplify.} \qquad \blacksquare$$

Self Check Simplify $\dfrac{2}{\dfrac{1}{x + 2} - 2}$.

Answer $\dfrac{2(x + 2)}{-2x - 3}$

## ■ SIMPLIFYING FRACTIONS WITH TERMS CONTAINING NEGATIVE EXPONENTS

Many fractions with terms containing negative exponents are complex fractions in disguise.

EXAMPLE 5 Simplify $\dfrac{x^{-1} + y^{-2}}{x^{-2} - y^{-1}}$.

*Solution* Write the fraction as a complex fraction and simplify:

$$\dfrac{x^{-1} + y^{-2}}{x^{-2} - y^{-1}} = \dfrac{\dfrac{1}{x} + \dfrac{1}{y^2}}{\dfrac{1}{x^2} - \dfrac{1}{y}}$$

$$= \frac{x^2y^2\left(\dfrac{1}{x} + \dfrac{1}{y^2}\right)}{x^2y^2\left(\dfrac{1}{x^2} - \dfrac{1}{y}\right)}$$ Multiply numerator and denominator by $x^2y^2$.

$$= \frac{xy^2 + x^2}{y^2 - x^2y}$$ Remove parentheses.

$$= \frac{x(y^2 + x)}{y(y - x^2)}$$ Attempt to simplify the fraction by factoring the numerator and denominator.

The result cannot be simplified. ∎

**Self Check** Simplify $\dfrac{x^{-2} - y^{-1}}{x^{-1} + y^{-2}}$.

**Answer** $\dfrac{y(y - x^2)}{x(y^2 + x)}$

Orals *Simplify each complex fraction.*

**1.** $\dfrac{\frac{2}{3}}{\frac{1}{2}}$ **2.** $\dfrac{2}{\frac{1}{2}}$ **3.** $\dfrac{\frac{1}{2}}{2}$ **4.** $\dfrac{1 + \frac{1}{2}}{\frac{1}{2}}$

# EXERCISE 6.6

***REVIEW*** *Write each expression as an expression involving only one exponent.*

**1.** $t^3t^4t^2$ **2.** $(a^0a^2)^3$ **3.** $-2r(r^3)^2$ **4.** $(s^3)^2(s^4)^0$

*Write each expression without using parentheses or negative exponents.*

**5.** $\left(\dfrac{3r}{4r^3}\right)^4$ **6.** $\left(\dfrac{12y^{-3}}{3y^2}\right)^{-2}$ **7.** $\left(\dfrac{6r^{-2}}{2r^3}\right)^{-2}$ **8.** $\left(\dfrac{4x^3}{5x^{-3}}\right)^{-2}$

***VOCABULARY AND CONCEPTS*** *Fill in each blank to make a true statement.*

**9.** If a fraction has a fraction in its numerator or denominator, it is called a _____.

**10.** The denominator of the complex fraction $\dfrac{\frac{3}{x} + \frac{x}{y}}{\frac{1}{x} + 2}$ is _____.

**11.** In Method 1, we write the numerator and denominator of a complex fraction as _____ fractions and then _____.

**12.** In Method 2, we multiply the numerator and denominator of the complex fraction by the _____ of the fractions in its numerator and denominator.

***PRACTICE*** *In Exercises 13–46, simplify each complex fraction.*

13. $\dfrac{\dfrac{2}{3}}{\dfrac{3}{4}}$

14. $\dfrac{\dfrac{3}{5}}{\dfrac{2}{7}}$

15. $\dfrac{\dfrac{4}{5}}{\dfrac{32}{15}}$

16. $\dfrac{\dfrac{7}{8}}{\dfrac{49}{4}}$

17. $\dfrac{\dfrac{2}{3}+1}{\dfrac{1}{3}+1}$

18. $\dfrac{\dfrac{3}{5}-2}{\dfrac{2}{5}-2}$

19. $\dfrac{\dfrac{1}{2}+\dfrac{3}{4}}{\dfrac{3}{2}+\dfrac{1}{4}}$

20. $\dfrac{\dfrac{2}{3}-\dfrac{5}{2}}{\dfrac{2}{3}-\dfrac{3}{2}}$

21. $\dfrac{\dfrac{x}{y}}{\dfrac{1}{x}}$

22. $\dfrac{\dfrac{y}{x}}{\dfrac{x}{xy}}$

23. $\dfrac{\dfrac{5t^2}{9x^2}}{\dfrac{3t}{x^2t}}$

24. $\dfrac{\dfrac{5w^2}{4tz}}{\dfrac{15wt}{z^2}}$

25. $\dfrac{\dfrac{1}{x}-3}{\dfrac{5}{x}+2}$

26. $\dfrac{\dfrac{1}{y}+3}{\dfrac{3}{y}-2}$

27. $\dfrac{\dfrac{2}{x}+2}{\dfrac{4}{x}+2}$

28. $\dfrac{\dfrac{3}{x}-3}{\dfrac{9}{x}-3}$

29. $\dfrac{\dfrac{3y}{x}-y}{y-\dfrac{y}{x}}$

30. $\dfrac{\dfrac{y}{x}+3y}{y+\dfrac{2y}{x}}$

31. $\dfrac{\dfrac{1}{x+1}}{1+\dfrac{1}{x+1}}$

32. $\dfrac{\dfrac{1}{x-1}}{1-\dfrac{1}{x-1}}$

33. $\dfrac{\dfrac{x}{x+2}}{\dfrac{x}{x+2}+x}$

34. $\dfrac{\dfrac{2}{x-2}}{\dfrac{2}{x-2}-1}$

35. $\dfrac{1}{\dfrac{1}{x}+\dfrac{1}{y}}$

36. $\dfrac{1}{\dfrac{b}{a}-\dfrac{a}{b}}$

37. $\dfrac{\dfrac{2}{x}}{\dfrac{2}{y}-\dfrac{4}{x}}$

38. $\dfrac{\dfrac{2y}{3}}{\dfrac{2y}{3}-\dfrac{8}{y}}$

39. $\dfrac{3+\dfrac{3}{x-1}}{3-\dfrac{3}{x}}$

40. $\dfrac{2-\dfrac{2}{x+1}}{2+\dfrac{2}{x}}$

41. $\dfrac{\dfrac{3}{x}+\dfrac{4}{x+1}}{\dfrac{2}{x+1}-\dfrac{3}{x}}$

42. $\dfrac{\dfrac{5}{y-3}-\dfrac{2}{y}}{\dfrac{1}{y}+\dfrac{2}{y-3}}$

43. $\dfrac{\dfrac{2}{x}-\dfrac{3}{x+1}}{\dfrac{2}{x+1}-\dfrac{3}{x}}$

44. $\dfrac{\dfrac{5}{y}+\dfrac{4}{y+1}}{\dfrac{4}{y}-\dfrac{5}{y+1}}$

45. $\dfrac{\dfrac{1}{y^2+y}-\dfrac{1}{xy+x}}{\dfrac{1}{xy+x}-\dfrac{1}{y^2+y}}$

46. $\dfrac{\dfrac{2}{b^2-1}-\dfrac{3}{ab-a}}{\dfrac{3}{ab-a}-\dfrac{2}{b^2-1}}$

*In Exercises 47–56, simplify each fraction.*

**47.** $\dfrac{x^{-2}}{y^{-1}}$

**48.** $\dfrac{a^{-4}}{b^{-2}}$

**49.** $\dfrac{1 + x^{-1}}{x^{-1} - 1}$

**50.** $\dfrac{y^{-2} + 1}{y^{-2} - 1}$

**51.** $\dfrac{a^{-2} + a}{a + 1}$

**52.** $\dfrac{t - t^{-2}}{1 - t^{-1}}$

**53.** $\dfrac{2x^{-1} + 4x^{-2}}{2x^{-2} + x^{-1}}$

**54.** $\dfrac{x^{-2} - 3x^{-3}}{3x^{-2} - 9x^{-3}}$

**55.** $\dfrac{1 - 25y^{-2}}{1 + 10y^{-1} + 25y^{-2}}$

**56.** $\dfrac{1 - 9x^{-2}}{1 - 6x^{-1} + 9x^{-2}}$

## WRITING

**57.** Explain how to use Method 1 to simplify

$$\dfrac{1 + \dfrac{1}{x}}{3 - \dfrac{1}{x}}$$

**58.** Explain how to use Method 2 to simplify the expression in Exercise 57.

## SOMETHING TO THINK ABOUT

**59.** Simplify these four complex fractions:

$$\dfrac{1}{1 + 1}, \quad \dfrac{1}{1 + \dfrac{1}{2}}, \quad \dfrac{1}{1 + \dfrac{1}{1 + \dfrac{1}{2}}}, \quad \text{and} \quad \dfrac{1}{1 + \dfrac{1}{1 + \dfrac{1}{1 + \dfrac{1}{2}}}}$$

**60.** In Exercise 59, what is the pattern in the numerators of the four answers? What would be the next answer?

# 6.7   Solving Equations That Contain Fractions

■ SOLVING EQUATIONS THAT CONTAIN FRACTIONS  ■ EXTRANEOUS SOLUTIONS  ■ FORMULAS

**Getting Ready**   *Simplify.*

**1.** $3\left(x + \dfrac{1}{3}\right)$

**2.** $8\left(x - \dfrac{1}{8}\right)$

**3.** $x\left(\dfrac{3}{x} + 2\right)$

**4.** $3y\left(\dfrac{1}{3} - \dfrac{2}{y}\right)$

**5.** $6x\left(\dfrac{5}{2x} + \dfrac{2}{3x}\right)$

**6.** $9x\left(\dfrac{7}{9} + \dfrac{2}{3x}\right)$

**7.** $(y - 1)\left(\dfrac{1}{y - 1} + 1\right)$

**8.** $(x + 2)\left(3 - \dfrac{1}{x + 2}\right)$

### ■ SOLVING EQUATIONS THAT CONTAIN FRACTIONS

To solve equations containing fractions, it is usually best to eliminate those fractions. To do so, we multiply both sides of the equation by the LCD of the fractions that appear in the equation. For example, to solve $\frac{x}{3} + 1 = \frac{x}{6}$, we multiply both sides of the equation by 6:

$$\frac{x}{3} + 1 = \frac{x}{6}$$

$$6\left(\frac{x}{3} + 1\right) = 6\left(\frac{x}{6}\right)$$

We then use the distributive property to remove parentheses, simplify, and solve the resulting equation for $x$.

$$6 \cdot \frac{x}{3} + 6 \cdot 1 = 6 \cdot \frac{x}{6}$$

$$2x + 6 = x$$

$$x + 6 = 0 \qquad \text{Subtract } x \text{ from both sides.}$$

$$x = -6 \qquad \text{Subtract 6 from both sides.}$$

$$\textit{Check: } \frac{x}{3} + 1 = \frac{x}{6}$$

$$\frac{-6}{3} + 1 \stackrel{?}{=} \frac{-6}{6} \qquad \text{Substitute } -6 \text{ for } x.$$

$$-2 + 1 \stackrel{?}{=} -1 \qquad \text{Simplify.}$$

$$-1 = -1$$

**EXAMPLE 1**  Solve $\dfrac{4}{x} + 1 = \dfrac{6}{x}$.

*Solution*  To clear the equation of fractions, we multiply both sides by the LCD of $\frac{4}{x}$ and $\frac{6}{x}$, which is $x$.

$$\frac{4}{x} + 1 = \frac{6}{x}$$

$$x\left(\frac{4}{x} + 1\right) = x\left(\frac{6}{x}\right) \qquad \text{Multiply both sides by } x.$$

$$x \cdot \frac{4}{x} + x \cdot 1 = x \cdot \frac{6}{x} \qquad \text{Remove parentheses.}$$

$$4 + x = 6 \qquad \text{Simplify.}$$

$$x = 2 \qquad \text{Subtract 4 from both sides.}$$

*Check:* $\dfrac{4}{x} + 1 = \dfrac{6}{x}$

$\dfrac{4}{2} + 1 \stackrel{?}{=} \dfrac{6}{2}$    Substitute 2 for $x$.

$2 + 1 \stackrel{?}{=} 3$    Simplify.

$3 = 3$    ■

**Self Check**    Solve $\frac{6}{x} - 1 = \frac{3}{x}$.

**Answer**    3

## ■ EXTRANEOUS SOLUTIONS

If we multiply both sides of an equation by an expression that involves a variable, as we did in Example 1, we *must* check the apparent solutions. The next example shows why.

**EXAMPLE 2**    Solve $\dfrac{x+3}{x-1} = \dfrac{4}{x-1}$.

*Solution*    To clear the equation of fractions, we multiply both sides by $x - 1$, the LCD of the fractions contained in the equation.

$$\dfrac{x+3}{x-1} = \dfrac{4}{x-1}$$

$$(x-1)\dfrac{x+3}{x-1} = (x-1)\dfrac{4}{x-1}$$    Multiply both sides by $x - 1$.

$$x + 3 = 4$$    Simplify.

$$x = 1$$    Subtract 3 from both sides.

Because both sides were multiplied by an expression containing a variable, we must check the apparent solution.

$$\dfrac{x+3}{x-1} = \dfrac{4}{x-1}$$

$$\dfrac{1+3}{1-1} \stackrel{?}{=} \dfrac{4}{1-1}$$    Substitute 1 for $x$.

$$\dfrac{4}{0} = \dfrac{4}{0}$$    Division by 0 is undefined.

Since zeros appear in the denominators, the fractions are undefined. Thus, 1 is a false solution, and the equation has no solutions. Such false solutions are often called **extraneous solutions**.    ■

**Self Check**

Solve $\frac{x+5}{x-2} = \frac{7}{x-2}$.

*Answer*

2 is extraneous.

**EXAMPLE 3**

Solve $\dfrac{3x+1}{x+1} - 2 = \dfrac{3(x-3)}{x+1}$.

*Solution*

To clear the equation of fractions, we multiply both sides by $x+1$, the LCD of the fractions contained in the equation.

$$\frac{3x+1}{x+1} - 2 = \frac{3(x-3)}{x+1}$$

$$(x+1)\left[\frac{3x+1}{x+1} - 2\right] = (x+1)\left[\frac{3(x-3)}{x+1}\right]$$

$$3x + 1 - 2(x+1) = 3(x-3) \qquad \text{Use the distributive property to remove parentheses.}$$

$$3x + 1 - 2x - 2 = 3x - 9 \qquad \text{Remove parentheses.}$$

$$x - 1 = 3x - 9 \qquad \text{Combine like terms.}$$

$$-2x = -8 \qquad \text{On both sides, subtract } 3x \text{ and add 1.}$$

$$x = 4 \qquad \text{Divide both sides by } -2.$$

*Check:* $\dfrac{3x+1}{x+1} - 2 = \dfrac{3(x-3)}{x+1}$

$$\frac{3(4)+1}{4+1} - 2 \stackrel{?}{=} \frac{3(4-3)}{4+1} \qquad \text{Substitute 4 for } x.$$

$$\frac{13}{5} - \frac{10}{5} \stackrel{?}{=} \frac{3(1)}{5}$$

$$\frac{3}{5} = \frac{3}{5}$$

**Self Check**

Solve $\frac{12}{x+1} - 5 = \frac{2}{x+1}$.

*Answer*

1

Many times, we will have to factor a denominator to find the LCD.

**EXAMPLE 4**

Solve $\dfrac{x+2}{x+3} + \dfrac{1}{x^2+2x-3} = 1$.

*Solution*

To find the LCD, we must factor the second denominator.

$$\frac{x+2}{x+3} + \frac{1}{x^2+2x-3} = 1$$

$$\frac{x+2}{x+3} + \frac{1}{(x+3)(x-1)} = 1 \qquad \text{Factor } x^2+2x-3.$$

To clear the equation of fractions, we multiply both sides by $(x + 3)(x - 1)$, the LCD of the fractions contained in the equation.

$$(x + 3)(x - 1)\left[\frac{x + 2}{x + 3} + \frac{1}{(x + 3)(x - 1)}\right] = (x + 3)(x - 1)1 \qquad \text{Multiply both sides by } (x + 3)(x - 1).$$

$$(x + 3)(x - 1)\frac{x + 2}{x + 3} + (x + 3)(x - 1)\frac{1}{(x + 3)(x - 1)} = (x + 3)(x - 1)1 \qquad \text{Remove brackets.}$$

$$(x - 1)(x + 2) + 1 = (x + 3)(x - 1) \qquad \text{Simplify.}$$

$$x^2 + x - 2 + 1 = x^2 + 2x - 3 \qquad \text{Remove parentheses.}$$

$$x - 2 + 1 = 2x - 3 \qquad \text{Subtract } x^2 \text{ from both sides.}$$

$$x - 1 = 2x - 3 \qquad \text{Combine like terms.}$$

$$-x - 1 = -3 \qquad \text{Subtract } 2x \text{ from both sides.}$$

$$-x = -2 \qquad \text{Add 1 to both sides.}$$

$$x = 2 \qquad \text{Divide both sides by } -1.$$

Verify that 2 is a solution of the given equation.     ■

**EXAMPLE 5**   Solve $\dfrac{4}{5} + y = \dfrac{4y - 50}{5y - 25}$.

*Solution*

$$\frac{4}{5} + y = \frac{4y - 50}{5y - 25}$$

$$\frac{4}{5} + y = \frac{4y - 50}{5(y - 5)} \qquad \text{Factor } 5y - 25.$$

$$5(y - 5)\left[\frac{4}{5} + y\right] = 5(y - 5)\left[\frac{4y - 50}{5(y - 5)}\right] \qquad \begin{array}{l}\text{Multiply both sides by}\\ 5(y - 5).\end{array}$$

$$4(y - 5) + 5y(y - 5) = 4y - 50 \qquad \text{Remove brackets.}$$

$$4y - 20 + 5y^2 - 25y = 4y - 50 \qquad \text{Remove parentheses.}$$

$$5y^2 - 25y - 20 = -50 \qquad \begin{array}{l}\text{Subtract } 4y \text{ from both}\\ \text{sides and rearrange terms.}\end{array}$$

$$5y^2 - 25y + 30 = 0 \qquad \text{Add 50 to both sides.}$$

$$y^2 - 5y + 6 = 0 \qquad \text{Divide both sides by 5.}$$

$$(y - 3)(y - 2) = 0 \qquad \text{Factor } y^2 - 5y + 6.$$

$$y - 3 = 0 \quad \text{or} \quad y - 2 = 0 \qquad \text{Set each factor equal to 0.}$$

$$y = 3 \qquad \qquad y = 2$$

Verify that 3 and 2 both satisfy the original equation.     ■

*Self Check*   Solve $\frac{x - 6}{3x - 9} - \frac{1}{3} = \frac{x}{2}$.

*Answer*   1, 2

### ■ FORMULAS

Many formulas are equations that contain fractions.

**EXAMPLE 6**   The formula $\frac{1}{r} = \frac{1}{r_1} + \frac{1}{r_2}$ is used in electronics to calculate parallel resistances. Solve the formula for $r$.

*Solution*   Clear the equation of fractions by multiplying both sides by the LCD, which is $rr_1r_2$.

$$\frac{1}{r} = \frac{1}{r_1} + \frac{1}{r_2}$$

$$rr_1r_2\left(\frac{1}{r}\right) = rr_1r_2\left(\frac{1}{r_1} + \frac{1}{r_2}\right) \qquad \text{Multiply both sides by } rr_1r_2.$$

$$\frac{rr_1r_2}{r} = \frac{rr_1r_2}{r_1} + \frac{rr_1r_2}{r_2} \qquad \text{Remove parentheses.}$$

$$r_1r_2 = rr_2 + rr_1 \qquad \text{Simplify.}$$

$$r_1r_2 = r(r_2 + r_1) \qquad \text{Factor out an } r.$$

$$\frac{r_1r_2}{r_2 + r_1} = r \qquad \text{Divide both sides by } r_2 + r_1.$$

or

$$r = \frac{r_1r_2}{r_2 + r_1}$$    ■

*Self Check*   Solve the formula in Example 6 for $r_1$.

*Answer*   $r_1 = \dfrac{rr_2}{r_2 - r}$

*Orals*   *Indicate your first step in solving each equation.*

**1.** $\dfrac{x-3}{5} = \dfrac{x}{2}$     **2.** $\dfrac{1}{x-1} = \dfrac{8}{x}$

**3.** $\dfrac{y}{9} + 5 = \dfrac{y+1}{3}$     **4.** $\dfrac{5x-8}{3} + 3x = \dfrac{x}{5}$

## EXERCISE 6.7

***REVIEW***   *Factor each expression.*

**1.** $x^2 + 4x$          **2.** $x^2 - 16y^2$          **3.** $2x^2 + x - 3$

**4.** $6a^2 - 5a - 6$     **5.** $x^4 - 16$           **6.** $4x^2 + 10x - 6$

***VOCABULARY AND CONCEPTS***   *Fill in each blank to make a true statement.*

**7.** False solutions that result from multiplying both sides of an equation by a variable are called _____ solutions.

**8.** If the product of two numbers is 1, the numbers are called _____.

**9.** To clear an equation of fractions, we multiply both sides by the _____ of the fractions in the equation.

**10.** If you multiply both sides of an equation by an expression that involves a variable, you must _____ the solution.

**11.** To clear the equation $\frac{1}{x} + \frac{2}{y} = 5$ of fractions, we multiply both sides by ___.

**12.** To clear the equation $\frac{x}{x-2} - \frac{x}{x-1} = 5$ of fractions, we multiply both sides by _____.

***PRACTICE***   *In Exercises 13–70, solve each equation and check the solution. If an equation has no solution, so indicate.*

**13.** $\dfrac{x}{2} + 4 = \dfrac{3x}{2}$

**14.** $\dfrac{y}{3} + 6 = \dfrac{4y}{3}$

**15.** $\dfrac{2y}{5} - 8 = \dfrac{4y}{5}$

**16.** $\dfrac{3x}{4} - 6 = \dfrac{x}{4}$

**17.** $\dfrac{x}{3} + 1 = \dfrac{x}{2}$

**18.** $\dfrac{x}{2} - 3 = \dfrac{x}{5}$

**19.** $\dfrac{x}{5} - \dfrac{x}{3} = -8$

**20.** $\dfrac{2}{3} + \dfrac{x}{4} = 7$

**21.** $\dfrac{3a}{2} + \dfrac{a}{3} = -22$

**22.** $\dfrac{x}{2} + x = \dfrac{9}{2}$

**23.** $\dfrac{x-3}{3} + 2x = -1$

**24.** $\dfrac{x+2}{2} - 3x = x + 8$

**25.** $\dfrac{z-3}{2} = z + 2$

**26.** $\dfrac{b+2}{3} = b - 2$

**27.** $\dfrac{5(x+1)}{8} = x + 1$

**28.** $\dfrac{3(x-1)}{2} + 2 = x$

**29.** $\dfrac{c-4}{4} = \dfrac{c+4}{8}$

**30.** $\dfrac{t+3}{2} = \dfrac{t-3}{3}$

**31.** $\dfrac{x+1}{3} + \dfrac{x-1}{5} = \dfrac{2}{15}$

**32.** $\dfrac{y-5}{7} + \dfrac{y-7}{5} = \dfrac{-2}{5}$

**33.** $\dfrac{3x-1}{6} - \dfrac{x+3}{2} = \dfrac{3x+4}{3}$

**34.** $\dfrac{2x+3}{3} + \dfrac{3x-4}{6} = \dfrac{x-2}{2}$

**35.** $\dfrac{3}{x} + 2 = 3$

**36.** $\dfrac{2}{x} + 9 = 11$

**37.** $\dfrac{5}{a} - \dfrac{4}{a} = 8 + \dfrac{1}{a}$

**38.** $\dfrac{11}{b} + \dfrac{13}{b} = 12$

**39.** $\dfrac{2}{y+1} + 5 = \dfrac{12}{y+1}$

**40.** $\dfrac{1}{t-3} = \dfrac{-2}{t-3} + 1$

**41.** $\dfrac{1}{x-1} + \dfrac{3}{x-1} = 1$

**42.** $\dfrac{3}{p+6} - 2 = \dfrac{7}{p+6}$

**43.** $\dfrac{a^2}{a+2} - \dfrac{4}{a+2} = a$

**44.** $\dfrac{z^2}{z+1} + 2 = \dfrac{1}{z+1}$

**45.** $\dfrac{x}{x-5} - \dfrac{5}{x-5} = 3$

**46.** $\dfrac{3}{y-2} + 1 = \dfrac{3}{y-2}$

**47.** $\dfrac{3r}{2} - \dfrac{3}{r} = \dfrac{3r}{2} + 3$

**48.** $\dfrac{2p}{3} - \dfrac{1}{p} = \dfrac{2p-1}{3}$

**49.** $\dfrac{1}{3} + \dfrac{2}{x-3} = 1$

**50.** $\dfrac{3}{5} + \dfrac{7}{x+2} = 2$

**51.** $\dfrac{u}{u-1} + \dfrac{1}{u} = \dfrac{u^2+1}{u^2-u}$

**52.** $\dfrac{v}{v+2} + \dfrac{1}{v-1} = 1$

**53.** $\dfrac{3}{x-2} + \dfrac{1}{x} = \dfrac{2(3x+2)}{x^2-2x}$

**54.** $\dfrac{5}{x} + \dfrac{3}{x+2} = \dfrac{-6}{x(x+2)}$

**55.** $\dfrac{7}{q^2-q-2} + \dfrac{1}{q+1} = \dfrac{3}{q-2}$

**56.** $\dfrac{-5}{s^2+s-2} + \dfrac{3}{s+2} = \dfrac{1}{s-1}$

**57.** $\dfrac{3y}{3y-6} + \dfrac{8}{y^2-4} = \dfrac{2y}{2y+4}$

**58.** $\dfrac{x-3}{4x-4} + \dfrac{1}{9} = \dfrac{x-5}{6x-6}$

**59.** $y + \dfrac{2}{3} = \dfrac{2y-12}{3y-9}$

**60.** $y + \dfrac{3}{4} = \dfrac{3y-50}{4y-24}$

**61.** $\dfrac{5}{4y+12} - \dfrac{3}{4} = \dfrac{5}{4y+12} - \dfrac{y}{4}$

**62.** $\dfrac{3}{5x-20} + \dfrac{4}{5} = \dfrac{3}{5x-20} - \dfrac{x}{5}$

**63.** $\dfrac{x}{x-1} - \dfrac{12}{x^2-x} = \dfrac{-1}{x-1}$

**64.** $1 - \dfrac{3}{b} = \dfrac{-8b}{b^2+3b}$

**65.** $\dfrac{z-4}{z-3} = \dfrac{z+2}{z+1}$

**66.** $\dfrac{a+2}{a+8} = \dfrac{a-3}{a-2}$

**67.** $\dfrac{n}{n^2-9} + \dfrac{n+8}{n+3} = \dfrac{n-8}{n-3}$

**68.** $\dfrac{x-3}{x-2} - \dfrac{1}{x} = \dfrac{x-3}{x}$

**69.** $\dfrac{b+2}{b+3} + 1 = \dfrac{-7}{b-5}$

**70.** $\dfrac{x-4}{x-3} + \dfrac{x-2}{x-3} = x-3$

**71.** Solve the formula $\dfrac{1}{a} + \dfrac{1}{b} = 1$ for $a$.

**72.** Solve the formula $\dfrac{1}{a} - \dfrac{1}{b} = 1$ for $b$.

**73. Optics**   The local length $f$ of a lens is given by the formula

$$\frac{1}{f} = \frac{1}{d_1} + \frac{1}{d_2}$$

where $d_1$ is the distance from the object to the lens and $d_2$ is the distance from the lens to the image.

Solve the formula for $f$.

**74.** Solve the formula in Exercise 73 for $d_1$.

## WRITING

**75.** Explain how you would decide what to do first when you solve an equation that involves fractions.

**76.** Explain why it is important to check your solutions to an equation that contains fractions with variables in the denominator.

## SOMETHING TO THINK ABOUT

**77.** What number is equal to its own reciprocal?

**78.** Solve $x^{-2} + x^{-1} = 0$.

# 6.8 Applications of Equations That Contain Fractions

■ PROBLEM SOLVING

Getting Ready
1. If it takes 5 hours to fill a pool, what part could be filled in 1 hour?
2. $x is invested at 5% annual interest. Write an expression for the interest earned in one year.
3. Write an expression for the amount of an investment that earns $y interest in one year at 5%.
4. Express how long it takes to travel $y$ miles at 52 mph.

## ■ PROBLEM SOLVING

EXAMPLE 1

**Number problem** If the same number is added to both the numerator and denominator of the fraction $\frac{3}{5}$, the result is $\frac{4}{5}$. Find the number.

*Analyze the problem*
We are asked to find a number. If we add it to both the numerator and denominator of a fraction, we will get $\frac{4}{5}$.

*Form an equation*
Let $n$ represent the unknown number and add $n$ to both the numerator and denominator of $\frac{3}{5}$. Then set the result equal to $\frac{4}{5}$ to get the equation

$$\frac{3+n}{5+n} = \frac{4}{5}$$

*Solve the equation*
To solve the equation, we proceed as follows:

$$\frac{3+n}{5+n} = \frac{4}{5}$$

$$5(5+n)\frac{3+n}{5+n} = 5(5+n)\frac{4}{5} \qquad \text{Multiply both sides by } 5(5+n).$$

$$5(3+n) = (5+n)4 \qquad \text{Simplify.}$$

$$15 + 5n = 20 + 4n \qquad \text{Use the distributive property to remove parentheses.}$$

$$5n = 5 + 4n \qquad \text{Subtract 15 from both sides.}$$

$$n = 5 \qquad \text{Subtract } 4n \text{ from both sides.}$$

*State the conclusion*
The number is 5.

*Check the result*
Add 5 to both the numerator and denominator of $\frac{3}{5}$ and get

$$\frac{3+5}{5+5} = \frac{8}{10} = \frac{4}{5}$$

The result checks.

■

**EXAMPLE 2**    **Draining an oil tank**    An inlet pipe can fill an oil tank in 7 days, and a second inlet pipe can fill the same tank in 9 days. If both pipes are used, how long will it take to fill the tank?

*Analyze the problem*    The key is to note what each pipe can do in 1 day. If you add what the first pipe can do in 1 day to what the second pipe can do in 1 day, the sum is what they can do together in 1 day. Since the first pipe can fill the tank in 7 days, it can do $\frac{1}{7}$ of the job in 1 day. Since the second pipe can fill the tank in 9 days, it can do $\frac{1}{9}$ of the job in 1 day. If it takes $x$ days for both pipes to fill the tank, together they can do $\frac{1}{x}$ of the job in 1 day.

*Form an equation*    Let $x$ represent the number of days it will take to fill the tank if both inlet pipes are used. Then form the equation

| What the first inlet pipe can do in 1 day | + | What the second inlet pipe can do in 1 day | = | what they can do together in 1 day. |
|:---:|:---:|:---:|:---:|:---:|
| $\dfrac{1}{7}$ | + | $\dfrac{1}{9}$ | = | $\dfrac{1}{x}$ |

*Solve the equation*    To solve the equation, we proceed as follows:

$$\frac{1}{7} + \frac{1}{9} = \frac{1}{x}$$

$$63x\left(\frac{1}{7} + \frac{1}{9}\right) = 63x\left(\frac{1}{x}\right) \qquad \text{Multiply both sides by } 63x.$$

$$9x + 7x = 63 \qquad \text{Use the distributive property to remove parentheses and simplify.}$$

$$16x = 63 \qquad \text{Combine like terms.}$$

$$x = \frac{63}{16} \qquad \text{Divide both sides by 16.}$$

*State the conclusion*    It will take $\frac{63}{16}$ or $3\frac{15}{16}$ days for both inlet pipes to fill the tank.

*Check the result*    In $\frac{63}{16}$ days, the first pipe fills $\frac{1}{7}\left(\frac{63}{16}\right)$ of the tank, and the second pipe fills $\frac{1}{9}\left(\frac{63}{16}\right)$ of the tank. The sum of these efforts, $\frac{9}{16} + \frac{7}{16}$, is equal to one full tank.    ■

**EXAMPLE 3**    **Track and field**    A coach can run 10 miles in the same amount of time that his best student athlete can run 12 miles. If the student can run 1 mph faster than the coach, how fast can the student run?

*Analyze the problem*    This is a uniform motion problem, which is based on the formula $d = rt$, where $d$ is the distance traveled, $r$ is the rate, and $t$ is the time. If we solve this formula for $t$, we obtain

$$t = \frac{d}{r}$$

If the coach runs 10 miles at some unknown rate of $r$ mph, it will take $\frac{10}{r}$ hours. If the student runs 12 miles at some unknown rate of $(r + 1)$ mph, it will take $\frac{12}{r+1}$ hours. We can organize the information of the problem as in Figure 6-3.

| | $d$ = | $r$ · | $t$ |
|---|---|---|---|
| **Student** | 12 | $r + 1$ | $\dfrac{12}{r+1}$ |
| **Coach** | 10 | $r$ | $\dfrac{10}{r}$ |

FIGURE 6-3

Because the times are given to be equal, we know that $\dfrac{12}{r+1} = \dfrac{10}{r}$.

***Form an equation***   Let $r$ be the rate that the coach can run. Then $r + 1$ is the rate that the student can run. We can form the equation

| The time it takes the student to run 12 miles | = | the time it takes the coach to run 10 miles. |
|---|---|---|
| $\dfrac{12}{r+1}$ | = | $\dfrac{10}{r}$ |

***Solve the equation***   We can solve the equation as follows:

$$\frac{12}{r+1} = \frac{10}{r}$$

$$r(r+1)\frac{12}{r+1} = r(r+1)\frac{10}{r} \qquad \text{Multiply both sides by } r(r+1).$$

$$12r = 10(r+1) \qquad \text{Simplify.}$$

$$12r = 10r + 10 \qquad \text{Use the distributive property to remove parentheses.}$$

$$2r = 10 \qquad \text{Subtract } 10r \text{ from both sides.}$$

$$r = 5 \qquad \text{Divide both sides by 2.}$$

***State the conclusion***   The coach can run 5 mph. The student, running 1 mph faster, can run 6 mph.

***Check the results***   Verify that this result checks.   ∎

**EXAMPLE 4**

**Comparing investments**   At one bank, a sum of money invested for one year will earn $96 interest. If invested in bonds, that same money would earn $108, because the interest rate paid by the bonds is 1% greater than that paid by the bank. Find the bank's rate of interest.

*Analyze the problem*    This interest problem is based on the formula $i = pr$, where $i$ is the interest, $p$ is the principal (the amount invested), and $r$ is the annual rate of interest. If we solve this formula for $p$, we obtain

$$p = \frac{i}{r}$$

If we let $r$ represent the bank's rate of interest, then $r + .01$ represents the rate paid by the bonds. If an investment at a bank earns $96 interest at some unknown rate $r$, the principal invested is $\frac{96}{r}$. If an investment in bonds earns $108 interest at some unknown rate $(r + .01)$, the principal invested is $\frac{108}{r+.01}$. We can organize the information of the problem as in Figure 6-4.

| | Interest = | Principal · | Rate |
|---|---|---|---|
| **Bank** | 96 | $\dfrac{96}{r}$ | $r$ |
| **Bonds** | 108 | $\dfrac{108}{r+.01}$ | $r + .01$ |

FIGURE 6-4

*Form an equation*    Because the same principal would be invested in either account, we can set up the following equation:

$$\frac{96}{r} = \frac{108}{r+.01}$$

*Solve the equation*    We can solve the equation as follows:

$$\frac{96}{r} = \frac{108}{r+.01}$$

$$r(r+.01)\cdot\frac{96}{r} = \frac{108}{r+.01}\cdot r(r+.01) \qquad \text{Multiply both sides by } r(r+.01).$$

$$96(r+.01) = 108r$$

$$96r + .96 = 108r \qquad\qquad \text{Remove parentheses.}$$

$$.96 = 12r \qquad\qquad \text{Subtract } 96r \text{ from both sides.}$$

$$.08 = r \qquad\qquad \text{Divide both sides by 12.}$$

*State the conclusion*    The bank's interest rate is .08, or 8%. The bonds pay 9% interest, a rate 1% greater than that paid by the bank.

*Check the results*    Verify that these rates check.    ∎

Orals    **1.** What is the formula that relates the principal $p$ that is invested, the earned interest $i$, and the rate $r$ for 1 year?

**2.** What is the formula that relates the distance $d$ traveled at a speed $r$, for a time $t$?

**3.** What is the formula that relates the cost $C$ of purchasing $q$ items that cost $\$d$ each?

## EXERCISE 6.8

**REVIEW**  *Solve each equation.*

**1.** $x^2 - 5x - 6 = 0$    **2.** $x^2 - 25 = 0$    **3.** $(t + 2)(t^2 + 7t + 12) = 0$    **4.** $2(y - 4) = -y^2$

**5.** $y^3 - y^2 = 0$    **6.** $5a^3 - 125a = 0$    **7.** $(x^2 - 1)(x^2 - 4) = 0$    **8.** $6t^3 + 35t^2 = 6t$

### VOCABULARY AND CONCEPTS

**9.** List the five steps used in problem solving.

**10.** Write 6% as a decimal.

### PRACTICE

**11. Number problem**  If the denominator of $\frac{3}{4}$ is increased by a number and the numerator is doubled, the result is 1. Find the number.

**12. Number problem**  If a number is added to the numerator of $\frac{7}{8}$ and the same number is subtracted from the denominator, the result is 2. Find the number.

**13. Number problem**  If a number is added to the numerator of $\frac{3}{4}$ and twice as much is added to the denominator, the result is $\frac{4}{7}$. Find the number.

**14. Number problem**  If a number is added to the numerator of $\frac{5}{7}$ and twice as much is subtracted from the denominator, the result is 8. Find the number.

**15. Number problem**  The sum of a number and its reciprocal is $\frac{13}{6}$. Find the numbers.

**16. Number problem**  The sum of the reciprocals of two consecutive even integers is $\frac{7}{24}$. Find the integers.

**17. Filling a pool**  An inlet pipe can fill an empty swimming pool in 5 hours, and another inlet pipe can fill the pool in 4 hours. How long will it take both pipes to fill the pool?

**18. Filling a pool**  One inlet pipe can fill an empty pool in 4 hours, and a drain can empty the pool in 8 hours. How long will it take the pipe to fill the pool if the drain is left open?

**19. Roofing a house**  A homeowner estimates that it will take 7 days to roof his house. A professional roofer estimates that he could roof the house in 4 days. How long will it take if the homeowner helps the roofer?

**20. Sewage treatment**  A sludge pool is filled by two inlet pipes. One pipe can fill the pool in 15 days and the other pipe can fill it in 21 days. However, if no sewage is added, waste removal will empty the pool in 36 days. How long will it take the two inlet pipes to fill an empty pool?

**21. Touring**  A tourist can bicycle 28 miles in the same time as he can walk 8 miles. If he can ride 10 mph faster than he can walk, how much time should he allow to walk a 30-mile trail? See Illustration 1. (*Hint:* How fast can he walk?)

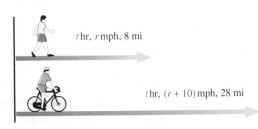

$t$ hr, $r$ mph, 8 mi

$t$ hr, $(r + 10)$ mph, 28 mi

ILLUSTRATION 1

**22. Comparing travel**   A plane can fly 300 miles in the same time as it takes a car to go 120 miles. If the car travels 90 mph slower than the plane, find the speed of the plane.

**23. Boating**   A boat that can travel 18 mph in still water can travel 22 miles downstream in the same amount of time that it can travel 14 miles upstream. Find the speed of the current in the river. (See Illustration 2.)

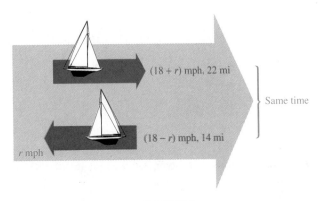

(18 + r) mph, 22 mi

(18 − r) mph, 14 mi

r mph

Same time

ILLUSTRATION 2

**24. Wind speed**   A plane can fly 300 miles downwind in the same amount of time as it can travel 210 miles upwind. Find the velocity of the wind if the plane can fly 255 mph in still air.

**25. Comparing investments**   Two certificates of deposit pay interest at rates that differ by 1%. Money invested for one year in the first CD earns $175 interest. The same principal invested in the other CD earns $200. Find the two rates of interest.

**26. Comparing interest rates**   Two bond funds pay interest at rates that differ by 2%. Money invested for one year in the first fund earns $315 interest. The same amount invested in the other fund earns $385. Find the lower rate of interest.

**27. Sharing costs**   Some office workers bought a $35 gift for their boss. If there had been two more employees to contribute, everyone's cost would have been $2 less. How many workers contributed to the gift?

**28. Sales**   A dealer bought some radios for a total of $1,200. She gave away 6 radios as gifts, sold each of the rest for $10 more than she paid for each radio, and broke even. How many radios did she buy?

**29. Sales**   A bookstore can purchase several calculators for a total cost of $120. If each calculator cost $1 less, the bookstore could purchase 10 additional calculators at the same total cost. How many calculators can be purchased at the regular price?

**30. Furnace repair**   A repairman purchased several furnace-blower motors for a total cost of $210. If his cost per motor had been $5 less, he could have purchased 1 additional motor. How many motors did he buy at the regular rate?

**31. River tours**   A river boat tour begins by going 60 miles upstream against a 5 mph current. There, the boat turns around and returns with the current. What still-water speed should the captain use to complete the tour in 5 hours?

**32. Travel time**   A company president flew 680 miles in a corporate jet but returned in a smaller plane that could fly only half as fast. If the total travel time was 6 hours, find the speeds of the planes.

*WRITING*

**33.** The key to solving shared work problems is to ask, "How much of the job could be done in 1 unit of time?" Explain.

**34.** It is difficult to check the solution of a shared work problem. Explain how you could decide if the answer is at least reasonable.

*SOMETHING TO THINK ABOUT*

**35.** Create a problem, involving either investment income or shared work, that can be solved by an equation that contains fractions.

**36.** Solve the problem you created in Exercise 35.

■ ■ ■ ■ ■ ■ ■ ■ ■

**MATHEMATICS IN ARCHITECTURE**

The area $A$ of either room discussed at the beginning of the chapter is given by $A = lw$, where $l$ and $w$ are the length and the width. We solve this equation for $l$:

$$l = \frac{A}{w}$$

Because the area of the living room is 288 ft$^2$ and its width is $(w + 4)$ ft, its length is $\frac{288}{w+4}$ ft. Similarly, the length of the dining room is $\frac{168}{w}$ ft. We summarize this information in a table.

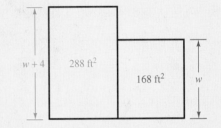

|  | Area | = | Length | · | Width |
|---|---|---|---|---|---|
| **Living room** | 288 | | $\dfrac{288}{w+4}$ | | $w + 4$ |
| **Dining room** | 168 | | $\dfrac{168}{w}$ | | $w$ |

| The length of the living room | + | the length of the dining room | = | the total length. |
|---|---|---|---|---|

$$\frac{288}{w+4} + \frac{168}{w} = 32$$

$$w(w+4)\left(\frac{288}{w+4} + \frac{168}{w}\right) = 32w(w+4) \qquad \text{Clear the equation of fractions.}$$

$$288w + 168w + 672 = 32w^2 + 128w \qquad \text{Remove parentheses.}$$

$$0 = 32w^2 - 328w - 672 \qquad \text{Subtract } 456w \text{ and } 672 \text{ from both sides.}$$

$$0 = 4w^2 - 41w - 84 \qquad \text{Divide both sides by 8.}$$

$$0 = (w - 12)(4w + 7) \qquad \text{Factor.}$$

$$w - 12 = 0 \quad \text{or} \quad 4w + 7 = 0 \qquad \text{Set each factor equal to 0.}$$

$$w = 12 \qquad\qquad w = -\frac{7}{4} \qquad \text{Solve each equation.}$$

Because the width cannot be negative, we discard the negative root. The architect must plan for a 12-foot-wide dining room.

■ ■ ■ ■ ■ ■ ■ ■ ■ ■ **PROJECTS**

1. If the sides of two similar triangles are in the ratio of 1 to 1, the triangles are said to be **congruent.** Congruent triangles have the same shape and the same size (area).

   **a.** Draw several triangles with sides of length 1, 1.5, and 2 inches. Are the triangles all congruent? What general rule could you make?

   **b.** Draw several triangles with the dimensions shown in Illustration 1. Are the triangles all congruent? What general rule could you make?

   **c.** Draw several triangles with the dimensions shown in Illustration 2. Are the triangles all congruent? What general rule could you make?

| ILLUSTRATION 1 | ILLUSTRATION 2 |

   **d.** If three angles of one triangle have the same measure as three angles of a second triangle, are the triangles congruent? Explain your answer.

2. Our solar system consists of nine planets and their moons, and some assorted asteroids, comets, and other debris, all orbiting the sun. If the sizes of the planets and their distances from the sun were reduced proportionally so that the sun was the size of an orange, earth would be a grain of sand, and the farthest planet, Pluto, would be half a mile away.

   The diameters of the planets and their distances from the sun are given in Table 6-2.

|  | Diameter (km) | Distance from sun (AU)* |
|---|---|---|
| Sun | $1.5 \times 10^6$ | 0 |
| Mercury | $4.9 \times 10^3$ | 0.39 |
| Venus | $1.2 \times 10^4$ | 0.72 |
| Earth | $1.3 \times 10^4$ | 1.0 |
| Mars | $6.8 \times 10^3$ | 1.5 |
| Jupiter | $1.4 \times 10^5$ | 5.2 |
| Saturn | $1.2 \times 10^5$ | 9.5 |
| Uranus | $5.1 \times 10^5$ | 19 |
| Neptune | $4.9 \times 10^5$ | 30 |
| Pluto | $2.3 \times 10^3$ | 39 |

*One AU (astronomical unit) is the distance from the earth to the sun, about 93 million miles.

TABLE 6-2

a. Use the information in Table 6-2 to draw a scale diagram of the relative *positions* of the sun and the planets. You will need a large sheet of paper, or perhaps the classroom chalkboard. From your diagram, which planets do you think are called the *inner planets,* and which are the *outer planets?*

b. Draw a scale diagram that shows the relative *sizes* of the sun and planets.

c. What difficulty would you have in drawing a scale diagram that shows both relative sizes and distances? Could you draw a scale diagram if you disregarded the enormous size of the sun? Write your observations in a brief paragraph.

# CHAPTER SUMMARY

## CONCEPTS

## REVIEW EXERCISES

### SECTION 6.1 — *Ratios*

A **ratio** is the comparison of two numbers by their indicated quotient.

**1.** Write each ratio as a fraction in lowest terms.
  **a.** 3 to 6
  **b.** $12x$ to $15x$
  **c.** 2 feet to 1 yard
  **d.** 5 pints to 3 quarts

The **unit cost** of an item is the ratio of its cost to its quantity.

**2.** If three pounds of coffee cost $8.79, find the unit cost (the cost per pound).

*Rates* are ratios that are used to compare quantities with different units.

**3.** If a factory used 2,275 kwh of electricity in February, what was the rate of energy consumption in kwh per week?

### SECTION 6.2 — *Proportions and Similar Triangles*

A **proportion** is a statement that two ratios are equal.

**4.** Determine whether the following equations are proportions.
  **a.** $\dfrac{4}{7} = \dfrac{20}{34}$
  **b.** $\dfrac{5}{7} = \dfrac{30}{42}$

In any proportion, the product of the extremes is equal to the product of the means.

**5.** Solve each proportion.
  **a.** $\dfrac{3}{x} = \dfrac{6}{9}$
  **b.** $\dfrac{x}{3} = \dfrac{x}{5}$
  **c.** $\dfrac{x-2}{5} = \dfrac{x}{7}$
  **d.** $\dfrac{4x-1}{18} = \dfrac{x}{6}$

The measures of corresponding sides of similar triangles are in proportion.

**6.** A telephone pole casts a shadow 12 feet long at the same time that a man 6 feet tall casts a shadow of 3.6 feet. How tall is the pole?

| SECTION 6.3 | *Simplifying Fractions* |
|---|---|

If $b$ and $c$ are not 0, then

$$\frac{a}{b} = \frac{a \cdot c}{b \cdot c}$$

$$\frac{a}{1} = a$$

$$\frac{a}{0} \text{ is undefined.}$$

**7.** Write each fraction in lowest terms. If a fraction is already in lowest terms, so indicate.

**a.** $\dfrac{10}{25}$      **b.** $-\dfrac{12}{18}$

**c.** $-\dfrac{51}{153}$      **d.** $\dfrac{105}{45}$

**e.** $\dfrac{3x^2}{6x^3}$      **f.** $\dfrac{5xy^2}{2x^2y^2}$

**g.** $\dfrac{x^2}{x^2 + x}$      **h.** $\dfrac{x + 2}{x^2 + 2x}$

**i.** $\dfrac{6xy}{3xy}$      **j.** $\dfrac{8x^2y}{2x(4xy)}$

**k.** $\dfrac{3p - 2}{2 - 3p}$      **l.** $\dfrac{x^2 - x - 56}{x^2 - 5x - 24}$

**m.** $\dfrac{2x^2 - 16x}{2x^2 - 18x + 16}$      **n.** $\dfrac{x^2 + x - 2}{x^2 - x - 2}$

| SECTION 6.4 | *Multiplying and Dividing Fractions* |
|---|---|

$$\frac{a}{b} \cdot \frac{c}{d} = \frac{a \cdot c}{b \cdot d} \quad (b, d \neq 0)$$

**8.** Do each multiplication and simplify.

**a.** $\dfrac{3xy}{2x} \cdot \dfrac{4x}{2y^2}$      **b.** $\dfrac{3x}{x^2 - x} \cdot \dfrac{2x - 2}{x^2}$

**c.** $\dfrac{x^2 + 3x + 2}{x^2 + 2x} \cdot \dfrac{x}{x + 1}$      **d.** $\dfrac{x^2 + x}{3x - 15} \cdot \dfrac{6x - 30}{x^2 + 2x + 1}$

$$\frac{a}{b} \div \frac{c}{d} = \frac{a}{b} \cdot \frac{d}{c} \quad (b, c, d \neq 0)$$

**9.** Do each division and simplify.

**a.** $\dfrac{3x^2}{5x^2y} \div \dfrac{6x}{15xy^2}$      **b.** $\dfrac{x^2 + 5x}{x^2 + 4x - 5} \div \dfrac{x^2}{x - 1}$

**c.** $\dfrac{x^2 - x - 6}{2x - 1} \div \dfrac{x^2 - 2x - 3}{2x^2 + x - 1}$

**d.** $\dfrac{x^2 - 3x}{x^2 - x - 6} \div \dfrac{x^2 - x}{x^2 + x - 2}$

**e.** $\dfrac{x^2 + 4x + 4}{x^2 + x - 6} \left( \dfrac{x - 2}{x - 1} \div \dfrac{x + 2}{x^2 + 2x - 3} \right)$

## SECTION 6.5 — Adding and Subtracting Fractions

$\dfrac{a}{d} + \dfrac{b}{d} = \dfrac{a+b}{d}$  $(d \neq 0)$

$\dfrac{a}{d} - \dfrac{b}{d} = \dfrac{a-b}{d}$  $(d \neq 0)$

To add or subtract fractions with unlike denominators, first find the LCD of the fractions. Then express each fraction in equivalent form with a common denominator. Finally, add or subtract the fractions.

**10.** Do each operation. Simplify all answers.

**a.** $\dfrac{x}{x+y} + \dfrac{y}{x+y}$

**b.** $\dfrac{3x}{x-7} - \dfrac{x-2}{x-7}$

**c.** $\dfrac{x}{x-1} + \dfrac{1}{x}$

**d.** $\dfrac{1}{7} - \dfrac{1}{x}$

**e.** $\dfrac{3}{x+1} - \dfrac{2}{x}$

**f.** $\dfrac{x+2}{2x} - \dfrac{2-x}{x^2}$

**g.** $\dfrac{x}{x+2} + \dfrac{3}{x} - \dfrac{4}{x^2+2x}$

**h.** $\dfrac{2}{x-1} - \dfrac{3}{x+1} + \dfrac{x-5}{x^2-1}$

## SECTION 6.6 — Complex Fractions

To simplify a complex fraction, use either of these methods:

**1.** Write the numerator and denominator of the complex fraction as single fractions, do the division of the fractions, and simplify.

**2.** Multiply both the numerator and the denominator of the complex fraction by the LCD of the fractions that appear in the numerator and the denominator; then simplify.

**11.** Simplify each complex fraction.

**a.** $\dfrac{\dfrac{3}{2}}{\dfrac{2}{3}}$

**b.** $\dfrac{\dfrac{3}{2}+1}{\dfrac{2}{3}+1}$

**c.** $\dfrac{\dfrac{1}{x}+1}{\dfrac{1}{x}-1}$

**d.** $\dfrac{1+\dfrac{3}{x}}{2-\dfrac{1}{x^2}}$

**e.** $\dfrac{\dfrac{2}{x-1}+\dfrac{x-1}{x+1}}{\dfrac{1}{x^2-1}}$

**f.** $\dfrac{\dfrac{a}{b}+c}{\dfrac{b}{a}+c}$

## Solving Equations That Contain Fractions

To solve an equation that contains fractions, change it to another equation without fractions. Do so by multiplying both sides by the LCD of the fractions. Check all solutions.

**12.** Solve each equation and check all answers.

a. $\dfrac{3}{x} = \dfrac{2}{x - 1}$        b. $\dfrac{5}{x + 4} = \dfrac{3}{x + 2}$

c. $\dfrac{2}{3x} + \dfrac{1}{x} = \dfrac{5}{9}$        d. $\dfrac{2x}{x + 4} = \dfrac{3}{x - 1}$

e. $\dfrac{2}{x - 1} + \dfrac{3}{x + 4} = \dfrac{-5}{x^2 + 3x - 4}$

f. $\dfrac{4}{x + 2} - \dfrac{3}{x + 3} = \dfrac{6}{x^2 + 5x + 6}$

**13.** Solve for $r_1$: $\dfrac{1}{r} = \dfrac{1}{r_1} + \dfrac{1}{r_2}$.

## Applications of Equations That Contain Fractions

**14.** The efficiency $E$ of a Carnot engine is given by the formula

$$E = 1 - \frac{T_2}{T_1}$$

Solve the formula for $T_1$.

**15.** Radioactive tracers are used for diagnostic work in nuclear medicine. The **effective half-life** $H$ of a radioactive material in a biological organism is given by the formula

$$H = \frac{RB}{R + B}$$

where $R$ is the radioactive half-life and $B$ is the biological half-life of the tracer. Solve the formula for $R$.

**16. Pumping a basement** If one pump can empty a flooded basement in 18 hours and a second pump can empty the basement in 20 hours, how long will it take to empty the basement when both pumps are used?

**17. Painting houses** If a homeowner can paint a house in 14 days and a professional painter can paint it in 10 days, how long will it take if they work together?

**18. Exercise**   A jogger can bicycle 30 miles in the same time as he can jog 10 miles. If he can ride 10 miles per hour faster than he can jog, how fast can he jog?

**19. Wind speed**   A plane can fly 400 miles downwind in the same amount of time as it can travel 320 miles upwind. If the plane can fly at 360 miles per hour in still air, find the velocity of the wind.

## ■ Chapter Test

**1.** Express as a ratio in lowest terms: 6 feet to 3 yards.

**2.** Is the equation $\dfrac{3xy}{5xy} = \dfrac{3xt}{5xt}$ a proportion?

**3.** Solve the proportion for $y$: $\dfrac{y}{y-1} = \dfrac{y-2}{y}$.

**4.** A tree casts a shadow that is 30 feet long when a 6-foot-tall man casts a shadow that is 4 feet long. How tall is the tree?

**5.** Simplify $\dfrac{48x^2y}{54xy^2}$.

**6.** Simplify $\dfrac{2x^2 - x - 3}{4x^2 - 9}$.

**7.** Simplify $\dfrac{3(x+2) - 3}{2x - 4 - (x-5)}$.

**8.** Multiply and simplify $\dfrac{12x^2y}{15xyz} \cdot \dfrac{25y^2z}{16xt}$.

**9.** Multiply and simplify $\dfrac{x^2 + 3x + 2}{3x + 9} \cdot \dfrac{x+3}{x^2 - 4}$.

**10.** Divide and simplify $\dfrac{8x^2y}{25xt} \div \dfrac{16x^2y^3}{30xyt^3}$.

**11.** Divide and simplify $\dfrac{x^2 - x}{3x^2 + 6x} \div \dfrac{3x - 3}{3x^3 + 6x^2}$.

**12.** Simplify $\dfrac{x^2 + xy}{x - y} \cdot \dfrac{x^2 - y^2}{x^2 - 2x} \div \dfrac{x^2 + 2xy + y^2}{x^2 - 4}$.

**13.** Add $\dfrac{5x - 4}{x - 1} + \dfrac{5x + 3}{x - 1}$.

**14.** Subtract $\dfrac{3y + 7}{2y + 3} - \dfrac{3(y - 2)}{2y + 3}$.

**15.** Add $\dfrac{x + 1}{x} + \dfrac{x - 1}{x + 1}$.

**16.** Subtract $\dfrac{5x}{x - 2} - 3$.

**17.** Simplify $\dfrac{\dfrac{8x^2}{xy^3}}{\dfrac{4y^3}{x^2y^3}}$.

**18.** Simplify $\dfrac{1 + \dfrac{y}{x}}{\dfrac{y}{x} - 1}$.

**19.** Solve for $x$: $\dfrac{x}{10} - \dfrac{1}{2} = \dfrac{x}{5}$.

**20.** Solve for $x$: $3x - \dfrac{2(x + 3)}{3} = 16 - \dfrac{x + 2}{2}$.

**21.** Solve for $x$: $\dfrac{7}{x + 4} - \dfrac{1}{2} = \dfrac{3}{x + 4}$.

**22.** Solve for $B$: $H = \dfrac{RB}{R + B}$.

**23. Cleaning highways** One highway worker could pick up all the trash on a strip of highway in 7 hours, and his helper could pick up the trash in 9 hours. How long will it take them if they work together?

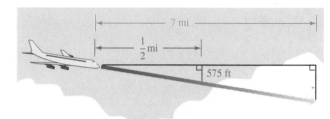

**24. Boating** A boat can motor 28 miles downstream in the same amount of time as it can motor 18 miles upstream. Find the speed of the current if the boat can motor at 23 mph in still water.

ILLUSTRATION 1

**25. Flight path** A plane drops 575 feet as it flies a horizontal distance of $\frac{1}{2}$ mile. How much altitude will it lose as it flies a horizontal distance of 7 miles? (See Illustration 1.)

## ■ Cumulative Review Exercises

*In Exercises 1–4, simplify each expression.*

**1.** $x^2 x^5$

**2.** $(x^2)^5$

**3.** $\dfrac{x^5}{x^2}$

**4.** $(3x^5)^0$

*In Exercises 5–8, simplify each expression.*

**5.** $(3x^2 - 2x) + (6x^3 - 3x^2 - 1)$

**6.** $(4x^3 - 2x) - (2x^3 - 2x^2 - 3x + 1)$

**7.** $3(5x^2 - 4x + 3) + 2(-x^2 + 2x - 4)$

**8.** $4(3x^2 - 4x - 1) - 2(-2x^2 + 4x - 3)$

*In Exercises 9–12, do each multiplication.*

**9.** $(3x^3 y^2)(-4x^2 y^3)$

**10.** $-5x^2(7x^3 - 2x^2 - 2)$

**11.** $(3x + 1)(2x + 4)$

**12.** $(5x - 4y)(3x + 2y)$

*In Exercises 13–14, do each division.*

**13.** $x + 3 \overline{) x^2 + 7x + 12}$

**14.** $2x - 3 \overline{) 2x^3 - x^2 - x - 3}$

*In Exercises 15–24, factor each expression.*

**15.** $3x^2 y - 6xy^2$

**16.** $3(a + b) + x(a + b)$

**17.** $2a + 2b + ab + b^2$

**18.** $25p^4 - 16q^2$

**19.** $x^2 - 11x - 12$

**20.** $x^2 - xy - 6y^2$

**21.** $6a^2 - 7a - 20$

**22.** $8m^2 - 10mn - 3n^2$

**23.** $p^3 - 27q^3$

**24.** $8r^3 + 64s^3$

*In Exercises 25–30, solve each equation.*

**25.** $\dfrac{4}{5}x + 6 = 18$

**26.** $5 - \dfrac{x+2}{3} = 7 - x$

**27.** $6x^2 - x - 2 = 0$

**28.** $5x^2 = 10x$

**29.** $x^2 + 3x + 2 = 0$

**30.** $2y^2 + 5y - 12 = 0$

*In Exercises 31–34, solve each inequality and graph the solution set.*

**31.** $5x - 3 > 7$

**32.** $7x - 9 < 5$

**33.** $-2 < -x + 3 < 5$

**34.** $0 \le \dfrac{4-x}{3} \le 2$

*In Exercises 35–36, graph each equation.*

**35.** Graph the equation $4x - 3y = 12$.

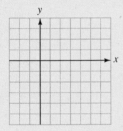

**36.** Graph the equation $3x + 4y = 4y + 12$.

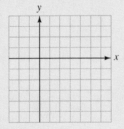

*In Exercises 37–38, solve each system of equations.*

**37.** $\begin{cases} x + y = 1 \\ x - y = 7 \end{cases}$

**38.** $\begin{cases} 4x + 9y = 8 \\ 2x - 6y = -3 \end{cases}$

*In Exercises 39–40, solve each system by graphing.*

**39.** $\begin{cases} x - y = 4 \\ 2x + y = 5 \end{cases}$

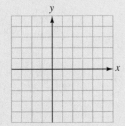

**40.** $\begin{cases} 3x + 2y \ge 6 \\ x + 3y \le 6 \end{cases}$

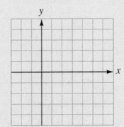

*In Exercises 41–44, $y = f(x) = 2x^2 - 3$. Find each value.*

**41.** $f(0)$          **42.** $f(3)$          **43.** $f(-2)$          **44.** $f(2x)$

*In Exercises 45–46, simplify each fraction.*

**45.** $\dfrac{x^2 + 2x + 1}{x^2 - 1}$                        **46.** $\dfrac{x^2 + 2x - 15}{x^2 + 3x - 10}$

*In Exercises 47–52, do the operation(s) and simplify when possible.*

**47.** $\dfrac{x^2 + x - 6}{5x - 5} \cdot \dfrac{5x - 10}{x + 3}$          **48.** $\dfrac{p^2 - p - 6}{3p - 9} \div \dfrac{p^2 + 6p + 9}{p^2 - 9}$

**49.** $\dfrac{3x}{x + 2} + \dfrac{5x}{x + 2} - \dfrac{7x - 2}{x + 2}$          **50.** $\dfrac{x - 1}{x + 1} + \dfrac{x + 1}{x - 1}$

**51.** $\dfrac{a + 1}{2a + 4} - \dfrac{a^2}{2a^2 - 8}$          **52.** $\dfrac{\dfrac{1}{x} + \dfrac{1}{y}}{\dfrac{1}{x} - \dfrac{1}{y}}$

# 7

# Roots and Radical Expressions

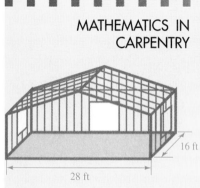

MATHEMATICS IN
CARPENTRY

To span the 16-foot-by-28-foot room shown in the illustration, a carpenter will use a scissors truss, with the ridge of the vaulted ceiling at the center of the room. The house plans call for the outside walls to be 8 feet high and the ridge of the room to be 12 feet high.

How many 4-foot-by-8-foot sheets of plaster board will be needed to drywall the entire ceiling?

After reading this chapter, you will be able to answer this question.

16 ft

28 ft

## 7.1  Square Roots and the Pythagorean Theorem

■ SQUARE ROOTS ■ USING A CALCULATOR TO FIND SQUARE ROOTS ■ USING A TABLE TO FIND SQUARE ROOTS ■ THE SQUARE ROOT FUNCTION ■ RIGHT TRIANGLES ■ THE PYTHAGOREAN THEOREM

Getting Ready    *Find each value.*

**1.** $3^2$          **2.** $4^2$          **3.** $2^3$          **4.** $5^3$

**5.** $3^4$          **6.** $4^4$          **7.** $(-3)^3$          **8.** $(-2)^5$

### ■ SQUARE ROOTS

5 in.

5 in.        5 in.

5 in.

FIGURE 7-1

To find the area $A$ of the square shown in Figure 7-1, we multiply its length by its width.

$$A = 5 \cdot 5$$
$$= 5^2$$
$$= 25$$

The area is 25 square inches.

We have seen that the product $5 \cdot 5$ can be denoted by the exponential expression $5^2$, where 5 is raised to the second power. Whenever we raise a number to the second power, we are squaring it, or finding its **square.** This example illustrates that the formula for the area of a square with a side of length $s$ is $A = s^2$.

Here are some more squares of numbers.

- The square of 3 is 9, because $3^2 = 9$.
- The square of $-3$ is 9, because $(-3)^2 = 9$.
- The square of 12 is 144, because $12^2 = 144$.
- The square of $-12$ is 144, because $(-12)^2 = 144$.

- The square of $\frac{1}{3}$ is $\frac{1}{9}$, because $\left(\frac{1}{3}\right)^2 = \frac{1}{9}$.
- The square of $-\frac{1}{3}$ is $\frac{1}{9}$, because $\left(-\frac{1}{3}\right)^2 = \frac{1}{9}$.
- The square of 0 is 0, because $0^2 = 0$.

In this section, we will reverse the squaring process and find **square roots** of numbers.

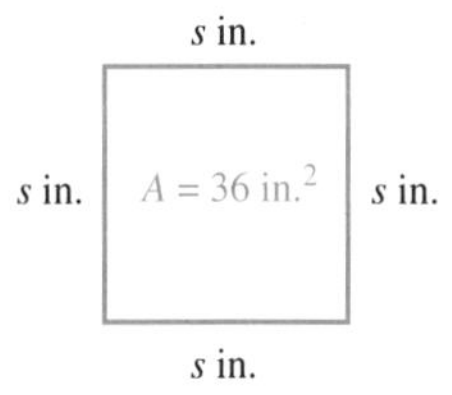

Suppose we know that the area of the square shown in Figure 7-2 is 36 square inches. To find the length of each side, we substitute 36 for $A$ in the formula $A = s^2$ and solve for $s$.

$s$ in.

$s$ in. | $A = 36$ in.$^2$ | $s$ in.

$s$ in.

FIGURE 7-2

$$A = s^2$$
$$36 = s^2$$

To solve for $s$, we must find a positive number whose square is 36. Since 6 is such a number, the sides of the square are 6 inches long. The number 6 is called a *square root* of 36, because $6^2 = 36$.

Here are more examples of square roots.

- 3 is a square root of 9, because $3^2 = 9$.
- $-3$ is a square root of 9, because $(-3)^2 = 9$.
- 12 is a square root of 144, because $12^2 = 144$.
- $-12$ is a square root of 144, because $(-12)^2 = 144$.
- $\frac{1}{3}$ is a square root of $\frac{1}{9}$, because $\left(\frac{1}{3}\right)^2 = \frac{1}{9}$.
- $-\frac{1}{3}$ is a square root of $\frac{1}{9}$, because $\left(-\frac{1}{3}\right)^2 = \frac{1}{9}$.
- 0 is a square root of 0, because $0^2 = 0$.

In general, the following is true.

### Square Roots
The number $b$ is a **square root of $a$** if $b^2 = a$.

All positive numbers have two square roots—one that is positive and one that is negative. The two square roots of 9 are 3 and $-3$, and the two square roots of 144 are 12 and $-12$. The number 0 is the only number that has just one square root, which is 0.

The symbol $\sqrt{\phantom{x}}$, called a **radical sign,** is used to represent the positive (or *principal*) square root of a number.

### Principal Square Roots
If $a > 0$, the expression $\sqrt{a}$ represents the **principal** (or positive) square root of $a$.

The principal square root of 0 is 0: $\sqrt{0} = 0$.

The expression under a radical sign is called a **radicand.**

The principal square root of a positive number is always positive. Although 3 and $-3$ are both square roots of 9, only 3 is the principal square root. The symbol $\sqrt{9}$ represents 3. To represent $-3$, we place a $-$ sign in front of the radical:

$$\sqrt{9} = 3 \qquad \text{and} \qquad -\sqrt{9} = -3$$

Likewise,

$$\sqrt{144} = 12 \qquad \text{and} \qquad -\sqrt{144} = -12$$

 a,g,k

**EXAMPLE 1**

Find each square root.

**a.** $\sqrt{0} = 0$          **b.** $\sqrt{1} = 1$

**c.** $-\sqrt{4} = -2$          **d.** $-\sqrt{81} = -9$

**e.** $\sqrt{225} = 15$          **f.** $\sqrt{169} = 13$

**g.** $-\sqrt{625} = -25$          **h.** $-\sqrt{900} = -30$

**i.** $\sqrt{576} = 24$          **j.** $\sqrt{1,600} = 40$

**k.** $\sqrt{\dfrac{1}{4}} = \dfrac{1}{2}$          **l.** $\sqrt{\dfrac{4}{9}} = \dfrac{2}{3}$          ∎

**Self Check**

Find each square root:   **a.** $\sqrt{121}$,   **b.** $-\sqrt{49}$,   **c.** $\sqrt{64}$,   **d.** $\sqrt{256}$,

**e.** $\sqrt{\dfrac{1}{25}}$,   and   **f.** $\sqrt{\dfrac{9}{49}}$.

**Answers**   **a.** 11,   **b.** $-7$,   **c.** 8,   **d.** 16,   **e.** $\frac{1}{5}$,   **f.** $\frac{3}{7}$

Square roots of certain numbers like 7 are hard to compute by hand. However, we can find $\sqrt{7}$ with a calculator or with a table of square roots.

## ■ USING A CALCULATOR TO FIND SQUARE ROOTS

To find the principal square root of 7, we enter 7 into a calculator and press the $\boxed{\sqrt{x}}$ key. The approximate value of $\sqrt{7}$ will appear on the calculator's display.

$$\sqrt{7} \approx 2.645751311 \qquad \text{Read} \approx \text{as "is approximately equal to."}$$

## ■ USING A TABLE TO FIND SQUARE ROOTS

To find the principal square root of 7, we can look in a table of square roots. (See Figure 7-3.) In the left column, headed by $n$, we locate the number 7. The column headed $\sqrt{n}$ contains the approximate value of $\sqrt{7}$.

$$\sqrt{7} \approx 2.646$$

| $n$ | $n^2$ | $\sqrt{n}$ | $n^3$ | $\sqrt[3]{n}$ |
|---|---|---|---|---|
| 5 | 25 | 2.236 | 125 | 1.710 |
| 6 | 36 | 2.449 | 216 | 1.817 |
| 7 | 49 | 2.646 | 343 | 1.913 |
| 8 | 64 | 2.828 | 512 | 2.000 |

FIGURE 7-3

Numbers such as 4, 9, 16, and 49 are called **integer squares,** because each one is the square of an integer. The square root of any integer square is an integer, and therefore a rational number:

$$\sqrt{4} = 2 \qquad \sqrt{9} = 3 \qquad \sqrt{16} = 4 \qquad \text{and} \qquad \sqrt{49} = 7$$

Square roots of positive integers that are not integer squares are called **irrational numbers.** For example, $\sqrt{7}$ is an irrational number. Recall that the set of rational numbers and the set of irrational numbers together make up the set of real numbers.

**WARNING!** Square roots of negative numbers are not real numbers. For example, $\sqrt{-4}$ is nonreal, because the square of no real number is $-4$. The number $\sqrt{-4}$ is an example from a set of numbers called **imaginary numbers,** which are discussed in later courses. Remember: *The square root of a negative number is not a real number.*

In this chapter, we will assume that *all radicands under square root symbols are either positive or 0.* Thus, in this chapter, all square roots will be real numbers.

## ■ THE SQUARE ROOT FUNCTION

Since there is one principal square root for every nonnegative real number $x$, the equation $y = f(x) = \sqrt{x}$ determines a square root function. To graph this function, we make a table of values and plot each pair of points. The graph appears in Figure 7-4.

The square root function has many real-life applications.

| $x$ | $f(x)$ | $(x, f(x))$ |
|-----|--------|-------------|
| 0   | 0      | $(0, 0)$    |
| 1   | 1      | $(1, 1)$    |
| 4   | 2      | $(4, 2)$    |
| 9   | 3      | $(9, 3)$    |
| 16  | 4      | $(16, 4)$   |

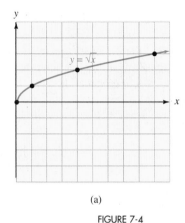

(a)

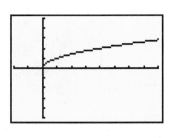

(b)

FIGURE 7-4

**EXAMPLE 2**

The *period* of a pendulum is the time required for the pendulum to swing back and forth to complete one cycle. (See Figure 7-5.) The period $t$ (in seconds) is a function of the pendulum's length $l$, which is defined by

$$t = f(l) = 1.11\sqrt{l}$$

Find the period of a pendulum that is 5 feet long.

*Solution*    We substitute 5 for $l$ in the formula and simplify.

$$t = 1.11\sqrt{l}$$
$$t = 1.11\sqrt{5}$$
$$\approx 2.482035455$$

The period is approximately 2.5 seconds.

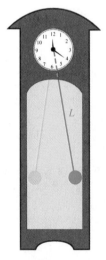

FIGURE 7-5    ∎

*Self Check*

To the nearest tenth, find the period of a pendulum that is 3 feet long.

*Answer*

1.9 sec

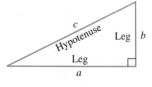

FIGURE 7-6

■ **RIGHT TRIANGLES**

A triangle that contains a 90° angle is called a **right triangle.** The longest side of a right triangle is the **hypotenuse,** which is the side opposite the right angle. The remaining two sides are the **legs** of the triangle. In the right triangle shown in Figure 7-6, side $c$ is the hypotenuse, and sides $a$ and $b$ are the legs.

## ■ THE PYTHAGOREAN THEOREM

The **Pythagorean theorem** provides a formula relating the lengths of the three sides of a right triangle.

> **The Pythagorean Theorem**
> If the length of the hypotenuse of a right triangle is $c$ and the lengths of the two legs are $a$ and $b$, then
> $$c^2 = a^2 + b^2$$

The Pythagorean theorem is useful because equal positive numbers have equal positive square roots.

> **Square Root Property of Equality**
> If $a$ and $b$ are positive numbers, then
> If $a = b$, then $\sqrt{a} = \sqrt{b}$.

Since the lengths of the sides of a triangle are positive numbers, we can use the square root property of equality and the Pythagorean theorem to find the length of an unknown side of a right triangle when we are given the lengths of the other two sides.

**EXAMPLE 3**

**Building a high-ropes adventure course**   The builder of a high-ropes course wants to stabilize the pole shown in Figure 7-7 by attaching a cable from a ground anchor 20 feet from its base to a point 15 feet up the pole. How long will the cable be?

*Solution*   We can use the Pythagorean theorem, with $a = 20$ and $b = 15$.

| | |
|---|---|
| $c^2 = a^2 + b^2$ | |
| $c^2 = 20^2 + 15^2$ | Substitute 20 for $a$ and 15 for $b$. |
| $c^2 = 400 + 225$ | $20^2 = 400$ and $15^2 = 225$. |
| $c^2 = 625$ | $400 + 225 = 625$. |
| $\sqrt{c^2} = \sqrt{625}$ | Take the positive square root of both sides. |
| $c = 25$ | $\sqrt{625} = 25$ and $\sqrt{c^2} = c$, because $c \cdot c = c^2$. |

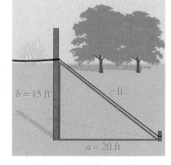

FIGURE 7-7

The cable will be 25 feet long.

■

■ ■ ■ ■ ■ ■ ■ ■ ■ ■ PERSPECTIVE

Because of the Pythagorean theorem, the ancient Greeks knew that the lengths of the sides of a right triangle could be natural numbers. For example, a right triangle could have sides of lengths 3, 4, and 5, because $3^2 + 4^2 = 5^2$. (Check it: $9 + 16 = 25$.) Similarly, a right triangle could have sides of 5, 12, and 13, because $5^2 + 12^2 = 13^2$. Natural numbers $a$, $b$, and $c$ that satisfy the equation $a^2 + b^2 = c^2$ are called **Pythagorean triples.** The triples 3, 4, 5 and 5, 12, 13 are two of infinitely many possibilities.

In 1637, the French mathematician Pierre de Fermat wrote a note in the margin of a book: There are no natural-number solutions $a$, $b$, and $c$ to the equation $a^n + b^n = c^n$ if $n$ is greater than 2. Fermat also mentioned that he had found a marvelous proof which wouldn't fit in the margin. Mathematicians have been trying to prove Fermat's last theorem ever since. Until recently, they had little success.

Princeton University mathematician Dr. Andrew Wiles first learned of Fermat's last theorem when he was 10 years old. He was so intrigued by the problem that he decided that he would study mathemat-

Dr. Andrew Wiles, Princeton University

ics. Dr. Wiles worked on the problem, isolating himself in a barren attic room in his Princeton home. "The problem was on my mind all the time," said Dr. Wiles. "When you are really desperate to find an answer, you can't let go." After seven years of concentrated work, Dr. Wiles announced an apparent solution in June of 1993.

EXAMPLE 4    **Saving cable**   The builder of a high-ropes course wants to use a 25-foot cable to stabilize the pole shown in Figure 7-8. To be safe, the ground anchor must be greater than 16 feet from the base of the pole. Is the cable long enough to use?

*Solution*    We can use the Pythagorean theorem, with $b = 16$ and $c = 25$.

$$c^2 = a^2 + b^2$$

$$25^2 = a^2 + 16^2$$    Substitute 25 for $c$ and 16 for $b$.

$$625 = a^2 + 256$$    $25^2 = 625$ and $16^2 = 256$.

$$369 = a^2$$    Subtract 256 from both sides.

$$\sqrt{369} = \sqrt{a^2}$$    Take the positive square root of both sides.

$$a \approx 19.20937271$$    Use a calculator.

FIGURE 7-8

Since the anchor can be more than 16 feet from the base, the cable is long enough.

■

**EXAMPLE 5**

**Reach of a ladder**   A 26-foot ladder rests against the side of a building. If the base of the ladder is 10 feet from the wall, how far up the side of the building will the ladder reach?

*Analyze the problem*   The wall, the ground, and the ladder form a right triangle, as shown in Figure 7-9. In this triangle, the hypotenuse is 26 feet, and one of the legs is the base-to-wall distance of 10 feet. We can let $d$ represent the other leg, which is the distance that the ladder will reach up the wall.

*Form an equation*   We can form the equation

| The hypotenuse squared | is | one leg squared | + | the other leg squared. |
|:---:|:---:|:---:|:---:|:---:|
| $26^2$ | $=$ | $10^2$ | $+$ | $d^2$ |

*Solve the equation*

$$26^2 = 10^2 + d^2$$
$$676 = 100 + d^2 \qquad \text{$26^2 = 676$ and}$$
$$\qquad\qquad\qquad\qquad 10^2 = 100.$$
$$676 - 100 = d^2 \qquad \text{Subtract 100 from}$$
$$\qquad\qquad\qquad\qquad \text{both sides.}$$
$$576 = d^2 \qquad\quad 676 - 100 = 576.$$
$$\sqrt{576} = \sqrt{d^2} \qquad \text{Take the square}$$
$$\qquad\qquad\qquad \text{root of both sides.}$$
$$24 = d \qquad\qquad \sqrt{576} = 24 \text{ and}$$
$$\qquad\qquad\qquad\quad \sqrt{d^2} = d, \text{ because}$$
$$\qquad\qquad\qquad\quad d \cdot d = d^2.$$

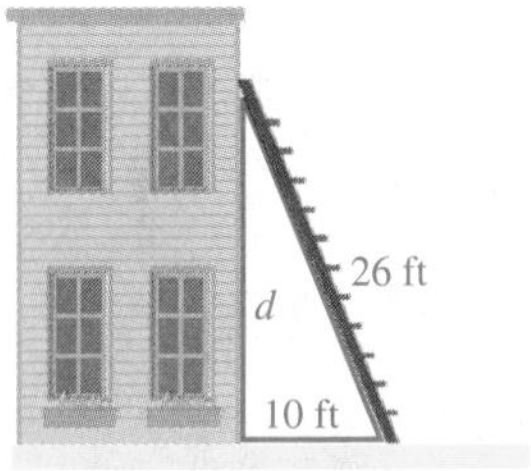

FIGURE 7-9

*State the conclusion*   The ladder will reach 24 feet up the side of the building.   ∎

**EXAMPLE 6**

**Measuring distance**   The gable end of the roof shown in Figure 7-10 is an isosceles right triangle with a span of 48 feet. Find the distance from the eaves to the peak.

*Analyze the problem*   The two equal sides of the isosceles right triangle are the two legs of the right triangle, and the span of 48 is the length of the hypotenuse. We can let $x$ represent the length of each leg, which is the distance from eaves to peak.

*Form an equation*   We can form the equation

| The hypotenuse squared | is | one leg squared | + | the other leg squared. |
|:---:|:---:|:---:|:---:|:---:|
| $48^2$ | $=$ | $x^2$ | $+$ | $x^2$ |

*Solve the equation*

$$48^2 = x^2 + x^2$$

$$2,304 = 2x^2$$

$$1,152 = x^2$$

$$\sqrt{1,152} = \sqrt{x^2}$$

$$33.9411255 \approx x$$

$48^2 = 2,304$ and $x^2 + x^2 = 2x^2.$

Divide both sides by 2.

Take the square root of both sides.

Use a calculator to find the approximate value of $\sqrt{1,152}$.

FIGURE 7-10

*State the conclusion*   The eaves-to-peak distance of the roof is approximately 34 feet. ■

Orals   *Find each root. Assume $x > 0$.*

**1.** $\sqrt{25}$       **2.** $\sqrt{4}$       **3.** $\sqrt{\dfrac{1}{9}}$       **4.** $\sqrt{\dfrac{25}{49}}$

**5.** $\sqrt{4x^2}$      **6.** $\sqrt{36x^4}$      **7.** $\sqrt{81y^6}$      **8.** $\sqrt{100z^4}$

## EXERCISE 7.1

**REVIEW**   *Graph each equation or inequality.*

**1.** $x = 3$       **2.** $y = -3$       **3.** $-2x + y = 4$       **4.** $4x - y > 4$

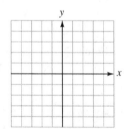   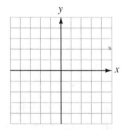   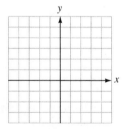   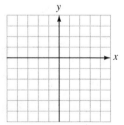

**VOCABULARY AND CONCEPTS**   *Fill in each blank to make a true statement.*

**5.** $b$ is a square root of $a$ if _____.

**6.** The symbol $\sqrt{\phantom{x}}$ is called a _____.

**7.** The principal square root of a positive number is _____.

**8.** The number under the radical sign is called the _____.

**9.** If a triangle has a right angle, it is called a _____ triangle.

**10.** The longest side of a _____ triangle is called the _____, and the other two sides are called *legs*.

**11.** The number 25 has _____ square roots. They are _____ and _____.

**12.** $\sqrt{-11}$ is not a _____ number.

**13.** The formula $A = s^2$ gives the area of a _____.

**14.** The principal square root of 0 is _____.

**15.** If the length of the _____ of a right triangle is $c$ and the legs are $a$ and $b$, then $c^2 =$ _____.

**16.** If $a$ and $b$ are positive numbers and if $a = b$, then _____.

**PRACTICE** *In Exercises 17–40, find each value.*

**17.** $\sqrt{9}$

**18.** $\sqrt{16}$

**19.** $\sqrt{49}$

**20.** $\sqrt{100}$

**21.** $\sqrt{36}$

**22.** $\sqrt{4}$

**23.** $\sqrt{\dfrac{1}{81}}$

**24.** $\sqrt{\dfrac{1}{121}}$

**25.** $-\sqrt{25}$

**26.** $-\sqrt{49}$

**27.** $-\sqrt{81}$

**28.** $-\sqrt{36}$

**29.** $\sqrt{196}$

**30.** $\sqrt{169}$

**31.** $\sqrt{\dfrac{9}{256}}$

**32.** $\sqrt{\dfrac{49}{225}}$

**33.** $-\sqrt{289}$

**34.** $\sqrt{400}$

**35.** $\sqrt{10,000}$

**36.** $-\sqrt{2,500}$

**37.** $\sqrt{324}$

**38.** $-\sqrt{625}$

**39.** $-\sqrt{3,600}$

**40.** $\sqrt{1,600}$

*In Exercises 41–64, use a calculator to find each square root to three decimal places.*

**41.** $\sqrt{2}$

**42.** $\sqrt{3}$

**43.** $\sqrt{5}$

**44.** $\sqrt{10}$

**45.** $\sqrt{6}$

**46.** $\sqrt{8}$

**47.** $\sqrt{11}$

**48.** $\sqrt{17}$

**49.** $\sqrt{23}$

**50.** $\sqrt{53}$

**51.** $\sqrt{95}$

**52.** $\sqrt{99}$

**53.** $\sqrt{6,428}$

**54.** $\sqrt{4,444}$

**55.** $-\sqrt{9,876}$

**56.** $-\sqrt{3,619}$

**57.** $\sqrt{21.35}$

**58.** $\sqrt{13.78}$

**59.** $\sqrt{0.3588}$

**60.** $\sqrt{0.9999}$

**61.** $\sqrt{0.9925}$

**62.** $\sqrt{0.12345}$

**63.** $-\sqrt{0.8372}$

**64.** $-\sqrt{0.4279}$

*In Exercises 65–72, tell whether each number is rational, irrational, or imaginary.*

**65.** $\sqrt{9}$

**66.** $\sqrt{17}$

**67.** $\sqrt{49}$

**68.** $\sqrt{-49}$

**69.** $-\sqrt{5}$

**70.** $\sqrt{0}$

**71.** $\sqrt{-100}$

**72.** $-\sqrt{225}$

*In Exercises 73–76, graph each function. Check your work with a graphing calculator.*

**73.** $f(x) = 1 + \sqrt{x}$

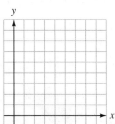

**74.** $f(x) = -1 + \sqrt{x}$

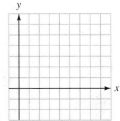

**75.** $f(x) = -\sqrt{x}$

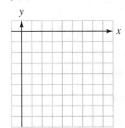

**76.** $f(x) = 1 - \sqrt{x}$

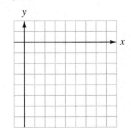

*In Exercises 77–84, refer to the right triangle in Illustration 1. Find the length of the unknown side.*

**77.** $a = 4$ and $b = 3$. Find $c$.

**78.** $a = 6$ and $b = 8$. Find $c$.

**79.** $a = 5$ and $b = 12$. Find $c$.

**80.** $a = 15$ and $c = 17$. Find $b$.

**81.** $a = 21$ and $c = 29$. Find $b$.

**82.** $b = 16$ and $c = 34$. Find $a$.

**83.** $b = 45$ and $c = 53$. Find $a$.

**84.** $a = 14$ and $c = 50$. Find $b$.

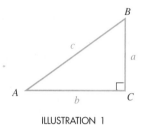

ILLUSTRATION 1

**APPLICATIONS** 🖩 *Use a calculator to help solve each problem. If an answer is not exact, give it to the nearest tenth.*

**85. Adjusting a ladder**   A 20-foot ladder reaches a window 16 feet above the ground. How far from the wall is the base of the ladder?

**86. Length of guy wires**   A 20-foot-tall tower is secured by three guy wires fastened at the top and to anchors 15 feet from the base of the tower. How long is each guy wire?

**87. Height of a pole**   A 34-foot-long wire reaches from the top of a telephone pole to a point on the ground 16 feet from the base of the pole. Find the height of the pole.

**88. Length of a path**   A rectangular garden has sides of 28 and 45 feet. Find the length of a path that extends from one corner to the opposite corner.

**89. Baseball**   A baseball diamond is a square, with each side 90 feet long. (See Illustration 2.) How far is it from home plate to second base?

**90. Television**   The size of the television screen shown in Illustration 3 is the diagonal measure of its rectangular screen. How large is the screen if it is 21 inches wide and 17 inches high?

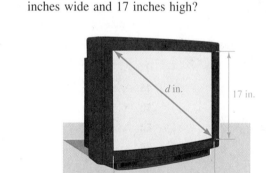

ILLUSTRATION 3

**91. Finding location**   A woman drives 4.2 miles east and then 4.0 miles north. How far is she from her starting point?

**92. Taking a shortcut**   Instead of walking on the sidewalk, students take a diagonal shortcut across the vacant lot shown in Illustration 4. How much distance do they save?

ILLUSTRATION 2

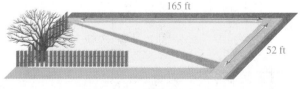

ILLUSTRATION 4

**93. Carpentry** A square-headed bolt is countersunk into a circular hole drilled in a wooden beam. The corners of the bolt must have $\frac{3}{8}$ inch clearance, as shown in Illustration 5. Find the diameter of the hole.

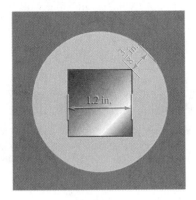

ILLUSTRATION 5

**94. Designing a tunnel** The entrance to a one-way tunnel is a rectangle with a semicircular roof. Its dimensions are given in Illustration 6. How tall can a 10-foot-wide truck be without getting stuck in the tunnel?

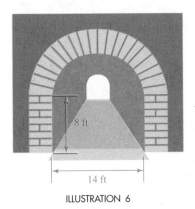

ILLUSTRATION 6

**95. Football** On first and ten, a coach tells his tight end to go out 6 yards, cut 45° to the right, and run 5 yards. (See Illustration 7.) The tight end follows instructions, catches a pass, and is tackled immediately. Does he gain the necessary 10 yards for a first down?

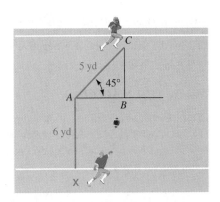

ILLUSTRATION 7

**96. Geometry** The legs of a right triangle are equal, and the hypotenuse is $2\sqrt{2}$ units long. Find the length of each leg.

**97. Geometry** The sides of a square are 3 feet long. Find the length of each diagonal of the square.

**98. Perimeter of a square** The diagonal of a square is 3 feet long. Find its perimeter.

**99. Altitude of a triangle** Find the altitude of the isosceles triangle shown in Illustration 8.

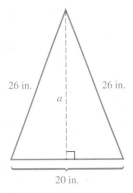

ILLUSTRATION 8

**100. Geometry** The square in Illustration 9 is inscribed in a circle. The sides of the square are 6 inches long. Find the area of the circle.

ILLUSTRATION 9

*WRITING*   *Write a paragraph using your own words.*

**101.** Explain why the square root of a negative number cannot be a real number.

**102.** Explain the Pythagorean theorem.

*SOMETHING TO THINK ABOUT*

**103.** To generate Pythagorean triples, pick natural numbers for $x$ and $y$ ($x > y$). Let $a = 2xy$ and $b = x^2 - y^2$, and $c = x^2 + y^2$. Why do you always get a Pythagorean triple?

**104.** Can you find a Pythagorean triple with $b = 10$? Explain.

# 7.2   *n*th Roots and Radicands That Contain Variables

■ CUBE ROOTS   ■ USING A CALCULATOR TO FIND CUBE ROOTS   ■ USING A TABLE TO FIND CUBE ROOTS   ■ THE CUBE ROOT FUNCTION   ■ *n*TH ROOTS   ■ RADICANDS THAT CONTAIN VARIABLES

Getting Ready   *Find each value.*

**1.** $2^3$  **2.** $4^3$  **3.** $(-5)^3$  **4.** $\left(-\dfrac{1}{2}\right)^3$

**5.** $3^4$  **6.** $\left(-\dfrac{1}{2}\right)^4$  **7.** $2^5$  **8.** $2^6$

## ■ CUBE ROOTS

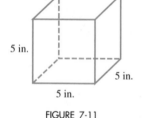

5 in.

5 in.

5 in.

FIGURE 7-11

To find the volume $V$ of the cube shown in Figure 7-11, we multiply its length, width, and height.

$$V = l \cdot w \cdot h$$
$$V = 5 \cdot 5 \cdot 5$$
$$= 5^3$$
$$= 125$$

The volume is 125 cubic inches.

We have seen that the product $5 \cdot 5 \cdot 5$ can be denoted by the exponential expression $5^3$, where 5 is raised to the third power. Whenever we raise a number to the third power, we are cubing it, or finding its **cube.** This example illustrates that the formula for the volume of a cube with each side of length $s$ is $V = s^3$.

Here are some more cubes of numbers.

- The cube of 3 is 27, because $3^3 = 27$.
- The cube of $-3$ is $-27$, because $(-3)^3 = -27$.
- The cube of 12 is 1,728, because $12^3 = 1,728$.

- The cube of $-12$ is $-1,728$, because $(-12)^3 = -1,728$.
- The cube of $\frac{1}{3}$ is $\frac{1}{27}$, because $\left(\frac{1}{3}\right)^3 = \frac{1}{27}$.
- The cube of $-\frac{1}{3}$ is $-\frac{1}{27}$, because $\left(-\frac{1}{3}\right)^3 = -\frac{1}{27}$.
- The cube of 0 is 0, because $0^3 = 0$.

In this section, we will reverse the cubing process and find **cube roots** of numbers. We will also consider fourth roots, fifth roots, and so on.

Suppose we know that the volume of the cube shown in Figure 7-12 is 216 cubic inches. To find the length of each side, we substitute 216 for $V$ in the formula $V = s^3$ and solve for $s$.

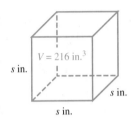

$$V = s^3$$
$$216 = s^3$$

To solve for $s$, we must find a number whose cube is 216. Since 6 is such a number, the sides of the cube are 6 inches long. The number 6 is called a *cube root* of 216, because $6^3 = 216$.

Here are more examples of cube roots.

- 3 is a cube root of 27, because $3^3 = 27$.
- $-3$ is a cube root of $-27$, because $(-3)^3 = -27$.
- 12 is a cube root of 1,728, because $12^3 = 1,728$.
- $-12$ is a cube root of $-1,728$, because $(-12)^3 = -1,728$.
- $\frac{1}{3}$ is a cube root of $\frac{1}{27}$, because $\left(\frac{1}{3}\right)^3 = \frac{1}{27}$.
- $-\frac{1}{3}$ is a cube root of $-\frac{1}{27}$, because $\left(-\frac{1}{3}\right)^3 = -\frac{1}{27}$.
- 0 is a cube root of 0, because $0^3 = 0$.

In general, the following is true.

## Cube Roots

The number $b$ is a **cube root of $a$** if $b^3 = a$.

All real numbers have one real cube root. As the previous examples show, a positive number has a positive cube root, a negative number has a negative cube root, and the cube root of 0 is 0.

## Cube Root Notation

The **cube root of $a$** is denoted by $\sqrt[3]{a}$. By definition,

$$\sqrt[3]{a} = b \quad \text{if} \quad b^3 = a$$

**EXAMPLE 1**     Find each cube root.

a. $\sqrt[3]{8} = 2$, because $2^3 = 8$.

b. $\sqrt[3]{343} = 7$, because $7^3 = 343$.

c. $\sqrt[3]{-8} = -2$, because $(-2)^3 = -8$.

d. $\sqrt[3]{-125} = -5$, because $(-5)^3 = -125$.   ∎

**Self Check**     Find each cube root:   a. $\sqrt[3]{64}$,   b. $\sqrt[3]{-64}$,   and   c. $\sqrt[3]{216}$.

**Answers**     a. 4,   b. −4,   c. 6

 b

**EXAMPLE 2**     Find each cube root.

a. $\sqrt[3]{\dfrac{1}{8}} = \dfrac{1}{2}$, because $\left(\dfrac{1}{2}\right)^3 = \dfrac{1}{2} \cdot \dfrac{1}{2} \cdot \dfrac{1}{2} = \dfrac{1}{8}$.

b. $\sqrt[3]{-\dfrac{125}{27}} = -\dfrac{5}{3}$, because $\left(-\dfrac{5}{3}\right)\left(-\dfrac{5}{3}\right)\left(-\dfrac{5}{3}\right) = -\dfrac{125}{27}$.   ∎

**Self Check**     Find each cube root:   a. $\sqrt[3]{\dfrac{1}{27}}$   and   b. $\sqrt[3]{-\dfrac{8}{125}}$.

**Answers**     a. $\frac{1}{3}$,   b. $-\frac{2}{5}$

Cube roots of numbers such as 7 are hard to compute by hand. However, we can find $\sqrt[3]{7}$ with a calculator or with a table of cube roots.

### ■ USING A CALCULATOR TO FIND CUBE ROOTS

To find $\sqrt[3]{7}$, we enter 7 into a calculator, press the $\boxed{\sqrt[x]{y}}$ key, enter 3, and press the $\boxed{=}$ key. The approximate value of $\sqrt[3]{7}$ will appear on the calculator's display.

$$\sqrt[3]{7} \approx 1.912931183$$

If your calculator does not have a $\boxed{\sqrt[x]{y}}$ key, you can use the $\boxed{y^x}$ key. We will see later that $\sqrt[3]{7} = 7^{1/3}$. To find the value of $7^{1/3}$, we enter 7 into the calculator and press these keys.

7 $\boxed{y^x}$ $\boxed{(}$ $\boxed{1}$ $\boxed{\div}$ $\boxed{3}$ $\boxed{)}$ $\boxed{=}$

The display will read $\mathtt{1.912931183}$.

### ■ USING A TABLE TO FIND CUBE ROOTS

To find the cube root of 7, we can look in a table of cube roots. (See Figure 7-13.) In the left column, headed by *n*, we locate the number 7. The column headed $\sqrt[3]{n}$ contains the approximate value of $\sqrt[3]{7}$.

$$\sqrt[3]{7} \approx 1.913$$

| *n* | $n^2$ | $\sqrt{n}$ | $n^3$ | $\sqrt[3]{n}$ |
|---|---|---|---|---|
| 5 | 25 | 2.236 | 125 | 1.710 |
| 6 | 36 | 2.449 | 216 | 1.817 |
| 7 | 49 | 2.646 | 343 | 1.913 |
| 8 | 64 | 2.828 | 512 | 2.000 |

FIGURE 7-13

Numbers such as 8, 27, 64, and 125 are called **integer cubes**, because each one is the cube of an integer. The cube root of any integer cube is an integer, and therefore a rational number:

$$\sqrt[3]{8} = 2 \qquad \sqrt[3]{27} = 3 \qquad \sqrt[3]{64} = 4 \qquad \text{and} \qquad \sqrt[3]{125} = 5$$

Cube roots of integers that are not integer cubes are irrational numbers. For example, $\sqrt[3]{7}$ and $\sqrt[3]{10}$ are irrational numbers.

### ■ THE CUBE ROOT FUNCTION

Since every real number has one real-number cube root, there is a cube root function $y = f(x) = \sqrt[3]{x}$. To graph this function, we substitute numbers for $x$, compute $f(x)$, plot the resulting ordered pairs, and connect them with a smooth curve, as shown in Figure 7-14.

| $x$ | $f(x)$ | $(x, f(x))$ |
|---|---|---|
| $-8$ | $-2$ | $(-8, -2)$ |
| $-1$ | $-1$ | $(-1, -1)$ |
| $0$ | $0$ | $(0, 0)$ |
| $1$ | $1$ | $(1, 1)$ |
| $8$ | $2$ | $(8, 2)$ |

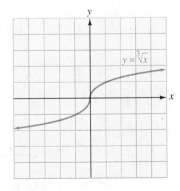

(a)

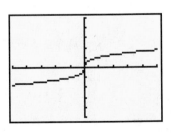

(b)

FIGURE 7-14

### ■ *n*TH ROOTS

Just as there are square roots and cube roots, there are also fourth roots, fifth roots, sixth roots, and so on. In general,

> **nth Root of *a***
> The **nth root of *a*** is denoted by $\sqrt[n]{a}$, and
> $$\sqrt[n]{a} = b \qquad \text{if} \qquad b^n = a$$
> The number *n* is called the **index** of the radical.
>     If *n* is an even natural number, *a* must be positive or 0.

In the square root symbol $\sqrt{\phantom{x}}$, the unwritten index is understood to be 2.

$$\sqrt{a} = \sqrt[2]{a}$$

 c,d

**EXAMPLE 3**    Find each root.

**a.** $\sqrt[4]{81} = 3$, because $3^4 = 81$.

**b.** $\sqrt[5]{32} = 2$, because $2^5 = 32$.

**c.** $\sqrt[5]{-32} = -2$, because $(-2)^5 = -32$.

**d.** $\sqrt[4]{-81}$ is not a real number, because no real number raised to the fourth power is $-81$. ■

*Self Check*    Find each root:   **a.** $\sqrt[4]{16}$,   **b.** $\sqrt[5]{243}$,   and   **c.** $\sqrt[5]{-1{,}024}$.
*Answers*    **a.** 2,   **b.** 3,   **c.** −4

**EXAMPLE 4**    Find each root.

**a.** $\sqrt[4]{\dfrac{1}{81}} = \dfrac{1}{3}$, because $\left(\dfrac{1}{3}\right)^4 = \dfrac{1}{81}$.

**b.** $\sqrt[5]{-\dfrac{32}{243}} = -\dfrac{2}{3}$, because $\left(-\dfrac{2}{3}\right)^5 = -\dfrac{32}{243}$. ■

*Self Check*    Find each root:   **a.** $\sqrt[4]{\frac{1}{16}}$   and   **b.** $\sqrt[5]{-\frac{243}{32}}$.
*Answers*    **a.** $\frac{1}{2}$,   **b.** $-\frac{3}{2}$

### ■ RADICANDS THAT CONTAIN VARIABLES

When *n* is even and $x \geq 0$, we say that the radical $\sqrt[n]{x}$ represents an **even root.** We can find even roots of many quantities that contain variables, provided that these variables represent positive numbers or 0.

EXAMPLE 5    Assume that each variable represents a positive number and find each root.

**a.** $\sqrt{x^2} = x$, because $x^2 = x^2$.

**b.** $\sqrt{x^4} = x^2$, because $(x^2)^2 = x^4$.

**c.** $\sqrt{x^4 y^2} = x^2 y$, because $(x^2 y)^2 = x^4 y^2$.

**d.** $\sqrt[4]{x^{12} y^8} = x^3 y^2$, because $(x^3 y^2)^4 = x^{12} y^8$.

**e.** $\sqrt[6]{64 p^{18} q^{12}} = 2 p^3 q^2$, because $(2 p^3 q^2)^6 = 64 p^{18} q^{12}$. ■

Self Check    Find each root:  **a.** $\sqrt{a^4 b^2}$,  **b.** $\sqrt{16 m^6 n^8}$,  and  **c.** $\sqrt[4]{16 x^4 y^8}$.

Answers    **a.** $a^2 b$,  **b.** $4 m^3 n^4$,  **c.** $2 x y^2$

When *n* is odd, we say that the radical $\sqrt[n]{x}$ represents an **odd root.**

EXAMPLE 6    Find each root.

**a.** $\sqrt[3]{x^6 y^3} = x^2 y$, because $(x^2 y)^3 = x^6 y^3$.

**b.** $\sqrt[5]{32 x^{10} y^5} = 2 x^2 y$, because $(2 x^2 y)^5 = 32 x^{10} y^5$. ■

Self Check    Find each root:  **a.** $\sqrt[3]{-27 p^6}$  and  **b.** $\sqrt[5]{\frac{1}{32} m^{10} n^{15}}$.

Answers    **a.** $-3 p^2$,  **b.** $\frac{1}{2} m^2 n^3$

■ ■ ■ ■ ■ ■ ■ ■ ■    **Radius of a Water Tank**

CALCULATORS    Engineers want to design a spherical tank that will hold 33,500 cubic feet of water. They know that the formula for the radius *r* of a sphere with volume *V* is given by the formula

$$ r = \sqrt[3]{\frac{3V}{4\pi}} \qquad \text{where } \pi = 3.14159 \dots $$

To use a calculator to find the radius *r*, they will substitute 33,500 for *V* and enter these numbers and press these keys.

**Keystrokes**

3  ×  33,500  ÷  (  4  ×  π  )  =  $\sqrt[x]{y}$  3  =

The display will read **19.99794636** . So the engineers should design a tank with a radius of 20 feet.

Orals *Simplify each radical.*

1. $\sqrt[3]{64}$
2. $\sqrt[3]{\dfrac{1}{64}}$
3. $\sqrt[3]{8}$
4. $\sqrt[3]{\dfrac{1}{8}}$

5. $\sqrt[4]{1}$
6. $\sqrt[4]{256}$
7. $\sqrt[5]{32}$
8. $\sqrt[5]{\dfrac{1}{32}}$

## EXERCISE 7.2

**REVIEW** *If $y = f(x) = 2x^2 - x - 1$, find each value.*

1. $f(0)$
2. $f(2)$
3. $f(-2)$
4. $f(-t)$

*Factor each expression.*

5. $x^2 - 16y^2$
6. $x^2 - 3x - 10$
7. $ax + ay + bx + by$
8. $2ax^2 + 2ax - 40a$

**VOCABULARY AND CONCEPTS** *Fill in each blank to make a true statement.*

9. If $p^3 = q$, $p$ is called the _____ of $q$.
10. If $p^4 = q$, $p$ is called a _____ of $q$.
11. In the notation $\sqrt[n]{x}$, $n$ is called the _____, and $x$ is called the _____.
12. If the index of a radical is an even number, the root is called an _____ root.
13. The formula for the volume of a cube with sides $s$ units long is _____.
14. $\sqrt[n]{a} = b$ if _____.
15. $\sqrt[3]{-216} = -6$, because _____.
16. $\sqrt[5]{\dfrac{32x^5}{243}} = \dfrac{2x}{3}$, because _____.

**PRACTICE** *In Exercises 17–32, find each value.*

17. $\sqrt[3]{1}$
18. $\sqrt[3]{8}$
19. $\sqrt[3]{27}$
20. $\sqrt[3]{0}$
21. $\sqrt[3]{-8}$
22. $\sqrt[3]{-1}$
23. $\sqrt[3]{-64}$
24. $\sqrt[3]{-27}$
25. $\sqrt[3]{125}$
26. $\sqrt[3]{1,000}$
27. $-\sqrt[3]{-1}$
28. $-\sqrt[3]{-27}$
29. $-\sqrt[3]{64}$
30. $-\sqrt[3]{343}$
31. $\sqrt[3]{729}$
32. $\sqrt[3]{512}$

*In Exercises 33–36, use a calculator to find each cube root. Give each answer to the nearest hundredth.*

33. $\sqrt[3]{32,100}$
34. $\sqrt[3]{-25,713}$
35. $\sqrt[3]{-0.11324}$
36. $\sqrt[3]{0.875}$

*In Exercises 37–40, graph each function.*

**37.** $f(x) = -\sqrt[3]{x}$

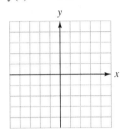

**38.** $f(x) = 1 + \sqrt[3]{x}$

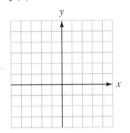

**39.** $f(x) = \sqrt[3]{x} - 2$

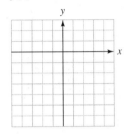

**40.** $f(x) = \sqrt[3]{x} + 2$

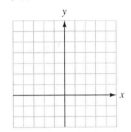

*In Exercises 41–48, find each value.*

**41.** $\sqrt[4]{16}$

**42.** $\sqrt[4]{81}$

**43.** $-\sqrt[5]{32}$

**44.** $-\sqrt[5]{243}$

**45.** $\sqrt[6]{1}$

**46.** $\sqrt[6]{0}$

**47.** $\sqrt[5]{-32}$

**48.** $\sqrt[7]{-1}$

*In Exercises 49–52, use a calculator to find each root. Give each answer to the nearest hundredth.*

**49.** $\sqrt[4]{125}$

**50.** $\sqrt[5]{12,450}$

**51.** $\sqrt[5]{-6,000}$

**52.** $\sqrt[6]{0.5}$

*In Exercises 53–76, write each expression without a radical sign. All variables represent positive numbers.*

**53.** $\sqrt{x^2y^2}$

**54.** $\sqrt{x^2y^4}$

**55.** $\sqrt{x^4z^4}$

**56.** $\sqrt{y^6z^8}$

**57.** $-\sqrt{x^4y^2}$

**58.** $-\sqrt{x^6y^4}$

**59.** $\sqrt{4z^2}$

**60.** $\sqrt{9t^6}$

**61.** $-\sqrt{9x^4y^2}$

**62.** $-\sqrt{16x^2y^4}$

**63.** $\sqrt{x^2y^2z^2}$

**64.** $\sqrt{x^4y^6z^8}$

**65.** $-\sqrt{x^2y^2z^4}$

**66.** $-\sqrt{a^8b^6c^2}$

**67.** $-\sqrt{25x^4z^{12}}$

**68.** $-\sqrt{100a^6b^4}$

**69.** $\sqrt{36z^{36}}$

**70.** $\sqrt{64y^{64}}$

**71.** $-\sqrt{16z^2}$

**72.** $-\sqrt{729x^8y^2}$

**73.** $\sqrt[3]{27y^3z^6}$

**74.** $\sqrt[3]{64x^3y^6z^9}$

**75.** $\sqrt[3]{-8p^6q^3}$

**76.** $\sqrt[3]{-r^{12}s^3t^6}$

**APPLICATIONS**   *Use a calculator to help solve each problem. Give each answer to the nearest hundredth.*

**77. Packaging**   If a cubical box has a volume of 2 cubic feet, how long is each side?

**78. Hot air balloons**   If a hot air balloon is in the shape of a sphere and has a volume of 15,000 cubic feet, how long is its radius?

**79. Windmills**   The power generated by a certain windmill is related to the speed of the wind by the formula

$$S = \sqrt[3]{\frac{P}{0.02}}$$

where $S$ is the speed of the wind (in mph) and $P$ is the power (in watts). Find the speed of the wind when the windmill is producing 400 watts of power.

**80. Astronomy**   Johannes Kepler discovered that a planet's mean distance $R$ from the sun (in astronomical units) is related to its period $T$ (in years) by the formula

$$R = \sqrt[3]{\frac{T^2}{k}}$$

Find $R$ when $T = 1.881$ and $k = 1.002$.

**81. Geometric sequences** The common ratio of a geometric sequence with four terms is given by the formula

$$r = \sqrt[3]{\frac{l}{a}}$$

where $a$ is the first term and $l$ is the last term. Find the common ratio of a geometric sequence that has a first term of 3 and a last term of 192.

**82. Business** The interest rate $i$ after five compoundings is given by the formula

$$\sqrt[5]{\frac{FV}{PV}} - 1 = i$$

where $FV$ is the future value and $PV$ is the present value. Find the interest rate $i$ if an investment of $1,000 grows to $1,338.23.

*WRITING*

**83.** Explain why a negative number can have a real number for its cube root.

**84.** Explain why a negative number cannot have a real number for its fourth root.

*SOMETHING TO THINK ABOUT*

**85.** Is $\sqrt{x^2 - 4x + 4} = x - 2$? What are the exceptions?

**86.** When is $\sqrt{x^2} \neq x$?

---

## 7.3 Solving Equations Containing Radicals; the Distance Formula

■ THE SQUARING PROPERTY OF EQUALITY ■ SOLVING EQUATIONS CONTAINING ONE SQUARE ROOT ■ SOLVING EQUATIONS CONTAINING TWO SQUARE ROOTS ■ SOLVING EQUATIONS CONTAINING CUBE ROOTS ■ THE DISTANCE FORMULA

Getting Ready    *In each set of numbers, verify that $a^2 + b^2 = c^2$.*

**1.** $a = 3, b = 4, c = 5$

**2.** $a = 6, b = 8, c = 10$

**3.** $a = 5, b = 12, c = 13$

**4.** $a = 9, b = 12, c = 15$

### ■ THE SQUARING PROPERTY OF EQUALITY

Before solving equations containing radicals, we note that if two numbers are equal, their squares are equal.

> **Squaring Property of Equality**
> If $a = b$, then $a^2 = b^2$.

If we square both sides of an equation, the resulting equation may or may not have the same solutions as the original one. For example, if we square both sides of the equation

**1.** $x = 2$, with the solution 2,

we obtain $(x)^2 = 2^2$, which simplifies as

**2.** $x^2 = 4$, with solutions of 2 and $-2$.     $2^2 = 4$ and $(-2)^2 = 4.$

Equations 1 and 2 are not equivalent, because they have different solution sets. The solution $-2$ of Equation 2 does not satisfy Equation 1. Because squaring both sides of an equation can produce an equation with solutions that don't satisfy the original one, we must always check each suspected solution in the original equation.

## ■ SOLVING EQUATIONS CONTAINING ONE SQUARE ROOT

To solve an equation containing square roots, we follow these steps.

> **Solving Equations Containing Square Roots**
> **1.** Whenever possible, isolate a single radical on one side of the equation.
> **2.** Square both sides of the equation and solve the resulting equation.
> **3.** Check the solution in the original equation. This step is required.

To solve the equation $\sqrt{x + 2} = 3$, we note that the radical is already isolated on one side. We proceed to Step 2 and square both sides to eliminate the radical. Since this might produce an equation with more solutions than the original one, we must check each solution.

**EXAMPLE 1**     Solve $\sqrt{x + 2} = 3$.

*Solution*     We square both sides to eliminate the radical and proceed as follows:

$$\sqrt{x + 2} = 3$$
$$\left(\sqrt{x + 2}\right)^2 = (3)^2 \qquad \text{Square both sides.}$$
$$x + 2 = 9 \qquad \left(\sqrt{x + 2}\right)^2 = x + 2 \text{ and } 3^2 = 9.$$
$$x = 7 \qquad \text{Subtract 2 from both sides.}$$

We check by substituting 7 for $x$ in the original equation.

$$\sqrt{x + 2} = 3$$
$$\sqrt{7 + 2} \stackrel{?}{=} 3 \qquad \text{Substitute 7 for } x.$$
$$\sqrt{9} \stackrel{?}{=} 3$$
$$3 = 3$$

The solution checks. Since no solutions are lost in this process, 7 is the only solution of the original equation.  ■

*Self Check*     Solve $\sqrt{x - 4} = 9$.

*Answer*     85

**EXAMPLE 2**   Solve $\sqrt{x+1}+5=3$.

*Solution*   We isolate the radical on one side and proceed as follows:

$$\sqrt{x+1}+5=3$$
$$\sqrt{x+1}=-2 \qquad \text{Subtract 5 from both sides.}$$
$$\left(\sqrt{x+1}\right)^2=(-2)^2 \qquad \text{Square both sides.}$$
$$x+1=4 \qquad \left(\sqrt{x+1}\right)^2=x+1 \text{ and } (-2)^2=4.$$
$$x=3 \qquad \text{Subtract 1 from both sides.}$$

We check by substituting 3 for $x$ in the original equation.

$$\sqrt{x+1}+5=3$$
$$\sqrt{3+1}+5 \overset{?}{=} 3 \qquad \text{Substitute 3 for } x.$$
$$\sqrt{4}+5 \overset{?}{=} 3$$
$$2+5 \overset{?}{=} 3$$
$$7 \neq 3$$

Since $7 \neq 3$, 3 is not a solution. This equation has no solution.   ∎

**Self Check**   Solve $\sqrt{x-2}-2=5$.
*Answer*   51

Example 2 shows that squaring both sides of an equation can lead to false solutions, called **extraneous solutions.** Such solutions do not satisfy the original equation and must be discarded.

**EXAMPLE 3**   The distance $d$ (in feet) that an object will fall in $t$ seconds is given by the formula $t=\sqrt{\frac{d}{16}}$. To find the height of the bridge shown in Figure 7-15, a man drops a stone into the water. If it takes the stone 3 seconds to hit the water, how high is the bridge?

FIGURE 7-15

*Solution*    We substitute 3 for $t$ in the formula and solve for $d$.

$$t = \sqrt{\frac{d}{16}}$$

$$3 = \sqrt{\frac{d}{16}} \qquad \text{Substitute 3 for } t.$$

$$(3)^2 = \left(\sqrt{\frac{d}{16}}\right)^2 \qquad \text{Square both sides.}$$

$$9 = \frac{d}{16} \qquad 3^2 = 9 \text{ and } \left(\sqrt{\frac{d}{16}}\right)^2 = \frac{d}{16}.$$

$$144 = d \qquad \text{Multiply both sides by 16.}$$

The bridge is 144 feet above the water. ■

*Self Check*    In Example 3, if it takes 4 seconds for the stone to hit the water, how high is the bridge?

*Answer*    256 ft

**EXAMPLE 4**    Solve $x = \sqrt{2x + 10} - 1$.

*Solution*    We add 1 to both sides to isolate the radical on the right-hand side and then square both sides to eliminate the radical.

$$x = \sqrt{2x + 10} - 1$$

$$x + 1 = \sqrt{2x + 10} \qquad \text{Add 1 to both sides to isolate the radical.}$$

$$(x + 1)^2 = \left(\sqrt{2x + 10}\right)^2 \qquad \text{Square both sides.}$$

$$x^2 + 2x + 1 = 2x + 10 \qquad \text{Remove parentheses.}$$

$$x^2 - 9 = 0 \qquad \text{Subtract } 2x \text{ and 10 from both sides.}$$

$$(x - 3)(x + 3) = 0 \qquad \text{Factor.}$$

$$x - 3 = 0 \quad \text{or} \quad x + 3 = 0 \qquad \text{Set each factor equal to 0.}$$

$$x = 3 \qquad\qquad x = -3 \qquad \text{Solve each linear equation.}$$

We check each possible solution.

| **For $x = 3$** | **For $x = -3$** |
|---|---|
| $x = \sqrt{2x + 10} - 1$ | $x = \sqrt{2x + 10} - 1$ |
| $3 \stackrel{?}{=} \sqrt{2(3) + 10} - 1$ | $-3 \stackrel{?}{=} \sqrt{2(-3) + 10} - 1$ |
| $3 \stackrel{?}{=} \sqrt{16} - 1$ | $-3 \stackrel{?}{=} \sqrt{4} - 1$ |
| $3 \stackrel{?}{=} 4 - 1$ | $-3 \stackrel{?}{=} 2 - 1$ |
| $3 = 3$ | $-3 \neq 1$ |

Since 3 is the only number that checks, it is the only solution. The false solution $-3$ is extraneous. ■

**Self Check**

**Answer**

Solve $x = \sqrt{3x + 1} + 1$.

5

### ■ SOLVING EQUATIONS CONTAINING TWO SQUARE ROOTS

The next example contains two square roots.

**EXAMPLE 5**

Solve $\sqrt{x + 12} = 3\sqrt{x + 4}$.

*Solution*    We square both sides to eliminate the radicals.

$$\sqrt{x + 12} = 3\sqrt{x + 4}$$
$$\left(\sqrt{x + 12}\right)^2 = \left(3\sqrt{x + 4}\right)^2 \qquad \text{Square both sides.}$$
$$x + 12 = 9(x + 4) \qquad \text{Simplify.}$$
$$x + 12 = 9x + 36 \qquad \text{Remove parentheses.}$$
$$-8x = 24 \qquad \text{Subtract } 9x \text{ and } 12 \text{ from both sides.}$$
$$x = -3 \qquad \text{Divide both sides by } -8.$$

We check the solution by substituting $-3$ for $x$ in the original equation.

$$\sqrt{x + 12} = 3\sqrt{x + 4}$$
$$\sqrt{-3 + 12} \overset{?}{=} 3\sqrt{-3 + 4} \qquad \text{Substitute } -3 \text{ for } x.$$
$$\sqrt{9} \overset{?}{=} 3\sqrt{1}$$
$$3 = 3 \qquad\qquad\qquad\qquad ■$$

**Self Check**

**Answer**

Solve $\sqrt{x - 4} = 2\sqrt{x - 16}$.

20

### ■ SOLVING EQUATIONS CONTAINING CUBE ROOTS

To solve an equation involving a cube root, we cube both sides of the equation.

**EXAMPLE 6**

Solve $\sqrt[3]{2x + 10} = 2$.

*Solution*    We cube both sides and proceed as follows:

$$\sqrt[3]{2x + 10} = 2$$
$$\left(\sqrt[3]{2x + 10}\right)^3 = (2)^3 \qquad \text{Cube both sides.}$$
$$2x + 10 = 8 \qquad \text{Simplify.}$$
$$2x = -2 \qquad \text{Subtract 10 from both sides.}$$
$$x = -1 \qquad \text{Divide both sides by 2.}$$

Check the result.                                                         ■

| | |
|---|---|
| *Self Check* | Solve $\sqrt[3]{3x-3} = 3$. |
| *Answer* | 10 |

### ■ THE DISTANCE FORMULA

We can use the Pythagorean theorem to derive a formula for finding the distance between two points $P(x_1, y_1)$ and $Q(x_2, y_2)$ on a rectangular coordinate system. The distance $d$ between points $P$ and $Q$ is the length of the hypotenuse of the triangle in Figure 7-16. The two legs have lengths $x_2 - x_1$ and $y_2 - y_1$.

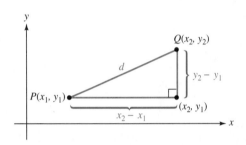

FIGURE 7-16

By the Pythagorean theorem, we have

**1.**  $d^2 = (x_2 - x_1)^2 + (y_2 - y_1)^2$

We can take the square root of both sides of Equation 1 to get the **distance formula.**

$$d = \sqrt{(x_2 - x_1)^2 + (y_2 - y_1)^2}$$

**The Distance Formula**
The distance $d$ between points $P(x_1, y_1)$ and $Q(x_2, y_2)$ is given by the formula
$$d = \sqrt{(x_2 - x_1)^2 + (y_2 - y_1)^2}$$

**EXAMPLE 7**    Find the distance between points $P(1, 5)$ and $Q(4, 9)$. (See Figure 7-17.)

*Solution*    We can use the distance formula by substituting 1 for $x_1$, 5 for $y_1$, 4 for $x_2$, and 9 for $y_2$.

$$\begin{aligned} d &= \sqrt{(x_2 - x_1)^2 + (y_2 - y_1)^2} \\ &= \sqrt{(4-1)^2 + (9-5)^2} \\ &= \sqrt{3^2 + 4^2} \\ &= \sqrt{9 + 16} \\ &= \sqrt{25} \\ &= 5 \end{aligned}$$

The distance between points $P$ and $Q$ is 5 units.    ■

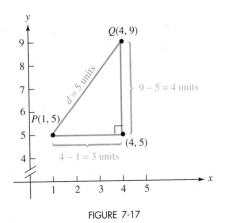

FIGURE 7-17

**Self Check**  Find the distance between $P(-2, 1)$ and $Q(4, 9)$.

*Answer*  10

**EXAMPLE 8**  **Building a freeway**  In a city, streets run north and south, and avenues run east and west. Streets and avenues are 750 feet apart. The city plans to construct a freeway from the intersection of 21st Street and 4th Avenue to the intersection of 111th Street and 60th Avenue. How long will it be?

*Solution*  We can represent the roads of the city by the coordinate system shown in Figure 7-18, where each unit on each axis represents 750 feet. We represent the end of the freeway at 21st Street and 4th Avenue by the point $(x_1, y_1) = (21, 4)$. The other end is $(x_2, y_2) = (111, 60)$.

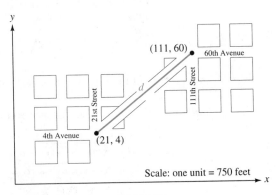

FIGURE 7-18

We can now use the distance formula to find the length of the freeway in blocks.

$$d = \sqrt{(x_2 - x_1)^2 + (y_2 - y_1)^2}$$
$$= \sqrt{(111 - 21)^2 + (60 - 4)^2}$$
$$= \sqrt{90^2 + 56^2}$$
$$= \sqrt{8,100 + 3,136}$$
$$= \sqrt{11,236}$$
$$= 106 \qquad \qquad \text{Use a calculator to find the square root.}$$

Because each block represents 750 feet, the length of the freeway will be $106 \cdot 750 = 79,500$ feet. Since there are 5,280 feet in 1 mile, we can divide 79,500 by 5,280 to convert 79,500 feet to 15.056818 miles. The freeway will be about 15 miles long. ■

Orals  *Solve each equation and check the solution.*

**1.** $\sqrt{x} = 4$                      **2.** $\sqrt{x - 1} = 2$

**3.** $\sqrt{x + 1} = 2$              **4.** $\sqrt{2x} = 1$

**5.** $\sqrt{\dfrac{x}{2}} = 1$              **6.** $\sqrt{2x - 1} = 1$

# EXERCISE 7.3

**REVIEW**  *Solve each system.*

**1.** $\begin{cases} x + y = 5 \\ x - y = -1 \end{cases}$          **2.** $\begin{cases} 2x + y = 0 \\ x + 3y = 5 \end{cases}$

**3.** $\begin{cases} 2x + 3y = 0 \\ 3x - 2y = 13 \end{cases}$          **4.** $\begin{cases} 3x - 4y = 11 \\ 4x + y = -17 \end{cases}$

**VOCABULARY AND CONCEPTS**  *Fill in each blank to make a true statement.*

**5.** A false solution that occurs because you square both sides of an equation is called an _____ solution.

**6.** The squaring property of equality states that if two numbers are _____, their _____ are equal.

**7.** If $a = b$, then $a^2 =$ ___.

**8.** The distance formula states that $d =$

_____.

*In Exercises 9–10, tell what is wrong with each solution.*

**9.** $\sqrt{x - 2} = 3$
$$x - 2 = 3$$
$$x = 5$$

**10.** $2 = \sqrt{x - 9}$
$$4 = x - 9$$
$$-5 = x$$
$$x = -5$$

***PRACTICE*** *In Exercises 11–58, solve each equation. Check all solutions. If an equation has no solutions, write "none."*

**11.** $\sqrt{x} = 3$

**12.** $\sqrt{x} = 5$

**13.** $\sqrt{x} = 7$

**14.** $\sqrt{x} = 2$

**15.** $\sqrt{x} = -4$

**16.** $\sqrt{x} = -1$

**17.** $\sqrt{x + 3} = 2$

**18.** $\sqrt{x - 2} = 3$

**19.** $\sqrt{x - 5} = 5$

**20.** $\sqrt{x + 8} = 12$

**21.** $\sqrt{3 - x} = -2$

**22.** $\sqrt{5 - x} = 10$

**23.** $\sqrt{6 + 2x} = 4$

**24.** $\sqrt{7 + x} = -4$

**25.** $\sqrt{5x - 5} = 5$

**26.** $\sqrt{6x + 19} = 7$

**27.** $\sqrt{4x - 3} = 3$

**28.** $\sqrt{11x - 2} = 3$

**29.** $\sqrt{13x + 14} = 1$

**30.** $\sqrt{8x + 9} = 1$

**31.** $\sqrt{x + 3} + 5 = 12$

**32.** $\sqrt{x - 5} - 3 = 4$

**33.** $\sqrt{2x + 10} + 3 = 5$

**34.** $\sqrt{3x + 4} + 7 = 12$

**35.** $\sqrt{5x + 9} + 4 = 7$

**36.** $\sqrt{9x + 25} - 2 = 3$

**37.** $\sqrt{7 - 5x} + 4 = 3$

**38.** $\sqrt{7 + 6x} - 4 = -3$

**39.** $\sqrt{x + 1} = x - 1$

**40.** $\sqrt{x + 4} = x - 2$

**41.** $\sqrt{x + 1} = x + 1$

**42.** $\sqrt{x + 9} = x + 7$

**43.** $\sqrt{7x + 2} - 2x = 0$

**44.** $\sqrt{3x + 3} + 5 = x$

**45.** $x - 1 = \sqrt{x - 1}$

**46.** $x - 2 = \sqrt{x + 10}$

**47.** $x = \sqrt{3 - x} + 3$

**48.** $x = \sqrt{x - 4} + 4$

**49.** $\sqrt{3x + 3} = 3\sqrt{x - 1}$

**50.** $2\sqrt{4x + 5} = 5\sqrt{x + 4}$

**51.** $2\sqrt{3x + 4} = \sqrt{5x + 9}$

**52.** $\sqrt{10 - 3x} = \sqrt{2x + 20}$

**53.** $\sqrt{3x + 6} = 2\sqrt{2x - 11}$

**54.** $2\sqrt{9x + 16} = \sqrt{3x + 64}$

**55.** $\sqrt[3]{x - 1} = 4$

**56.** $\sqrt[3]{2x + 5} = 3$

**57.** $\sqrt[3]{\dfrac{1}{2}x - 3} = 2$

**58.** $\sqrt[4]{x + 4} = 1$

*In Exercises 59–66, find the distance between points P and Q.*

**59.** $P(3, -4)$ and $Q(0, 0)$

**60.** $P(0, 0)$ and $Q(-6, 8)$

**61.** $P(2, 4)$ and $Q(5, 8)$

**62.** $P(5, 9)$ and $Q(8, 13)$

**63.** $P(-2, -8)$ and $Q(3, 4)$

**64.** $P(-5, -2)$ and $Q(7, 3)$

**65.** $P(6, 8)$ and $Q(12, 16)$

**66.** $P(10, 4)$ and $Q(2, -2)$

**APPLICATIONS** ▦ *Use a calculator to help solve each problem.*

**67. Falling objects** The distance $s$ (in feet) that an object will fall in $t$ seconds is given by the formula

$$t = \frac{\sqrt{s}}{4}$$

How deep is a shaft if a stone dropped down it hits bottom in 4 seconds?

**68. Falling objects** How deep would the shaft in Exercise 67 be if the stone hit bottom in 3 seconds?

**69. Horology** The time $t$ (in seconds) required for a pendulum to swing through one cycle is given by the formula $t = 1.11\sqrt{L}$. Find the length $L$ of a pendulum that completes one cycle in $\frac{3}{2}$ seconds.

**70. Foucault pendulum**    A long pendulum in Chicago's Museum of Science and Industry completes one cycle in 8.91 seconds. (See Illustration 1.) How long is it?

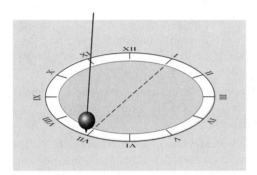

ILLUSTRATION 1

**71. Electronics**    The current $I$ (in amperes), the resistance $R$ (in ohms), and the power $P$ (in watts) are related by the formula

$$I = \sqrt{\frac{P}{R}}$$

Find the power used by an electrical appliance that draws 7 amps when the resistance is 20 ohms.

**72. Electronics**    Find the resistance of a 500-watt space heater that draws 7 amperes. (See Exercise 71.)

**73. Road safety**    The formula $s = k\sqrt{d}$ relates the speed $s$ (in mph) of a car and the distance $d$ of the skid when a driver hits the brakes. For wet pavement, $k = 3.24$. How far will a car skid if it is going 56 mph?

**74. Road safety**    How far will the car in Exercise 73 skid if it is going 80 mph?

**75. Road safety**    How far will the car in Exercise 73 skid if it is going 50 mph on dry pavement? On dry pavement, $k = 5.34$.

**76. Road safety**    How far will the car in Exercise 73 skid if it is going 80 mph on dry pavement? On dry pavement, $k = 5.34$.

**77. Satellite orbits**    The orbital speed $s$ of an earth satellite is related to its distance $r$ from the earth's center by the formula

$$s = \frac{2.029 \times 10^7}{\sqrt{r}}$$

If the satellite's orbital speed is $7 \times 10^3$ meters per second, find its altitude $a$ above the earth's surface. (See Illustration 2.)

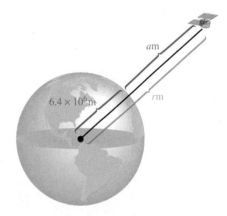

ILLUSTRATION 2

**78. Highway design**    A curve banked at 8° will accommodate traffic traveling $s$ miles per hour if the radius of the curve is $r$ feet, according to the equation $s = 1.45\sqrt{r}$. If highway engineers expect 65-mph traffic, what radius should they specify? (See Illustration 3.)

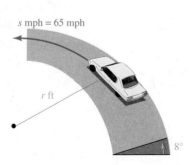

ILLUSTRATION 3

**79. Relativity** Einstein's theory of relativity predicts that an object moving at speed $v$ will be shortened in the direction of its motion by a factor $f$ given by

$$f = \sqrt{1 - \frac{v^2}{c^2}}$$

where $c$ is the speed of light. Solve this formula for $v^2$.

**80. Relativity** Einstein's theory of relativity predicts that a clock moving at speed $v$ will run slower by a factor $f$ given by

$$f = \frac{1}{\sqrt{1 - \frac{v^2}{c^2}}}$$

where $c$ is the speed of light. Solve this equation for $v^2$.

**81. Windmills** The power produced by a certain windmill is related to the speed of the wind by the formula

$$s = \sqrt[3]{\frac{P}{0.02}}$$

where $P$ is the power (in watts) and $s$ is the speed of the wind (in mph). How much power will the windmill produce if the wind is blowing at 30 mph?

**82. Windmills** If the wind is blowing at 20 mph, how much power is the windmill in Exercise 81 producing?

**83. Road construction** If the freeway of Example 8 were to go from 21st Street and 4th Avenue to 120th Street and 70th Avenue, how long would it be?

**84. Road construction** If the freeway of Example 8 were to go from 10th Street and 3rd Avenue to 100th Street and 60th Avenue, how long would it be?

*WRITING*

**85.** Explain why a check is necessary when solving radical equations.

**86.** How would you know, without solving it, that the equation $\sqrt{x + 2} = -4$ has no solutions?

*SOMETHING TO THINK ABOUT*

**87.** Solve the equation $\sqrt[4]{3x + 4} = 5$.

**88.** Solve the equation $\sqrt{\sqrt{x + 2}} = 3$.

# 7.4  Simplifying Radical Expressions

■ THE MULTIPLICATION PROPERTY OF RADICALS  ■ SIMPLIFYING RADICALS  ■ THE DIVISION PROPERTY OF RADICALS  ■ SIMPLIFYING CUBE ROOTS

Getting Ready  *Simplify each radical. Assume that all variables represent positive numbers.*

**1.** $\sqrt{100}$     **2.** $\sqrt{4}$     **3.** $\sqrt{25}$     **4.** $\sqrt{144}$

**5.** $\sqrt{9x^2}$     **6.** $\sqrt{16x^4}$     **7.** $\sqrt[3]{27x^3y^6}$     **8.** $\sqrt[3]{-8x^6y^9}$

## THE MULTIPLICATION PROPERTY OF RADICALS

We introduce the first of two properties of radicals with the following examples:

$$\sqrt{4 \cdot 25} = \sqrt{100} \qquad\qquad \sqrt{4}\sqrt{25} = 2 \cdot 5$$
$$= 10 \qquad\qquad\qquad\qquad = 10$$

In each case, the answer is 10. Thus, $\sqrt{4 \cdot 25} = \sqrt{4}\sqrt{25}$. Likewise,

$$\sqrt{9 \cdot 16} = \sqrt{144} \qquad\qquad \sqrt{9}\sqrt{16} = 3 \cdot 4$$
$$= 12 \qquad\qquad\qquad\qquad = 12$$

In each case, the answer is 12. Thus, $\sqrt{9 \cdot 16} = \sqrt{9}\sqrt{16}$. These results suggest the **multiplication property of radicals.**

> **Multiplication Property of Radicals**
> If $a$ and $b$ are positive or 0, then
> $$\sqrt{ab} = \sqrt{a}\sqrt{b}$$

In words, *the square root of the product of two nonnegative numbers is equal to the product of their square roots.*

## SIMPLIFYING RADICALS

A radical is in simplified form when each of the following statements is true.

> **Simplified Form of a Square Root Radical**
> **1.** Except for 1, the radicand has no perfect square factors.
> **2.** No fraction appears in a radicand.
> **3.** No radical appears in the denominator of a fraction.

We can use the multiplication property of radicals to simplify radicals that have perfect square factors. For example, we can simplify $\sqrt{12}$ as follows:

$$\sqrt{12} = \sqrt{4 \cdot 3} \qquad \text{Factor 12 as } 4 \cdot 3, \text{ where 4 is a perfect square.}$$
$$= \sqrt{4}\sqrt{3} \qquad \text{Use the multiplication property of radicals:}$$
$$\sqrt{a \cdot b} = \sqrt{a}\sqrt{b}.$$
$$= 2\sqrt{3} \qquad \text{Write } \sqrt{4} \text{ as 2.}$$

To simplify more difficult radicals, we need to know the integers that are perfect squares. For example, 81 is a perfect square, because $9^2 = 81$. The first 20 integer squares are

1, 4, 9, 16, 25, 36, 49, 64, 81, 100, 121, 144, 169, 196, 225, 256, 289, 324, 361, 400

Expressions with variables can also be perfect squares. For example, $9x^4y^2$ is a perfect square, because

$$9x^4y^2 = (3x^2y)^2$$

**EXAMPLE 1**    Simplify $\sqrt{72x^3}$   ($x \geq 0$).

*Solution*    We factor $72x^3$ into two factors, one of which is the greatest perfect square that divides $72x^3$. Since

- 36 is the greatest perfect square that divides 72, and
- $x^2$ is the greatest perfect square that divides $x^3$,

the greatest perfect square that divides $72x^3$ is $36x^2$. We can now use the multiplication property of radicals and simplify to get

$$\sqrt{72x^3} = \sqrt{36x^2 \cdot 2x}$$
$$= \sqrt{36x^2}\sqrt{2x} \qquad \text{The square root of a product is equal to the product of the square roots.}$$
$$= 6x\sqrt{2x} \qquad \sqrt{36x^2} = 6x. \qquad \blacksquare$$

**Self Check**    Simplify $\sqrt{50y^3}$.
*Answer*    $5y\sqrt{2y}$

**EXAMPLE 2**    Simplify $\sqrt{45x^2y^3}$   ($x \geq 0$, $y \geq 0$).

*Solution*    We look for the greatest perfect square that divides $45x^2y^3$. Because

- 9 is the greatest perfect square that divides 45,
- $x^2$ is the greatest perfect square that divides $x^2$, and
- $y^2$ is the greatest perfect square that divides $y^3$,

the factor $9x^2y^2$ is the greatest perfect square that divides $45x^2y^3$.
We can now use the multiplication property of radicals and simplify to get

$$\sqrt{45x^2y^3} = \sqrt{9x^2y^2 \cdot 5y}$$
$$= \sqrt{9x^2y^2}\sqrt{5y} \qquad \text{The square root of a product is equal to the product of square roots.}$$
$$= 3xy\sqrt{5y} \qquad \sqrt{9x^2y^2} = 3xy. \qquad \blacksquare$$

**Self Check**    Simplify $\sqrt{63a^3b^2}$.
*Answer*    $3ab\sqrt{7a}$

**EXAMPLE 3**    Simplify $3a\sqrt{288a^5b^7}$    $(a \geq 0,\ b \geq 0)$.

*Solution*    We look for the greatest perfect square that divides $288a^5b^7$. Because

- 144 is the greatest perfect square that divides 288,
- $a^4$ is the greatest perfect square that divides $a^5$, and
- $b^6$ is the greatest perfect square that divides $b^7$,

the factor $144a^4b^6$ is the greatest perfect square that divides $288a^5b^7$.
We can now use the multiplication property of radicals and simplify to get

$$
\begin{aligned}
3a\sqrt{288a^5b^7} &= 3a\sqrt{144a^4b^6 \cdot 2ab} \\
&= 3a\sqrt{144a^4b^6}\sqrt{2ab} \quad \text{\small The square root of a product is equal} \\
&\qquad\qquad\qquad\qquad\qquad \text{\small to the product of the square roots.} \\
&= 3a\left(12a^2b^3\sqrt{2ab}\right) \quad \text{\small } \sqrt{144a^4b^6} = 12a^2b^3. \\
&= 36a^3b^3\sqrt{2ab}
\end{aligned}
$$

&#9632;

*Self Check*
*Answer*

Simplify $5p\sqrt{300p^3q^9}$.
$50p^2q^4\sqrt{3pq}$

&#9632; THE DIVISION PROPERTY OF RADICALS

To find the second property of radicals, we consider these examples.

$$
\sqrt{\frac{100}{25}} = \sqrt{4} \qquad \text{and} \qquad \frac{\sqrt{100}}{\sqrt{25}} = \frac{10}{5}
$$
$$
\qquad\qquad = 2 \qquad\qquad\qquad\qquad\qquad = 2
$$

Since the answer is 2 in each case, $\sqrt{\frac{100}{25}} = \frac{\sqrt{100}}{\sqrt{25}}$. Likewise,

$$
\sqrt{\frac{36}{4}} = \sqrt{9} \qquad\qquad \frac{\sqrt{36}}{\sqrt{4}} = \frac{6}{2}
$$
$$
\qquad\qquad = 3 \qquad\qquad\qquad\qquad = 3
$$

Since the answer is 3 in each case, $\sqrt{\frac{36}{4}} = \frac{\sqrt{36}}{\sqrt{4}}$. These results suggest the **division property of radicals.**

**Division Property of Radicals**
If $a \geq 0$ and $b > 0$, then
$$
\sqrt{\frac{a}{b}} = \frac{\sqrt{a}}{\sqrt{b}}
$$

In words, *the square root of the quotient of two numbers is the quotient of their square roots.*

We can use the division property of radicals to simplify radicals that have fractions in their radicands. For example,

$$\sqrt{\frac{59}{49}} = \frac{\sqrt{59}}{\sqrt{49}}$$
$$= \frac{\sqrt{59}}{7} \qquad \sqrt{49} = 7.$$

**EXAMPLE 4**  Simplify $\sqrt{\dfrac{108}{25}}$.

*Solution*  $\sqrt{\dfrac{108}{25}} = \dfrac{\sqrt{108}}{\sqrt{25}}$   The square root of a quotient is equal to the quotient of the square roots.

$= \dfrac{\sqrt{36 \cdot 3}}{5}$   Factor 108 using the factorization involving 36, the largest perfect square factor of 108, and write $\sqrt{25}$ as 5.

$= \dfrac{\sqrt{36}\sqrt{3}}{5}$   The square root of a product is equal to the product of the square roots.

$= \dfrac{6\sqrt{3}}{5}$   $\sqrt{36} = 6.$   ∎

*Self Check*  Simplify $\sqrt{\frac{20}{49}}$.

*Answer*  $\dfrac{2\sqrt{5}}{7}$

**EXAMPLE 5**  Simplify $\sqrt{\dfrac{44x^3}{9xy^2}}$   $(x > 0, y > 0)$.

*Solution*  $\sqrt{\dfrac{44x^3}{9xy^2}} = \sqrt{\dfrac{44x^2}{9y^2}}$   Simplify the fraction by dividing out the common factor of $x$.

$= \dfrac{\sqrt{44x^2}}{\sqrt{9y^2}}$   The square root of a quotient is equal to the quotient of the square roots.

$= \dfrac{\sqrt{4x^2}\sqrt{11}}{\sqrt{9y^2}}$   The square root of a product is equal to the product of the square roots: $\sqrt{44x^2} = \sqrt{4x^2 \cdot 11}$ $= \sqrt{4x^2}\sqrt{11}.$

$= \dfrac{2x\sqrt{11}}{3y}$   $\sqrt{4x^2} = 2x$ and $\sqrt{9y^2} = 3y.$   ∎

Self Check   Simplify $\sqrt{\dfrac{99b^3}{16a^2b}}$. Assume that all variables are positive numbers.

Answer   $\dfrac{3b\sqrt{11}}{4a}$

■ SIMPLIFYING CUBE ROOTS

The multiplication and division properties of radicals are also true for cube roots and higher. To simplify a cube root, we must know the following perfect cube integers:

1, 8, 27, 64, 125, 216, 343, 512, 729, 1,000

Expressions with variables can also be perfect cubes. For example, $27x^6y^3$ is a perfect cube, because

$$27x^6y^3 = (3x^2y)^3$$

**EXAMPLE 6**   Simplify   **a.** $\sqrt[3]{16x^3y^4}$   and   **b.** $\sqrt[3]{\dfrac{64n^4}{27m^3}}$.

*Solution*   **a.** We look for the greatest perfect cube that divides $16x^3y^4$. Because

- 8 is the greatest perfect cube that divides 16,
- $x^3$ is the greatest perfect cube that divides $x^3$, and
- $y^3$ is the greatest perfect cube that divides $y^4$,

the greatest perfect cube factor that divides $16x^3y^4$ is $8x^3y^3$.
We can now use the multiplication property of radicals to get

$$\sqrt[3]{16x^3y^4} = \sqrt[3]{8x^3y^3 \cdot 2y}$$
$$= \sqrt[3]{8x^3y^3}\sqrt[3]{2y} \qquad \text{The cube root of a product is equal to the product of the cube roots.}$$
$$= 2xy\sqrt[3]{2y} \qquad \sqrt[3]{8x^3y^3} = 2xy.$$

**b.** $\sqrt[3]{\dfrac{64n^4}{27m^3}} = \dfrac{\sqrt[3]{64n^4}}{\sqrt[3]{27m^3}}$   The cube root of a quotient is equal to the quotient of the cube roots.

$$= \dfrac{\sqrt[3]{64n^3}\sqrt[3]{n}}{3m} \qquad \text{Use the multiplication property of radicals, and write } \sqrt[3]{27m^3} \text{ as } 3m.$$

$$= \dfrac{4n\sqrt[3]{n}}{3m} \qquad \sqrt[3]{64n^3} = 4n.$$

■

Self Check   Simplify  **a.** $\sqrt[3]{54a^3b^5}$  and  **b.** $\sqrt[3]{\dfrac{27q^5}{64p^3}}$.

Answers   **a.** $3ab\sqrt[3]{2b^2}$,  **b.** $\dfrac{3q\sqrt[3]{q^2}}{4p}$

**WARNING!**  Note that $\sqrt{a+b} \neq \sqrt{a} + \sqrt{b}$ and $\sqrt{a-b} \neq \sqrt{a} - \sqrt{b}$.
To see that this is true, we consider these correct simplifications:

$$\sqrt{9+16} = \sqrt{25} = 5 \quad \text{and} \quad \sqrt{25-16} = \sqrt{9} = 3$$

It is incorrect to write

$$\sqrt{9+16} = \sqrt{9} + \sqrt{16} \qquad\qquad \sqrt{25-16} = \sqrt{25} - \sqrt{16}$$
$$= 3 + 4 \qquad\qquad\qquad\qquad\qquad = 5 - 4$$
$$= 7 \qquad\qquad\qquad\qquad\qquad\qquad = 1$$

Orals   *Simplify each radical ($x > 0$, $a > 0$).*

**1.** $\sqrt{8}$

**2.** $\sqrt{12}$

**3.** $\dfrac{\sqrt{5}}{\sqrt{9}}$

**4.** $\dfrac{\sqrt{7}}{\sqrt{25}}$

**5.** $\sqrt{4x^2}$

**6.** $\sqrt{9a^4}$

## EXERCISE 7.4

**REVIEW**   *Simplify each fraction.*

**1.** $\dfrac{5xy^2z^3}{10x^2y^2z^4}$

**2.** $\dfrac{35a^3b^2c}{63a^2b^3c^2}$

**3.** $\dfrac{a^2 - a - 2}{a^2 + a - 6}$

**4.** $\dfrac{y^2 + 3y - 18}{y^2 - 9}$

**VOCABULARY AND CONCEPTS**   *Fill in each blank to make a true statement.*

**5.** Squares of integers such as 4, 9, and 16 are called _____ squares.

**6.** Cubes of integers such as 8, 27, and 64 are called perfect _____.

**7.** $\sqrt{ab} =$ _____

**8.** $\sqrt{\dfrac{a}{b}} =$ _____  ($b \neq 0$)

*Tell what is wrong with each solution.*

**9.** $\sqrt{13} = \sqrt{9+4}$
   $= \sqrt{9} + \sqrt{4}$
   $= 3 + 2$
   $= 5$

**10.** $\sqrt{7} = \sqrt{16-9}$
   $= \sqrt{16} - \sqrt{9}$
   $= 4 - 3$
   $= 1$

**PRACTICE** *In Exercises 11–58, simplify each radical. Assume that all variables represent positive numbers.*

11. $\sqrt{20}$      12. $\sqrt{18}$      13. $\sqrt{50}$      14. $\sqrt{75}$

15. $\sqrt{45}$      16. $\sqrt{54}$      17. $\sqrt{98}$      18. $\sqrt{27}$

19. $\sqrt{48}$      20. $\sqrt{128}$     21. $\sqrt{200}$     22. $\sqrt{300}$

23. $\sqrt{192}$     24. $\sqrt{250}$     25. $\sqrt{88}$      26. $\sqrt{275}$

27. $\sqrt{324}$     28. $\sqrt{405}$     29. $\sqrt{147}$     30. $\sqrt{722}$

31. $\sqrt{180}$     32. $\sqrt{320}$     33. $\sqrt{432}$     34. $\sqrt{720}$

35. $4\sqrt{288}$    36. $2\sqrt{800}$    37. $-7\sqrt{1,000}$  38. $-3\sqrt{252}$

39. $2\sqrt{245}$    40. $3\sqrt{196}$    41. $-5\sqrt{162}$   42. $-4\sqrt{243}$

43. $\sqrt{25x}$     44. $\sqrt{36y}$     45. $\sqrt{a^2b}$     46. $\sqrt{rs^2}$

47. $\sqrt{9x^2y}$                        48. $\sqrt{16xy^2}$

49. $\dfrac{1}{5}x^2y\sqrt{50x^2y^2}$      50. $\dfrac{1}{5}x^5y\sqrt{75x^3y^2}$

51. $12x\sqrt{16x^2y^3}$                   52. $-4x^5y^3\sqrt{36x^3y^3}$

53. $-3xyz\sqrt{18x^3y^5}$                 54. $15xy^2\sqrt{72x^2y^3}$

55. $\dfrac{3}{4}\sqrt{192a^3b^5}$         56. $-\dfrac{2}{9}\sqrt{162r^3s^3t}$

57. $-\dfrac{2}{5}\sqrt{80mn^2}$           58. $\dfrac{5}{6}\sqrt{180ab^2c}$

*In Exercises 59–74, write each quotient as the quotient of two radicals and simplify.*

59. $\sqrt{\dfrac{25}{9}}$     60. $\sqrt{\dfrac{36}{49}}$    61. $\sqrt{\dfrac{81}{64}}$    62. $\sqrt{\dfrac{121}{144}}$

63. $\sqrt{\dfrac{26}{25}}$    64. $\sqrt{\dfrac{17}{169}}$   65. $\sqrt{\dfrac{20}{49}}$    66. $\sqrt{\dfrac{50}{9}}$

67. $\sqrt{\dfrac{48}{81}}$    68. $\sqrt{\dfrac{27}{64}}$    69. $\sqrt{\dfrac{32}{25}}$    70. $\sqrt{\dfrac{75}{16}}$

71. $\sqrt{\dfrac{125}{121}}$  72. $\sqrt{\dfrac{250}{49}}$   73. $\sqrt{\dfrac{245}{36}}$   74. $\sqrt{\dfrac{500}{81}}$

*In Exercises 75–82, simplify each expression. All variables represent positive numbers.*

75. $\sqrt{\dfrac{72x^3}{y^2}}$    76. $\sqrt{\dfrac{108a^3b^2}{c^2d^4}}$    77. $\sqrt{\dfrac{125m^2n^5}{64n}}$    78. $\sqrt{\dfrac{72p^5q^7}{16pq^3}}$

79. $\sqrt{\dfrac{128m^3n^5}{36mn^7}}$    80. $\sqrt{\dfrac{75p^3q^2}{9p^5q^4}}$    81. $\sqrt{\dfrac{12r^7s^6t}{81r^5s^2t}}$    82. $\sqrt{\dfrac{36m^2n^9}{100mn^3}}$

*In Exercises 83–96, simplify each cube root.*

**83.** $\sqrt[3]{8x^3}$

**84.** $\sqrt[3]{27x^3y^3}$

**85.** $\sqrt[3]{-64x^5}$

**86.** $\sqrt[3]{-16x^4y^3}$

**87.** $\sqrt[3]{54x^3y^4z^6}$

**88.** $\sqrt[3]{-24x^5y^5z^4}$

**89.** $\sqrt[3]{-81x^2y^3z^4}$

**90.** $\sqrt[3]{1,600xy^2z^3}$

**91.** $\sqrt[3]{\dfrac{27m^3}{8n^6}}$

**92.** $\sqrt[3]{\dfrac{125t^9}{27s^6}}$

**93.** $\sqrt[3]{\dfrac{16r^4s^5}{1,000t^3}}$

**94.** $\sqrt[3]{\dfrac{54m^4n^3}{r^3s^6}}$

**95.** $\sqrt[3]{\dfrac{250a^3b^4}{16b}}$

**96.** $\sqrt[3]{\dfrac{81p^5q^3}{1,000p^2q^6}}$

## WRITING

**97.** Explain the multiplication property of radicals.

**98.** Explain the division property of radicals.

## SOMETHING TO THINK ABOUT

**99.** Find the errors.

$$\left(\sqrt{a+b}\right)^2 = \left(\sqrt{a}+\sqrt{b}\right)^2$$
$$= \left(\sqrt{a}\right)^2 + \left(\sqrt{b}\right)^2$$
$$= a + b$$

In spite of the errors, is the conclusion correct?

**100.** Use scientific notation to simplify $\sqrt{0.00000004}$.

## 7.5 Adding and Subtracting Radical Expressions

■ COMBINING LIKE RADICALS  ■ COMBINING EXPRESSIONS WITH HIGHER-ORDER RADICALS

Getting Ready   *Combine like terms.*

**1.** $3x + 4x$   **2.** $5y - 2y$   **3.** $7xy - 2xy$   **4.** $7t^2 + 2t^2$

*Simplify each radical ($x \geq 0$, $y \geq 0$).*

**5.** $\sqrt{20}$   **6.** $\sqrt{45}$   **7.** $\sqrt{8x^2y}$   **8.** $\sqrt{18x^2y}$

## ■ COMBINING LIKE RADICALS

When adding monomials, we can combine **like terms.** For example,

$$3x + 5x = (3+5)x \qquad \text{Use the distributive property.}$$
$$= 8x$$

It is often possible to combine terms that contain *like radicals.*

### Like Radicals

Radicals are called **like radicals** when they have the same index and the same radicand.

Since the terms $3\sqrt{2}$ and $5\sqrt{2}$ contain like radicals, they are like terms and can be combined.

$$3\sqrt{2} + 5\sqrt{2} = (3 + 5)\sqrt{2} \qquad \text{Use the distributive property.}$$
$$= 8\sqrt{2}$$

Likewise,

$$5x\sqrt{3y} - 2x\sqrt{3y} = (5x - 2x)\sqrt{3y} \qquad \text{Use the distributive property.}$$
$$= 3x\sqrt{3y}$$

Radicals such as $3\sqrt{18}$ and $5\sqrt{8}$ can be simplified so that they contain like radicals. They can then be combined.

**EXAMPLE 1**   Simplify $3\sqrt{18} + 5\sqrt{8}$.

*Solution*   The radical $\sqrt{18}$ is not in simplified form, because 18 has a perfect square factor of 9. The radical $\sqrt{8}$ is not in simplified form either, because 8 has a perfect square factor of 4. To simplify the radicals and add them, we proceed as follows.

$$3\sqrt{18} + 5\sqrt{8} = 3\sqrt{9 \cdot 2} + 5\sqrt{4 \cdot 2} \qquad \text{Factor 18 and 8. Look for perfect square factors.}$$

$$= 3\sqrt{9}\sqrt{2} + 5\sqrt{4}\sqrt{2} \qquad \text{The square root of a product is equal to the product of the square roots.}$$

$$= 3(3)\sqrt{2} + 5(2)\sqrt{2} \qquad \sqrt{9} = 3 \text{ and } \sqrt{4} = 2.$$
$$= 9\sqrt{2} + 10\sqrt{2} \qquad 3(3) = 9 \text{ and } 5(2) = 10.$$
$$= 19\sqrt{2} \qquad \text{Combine like terms.} \qquad ■$$

*Self Check*   Simplify $2\sqrt{50} + \sqrt{32}$.

*Answer*   $14\sqrt{2}$

**EXAMPLE 2**    Simplify $\sqrt{20} + \sqrt{45} + 3\sqrt{5}$.

*Solution*    We simplify the first two radicals and combine like terms.

$$\sqrt{20} + \sqrt{45} + 3\sqrt{5}$$
$$= \sqrt{4 \cdot 5} + \sqrt{9 \cdot 5} + 3\sqrt{5} \qquad \text{Factor.}$$
$$= \sqrt{4}\sqrt{5} + \sqrt{9}\sqrt{5} + 3\sqrt{5} \qquad \begin{array}{l}\text{The square root of a product is equal}\\ \text{to the product of the square roots.}\end{array}$$
$$= 2\sqrt{5} + 3\sqrt{5} + 3\sqrt{5} \qquad \sqrt{4} = 2 \text{ and } \sqrt{9} = 3.$$
$$= 8\sqrt{5} \qquad \text{Combine like terms.} \qquad \blacksquare$$

**Self Check**    Simplify $\sqrt{12} + \sqrt{27} + \sqrt{75}$.
**Answer**    $10\sqrt{3}$

**EXAMPLE 3**    Simplify $\sqrt{8x^2y} + \sqrt{18x^2y}$   $(x > 0, y > 0)$.

*Solution*    We simplify each radical and combine like terms.

$$\sqrt{8x^2y} + \sqrt{18x^2y}$$
$$= \sqrt{4 \cdot 2x^2y} + \sqrt{9 \cdot 2x^2y} \qquad \text{Factor.}$$
$$= \sqrt{4x^2}\sqrt{2y} + \sqrt{9x^2}\sqrt{2y} \qquad \begin{array}{l}\text{The square root of a product is equal}\\ \text{to the product of the square roots.}\end{array}$$
$$= 2x\sqrt{2y} + 3x\sqrt{2y} \qquad \sqrt{4x^2} = 2x \text{ and } \sqrt{9x^2} = 3x.$$
$$= 5x\sqrt{2y} \qquad \text{Combine like terms.} \qquad \blacksquare$$

**Self Check**    Simplify $\sqrt{12xy^2} + \sqrt{27xy^2}$.
**Answer**    $5y\sqrt{3x}$

**EXAMPLE 4**    Simplify $\sqrt{28x^2y} - 2\sqrt{63y^3}$   $(x > 0, y > 0)$.

*Solution*    We simplify each radical and combine like terms.

$$\sqrt{28x^2y} - 2\sqrt{63y^3}$$
$$= \sqrt{4 \cdot 7x^2y} - 2\sqrt{9 \cdot 7y^2y} \qquad \text{Factor.}$$
$$= \sqrt{4x^2}\sqrt{7y} - 2\sqrt{9y^2}\sqrt{7y} \qquad \begin{array}{l}\text{The square root of a product is equal}\\ \text{to the product of the square roots.}\end{array}$$
$$= 2x\sqrt{7y} - 2 \cdot 3y\sqrt{7y} \qquad \sqrt{4x^2} = 2x \text{ and } \sqrt{9y^2} = 3y.$$
$$= 2x\sqrt{7y} - 6y\sqrt{7y} \qquad \text{Simplify.}$$

Since the variables in the terms are different, the expression does not simplify further. $\qquad \blacksquare$

Self Check

Answer

Simplify $\sqrt{20xy^2} - \sqrt{80x^3}$.

$2y\sqrt{5x} - 4x\sqrt{5x}$

**EXAMPLE 5** Simplify $\sqrt{27xy} + \sqrt{20xy}$ $(x > 0, y > 0)$.

Solution

$\sqrt{27xy} + \sqrt{20xy}$

$= \sqrt{9 \cdot 3xy} + \sqrt{4 \cdot 5xy}$      Factor.

$= \sqrt{9}\sqrt{3xy} + \sqrt{4}\sqrt{5xy}$      The square root of a product is equal to the product of the square roots.

$= 3\sqrt{3xy} + 2\sqrt{5xy}$      $\sqrt{9} = 3$ and $\sqrt{4} = 2$.

Since the terms have unlike radicals, the expression does not simplify further. ■

Self Check

Answer

Simplify $\sqrt{75ab} + \sqrt{72ab}$.

$5\sqrt{3ab} + 6\sqrt{2ab}$

**EXAMPLE 6** Simplify $\sqrt{8x} + \sqrt{3y} - \sqrt{50x} + \sqrt{27y}$ $(x > 0, y > 0)$.

Solution We simplify the radicals and combine like terms, when possible.

$\sqrt{8x} + \sqrt{3y} - \sqrt{50x} + \sqrt{27y}$

$= \sqrt{4 \cdot 2x} + \sqrt{3y} - \sqrt{25 \cdot 2x} + \sqrt{9 \cdot 3y}$

$= \sqrt{4}\sqrt{2x} + \sqrt{3y} - \sqrt{25}\sqrt{2x} + \sqrt{9}\sqrt{3y}$

$= 2\sqrt{2x} + \sqrt{3y} - 5\sqrt{2x} + 3\sqrt{3y}$

$= -3\sqrt{2x} + 4\sqrt{3y}$ ■

Self Check

Answer

Simplify $\sqrt{32x} - \sqrt{5y} - \sqrt{200x} + \sqrt{125y}$.

$-6\sqrt{2x} + 4\sqrt{5y}$

## ■ COMBINING EXPRESSIONS WITH HIGHER-ORDER RADICALS

It is often possible to combine terms containing like radicals other than square roots.

**EXAMPLE 7** Simplify $\sqrt[3]{81x^4} - x\sqrt[3]{24x}$.

Solution We simplify each radical and combine like terms, when possible.

$\sqrt[3]{81x^4} - x\sqrt[3]{24x} = \sqrt[3]{27x^3 \cdot 3x} - x\sqrt[3]{8 \cdot 3x}$

$= \sqrt[3]{27x^3}\sqrt[3]{3x} - x\sqrt[3]{8}\sqrt[3]{3x}$

$= 3x\sqrt[3]{3x} - 2x\sqrt[3]{3x}$

$= x\sqrt[3]{3x}$ ■

| Self Check | Simplify $\sqrt[3]{24a^4} + a\sqrt[3]{81a}$. |
|---|---|
| Answer | $5a\sqrt[3]{3a}$ |

Orals   *Combine like radicals.*

**1.** $3\sqrt{5} + 2\sqrt{5}$

**2.** $6\sqrt{7} - \sqrt{7}$

**3.** $4\sqrt{xy} - 2\sqrt{xy}$

**4.** $\sqrt{5yz} + 3\sqrt{5yz}$

**5.** $\sqrt{8} + \sqrt{2}$

**6.** $\sqrt{3} - \sqrt{12}$

## EXERCISE 7.5

***REVIEW***   *Solve each proportion.*

**1.** $\dfrac{a-2}{8} = \dfrac{a+10}{24}$

**2.** $\dfrac{6}{t+12} = \dfrac{18}{4t}$

**3.** $\dfrac{-2}{x+14} = \dfrac{6}{x-6}$

**4.** $\dfrac{y-4}{4} = \dfrac{y+2}{12}$

***VOCABULARY AND CONCEPTS***   *Fill in each blank to make a true statement.*

**5.** _____ have the same variables with the same exponents.

**6.** _____ have the same index and the same radicand.

*In Exercises 7–10, tell whether the terms contain like radicals.*

**7.** $5\sqrt{2}$ and $2\sqrt{3}$

**8.** $7\sqrt{3x}$ and $3\sqrt{3x}$

**9.** $125\sqrt[3]{13a}$ and $-\sqrt[3]{13a}$

**10.** $-17\sqrt[4]{5x}$ and $25\sqrt[3]{5x}$

**11.** What is wrong with the following work?
$$7\sqrt{5} - 3\sqrt{2} = 4\sqrt{3}$$
Use your calculator to show that the left-hand side is not equal to the right-hand side.

**12.** What is wrong with the following work?
$$12\sqrt{7} + 20\sqrt{11} = 32\sqrt{18}$$
Use your calculator to show that the left-hand side is not equal to the right-hand side.

***PRACTICE***   *In Exercises 13–36, find each sum.*

**13.** $\sqrt{12} + \sqrt{27}$

**14.** $\sqrt{20} + \sqrt{45}$

**15.** $\sqrt{48} + \sqrt{75}$

**16.** $\sqrt{48} + \sqrt{108}$

**17.** $\sqrt{45} + \sqrt{80}$

**18.** $\sqrt{80} + \sqrt{125}$

**19.** $\sqrt{125} + \sqrt{245}$

**20.** $\sqrt{36} + \sqrt{196}$

**21.** $\sqrt{20} + \sqrt{180}$

**22.** $\sqrt{80} + \sqrt{245}$

**23.** $\sqrt{160} + \sqrt{360}$

**24.** $\sqrt{12} + \sqrt{147}$

**25.** $3\sqrt{45} + 4\sqrt{245}$

**26.** $2\sqrt{28} + 7\sqrt{63}$

**27.** $2\sqrt{28} + 2\sqrt{112}$

**28.** $4\sqrt{63} + 6\sqrt{112}$

**29.** $5\sqrt{32} + 3\sqrt{72}$

**30.** $3\sqrt{72} + 2\sqrt{128}$

**31.** $3\sqrt{98} + 8\sqrt{128}$

**32.** $5\sqrt{90} + 7\sqrt{250}$

**33.** $\sqrt{20} + \sqrt{45} + \sqrt{80}$

**34.** $\sqrt{48} + \sqrt{27} + \sqrt{75}$

**35.** $\sqrt{24} + \sqrt{150} + \sqrt{240}$

**36.** $\sqrt{28} + \sqrt{63} + \sqrt{112}$

*In Exercises 37–56, find each difference, when possible.*

**37.** $\sqrt{18} - \sqrt{8}$

**38.** $\sqrt{32} - \sqrt{18}$

**39.** $\sqrt{9} - \sqrt{50}$

**40.** $\sqrt{50} - \sqrt{32}$

**41.** $\sqrt{72} - \sqrt{32}$

**42.** $\sqrt{98} - \sqrt{72}$

**43.** $\sqrt{12} - \sqrt{48}$

**44.** $\sqrt{48} - \sqrt{75}$

**45.** $\sqrt{108} - \sqrt{75}$

**46.** $\sqrt{147} - \sqrt{48}$

**47.** $\sqrt{1,000} - \sqrt{360}$

**48.** $\sqrt{180} - \sqrt{125}$

**49.** $2\sqrt{80} - 3\sqrt{125}$

**50.** $3\sqrt{245} - 2\sqrt{180}$

**51.** $8\sqrt{96} - 5\sqrt{24}$

**52.** $3\sqrt{216} - 3\sqrt{150}$

**53.** $\sqrt{288} - 3\sqrt{200}$

**54.** $\sqrt{392} - 2\sqrt{128}$

**55.** $5\sqrt{250} - 3\sqrt{160}$

**56.** $4\sqrt{490} - 3\sqrt{360}$

*In Exercises 57–68, simplify each expression.*

**57.** $\sqrt{12} + \sqrt{18} - \sqrt{27}$

**58.** $\sqrt{8} - \sqrt{50} + \sqrt{72}$

**59.** $\sqrt{200} - \sqrt{75} + \sqrt{48}$

**60.** $\sqrt{20} + \sqrt{80} - \sqrt{125}$

**61.** $\sqrt{24} - \sqrt{150} - \sqrt{54}$

**62.** $\sqrt{98} - \sqrt{300} + \sqrt{800}$

**63.** $\sqrt{200} + \sqrt{300} - \sqrt{75}$

**64.** $\sqrt{175} + \sqrt{125} - \sqrt{28}$

**65.** $\sqrt{48} - \sqrt{8} + \sqrt{27} - \sqrt{32}$

**66.** $\sqrt{162} + \sqrt{50} - \sqrt{75} - \sqrt{108}$

**67.** $\sqrt{147} + \sqrt{216} - \sqrt{108} - \sqrt{27}$

**68.** $\sqrt{180} - \sqrt{112} + \sqrt{45} - \sqrt{700}$

*In Exercises 69–82, simplify each expression. All variables represent positive numbers.*

**69.** $\sqrt{2x^2} + \sqrt{8x^2}$

**70.** $\sqrt{3y^2} - \sqrt{12y^2}$

**71.** $\sqrt{2x^3} + \sqrt{8x^3}$

**72.** $\sqrt{3y^3} - \sqrt{12y^3}$

**73.** $\sqrt{18x^2y} - \sqrt{27x^2y}$

**74.** $\sqrt{49xy} + \sqrt{xy}$

**75.** $\sqrt{32x^5} - \sqrt{18x^5}$

**76.** $\sqrt{27xy^3} - \sqrt{48xy^3}$

**77.** $3\sqrt{54x^2} + 5\sqrt{24x^2}$

**78.** $3\sqrt{24x^4y^3} + 2\sqrt{54x^4y^3}$

**79.** $y\sqrt{490y} - 2\sqrt{360y^3}$

**80.** $3\sqrt{20x} + 2\sqrt{63y}$

**81.** $\sqrt{20x^3y} + \sqrt{45x^5y^3} - \sqrt{80x^7y^5}$

**82.** $x\sqrt{48xy^2} - y\sqrt{27x^3} + \sqrt{75x^3y^2}$

*In Exercises 83–94, simplify each expression.*

**83.** $\sqrt[3]{16} + \sqrt[3]{54}$

**84.** $\sqrt[3]{24} - \sqrt[3]{81}$

**85.** $\sqrt[3]{81} - \sqrt[3]{24}$

**86.** $\sqrt[3]{32} + \sqrt[3]{108}$

**87.** $\sqrt[3]{40} + \sqrt[3]{125}$

**88.** $\sqrt[3]{3,000} - \sqrt[3]{192}$

**89.** $\sqrt[3]{x^4} - \sqrt[3]{x^7}$

**90.** $\sqrt[3]{8x^5} + \sqrt[3]{27x^8}$

**91.** $\sqrt[3]{192x^4y^5} - \sqrt[3]{24x^4y^5}$

**92.** $\sqrt[3]{24a^5b^4} + \sqrt[3]{81a^5b^4}$

**93.** $\sqrt[3]{135x^7y^4} - \sqrt[3]{40x^7y^4}$

**94.** $\sqrt[3]{56a^4b^5} + \sqrt[3]{7a^4b^5}$

*WRITING*

**95.** Explain why $\sqrt{3x} + \sqrt{2y}$ cannot be combined.

**96.** Explain why $\sqrt{4x}$ and $\sqrt[3]{4x}$ cannot be combined.

*SOMETHING TO THINK ABOUT*

**97.** What is wrong with the following work?

$$\sqrt{27} - \sqrt{75} = \sqrt{9}\sqrt{3} - \sqrt{25}\sqrt{3}$$
$$= 3\sqrt{3} - 5\sqrt{3}$$
$$= -2$$

**98.** What is wrong with the following work?

$$\sqrt{8} + \sqrt{12} = \sqrt{4}\sqrt{2} + \sqrt{4}\sqrt{3}$$
$$= \sqrt{4}(2) + \sqrt{4}(3)$$
$$= 2\sqrt{4} + 3\sqrt{4}$$
$$= 5\sqrt{4}$$

# 7.6 Multiplying and Dividing Radical Expressions

■ MULTIPLYING RADICAL EXPRESSIONS  ■ MULTIPLYING MONOMIALS CONTAINING RADICALS
■ MULTIPLYING POLYNOMIALS CONTAINING RADICALS BY MONOMIALS CONTAINING RADICALS
■ MULTIPLYING BINOMIALS CONTAINING RADICALS  ■ DIVIDING RADICAL EXPRESSIONS
■ RATIONALIZING MONOMIAL DENOMINATORS  ■ RATIONALIZING BINOMIAL DENOMINATORS

Getting Ready  *Do each operation and simplify, when possible.*

**1.** $x^2x^3$  **2.** $y^3y^4$  **3.** $\dfrac{x^5}{x^2}$  **4.** $\dfrac{y^8}{y^5}$

**5.** $x(x + 2)$  **6.** $2y^3(3y^2 - 4y)$
**7.** $(x + 2)(x - 3)$  **8.** $(2x + 3y)(3x + 2y)$

## ■ MULTIPLYING RADICAL EXPRESSIONS

Recall that the *product of the square roots of two nonnegative numbers is equal to the square root of the product of those numbers.* For example,

$$\sqrt{2}\sqrt{8} = \sqrt{2 \cdot 8} \qquad \sqrt{3}\sqrt{27} = \sqrt{3 \cdot 27} \qquad \sqrt{x}\sqrt{x^3} = \sqrt{x \cdot x^3}$$
$$= \sqrt{16} \qquad\qquad = \sqrt{81} \qquad\qquad = \sqrt{x^4}$$
$$= 4 \qquad\qquad\quad = 9 \qquad\qquad\quad = x^2$$

Likewise, the *product of the cube roots of two numbers is equal to the cube root of the product of those numbers.* For example,

$$\sqrt[3]{2} \cdot \sqrt[3]{4} = \sqrt[3]{2 \cdot 4} \qquad \sqrt[3]{4} \cdot \sqrt[3]{16} = \sqrt[3]{4 \cdot 16} \qquad \sqrt[3]{3x^2} \cdot \sqrt[3]{9x} = \sqrt[3]{3x^2 \cdot 9x}$$
$$= \sqrt[3]{8} \qquad\qquad = \sqrt[3]{64} \qquad\qquad = \sqrt[3]{27x^3}$$
$$= 2 \qquad\qquad\quad = 4 \qquad\qquad\quad = 3x$$

### ■ MULTIPLYING MONOMIALS CONTAINING RADICALS

To multiply monomials containing radical expressions, we multiply the coefficients and multiply the radicals separately and then simplify the result, when possible.

**EXAMPLE 1**  Multiply **a.** $3\sqrt{6}$ by $4\sqrt{3}$ and **b.** $-2\sqrt[3]{7x}$ by $6\sqrt[3]{49x^2}$.

*Solution*  The commutative and associative properties enable us to multiply the integers and the radicals separately.

**a.** $3\sqrt{6} \cdot 4\sqrt{3} = 3(4)\sqrt{6}\sqrt{3}$
$\qquad = 12\sqrt{18}$
$\qquad = 12\sqrt{9}\sqrt{2}$
$\qquad = 12(3)\sqrt{2}$
$\qquad = 36\sqrt{2}$

**b.** $-2\sqrt[3]{7x} \cdot 6\sqrt[3]{49x^2} = -2(6)\sqrt[3]{7x}\sqrt[3]{49x^2}$
$\qquad = -12\sqrt[3]{7 \cdot 49 \cdot x \cdot x^2}$
$\qquad = -12\sqrt[3]{343x^3}$
$\qquad = -12(7x)$
$\qquad = -84x$  ■

**Self Check**  Multiply **a.** $(2\sqrt{2x})(3\sqrt{3x})$ and **b.** $(5\sqrt[3]{2})(-2\sqrt[3]{4})$.
**Answers**  **a.** $6x\sqrt{6}$,  **b.** $-20$

### ■ MULTIPLYING POLYNOMIALS CONTAINING RADICALS BY MONOMIALS CONTAINING RADICALS

Recall that to multiply a polynomial by a monomial, we use the distributive property to remove parentheses and combine like terms.

**EXAMPLE 2**  Multiply **a.** $\sqrt{2x}(\sqrt{6x} + \sqrt{8x})$ and **b.** $\sqrt[3]{3}(\sqrt[3]{9} - 2)$.

*Solution*  **a.** $\sqrt{2x}(\sqrt{6x} + \sqrt{8x}) = \sqrt{2x}\sqrt{6x} + \sqrt{2x}\sqrt{8x}$ 　　Use the distributive property to remove parentheses.

$\qquad\qquad = \sqrt{12x^2} + \sqrt{16x^2}$ 　　The product of two square roots is equal to the square root of the product.

$\qquad\qquad = \sqrt{4 \cdot 3 \cdot x^2} + \sqrt{16x^2}$ 　　Factor 12.

$\qquad\qquad = \sqrt{4x^2}\sqrt{3} + \sqrt{16x^2}$ 　　The square root of a product is equal to the product of the square roots.

$\qquad\qquad = 2x\sqrt{3} + 4x$ 　　$\sqrt{4x^2} = 2x$ and $\sqrt{16x^2} = 4x$.

**b.** $\sqrt[3]{3}(\sqrt[3]{9} - 2) = \sqrt[3]{3}\sqrt[3]{9} - \sqrt[3]{3} \cdot 2$ 　　Use the distributive property to remove parentheses.

$\qquad\qquad = \sqrt[3]{27} - 2\sqrt[3]{3}$ 　　The product of two cube roots is equal to the cube root of the product.

$\qquad\qquad = 3 - 2\sqrt[3]{3}$ 　　$\sqrt[3]{27} = 3$.  ■

Self Check     Multiply   **a.** $\sqrt{3}\left(\sqrt{6} - \sqrt{3}\right)$   and   **b.** $\sqrt[3]{2x}\left(3 - \sqrt[3]{4x^2}\right)$.
Answers      **a.** $3\sqrt{2} - 3$,   **b.** $3\sqrt[3]{2x} - 2x$

■   MULTIPLYING BINOMIALS CONTAINING RADICALS

To multiply two binomials, we multiply each term of one binomial by each term of the other binomial and simplify.

EXAMPLE 3     Multiply $\left(\sqrt{3} + \sqrt{2}\right)\left(\sqrt{3} - \sqrt{2}\right)$.

Solution     We can find the product with the FOIL method.

$$\left(\sqrt{3} + \sqrt{2}\right)\left(\sqrt{3} - \sqrt{2}\right)$$

$$= \sqrt{3}\sqrt{3} - \sqrt{3}\sqrt{2} + \sqrt{2}\sqrt{3} - \sqrt{2}\sqrt{2} \qquad \text{Use the FOIL method.}$$

$$= 3 - 2 \qquad\qquad\qquad \text{Combine like terms}$$
$$\qquad\qquad\qquad\qquad\qquad\qquad \text{and simplify.}$$

$$= 1 \qquad\qquad\qquad\qquad \text{Simplify.} \qquad ■$$

Self Check     Multiply $\left(\sqrt{5} + \sqrt{3}\right)\left(\sqrt{5} - \sqrt{3}\right)$.
Answer      2

EXAMPLE 4     Multiply $\left(\sqrt{3x} + 1\right)\left(\sqrt{3x} + 2\right)$.

Solution     $\left(\sqrt{3x} + 1\right)\left(\sqrt{3x} + 2\right)$

$$= \sqrt{3x}\sqrt{3x} + 2\sqrt{3x} + \sqrt{3x} + 2 \qquad \text{Use the FOIL method.}$$
$$= 3x + 3\sqrt{3x} + 2 \qquad\qquad\qquad \text{Combine like terms and simplify.}$$
$$\qquad\qquad\qquad\qquad\qquad\qquad\qquad\qquad\qquad\qquad ■$$

Self Check     Multiply $\left(\sqrt{5a} - 2\right)\left(\sqrt{5a} + 3\right)$.
Answer      $5a + \sqrt{5a} - 6$

EXAMPLE 5     Multiply $\left(\sqrt[3]{4x} - 3\right)\left(\sqrt[3]{2x^2} + 1\right)$.

*Solution*

$$\left(\sqrt[3]{4x} - 3\right)\left(\sqrt[3]{2x^2} + 1\right)$$

$$= \sqrt[3]{4x}\sqrt[3]{2x^2} + 1 \cdot \sqrt[3]{4x} - 3\sqrt[3]{2x^2} - 3 \qquad \text{Use the FOIL method.}$$

$$= \sqrt[3]{8x^3} + \sqrt[3]{4x} - 3\sqrt[3]{2x^2} - 3 \qquad \begin{array}{l}\text{The product of two cube} \\ \text{roots is equal to the cube} \\ \text{root of the product.}\end{array}$$

$$= 2x + \sqrt[3]{4x} - 3\sqrt[3]{2x^2} - 3 \qquad \text{Simplify.} \qquad \blacksquare$$

*Self Check*

*Answer*

Multiply $\left(\sqrt[3]{3x} + 1\right)\left(\sqrt[3]{9x^2} - 2\right)$.

$3x - 2\sqrt[3]{3x} + \sqrt[3]{9x^2} - 2$

## ■ DIVIDING RADICAL EXPRESSIONS

To divide radical expressions, we use the division property of radicals. For example, to divide $\sqrt{108}$ by $\sqrt{36}$, we proceed as follows:

$$\frac{\sqrt{108}}{\sqrt{36}} = \sqrt{\frac{108}{36}} \qquad \begin{array}{l}\text{The quotient of two square roots is the square root} \\ \text{of the quotient.}\end{array}$$

$$= \sqrt{3}$$

**EXAMPLE 6** Divide $\dfrac{\sqrt{22a^2b^7}}{\sqrt{99a^4b^3}}$ $(a > 0, b > 0)$.

*Solution*

$$\frac{\sqrt{22a^2b^7}}{\sqrt{99a^4b^3}} = \sqrt{\frac{22a^2b^7}{99a^4b^3}}$$

$$= \sqrt{\frac{2b^4}{9a^2}} \qquad \text{Simplify the radicand.}$$

$$= \frac{\sqrt{2b^4}}{\sqrt{9a^2}} \qquad \begin{array}{l}\text{The square root of a quotient is equal to the} \\ \text{quotient of the square roots.}\end{array}$$

$$= \frac{\sqrt{b^4}\sqrt{2}}{\sqrt{9a^2}} \qquad \begin{array}{l}\text{The square root of a product is equal to the product} \\ \text{of the square roots.}\end{array}$$

$$= \frac{b^2\sqrt{2}}{3a} \qquad \text{Simplify the radicals.} \qquad \blacksquare$$

*Self Check*

*Answer*

Divide $\dfrac{\sqrt{44x^3y^9}}{\sqrt{99x^5y^5}}$.

$\dfrac{2y^2}{3x}$

### ■ RATIONALIZING MONOMIAL DENOMINATORS

We can **rationalize the denominator** to simplify fractions that have radicals in their denominators. For example, we can eliminate the radical in the denominator of $\frac{1}{\sqrt{2}}$ by multiplying both the numerator and the denominator by $\sqrt{2}$. We note that $(\sqrt{2})(\sqrt{2})$ is the rational number 2.

$$\frac{1}{\sqrt{2}} = \frac{1\sqrt{2}}{\sqrt{2}\sqrt{2}} \qquad \text{Multiply both numerator and denominator by } \sqrt{2}.$$

$$= \frac{\sqrt{2}}{2} \qquad 1 \cdot \sqrt{2} = \sqrt{2} \text{ and } \sqrt{2}\sqrt{2} = 2.$$

**EXAMPLE 7**    Rationalize each denominator:   **a.** $\dfrac{3}{\sqrt{3}}$   and   **b.** $\dfrac{2}{\sqrt[3]{3}}$.

*Solution*    **a.** We multiply the numerator and denominator by $\sqrt{3}$ and simplify.

$$\frac{3}{\sqrt{3}} = \frac{3\sqrt{3}}{\sqrt{3}\sqrt{3}} \qquad \text{Multiply the numerator and denominator by } \sqrt{3}.$$

$$= \frac{3\sqrt{3}}{3} \qquad \sqrt{3}\sqrt{3} = 3.$$

$$= \sqrt{3} \qquad \frac{3}{3} = 1.$$

**b.** Since $\sqrt[3]{3}\sqrt[3]{9} = \sqrt[3]{27}$ and 27 is a perfect integer cube, we multiply the numerator and denominator by $\sqrt[3]{9}$ and simplify.

$$\frac{2}{\sqrt[3]{3}} = \frac{2\sqrt[3]{9}}{\sqrt[3]{3}\sqrt[3]{9}} = \frac{2\sqrt[3]{9}}{\sqrt[3]{27}}$$

$$= \frac{2\sqrt[3]{9}}{3} \qquad ■$$

*Self Check*    Rationalize each denominator:   **a.** $\dfrac{2}{\sqrt{5}}$   and   **b.** $\dfrac{5}{\sqrt[3]{5}}$.

*Answers*    **a.** $\dfrac{2\sqrt{5}}{5}$,   **b.** $\sqrt[3]{25}$

**EXAMPLE 8**    Divide $\dfrac{5}{\sqrt{20x}}$   $(x > 0)$.

*Solution*    To rationalize the denominator, we do not need to multiply numerator and denominator by $\sqrt{20x}$. To keep the numbers small, we can multiply by $\sqrt{5x}$, because $5x \cdot 20x = 100x^2$, which is a perfect square.

$$\frac{5}{\sqrt{20x}} = \frac{5\sqrt{5x}}{\sqrt{20x}\sqrt{5x}}$$ Multiply numerator and denominator by $\sqrt{5x}$.

$$= \frac{5\sqrt{5x}}{\sqrt{100x^2}}$$ The product of two square roots is equal to the square root of the product.

$$= \frac{5\sqrt{5x}}{10x}$$ $\sqrt{100x^2} = 10x$.

$$= \frac{\sqrt{5x}}{2x}$$ $\frac{5}{10} = \frac{1}{2}$. ∎

**Self Check**  Divide $\dfrac{6}{\sqrt{30y}}$.

*Answer*  $\dfrac{\sqrt{30y}}{5y}$

**EXAMPLE 9**  Divide $\sqrt{\dfrac{3x^3y^2}{27xy^3}}$  $(x > 0, y > 0)$.

*Solution*  $\sqrt{\dfrac{3x^3y^2}{27xy^3}} = \sqrt{\dfrac{x^2}{9y}}$  Simplify the fraction within the radical.

$$= \sqrt{\dfrac{x^2 \cdot y}{9y \cdot y}}$$  Multiply both numerator and denominator by $y$.

$$= \dfrac{\sqrt{x^2y}}{\sqrt{9y^2}}$$  The square root of a quotient is the quotient of the square roots.

$$= \dfrac{x\sqrt{y}}{3y}$$  Simplify. ∎

**Self Check**  Divide $\sqrt{\dfrac{3a^5b}{108ab^2}}$.

*Answer*  $\dfrac{a^2\sqrt{b}}{6b}$

## ■ RATIONALIZING BINOMIAL DENOMINATORS

Since the denominators of many fractions such as $\frac{2}{\sqrt{3}-1}$ contain radicals, they are not in simplified form. Because the denominator is a *binomial*, multiplying the denominator by $\sqrt{3}$ will not make it a rational number. The key to rationalizing the denominator is to multiply the numerator and denominator by $\sqrt{3} + 1$, because the product $(\sqrt{3} + 1)(\sqrt{3} - 1)$ has no radicals. Radical expressions such as $\sqrt{3} + 1$ and $\sqrt{3} - 1$ are called **conjugates** of each other.

**EXAMPLE 10**  Divide $\dfrac{2}{\sqrt{3} - 1}$.

*Solution*  Multiply the numerator and denominator by the conjugate of the denominator.

$$\frac{2}{\sqrt{3} - 1} = \frac{2(\sqrt{3} + 1)}{(\sqrt{3} - 1)(\sqrt{3} + 1)} \qquad \text{Multiply numerator and denominator by the conjugate of the denominator.}$$

$$= \frac{2(\sqrt{3} + 1)}{3 - 1} \qquad \text{Multiply the binomials in the denominator.}$$

$$= \frac{2(\sqrt{3} + 1)}{2} \qquad \text{Simplify.}$$

$$= \sqrt{3} + 1 \qquad \text{Divide out the common factor of 2.} \qquad \blacksquare$$

**Self Check**  Divide $\dfrac{3}{\sqrt{2} + 1}$.

**Answer**  $3(\sqrt{2} - 1)$

**EXAMPLE 11**  Divide $\dfrac{\sqrt{x} + 1}{\sqrt{x} - 1}$.

*Solution*  We multiply the numerator and denominator by the conjugate of the denominator, which is $\sqrt{x} + 1$.

$$\frac{\sqrt{x} + 1}{\sqrt{x} - 1} = \frac{(\sqrt{x} + 1)(\sqrt{x} + 1)}{(\sqrt{x} - 1)(\sqrt{x} + 1)} \qquad \text{Multiply numerator and denominator by } \sqrt{x} + 1.$$

$$= \frac{\sqrt{x}\sqrt{x} + \sqrt{x}(1) + 1(\sqrt{x}) + 1}{\sqrt{x}\sqrt{x} + \sqrt{x}(1) - 1(\sqrt{x}) - 1} \qquad \text{Multiply the binomials.}$$

$$= \frac{x + 2\sqrt{x} + 1}{x - 1} \qquad \text{Simplify.} \qquad \blacksquare$$

**Self Check**  Divide $\dfrac{\sqrt{x} - 1}{\sqrt{x} + 1}$.

**Answer**  $\dfrac{x - 2\sqrt{x} + 1}{x - 1}$

**Orals**   *Do each multiplication. All variables represent positive numbers.*

**1.** $\sqrt{5}\sqrt{5}$                                    **2.** $\sqrt{2}\sqrt{50}$

**3.** $\sqrt{x}(\sqrt{x} + 2)$                        **4.** $\sqrt{y}(4 - \sqrt{y})$

**5.** $(\sqrt{x} + 1)(\sqrt{x} - 1)$              **6.** $(1 + \sqrt{z})(1 - \sqrt{z})$

*Rationalize each denominator.*

**7.** $\dfrac{1}{\sqrt{5}}$                                    **8.** $\dfrac{x}{\sqrt{x}}$

## EXERCISE 7.6

**REVIEW**   *Factor each polynomial.*

**1.** $x^2 - 4x - 21$

**2.** $y^2 + 6y - 27$

**3.** $6x^2y - 15xy$

**4.** $x^3 + 8$

**VOCABULARY AND CONCEPTS**   *Fill in each blank to make a true statement.*

**5.** The symbol $\sqrt{\phantom{x}}$ is called a _____ sign.

**6.** The method of changing a radical denominator of a fraction into a rational number is called _____ the denominator.

**7.** To change the radicand in $\sqrt{11}$ into a perfect integer square, we multiply it by _____.

**8.** To change the radicand in $\sqrt[3]{11}$ into a perfect integer cube, we multiply it by _____.

**9.** To rationalize the denominator of $\frac{x}{\sqrt{7}}$, we multiply the numerator and denominator by _____.

**10.** To rationalize the denominator of $\frac{x}{\sqrt{x} + 1}$, we multiply the numerator and denominator by _____.

**PRACTICE**   *In Exercises 11–34, do each multiplication. All variables represent positive numbers.*

**11.** $\sqrt{3}\sqrt{3}$          **12.** $\sqrt{7}\sqrt{7}$          **13.** $\sqrt{2}\sqrt{8}$          **14.** $\sqrt{27}\sqrt{3}$

**15.** $\sqrt{16}\sqrt{4}$          **16.** $\sqrt{32}\sqrt{2}$          **17.** $\sqrt[3]{8}\sqrt[3]{8}$          **18.** $\sqrt[3]{4}\sqrt[3]{250}$

**19.** $\sqrt{x^3}\sqrt{x^3}$          **20.** $\sqrt{a^7}\sqrt{a^3}$          **21.** $\sqrt{b^8}\sqrt{b^6}$          **22.** $\sqrt{y^4}\sqrt{y^8}$

**23.** $(2\sqrt{5})(2\sqrt{3})$

**24.** $(4\sqrt{3})(2\sqrt{2})$

**25.** $(-5\sqrt{6})(4\sqrt{3})$

**26.** $(6\sqrt{3})(-7\sqrt{3})$

**27.** $(2\sqrt[3]{4})(3\sqrt[3]{16})$

**28.** $(-3\sqrt[3]{100})(\sqrt[3]{10})$

**29.** $(4\sqrt{x})(-2\sqrt{x})$

**30.** $(3\sqrt{y})(15\sqrt{y})$

**31.** $(-14\sqrt{50x})(-5\sqrt{20x})$

**32.** $(12\sqrt{24y})(-16\sqrt{2y})$

**33.** $\sqrt{8x}\sqrt{2x^3y}$

**34.** $\sqrt{27y}\sqrt{3y^3}$

*In Exercises 35–52, do each multiplication. All variables represent positive numbers.*

**35.** $\sqrt{2}(\sqrt{2}+1)$ 　　　　　　　　　　**36.** $\sqrt{3}(\sqrt{3}-2)$

**37.** $\sqrt{3}(\sqrt{27}-1)$ 　　　　　　　　　**38.** $\sqrt{2}(\sqrt{8}-1)$

**39.** $\sqrt{7}(\sqrt{7}-3)$ 　　　　　　　　　**40.** $\sqrt{5}(\sqrt{5}+2)$

**41.** $\sqrt{5}(3-\sqrt{5})$ 　　**42.** $\sqrt{7}(2+\sqrt{7})$ 　　**43.** $\sqrt{3}(\sqrt{6}+1)$ 　　**44.** $\sqrt{2}(\sqrt{6}-2)$

**45.** $\sqrt[3]{7}(\sqrt[3]{49}-2)$ 　　**46.** $\sqrt[3]{5}(\sqrt[3]{25}+3)$ 　　**47.** $\sqrt{x}(\sqrt{3x}-2)$ 　　**48.** $\sqrt{y}(\sqrt{y}+5)$

**49.** $2\sqrt{x}(\sqrt{9x}+3)$ 　　**50.** $3\sqrt{z}(\sqrt{4z}-\sqrt{z})$ 　　**51.** $3\sqrt{x}(2+\sqrt{x})$ 　　**52.** $5\sqrt{y}(5-\sqrt{5y})$

*In Exercises 53–66, do each multiplication. All variables represent positive numbers.*

**53.** $(\sqrt{2}+1)(\sqrt{2}-1)$ 　　　　　　　**54.** $(\sqrt{3}-1)(\sqrt{3}+1)$

**55.** $(\sqrt{5}+2)(\sqrt{5}-2)$ 　　　　　　　**56.** $(\sqrt{7}+5)(\sqrt{7}-5)$

**57.** $(\sqrt[3]{2}+1)(\sqrt[3]{2}+1)$ 　　　　　**58.** $(\sqrt[3]{5}-2)(\sqrt[3]{5}-2)$

**59.** $(\sqrt{7}-x)(\sqrt{7}+x)$ 　　　　　　　**60.** $(\sqrt{2}-\sqrt{x})(\sqrt{x}+\sqrt{2})$

**61.** $(\sqrt{6x}+\sqrt{7})(\sqrt{6x}-\sqrt{7})$ 　　**62.** $(\sqrt{8y}+\sqrt{2z})(\sqrt{8y}-\sqrt{2z})$

**63.** $(\sqrt{2x}+3)(\sqrt{8x}-6)$ 　　　　　　**64.** $(\sqrt{5y}-3)(\sqrt{20y}+6)$

**65.** $(\sqrt{8xy}+1)(\sqrt{8xy}+1)$ 　　　　　**66.** $(\sqrt{5x}+3\sqrt{y})(\sqrt{5x}-3\sqrt{y})$

*In Exercises 67–78, simplify each expression. Assume that all variables represent positive numbers.*

**67.** $\dfrac{\sqrt{12x^3}}{\sqrt{27x}}$ 　　**68.** $\dfrac{\sqrt{32}}{\sqrt{98x^2}}$ 　　**69.** $\dfrac{\sqrt{18xy^2}}{\sqrt{25x}}$ 　　**70.** $\dfrac{\sqrt{27y^3}}{\sqrt{75x^2y}}$

**71.** $\dfrac{\sqrt{196xy^3}}{\sqrt{49x^3y}}$ 　　**72.** $\dfrac{\sqrt{50xyz^4}}{\sqrt{98xyz^2}}$ 　　**73.** $\dfrac{\sqrt[3]{16x^6}}{\sqrt[3]{54x^3}}$ 　　**74.** $\dfrac{\sqrt[3]{128a^6b^3}}{\sqrt[3]{16a^3b^6}}$

**75.** $\dfrac{\sqrt{3x^2y^3}}{\sqrt{27x}}$ 　　**76.** $\dfrac{\sqrt{44x^2y^5}}{\sqrt{99x^4y}}$ 　　**77.** $\dfrac{\sqrt{5x}\sqrt{10y^2}}{\sqrt{x^3y}}$ 　　**78.** $\dfrac{\sqrt{7y}\sqrt{14x}}{\sqrt{8xy}}$

*In Exercises 79–106, do each division by rationalizing the denominator and simplifying. All variables represent positive numbers.*

**79.** $\dfrac{1}{\sqrt{3}}$ 　　**80.** $\dfrac{1}{\sqrt{5}}$ 　　**81.** $\dfrac{2}{\sqrt{7}}$ 　　**82.** $\dfrac{3}{\sqrt{11}}$

**83.** $\dfrac{5}{\sqrt[3]{5}}$ 　　**84.** $\dfrac{7}{\sqrt[3]{7}}$ 　　**85.** $\dfrac{9}{\sqrt{27}}$ 　　**86.** $\dfrac{4}{\sqrt{20}}$

**87.** $\dfrac{3}{\sqrt{32}}$ 　　**88.** $\dfrac{5}{\sqrt{18}}$ 　　**89.** $\dfrac{4}{\sqrt[3]{4}}$ 　　**90.** $\dfrac{7}{\sqrt[3]{10}}$

**91.** $\dfrac{\sqrt{5}}{\sqrt{3}}$

**92.** $\dfrac{\sqrt{3}}{\sqrt{5}}$

**93.** $\dfrac{10}{\sqrt{x}}$

**94.** $\dfrac{12}{\sqrt{y}}$

**95.** $\dfrac{\sqrt{9}}{\sqrt{2x}}$

**96.** $\dfrac{\sqrt{4}}{\sqrt{3z}}$

**97.** $\dfrac{\sqrt{2x}}{\sqrt{9y}}$

**98.** $\dfrac{\sqrt{3xy}}{\sqrt{4x}}$

**99.** $\dfrac{\sqrt[3]{5}}{\sqrt[3]{2}}$

**100.** $\dfrac{\sqrt[3]{2}}{\sqrt[3]{5}}$

**101.** $\dfrac{\sqrt[3]{2x^2}}{\sqrt[3]{2x}}$

**102.** $\dfrac{\sqrt[3]{3y^4}}{\sqrt[3]{3y}}$

**103.** $\dfrac{2}{\sqrt[3]{4x^2y}}$

**104.** $\dfrac{3}{\sqrt[3]{9xy^2}}$

**105.** $\dfrac{-5}{\sqrt[3]{25a^2b^2}}$

**106.** $\dfrac{-4}{\sqrt[3]{4ab^2c^2}}$

*In Exercises 107–126, do each division by rationalizing the denominator and simplifying. All variables represent positive numbers.*

**107.** $\dfrac{3}{\sqrt{3}-1}$

**108.** $\dfrac{3}{\sqrt{5}-2}$

**109.** $\dfrac{3}{\sqrt{7}+2}$

**110.** $\dfrac{5}{\sqrt{8}+3}$

**111.** $\dfrac{12}{3-\sqrt{3}}$

**112.** $\dfrac{10}{5-\sqrt{5}}$

**113.** $\dfrac{\sqrt{2}}{\sqrt{2}+1}$

**114.** $\dfrac{\sqrt{3}}{\sqrt{3}-1}$

**115.** $\dfrac{-\sqrt{3}}{\sqrt{3}+1}$

**116.** $\dfrac{-\sqrt{2}}{\sqrt{2}-1}$

**117.** $\dfrac{5}{\sqrt{3}+\sqrt{2}}$

**118.** $\dfrac{3}{\sqrt{3}-\sqrt{2}}$

**119.** $\dfrac{\sqrt{x}+2}{\sqrt{x}-2}$

**120.** $\dfrac{\sqrt{x}-3}{\sqrt{x}+3}$

**121.** $\dfrac{\sqrt{3x}-2}{\sqrt{3x}+2}$

**122.** $\dfrac{\sqrt{5x}+3}{\sqrt{5x}-3}$

**123.** $\dfrac{\sqrt{3x}-1}{\sqrt{3x}+1}$

**124.** $\dfrac{\sqrt{2x}+5}{\sqrt{2x}+3}$

**125.** $\dfrac{\sqrt{3y}+3}{\sqrt{3y}-2}$

**126.** $\dfrac{\sqrt{5x}-1}{\sqrt{5x}+2}$

## WRITING

**127.** How do you know when a radical has been simplified?

**128.** Explain how to rationalize a denominator.

## SOMETHING TO THINK ABOUT

**129.** How would you make the numerator of $\dfrac{\sqrt{3}}{2}$ a rational number?

**130.** Rationalize the numerator of $\dfrac{\sqrt{5}+2}{5}$.

# 7.7 Rational Exponents

■ FRACTIONAL EXPONENTS WITH NUMERATORS OF 1  ■ FRACTIONAL EXPONENTS WITH
NUMERATORS OTHER THAN 1  ■ RULES OF EXPONENTS

**Getting Ready**   *Simplify each expression.*

**1.** $x^3x^2$      **2.** $(x^4)^3$      **3.** $\dfrac{x^7}{x^2}$      **4.** $(xy)^0$

**5.** $(ab)^{-1}$      **6.** $(a^3b^2)^4$      **7.** $\left(\dfrac{a^3}{b^2}\right)^3$      **8.** $(a^3a^2a)^4$

## ■ FRACTIONAL EXPONENTS WITH NUMERATORS OF 1

We have seen that a positive integer exponent indicates the number of times that a base is to be used as a factor in a product. For example, $x^4$ means that $x$ is to be used as a factor four times.

$$\overbrace{x^4 = x \cdot x \cdot x \cdot x}^{4 \text{ factors of } x}$$

Furthermore, we recall the following rules of exponents.

### Rules of Exponents
If $m$ and $n$ are natural numbers and there are no divisions by 0, then

$$x^m x^n = x^{m+n} \qquad (x^m)^n = x^{m \cdot n} \qquad (xy)^n = x^n y^n \qquad \left(\dfrac{x}{y}\right)^n = \dfrac{x^n}{y^n}$$

$$x^0 = 1 \qquad x^{-1} = \dfrac{1}{x} \qquad \dfrac{x^m}{x^n} = x^{m-n}$$

In this section, we will extend the definition and rules of exponents to cover fractional exponents. To give meaning to rational (fractional) exponents, we consider $\sqrt{7}$. Because $\sqrt{7}$ is the positive number whose square is 7, we have

$$\left(\sqrt{7}\right)^2 = 7$$

We now consider the symbol $7^{1/2}$. If fractional exponents are to follow the same rules as integer exponents, the square of $7^{1/2}$ must be 7, because

$$(7^{1/2})^2 = 7^{(1/2)2} \qquad \text{Keep the base and multiply the exponents.}$$
$$= 7^1 \qquad \tfrac{1}{2} \cdot 2 = 1.$$
$$= 7$$

Since $(7^{1/2})^2$ and $(\sqrt{7})^2$ are both equal to 7, we define $7^{1/2}$ to be $\sqrt{7}$. Similarly, we make these definitions.

$$7^{1/3} = \sqrt[3]{7}$$
$$7^{1/7} = \sqrt[7]{7}$$

and so on.

---

**Rational Exponents**

If $n$ is a positive integer greater than 1 and $\sqrt[n]{x}$ is a real number, then

$$x^{1/n} = \sqrt[n]{x}$$

---

**EXAMPLE 1**  Simplify  **a.** $64^{1/2}$,  **b.** $64^{1/3}$,  **c.** $(-64)^{1/3}$,  and  **d.** $64^{1/6}$.

*Solution*  **a.** $64^{1/2} = \sqrt{64} = 8$  **b.** $64^{1/3} = \sqrt[3]{64} = 4$

**c.** $(-64)^{1/3} = \sqrt[3]{-64} = -4$  **d.** $64^{1/6} = \sqrt[6]{64} = 2$  ∎

---

**Self Check**  Simplify  **a.** $81^{1/2}$,  **b.** $125^{1/3}$,  and  **c.** $(-32)^{1/5}$.

*Answers*  **a.** 9,  **b.** 5,  **c.** $-2$

---

### ■ FRACTIONAL EXPONENTS WITH NUMERATORS OTHER THAN 1

We can extend the definition of $x^{1/n}$ to cover fractional exponents for which the numerator is not 1. For example, because $4^{3/2}$ can be written as $(4^{1/2})^3$, we have

$$4^{3/2} = (4^{1/2})^3 = (\sqrt{4})^3 = 2^3 = 8$$

Because $4^{3/2}$ can also be written as $(4^3)^{1/2}$, we have

$$4^{3/2} = (4^3)^{1/2} = 64^{1/2} = \sqrt{64} = 8$$

In general, $x^{m/n}$ can be written as $(x^{1/n})^m$ or as $(x^m)^{1/n}$. Since $(x^{1/n})^m = (\sqrt[n]{x})^m$ and $(x^m)^{1/n} = \sqrt[n]{x^m}$, we make the following definition.

---

**Changing from Rational Exponents to Radicals**

If $m$ and $n$ are positive integers, $x$ is nonnegative, and the fraction $m/n$ cannot be simplified, then

$$x^{m/n} = \sqrt[n]{x^m} = (\sqrt[n]{x})^m$$

---

b

**EXAMPLE 2**   Simplify  **a.** $8^{2/3}$  and  **b.** $(-27)^{4/3}$.

*Solution*   **a.** $8^{2/3} = \left(\sqrt[3]{8}\right)^2$     or     $8^{2/3} = \sqrt[3]{8^2}$
  $= 2^2$                 $= \sqrt[3]{64}$
  $= 4$                   $= 4$

  **b.** $(-27)^{4/3} = \left(\sqrt[3]{-27}\right)^4$     or     $(-27)^{4/3} = \sqrt[3]{(-27)^4}$
    $= (-3)^4$                 $= \sqrt[3]{531,441}$
    $= 81$                     $= 81$ ■

**Self Check**   Simplify  **a.** $16^{3/2}$  and  **b.** $(-8)^{4/3}$.
**Answers**   **a.** 64,  **b.** 16

The work in Example 2 suggests that in order to avoid large numbers, it is usually easier to take the root of the base first.

**EXAMPLE 3**   Simplify  **a.** $125^{4/3}$,  **b.** $9^{5/2}$,  **c.** $-25^{3/2}$,  and  **d.** $(-27)^{2/3}$.

*Solution*   **a.** $125^{4/3} = \left(\sqrt[3]{125}\right)^4$          **b.** $9^{5/2} = \left(\sqrt{9}\right)^5$
    $= (5)^4$                      $= (3)^5$
    $= 625$                        $= 243$

  **c.** $-25^{3/2} = -\left(\sqrt{25}\right)^3$          **d.** $(-27)^{2/3} = \left(\sqrt[3]{-27}\right)^2$
    $= -(5)^3$                      $= (-3)^2$
    $= -125$                        $= 9$ ■

**Self Check**   Simplify  **a.** $100^{3/2}$  and  **b.** $(-8)^{2/3}$.
**Answers**   **a.** 1,000,  **b.** 4

## ■ RULES OF EXPONENTS

Because of the definition of $x^{1/n}$, the familiar rules of exponents are valid for rational exponents. The following example illustrates the use of each rule.

f,g

**EXAMPLE 4**   Simplify each expression.

|  | *Problem* | *Rule* |
|---|---|---|
| **a.** | $4^{2/5}4^{1/5} = 4^{2/5+1/5} = 4^{3/5}$ | $x^m x^n = x^{m+n}$ |
| **b.** | $(5^{2/3})^{1/2} = 5^{(2/3)(1/2)} = 5^{1/3}$ | $(x^m)^n = x^{m \cdot n}$ |
| **c.** | $(3x)^{2/3} = 3^{2/3}x^{2/3}$ | $(xy)^m = x^m y^m$ |

**d.** $\dfrac{4^{3/5}}{4^{2/5}} = 4^{3/5-2/5} = 4^{1/5}$          $\dfrac{x^m}{x^n} = x^{m-n}$

**e.** $\left(\dfrac{3}{2}\right)^{2/5} = \dfrac{3^{2/5}}{2^{2/5}}$          $\left(\dfrac{x}{y}\right)^n = \dfrac{x^n}{y^n}$

**f.** $4^{-2/3} = \dfrac{1}{4^{2/3}}$          $x^{-n} = \dfrac{1}{x^n}$

**g.** $5^0 = 1$          $x^0 = 1 \quad (x \neq 0)$          ■

**Self Check**     Simplify   **a.** $5^{1/3}5^{1/3}$,   **b.** $(5^{1/3})^4$,   **c.** $(3x)^{1/5}$,   **d.** $\dfrac{5^{3/7}}{5^{2/7}}$,   **e.** $\left(\dfrac{2}{3}\right)^{2/3}$,
**f.** $5^{-2/7}$,   and   **g.** $12^0$.

**Answers**     **a.** $5^{2/3}$,   **b.** $5^{4/3}$,   **c.** $3^{1/5}x^{1/5}$,   **d.** $5^{1/7}$,   **e.** $\dfrac{2^{2/3}}{3^{2/3}}$,   **f.** $\dfrac{1}{5^{2/7}}$,   **g.** 1

We can often use the rules of exponents to simplify expressions containing rational exponents.

**EXAMPLE 5**     Simplify   **a.** $64^{-2/3}$,   **b.** $(x^2)^{1/2}$,   **c.** $(x^6y^4)^{1/2}$,   and   **d.** $(27x^{12})^{-1/3}$   $(x > 0$ and $y > 0)$.

*Solution*   **a.** $64^{-2/3} = \dfrac{1}{64^{2/3}}$          **b.** $(x^2)^{1/2} = x^{2(1/2)}$

$\qquad\qquad = \dfrac{1}{(64^{1/3})^2}$          $\qquad\qquad = x^1$

$\qquad\qquad = \dfrac{1}{4^2}$          $\qquad\qquad = x$          /

$\qquad\qquad = \dfrac{1}{16}$

**c.** $(x^6y^4)^{1/2} = x^{6(1/2)}y^{4(1/2)}$          **d.** $(27x^{12})^{-1/3} = \dfrac{1}{(27x^{12})^{1/3}}$

$\qquad\qquad = x^3y^2$          $\qquad\qquad = \dfrac{1}{27^{1/3}x^{12(1/3)}}$

$\qquad\qquad\qquad\qquad\qquad = \dfrac{1}{3x^4}$          ■

**Self Check**     Simplify   **a.** $25^{-3/2}$,   **b.** $(x^3)^{1/3}$,   and   **c.** $(x^6y^9)^{-2/3}$.

**Answers**     **a.** $\dfrac{1}{125}$,   **b.** $x$,   **c.** $\dfrac{1}{x^4y^6}$

b   **EXAMPLE 6**   Simplify  **a.** $x^{1/3}x^{1/2}$,  **b.** $\dfrac{3x^{2/3}}{6x^{1/5}}$,  and  **c.** $\dfrac{2x^{-1/2}}{x^{3/4}}$  $(x > 0)$.

*Solution*   **a.** $x^{1/3}x^{1/2} = x^{2/6}x^{3/6}$   Get a common denominator in the fractional exponents.

$\qquad\qquad\quad = x^{5/6}$   Keep the base and add the exponents.

**b.** $\dfrac{3x^{2/3}}{6x^{1/5}} = \dfrac{3x^{10/15}}{6x^{3/15}}$   Get a common denominator in the fractional exponents.

$\qquad\qquad = \dfrac{1}{2}x^{10/15 - 3/15}$   Simplify $\tfrac{3}{6}$, keep the base, and subtract the exponents.

$\qquad\qquad = \dfrac{1}{2}x^{7/15}$

**c.** $\dfrac{2x^{-1/2}}{x^{3/4}} = \dfrac{2x^{-2/4}}{x^{3/4}}$   Get a common denominator in the fractional exponents.

$\qquad\qquad = 2x^{-2/4 - 3/4}$   Keep the base and subtract the exponents.

$\qquad\qquad = 2x^{-5/4}$   Simplify.

$\qquad\qquad = \dfrac{2}{x^{5/4}}$   $x^{-5/4} = \frac{1}{x^{5/4}}$.   ∎

*Self Check*   Simplify  **a.** $x^{2/3}x^{1/2}$  and  **b.** $\dfrac{x^{2/3}}{2x^{1/4}}$.

*Answers*   **a.** $x^{7/6}$,  **b.** $\frac{1}{2}x^{5/12}$

Orals   *Find each value.*

**1.** $16^{1/2}$      **2.** $25^{1/2}$      **3.** $27^{1/3}$      **4.** $81^{1/4}$

**5.** $8^{2/3}$      **6.** $32^{3/5}$      **7.** $9^{-1/2}$      **8.** $64^{-1/3}$

# EXERCISE 7.7

**REVIEW**   *Factor each expression.*

**1.** $3z^2 - 15tz + 12t^2$

**2.** $a^4 - b^4$

*Solve each equation.*

**3.** $\dfrac{x-5}{7} + \dfrac{2}{5} = \dfrac{7-x}{5}$

**4.** $\dfrac{t}{t+2} - 1 = \dfrac{1}{1-t}$

***VOCABULARY AND CONCEPTS*** *Fill in each blank to make a true statement.*

**5.** A fractional exponent is also called a _____ exponent.

**6.** In the expression $(2x)^{1/3}$, $2x$ is called the _____, and the exponent is _.

**7.** $x^m x^n =$ _____

**8.** $(x^m)^n =$ _____

**9.** $\left(\dfrac{x}{y}\right)^n =$ __

**10.** $x^0 =$ __

**11.** $x^{-1} =$ __

**12.** $\dfrac{x^m}{x^n} =$ _____

***PRACTICE*** *In Exercises 13–36, simplify each expression.*

**13.** $81^{1/2}$

**14.** $100^{1/2}$

**15.** $-144^{1/2}$

**16.** $-400^{1/2}$

**17.** $\left(\dfrac{1}{4}\right)^{1/2}$

**18.** $\left(\dfrac{1}{25}\right)^{1/2}$

**19.** $\left(\dfrac{4}{49}\right)^{1/2}$

**20.** $\left(\dfrac{9}{64}\right)^{1/2}$

**21.** $27^{1/3}$

**22.** $8^{1/3}$

**23.** $-125^{1/3}$

**24.** $-1{,}000^{1/3}$

**25.** $(-8)^{1/3}$

**26.** $(-125)^{1/3}$

**27.** $\left(\dfrac{1}{64}\right)^{1/3}$

**28.** $\left(\dfrac{1}{1{,}000}\right)^{1/3}$

**29.** $\left(\dfrac{27}{64}\right)^{1/3}$

**30.** $\left(\dfrac{64}{125}\right)^{1/3}$

**31.** $16^{1/4}$

**32.** $81^{1/4}$

**33.** $32^{1/5}$

**34.** $-32^{1/5}$

**35.** $-243^{1/5}$

**36.** $\left(-\dfrac{1}{32}\right)^{1/5}$

*In Exercises 37–56, simplify each expression.*

**37.** $81^{3/2}$

**38.** $16^{3/2}$

**39.** $25^{3/2}$

**40.** $4^{5/2}$

**41.** $125^{2/3}$

**42.** $8^{4/3}$

**43.** $1{,}000^{2/3}$

**44.** $27^{2/3}$

**45.** $(-8)^{2/3}$

**46.** $(-125)^{2/3}$

**47.** $32^{3/5}$

**48.** $-243^{3/5}$

**49.** $81^{3/4}$

**50.** $256^{3/4}$

**51.** $(-32)^{3/5}$

**52.** $243^{2/5}$

**53.** $\left(\dfrac{8}{27}\right)^{2/3}$

**54.** $\left(\dfrac{27}{64}\right)^{2/3}$

**55.** $\left(\dfrac{16}{625}\right)^{3/4}$

**56.** $\left(\dfrac{49}{64}\right)^{3/2}$

*In Exercises 57–80, simplify each expression. Write each answer without using negative exponents.*

**57.** $6^{3/5}6^{2/5}$

**58.** $3^{4/7}3^{3/7}$

**59.** $5^{2/3}5^{4/3}$

**60.** $2^{7/8}2^{9/8}$

**61.** $(7^{2/5})^{5/2}$

**62.** $(8^{1/3})^3$

**63.** $(5^{2/7})^7$

**64.** $(3^{3/8})^8$

**65.** $\dfrac{8^{3/2}}{8^{1/2}}$

**66.** $\dfrac{11^{9/7}}{11^{2/7}}$

**67.** $\dfrac{5^{11/3}}{5^{2/3}}$

**68.** $\dfrac{27^{13/15}}{27^{8/15}}$

**69.** $(2^{1/2}3^{1/2})^2$

**70.** $(3^{2/3}5^{1/3})^3$

**71.** $(4^{3/4}3^{1/4})^4$

**72.** $(2^{1/5}3^{2/5})^5$

**73.** $4^{-1/2}$

**74.** $8^{-1/3}$

**75.** $27^{-2/3}$

**76.** $36^{-3/2}$

**77.** $16^{-3/2}$

**78.** $100^{-5/2}$

**79.** $(-27)^{-4/3}$

**80.** $(-8)^{-4/3}$

*In Exercises 81–100, simplify each expression. Assume that all variables represent positive numbers.*

**81.** $(x^{1/2})^2$

**82.** $(x^9)^{1/3}$

**83.** $(x^{12})^{1/6}$

**84.** $(x^{18})^{1/9}$

**85.** $(x^{18})^{2/9}$

**86.** $(x^{12})^{3/4}$

**87.** $x^{5/6}x^{7/6}$

**88.** $x^{2/3}x^{7/3}$

**89.** $y^{4/7}y^{10/7}$

**90.** $y^{5/11}y^{6/11}$

**91.** $\dfrac{x^{3/5}}{x^{1/5}}$

**92.** $\dfrac{x^{4/3}}{x^{2/3}}$

**93.** $\dfrac{x^{1/7}x^{3/7}}{x^{2/7}}$

**94.** $\dfrac{x^{5/6}x^{5/6}}{x^{7/6}}$

**95.** $\left(\dfrac{x^{3/5}}{x^{2/5}}\right)^5$

**96.** $\left(\dfrac{x^{2/9}}{x^{1/9}}\right)^9$

**97.** $\left(\dfrac{y^{2/7}y^{3/7}}{y^{4/7}}\right)^{49}$

**98.** $\left(\dfrac{z^{3/5}z^{6/5}}{z^{2/5}}\right)^5$

**99.** $\left(\dfrac{y^{5/6}y^{7/6}}{y^{1/3}y}\right)^3$

**100.** $\left(\dfrac{t^{4/9}t^{5/9}}{t^{1/9}t^{2/9}}\right)^9$

*In Exercises 101–110, simplify each expression. Assume that all variables represent positive numbers.*

**101.** $x^{2/3}x^{3/4}$

**102.** $a^{3/5}a^{1/2}$

**103.** $(b^{1/2})^{3/5}$

**104.** $(x^{2/5})^{4/7}$

**105.** $\dfrac{t^{2/3}}{t^{2/5}}$

**106.** $\dfrac{p^{3/4}}{p^{1/3}}$

**107.** $\dfrac{x^{4/5}x^{1/3}}{x^{2/15}}$

**108.** $\dfrac{y^{2/3}y^{3/5}}{y^{1/5}}$

**109.** $\dfrac{a^{2/5}a^{1/5}}{a^{-1/3}}$

**110.** $\dfrac{q^{3/4}q^{4/5}}{q^{-2/3}}$

**WRITING**

**111.** Is $(-4)^{1/2}$ a real number? Explain.

**112.** Is $(-8)^{1/3}$ a real number? Explain.

**SOMETHING TO THINK ABOUT**   *If $x > y$, which is the larger number in each pair?*

**113.** $2^x$, $2^y$

**114.** $\left(\dfrac{1}{2}\right)^x$, $\left(\dfrac{1}{2}\right)^y$

■ ■ ■ ■ ■ ■ ■ ■ ■

**MATHEMATICS IN CARPENTRY**

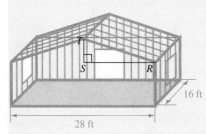

16 ft

28 ft

To find the area of the ceiling of the room discussed at the beginning of the chapter, we must find the length of the rafter $RT$ in right triangle $RST$. Since the ridge is at the center of the room, the length of $RS$ is 14 ft. Since the height of the ridge is 12 feet and the height of the outside wall is 8 feet, the length of $ST$ is 4 feet. We can use the Pythagorean theorem to find the length of $RT$.

$$(RT)^2 = (RS)^2 + (ST)^2$$
$$(RT)^2 = 14^2 + 4^2$$
$$= 212$$
$$RT = \sqrt{212} \qquad \text{Take the square root of both sides.}$$

Since the area $A$ of the ceiling is the sum of its two rectangular parts, we have

$$A = 2\left(16 \cdot \sqrt{212}\right) \approx 465.9270329$$

To find the number of sheets of plaster board needed to drywall the ceiling, we divide the area of the ceiling by the area of one 4-by-8-foot sheet of plaster board.

$$465.9270329 \div 32 \approx 14.56021978$$

The carpenter will need at least 15 sheets of plaster board.

# ■ ■ ■ ■ ■ ■ ■ ■ ■ ■  PROJECT

The Italian mathematician and physicist Galileo Galilei (1564–1642) is best known as the inventor of the telescope and for his discovery of four of the moons of Jupiter, still known as the *Galilean satellites.* Less known is his discovery that a pendulum could be used to keep accurate time. While praying one day in the cathedral, Galileo noticed a suspended candle left swinging after it had been lit. Using his own pulse as a timer, Galileo discovered that the time for one swing remained unchanged as the swings themselves became smaller. By more experimenting, Galileo discovered the relationship between the length of a pendulum and its **period,** the time it takes to complete one swing.

You can discover the relationship, too. You will need a stopwatch, a calculator, a meter stick, and a pendulum. To make the pendulum, try tying a length of string to a rubber band and wrapping the band tightly around a small rock. By tying the free end of the string to a support, you can change the length of the pendulum.

- Start the pendulum swinging and use the stopwatch to determine its period—the time it takes to go from left to right and back to left again. (You might time 10 complete swings and then divide by 10.) Do this for pendulums of at least eight different lengths. Let $t$ be the period (measured in seconds) and let $l$ be the length (measured in centimeters), and record your results in the table of Illustration 1.

- Plot the points $(l, t)$ on the axes in the figure. Does the graph appear to be a line?

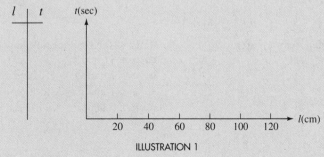

ILLUSTRATION 1

*(continued)*

■ ■ ■ ■ ■ ■ ■ ■ ■ ■ **PROJECT** *(continued)*

- The pendulum's period $t$ and length $l$ are related by the formula $t = a\sqrt{l}$ for some number $a$. From your experimental data, find the approximate value of $a$.
- Use your formula to predict the period of a pendulum 2 meters long.
- Time the period of a 2-meter pendulum. How close was your prediction?

# C H A P T E R   S U M M A R Y

CONCEPTS                     REVIEW EXERCISES

| SECTION 7.1 | *Square Roots and the Pythagorean Theorem* |

The number $b$ is a **square root** of $a$ if $b^2 = a$.

The **principal square root** of a positive number $a$, denoted by $\sqrt{a}$, is the positive square root of $a$.

**1.** Find each square root.
   **a.** $\sqrt{25}$  **b.** $\sqrt{64}$  **c.** $-\sqrt{144}$  **d.** $-\sqrt{289}$
   **e.** $\sqrt{256}$  **f.** $-\sqrt{64}$  **g.** $\sqrt{169}$  **h.** $-\sqrt{225}$

**2.** Use a calculator to find each root to three decimal places.
   **a.** $\sqrt{21}$  **b.** $-\sqrt{15}$
   **c.** $-\sqrt{57.3}$  **d.** $\sqrt{751.9}$

**3.** Graph each function.
   **a.** $f(x) = \sqrt{x}$  **b.** $f(x) = 2 - \sqrt{x}$

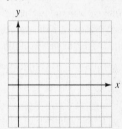 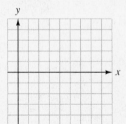

**The Pythagorean theorem:**
$$a^2 + b^2 = c^2$$

**4.** Refer to the right triangle shown in Illustration 1.
   **a.** $a = 21$ and $b = 28$. Find $c$.
   **b.** $a = 25$ and $c = 65$. Find $b$.
   **c.** $a = 1$ and $c = \sqrt{2}$. Find $b$.
   **d.** $b = 5$ and $c = 7$. Find $a$.

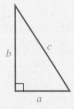

ILLUSTRATION 1

**5. Installing windows**  The window frame shown in Illustration 2 is 32 inches by 60 inches. It is to be shipped with a temporary brace attached diagonally. Find the length of the brace.

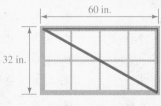

60 in.

32 in.

ILLUSTRATION 2

**6. Height of a mast**  A 53-foot rope runs from the top of the mast shown in Illustration 3 to a point 28 feet from its base. Find the height of the mast.

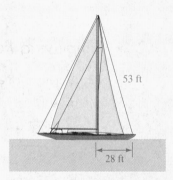

53 ft

28 ft

ILLUSTRATION 3

| SECTION 7.2 | *nth Roots and Radicands That Contain Variables* |
|---|---|

The number $b$ is a **cube root** of $a$ if $b^3 = a$.

**7.** Find each root.
  **a.** $\sqrt[3]{-27}$    **b.** $-\sqrt[3]{125}$    **c.** $\sqrt[4]{81}$    **d.** $\sqrt[5]{32}$

The number $b$ is an ***n*th root** of $a$ if $b^n = a$.

**8.** Use a calculator to find each root to three decimal places.
  **a.** $\sqrt[3]{54.3}$    **b.** $\sqrt[3]{0.003}$
  **c.** $\sqrt[3]{-0.055}$    **d.** $\sqrt[3]{-63,777}$

**9.** Find each root.
  **a.** $\sqrt{x^6}$    **b.** $\sqrt{16x^4y^2}$
  **c.** $\sqrt[3]{27x^3}$    **d.** $\sqrt[3]{1,000a^6b^3}$

| | |
|---|---|
| **SECTION 7.3** | *Solving Equations Containing Radicals; the Distance Formula* |

If $a = b$, then $a^2 = b^2$.

**10.** Solve each equation and check all solutions.

    **a.** $\sqrt{x + 3} = 3$         **b.** $\sqrt{2x + 10} = 2$

    **c.** $\sqrt{3x + 4} = -2\sqrt{x}$     **d.** $\sqrt{2(x + 4)} - \sqrt{4x} = 0$

    **e.** $\sqrt{x + 5} = x - 1$

    **f.** $\sqrt{2x + 9} = x - 3$

    **g.** $\sqrt{2x + 5} - 1 = x$

    **h.** $\sqrt{4a + 13} + 2 = a$

**The distance formula:**

$$d = \sqrt{(x_2 - x_1)^2 + (y_2 - y_1)^2}$$

**11.** Find the distance between the points.

    **a.** $(-7, 12), (-4, 8)$        **b.** $(-15, -3), (-10, -15)$

    **c.** $(1, 1), (-1, 1)$            **d.** $(-10, 11), (10, -10)$

| | |
|---|---|
| **SECTION 7.4** | *Simplifying Radical Expressions* |

$\sqrt{ab} = \sqrt{a}\sqrt{b}$

**12.** Simplify each expression. All variables represent positive numbers.

    **a.** $\sqrt{32}$            **b.** $\sqrt{50}$

    **c.** $\sqrt{500}$          **d.** $\sqrt{112}$

    **e.** $\sqrt{80x^2}$         **f.** $\sqrt{63y^2}$

    **g.** $-\sqrt{250t^3}$      **h.** $-\sqrt{700z^5}$

    **i.** $\sqrt{200x^2y}$      **j.** $\sqrt{75y^2z}$

    **k.** $\sqrt[3]{8x^2y^3}$      **l.** $\sqrt[3]{250x^4y^3}$

$\sqrt{\dfrac{a}{b}} = \dfrac{\sqrt{a}}{\sqrt{b}}$   $(b \neq 0)$

**13.** Simplify each expression. All variables represent positive numbers.

    **a.** $\sqrt{\dfrac{16}{25}}$        **b.** $\sqrt{\dfrac{100}{49}}$

    **c.** $\sqrt[3]{\dfrac{1,000}{27}}$      **d.** $\sqrt[3]{\dfrac{16}{64}}$

    **e.** $\sqrt{\dfrac{60}{49}}$       **f.** $\sqrt{\dfrac{80}{225}}$

    **g.** $\sqrt{\dfrac{242x^4}{169x^2}}$    **h.** $\sqrt{\dfrac{450a^6}{196a^2}}$

| **SECTION 7.5** | *Adding and Subtracting Radical Expressions* |
|---|---|

| Radical expressions can be added or subtracted if they contain **like radicals**. | **14.** Do the operations. All variables represent positive numbers. |
|---|---|

**a.** $\sqrt{2} + \sqrt{8} - \sqrt{18}$      **b.** $\sqrt{3} + \sqrt{27} - \sqrt{12}$

**c.** $3\sqrt{5} + 5\sqrt{45}$      **d.** $5\sqrt{28} - 3\sqrt{63}$

**e.** $3\sqrt{2x^2y} + 2x\sqrt{2y}$

**f.** $3y\sqrt{5xy^3} - y^2\sqrt{20xy}$

**g.** $\sqrt[3]{16} + \sqrt[3]{54}$

**h.** $\sqrt[3]{2,000x^3} - \sqrt[3]{128x^3}$

| **SECTION 7.6** | *Multiplying and Dividing Radical Expressions* |
|---|---|

**15.** Do the operations.

**a.** $\left(3\sqrt{2}\right)\left(-2\sqrt{3}\right)$      **b.** $\left(-5\sqrt{x}\right)\left(-2\sqrt{x}\right)$

**c.** $\left(3\sqrt{3x}\right)\left(4\sqrt{6x}\right)$      **d.** $\left(-2\sqrt{27y^3}\right)\left(y\sqrt{2y}\right)$

**e.** $\left(\sqrt[3]{4}\right)\left(2\sqrt[3]{4}\right)$      **f.** $\left(-2\sqrt[3]{32x^2}\right)\left(3\sqrt[3]{2x^2}\right)$

**g.** $\sqrt{2}\left(\sqrt{8} - \sqrt{18}\right)$

**h.** $\sqrt{6y}\left(\sqrt{2y} + \sqrt{75}\right)$

**i.** $\left(\sqrt{3} + \sqrt{5}\right)\left(\sqrt{3} - \sqrt{5}\right)$

**j.** $\left(\sqrt{15} + 3x\right)\left(\sqrt{15} + 3x\right)$

**k.** $\left(\sqrt[3]{3} + 2\right)\left(\sqrt[3]{3} - 1\right)$

**l.** $\left(\sqrt[3]{5} - 1\right)\left(\sqrt[3]{5} + 1\right)$

| If a radical appears as a monomial in the denominator of a fraction, **rationalize the denominator** by multiplying the numerator and denominator by some appropriate radical. | **16.** Rationalize each denominator. |
|---|---|

**a.** $\dfrac{1}{\sqrt{7}}$      **b.** $\dfrac{3}{\sqrt{18}}$

**c.** $\dfrac{8}{\sqrt[3]{16}}$      **d.** $\dfrac{10}{\sqrt[3]{32}}$

| If the denominator of a fraction contains radicals within a binomial, multiply numerator and denominator by the conjugate of the denominator. | **17.** Rationalize each denominator. |
|---|---|

**a.** $\dfrac{7}{\sqrt{2} + 1}$      **b.** $\dfrac{3}{\sqrt{3} - 1}$

**c.** $\dfrac{2\sqrt{5}}{\sqrt{5} + \sqrt{3}}$      **d.** $\dfrac{\sqrt{7x} + \sqrt{x}}{\sqrt{7x} - \sqrt{x}}$

| SECTION 7.7 | *Rational Exponents* |

$x^{1/n} = \sqrt[n]{x}$

$x^{m/n} = \sqrt[n]{x^m} = \left(\sqrt[n]{x}\right)^m$

**18.** Simplify each expression. Write answers without using negative exponents.

**a.** $49^{1/2}$

**b.** $(-1,000)^{1/3}$

**c.** $36^{3/2}$

**d.** $\left(\dfrac{4}{9}\right)^{5/2}$

**e.** $8^{2/3}8^{4/3}$

**f.** $\dfrac{5^{17/7}}{5^{3/7}}$

**g.** $\dfrac{x^{4/5}x^{3/5}}{(x^{2/5})^3}$

**h.** $\left(\dfrac{r^{1/3}r^{2/3}}{r^{4/3}}\right)^3$

**i.** $6^{5/3}6^{-2/3}$

**j.** $\dfrac{5^{2/3}}{5^{-1/3}}$

**k.** $\dfrac{x^{2/5}x^{1/5}}{x^{-2/5}}$

**l.** $(a^4b^8)^{-1/2}$

**m.** $x^{1/3}x^{2/5}$

**n.** $\dfrac{t^{3/4}}{t^{2/3}}$

**o.** $\dfrac{x^{-4/5}x^{1/3}}{x^{1/3}}$

**p.** $\dfrac{r^{1/4}r^{1/3}}{r^{5/6}}$

## ■ Chapter Test

*In Problems 1–4, write each expression without a radical sign. Assume $x > 0$.*

**1.** $\sqrt{100}$

**2.** $-\sqrt{400}$

**3.** $\sqrt[3]{-27}$

**4.** $\sqrt{3x}\sqrt{27x}$

**5.** Find the length of the hypotenuse of a right triangle with legs of 5 inches and 12 inches.

**6.** A 26-foot ladder reaches a point on a wall 24 feet above the ground. How far from the wall is the ladder's base?

*In Problems 7–12, simplify each expression. Assume $x > 0$ and $y > 0$.*

**7.** $\sqrt{8x^2}$

**8.** $\sqrt{54x^3y}$

**9.** $\sqrt{\dfrac{320}{10}}$

**10.** $\sqrt{\dfrac{18x^2y^3}{2xy}}$

**11.** $\sqrt[3]{x^6y^6}$

**12.** $\sqrt[4]{\dfrac{16x^8}{y^4}}$

*In Problems 13–18, solve each equation.*

**13.** $\sqrt{x} + 3 = 9$

**14.** $\sqrt{x - 2} - 2 = 6$

**15.** $\sqrt{3x + 9} = 2\sqrt{x + 1}$

**16.** $3\sqrt{x - 3} = \sqrt{2x + 8}$

**17.** $\sqrt{3x + 1} = x - 1$

**18.** $\sqrt[3]{x - 2} = 3$

**19.** Find the distance between the points $(1, 4)$ and $(7, 12)$.

**20.** Find the distance between points $(-2, -3)$ and $(-5, 1)$.

*In Problems 21–26, do each operation and simplify.*

**21.** $\sqrt{12} + \sqrt{27}$

**22.** $\sqrt{8x^3} - x\sqrt{18x}$

**23.** $\left(-2\sqrt{8x}\right)\left(3\sqrt{12x}\right)$

**24.** $\sqrt{3}\left(\sqrt{8} + \sqrt{6}\right)$

**25.** $\left(\sqrt{2} + \sqrt{3}\right)\left(\sqrt{2} - \sqrt{3}\right)$

**26.** $\left(2\sqrt{x} + 2\right)\left(\sqrt{x} - 3\right)$

*In Problems 27–30, rationalize each denominator.*

**27.** $\dfrac{2}{\sqrt{2}}$

**28.** $\sqrt{\dfrac{3xy^3}{48x^2}}$

**29.** $\dfrac{2}{\sqrt{5} - 2}$

**30.** $\dfrac{\sqrt{3x}}{\sqrt{x} + 2}$

*In Problems 31–36, simplify each expression and write all answers without using negative exponents. All variables represent positive numbers.*

**31.** $121^{1/2}$

**32.** $27^{-4/3}$

**33.** $(y^{15})^{2/5}$

**34.** $\left(\dfrac{a^{5/3}a^{4/3}}{(a^{1/3})^2 a^{2/3}}\right)^6$

**35.** $p^{2/3}p^{3/4}$

**36.** $\dfrac{x^{2/3}x^{-4/5}}{x^{2/15}}$

# 8 Writing Equations of Lines; Variation

MATHEMATICS
IN REAL ESTATE

An investor bought an office building for $465,000, excluding the value of the land. At that time, a real estate appraiser estimated that the building would retain 80% of its value after 40 years. For tax purposes, the investor used linear depreciation to depreciate the building over a period of 40 years. If the investor sells the building for $400,000 after 35 years, find the taxable capital gain.

After you have read this chapter, you will be able to answer this question.

Recall that in Chapter 3 we introduced the rectangular coordinate system and learned how to plot ordered pairs of real numbers $(x, y)$. We then started with linear equations and learned how to construct their graphs. In this chapter, we will start with straight-line graphs and learn how to write their equations. We will then apply this skill to solve many applications problems.

## 8.1   The Slope of a Nonvertical Line

■ SLOPE OF A LINE   ■ INTERPRETATION OF SLOPE   ■ SLOPES OF HORIZONTAL AND VERTICAL LINES
■ SLOPES OF PARALLEL LINES   ■ SLOPES OF PERPENDICULAR LINES

Getting Ready    *Simplify each expression.*

**1.** $\dfrac{6-2}{12-8}$      **2.** $\dfrac{12-3}{11-8}$      **3.** $\dfrac{4-16}{6-2}$      **4.** $\dfrac{2-9}{21-7}$

### ■ SLOPE OF A LINE

A service offered by an internet company costs $2 per month plus $3 for each hour of connect time. The table shown in Figure 8-1(a) gives the cost $y$ for different hours $x$ of connect time. If we construct a graph from this data, we get the line shown in Figure 8-1(b).

**Hours of connect time**

| $x$ | 0 | 1 | 2 | 3 | 4 | 5 |
|-----|---|---|---|----|----|----|
| $y$ | 2 | 5 | 8 | 11 | 14 | 17 |

Cost

(a)

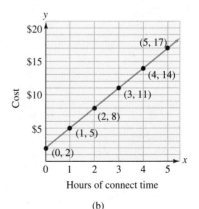

(b)

FIGURE 8-1

From the graph, we can see that if $x$ changes from 0 to 1, $y$ changes from 2 to 5. As $x$ changes from 1 to 2, $y$ changes from 5 to 8, and so on. The ratio of the change in $y$ divided by the change in $x$ is the constant 3.

$$\frac{\text{Change in } y}{\text{Change in } x} = \frac{5-2}{1-0} = \frac{8-5}{2-1} = \frac{11-8}{3-2} = \frac{14-11}{4-3} = \frac{17-14}{5-4} = \frac{3}{1} = 3$$

The ratio of the change in $y$ divided by the change in $x$ between any two points on any line is always a constant. This constant rate of change is called the **slope** of the line.

To distinguish between the coordinates of points $P$ and $Q$ in Figure 8-2, we use **subscript notation**. Point $P$ is denoted as $P(x_1, y_1)$ and is read as "point $P$ with coordinates of $x$ sub 1 and $y$ sub 1." Point $Q$ is denoted as $Q(x_2, y_2)$ and is read as "point $Q$ with coordinates of $x$ sub 2 and $y$ sub 2."

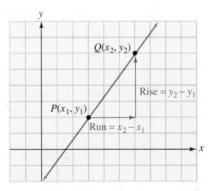

FIGURE 8-2

As a point on the line in Figure 8-2 moves from $P$ to $Q$, its $y$-coordinate changes by the amount $y_2 - y_1$, and its $x$-coordinate changes by $x_2 - x_1$. The change in $y$ is often called the **rise** of the line between points $P$ and $Q$, and the change in $x$ is often called the **run**.

### The Slope of a Nonvertical Line

The **slope of the nonvertical line** passing through points $P(x_1, y_1)$ and $Q(x_2, y_2)$ is

$$m = \frac{\text{change in } y}{\text{change in } x} = \frac{\text{rise}}{\text{run}} = \frac{y_2 - y_1}{x_2 - x_1} \quad (x_2 \neq x_1)$$

EXAMPLE 1

*Solution*

Find the slope of the line shown in Figure 8-3.

We can let $P(x_1, y_1) = P(-3, 2)$ and $Q(x_2, y_2) = Q(2, -5)$. Then $x_1 = -3$, $y_1 = 2$, $x_2 = 2$, and $y_2 = -5$. To find the slope, we substitute these values into the formula for slope and simplify.

$$m = \frac{\text{change in } y}{\text{change in } x}$$

$$= \frac{y_2 - y_1}{x_2 - x_1}$$

$$= \frac{-5 - 2}{2 - (-3)} \qquad \text{Substitute } -5 \text{ for } y_2, 2 \text{ for } y_1, 2 \text{ for } x_2, \text{ and } -3 \text{ for } x_1.$$

$$= \frac{-7}{5}$$

$$= -\frac{7}{5}$$

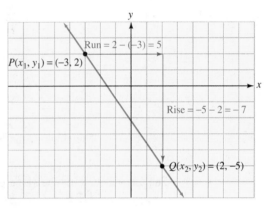

FIGURE 8-3

The slope of the line is $-\frac{7}{5}$. We would obtain the same result if we had let $P(x_1, y_1) = P(2, -5)$ and $Q(x_2, y_2) = Q(-3, 2)$. ∎

Self Check

Answer

Find the slope of the line passing through points $P(-2, 4)$ and $Q(3, -4)$.

$-\frac{8}{5}$

**WARNING!** When calculating slope, always subtract the $y$ values and the $x$ values in the same order.

$$m = \frac{y_2 - y_1}{x_2 - x_1} \qquad \text{or} \qquad m = \frac{y_1 - y_2}{x_1 - x_2}$$

However,

$$m \neq \frac{y_2 - y_1}{x_1 - x_2} \qquad \text{and} \qquad m \neq \frac{y_1 - y_2}{x_2 - x_1}$$

**EXAMPLE 2** Find the slope of the line determined by $3x - 4y = 12$.

*Solution* We first find the coordinates of two points on the line.

- If $x = 0$, then $y = -3$, and the point $(0, -3)$ is on the line.
- If $y = 0$, then $x = 4$, and the point $(4, 0)$ is on the line.

We then refer to Figure 8-4 and find the slope of the line between $P(0, -3)$ and $Q(4, 0)$.

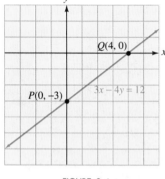

FIGURE 8-4

$$m = \frac{\text{change in } y}{\text{change in } x}$$

$$= \frac{y_2 - y_1}{x_2 - x_1}$$

$$= \frac{0 - (-3)}{4 - 0} \qquad \text{Substitute 0 for } y_2, -3 \text{ for } y_1, 4 \text{ for } x_2, \text{ and 0 for } x_1.$$

$$= \frac{3}{4}$$

The slope of the line is $\frac{3}{4}$.  ∎

**Self Check** Find the slope of the line determined by $4x + 3y = 12$.

**Answer** $-\frac{4}{3}$

## ▓ INTERPRETATION OF SLOPE

Many problems involve equations of lines and their slopes.

**EXAMPLE 3** **Cost of carpet** If carpet costs \$25 per square yard, the total cost $c$ of $n$ square yards is the price per square yard times the number of square yards purchased:

| $c$ | $=$ | the cost per square yard | $\cdot$ | the number of square yards. |
|---|---|---|---|---|
| $c$ | $=$ | 25 | $\cdot$ | $n$ |

Graph the equation $c = 25n$ and interpret the slope of the line.

*Solution* We can graph the equation on a coordinate system with a vertical $c$-axis and a horizontal $n$-axis. Figure 8-5 shows a table of ordered pairs and the graph.

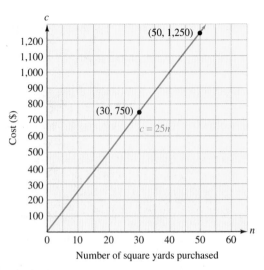

FIGURE 8-5

| | $c = 25n$ | |
| --- | --- | --- |
| $x$ | $y$ | $(x, y)$ |
| 10 | 250 | (10, 250) |
| 20 | 500 | (20, 500) |
| 30 | 750 | (30, 750) |
| 40 | 1,000 | (40, 1,000) |
| 50 | 1,250 | (50, 1,250) |

If we pick the points (30, 750) and (50, 1,250) to find the slope, we have

$$m = \frac{\text{change in } c}{\text{change in } n}$$

$$= \frac{c_2 - c_1}{n_2 - n_1}$$

$$= \frac{1{,}250 - 750}{50 - 30} \qquad \text{Substitute 1,250 for } c_2, \text{ 750 for } c_1, \text{ 50 for } n_2,$$
$$\text{and 30 for } n_1.$$

$$= \frac{500}{20}$$

$$= 25$$

The slope of 25 (in dollars/square yard) is the ratio of the change in the cost to the change in the number of square yards purchased. The slope is the cost in dollars per square yard. ∎

**EXAMPLE 4**

**Rate of descent**    It takes a skier 25 minutes to complete the course shown in Figure 8-6. Find his average rate of descent in feet per minute.

*Solution*    To find the average rate of descent, we must find the ratio of the change in altitude to the change in time. To find this ratio, we calculate the slope of the line passing through the points (0, 12,000) and (25, 8,500).

$$\text{Average rate of descent} = \frac{12{,}000 - 8{,}500}{0 - 25}$$

$$= \frac{3{,}500}{-25}$$

$$= -140$$

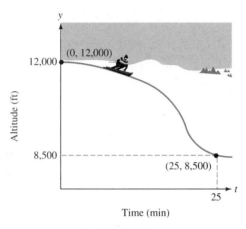

FIGURE 8-6

The average rate of descent is $-140$ ft/min. ∎

### ■ SLOPES OF HORIZONTAL AND VERTICAL LINES

If $P(x_1, y_1)$ and $Q(x_2, y_2)$ are points on the horizontal line shown in Figure 8-7(a), then $y_1 = y_2$, and the numerator of the fraction

$$\frac{y_2 - y_1}{x_2 - x_1} \qquad \text{On a horizontal line, } x_2 \neq x_1.$$

is 0. Thus, the value of the fraction is 0, and the slope of the horizontal line is 0.

If $P(x_1, y_1)$ and $Q(x_2, y_2)$ are two points on the vertical line shown in Figure 8-7(b), then $x_1 = x_2$, and the denominator of the fraction

$$\frac{y_2 - y_1}{x_2 - x_1} \qquad \text{On a vertical line, } y_2 \neq y_1.$$

is 0. Since the denominator of a fraction cannot be 0, a vertical line has no defined slope.

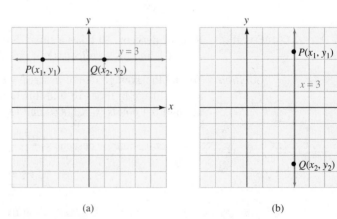

(a)                                                    (b)

FIGURE 8-7

### Slopes of Horizontal and Vertical Lines

All horizontal lines (lines with equations of the form $y = b$) have a slope of 0.

Vertical lines (lines with equations of the form $x = a$) have no defined slope.

If a line rises as we follow it from left to right, as in Figure 8-8(a), its slope is positive. If a line drops as we follow it from left to right, as in Figure 8-8(b), its slope is negative. If a line is horizontal, as in Figure 8-8(c), its slope is 0. If a line is vertical, as in Figure 8-8(d), it has no defined slope.

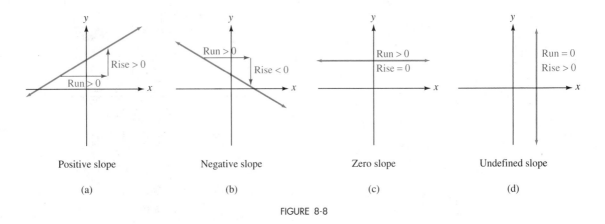

Positive slope    Negative slope    Zero slope    Undefined slope

(a)    (b)    (c)    (d)

FIGURE 8-8

### ■ SLOPES OF PARALLEL LINES

To see a relationship between parallel lines and their slopes, we refer to the parallel lines $l_1$ and $l_2$ shown in Figure 8-9, with slopes of $m_1$ and $m_2$, respectively. Because right triangles $ABC$ and $DEF$ are similar, it follows that

$$m_1 = \frac{\Delta y \text{ of } l_1}{\Delta x \text{ of } l_1}$$

$$= \frac{\Delta y \text{ of } l_2}{\Delta x \text{ of } l_2}$$

$$= m_2$$

Read $\Delta y$ as "the change in $y$."
Read $\Delta x$ as "the change in $x$."

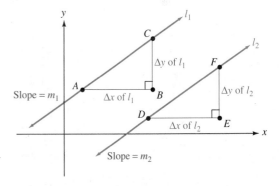

FIGURE 8-9

This shows that if two nonvertical lines are parallel, they have the same slope. It is also true that when two lines have the same slope, they are parallel.

### Slopes of Parallel Lines

Nonvertical parallel lines have the same slope, and lines having the same slope are parallel.

Since vertical lines are parallel, lines with no defined slope are parallel.

**EXAMPLE 5**     The lines in Figure 8-10 are parallel. Find $y$.

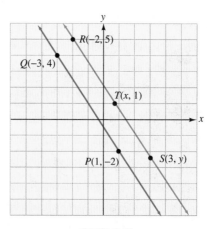

FIGURE 8-10

*Solution*     Since the lines are parallel, they have equal slopes. To find $y$, we find the slope of each line, set them equal, and solve the resulting equation.

| *Slope of PQ* | *Slope of RS* | |
|---|---|---|
| $\dfrac{-2-4}{1-(-3)}$ | $=\dfrac{y-5}{3-(-2)}$ | |
| $\dfrac{-6}{4}$ | $=\dfrac{y-5}{5}$ | |
| $-30 = 4(y-5)$ | | Multiply both sides by 20. |
| $-30 = 4y - 20$ | | Use the distributive property. |
| $-10 = 4y$ | | Add 20 to both sides. |
| $-\dfrac{5}{2} = y$ | | Divide both sides by 4 and simplify. |

Thus, $y = -\frac{5}{2}$.     ■

**Self Check**     In Figure 8-10, find $x$.

**Answer**     $\frac{2}{3}$

### ■ SLOPES OF PERPENDICULAR LINES

Two real numbers $a$ and $b$ are called **negative reciprocals** if $ab = -1$. For example,

$$-\frac{4}{3} \quad \text{and} \quad \frac{3}{4}$$

are negative reciprocals, because $-\frac{4}{3}\left(\frac{3}{4}\right) = -1$.

The following theorem relates perpendicular lines and their slopes.

**Slopes of Perpendicular Lines**

If two nonvertical lines are perpendicular, their slopes are negative reciprocals.

If the slopes of two lines are negative reciprocals, the lines are perpendicular.

Because a horizontal line is perpendicular to a vertical line, a line with a slope of 0 is perpendicular to a line with no defined slope.

**EXAMPLE 6**   Are the lines shown in Figure 8-11 perpendicular?

*Solution*   We find the slopes of the lines to see whether they are negative reciprocals.

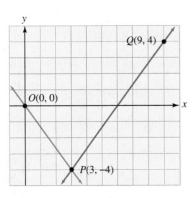

FIGURE 8-11

$$\text{Slope of } OP = \frac{\Delta y}{\Delta x} \quad \text{and} \quad \text{Slope of } PQ = \frac{\Delta y}{\Delta x}$$

$$= \frac{y_2 - y_1}{x_2 - x_1} \qquad\qquad = \frac{y_2 - y_1}{x_2 - x_1}$$

$$= \frac{-4 - 0}{3 - 0} \qquad\qquad = \frac{4 - (-4)}{9 - 3}$$

$$= -\frac{4}{3} \qquad\qquad\qquad = \frac{8}{6}$$

$$\qquad\qquad\qquad\qquad\qquad = \frac{4}{3}$$

Since their slopes are not negative reciprocals, the lines are not perpendicular. ■

*Self Check*   In Figure 8-11, is $PQ$ perpendicular to a line passing through points $P$ and $R(0, -1)$?

*Answer*   no

Orals   *Find the slope of the line passing through*

**1.** $(0, 0)$, $(1, 3)$                    **2.** $(0, 0)$, $(3, 6)$

**3.** Are lines with slopes of $-2$ and $\dfrac{8}{-4}$ parallel?

**4.** Find the negative reciprocal of $-0.2$.

**5.** Are lines with slopes of $-2$ and $\frac{1}{2}$ perpendicular?

## EXERCISE 8.1

**REVIEW**   *Simplify each expression. Write all answers without negative exponents.*

**1.** $(x^3 y^2)^3$

**2.** $\left(\dfrac{x^5}{x^3}\right)^3$

**3.** $(x^{-3} y^2)^{-4}$

**4.** $\left(\dfrac{x^{-6}}{y^3}\right)^{-4}$

**5.** $\left(\dfrac{3x^2 y^3}{8}\right)^0$

**6.** $\left(\dfrac{x^3 x^{-7} y^{-6}}{x^4 y^{-3} y^{-2}}\right)^{-2}$

**VOCABULARY AND CONCEPTS**   *Fill in each blank to make a true statement.*

**7.** The slope of a line is the change in ___ divided by the change in ___.

**8.** The point $(x_1, y_1)$ is read as "$x$ ___ 1 $y$ ___ 1."

**9.** Slope is sometimes defined as rise over ____.

**10.** The slope of a _____ line is 0.

**11.** The slope of a vertical line is _____.

**12.** Slopes of parallel lines are ____.

**13.** Slopes of _____ lines are negative reciprocals.

**14.** If a line rises as $x$ gets larger, the line has a _____ slope.

**PRACTICE**   *In Exercises 15–26, find the slope of the line that passes through the given points, if possible.*

**15.** $(0, 0)$, $(3, 9)$

**16.** $(9, 6)$, $(0, 0)$

**17.** $(-1, 8)$, $(6, 1)$

**18.** $(-5, -8)$, $(3, 8)$

**19.** $(3, -1)$, $(-6, 2)$

**20.** $(0, -8)$, $(-5, 0)$

**21.** $(7, 5)$, $(-9, 5)$

**22.** $(2, -8)$, $(3, -8)$

**23.** $(-7, -5)$, $(-7, -2)$

**24.** $(3, -5)$, $(3, 14)$

**25.** $(a, b)$, $(b, a)$

**26.** $(a, b)$, $(-b, -a)$

*In Exercises 27–34, find the slope of the line determined by each equation.*

**27.** $3x + 2y = 12$

**28.** $2x - y = 6$

**29.** $3x = 4y - 2$

**30.** $x = y$

**31.** $y = \dfrac{x - 4}{2}$

**32.** $x = \dfrac{3 - y}{4}$

**33.** $4y = 3(y + 2)$

**34.** $x + y = \dfrac{2 - 3y}{3}$

*In Exercises 35–40, tell whether the slope of the line in each graph is positive, negative, 0, or undefined.*

**35.**

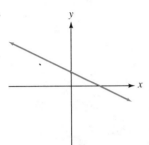

**36.**

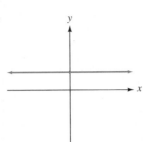

**37.**

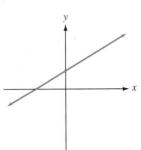

**38.**

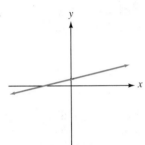

**39.**

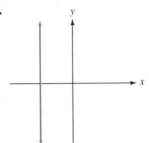

**40.**

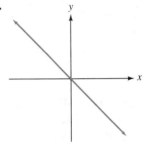

*In Exercises 41–46, tell whether the lines with the given slopes are parallel, perpendicular, or neither.*

**41.** $m_1 = 3, m_2 = -\dfrac{1}{3}$

**42.** $m_1 = \dfrac{1}{4}, m_2 = 4$

**43.** $m_1 = 4, m_2 = 0.25$

**44.** $m_1 = -5, m_2 = \dfrac{1}{-0.2}$

**45.** $m_1 = \dfrac{a}{b}, m_2 = \left(\dfrac{b}{a}\right)^{-1}$

**46.** $m_1 = \dfrac{c}{d}, m_2 = \dfrac{d}{c}$

*In Exercises 47–52, tell whether the line PQ is parallel or perpendicular (or neither) to a line with a slope of −2.*

**47.** $P(3, 4), Q(4, 2)$

**48.** $P(6, 4), Q(8, 5)$

**49.** $P(-2, 1), Q(6, 5)$

**50.** $P(3, 4), Q(-3, -5)$

**51.** $P(5, 4), Q(6, 6)$

**52.** $P(-2, 3), Q(4, -9)$

*In Exercises 53–58, find the slopes of lines PQ and PR and tell whether the points P, Q, and R lie on the same line. (Hint: Two lines with the same slope and a point in common must be the same line.)*

**53.** $P(-2, 4), Q(4, 8), R(8, 12)$

**54.** $P(6, 10), Q(0, 6), R(3, 8)$

**55.** $P(-4, 10), Q(-6, 0), R(-1, 5)$

**56.** $P(-10, -13), Q(-8, -10), R(-12, -16)$

**57.** $P(-2, 4), Q(0, 8), R(2, 12)$

**58.** $P(8, -4), Q(0, -12), R(8, -20)$

**59.** Find the equation of the *x*-axis and its slope.

**60.** Find the equation of the *y*-axis and its slope, if any.

*APPLICATIONS*

**61. Grade of a road** If the vertical rise of the road shown in Illustration 1 is 24 feet for a horizontal run of 1 mile, find the slope of the road. (*Hint:* 1 mile = 5,280 feet.)

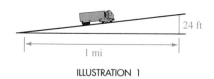

ILLUSTRATION 1

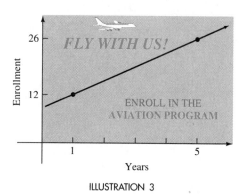

ILLUSTRATION 3

**62. Pitch of a roof** If the rise of the roof shown in Illustration 2 is 5 feet for a run of 12 feet, find the pitch of the roof.

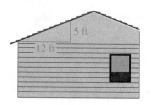

ILLUSTRATION 2

**63. Slope of a ramp** If a ramp rises 4 feet over a run of 12 feet, find its slope.

**64. Slope of a ladder** A 26-foot ladder leans against a building, with its base 10 feet from the building. Find the slope of the ladder.

**65. Rate of growth** When a college started an aviation program, the administration agreed to predict enrollments using a straight-line method. If the enrollment during the first year was 12, and the enrollment during the fifth year was 26, find the rate of growth per year (the slope of the line). See Illustration 3.

**66. Rate of growth** A small business predicts sales according to a straight-line method. If sales were $50,000 in the first year and $110,000 in the third year, find the rate of growth in sales per year (the slope of the line).

**67. Rate of decrease** The price of computer equipment has been dropping steadily for the past ten years. If a desktop PC cost $6,700 ten years ago, and the same computing power cost $2,200 three years ago, find the rate of decrease per year. (Assume a straight-line model).

**68. Hospital costs** Illustration 4 shows the changing mean daily cost for a hospital room. Find the rate of change per year of the portion of the room cost that was absorbed by the hospital between 1980 and 1990.

|      | Cost passed on to the patient | Total cost to the hospital |
|------|:-----------------------------:|:--------------------------:|
| 1980 | $130                          | $245                       |
| 1985 | 214                           | 459                        |
| 1990 | 295                           | 670                        |

ILLUSTRATION 4

*WRITING*

**69.** Explain why a vertical line has no defined slope.

**70.** Explain how to determine from their slopes whether two lines are parallel, perpendicular, or neither.

*SOMETHING TO THINK ABOUT*

**71.** The points $(3, a)$, $(5, 7)$, and $(7, 10)$ lie on a line. Find $a$.

**72.** The line passing through points $A(1, 3)$ and $B(-2, 7)$ is perpendicular to the line passing through points $C(4, b)$ and $D(8, -1)$. Find $b$.

## 8.2  Writing Equations of Lines

■ POINT–SLOPE FORM OF THE EQUATION OF A LINE  ■ SLOPE–INTERCEPT FORM OF THE EQUATION OF A LINE  ■ USING SLOPE AS AN AID IN GRAPHING  ■ GENERAL FORM OF THE EQUATION OF A LINE  ■ APPLICATIONS

Getting Ready    *Solve each equation for y.*

**1.** $2x + y = 12$                           **2.** $2x - 4y = 9$

### ■ POINT–SLOPE FORM OF THE EQUATION OF A LINE

Suppose that line $l$ shown in Figure 8-12 has a slope of $m$ and passes through the point $P(x_1, y_1)$. If $Q(x, y)$ is a second point on line $l$, we have

$$m = \frac{y - y_1}{x - x_1}$$

or, after we multiply both sides by $x - x_1$,

**1.** $\quad y - y_1 = m(x - x_1)$

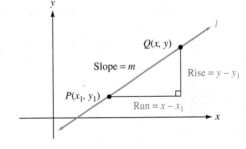

FIGURE 8-12

Because Equation 1 displays the coordinates of the point $(x_1, y_1)$ on the line and the slope $m$ of the line, it is called the **point–slope form** of the equation of a line.

> **Point–Slope Form of the Equation of a Line**
> The equation of the line passing through $P(x_1, y_1)$ and with slope $m$ is
> $$y - y_1 = m(x - x_1)$$

**EXAMPLE 1**  Write the equation of the line that has a slope of $\frac{2}{3}$ and passes through $P(-4, 2)$.

*Solution*  We substitute $\frac{2}{3}$ for $m$, $-4$ for $x_1$, and 2 for $y_1$ into the point–slope form and simplify.

$$y - y_1 = m(x - x_1)$$

$$y - 2 = \frac{2}{3}\,[x - (-4)] \qquad \text{Substitute } \tfrac{2}{3} \text{ for } m, -4 \text{ for } x_1, \text{ and 2 for } y_1.$$

$$y - 2 = \frac{2}{3}\,(x + 4) \qquad\qquad -(-4) = 4.$$

$$y - 2 = \frac{2}{3}x + \frac{8}{3} \qquad\qquad \text{Use the distributive property to remove parentheses.}$$

$$y = \frac{2}{3}x + \frac{14}{3} \qquad\qquad \text{Add 2 to both sides and simplify.}$$

The equation is $y = \frac{2}{3}x + \frac{14}{3}$. ■

**Self Check**  Write the equation of the line that has a slope of $\frac{2}{3}$ and passes through $(4, -5)$.
**Answer**  $y = \frac{2}{3}x - \frac{23}{3}$

In the next example, we are given a graph and asked to write its equation.

**EXAMPLE 2**  Write the equation of the line shown in Figure 8-13.

*Solution*  First we find the slope of the line by substituting $-6$ for $y_2$, 4 for $y_1$, 8 for $x_2$, and $-5$ for $x_1$.

$$m = \frac{y_2 - y_1}{x_2 - x_1}$$

$$= \frac{-6 - 4}{8 - (-5)}$$

$$= -\frac{10}{13}$$

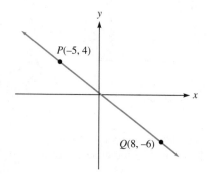

FIGURE 8-13

Because the line passes through both $P$ and $Q$, we can choose either point and substitute its coordinates into the point–slope form. If we choose $P(-5, 4)$, we substitute $-5$ for $x_1$, 4 for $y_1$, and $-\frac{10}{13}$ for $m$ and proceed as follows.

$$y - y_1 = m(x - x_1)$$

$$y - 4 = -\frac{10}{13}\,[x - (-5)] \qquad \text{Substitute } -\tfrac{10}{13} \text{ for } m, -5 \text{ for } x_1, \text{ and } 4 \text{ for } y_1.$$

$$y - 4 = -\frac{10}{13}\,(x + 5) \qquad -(-5) = 5.$$

$$y - 4 = -\frac{10}{13}x - \frac{50}{13} \qquad \text{Use the distributive property to remove parentheses}$$

$$y = -\frac{10}{13}x + \frac{2}{13} \qquad \text{Add 4 to both sides and simplify.}$$

The equation is $y = -\frac{10}{13}x + \frac{2}{13}$.    ∎

**Self Check**

Write the equation of the line passing through $(-2, 5)$ and $(5, -2)$.

**Answer**

$y = -x + 3$

## SLOPE–INTERCEPT FORM OF THE EQUATION OF A LINE

Since the $y$-intercept of the line shown in Figure 8-14 is the point $(0, b)$, we can write the equation of the line by substituting 0 for $x_1$ and $b$ for $y_1$ in the point–slope form and simplifying.

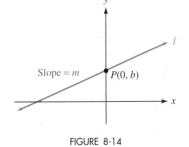

FIGURE 8-14

$$y - y_1 = m(x - x_1)$$

$$y - b = m(x - 0)$$

$$y - b = mx$$

**2.**    $$y = mx + b$$

Because Equation 2 displays the slope $m$ and the $y$-coordinate $b$ of the $y$-intercept, it is called the **slope–intercept form** of the equation of a line.

> **Slope–Intercept Form of the Equation of a Line**
> The equation of the line with slope $m$ and $y$-intercept $(0, b)$ is
> $$y = mx + b$$

**EXAMPLE 3**

Use the slope–intercept form to write the equation of the line that has a slope of 5 and passes through the point $P(-2, 9)$.

**Solution**

Since we are given that $m = 5$ and that the pair $(-2, 9)$ satisfies the equation, we can substitute $-2$ for $x$, 9 for $y$, and 5 for $m$ in the equation $y = mx + b$ and solve for $b$.

$$y = mx + b$$

$$9 = 5(-2) + b \qquad \text{Substitute 9 for } y, \text{ 5 for } m, \text{ and } -2 \text{ for } x.$$

$$9 = -10 + b \qquad \text{Simplify.}$$

$$19 = b \qquad \text{Add 10 to both sides.}$$

Because $m = 5$ and $b = 19$, the equation is $y = 5x + 19$.    ∎

<table>
<tr><td>

*Self Check*

*Answer*

</td><td>

Write the equation of a line that has a slope of $-2$ and passes through $(2, 3)$.

$y = -2x + 7$

</td></tr>
</table>

## ■ USING SLOPE AS AN AID IN GRAPHING

It is easy to graph a linear equation when it is written in slope–intercept form. For example, to graph $y = \frac{4}{3}x - 2$, we note that $b = -2$ and that the $y$-intercept is $(0, b) = (0, -2)$. (See Figure 8-15.)

    Because the slope of the line is $\frac{4}{3}$, we can locate another point $Q$ on the line by starting at point $P$ and counting 3 units to the right and 4 units up. The change in $x$ from point $P$ to point $Q$ is 3, and the corresponding change in $y$ is 4. The line joining points $P$ and $Q$ is the graph of the equation.

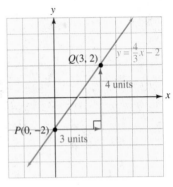

FIGURE 8-15

EXAMPLE 4

Find the slope and the $y$-intercept of the line with the equation $2(x - 3) = -3(y + 5)$. Then graph the line.

*Solution*

We write the equation in the form $y = mx + b$ to find the slope $m$ and the $y$-intercept $(0, b)$.

$$2(x - 3) = -3(y + 5)$$

$$2x - 6 = -3y - 15 \qquad \text{Use the distributive property to remove parentheses.}$$

$$2x + 3y - 6 = -15 \qquad \text{Add } 3y \text{ to both sides.}$$

$$3y - 6 = -2x - 15 \qquad \text{Subtract } 2x \text{ from both sides.}$$

$$3y = -2x - 9 \qquad \text{Add 6 to both sides.}$$

$$y = -\frac{2}{3}x - 3 \qquad \text{Divide both sides by 3.}$$

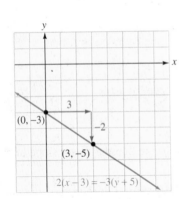

FIGURE 8-16

The slope is $-\frac{2}{3}$, and the $y$-intercept is $(0, -3)$. To draw the graph, we plot the $y$-intercept $(0, -3)$ and then locate a second point on the line by moving 3 units to the right and 2 units down. We draw a line through the two points to obtain the graph shown in Figure 8-16. ■

<table>
<tr><td>

*Self Check*

*Answer*

</td><td>

Find the slope and the $y$-intercept of the line with the equation $-3(x + 2) = 2(y - 3)$.

$y = -\frac{3}{2}x$

</td></tr>
</table>

**EXAMPLE 5**    Show that the lines represented by $4x + 8y = 16$ and $x = 6 - 2y$ are parallel.

*Solution*    We solve each equation for $y$ to see whether the lines are distinct and whether their slopes are equal.

$$4x + 8y = 16 \qquad\qquad\qquad x = 6 - 2y$$
$$8y = -4x + 16 \qquad\qquad\qquad 2y = -x + 6$$
$$y = -\frac{1}{2}x + 2 \qquad\qquad\qquad y = -\frac{1}{2}x + 3$$

Since the values of $b$ in these equations are different, the lines are distinct. Since the slope of each line is $-\frac{1}{2}$, the lines are parallel. ■

**Self Check**    Are the lines represented by $3x + y = 2$ and $2y = -6x - 3$ parallel?

*Answer*    yes

**EXAMPLE 6**    Show that the lines represented by $4x + 8y = 16$ and $4x - 2y = 22$ are perpendicular.

*Solution*    We solve each equation for $y$ to see whether the slopes of their straight-line graphs are negative reciprocals.

$$4x + 8y = 16 \qquad\qquad\qquad 4x - 2y = 22$$
$$8y = -4x + 16 \qquad\qquad\qquad -2y = -4x + 22$$
$$y = -\frac{1}{2}x + 2 \qquad\qquad\qquad y = 2x - 11$$

Since the slopes of $-\frac{1}{2}$ and $2$ are negative reciprocals, the lines are perpendicular. ■

**Self Check**    Are the lines represented by $y = 2x + 3$ and $2x + y = 7$ perpendicular?

*Answer*    no

**EXAMPLE 7**    Write the equation of the line passing through $P(-3, 2)$ and parallel to the line $y = 8x - 5$.

*Solution*    Since the equation is in slope–intercept form and the slope of the line given by $y = 8x - 5$ is the coefficient of $x$, the slope is 8. Since the desired equation is to have a graph that is parallel to the graph of $y = 8x - 5$, its slope must also be 8.

We substitute $-3$ for $x_1$, 2 for $y_1$, and 8 for $m$ in the point–slope form and simplify.

$$y - y_1 = m(x - x_1)$$
$$y - 2 = 8[x - (-3)] \qquad \text{Substitute 2 for } y_1, 8 \text{ for } m, \text{ and } -3 \text{ for } x_1.$$
$$y - 2 = 8(x + 3) \qquad -(-3) = 3.$$
$$y - 2 = 8x + 24 \qquad \text{Use the distributive property to remove parentheses.}$$
$$y = 8x + 26 \qquad \text{Add 2 to both sides.}$$

The equation of the desired line is $y = 8x + 26$. ■

**Self Check**  Write the equation of the line passing through $(0, 0)$ and parallel to the line $y = 8x - 3$.

**Answer**  $y = 8x$

**EXAMPLE 8**  Write the equation of the line that passes through $P(-2, 5)$ and is perpendicular to the line $y = 8x - 3$.

**Solution**  Since the slope of the given line is 8, the slope of the desired line must be $-\frac{1}{8}$, which is the negative reciprocal of 8.

We substitute $-2$ for $x_1$, 5 for $y_1$, and $-\frac{1}{8}$ for $m$ into the point–slope form and simplify.

$$y - y_1 = m(x - x_1)$$
$$y - 5 = -\frac{1}{8}[x - (-2)] \qquad \text{Substitute 5 for } y_1, -\frac{1}{8} \text{ for } m, \text{ and } -2 \text{ for } x_1.$$
$$y - 5 = -\frac{1}{8}(x + 2) \qquad -(-2) = 2.$$
$$8y - 40 = -(x + 2) \qquad \text{Multiply both sides by 8.}$$
$$8y - 40 = -x - 2 \qquad \text{Use the distributive property to remove parentheses.}$$
$$x + 8y - 40 = -2 \qquad \text{Add } x \text{ to both sides.}$$
$$x + 8y = 38 \qquad \text{Add 40 to both sides.}$$

The equation of the line is $x + 8y = 38$. ■

**Self Check**  Write the equation of the line that passes through $(0, 0)$ and is perpendicular to the line $y = 8x - 3$.

**Answer**  $y = -\frac{1}{8}x$

## ■ GENERAL FORM OF THE EQUATION OF A LINE

The final equation in Example 8 is written in the form $Ax + By = C$, where $A$, $B$, and $C$ are constants. When an equation is written in this form, it is said to be written in **general form.**

**WARNING!** When writing equations in general form, it is customary to clear the equation of fractions and make $A$ positive.

It is also customary to make $A$, $B$, and $C$ as small as possible. For example, the equation $6x + 12y = 24$ can be changed to $x + 2y = 4$ by dividing both sides by 6.

**Finding the Slope and y-Intercept from the General Form**
If $A$, $B$, and $C$ are real numbers and $B \neq 0$, the graph of the equation

$$Ax + By = C$$

is a nonvertical line with slope of $-\dfrac{A}{B}$ and a $y$-intercept of $\left(0, \dfrac{C}{B}\right)$.

You will be asked to justify the previous results in the exercises. You will also be asked to show that if $B = 0$, the equation $Ax + By = C$ represents a vertical line with $x$-intercept of $\left(\frac{C}{A}, 0\right)$.

**EXAMPLE 9**

Show that the lines represented by $6x + 5y = 7$ and $5x - 6y = 12$ are perpendicular.

*Solution*

To show that the lines are perpendicular, we will show that their slopes are negative reciprocals. Since the equation $6x + 5y = 7$ is written in general form, with $A = 6$, $B = 5$, and $C = 7$, the slope of its graph is

$$m_1 = -\frac{A}{B} = -\frac{6}{5}$$

Since the equation $5x - 6y = 12$ is also written in general form, with $A = 5$, $B = -6$, and $C = 12$, the slope of its graph is

$$m_2 = -\frac{A}{B} = -\frac{5}{-6} = \frac{5}{6}$$

Since the slopes are negative reciprocals, the lines are perpendicular. ■

*Self Check*

*Answer*

Are the lines represented by $5x - 3y = 12$ and $3x - 5y = 5$ perpendicular?

no

We summarize the various forms for the equation of a line as follows.

**General form** of a linear equation — $Ax + By = C$

$A$ and $B$ cannot both be 0, $A \geq 0$, and $A$, $B$, and $C$ are integers, when possible.

**Slope–intercept form** of a linear equation — $y = mx + b$

The slope is $m$, and the $y$-intercept is $(0, b)$.

**Point–slope form** of a linear equation — $y - y_1 = m(x - x_1)$

The slope is $m$, and the line passes through $(x_1, y_1)$.

**A horizontal line** — $y = b$

The slope is 0, and the $y$-intercept is $(0, b)$.

**A vertical line** — $x = a$

There is no defined slope, and the $x$-intercept is $(a, 0)$.

## ■ APPLICATIONS

**EXAMPLE 10**

**Water billing**   A linear equation can be used to represent a water department's monthly charge for water usage in relation to the number of gallons used. If a customer is charged \$12 for using 1,000 gallons and \$16 for using 1,800 gallons, find the charge for 2,000 gallons.

*Solution*   When a customer uses 1,000 gallons, the charge is \$12. When 1,800 gallons are used, the charge is \$16. Since $y$ is related to $x$ by a linear equation, the points $P(1,000, 12)$ and $Q(1,800, 16)$ will lie on the line shown in Figure 8-17.

To write the equation of the line passing through $P$ and $Q$, we first find the slope of the line passing through those points:

$$m = \frac{y_2 - y_1}{x_2 - x_1}$$

$$= \frac{16 - 12}{1,800 - 1,000}$$

$$= \frac{4}{800}$$

$$= \frac{1}{200}$$

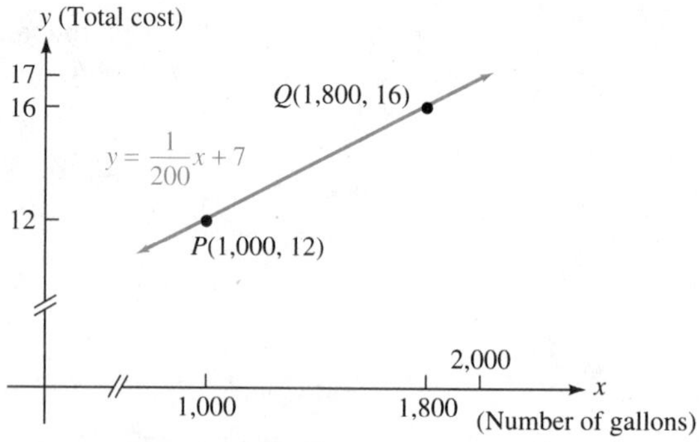

FIGURE 8-17

We then substitute $\frac{1}{200}$ for $m$ and the coordinates of one of the known points (say, $P(1,000, 12)$) into the point–slope form of the equation of a line and proceed as follows:

$$y - y_1 = m(x - x_1)$$

$$y - 12 = \frac{1}{200}(x - 1,000)$$

$$y - 12 = \frac{1}{200}x - \frac{1}{200}(1,000)$$

$$y - 12 = \frac{1}{200}x - 5$$

**1.** $\qquad y = \frac{1}{200}x + 7 \qquad\qquad$ Add 12 to both sides.

To find the charge for 2,000 gallons, we substitute 2,000 for $x$ in Equation 1 and find $y$.

$$y = \frac{1}{200}(2,000) + 7$$

$$y = 10 + 7$$

$$y = 17$$

The charge for 2,000 gallons of water is $17. ∎

As machinery wears out, it is worth less. Accountants often estimate the decreasing value of aging equipment with **linear depreciation**, a method based on the equation of a line.

**EXAMPLE 11**

**Linear depreciation**   A company buys a $12,500 computer with an estimated life of 6 years. The computer can then be sold as scrap for an estimated *salvage value* of $500. If $y$ represents the value of the computer after $x$ years of use and $y$ and $x$ are related by the equation of a line,

**a.** Find the equation of the line.

**b.** Find the value of the computer after 2 years.

**c.** Find the economic meaning of the $y$-intercept of the line.

**d.** Find the economic meaning of the slope of the line.

*Solution*   **a.** To find the equation of the line, we calculate its slope and then use the point–slope form to find its equation.

When the computer is new, its age $x$ is 0 and its value $y$ is the purchase price of $12,500. When it is six years old, $x = 6$ and $y = 500$, its salvage value. Since the line passes through the points (0, 12,500) and (6, 500) as shown in Figure 8-18, the slope of the line is

$$m = \frac{y_2 - y_1}{x_2 - x_1}$$
$$= \frac{500 - 12,500}{6 - 0}$$
$$= \frac{-12,000}{6}$$
$$= -2,000$$

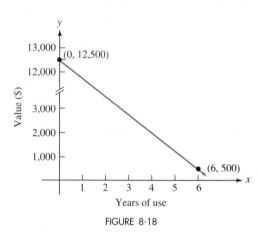

FIGURE 8-18

To find the equation of the line, we substitute $-2,000$ for $m$, 0 for $x_1$, and 12,500 for $y_1$ in the point–slope form and simplify.

$$y - y_1 = m(x - x_1)$$
$$y - 12,500 = -2,000(x - 0)$$
$$y = -2,000x + 12,500$$

The current value of the computer is related to its age by the equation $y = -2,000x + 12,500$.

**b.** To find the value of the computer after 2 years, we substitute 2 for $x$ in the equation $y = -2,000x + 12,500$.

$$y = -2,000x + 12,500$$
$$y = -2,000(2) + 12,500$$
$$= -4,000 + 12,500$$
$$= 8,500$$

In 2 years, the computer will be worth $8,500.

**c.** Since the $y$-intercept of the graph is (0, 12,500), the $y$-coordinate of the $y$-intercept is the computer's original purchase price.

**d.** Each year, the value decreases by $2,000, because the slope of the line is $-2,000$. The slope of the depreciation line is called the **annual depreciation rate.** ■

Orals *Write the point–slope form of the equation of a line with m = 2, passing through the given point.*

**1.** $(2, 3)$ **2.** $(-3, 8)$

*Write the equation of a line with m = −3 and y-intercept of*

**3.** $(0, 5)$ **4.** $(0, -7)$

*Tell whether the lines are parallel, perpendicular, or neither.*

**5.** $y = 3x - 4$, $y = 3x + 5$ **6.** $y = -3x + 7$, $x = 3y - 1$

## EXERCISE 8.2

**REVIEW** *Solve each equation.*

**1.** $3(x + 2) + x = 5x$

**2.** $12b + 6(3 - b) = b + 3$

**3.** $\dfrac{5(2 - x)}{3} - 1 = x + 5$

**4.** $\dfrac{r - 13}{3} = \dfrac{r + 2}{6} - 2$

**5. Mixing alloys** In 60 ounces of alloy for watch cases, there are 20 ounces of gold. How much copper must be added to the alloy so that a watch case weighing 4 ounces, made from the new alloy, will contain exactly one ounce of gold?

**6. Mixing coffee** To make a mixture of 80 pounds of coffee worth $272, a grocer mixes coffee worth $3.25 a pound with coffee worth $3.85 a pound. How many pounds of the cheaper coffee should the grocer use?

**VOCABULARY AND CONCEPTS** *Fill in each blank to make a true statement.*

**7.** Write the point–slope form of the equation of a line. _____.

**8.** Write the slope–intercept form of the equation of a line. _____

**9.** In the graph of the equation $y = -3x + 7$, _____ is the slope, and __ is the $y$-coordinate of the $y$-intercept.

**10.** In the graph of the equation $y - 3 = 7(x + 2)$, __ is the slope, and the line passes through _____.

**11.** Write the general form of the equation of a line. _____

**12.** The slope of the line represented by $3x + 2y = 7$ is ____.

**13.** $y = b$ is the equation of a _____ line.

**14.** $x = a$ is the equation of a _____ line.

**PRACTICE** *In Exercises 15–18, use point–slope form to write the equation of the line with the given properties. Write each equation in general form.*

**15.** $m = 8$, passing through $P(0, 7)$

**16.** $m = -5$, passing through $P(0, -4)$

**17.** $m = -3$, passing through $P(4, 0)$

**18.** $m = 7$, passing through $P(-3, 0)$

*In Exercises 19–20, use point–slope form to write the equation of each line. Write the equation in general form.*

**19.**

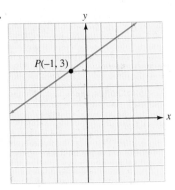

**20.**

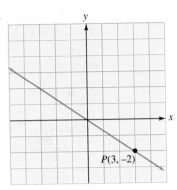

*In Exercises 21–24, use point–slope form to write the equation of the line passing through the two given points. Write each equation in slope–intercept form.*

**21.** $P(0, 0)$, $Q(8, 8)$

**23.** $P(3, 4)$, $Q(0, 5)$

**22.** $P(-5, -5)$, $Q(4, 4)$

**24.** $P(6, 0)$, $Q(8, 6)$

*In Exercises 25–26, use point–slope form to write the equation of each line. Write each answer in slope–intercept form.*

**25.**

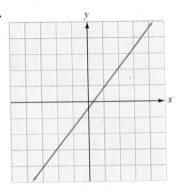

**26.**

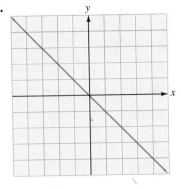

*In Exercises 27–34, use the slope–intercept form to write the equation of the line with the given properties. Write each equation in slope–intercept form.*

**27.** $m = 5$, $b = 19$

**29.** $m = -9$, passing through $P(7, 5)$

**31.** $m = 0$, passing through $P(2, 8)$

**28.** $m = -8$, $b = 13$

**30.** $m = 5$, passing through $P(-2, -5)$

**32.** $m = -9$, passing through the origin

**33.** Passing through $(6, 8)$ and $Q(4, 9)$

**34.** Passing through $P(-4, 5)$ and $Q\left(0, -\dfrac{7}{3}\right)$

*In Exercises 35–40, write each equation in slope–intercept form to find the slope and the y-intercept. Then use the slope and y-intercept to draw the line.*

**35.** $x - y = 1$

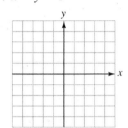

**36.** $x + y = 2$

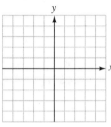

**37.** $2x = 3y - 6$

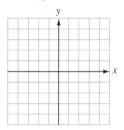

**38.** $5x = -4y + 10$

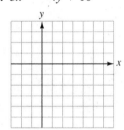

**39.** $3y = -2x + 18$

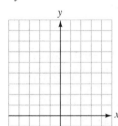

**40.** $-8x = 9y + 36$

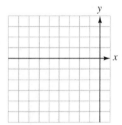

*In Exercises 41–46, find the slope and the y-intercept of the line determined by the given equation.*

**41.** $3x - 2y = 8$

**42.** $x - 2y = -6$

**43.** $-2x - 6y = 5$

**44.** $10x - 15y = 4$

**45.** $7x = 2y - 4$

**46.** $15x = -2y - 14$

*In Exercises 47–58, tell whether the graphs of each pair of equations are parallel, perpendicular, or neither.*

**47.** $y = 8x + 4$, $y = 8x - 7$

**48.** $y = 5x - 13$, $y = \dfrac{1}{5}x + 28$

**49.** $x + y = 6$, $y = x + 8$

**50.** $x = y + 5$, $y = x + 8$

**51.** $y = 3x + 9$, $2y = 6x - 10$

**52.** $2x + 3y = 11$, $3x - 2y = 15$

**53.** $x = 3y + 9$, $y = -3x + 2$

**54.** $3x + 6y = 7$, $y = \dfrac{1}{2}x$

**55.** $y = 8$, $x = 4$

**56.** $x = -3$, $x = -7$

**57.** $3x = y - 2$, $3(y - 3) + x = 0$

**58.** $2y = 8$, $x = y$

*In Exercises 59–64, write the equation of the line that passes through the given point and is parallel to the given line. Write the answer in slope–intercept form.*

**59.** $P(0, 0)$, $y = 4x - 9$

**60.** $P(0, 0)$, $x = -3y - 25$

**61.** $P(4, 13)$, $4x - y = 7$

**62.** $P(-2, -9)$, $y + 3x = -12$

**63.** $P\left(0, -\dfrac{26}{5}\right)$, $x = \dfrac{5}{4}y - 2$

**64.** $P(3, -9)$, $x = -\dfrac{3}{4}y + 5$

*In Exercises 65–70, write the equation of the line that passes through the given point and is perpendicular to the given line. Write the answer in slope–intercept form.*

**65.** $P(0, 0)$, $y = 4x - 9$

**66.** $P(0, 0)$, $x = -3y - 25$

**67.** $P(6, 4)$, $4x - y = 7$

**68.** $P(3, 6)$, $y + 3x = -12$

**69.** $P(-4, 8)$, $x = \dfrac{5}{4}y - 2$

**70.** $P(5, -2)$, $x = -\dfrac{3}{4}y + 5$

*In Exercises 71–74, use the method of Example 9 to determine whether the graphs determined by each pair of equations are parallel, perpendicular, or neither.*

**71.** $3x + 4y = 20$, $4x - 3y = 20$

**72.** $6x - 8y = 32$, $3x - 4y = 25$

**73.** $6x + 9y = 36$, $6x + 9y = 32$

**74.** $10x + 12y = 30$, $6x + 5y = 24$

**75.** Find the equation of the line perpendicular to the line $y = 5$ and passing through the midpoint of the segment joining $(4, 4)$ and $(-8, 10)$.

**76.** Find the equation of the line parallel to the line $y = -8$ and passing through the midpoint of the segment joining $(-8, 2)$ and $(-7, 8)$.

**77.** Find the equation of the line parallel to the line $x = 8$ and passing through the midpoint of the segment joining $(2, -6)$ and $(8, 10)$.

**78.** Find the equation of the line perpendicular to the line $x = 5$ and passing through the midpoint of the segment joining $(-12, 2)$ and $(14, -8)$.

**79.** Solve $Ax + By = C$ for $y$ and thereby show that the slope of its graph is $-\frac{A}{B}$ and its $y$-intercept is $\left(0, \frac{C}{B}\right)$.

**80.** Show that the $x$-intercept of the graph of $Ax + By = C$ is $\left(\frac{C}{A}, 0\right)$.

**APPLICATIONS**   *Assume straight-line depreciation or straight-line appreciation.*

**81. Finding a depreciation equation**   A taxicab was purchased for $24,300. Its salvage value at the end of its 7-year useful life is expected to be $1,900. Find the depreciation equation.

**82. Finding a depreciation equation**   A small business purchases the computer system shown in Illustration 1. It will be depreciated over a 4-year period, when its salvage value will be $300. Find the depreciation equation.

**83. Finding an appreciation equation**   An apartment building is purchased for $475,000. The owners expect the property to double in value in 10 years. Find the appreciation equation.

**84. Finding an appreciation equation**   A house purchased for $112,000 is expected to double in value in 12 years. Find its appreciation equation.

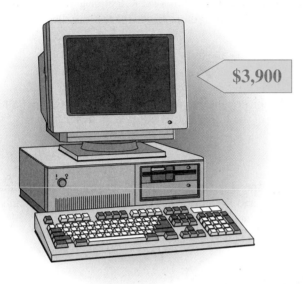

ILLUSTRATION 1

**85. Finding a depreciation equation** Find the depreciation equation for the TV in the want ad in Illustration 2.

> *For Sale*: 3-year-old 54-inch TV, $1,900 new. Asking $1,190. Call 875-5555. Ask for Mike.

ILLUSTRATION 2

**86. Depreciating a word processor** A word processor cost $555 when new and is expected to be worth $80 after 5 years. What will it be worth after 3 years?

**87. Finding salvage value** A copier cost $1,050 when new and will be depreciated at the rate of $120 per year. If the useful life of the copier is 8 years, find its salvage value.

**88. Finding annual rate of depreciation** A truck that cost $27,600 when new will have no salvage value after 12 years. Find its annual rate of depreciation.

**89. Finding the value of antiques** An antique table is expected to appreciate $40 each year. If the table will be worth $450 in 2 years, what will it be worth in 13 years?

**90. Finding the value of antiques** An antique clock is expected to be worth $350 after 2 years and $530 after 5 years. What will the clock be worth after 7 years?

**91. Finding the purchase price of real estate** A cottage that was purchased 3 years ago is now appraised at $47,700. If the property has been appreciating $3,500 per year, find its original purchase price.

**92. Charges for computer repair** A computer-repair company charges a fixed amount, plus an hourly rate, for a service call. Use the information in Illustration 3 to find the hourly rate.

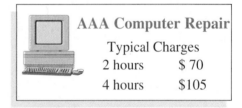

| AAA Computer Repair |  |
|---|---|
| Typical Charges | |
| 2 hours | $ 70 |
| 4 hours | $105 |

ILLUSTRATION 3

**93. Charges for automobile repair** An auto repair shop charges an hourly rate, plus the cost of parts. If the cost of labor for a $1\frac{1}{2}$-hour radiator repair job is $69, find the cost of labor for a 5-hour transmission overhaul.

**94. Finding printing charges** A printer charges a fixed setup cost, plus $1 for every 100 copies. If 700 copies cost $52, how much will it cost to print 1,000 copies?

**95. Predicting fires** A local fire department recognizes that city growth and the number of reported fires are related by a linear equation. City records show that 300 fires were reported in a year when the local population was 57,000 persons, and 325 fires were reported in a year when the population was 59,000 persons. How many fires can be expected when the population reaches 100,000 persons?

**96. Estimating the cost of rain gutter** A neighbor says that an installer of rain gutter charges $60, plus a dollar amount per foot. If the neighbor paid $435 for the installation of 250 feet of gutter, how much will it cost you to have 300 feet installed?

*WRITING*

**97.** Explain how to find the equation of a line passing through two given points.

**98.** In straight-line depreciation, explain why the slope of the line is called the *rate of depreciation*.

*SOMETHING TO THINK ABOUT*

**99.** If the graph of $y = ax + b$ passes through quadrants I, II, and IV, what can be known about the constants $a$ and $b$?

**100.** The graph of $Ax + By = C$ passes through quadrants I and IV only. What is known about the constants $A$, $B$, and $C$?

## 8.3  More on Functions

■ FUNCTIONS ■ THE ABSOLUTE VALUE FUNCTION ■ RATIONAL FUNCTIONS ■ FINDING DOMAINS AND RANGES FROM GRAPHS ■ THE VERTICAL LINE TEST ■ AN APPLICATION OF A RATIONAL FUNCTION

Getting Ready    *Let $y = x^2 - 2$. Find y when*

**1.** $x = 0$    **2.** $x = 2$    **3.** $x = -1$    **4.** $x = -2$

*Let $y = x^3 + 1$. Find y when*

**5.** $x = 0$    **6.** $x = -2$    **7.** $x = 1$    **8.** $x = \dfrac{1}{2}$

### ■ FUNCTIONS

In Section 4.4, we introduced the concept of function. Recall the following definition.

> **Function**
> Any equation in $x$ and $y$ where each value of $x$ (the input) determines one value of $y$ (the output) is called a **function**.
>
> The set of all input values is called the **domain** of the function, and the set of all output values $y$ is called the **range**.

We then introduced a special notation for functions called **function notation**.

### Function Notation

The notation $y = f(x)$ denotes that the variable $y$ is a function of $x$.

Recall that the notation $y = f(x)$ provides a way to denote the values of $y$ in a function that correspond to individual values of $x$. For example, $f(2)$ represents the $y$ value of $y = f(x)$ when $x = 2$. Since $y$ depends on $x$, $y$ is called the **dependent variable**, and $x$ is called the **independent variable**.

**EXAMPLE 1**    If $y = f(x) = x^2 - 3$, find   **a.** $f(2)$   and   **b.** $f(-3)$.

*Solution*    **a.** To find $f(2)$, we substitute 2 for $x$.

$$f(x) = x^2 - 3$$
$$f(2) = (2)^2 - 3$$
$$= 4 - 3$$
$$= 1$$

**b.** To find $f(-3)$, we substitute $-3$ for $x$.

$$f(x) = x^2 - 3$$
$$f(-3) = (-3)^2 - 3$$
$$= 9 - 3$$
$$= 6 \qquad\blacksquare$$

**Self Check**

**Answer**

If $y = f(x) = \frac{1}{2}x + 7$, find $f(-4)$.

5

Finally, we graphed several polynomial functions, including those shown in Figure 8-19.

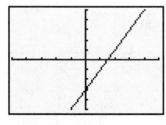

$y = f(x) = 2x - 3$
A linear function

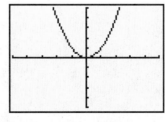

$y = f(x) = x^2$
A quadratic function

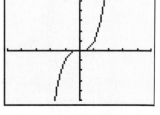

$y = f(x) = x^3$
A cubic function

FIGURE 8-19

In Chapter 7, we graphed the square root function and the cube root function, shown in Figure 8-20.

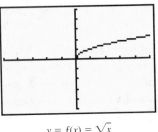

$y = f(x) = \sqrt{x}$
The square root function

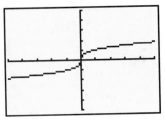

$y = f(x) = \sqrt[3]{x}$
The cube root function

FIGURE 8-20

## ■ THE ABSOLUTE VALUE FUNCTION

In this section, we discuss several more functions. The next example introduces the absolute value function.

**EXAMPLE 2**    Determine whether the equation $y = |x|$ determines a function. If so, determine its domain and range and draw its graph.

*Solution*    Since every real number $x$ has a single absolute value, the equation does define $y$ to be a function of $x$.

Since we can find the absolute value of any real number $x$, the domain is the set of all real numbers.

Since the absolute value of any real number is always positive or 0, the range is the set of nonnegative real numbers.

To construct the graph of the function $y = f(x) = |x|$, we make a table of values, plot the points, and join the points as in Figure 8-21.

$y = f(x) = |x|$

| $x$ | $y$ | $(x, y)$ |
|---|---|---|
| $-2$ | 2 | $(-2, 2)$ |
| $-1$ | 1 | $(-1, 1)$ |
| 0 | 0 | $(0, 0)$ |
| 1 | 1 | $(1, 1)$ |
| 2 | 2 | $(2, 2)$ |

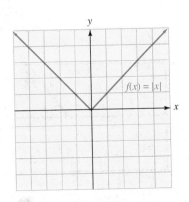

FIGURE 8-21

Self Check    Graph $y = f(x) = |x| + 1$ and compare the graph with the graph of $y = f(x) = |x|$.

Answer    the same graph, but 1 unit higher

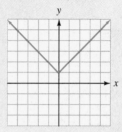

### ■ RATIONAL FUNCTIONS

If $y$ is equal to a polynomial divided by a polynomial, we call the resulting function a **rational function**. The simplest of these functions is considered in the next example.

**EXAMPLE 3**    Determine whether the equation $y = \frac{1}{x}$ represents a function. If so, determine its domain and range and draw its graph.

Solution    Since every value of $x$, except 0, determines one value of $y$, the equation does define $y$ to be a function of $x$.

Since the denominator of a fraction cannot be 0, the domain is the set of all real numbers $x$ except 0.

Since a fraction with a numerator of 1 cannot be 0, the range is the set of all real numbers except 0.

To construct the graph of the function $y = f(x) = \frac{1}{x}$, we make a table of values, plot the points, and join the points as in Figure 8-22.

$$y = f(x) = \frac{1}{x}$$

| $x$ | $y$ | $(x, y)$ |
|---|---|---|
| $-4$ | $-\frac{1}{4}$ | $\left(-4, -\frac{1}{4}\right)$ |
| $-2$ | $-\frac{1}{2}$ | $\left(-2, -\frac{1}{2}\right)$ |
| $-1$ | $-1$ | $(-1, -1)$ |
| $-\frac{1}{4}$ | $-4$ | $\left(-\frac{1}{4}, -4\right)$ |
| $\frac{1}{4}$ | $4$ | $\left(\frac{1}{4}, 4\right)$ |
| $1$ | $1$ | $(1, 1)$ |
| $2$ | $\frac{1}{2}$ | $\left(2, \frac{1}{2}\right)$ |
| $4$ | $\frac{1}{4}$ | $\left(4, \frac{1}{4}\right)$ |

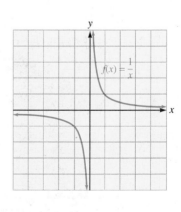

FIGURE 8-22

*Self Check*

*Answer*

Graph $y = f(x) = \frac{4}{x}$.

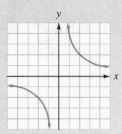

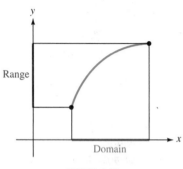

FIGURE 8-23

## ■ FINDING DOMAINS AND RANGES FROM GRAPHS

The graph of a function is the graph of the ordered pairs $(x, y)$ that define the function. For the graph shown in Figure 8-23, the domain is shown on the $x$-axis, and the range is shown on the $y$-axis. For any $x$ in the domain, there is one value of $y$ in the range.

**EXAMPLE 4**

Find the domain and range of the function defined by $y = f(x) = x^2 - 1$.

*Solution*

The graph of $y = f(x) = x^2 - 1$ is shown in Figure 8-24.

Since every number $x$ determines a corresponding value of $y$, the domain is the set of real numbers.

Since the values of $y$ are never less than $-1$, the range is the set of numbers greater than or equal to $-1$.

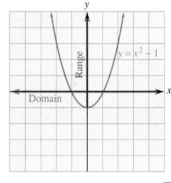

FIGURE 8-24 ■

*Self Check*

Find the domain and the range of the function whose graph is shown below.

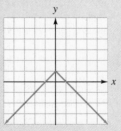

*Answers*

The domain is the set of all real numbers; the range is the set of all real numbers less than or equal to 1.

### ■ THE VERTICAL LINE TEST

The **vertical line test** can be used to determine whether the graph of an equation represents a function. If any vertical line intersects a graph more than once, the graph cannot represent a function, because to one number $x$, there would correspond more than one value of $y$.

The graph in Figure 8-25(a) represents a function, because every vertical line that intersects the graph does so exactly once. The graph in Figure 8-25(b) does not represent a function, because some vertical lines intersect the graph more than once.

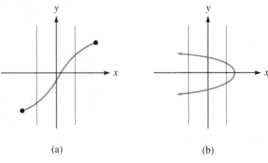

(a)                         (b)

FIGURE 8-25

EXAMPLE 5    Tell whether each graph represents a function.

**a.**            **b.**

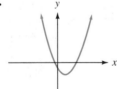    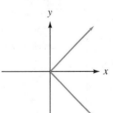

**c.**            **d.**

    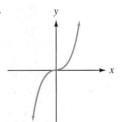

*Solution*    **a.** Since any vertical line that intersects the graph does so only once, the graph represents a function.

**b.** Since some vertical lines that intersect the graph do so twice, the graph does not represent a function.

**c.** Since some vertical lines that intersect the graph do so twice, the graph does not represent a function.

**d.** Since any vertical line that intersects the graph does so only once, the graph represents a function.    ■

Self Check

Tell whether each graph represents a function.

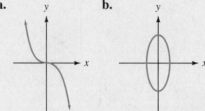

**a.** **b.**

Answers    **a.** function,   **b.** not a function

## ■ AN APPLICATION OF A RATIONAL FUNCTION

Suppose that the cost of telephone service is $5 per month plus 5¢ per call. If $n$ represents the number of calls made during one month, the cost $C$ of phone service that month will be $C = 0.05n + 5$. If we divide the total cost $C$ by the number of calls $n$, we will obtain the average cost per call, which we will denote as $\bar{c}$.

**1.**   $\bar{c} = \dfrac{C}{n} = \dfrac{0.05n + 5}{n}$    $\bar{c}$ is the average cost per call, $C$ is the total monthly cost, and $n$ is the number of phone calls made that month.

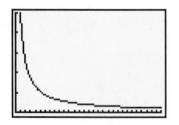

FIGURE 8-26

If we use a graphing calculator with settings of [0, 30] for $x$ and [0, 5] for $y$ to graph the function $\bar{c} = \frac{0.05n + 5}{n}$, we get the graph shown in Figure 8-26. Note that the graph of the function passes the vertical line test, as expected.

From the graph, we can see that the average cost per call decreases as the number of phone calls increases. Since the cost of each extra call is 5¢, the average cost can approach $0.05 per phone call but never drop below it. Thus, the graph of the function approaches the line $y = 0.05$ as $n$ increases without bound. When a graph approaches a line, we call the line an **asymptote**. The line $y = 0.05$ is a **horizontal asymptote** of the graph.

As the number of calls $n$ gets smaller and approaches 0, the graph approaches the $y$-axis but never touches it. The $y$-axis is a **vertical asymptote** of the graph.

**EXAMPLE 6**    Find the average cost per call when   **a.** 5 calls   and   **b.** 25 calls are made.

*Solution*   **a.** To find the average cost for 5 phone calls, we substitute 5 for $n$ in Equation 1 and simplify:

$$\bar{c} = f(5) = \frac{0.05(5) + 5}{5} = 1.05$$

The average cost per call for 5 calls is $1.05.

**b.** To find the average cost for 25 phone calls, we substitute 25 for $n$ in Equation 1 and simplify:

$$\bar{c} = f(25) = \frac{0.05(25) + 5}{25} = 0.25$$

The average cost per call for 25 phone calls is $0.25.   ■

*Self Check*   In Example 6, find the average cost per call when 100 phone calls are made.
*Answer*    $0.10

Orals    *If $f(x) = 2x + 1$, find*

**1.** $f(0)$          **2.** $f(1)$          **3.** $f(-1)$          **4.** $f(-2)$

# EXERCISE 8.3

***REVIEW***   *In Exercises 1–4, solve each equation.*

**1.** $\dfrac{y + 2}{2} = 4(y + 2)$

**2.** $\dfrac{3z - 1}{6} - \dfrac{3z + 4}{3} = \dfrac{z + 3}{2}$

**3.** $\dfrac{2}{x - 3} - 1 = -\dfrac{1}{3}$

**4.** $\dfrac{5}{x} + \dfrac{6}{x^2 + 2x} = \dfrac{-3}{x + 2}$

***VOCABULARY AND CONCEPTS***   *Fill in each blank to make a true statement.*

**5.** Any equation in $x$ and $y$ where each _____ $x$ determines one output $y$ is called a _____.

**6.** In a function, the set of all inputs is called the _____ of the function.

**7.** In a function, the set of all outputs is called the _____ of the function.

**8.** In the function $y = f(x)$, $y$ is called the _____ variable.

**9.** In the function $y = f(x)$, $x$ is called the _____ variable.

**10.** The function $y = f(x) = |x|$ is called the _____ value function.

**11.** If a vertical line intersects a graph more than once, the graph _____ represent a function.

**12.** The function $y = f(x) = \frac{5}{x}$ is an example of a _____ function.

***PRACTICE***   *In Exercises 13–20, find $f(3)$, $f(0)$, and $f(-1)$.*

**13.** $f(x) = 3x$

**14.** $f(x) = -4x$

**15.** $f(x) = 2x - 3$

**16.** $f(x) = 3x - 5$

**17.** $f(x) = 7 + 5x$

**18.** $f(x) = 3 + 3x$

**19.** $f(x) = 9 - 2x$

**20.** $f(x) = 12 + 3x$

*In Exercises 21–28, find $f(1)$, $f(-2)$, and $f(3)$.*

**21.** $f(x) = x^2$

**22.** $f(x) = x^2 - 2$

**23.** $f(x) = x^3 - 1$

**24.** $f(x) = x^3$

**25.** $f(x) = (x + 1)^2$

**26.** $f(x) = (x - 3)^2$

**27.** $f(x) = 2x^2 - x$

**28.** $f(x) = 5x^2 + 2x - 1$

*In Exercises 29–32, graph each function.*

**29.** $y = f(x) = |x| - 1$

**30.** $y = f(x) = -|x| + 2$

**31.** $y = f(x) = -\dfrac{9}{x}$

**32.** $y = f(x) = \dfrac{4}{x}$

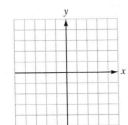

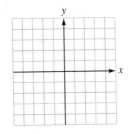

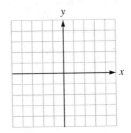

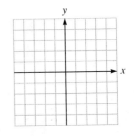

In Exercises 33–36, find the domain and range of each function.

**33.** $y = -x^2 + 2$

**34.** $y = (x - 2)^2$

**35.** $y = |x - 2| - 2$

**36.** $y = -|x + 1|$

*In Exercises 37–40, tell whether each graph represents a function.*

**37.**

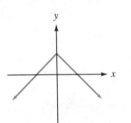

**38.**

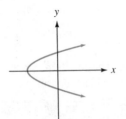

**39.**

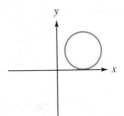

**40.**

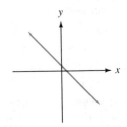

***APPLICATIONS*** *In Exercises 41–44, assume that long-distance service is $12 a month plus 10¢ per phone call.*

**41.** Write a function that will give the cost $C$ per month for making $n$ phone calls.

**42.** Write a function that will give the average cost $\bar{c}$ per phone call during the month.

**43.** Find the cost if 20 phone calls were made during the month

**44.** Find the average cost per call if 60 calls were made.

*In Exercises 45–48, assume that a cellular phone company charges $15 a month and 30¢ per phone call.*

**45.** Write a function that will give the cost $C$ per month for making $n$ phone calls.

**46.** Write a function that will give the average cost $\bar{c}$ per phone call during the month.

**47.** Find the cost if 45 phone calls were made during the month.

**48.** Find the average cost per call if 45 calls were made.

***WRITING***

**49.** Define the domain and range of a function.

**50.** Explain why the vertical line test works.

***SOMETHING TO THINK ABOUT*** *Let $f(x) = 2x + 1$ and $g(x) = x^2$. Assume $f(x) \neq 0$ and $g(x) \neq 0$.*

**51.** Is $f(x) + g(x) = g(x) + f(x)$?

**52.** Is $f(x) - g(x) = g(x) - f(x)$?

**53.** Is $f(x) \cdot g(x) = g(x) \cdot f(x)$?

**54.** Is $\dfrac{f(x)}{g(x)} = \dfrac{g(x)}{f(x)}$?

# 8.4 Variation

■ DIRECT VARIATION ■ INVERSE VARIATION ■ JOINT VARIATION ■ COMBINED VARIATION

Getting Ready    *Solve for k.*

**1.** $8 = 2k$     **2.** $8 = \dfrac{k}{2}$     **3.** $A = kbh$     **4.** $P = \dfrac{kT}{V}$

We now introduce some special terminology that scientists use to describe functions.

## ■ DIRECT VARIATION

**Direct Variation**
The words *y varies directly with x* mean that

   $y = kx$

for some constant $k$. The constant $k$ is called the **constant of variation**.

Since $k$ is a constant, the equation $y = kx$ is a linear function. Its graphs for three values of $k$ are shown in Figure 8-27.

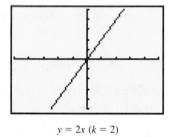

$y = 2x$ ($k = 2$)

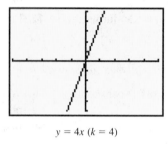

$y = 4x$ ($k = 4$)

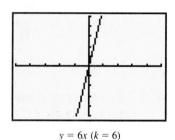

$y = 6x$ ($k = 6$)

FIGURE 8-27

The more force that is applied to a spring, the more it will stretch. Scientists call this fact Hooke's law: *The distance a spring will stretch varies directly with the force applied.* If $d$ represents distance and $f$ represents force, this relationship can be expressed by the equation

**1.**   $d = kf$

where $k$ is the constant of variation. If a spring stretches 5 inches when a weight of 2 pounds is attached, we can find the constant of variation by substituting 5 for $d$ and 2 for $f$ in Equation 1 and solving for $k$.

$$d = kf$$
$$5 = k(2)$$
$$\frac{5}{2} = k$$

To find the distance that the spring will stretch when a weight of 6 pounds is attached, we substitute $\frac{5}{2}$ for $k$ and 6 for $f$ in Equation 1 and solve for $d$.

$$d = kf$$
$$d = \frac{5}{2}(6)$$
$$d = 15$$

The spring will stretch 15 inches when a weight of 6 pounds is attached.

EXAMPLE 1    At a constant speed, the distance traveled varies directly with time. If a bus driver can drive 105 miles in 3 hours, how far would he drive in 5 hours?

*Solution*    We let $d$ represent a distance traveled and let $t$ represent time. We then translate the words *distance varies directly with time* into the equation

**2.**   $d = kt$

To find the constant of variation, $k$, we substitute 105 for $d$ and 3 for $t$ in Equation 2 and solve for $k$.

$$d = kt$$
$$105 = k(3)$$
$$35 = k \qquad \text{Divide both sides by 3.}$$

We can now substitute 35 for $k$ in Equation 2 to obtain Equation 3.

**3.** $\ d = 35t$

To find the distance traveled in 5 hours, we substitute 5 for $t$ in Equation 3.

$$d = 35t$$
$$d = 35(5)$$
$$d = 175$$

In 5 hours, the bus driver would travel 175 miles. ∎

**Self Check**
**Answer**

In Example 1, how far would the bus driver drive in 8 hours?
280 mi

## ■ INVERSE VARIATION

**Inverse Variation**
The words *y varies inversely with x* mean that
$$y = \frac{k}{x}$$
for some constant $k$. The constant $k$ is the **constant of variation**.

Since $k$ is a constant, the equation that defines inverse variation is a rational function. The first-quadrant graphs for three positive values of $k$ are shown in Figure 8-28.

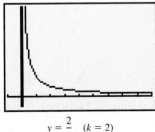

$y = \dfrac{2}{x}$ $(k = 2)$

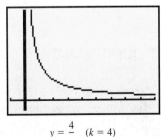

$y = \dfrac{4}{x}$ $(k = 4)$

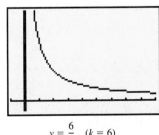

$y = \dfrac{6}{x}$ $(k = 6)$

FIGURE 8-28

Under constant temperature, the volume occupied by a gas varies inversely with its pressure. If $V$ represents volume and $p$ represents pressure, this relationship is expressed by the equation

4. $V = \dfrac{k}{p}$

**EXAMPLE 2**   A gas occupies a volume of 15 cubic inches when placed under 4 pounds per square inch of pressure. How much pressure is needed to compress the gas into a volume of 10 cubic inches?

*Solution*   To find the constant of variation, we substitute 15 for $V$ and 4 for $p$ in Equation 4 and solve for $k$.

$$V = \dfrac{k}{p}$$

$$15 = \dfrac{k}{4}$$

$$60 = k \qquad \text{Multiply both sides by 4.}$$

To find the pressure needed to compress the gas into a volume of 10 cubic inches, we substitute 60 for $k$ and 10 for $V$ in Equation 4 and solve for $p$.

$$V = \dfrac{k}{p}$$

$$10 = \dfrac{60}{p}$$

$$10p = 60 \qquad \text{Multiply both sides by } p.$$

$$p = 6 \qquad \text{Divide both sides by 10.}$$

It will take 6 pounds per square inch of pressure to compress the gas into a volume of 10 cubic inches.   ■

*Self Check*   In Example 2, how much pressure is needed to compress the gas into a volume of 8 cubic inches?

*Answer*   7.5 lb/in.$^2$

■ **JOINT VARIATION**

**Joint Variation**
The words *y varies jointly with x and z* mean that

$y = kxz$

for some constant $k$. The constant $k$ is the **constant of variation.**

The area of a rectangle depends on its length $l$ and its width $w$ by the formula

$$A = lw$$

We could say that the area of the rectangle varies jointly with its length and its width. In this example, the constant of variation is $k = 1$.

**EXAMPLE 3**   The area of a triangle varies jointly with the length of its base and its height. If a triangle with an area of 63 square inches has a base of 18 inches and a height of 7 inches, find the area of a triangle with a base of 12 inches and a height of 10 inches.

*Solution*   We let $A$ represent the area of the triangle, $b$ represent the length of the base, and $h$ represent the height. We translate the words *area varies jointly with the length of the base and the height* into the formula

**5.**   $A = kbh$

We are given that $A = 63$ when $b = 18$ and $h = 7$. To find $k$, we substitute these values into Equation 5 and solve for $k$.

$$A = kbh$$
$$63 = k(18)(7)$$
$$63 = k(126)$$
$$\frac{63}{126} = k \qquad \text{Divide both sides by 126.}$$
$$\frac{1}{2} = k \qquad \text{Simplify.}$$

Thus, $k = \frac{1}{2}$, and the formula for finding the area is

**6.**   $A = \dfrac{1}{2}bh$

To find the area of a triangle with a base of 12 inches and a height of 10 inches, we substitute 12 for $b$ and 10 for $h$ in Equation 6.

$$A = \frac{1}{2}bh$$
$$A = \frac{1}{2}(12)(10)$$
$$A = 60$$

The area is 60 square inches.   ■

*Self Check*   Find the area of a triangle with a base of 14 inches and a height of 8 inches.
*Answer*   56 in.$^2$

### ■ COMBINED VARIATION

Combined variation involves a combination of direct and inverse variation.

**EXAMPLE 4**    The pressure of a fixed amount of gas varies directly with its temperature and inversely with its volume. A sample of gas at a pressure of 1 atmosphere occupies a volume of 3 cubic meters when its temperature is 273 degrees Kelvin (about 0° Celsius). Find the pressure after the gas is heated to 364 K and compressed to 1 cubic meter.

*Solution*    We let $P$ represent the pressure of the gas, $T$ represent its temperature, and $V$ represent its volume. The words *the pressure varies directly with temperature and inversely with volume* translate into the equation

**7.**   $P = \dfrac{kT}{V}$

To find $k$, we substitute 1 for $P$, 273 for $T$, and 3 for $V$ in Equation 7.

$$P = \frac{kT}{V}$$

$$1 = \frac{k(273)}{3}$$

$$1 = 91k \qquad \tfrac{273}{3} = 91.$$

$$\frac{1}{91} = k$$

Since $k = \tfrac{1}{91}$, the formula is

$$P = \frac{1}{91} \cdot \frac{T}{V} \qquad \text{or} \qquad P = \frac{T}{91V}$$

To find the pressure under the new conditions, we substitute 364 for $T$ and 1 for $V$ in the previous equation and solve for $P$.

$$P = \frac{T}{91V}$$

$$P = \frac{364}{91(1)}$$

$$= 4$$

The pressure of the heated and compressed gas is 4 atmospheres.    ■

*Self Check*    In Example 4, find the pressure after the gas is heated to 373.1 K and compressed into 1 cubic meter.

*Answer*    4.1 atmospheres

Orals    *Tell whether each equation indicates direct variation, inverse variation, joint variation, or combined variation.*

**1.** $y = \dfrac{5}{x}$        **2.** $y = 3x$        **3.** $y = \dfrac{3x}{z}$        **4.** $y = 3xz$

**5.** $y = \dfrac{1}{2}x$        **6.** $y = \dfrac{2x}{z}$        **7.** $y = \dfrac{1}{2}xz$        **8.** $y = 4x$

## EXERCISE 8.4

***REVIEW***   *Remove parentheses and simplify.*

**1.** $2(x + 4) + 3(2x - 1)$

**2.** $-3(3x + 5) - 2(2x + 4)$

**3.** $3x(x^2 - 2) - 6x^2(x - 1)$

**4.** $-5a^2(a + 1) - 3a(a^2 + 4a - 3)$

***VOCABULARY AND CONCEPTS***   *Fill in each blank to make a true statement. Assume that $k$ is a constant.*

**5.** The equation $y = kx$ represents _____ variation.

**6.** The equation _____ represents inverse variation.

**7.** The equation $y = kx$ indicates a _____ function.

**8.** The equation $y = \frac{k}{x}$ indicates a _____ function.

**9.** The equation $y = kxz$ represents _____ variation.

**10.** The equation $y = \frac{kx}{z}$ means that $y$ varies _____ with $x$ and _____ with $z$.

*In Exercises 11–20, express each sentence as a formula.*

**11.** The distance $d$ a car can travel while moving at a constant speed varies directly with $n$, the number of gallons of gasoline it consumes.

**12.** A farmer's harvest $h$ varies directly with $a$, the number of acres he plants.

**13.** For a fixed area, the length $l$ of a rectangle varies inversely with its width $w$.

**14.** The value $v$ of a boat varies inversely with its age $a$.

**15.** The area $A$ of a circle varies directly with the square of its radius $r$.

**16.** The distance $s$ that a body falls varies directly with the square of the time $t$.

**17.** The distance $d$ traveled varies jointly with the speed $s$ and the time $t$.

**18.** The interest $i$ on a savings account varies jointly with the rate $r$ and the time $t$.

**19.** The current $I$ varies directly with the voltage $V$ and inversely with the resistance $R$.

**20.** The force of gravity $F$ varies directly with the product of the masses $m_1$ and $m_2$, and inversely with the square of the distance between them.

***PRACTICE***   *In Exercises 21–38, assume that all variables represent positive numbers.*

**21.** Assume that $y$ varies directly with $x$. If $y = 10$ when $x = 2$, find $y$ when $x = 7$.

**22.** Assume that $A$ varies directly with $z$. If $A = 30$ when $z = 5$, find $A$ when $z = 9$.

**23.** Assume that $r$ varies directly with $s$. If $r = 21$ when $s = 6$, find $r$ when $s = 12$.

**24.** Assume that $d$ varies directly with $t$. If $d = 15$ when $t = 3$, find $t$ when $d = 3$.

**25.** Assume that $s$ varies directly with $t^2$. If $s = 12$ when $t = 4$, find $s$ when $t = 30$.

**26.** Assume that $y$ varies directly with $x^3$. If $y = 16$ when $x = 2$, find $y$ when $x = 3$.

**27.** Assume that $y$ varies inversely with $x$. If $y = 8$ when $x = 1$, find $y$ when $x = 8$.

**28.** Assume that $V$ varies inversely with $p$. If $V = 30$ when $p = 5$, find $V$ when $p = 6$.

**29.** Assume that $r$ varies inversely with $s$. If $r = 40$ when $s = 10$, find $r$ when $s = 15$.

**30.** Assume that $J$ varies inversely with $v$. If $J = 90$ when $v = 5$, find $J$ when $v = 45$.

**31.** Assume that $y$ varies inversely with $x^2$. If $y = 6$ when $x = 4$, find $y$ when $x = 2$.

**32.** Assume that $i$ varies inversely with $d^2$. If $i = 6$ when $d = 3$, find $i$ when $d = 2$.

**33.** Assume that $y$ varies jointly with $r$ and $s$. If $y = 4$ when $r = 2$ and $s = 6$, find $y$ when $r = 3$ and $s = 4$.

**34.** Assume that $A$ varies jointly with $x$ and $y$. If $A = 18$ when $x = 3$ and $y = 3$, find $A$ when $x = 7$ and $y = 9$.

**35.** Assume that $D$ varies jointly with $p$ and $q$. If $D = 20$ when $p$ and $q$ are both 5, find $D$ when $p$ and $q$ are both 10.

**36.** Assume that $z$ varies jointly with $r$ and the square of $s$. If $z = 24$ when $r$ and $s$ are 2, find $z$ when $r = 3$ and $s = 4$.

**37.** Assume that $y$ varies directly with $a$ and inversely with $b$. If $y = 1$ when $a = 2$ and $b = 10$, find $y$ when $a = 7$ and $b = 14$.

**38.** Assume that $y$ varies directly with the square of $x$ and inversely with $z$. If $y = 1$ when $x = 2$ and $z = 10$, find $y$ when $x = 4$ and $z = 5$.

## APPLICATIONS

**39. Objects in free fall**   The distance traveled by an object in free fall varies directly with the square of the time that it falls. If the object falls 256 feet in 4 seconds, how far will it fall in 6 seconds?

**40. Traveling range**   The distance that a car can travel without refueling varies directly with the number of gallons of gasoline in the tank. If a car can go 360 miles on 12 gallons of gas, how far can it go on 7 gallons?

**41. Computing interest**   For a fixed rate and principal, the interest earned in a bank account paying simple interest varies directly with the length of time the principal is left on deposit. If an investment of $5,000 earns $700 in 2 years, how much will it earn in 7 years?

**42. Computing forces**   The force of gravity acting on an object varies directly with the mass of the object. The force on a mass of 5 kilograms is 49 newtons. What is the force acting on a mass of 12 kilograms?

**43. Commuting time**   The time it takes a car to travel a certain distance varies inversely with its rate of speed. If a certain trip takes 3 hours when the driver travels at 50 mph, how long will the trip take when the driver travels at 60 mph?

**44. Geometry**   For a fixed area, the length of a rectangle is inversely proportional to its width. A rectangle has a width of 12 feet and a length of 20 feet. If the length is increased to 24 feet, find the width of the rectangle.

**45. Computing pressures**   If the temperature of a gas is constant, the volume occupied varies inversely with the pressure. If a gas occupies a volume of 40 cubic meters under a pressure of 8 atmospheres, find the volume when the pressure is changed to 6 atmospheres.

**46. Computing depreciation**   Assume that the value of a machine varies inversely with its age. If a drill press is worth $300 when it is 2 years old, find its value when it is 6 years old. How much has the machine depreciated in those 4 years?

**47. Computing interest**   The interest earned on a fixed amount of money varies jointly with the annual interest rate and the time that the money is left on deposit. If an account earns $120 at 8% annual interest when left on deposit for 2 years, how much interest would be earned in 3 years at an annual rate of 12%?

**48. Cost of a well**   The cost of drilling a water well is jointly proportional to the length and diameter of the steel casing. If a 30-foot well using 4-inch casing costs $1,200, find the cost of a 35-foot well using 6-inch casing.

**49. Electronics**   The current in a circuit varies directly with the voltage and inversely with the resistance. If a current of 4 amperes flows when 36 volts is applied to a 9-ohm resistance, find the current when the voltage is 42 volts and the resistance is 11 ohms.

**50. Building construction**   The deflection of a beam is inversely proportional to its width and the cube of its depth. If the deflection is 2 inches when the width is 4 inches and the depth is 3 inches, find the deflection when the width is 3 inches and the depth is 4 inches.

## WRITING

**51.** Explain why the words *y varies jointly with x and z* mean the same as the words *y varies directly with the product of x and z*.

**52.** Explain the meaning of combined variation.

## SOMETHING TO THINK ABOUT

**53.** Can direct variation be defined as $\frac{y}{x} = k$, rather than $y = kx$?

**54.** Can inverse variation be defined as $xy = k$, rather than $y = \frac{k}{x}$?

■ ■ ■ ■ ■ ■ ■ ■

MATHEMATICS IN REAL ESTATE   When the office building discussed at the beginning of the chapter was purchased, its value was $465,000. Forty years later, it is estimated to be worth 80% of $465,000, or $372,000. Since the points (0, 465,000) and (40, 372,000) lie on the straight-line depreciation graph of the building, we can write the depreciation equation as follows:

$$m = \frac{372,000 - 465,000}{40 - 0} = -2,325$$

Since the y-intercept is (0, 465,000) and the slope is $m = -2,325$, we have

$$y = mx + b$$
$$y = -2,325x + 465,000$$

To find the value after 35 years, we substitute 35 for $x$ and solve for $y$ to get $383,625. Since the investor sold it for $400,000, the taxable capital gain is $400,000 - $383,625, or $16,375.

# PROJECT

Graphs are often used in newspapers and magazines to convey complex information at a glance. Unfortunately, it is easy to use graphs to convey misleading information. For example, the profit percents of a company for several years are given in the table and two graphs in Figure 8-29.

| Year | Profit |
|------|--------|
| 1989 | 6.2% |
| 1990 | 6.0% |
| 1991 | 6.2% |
| 1992 | 6.1% |
| 1993 | 6.3% |
| 1994 | 6.6% |

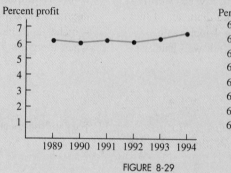 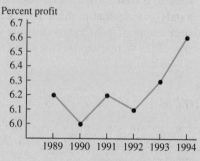

FIGURE 8-29

The first graph in the figure accurately indicates the company's steady performance over five years. But because the vertical axis of the second graph does not start at zero, the performance appears deceptively erratic.

As your college's head librarian, you spend much of your time writing reports, either trying to make the school library look good (for college promotional literature) or bad (to encourage greater funding). In 1989, the library held a collection of 17,000 volumes. Over the years, the library has acquired many new books and has retired several old books. The details appear in Table 8-1.

| Year | Volumes acquired | Volumes removed |
|------|------------------|-----------------|
| 1989 | 215 | 137 |
| 1990 | 217 | 145 |
| 1991 | 235 | 185 |
| 1992 | 257 | 210 |
| 1993 | 270 | 200 |
| 1994 | 275 | 180 |

TABLE 8-1

Using the data in the table,

- Draw a misleading graph that makes the library look good.
- Draw a misleading graph that makes the library look bad.
- Draw a graph that accurately reflects the library's condition.

# CHAPTER SUMMARY

CONCEPT

REVIEW EXERCISES

| **SECTION 8.1** |
|---|

*The Slope of a Nonvertical Line*

The slope of a line passing through $(x_1, y_1)$ and $(x_2, y_2)$ is given by the formula

$$m = \frac{y_2 - y_1}{x_2 - x_1} \quad (x_2 \neq x_1)$$

**1.** Find the slope of the line passing through the given points.
  **a.** $(1, 4), (2, 3)$        **b.** $(-1, 3), (3, -2)$
  **c.** $(-1, -1), (-3, 0)$     **d.** $(-8, 2), (3, 2)$

**2.** Find the slope of the line determined by each equation.
  **a.** $4x - 3y = 12$       **b.** $x = 4y + 2$

Horizontal lines have a slope of 0.

The slope of a vertical line is undefined.

**3.** Tell whether the slope of each line is positive, negative, 0, or undefined.
  **a.**                          **b.**

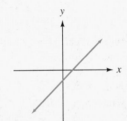

  **c.**                          **d.**

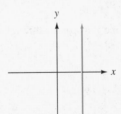 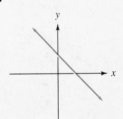

Parallel lines have the same slope.

The product of the slopes of perpendicular lines is −1.

**4.** Tell whether lines with the given slopes are parallel, perpendicular, or neither.
  **a.** 4 and $-\dfrac{1}{4}$           **b.** 0.2 and $\dfrac{1}{5}$

  **c.** 5 and $\dfrac{1}{5}$             **d.** $\dfrac{2}{4}$ and 0.5

  **e.** −5 and $\dfrac{1}{5}$          **f.** 0.25 and $\dfrac{1}{4}$

**5.** Find the slope of a roof if it rises 2 feet for every run of 8 feet.

**6.** Find the average rate of growth of a business if sales were $25,000 the first year and $66,000 the third year.

| | |
|---|---|
| **SECTION 8.2** | *Writing Equations of Lines* |

Point–slope form of a line:

$$y - y_1 = m(x - x_1)$$

**7.** Use the point–slope form of a linear equation to find the equation of each line. Write the equation in general form.

   **a.** $m = 3$ and passing through $(0, 0)$

   **b.** $m = -\dfrac{1}{3}$ and passing through $\left(1, \dfrac{2}{3}\right)$

   **c.** $m = \dfrac{1}{9}$ and passing through $(-27, -2)$

   **d.** $m = -\dfrac{3}{5}$ and passing through $\left(1, -\dfrac{1}{5}\right)$

Slope–intercept form of a line:

$$y = mx + b$$

**8.** Find the slope and the $y$-intercept of the line defined by each equation.

   **a.** $y = 5x + 2$                 **b.** $y = -\dfrac{x}{2} + 4$

   **c.** $y + 3 = 0$                **d.** $x + 3y = 1$

**9.** Use the slope–intercept form to find the equation of each line. Write the equation in general form.

   **a.** $m = -3$, $y$-intercept $(0, 2)$

   **b.** $m = 0$, $y$-intercept $(0, -7)$

   **c.** $m = 7$, $y$-intercept $(0, 0)$

   **d.** $m = \dfrac{1}{2}$, $y$-intercept $\left(0, -\dfrac{3}{2}\right)$

**10.** Graph the line that passes through the given point and has the given slope.

   **a.** $(-1, 3)$, $m = -2$            **b.** $(1, -2)$, $m = 2$

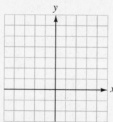 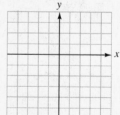

   **c.** $\left(0, \dfrac{1}{2}\right)$, $m = \dfrac{3}{2}$         **d.** $(-3, 0)$, $m = -\dfrac{5}{2}$

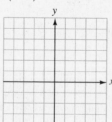 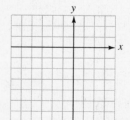

**11.** Tell whether the graphs of the equations are parallel, perpendicular, or neither.
  **a.** $y = 3x$, $x = 3y$
  **b.** $3x = y$, $x = -3y$
  **c.** $x + 2y = y - x$, $2x + y = 3$
  **d.** $3x + 2y = 7$, $2x - 3y = 8$

**12.** Write the equation of each line in general form.
  **a.** parallel to $y = 7x - 18$, passing through $(2, 5)$
  **b.** parallel to $3x + 2y = 7$, passing through $(-3, 5)$
  **c.** perpendicular to $2x - 5y = 12$, passing through $(0, 0)$
  **d.** perpendicular to $y = \dfrac{x}{3} + 17$, $y$-intercept at $(0, -4)$

**13.** Find the slope of each line.
  **a.** $4x + 5y = 10$          **b.** $3x - 5y = 17$

**14.** A company buys a fax machine for \$2,700 and will sell it for \$200 at the end of its useful life. If the company depreciates the machine linearly over a 5-year period, what will the machine be worth after 3 years?

| SECTION 8.3 | *More on Functions* |
|---|---|

Any equation in $x$ and $y$ where each value of $x$ determines one value of $y$ is called a **function.**

The set of all inputs into a function is called the **domain.** The set of all outputs is called the **range.**

**15.** Let $f(x) = x^2 - x + 1$ and find each value.
  **a.** $f(0)$          **b.** $f(-2)$
  **c.** $f(3)$          **d.** $f(0.5)$

**16.** Graph each function.
  **a.** $y = f(x) = |x| - 3$          **b.** $y = f(x) = \dfrac{-4}{x}$

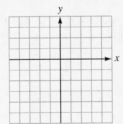
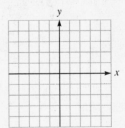

**17.** Find the domain and range of each function.
  **a.** $y = f(x) = |x| - 3$
  **b.** $y = f(x) = \dfrac{-4}{x}$

If any vertical line intersects a graph more than once, the graph does not represent a function.

**18.** Tell whether each graph represents a function.

**a.**

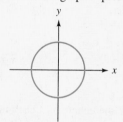

**b.**

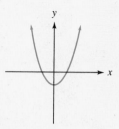

**19.** Assume that a health club membership costs $50 a month plus $5 per visit.
  **a.** Write a function that represents the monthly cost of membership.

  **b.** Find the monthly cost if the member makes 4 visits.

  **c.** Write a function that represents the average cost per visit.

  **d.** Find the average cost per visit if the member makes 8 visits.

---

## SECTION 8.4    *Variation*

**20.** Express each variation as an equation. Then find the requested value.

$y = kx$ represents **direct variation.**

  **a.** $s$ varies directly with the square of $t$. Find $s$ when $t = 10$ if $s = 64$ when $t = 4$.

$y = \dfrac{k}{x}$ represents **inverse variation.**

  **b.** $l$ varies inversely with $w$. Find the constant of variation if $l = 30$ when $w = 20$.

$y = kxz$ represents **joint variation.**

  **c.** $R$ varies jointly with $b$ and $c$. If $R = 72$ when $b = 4$ and $c = 24$, find $R$ when $b = 6$ and $c = 18$.

Direct and inverse variation are used together in **combined variation.**

  **d.** $s$ varies directly with $w$ and inversely with the square of $m$. If $s = \frac{7}{4}$ when $w$ and $m$ are 4, find $s$ when $w = 5$ and $m = 7$.

---

## ■ Chapter Test

**1.** Find the slope of the line passing through $(0, 0)$ and $(6, 8)$.

**2.** Find the slope of the line passing through $(-1, 3)$ and $(3, -1)$.

**3.** Find the slope of the line determined by $2x + y = 3$.

**4.** Find the $y$-intercept of the line determined by $2y - 7(x + 5) = 7$.

**5.** Find the slope of the line $y = 5$, if any.

**6.** Find the slope of the line $x = -2$, if any.

**7.** If two lines are parallel, their slopes are _____.

**8.** If two lines are perpendicular, the product of their slopes is ____.

*In Problems 9–10, tell whether lines with the given slopes are parallel, perpendicular, or neither.*

**9.** 0.5 and −2

**10.** $\sqrt{25}$ and 5

**11.** If a ramp rises 3 feet over a run of 12 feet, find the slope of the ramp.

**12.** If a business had sales of $50,000 the second year in business and $100,000 in sales the fifth year, find the annual rate of growth in sales.

**13.** Find the slope of a line parallel to a line with a slope of 2.

**14.** Find the slope of a line perpendicular to a line with a slope of 2.

**15.** In general form, write the equation of a line that has a slope of 7 and passes through the point $(-2, 5)$.

**16.** In general form, write the equation of a line that has a slope of $\frac{1}{2}$ and a $y$-intercept of $(0, 3)$.

**17.** Write the equation of a line that is parallel to the $y$-axis and passes through $(-3, 17)$.

**18.** Write the equation of a line that passes through $(3, -5)$ and is perpendicular to the line with the equation $y = \frac{1}{3}x + 11$.

*In Problems 19–20, suppose that $y = f(x) = 3x - 2$ and find each value.*

**19.** $f(2)$

**20.** $f(-3)$

*In Problems 21–22, graph each function and give its domain and range.*

**21.** $y = f(x) = -|x| + 4$

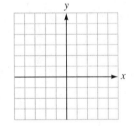

**22.** $y = f(x) = -\dfrac{1}{x}$

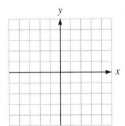

*In Problems 23–24, tell whether each graph represents a function.*

**23.**

**24.**

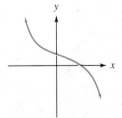

**25.** If $y$ varies directly with $x$ and $y = 32$ when $x = 8$, find $x$ when $y = 4$.

**26.** If $i$ varies inversely with the square of $d$, find the constant of variation if $i = 100$ when $d = 2$.

## ■ Cumulative Review Exercises

*In Exercises 1–4, simplify each expression.*

**1.** $(3x^2 + 2x) + (6x^3 - 3x^2 + 1)$

**2.** $(5x^3 - 2x) - (2x^3 - 3x^2 - 3x - 1)$

**3.** $3(6x^2 - 3x + 3) + 2(-x^2 + 2x - 5)$

**4.** $5(3x^2 - 4x - 1) - 2(-2x^2 + 4x + 3)$

*In Exercises 5–8, do each multiplication.*

**5.** $(5x^4y^3)(-3x^2y^3)$

**6.** $-3x^2(-5x^3 - 3x^2 + 2)$

**7.** $(2x + 3)(3x + 4)$

**8.** $(4x - 3y)(3x - 2y)$

*In Exercises 9–10, do each division.*

**9.** $x + 3 \overline{)x^2 - x - 12}$

**10.** $2x - 1 \overline{)2x^3 + x^2 - x - 3}$

*In Exercises 11–20, factor each expression.*

**11.** $2x^2y - 4xy^2$

**12.** $5(x + y) + a(x + y)$

**13.** $3a + 3b + ab + b^2$

**14.** $49p^4 - 16q^2$

**15.** $x^2 - 9x - 36$

**16.** $x^2 - 3xy - 10y^2$

**17.** $12a^2 + a - 20$

**18.** $10m^2 - 13mn - 3n^2$

**19.** $p^3 - 64q^3$

**20.** $2r^3 + 54s^3$

*In Exercises 21–24, solve each equation.*

**21.** $a^2 + 3a = -2$

**22.** $2b^2 - 12 = -5b$

**23.** $\dfrac{4}{a} = \dfrac{6}{a} - 1$

**24.** $\dfrac{a + 2}{a + 3} - 1 = \dfrac{-1}{a^2 + 2a - 3}$

*In Exercises 25–26, solve each proportion.*

**25.** $\dfrac{4 - a}{13} = \dfrac{11}{26}$

**26.** $\dfrac{3a - 2}{7} = \dfrac{a}{28}$

*In Exercises 27–28, graph each equation.*

**27.** $5x - 4y = 20$

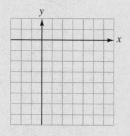

**28.** $y = -x^2$

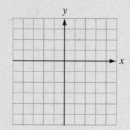

*In Exercises 29–30, find the slope of the line with the given properties.*

**29.** Passing through $(-2, 4)$ and $(6, 8)$

**30.** The equation of the line is $3x + 6y = 13$.

*In Exercises 31–32, write the equation of the line with the following properties.*

**31.** Slope of $\frac{2}{3}$, $y$-intercept of $(0, 5)$

**32.** Passing through $(-2, 4)$ and $(6, 10)$

*In Exercises 33–34, are the graphs of the lines parallel or perpendicular?*

**33.** $\begin{cases} 3x + 4y = 15 \\ 4x - 3y = 25 \end{cases}$

**34.** $\begin{cases} 3x + 4y = 15 \\ 6x = 15 - 8y \end{cases}$

*In Exercises 35–36, graph each inequality.*

**35.** $y \le -2x + 4$

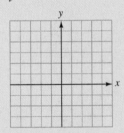

**36.** $3x + 4y \le 12$

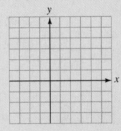

*In Exercises 37–40, $y = f(x) = 2x^2 - 3$. Find each value.*

**37.** $f(0)$

**38.** $f(3)$

**39.** $f(-2)$

**40.** $f(2x)$

**41.** Assume that $y$ varies directly with $x$. If $y = 4$ when $x = 10$, find $y$ when $x = 30$.

**42.** Assume that $y$ varies inversely with $x$. If $y = 8$ when $x = 2$, find $y$ when $x = 8$.

# 9

# Quadratic Equations and Graphing Quadratic Functions

**MATHEMATICS IN SALES**

American Appliance has found that it can sell more top-of-the-line refrigerators each month if it lowers the price. Over the years, the sales department has found that the store will sell $x$ refrigerators at a price of $\$\left(1,800 - \frac{3}{2}x\right)$. What price should the store charge to maximize its revenue?

At that price, how many refrigerators should the store order from the distributor each month?

After reading this chapter, you will be able to answer these questions.

## 9.1 Quadratic Equations

■ QUADRATIC EQUATIONS ■ SOLVING QUADRATIC EQUATIONS BY FACTORING ■ THE SQUARE ROOT METHOD ■ APPLICATIONS

Getting Ready  *Factor each polynomial.*

**1.** $4x^2 - 2x$  **2.** $x^2 - 9$  **3.** $x^2 + x - 6$  **4.** $2x^2 + 3x - 9$

*Find each square root.*

**5.** $\sqrt{25}$  **6.** $-\sqrt{36}$  **7.** $\sqrt{20}$  **8.** $-\sqrt{50}$

### ■ QUADRATIC EQUATIONS

Recall the definition of a quadratic equation.

> **Quadratic Equations**
> A **quadratic equation** in one variable is an equation of the form
> $$ax^2 + bx + c = 0 \quad (a \neq 0)$$
> where $a$, $b$, and $c$ are real numbers.

We have previously solved quadratic equations by factoring and using the zero-factor property.

> **Zero-Factor Property**
> Suppose that $a$ and $b$ represent real numbers. Then
> If $ab = 0$, then $a = 0$ or $b = 0$.

The zero-factor property states that when the product of two numbers is 0, at least one of the numbers is 0. For example, the equation $(x - 4)(x + 5) = 0$ indicates that a product is equal to 0. By the zero-factor property, one of the factors must be 0:

$$x - 4 = 0 \quad \text{or} \quad x + 5 = 0$$

We can solve each of these linear equations to get

$$x = 4 \quad \text{or} \quad x = -5$$

The equation $(x - 4)(x + 5) = 0$ has two solutions: 4 and $-5$.

## ■ SOLVING QUADRATIC EQUATIONS BY FACTORING

**Factoring Method**
To solve a quadratic equation by factoring, we
1. Write the equation in $ax^2 + bx + c = 0$ form (called **quadratic form**).
2. Factor the left-hand side of the equation.
3. Use the zero-factor property to set each factor equal to 0.
4. Solve each resulting linear equation.

We will review the factoring method in Examples 1 and 2.

**EXAMPLE 1**    Solve $6x^2 - 3x = 0$.

*Solution*    Since the equation is already in quadratic form, we begin by factoring the left-hand side of the equation.

$$6x^2 - 3x = 0$$
$$3x(2x - 1) = 0 \qquad \text{Factor out } 3x.$$

By the zero-factor property, we have

$$3x = 0 \quad \text{or} \quad 2x - 1 = 0$$

We can solve these linear equations to get

$$3x = 0 \quad \text{or} \quad 2x - 1 = 0$$
$$x = 0 \qquad \qquad 2x = 1$$
$$x = \frac{1}{2}$$

*Check:* **For $x = 0$**       or       **For $x = \dfrac{1}{2}$**

$$6x^2 - 3x = 0$$

$$6(0)^2 - 3(0) \overset{?}{=} 0$$

$$6(0) - 0 \overset{?}{=} 0$$

$$0 - 0 \overset{?}{=} 0$$

$$0 = 0$$

$$6x^2 - 3x = 0$$

$$6\left(\dfrac{1}{2}\right)^2 - 3\left(\dfrac{1}{2}\right) \overset{?}{=} 0$$

$$6\left(\dfrac{1}{4}\right) - \dfrac{3}{2} \overset{?}{=} 0$$

$$\dfrac{3}{2} - \dfrac{3}{2} \overset{?}{=} 0$$

$$0 = 0$$

Both solutions check.

**Self Check**

**Answers**

Solve $5x^2 + 10x = 0$.

$0, -2$

**WARNING!**   Don't make the following error.

$$6x^2 - 3x = 0$$

$$6x^2 = 3x \qquad \text{Add } 3x \text{ to both sides.}$$

$$2x = 1 \qquad \text{Divide both sides by } 3x.$$

$$x = \dfrac{1}{2} \qquad \text{Divide both sides by } 2.$$

By dividing by $3x$, the solution $x = 0$ is lost.

**EXAMPLE 2**

*Solution*

Solve $6x^2 - x = 2$.

To write the equation in quadratic form, we begin by subtracting 2 from both sides of the equation.

$$6x^2 - x = 2$$

$$6x^2 - x - 2 = 0 \qquad \text{Subtract 2 from both sides.}$$

$$(3x - 2)(2x + 1) = 0 \qquad \text{Factor the trinomial.}$$

$$3x - 2 = 0 \quad \text{or} \quad 2x + 1 = 0 \qquad \text{Set each factor equal to 0.}$$

$$3x = 2 \qquad\qquad 2x = -1 \qquad \text{Solve each linear equation.}$$

$$x = \dfrac{2}{3} \qquad\qquad x = -\dfrac{1}{2}$$

*Check:* **For $x = \dfrac{2}{3}$** | **For $x = -\dfrac{1}{2}$**

$$6x^2 - x = 2 \qquad\qquad 6x^2 - x = 2$$

$$6\left(\frac{2}{3}\right)^2 - \frac{2}{3} \overset{?}{=} 2 \qquad 6\left(-\frac{1}{2}\right)^2 - \left(-\frac{1}{2}\right) \overset{?}{=} 2$$

$$6\left(\frac{4}{9}\right) - \frac{6}{9} \overset{?}{=} 2 \qquad\qquad 6\left(\frac{1}{4}\right) + \frac{2}{4} \overset{?}{=} 2$$

$$\frac{24}{9} - \frac{6}{9} \overset{?}{=} 2 \qquad\qquad\qquad \frac{6}{4} + \frac{2}{4} \overset{?}{=} 2$$

$$\frac{18}{9} \overset{?}{=} 2 \qquad\qquad\qquad\qquad \frac{8}{4} \overset{?}{=} 2$$

$$2 = 2 \qquad\qquad\qquad\qquad\quad 2 = 2$$

| **Self Check** | Solve $x^2 - 6 = x$. |
|---|---|
| **Answers** | $3, -2$ |

The factoring method doesn't work with many quadratic equations. For example, the trinomial in the equation $x^2 + 5x + 1 = 0$ cannot be factored by using integer coefficients. To solve such equations, we need to develop other methods. The first of these is the square root method.

## ■ THE SQUARE ROOT METHOD

If $x^2 = c$ $(c \geq 0)$, then $x$ is a number whose square is $c$. Since $\left(\sqrt{c}\right)^2 = c$ and $\left(-\sqrt{c}\right)^2 = c$, the equation $x^2 = c$ has two solutions.

> **Square Root Method**
> If $c > 0$, the equation $x^2 = c$ has two solutions:
> $$x = \sqrt{c} \qquad \text{or} \qquad x = -\sqrt{c}$$

We can write the previous result with double-sign notation. The equation $x = \pm\sqrt{c}$ (read as "$x$ equals plus or minus $\sqrt{c}$") means that $x = \sqrt{c}$ or $x = -\sqrt{c}$.

**EXAMPLE 3**  Solve $x^2 = 16$.

*Solution*  The equation $x^2 = 16$ has two solutions:

$$x = \sqrt{16} \quad \text{or} \quad x = -\sqrt{16}$$
$$= 4 \qquad\quad | \qquad = -4$$

Using double-sign notation, we have $x = \pm 4$.

*Check:* **For $x = 4$** | **For $x = -4$**

$$x^2 = 16$$
$$4^2 \stackrel{?}{=} 16$$
$$16 = 16$$

$$x^2 = 16$$
$$(-4)^2 \stackrel{?}{=} 16$$
$$16 = 16$$

∎

**Self Check**

**Answers**

Solve $y^2 = 36$.

6, −6

The equation in Example 3 can also be solved by factoring.

$$x^2 = 16$$
$$x^2 - 16 = 0 \qquad \text{Subtract 16 from both sides.}$$
$$(x + 4)(x - 4) = 0 \qquad \text{Factor the difference of two squares.}$$
$$x + 4 = 0 \quad \text{or} \quad x - 4 = 0$$
$$x = -4 \qquad\qquad x = 4$$

**EXAMPLE 4**

Solve $3x^2 - 9 = 0$.

*Solution*

We can solve the equation by the square root method.

$$3x^2 - 9 = 0$$
$$3x^2 = 9 \qquad \text{Add 9 to both sides.}$$
$$x^2 = 3 \qquad \text{Divide both sides by 3.}$$

This equation has two solutions:

$$x = \sqrt{3} \quad \text{or} \quad x = -\sqrt{3}$$

*Check:* **For $x = \sqrt{3}$** | **For $x = -\sqrt{3}$**

$$3x^2 - 9 = 0$$
$$3(\sqrt{3})^2 - 9 \stackrel{?}{=} 0$$
$$3(3) - 9 \stackrel{?}{=} 0$$
$$9 - 9 \stackrel{?}{=} 0$$
$$0 = 0$$

$$3x^2 - 9 = 0$$
$$3(-\sqrt{3})^2 - 9 \stackrel{?}{=} 0$$
$$3(3) - 9 \stackrel{?}{=} 0$$
$$9 - 9 \stackrel{?}{=} 0$$
$$0 = 0$$

The solutions can be written as $x = \pm\sqrt{3}$. To the nearest tenth, the solutions are $\pm 1.7$. ∎

**Self Check**

**Answers**

Solve $2x^2 - 10 = 0$. Give the results to the nearest tenth.

$\pm 2.2$

**EXAMPLE 5**    Solve $(x + 1)^2 = 9$.

*Solution*    The two solutions are

$$x + 1 = \sqrt{9} \quad \text{or} \quad x + 1 = -\sqrt{9}$$
$$x + 1 = 3 \qquad\qquad x + 1 = -3$$
$$x = 2 \qquad\qquad\quad x = -4$$

*Check:* **For $x = 2$**            **For $x = -4$**

$$(x + 1)^2 = 9 \qquad\qquad (x + 1)^2 = 9$$
$$(2 + 1)^2 \stackrel{?}{=} 9 \qquad (-4 + 1)^2 \stackrel{?}{=} 9$$
$$3^2 \stackrel{?}{=} 9 \qquad\qquad (-3)^2 \stackrel{?}{=} 9$$
$$9 = 9 \qquad\qquad\quad 9 = 9$$

*Self Check*    Solve $(y + 2)^2 = 16$.
*Answers*    $2, -6$

**EXAMPLE 6**    Solve $(x - 2)^2 - 18 = 0$.

*Solution*    $(x - 2)^2 - 18 = 0$
$$(x - 2)^2 = 18 \qquad \text{Add 18 to both sides.}$$

The two solutions are

$$x - 2 = \sqrt{18} \qquad \text{or} \quad x - 2 = -\sqrt{18}$$
$$x = 2 + \sqrt{18} \qquad\qquad x = 2 - \sqrt{18}$$
$$x = 2 + 3\sqrt{2} \qquad\qquad x = 2 - 3\sqrt{2} \qquad \sqrt{18} = \sqrt{9}\,\sqrt{2} = 3\sqrt{2}.$$

*Check:* **For $x = 2 + 3\sqrt{2}$**                     **For $x = 2 - 3\sqrt{2}$**

$$(x - 2)^2 - 18 = 0 \qquad\qquad\qquad (x - 2)^2 - 18 = 0$$
$$\left(2 + 3\sqrt{2} - 2\right)^2 - 18 \stackrel{?}{=} 0 \qquad \left(2 - 3\sqrt{2} - 2\right)^2 - 18 \stackrel{?}{=} 0$$
$$\left(3\sqrt{2}\right)^2 - 18 \stackrel{?}{=} 0 \qquad\qquad \left(-3\sqrt{2}\right)^2 - 18 \stackrel{?}{=} 0$$
$$18 - 18 \stackrel{?}{=} 0 \qquad\qquad\qquad 18 - 18 \stackrel{?}{=} 0$$
$$0 = 0 \qquad\qquad\qquad\qquad 0 = 0$$

To the nearest tenth, the results are 6.2 and $-2.2$.

*Self Check*    Solve $(x + 2)^2 - 12 = 0$. Give the results to the nearest tenth.
*Answers*    $1.5, -5.5$

**EXAMPLE 7**     Solve $3x^2 - 4 = 2(x^2 + 2)$.

*Solution*
$$3x^2 - 4 = 2(x^2 + 2)$$
$$3x^2 - 4 = 2x^2 + 4 \qquad \text{Remove parentheses.}$$
$$3x^2 = 2x^2 + 8 \qquad \text{Add 4 to both sides.}$$
$$x^2 = 8 \qquad \text{Subtract } 2x^2 \text{ from both sides.}$$
$$x = \sqrt{8} \quad \text{or} \quad x = -\sqrt{8}$$
$$= 2\sqrt{2} \quad | \quad = -2\sqrt{2} \qquad \text{Simplify the radical.}$$

Both solutions check. To the nearest tenth, the results are $\pm 2.8$. ∎

*Self Check*     Solve $3x^2 = 2(x^2 + 3) + 6$. Give the results to the nearest tenth.

*Answers*     $\pm 3.5$

### ■ APPLICATIONS

The solutions of many problems involve quadratic equations.

**EXAMPLE 8**     **Integer problems**   The product of the first and third of three consecutive positive odd integers is 77. Find the integers.

*Analyze the problem*     Some consecutive positive odd integers are 3, 5, and 7 and 25, 27, and 29. These examples illustrate that to obtain the next consecutive positive odd integer, we must add 2. If $x$ represents a positive odd integer, the expressions $x$, $x + 2$, and $x + 4$ represent three consecutive positive odd integers.

*Form an equation*     We can let $x$ represent the first odd integer. Then $x + 2$ will represent the second odd integer, and $x + 4$ will represent the third. The product of the first and third is $x(x + 4)$, which we know to be equal to 77. Thus, we have the equation

$$x(x + 4) = 77$$

*Solve the equation*     We can solve the equation as follows.

$$x(x + 4) = 77$$
$$x^2 + 4x - 77 = 0 \qquad \text{Use the distributive property to remove parentheses.}$$
$$(x + 11)(x - 7) = 0 \qquad \text{Factor.}$$
$$x + 11 = 0 \qquad \text{or} \quad x - 7 = 0 \qquad \text{Set each factor equal to 0.}$$
$$x = -11 \quad | \quad x = 7$$

*State the conclusion*     Since we are looking for three *positive* odd integers, we ignore the solution of $-11$. Thus, the first of the integers is 7, the second is 9, and the third is 11.
        Check the result. ∎

**EXAMPLE 9**

**Geometry**   The oriental rug shown in Figure 9-1 is 3 feet longer than it is wide. Find the dimensions of the rug if its area is 180 ft².

*Analyze the problem*   We can let $w$ represent the width of the rug. Then $w + 3$ will represent its length. Since the area of a rectangle is given by the formula $A = lw$ (Area = length × width), the area of the rug is $(w + 3)w$, which is equal to 180.

*Form an equation*   This gives the equation

| The length of the rug | · | the width of the rug | = | the area of the rug. |
|---|---|---|---|---|
| $(w + 3)$ | · | $w$ | = | 180 |

*Solve the equation*   We can solve this equation as follows:

$$(w + 3)w = 180$$
$$w^2 + 3w = 180 \qquad \text{Use the distributive property to remove parentheses.}$$
$$w^2 + 3w - 180 = 0 \qquad \text{Subtract 180 from both sides.}$$
$$(w - 12)(w + 15) = 0 \qquad \text{Factor.}$$
$$w - 12 = 0 \quad \text{or} \quad w + 15 = 0$$
$$w = 12 \quad | \quad w = -15$$

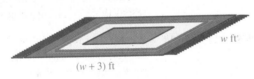

$(w + 3)$ ft

$w$ ft

FIGURE 9-1

*State the conclusion*   When $w = 12$, the length, $w + 3$, is 15. The dimensions of the rug are 12 feet by 15 feet. We discard the solution $w = -15$, because a rug cannot have a negative width. Check the result.   ■

**EXAMPLE 10**

**Ballistics**   If an object is thrown straight up into the air with an initial velocity of 112 feet per second, its height after $t$ seconds is given by the formula

$$h = 112t - 16t^2$$

where $h$ represents the height of the object in feet. After the object has been thrown, in how many seconds will it hit the ground?

*Solution*    When the object hits the ground, its height will be 0. Thus, we set $h$ equal to 0 and solve for $t$.

$$h = 112t - 16t^2$$
$$0 = 112t - 16t^2$$
$$0 = 16t(7 - t) \qquad \text{Factor out } 16t.$$
$$16t = 0 \quad \text{or} \quad 7 - t = 0 \qquad \text{Set each factor equal to 0.}$$
$$t = 0 \quad | \quad \phantom{xx} t = 7 \qquad \text{Solve each linear equation.}$$

When $t = 0$, the object's height above the ground is 0 feet, because it has not been released. When $t = 7$, the height is again 0 feet, and the object has hit the ground. The solution is 7 seconds.  ■

Orals    *Solve by factoring.*

**1.** $x^2 - 25 = 0$                                       **2.** $x^2 - 5x + 6 = 0$

*Solve by the square root method.*

**3.** $x^2 = 100$                                          **4.** $x^2 = 49$

## EXERCISE 9.1

**REVIEW**    *Write each expression without parentheses.*

**1.** $(y - 1)^2$             **2.** $(z + 2)^2$             **3.** $(x + y)^2$
**4.** $(a - b)^2$             **5.** $(2r - s)^2$           **6.** $(m + 3n)^2$

**VOCABULARY AND CONCEPTS**    *Fill in each blank to make a true statement.*

**7.** Any equation that can be written in the form $ax^2 + bx + c = 0 \ (a \neq 0)$ is called a _____ equation.

**8.** In the equation $3x^2 - 4x + 5 = 0$, $a =$ __, $b =$ _____, and $c =$ __.

**9.** If $ab = 0$, then $a =$ __ or $b =$ __.

**10.** In the quadratic equation $ax^2 + bx + c = 0$, $a \neq$ __.

**11.** The equation $x^2 = c \ (c > 0)$ has _____ solutions.

**12.** The solutions of $x^2 = c \ (c > 0)$ are _____ and _____.

**PRACTICE**    *In Exercises 13–20, solve each equation.*

**13.** $(x - 2)(x + 3) = 0$                             **14.** $(x - 3)(x - 2) = 0$
**15.** $(x - 4)(x + 1) = 0$                             **16.** $(x + 5)(x + 2) = 0$
**17.** $(2x - 5)(3x + 6) = 0$                        **18.** $(3x - 4)(x + 1) = 0$
**19.** $(x - 1)(x + 2)(x - 3) = 0$                  **20.** $(x + 2)(x + 3)(x - 4) = 0$

*In Exercises 21–32, use the factoring method to solve each equation.*

**21.** $x^2 - 9 = 0$      **22.** $x^2 + x = 0$      **23.** $3x^2 + 9x = 0$      **24.** $2x^2 - 8 = 0$

**25.** $x^2 - 5x + 6 = 0$      **26.** $x^2 + 7x + 12 = 0$      **27.** $3x^2 + x - 2 = 0$      **28.** $2x^2 - x - 6 = 0$

**29.** $6x^2 + 11x + 3 = 0$      **30.** $5x^2 + 13x - 6 = 0$      **31.** $10x^2 + x - 2 = 0$      **32.** $6x^2 + 37x + 6 = 0$

*In Exercises 33–44, use the square root method to solve each equation for x. Assume that a and b are positive.*

**33.** $x^2 = 1$      **34.** $x^2 = 4$      **35.** $x^2 = 9$      **36.** $x^2 = 32$

**37.** $x^2 = 20$      **38.** $x^2 = 0$      **39.** $3x^2 = 27$      **40.** $4x^2 = 64$

**41.** $4x^2 = 16$      **42.** $5x^2 = 125$      **43.** $x^2 = a$      **44.** $x^2 = 4b$

*In Exercises 45–54, use the square root method to solve each equation. Assume that a, b, c, and y are positive.*

**45.** $(x + 1)^2 = 25$      **46.** $(x - 1)^2 = 49$      **47.** $(x + 2)^2 = 81$      **48.** $(x + 3)^2 = 16$

**49.** $(x - 2)^2 = 8$      **50.** $(x + 2)^2 = 50$

**51.** $(x - a)^2 = 4a^2$      **52.** $(x + y)^2 = 9y^2$

**53.** $(x + b)^2 = 16c^2$      **54.** $(x - c)^2 = 25b^2$

*In Exercises 55–60, factor the trinomial square and use the square root method to solve each equation.*

**55.** $x^2 + 4x + 4 = 4$      **56.** $x^2 - 6x + 9 = 9$

**57.** $9x^2 - 12x + 4 = 16$      **58.** $4x^2 - 20x + 25 = 36$

**59.** $4x^2 + 4x + 1 = 20$      **60.** $9x^2 + 12x + 4 = 12$

*In Exercises 61–66, solve each equation.*

**61.** $6(x^2 - 1) = 4(x^2 + 3)$      **62.** $5(x^2 - 2) = 2(x^2 + 1)$

**63.** $8(x^2 - 6) = 4(x^2 + 13)$      **64.** $8(x^2 - 1) = 5(x^2 + 10) + 50$

**65.** $5(x + 1)^2 = (x + 1)^2 + 32$      **66.** $6(x - 4)^2 = 4(x - 4)^2 + 36$

## APPLICATIONS

**67. Finding consecutive integers** The product of two consecutive positive even integers is 48. Find the integers.

**68. Finding consecutive integers** The product of the first and second of three consecutive positive odd integers is 35. Find the sum of the three integers.

**69. Finding the sum of two squares** The sum of the squares of two consecutive even negative integers is 52. Find the integers.

**70. Integer problem** The sum of an integer and four times its reciprocal is 8.5. Find the integer.

**71. Geometric problem** A rectangular mural is 4 feet longer than it is wide. Find its dimensions if its area is 32 square feet.

**72. Geometric problem** The length of a 220-square-foot rectangular garden is 2 feet more than twice its width. Find its perimeter.

**73. Finding the height of a triangle**  The triangle shown in Illustration 1 has an area of 30 square inches. Find its height.

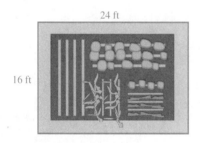

ILLUSTRATION 1

**74. Finding the base of a triangle**  Find the length of the base of the triangle shown in Illustration 1.

**75. Finding the dimensions of a garden**  The rectangular garden shown in Illustration 2 is surrounded by a walk of uniform width. Find the dimensions of the garden if its area is 180 square feet.

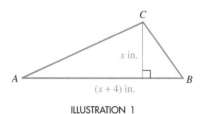

ILLUSTRATION 2

**76. Building a deck around a pool**  The owner of the pool in Illustration 3 wants to surround it with a deck of uniform width. If he can afford 368 square feet of decking, how wide can he make the deck?

**77. Falling objects**  An object will fall $s$ feet in $t$ seconds, where $s = 16t^2$. If a workman 1,454 feet above the ground at the top of the Sears Tower drops a hammer, how long will it take for the hammer to hit the ground?

ILLUSTRATION 3

**78. Falling coins**  A tourist drops a penny from the observation deck of the World Trade Center, 1,377 feet above the ground. How long will it take for the penny to hit the ground? (See Exercise 77.)

*In Exercises 79–81, the height h of a toy rocket in flight is given by the formula $h = -16t^2 + 144t$, where t is the time of the flight in seconds.*

**79. Height of a rocket**  Find the height of the rocket in 3 seconds.

**80. Flight of a rocket**  How long will it take for the rocket to hit the ground?

**81. Maximum height of a rocket**  If the maximum height of the rocket occurs halfway through its flight, how high will the rocket go?

*In Exercises 82–84, a gun is fired straight up with a muzzle velocity of 1,088 feet per second. The height h of the bullet is given by the formula $h = -16t^2 + 1,088t$, where t is the time in seconds.*

**82. Height of a bullet**  Find the height of the bullet after 10 seconds.

**83. Height of a bullet**  When will the bullet hit the ground?

**84. Maximum height of a bullet**  How high will the bullet go?

**WRITING**

**85.** Explain how to solve a quadratic equation by the factoring method.

**86.** Explain how to solve the equation $x^2 = 81$ by the square root method.

## *SOMETHING TO THINK ABOUT*

**87.** Find the error in the following solution.

$$6x^2 - 3x = 0$$

$$6x^2 = 3x \qquad \text{Add } 3x \text{ to both sides.}$$

$$2x = 1 \qquad \text{Divide by } 3x.$$

$$x = \frac{1}{2} \qquad \text{Divide by } 2.$$

**88.** What would happen if you solved $x^2 = c$ $(c < 0)$ by the square root method? Would the roots be real numbers?

# 9.2 Completing the Square

■ COMPLETING THE SQUARE ■ SOLVING EQUATIONS WITH LEAD COEFFICIENTS OF 1 ■ SOLVING EQUATIONS WITH LEAD COEFFICIENTS OTHER THAN 1

Getting Ready *Find one-half of each number and square it.*

**1.** 6      **2.** 10      **3.** 2      **4.** 5

**5.** $-8$      **6.** $-12$      **7.** $\dfrac{1}{2}$      **8.** $\dfrac{2}{3}$

## ■ COMPLETING THE SQUARE

When the polynomial in a quadratic equation doesn't factor easily, we can solve the equation by using a method called **completing the square**.

The method of completing the square is based on the following special products:

$$x^2 + 2bx + b^2 = (x + b)^2 \qquad \text{and} \qquad x^2 - 2bx + b^2 = (x - b)^2$$

The trinomials $x^2 + 2bx + b^2$ and $x^2 - 2bx + b^2$ are both trinomial squares, because each one factors as the square of a binomial. In each trinomial, if we take one-half of the coefficient of the $x$ and square it, we get the third term.

In $x^2 + 2bx + b^2$, if we take $\frac{1}{2}(2b)$, which is $b$, and square it, we get the third term $b^2$.

In $x^2 - 2bx + b^2$, if we take $\frac{1}{2}(-2b) = -b$ and square it, we get $(-b)^2 = b^2$, which is the third term.

To form a trinomial square from the binomial $x^2 + 12x$, we take one-half of the coefficient of $x$ (the 12), square it, and add it to $x^2 + 12x$.

$$x^2 + 12x + \left[ \frac{1}{2}(12) \right]^2 = x^2 + 12x + (6)^2$$

$$= x^2 + 12x + 36$$

This result is a trinomial square, because $x^2 + 12x + 36 = (x + 6)^2$.

C

**EXAMPLE 1**    Form trinomial squares using   **a.** $x^2 + 4x$,   **b.** $x^2 - 6x$,   and   **c.** $x^2 - 5x$.

*Solution*    **a.** $x^2 + 4x + \left[\dfrac{1}{2}(4)\right]^2 = x^2 + 4x + (2)^2$

$$= x^2 + 4x + 4 \qquad \text{This is } (x + 2)^2.$$

**b.** $x^2 - 6x + \left[\dfrac{1}{2}(-6)\right]^2 = x^2 - 6x + (-3)^2$

$$= x^2 - 6x + 9 \qquad \text{This is } (x - 3)^2.$$

**c.** $x^2 - 5x + \left[\dfrac{1}{2}(-5)\right]^2 = x^2 - 5x + \left(-\dfrac{5}{2}\right)^2$

$$= x^2 - 5x + \dfrac{25}{4} \qquad \text{This is } \left(x - \tfrac{5}{2}\right)^2.$$

In each case, note that $\frac{1}{2}$ of the coefficient of $x$ is the second term of the binomial factorization.    ■

*Self Check*    Add a constant to make each expression a trinomial square:   **a.** $y^2 + 6y$,
**b.** $y^2 - 8y$,   and   **c.** $y^2 + 3y$.

*Answers*    **a.** $y^2 + 6y + 9$,   **b.** $y^2 - 8y + 16$,   **c.** $y^2 + 3y + \frac{9}{4}$

## ■ SOLVING EQUATIONS WITH LEAD COEFFICIENTS OF 1

If the quadratic equation $ax^2 + bx + c = 0$ has a lead coefficient of 1 ($a = 1$), it is
easy to solve by completing the square.

**EXAMPLE 2**    Solve $x^2 - 4x - 13 = 0$. Give each answer to the nearest hundredth.

*Solution*    Since the coefficient of $x^2$ is 1, we can solve by completing the square as follows:

$$x^2 - 4x - 13 = 0$$
$$x^2 - 4x = 13 \qquad \text{Add 13 to both sides.}$$

We then find one-half of the coefficient of $x$, square it, and add the result to both
sides to make the left-hand side a trinomial square.

$$x^2 - 4x + \left[\dfrac{1}{2}(-4)\right]^2 = 13 + \left[\dfrac{1}{2}(-4)\right]^2$$

$$x^2 - 4x + 4 = 13 + 4 \qquad \text{Simplify.}$$
$$(x - 2)^2 = 17 \qquad \text{Factor } x^2 - 4x + 4 \text{ and simplify.}$$
$$x - 2 = \pm\sqrt{17} \qquad \text{Use the square root method to solve for } x - 2.$$
$$x = 2 \pm \sqrt{17} \qquad \text{Add 2 to both sides.}$$

Because of the $\pm$ sign, there are two solutions.

$$x = 2 + \sqrt{17} \qquad \text{or} \quad x = 2 - \sqrt{17}$$
$$\approx 2 + 4.123105626 \qquad \approx 2 - 4.123105626$$
$$\approx 6.12 \qquad\qquad\qquad \approx -2.12 \qquad\blacksquare$$

*Self Check*
*Answers*

Solve $x^2 + x - 3 = 0$. Give each answer to the nearest hundredth.

$1.30, -2.30$

## ■ SOLVING EQUATIONS WITH LEAD COEFFICIENTS OTHER THAN 1

If the quadratic equation $ax^2 + bx + c = 0$ ($a \neq 0$) has a lead coefficient other than 1, we can make the lead coefficient 1 by dividing both sides of the equation by $a$.

**EXAMPLE 3**

Solve $4x^2 + 4x - 3 = 0$.

*Solution*

We divide both sides by 4 so that the coefficient of $x^2$ is 1. We then proceed as follows.

$$4x^2 + 4x - 3 = 0$$

$$x^2 + x - \frac{3}{4} = 0 \qquad\qquad \text{Divide both sides by 4.}$$

$$x^2 + x = \frac{3}{4} \qquad\qquad \text{Add } \tfrac{3}{4} \text{ to both sides.}$$

$$x^2 + x + \left(\frac{1}{2}\right)^2 = \frac{3}{4} + \left(\frac{1}{2}\right)^2 \qquad \text{Add } \left(\tfrac{1}{2}\right)^2 \text{ to both sides to complete the square.}$$

$$\left(x + \frac{1}{2}\right)^2 = 1 \qquad\qquad \text{Factor and simplify.}$$

$$x + \frac{1}{2} = \pm 1 \qquad\qquad \text{Solve for } x + \tfrac{1}{2}.$$

$$x = -\frac{1}{2} \pm 1 \qquad\qquad \text{Add } -\tfrac{1}{2} \text{ to both sides.}$$

$$x = -\frac{1}{2} + 1 \quad \text{or} \quad x = -\frac{1}{2} - 1$$

$$= \frac{1}{2} \qquad\qquad\qquad = -\frac{3}{2}$$

Check each solution. Note that this equation can also be solved by factoring. ■

Self Check  Solve $2x^2 - 5x - 3 = 0$.

Answers  $3, -\frac{1}{2}$

The previous examples illustrate that to solve a quadratic equation of the form $ax^2 + bx + c = 0$ by completing the square, we follow these steps.

### Completing the Square

1. If the coefficient of $x^2$ is not 1, make it 1 by dividing both sides of the equation by the coefficient of $x^2$.

2. If necessary, add a number to both sides of the equation to get the constant term on the right-hand side.

3. Complete the square.

   **a.** Find half the coefficient of $x$ and square it.

   **b.** Add that square to both sides of the equation.

4. Factor the trinomial square and combine terms.

5. Solve the resulting quadratic equation.

6. Check each solution.

EXAMPLE 4    Solve $2x^2 - 2 = -4x$. Give the result to the nearest tenth.

Solution    We add $4x$ to both sides to determine whether the equation can be solved by factoring.

$$2x^2 + 4x - 2 = 0 \qquad \text{Add } 4x \text{ to both sides.}$$

**1.** $\qquad x^2 + 2x - 1 = 0 \qquad \text{Divide both sides by 2.}$

Since Equation 1 cannot be solved by factoring, we complete the square.

$$x^2 + 2x = 1 \qquad\qquad\qquad \text{Add 1 to both sides.}$$

$$x^2 + 2x + (1)^2 = 1 + (1)^2 \qquad \text{Add } 1^2 \text{ to both sides to complete the square.}$$

$$(x + 1)^2 = 2 \qquad\qquad\qquad \text{Factor and simplify.}$$

$$x + 1 = \pm \sqrt{2} \qquad\qquad \text{Solve for } x + 1.$$

$$x = -1 \pm \sqrt{2} \qquad\qquad \text{Subtract 1 from both sides.}$$

$$x = -1 + \sqrt{2} \quad \text{or} \quad x = -1 - \sqrt{2}$$

Both solutions check. To the nearest tenth, the solutions are $0.4$ and $-2.4$. ■

Self Check  Solve $3x^2 + 6x = 3$. Give the result to the nearest hundredth.

Answers  $0.41, -2.41$

■ ■ ■ ■ ■ ■ ■ ■ ■ **PERSPECTIVE**

Clay tablets that survive from the early period of the Babylonian civilization, 1800 to 1600 B.C., show that the Babylonians were accomplished mathematicians. These tablets were compiled for use by the Babylonian merchants. Many of these tablets contain multiplication tables and, for division, lists of reciprocals. For more abstract mathematical purposes, others provide tables of squares, cubes, square roots, and cube roots. Still others contain lists of problems and exercises. Some of these problems are practical, but many are puzzle problems, just for fun. Several

problems and their solutions indicate that the Babylonians knew the Pythagorean theorem centuries before the Greeks discovered it.

The Babylonians could also solve certain quadratic equations. For example, one problem from a Babylonian tablet asks, "What number added to its reciprocal is 5?" Today, we would translate this question into the equation $x + \frac{1}{x} = 5$. Can you show that this equation is equivalent to the quadratic equation $x^2 - 5x + 1 = 0$?

Orals  *What number must be added to each binomial to make a trinomial square?*

**1.** $x^2 + 4x$      **2.** $x^2 + 6x$      **3.** $x^2 - 8x$

**4.** $x^2 - 10x$      **5.** $x^2 + 5x$      **6.** $x^2 - 3x$

# EXERCISE 9.2

***REVIEW***  *Solve each equation.*

**1.** $\dfrac{3t(2t + 1)}{2} + 6 = 3t^2$          **2.** $20r^2 - 11r - 3 = 0$

**3.** $\dfrac{2}{3x} - \dfrac{5}{9} = -\dfrac{1}{x}$          **4.** $\sqrt{x + 12} = \sqrt{3x}$

***VOCABULARY AND CONCEPTS***  *Fill in each blank to make a true statement.*

**5.** If the polynomial in the equation $ax^2 + bx + c = 0$ doesn't factor, we can solve the equation by _____ the square.

**6.** Since $x^2 + 12x + 36 = (x + 6)^2$, we call the trinomial a trinomial _____.

**7.** $x^2 + 2bx + b^2 = $ _____

**8.** $x^2 - 2bx + b^2 = $ _____

**9.** To complete the square on $x^2 + 8x$, we add the _____ of one-half of __, which is 16.

**10.** To complete the square on $x^2 - 10x$, we add the square of _____ of $-10$, which is 25.

*In Exercises 11–22, complete the square to make a trinomial square.*

**11.** $x^2 + 2x$        **12.** $x^2 + 12x$        **13.** $x^2 - 4x$

**14.** $x^2 - 14x$        **15.** $x^2 + 7x$        **16.** $x^2 + 21x$

**17.** $a^2 - 3a$

**18.** $b^2 - 13b$

**19.** $b^2 + \dfrac{2}{3}b$

**20.** $a^2 + \dfrac{8}{5}a$

**21.** $c^2 - \dfrac{5}{2}c$

**22.** $c^2 - \dfrac{11}{3}c$

*In Exercises 23–40, solve each equation by completing the square.*

**23.** $x^2 + 6x + 8 = 0$

**24.** $x^2 + 8x + 12 = 0$

**25.** $x^2 - 8x + 12 = 0$

**26.** $x^2 - 4x + 3 = 0$

**27.** $x^2 - 2x - 15 = 0$

**28.** $x^2 - 2x - 8 = 0$

**29.** $x^2 - 7x + 12 = 0$

**30.** $x^2 - 7x + 10 = 0$

**31.** $x^2 + 5x - 6 = 0$

**32.** $x^2 = 14 - 5x$

**33.** $2x^2 = 4 - 2x$

**34.** $3x^2 + 9x + 6 = 0$

**35.** $3x^2 + 48 = -24x$

**36.** $3x^2 = 3x + 6$

**37.** $2x^2 = 3x + 2$

**38.** $3x^2 = 2 - 5x$

**39.** $4x^2 = 2 - 7x$

**40.** $2x^2 = 5x + 3$

*In Exercises 41–48, solve each equation.*

**41.** $x^2 + 4x + 1 = 0$

**42.** $x^2 + 6x + 2 = 0$

**43.** $x^2 - 2x - 4 = 0$

**44.** $x^2 - 4x - 2 = 0$

**45.** $x^2 = 4x + 3$

**46.** $x^2 = 6x - 3$

**47.** $2x^2 = 2 - 4x$

**48.** $3x^2 = 12 - 6x$

*In Exercises 49–52, write each equation in the form $ax^2 + bx + c = 0$ and solve it by completing the square.*

**49.** $2x(x + 3) = 8$

**50.** $3x(x - 2) = 9$

**51.** $6(x^2 - 1) = 5x$

**52.** $2(3x^2 - 2) = 5x$

### WRITING

**53.** Explain how to complete the square.

**54.** Explain why the coefficient of $x^2$ should be 1 before completing the square.

**SOMETHING TO THINK ABOUT**  *Consider this method of completing the square on x in the binomial $ax^2 + bx$: Multiply the binomial by 4a and then add $b^2$. Complete the square on x in each binomial.*

**55.** $2x^2 + 6x$

**56.** $3x^2 - 4x$

## 9.3  The Quadratic Formula

■ THE QUADRATIC FORMULA  ■ QUADRATIC EQUATIONS WITH NO REAL ROOTS  ■ APPLICATIONS

*Getting Ready*  *Evaluate $b^2 - 4ac$ when a, b, and c have the following values.*

**1.** $a = 1, b = 2, c = 3$

**2.** $a = 4, b = 3, c = 1$

**3.** $a = 1, b = 0, c = -2$

**4.** $a = 2, b = 4, c = 2$

### ■ THE QUADRATIC FORMULA

We can solve any quadratic equation by the method of completing the square, but the work is often tedious. In this section, we will develop a formula, called the *quadratic formula*, that will enable us to solve quadratic equations with much less effort.

We can solve the **general quadratic equation** $ax^2 + bx + c = 0$ ($a \ne 0$) by completing the square.

$$ax^2 + bx + c = 0$$

$$\frac{ax^2}{a} + \frac{bx}{a} + \frac{c}{a} = \frac{0}{a} \qquad \text{Divide both sides by } a.$$

$$x^2 + \frac{b}{a}x + \frac{c}{a} = 0 \qquad \text{Simplify: } \tfrac{a}{a} = 1; \tfrac{0}{a} = 0.$$

$$x^2 + \frac{b}{a}x \phantom{+\frac{c}{a}} = -\frac{c}{a} \qquad \text{Subtract } \tfrac{c}{a} \text{ from both sides.}$$

We can now complete the square on $x$ by adding $\left(\dfrac{1}{2} \cdot \dfrac{b}{a}\right)^2$, or $\dfrac{b^2}{4a^2}$, to both sides:

$$x^2 + \frac{b}{a}x + \frac{b^2}{4a^2} = \frac{b^2}{4a^2} - \frac{c}{a}$$

After factoring the trinomial on the left-hand side and adding the fractions on the right-hand side, we have

$$\left(x + \frac{b}{2a}\right)\left(x + \frac{b}{2a}\right) = \frac{b^2}{4a^2} - \frac{4ac}{4aa}$$

**1.** 
$$\left(x + \frac{b}{2a}\right)^2 = \frac{b^2 - 4ac}{4a^2}$$

Equation 1 can be solved by the square root method to obtain

$$x + \frac{b}{2a} = \sqrt{\frac{b^2 - 4ac}{4a^2}} \qquad \text{and} \qquad x + \frac{b}{2a} = -\sqrt{\frac{b^2 - 4ac}{4a^2}}$$

$$x + \frac{b}{2a} = \frac{\sqrt{b^2 - 4ac}}{\sqrt{4a^2}} \qquad\qquad x + \frac{b}{2a} = -\frac{\sqrt{b^2 - 4ac}}{\sqrt{4a^2}}$$

$$x + \frac{b}{2a} = \frac{\sqrt{b^2 - 4ac}}{2a} \qquad\qquad x + \frac{b}{2a} = -\frac{\sqrt{b^2 - 4ac}}{2a}$$

$$x = -\frac{b}{2a} + \frac{\sqrt{b^2 - 4ac}}{2a} \qquad\qquad x = -\frac{b}{2a} - \frac{\sqrt{b^2 - 4ac}}{2a}$$

$$x = \frac{-b + \sqrt{b^2 - 4ac}}{2a} \qquad\qquad x = \frac{-b - \sqrt{b^2 - 4ac}}{2a}$$

These solutions are usually written in one expression called the **quadratic formula**.

## Quadratic Formula

The solutions of the quadratic equation $ax^2 + bx + c = 0$ are

$$x = \frac{-b \pm \sqrt{b^2 - 4ac}}{2a} \quad (a \neq 0)$$

**WARNING!** When you write the quadratic formula, be careful to draw the fraction bar so that it underlines the complete numerator. Do not write

$$x = -b \pm \frac{\sqrt{b^2 - 4ac}}{2a}$$

**EXAMPLE 1**    Solve $x^2 + 5x + 6 = 0$.

*Solution*    In this equation, $a = 1$, $b = 5$, and $c = 6$. We substitute these values into the quadratic formula and simplify.

$$x = \frac{-b \pm \sqrt{b^2 - 4ac}}{2a}$$

$$= \frac{-5 \pm \sqrt{5^2 - 4(1)(6)}}{2(1)} \qquad \text{Substitute 1 for } a, \text{ 5 for } b, \text{ and 6 for } c.$$

$$= \frac{-5 \pm \sqrt{25 - 24}}{2}$$

$$= \frac{-5 \pm \sqrt{1}}{2}$$

$$= \frac{-5 \pm 1}{2}$$

Thus,

$$x = \frac{-5 + 1}{2} \quad \text{and} \quad x = \frac{-5 - 1}{2}$$

$$= \frac{-4}{2} \qquad\qquad\quad = \frac{-6}{2}$$

$$= -2 \qquad\qquad\quad = -3$$

Check both solutions.    ∎

*Self Check*    Solve $x^2 - 4x - 12 = 0$.

*Answers*    $6, -2$

**WARNING!**   Be sure to write a quadratic equation in quadratic form before identifying the values of $a$, $b$, and $c$.

**EXAMPLE 2**   Solve $2x^2 = 5x + 3$.

*Solution*   We begin by writing the equation in quadratic form.

$$2x^2 = 5x + 3$$
$$2x^2 - 5x - 3 = 0 \qquad \text{Add } -5x \text{ and } -3 \text{ to both sides.}$$

In this equation, $a = 2$, $b = -5$, and $c = -3$. We substitute these values into the quadratic formula and simplify.

$$x = \frac{-b \pm \sqrt{b^2 - 4ac}}{2a}$$

$$= \frac{-(-5) \pm \sqrt{(-5)^2 - 4(2)(-3)}}{2(2)} \qquad \text{Substitute 2 for } a, -5 \text{ for } b, \text{ and } -3 \text{ for } c.$$

$$= \frac{5 \pm \sqrt{25 + 24}}{4}$$

$$= \frac{5 \pm \sqrt{49}}{4}$$

$$= \frac{5 \pm 7}{4}$$

Thus,

$$x = \frac{5 + 7}{4} \quad \text{or} \quad x = \frac{5 - 7}{4}$$

$$= \frac{12}{4} \qquad\qquad = \frac{-2}{4}$$

$$= 3 \qquad\qquad\quad = -\frac{1}{2}$$

Check both solutions.   ◼

**Self Check**   Solve $4x^2 + 4x = 3$.

*Answers*   $\frac{1}{2}$, $-\frac{3}{2}$

**EXAMPLE 3**    Solve $3x^2 = 2x + 4$.

*Solution*    We begin by writing the equation in quadratic form.

$$3x^2 = 2x + 4$$
$$3x^2 - 2x - 4 = 0 \qquad \text{Add } -2x \text{ and } -4 \text{ to both sides.}$$

In this equation, $a = 3$, $b = -2$, and $c = -4$. We substitute these values into the quadratic formula and simplify.

$$x = \frac{-b \pm \sqrt{b^2 - 4ac}}{2a}$$

$$= \frac{-(-2) \pm \sqrt{(-2)^2 - 4(3)(-4)}}{2(3)} \qquad \text{Substitute 3 for } a, -2 \text{ for } b, \text{ and } -4 \text{ for } c.$$

$$= \frac{2 \pm \sqrt{4 + 48}}{6}$$

$$= \frac{2 \pm \sqrt{52}}{6}$$

$$= \frac{2 \pm 2\sqrt{13}}{6} \qquad \sqrt{52} = \sqrt{4 \cdot 13} = \sqrt{4}\sqrt{13} = 2\sqrt{13}.$$

$$= \frac{2\left(1 \pm \sqrt{13}\right)}{6} \qquad \text{Factor out 2.}$$

$$= \frac{1 \pm \sqrt{13}}{3} \qquad \text{Divide out the common factor of 2.}$$

Thus,

$$x = \frac{1}{3} + \frac{\sqrt{13}}{3} \quad \text{or} \quad x = \frac{1}{3} - \frac{\sqrt{13}}{3}$$

Both solutions check. To the nearest tenth, the solutions are 1.5 and $-0.9$.    ∎

*Self Check*    Solve $2x^2 - 5x - 5 = 0$. Give the results to the nearest tenth.
*Answers*    $3.3, -0.8$

## ■ QUADRATIC EQUATIONS WITH NO REAL ROOTS

The next example shows that some quadratic equations have no real-number solutions.

**EXAMPLE 4**   Solve $x^2 + 2x + 5 = 0$.

*Solution*   In this equation, $a = 1$, $b = 2$, and $c = 5$. We substitute these values into the quadratic formula.

$$x = \frac{-b \pm \sqrt{b^2 - 4ac}}{2a}$$

$$= \frac{-2 \pm \sqrt{2^2 - 4(1)(5)}}{2(1)} \qquad \text{Substitute 1 for } a, 2 \text{ for } b, \text{ and } 5 \text{ for } c.$$

$$= \frac{-2 \pm \sqrt{4 - 20}}{6}$$

$$= \frac{-2 \pm \sqrt{-16}}{6}$$

Since $\sqrt{-16}$ is not a real number, there are no real-number solutions.   ■

*Self Check*   Does the equation $2x^2 + 5x + 4 = 0$ have any real-number solutions?

*Answer*   no

■ **APPLICATIONS**

**EXAMPLE 5**   **Manufacturing**   A manufacturer of television parts received an order for 52-inch
picture tubes (measured along the diagonal). The tubes are to be rectangular in
shape and 4 inches wider than they are high. Find the dimensions of a tube.

*Analyze the problem*   We can let $h$ represent the height of a
picture tube, as shown in Figure 9-2.
Then $h + 4$ will represent the width. A
52-inch picture tube measures 52 inches
along its diagonal.

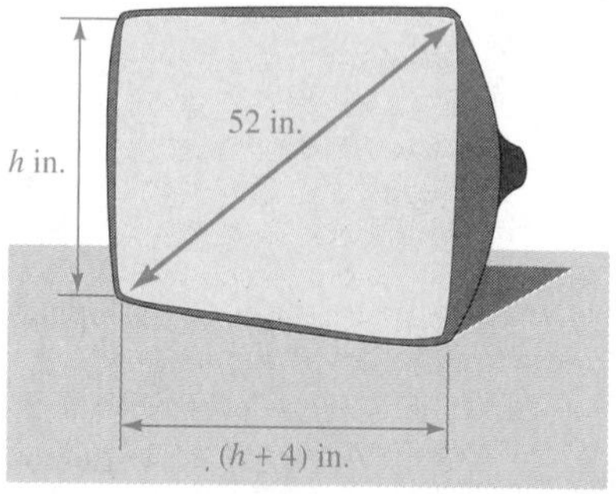

*Form an equation*   Since the sides of the tube and its diagonal form a right triangle, we can use the
Pythagorean theorem to form the equation

$$h^2 + (h + 4)^2 = 52^2$$

**FIGURE 9-2**

*Solve the equation*   which we can solve as follows.

$$h^2 + h^2 + 8h + 16 = 2{,}704 \qquad \text{Expand } (h + 4)^2; 52^2 = 2{,}704.$$

$$2h^2 + 8h - 2{,}688 = 0 \qquad \text{Subtract 2,704 from both sides.}$$

$$h^2 + 4h - 1{,}344 = 0 \qquad \text{Divide both sides by 2.}$$

We can solve this equation with the quadratic formula.

$$h = \frac{-b \pm \sqrt{b^2 - 4ac}}{2a}$$

$$h = \frac{-4 \pm \sqrt{(4)^2 - 4(1)(-1,344)}}{2(1)}$$

$$h = \frac{-4 \pm \sqrt{16 + 5,376}}{2}$$

$$h = \frac{-4 \pm \sqrt{5,392}}{2}$$

$$h \approx \frac{-4 \pm 73.430239}{2}$$

$$h \approx \frac{-4 + 73.430239}{2} \quad \text{or} \quad h \approx \frac{-4 - 73.430239}{2}$$

$$\approx \frac{69.430239}{2} \qquad\qquad \approx \frac{-77.430239}{2}$$

$$\approx 34.715119 \qquad\qquad \approx -38.715119$$

***State the conclusion***   The height of each tube will be approximately 34.7 inches, and the width will be approximately 34.7 + 4 or 38.7 inches. We discard the second solution, because the diagonal measure of a TV picture tube cannot be negative.

Check the result.   ∎

| EXAMPLE 6 |
|-----------|

**Finance**   If $P is invested at an annual rate of $r\%$, it will grow to an amount of $A in $n$ years according to the formula $A = P(1 + r)^n$. What interest rate is needed for a $5,000 investment to grow to $5,618 after 2 years?

***Analyze the problem***   Here we are given a formula to find the amount. We will substitute the values into the formula and find $r$.

***Form an equation***   We substitute 5,000 for $P$, 5,618 for $A$, and 2 for $n$ in the formula.

$$A = P(1 + r)^n$$
$$5,618 = 5,000(1 + r)^2$$

***Solve the equation***   We then solve the equation as follows.

$$5,618 = 5,000(1 + r)^2$$
$$5,618 = 5,000(1 + 2r + r^2) \qquad \text{Expand } (1 + r)^2.$$
$$5,618 = 5,000 + 10,000r + 5,000r^2 \qquad \text{Remove parentheses.}$$

$$5,000r^2 + 10,000r - 618 = 0 \qquad \text{Subtract 5,618 from both sides.}$$

We can use a calculator and solve this equation with the quadratic formula.

$$r = \frac{-b \pm \sqrt{b^2 - 4ac}}{2a}$$

$$r = \frac{-10,000 \pm \sqrt{10,000^2 - 4(5,000)(-618)}}{2(5,000)}$$

$$r = \frac{-10,000 \pm \sqrt{100,000,000 + 12,360,000}}{10,000}$$

$$r = \frac{-10,000 \pm \sqrt{112,360,000}}{10,000}$$

$$r = \frac{-10,000 \pm 10,600}{10,000}$$

$$r = \frac{-10,000 + 10,600}{10,000} \quad \text{or} \quad r = \frac{-10,000 - 10,600}{10,000}$$

$$= \frac{600}{10,000} \qquad\qquad = \frac{-20,600}{10,000}$$

$$= 6\% \qquad\qquad = -206\%$$

**State the conclusion**    The required rate is 6%. The rate of $-206\%$ has no meaning in this problem. Check the result.   ■

Orals    *Find the values of a, b, and c in each equation.*

**1.** $3x^2 + 4x - 12 = 0$          **2.** $-2x^2 - 4x + 10 = 0$

**3.** $5x^2 - x = 1$          **4.** $x^2 - 9 = -3x$

## EXERCISE 9.3

**REVIEW**    *Solve each formula for the indicated variable.*

**1.** $A = p + prt$, for $r$          **2.** $F = \dfrac{GMm}{d^2}$, for $M$

*Write the equation of the line with the given properties in general form.*

**3.** Slope of $\frac{3}{5}$ and passing through $(0, 12)$          **4.** Passes through $(6, 8)$ and the origin

*Simplify each expression. Assume that all variables represent positive numbers.*

**5.** $\sqrt{80}$          **6.** $12\sqrt{x^3 y^2}$

**7.** $\dfrac{x}{\sqrt{7x}}$          **8.** $\dfrac{\sqrt{x} + 2}{\sqrt{x} - 2}$

***VOCABULARY AND CONCEPTS***   *Fill in each blank to make a true statement.*

**9.** Write the general quadratic equation.

_____ $(a \neq 0)$

**10.** Write the quadratic formula.

_____

**11.** In the quadratic equation $ax^2 + bx + c = 0$, $a$ cannot equal __.

**12.** In the quadratic equation $3x^2 - 5 = 0$, $a = $ __, $b = $ __, and $c = $ __.

**13.** In the quadratic equation $-4x^2 + 8x = 0$, $a = $ ___, $b = $ __, and $c = $ __.

**14.** If $a$, $b$, and $c$ are three sides of a right triangle and $c$ is the hypotenuse, then $c^2 = $ _____.

***PRACTICE***   *In Exercises 15–26, change each equation into quadratic form, if necessary, and find the values of a, b, and c.* ***Do not solve the equation.***

**15.** $x^2 + 4x + 3 = 0$

**16.** $x^2 + x - 4 = 0$

**17.** $3x^2 - 2x + 7 = 0$

**18.** $4x^2 + 7x - 3 = 0$

**19.** $4y^2 = 2y - 1$

**20.** $2x = 3x^2 + 4$

**21.** $x(3x - 5) = 2$

**22.** $y(5y + 10) = 8$

**23.** $7(x^2 + 3) = -14x$

**24.** $5(a^2 + 5) = -4a$

**25.** $(2a + 3)(a - 2) = (a + 1)(a - 1)$

**26.** $(3a + 2)(a - 1) = (2a + 7)(a - 1)$

*In Exercises 27–54, use the quadratic formula to find all real solutions of each equation.*

**27.** $x^2 - 5x + 6 = 0$

**28.** $x^2 + 5x + 4 = 0$

**29.** $x^2 + 7x + 12 = 0$

**30.** $x^2 - x - 12 = 0$

**31.** $2x^2 - x - 1 = 0$

**32.** $2x^2 + 3x - 2 = 0$

**33.** $3x^2 + 5x + 2 = 0$

**34.** $3x^2 - 4x + 1 = 0$

**35.** $4x^2 + 4x - 3 = 0$

**36.** $4x^2 + 3x - 1 = 0$

**37.** $5x^2 - 8x - 4 = 0$

**38.** $6x^2 - 8x + 2 = 0$

**39.** $x^2 + 3x + 1 = 0$

**40.** $x^2 + 3x - 2 = 0$

**41.** $x^2 + 5x - 3 = 0$

**42.** $x^2 + 5x + 3 = 0$

**43.** $x^2 + 2x + 7 = 0$

**44.** $2x^2 - x + 2 = 0$

**45.** $2x^2 + x = 5$

**46.** $3x^2 - x = 1$

**47.** $x^2 + 1 = -4x$

**48.** $x^2 + 1 = -8x$

**49.** $x^2 + 5 = 2x$

**50.** $2x^2 + 3x = -3$

**51.** $x^2 = 1 - 2x$

**52.** $x^2 = 2 - 2x$

**53.** $3x^2 = 6x + 2$

**54.** $3x^2 = -8x - 2$

## *APPLICATIONS*

**55. Finding dimensions** The picture frame in Illustration 1 is 2 inches wider than it is high. Find its dimensions.

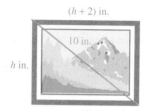

$(h + 2)$ in.

10 in.

$h$ in.

ILLUSTRATION 1

**56. Installing sidewalks** A 170-meter-long sidewalk from the mathematics building M to the student center C is shown in Illustration 2. However, students prefer to walk directly from M to C. How long are the two pieces of the existing sidewalk?

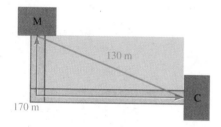

M

130 m

C

170 m

ILLUSTRATION 2

**57. Navigation** Two boats left port at the same time, one sailing east and one sailing south. If one boat sailed 10 nautical miles more than the other and they are 50 nautical miles apart, how far did each boat sail?

**58. Navigation** One plane heads west from an airport flying at 200 mph. One hour later, a second plane heads north from the same airport, flying at the same speed. When will the planes be 1,000 miles apart?

*In Exercises 59–60, use the formula $A = P(1 + r)^2$ to find the amount $A$ that $P$ will become when invested at an annual rate of r% for 2 years.*

**59. Investing** What interest rate is needed for $5,000 to grow to $5,724.50 in 2 years?

**60. Investing** What interest rate is needed for $7,000 to grow to $8,470 in 2 years?

**61. Manufacturing** An electronics firm has found that its revenue for manufacturing and selling $x$ television sets is given by the formula $R = -\frac{1}{6}x^2 + 450x$. How much revenue will be earned by manufacturing 600 television sets?

**62. Wholesale revenue** When a wholesaler sells $n$ CD players, the revenue $R$ is given by the formula $R = 150n - \frac{1}{2}n^2$. How many players would the wholesaler have to sell to recieve $11,250?

**63. Fabricating metal** A piece of tin, 12 inches on a side, is to have four equal squares cut from its corners, as in Illustration 3. If the edges are then to be folded up to make a box with a floor area of 64 square inches, find the depth of the box.

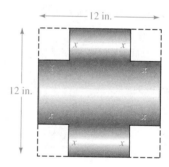

12 in.

$x$    $x$

12 in.

$x$    $x$

ILLUSTRATION 3

**64. Making gutters** A piece of sheet metal, 18 inches wide, is bent to form the gutter shown in Illustration 4. If the cross-sectional area is 36 square inches, find the depth of the gutter.

$x$

$x$

18 - 2$x$

ILLUSTRATION 4

**65. Filling a tank**   Two pipes are used to fill a water tank. The first pipe can fill the tank in 4 hours, and the two pipes together can fill the tank in 2 hours less time than the second pipe alone. How long would it take for the second pipe to fill the tank?

**66. Filling a pool**   A small hose requires 6 more hours to fill a swimming pool than a larger hose. If the two hoses can fill the pool in 4 hours, how long would it take the larger hose alone?

## WRITING

**67.** Explain how to use the quadratic formula.

**68.** Explain the meaning of the $\pm$ symbol.

**69.** Choose one of the previous application problems and list the steps you followed as you worked it.

**70.** The binomial $b^2 - 4ac$ is called the **discriminant**. From its value, you can predict whether the solutions of a given quadratic equation are real or nonreal numbers. Explain.

***SOMETHING TO THINK ABOUT***   *In Exercises 71–72, use these facts. The two solutions of the equation* $ax^2 + bx + c = 0$   $(a \neq 0)$ *are*

$$x_1 = \frac{-b + \sqrt{b^2 - 4ac}}{2a} \quad \text{and} \quad x_2 = \frac{-b - \sqrt{b^2 - 4ac}}{2a}$$

**71.** Show that $x_1 + x_2 = -\dfrac{b}{a}$.

**72.** Show that $x_1 x_2 = \dfrac{c}{a}$.

# 9.4 Graphing Quadratic Functions

■ GRAPHING QUADRATIC FUNCTIONS ■ FINDING THE VERTEX OF A PARABOLA ■ APPLICATIONS

Getting Ready    *If* $y = f(x) = 2x^2 - x + 2$, *find each value.*

**1.** $f(0)$         **2.** $f(1)$         **3.** $f(-1)$         **4.** $f(-2)$

*If* $x = -\dfrac{b}{2a}$ *find x when a and b have the following values.*

**5.** $a = 2, b = 8$         **6.** $a = 5, b = -20$

## ■ GRAPHING QUADRATIC FUNCTIONS

The function defined by the equation $y = mx + b$ is a linear function, because its right-hand side is a first-degree polynomial in the variable $x$. The function defined by $y = ax^2 + bx + c$ $(a \neq 0)$ is called a **quadratic function**, because its right-hand side is a second-degree polynomial in the variable $x$. In this section, we will discuss many quadratic functions.

A basic quadratic function is defined by the equation $y = x^2$. Recall that to graph this function, we find several ordered pairs $(x, y)$ that satisfy the equation, plot the pairs, and join the points with a smooth curve. A table of values and the graph appear in Figure 9-3. The graph of a quadratic function is called a **parabola**. The lowest point on the parabola is called its **vertex**. The vertex of the parabola shown in Figure 9-3 is the point $V(0, 0)$.

$y = x^2$

| $x$ | $y$ | $(x, y)$ |
|-----|-----|----------|
| $-3$ | 9 | $(-3, 9)$ |
| $-2$ | 4 | $(-2, 4)$ |
| $-1$ | 1 | $(-1, 1)$ |
| 0 | 0 | $(0, 0)$ |
| 1 | 1 | $(1, 1)$ |
| 2 | 4 | $(2, 4)$ |
| 3 | 9 | $(3, 9)$ |

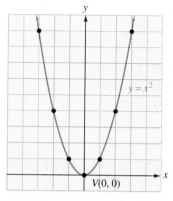

FIGURE 9-3

**EXAMPLE 1**    Graph $y = x^2 - 3$. Compare the graph with Figure 9-3 and tell what you notice.

*Solution*    To find ordered pairs $(x, y)$ that satisfy the equation, we pick several numbers $x$ and find the corresponding values of $y$. If we let $x = 3$, we have

$y = x^2 - 3$

$y = 3^2 - 3$        Substitute 3 for $x$.

$y = 6$

The ordered pair $(3, 6)$ and others satisfying the equation appear in the table shown in Figure 9-4. To graph the function, we plot the points and draw a smooth curve passing through them.

The resulting parabola is the graph of $y = x^2 - 3$. The vertex of the parabola is the point $V(0, -3)$.

Note that the graph of $y = x^2 - 3$ looks just like the graph of $y = x^2$, except that it is 3 units lower.

$y = x^2 - 3$

| $x$ | $y$ | $(x, y)$ |
|-----|-----|----------|
| 3 | 6 | $(3, 6)$ |
| 2 | 1 | $(2, 1)$ |
| 1 | $-2$ | $(1, -2)$ |
| 0 | $-3$ | $(0, -3)$ |
| $-1$ | $-2$ | $(-1, -2)$ |
| $-2$ | 1 | $(-2, 1)$ |
| $-3$ | 6 | $(-3, 6)$ |

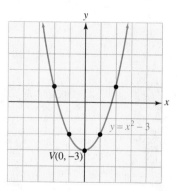

FIGURE 9-4

■

Self Check   Graph $y = x^2 + 2$. What do you notice?

Answer   The graph has the same shape as the graph of $y = x^2$, but it is 2 units higher.

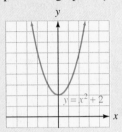

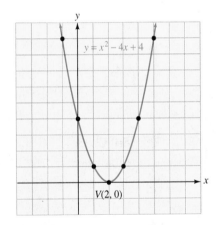

EXAMPLE 2   Graph $y = x^2 - 4x + 4$ and find its vertex.

Solution   To construct a table like the one shown in Figure 9-5, we pick several numbers $x$ and find the corresponding values of $y$. To graph the function, we plot the points and join them with a smooth curve.

$$y = x^2 - 4x + 4$$

| $x$ | $y$ | $(x, y)$ |
|-----|-----|----------|
| $-1$ | 9 | $(-1, 9)$ |
| 0 | 4 | $(0, 4)$ |
| 1 | 1 | $(1, 1)$ |
| 2 | 0 | $(2, 0)$ |
| 3 | 1 | $(3, 1)$ |
| 4 | 4 | $(4, 4)$ |
| 5 | 9 | $(5, 9)$ |

FIGURE 9-5

Since the graph is a parabola that opens upward, the vertex is the lowest point on the graph, the point $V(2, 0)$. ∎

Self Check   Graph $y = x^2 - 6x + 9$ and find its vertex.

Answer   The vertex is at $(3, 0)$.

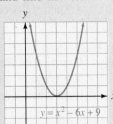

EXAMPLE 3    Graph $y = -x^2 + 2x - 1$ and find its vertex.

*Solution*    We construct the table shown in Figure 9-6, plot the points, and draw the graph.

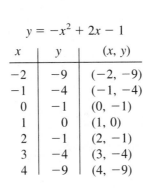

$$y = -x^2 + 2x - 1$$

| $x$ | $y$ | $(x, y)$ |
|-----|-----|----------|
| $-2$ | $-9$ | $(-2, -9)$ |
| $-1$ | $-4$ | $(-1, -4)$ |
| $0$ | $-1$ | $(0, -1)$ |
| $1$ | $0$ | $(1, 0)$ |
| $2$ | $-1$ | $(2, -1)$ |
| $3$ | $-4$ | $(3, -4)$ |
| $4$ | $-9$ | $(4, -9)$ |

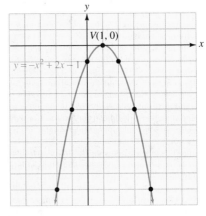

FIGURE 9-6

Since the parabola opens downward, its vertex is its highest point, the point $V(1, 0)$. ■

**Self Check**    Graph $y = -x^2 - 4x - 4$ and find its vertex.

*Answer*    The vertex is at $(-2, 0)$.

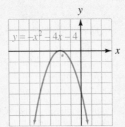

The results of these first three examples suggest the following fact.

**Graphs of Parabolas**
The graph of the equation $y = ax^2 + bx + c$ $(a \neq 0)$ is a parabola. It opens upward when $a > 0$, and it opens downward when $a < 0$.

## Graphing Quadratic Functions

**GRAPHING CALCULATORS**

We can use a graphing calculator to draw the graphs of Examples 1–3. If we use window values of $x = [-10, 10]$ and $y = [-10, 10]$, enter the quadratic polynomial, and press the GRAPH key, we will obtain the graphs shown in Figure 9-7.

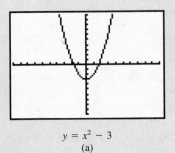

$y = x^2 - 3$
(a)

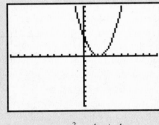

$y = x^2 - 4x + 4$
(b)

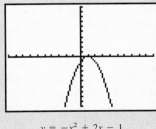

$y = -x^2 + 2x - 1$
(c)

FIGURE 9-7

## ■ FINDING THE VERTEX OF A PARABOLA

It is easier to graph a parabola when we know the coordinates of its vertex. We can find the coordinates of the vertex of the graph of

**1.** $y = x^2 - 6x + 8$

if we complete the square in the following way.

$y = x^2 - 6x + 8$

$y = x^2 - 6x + 9 - 9 + 8$     Add 9 to complete the square on $x^2 - 6x$ and then subtract 9.

$y = (x - 3)^2 - 1$     Factor $x^2 - 6x + 9$ and combine like terms.

Since $a > 0$ in Equation 1, the graph will be a parabola that opens upward. The vertex will be the lowest point on the parabola, and the $y$-coordinate of the vertex will be the smallest possible value of $y$. Because $(x - 3)^2 \geq 0$, the smallest value of $y$ occurs when $(x - 3)^2 = 0$ or when $x = 3$. To find the corresponding value of $y$, we substitute 3 for $x$ in the equation $y = (x - 3)^2 - 1$ and simplify.

$y = (x - 3)^2 - 1$

$y = (3 - 3)^2 - 1$     Substitute 3 for $x$.

$y = 0^2 - 1$

$y = -1$

The vertex of the parabola is the point $(3, -1)$. The graph appears in Figure 9-8.

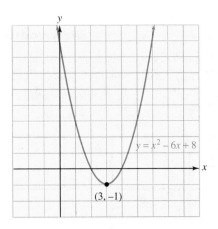

FIGURE 9-8

A generalization of this discussion leads to the following fact.

### Graphs of Parabolas with Vertex at ($h$, $k$)

The graph of an equation of the form

$$y = a(x - h)^2 + k$$

is a parabola with its vertex at the point with coordinates ($h$, $k$). The parabola opens upward if $a > 0$, and it opens downward if $a < 0$.

■ ■ ■ ■ ■ ■ ■ ■ ■ ■ **Finding the Vertex of a Parabola**

GRAPHING CALCULATORS
To find the vertex of the parabola determined by the equation $y = x^2 - 6x + 8$, we graph the function as in Figure 9-9(a). We then trace to move the cursor to the lowest point on the graph, as in Figure 9-9(b). If we zoom and trace again, we will obtain Figure 9-9(c). After repeated zooms, we will see that the vertex is the point with coordinates of $(3, -1)$.

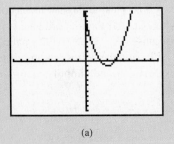

(a)

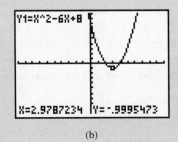

(b)

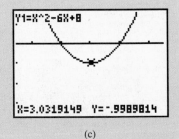

(c)

FIGURE 9-9

**EXAMPLE 4**    Find the vertex of the parabola determined by $y = -4(x - 3)^2 - 2$. Does the parabola open upward or downward?

*Solution*  Since $a = -4$ and $-4 < 0$, the parabola opens downward.

In the equation $y = a(x - h)^2 + k$, the coordinates of the vertex are given by the ordered pair $(h, k)$. In the equation $y = -4(x - 3)^2 - 2$, $h = 3$ and $k = -2$. Thus, the vertex is the point $(h, k) = (3, -2)$.  ∎

*Self Check*

*Answer*

Confirm the results of Example 4 by using a graphing calculator.

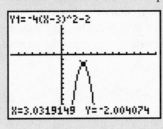

| EXAMPLE 5 |

Find the vertex of the parabola determined by $y = 5(x + 1)^2 + 4$. Does the parabola open upward or downward?

*Solution*  Since $a = 5$ and $5 > 0$, the parabola opens upward.

The equation $y = 5(x + 1)^2 + 4$ is equivalent to the equation

$$y = 5[x - (-1)]^2 + 4$$

Since $h = -1$ and $k = 4$, the vertex is the point $(h, k) = (-1, 4)$.  ∎

*Self Check*

*Answer*

Confirm the results of Example 5 by using a graphing calculator.

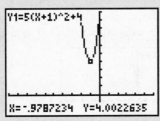

| EXAMPLE 6 |

Find the vertex of the parabola determined by $y = 2x^2 + 8x + 2$ and graph the parabola.

*Solution*  To make the coefficient of $x^2$ equal to 1, we factor 2 out of the binomial $2x^2 + 8x$ and proceed as follows:

1.  $y = 2x^2 + 8x + 2$

     $= 2(x^2 + 4x) + 2$          Factor 2 out of $2x^2 + 8x$.

     $= 2(x^2 + 4x + 4 - 4) + 2$    Complete the square on $x^2 + 4x$.

     $= 2[(x + 2)^2 - 4] + 2$       Factor $x^2 + 4x + 4$.

     $= 2(x + 2)^2 - 2 \cdot 4 + 2$     Distribute the multiplication by 2.

     $= 2(x + 2)^2 - 6$           Simplify and combine like terms.

or

$$y = 2[x - (-2)]^2 + (-6)$$

Since $h = -2$ and $k = -6$, the vertex of the parabola is the point $(-2, -6)$. Since $a = 2$, the parabola opens upward.

We can pick numbers $x$ on either side of $x = -2$ to construct the table shown in Figure 9-10. To find the $y$-intercept, we substitute 0 for $x$ in Equation 1 and solve for $y$: When $x = 0$, $y = 2$. Thus, the $y$-intercept is $(0, 2)$. We find some more ordered pairs, plot the points, and draw the parabola.

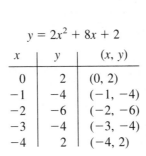

$$y = 2x^2 + 8x + 2$$

| $x$ | $y$ | $(x, y)$ |
|-----|-----|----------|
| 0 | 2 | $(0, 2)$ |
| $-1$ | $-4$ | $(-1, -4)$ |
| $-2$ | $-6$ | $(-2, -6)$ |
| $-3$ | $-4$ | $(-3, -4)$ |
| $-4$ | 2 | $(-4, 2)$ |

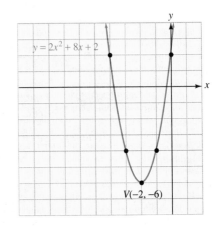

FIGURE 9-10 ∎

**Self Check**

**Answer**

Confirm the results of Example 6 by using a graphing calculator.

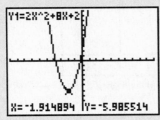

Much can be determined about the graph of $y = ax^2 + bx + c$ from the coefficients $a$, $b$, and $c$:

- The $y$-intercept $(0, y)$ is determined by the value of $y$ attained when $x = 0$: The $y$-intercept is $(0, c)$.

- The $x$-intercepts (if any) are determined by the numbers $x$ that make $y = 0$. To find them, we solve the quadratic equation $ax^2 + bx + c = 0$.

- By using the methods of Example 6, we can complete the square on $x$ in the equation $y = ax^2 + bx + c$ and find the $x$-coordinate of the vertex of the parabola, which is $x = -\frac{b}{2a}$.

We summarize these results as follows.

---

### Graphing the Parabola $y = ax^2 + bx + c$

The $y$-intercept is $(0, c)$.

The $x$-intercepts (if any) are determined by the solutions of $ax^2 + bx + c = 0$.

The $x$-coordinate of the vertex of the parabola $y = ax^2 + bx + c$ is $x = -\frac{b}{2a}$.

To find the $y$-coordinate of the vertex, substitute $-\frac{b}{2a}$ for $x$ into the equation $y = ax^2 + bx + c$ and solve for $y$.

---

**EXAMPLE 7**    Graph $y = x^2 - 2x - 3$.

*Solution*    The equation is in the form $y = ax^2 + bx + c$, with $a = 1$, $b = -2$, and $c = -3$. Since $a > 0$, the parabola opens upward. To find the $x$-coordinate of the vertex, we substitute the values for $a$ and $b$ into the formula $x = -\frac{b}{2a}$.

$$x = -\frac{b}{2a}$$

$$x = -\frac{-2}{2(1)}$$

$$= 1$$

The $x$-coordinate of the vertex is $x = 1$. To find the $y$-coordinate, we substitute 1 for $x$ in the equation and solve for $y$.

$$y = x^2 - 2x - 3$$
$$y = 1^2 - 2 \cdot 1 - 3$$
$$= 1 - 2 - 3$$
$$= -4$$

The vertex of the parabola is the point $(1, -4)$.

To graph the parabola, we find several other points with coordinates that satisfy the equation. One easy point to find is the $y$-intercept. It is the value of $y$ when $x = 0$. Thus, the parabola passes through the point $(0, -3)$.

To find the $x$-intercepts of the graph, we set $y$ equal to 0 and solve the resulting quadratic equation:

$$y = x^2 - 2x - 3$$
$$0 = x^2 - 2x - 3$$
$$0 = (x - 3)(x + 1) \qquad \text{Factor.}$$
$$x - 3 = 0 \quad \text{or} \quad x + 1 = 0 \qquad \text{Set each factor equal to 0.}$$
$$x = 3 \qquad\qquad x = -1$$

Since the $x$-intercepts of the graph are $(3, 0)$ and $(-1, 0)$, the graph passes through these points. The graph appears in Figure 9-11. ■

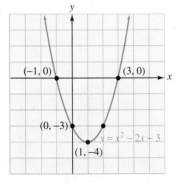

FIGURE 9-11

Self Check
Answer

Confirm the results of Example 7 by using a graphing calculator.

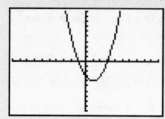

■ APPLICATIONS

EXAMPLE 8

**Finding maximum revenue**   An electronics firm manufactures a certain type of radio. Over the past 10 years, the firm has learned that it can sell $x$ radios at a price of $\left(200 - \frac{1}{5}x\right)$ dollars. How many radios should the firm manufacture and sell to maximize its revenue? Find the maximum revenue.

*Solution*    The revenue obtained is the product of the number of radios that the firm sells $(x)$ and the price of each radio $\left(200 - \frac{1}{5}x\right)$. Thus, the revenue $R$ is given by the formula

$$R = x\left(200 - \frac{1}{5}x\right) \qquad \text{or} \qquad R = -\frac{1}{5}x^2 + 200x$$

Since the graph of this function is a parabola that opens downward, the maximum value of $R$ will be the value of $R$ determined by the vertex of the parabola. Because the $x$-coordinate of the vertex is at $x = \frac{-b}{2a}$, we have

$$x = \frac{-b}{2a}$$

$$= \frac{-200}{2\left(-\dfrac{1}{5}\right)}$$

$$= \frac{-200}{-\dfrac{2}{5}}$$

$$= (-200)\left(-\frac{5}{2}\right)$$

$$= 500$$

If the firm manufactures 500 radios, the maximum revenue will be

$$R = -\frac{1}{5}x^2 + 200x$$

$$= -\frac{1}{5}(500)^2 + 200(500)$$

$$= 50,000$$

The firm should manufacture 500 radios to get a maximum revenue of $50,000.   ■

■ ■ ■ ■ ■ ■ ■ ■ ■

GRAPHING CALCULATORS

## Solving Quadratic Equations

We can use graphing methods to solve quadratic equations. For example, the solutions of the equation $x^2 - x - 3 = 0$ are the values of $x$ that will make $y = 0$ in the quadratic function $y = x^2 - x - 3$. To find these numbers, we graph the quadratic function and read the $x$-intercepts of the graph.

If we use window values of $x = [-10, 10]$ and $y = [-10, 10]$ and graph the function $y = x^2 - x - 3$, we will obtain the graph shown in Figure 9-12(a). We then trace to move the cursor close to the positive $x$-intercept until we have the $x$-coordinate shown in Figure 9-12(b).

To obtain better results, we zoom and trace again to obtain the graph in Figure 9-12(c). We can now see that the positive $x$-intercept is approximately $(2.3, 0)$, and a solution to the equation is approximately 2.3. For better results, we would do additional zooms.

Similar steps will show that the negative $x$-intercept is approximately $(-1.3, 0)$ and that the second solution of the equation is $x = -1.3$.

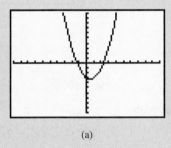
(a)

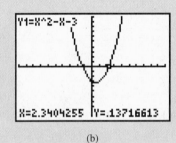
(b)

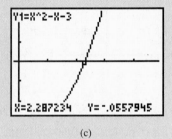
(c)

FIGURE 9-12

Orals *The graph of each equation is a parabola. Does it open up or down?*

**1.** $y = 3x^2 - 2x + 4$　　　　　　**2.** $y = 2x^2 + x - 5$

**3.** $y = -x^2 - 3x - 5$　　　　　　**4.** $y = -3x^2 + 4x - 1$

**5.** $y - 4 = -\dfrac{3}{2}x^2 + x$　　　　**6.** $y + 5 = \dfrac{3}{7}x^2 - x$

## EXERCISE 9.4

***REVIEW*** *Simplify each expression. Assume that all variables represent positive values.*

**1.** $\sqrt{12} + \sqrt{27}$　　　　　　　　**2.** $3\sqrt{6y}\left(-4\sqrt{3y}\right)$

**3.** $\left(\sqrt{3} + 1\right)\left(\sqrt{3} - 1\right)$　　　　　**4.** $\dfrac{\sqrt{x} + 2}{\sqrt{x} - 2}$

***VOCABULARY AND CONCEPTS*** *Fill in each blank to make a true statement.*

**5.** A function defined by the equation $y = ax^2 + bx + c$ ($a \neq 0$) is called a _____ function.

**6.** The lowest (or highest) point on a parabola is called the _____ of the parabola.

**7.** The point where a parabola intersects the $y$-axis is called the _____.

**8.** An $x$-intercept is a point where a parabola intersects the _____.

**9.** The graph of $y = ax^2 + bx + c$ ($a \neq 0$) opens upward when _____ and downward when _____.

**10.** The vertex of the parabolic graph of $y = a(x - h)^2 + k$ is the point _____.

**11.** The $y$-intercept of the graph of $y = ax^2 + bx + c$ is the point _____.

**12.** The $x$-coordinate of the vertex of the graph of $y = ax^2 + bx + c$ is $x =$ _____.

***PRACTICE*** *In Exercises 13–16, graph each equation and compare the graph with the graph of* $y = x^2$. *Check your work with a graphing calculator.*

**13.** $y = x^2 + 1$

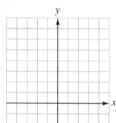

**14.** $y = x^2 - 4$

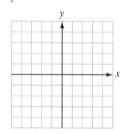

**15.** $y = -x^2$

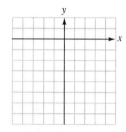

**16.** $y = -(x - 1)^2$

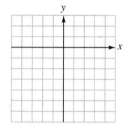

*In Exercises 17–24, graph each equation. Check your work with a graphing calculator.*

**17.** $y = x^2 + x$

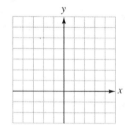

**18.** $y = x^2 - 2x$

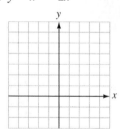

**19.** $y = -x^2 - 4x$

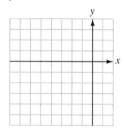

**20.** $y = -x^2 + 2x$

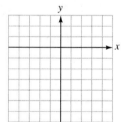

**21.** $y = x^2 + 4x + 4$

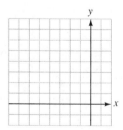

**22.** $y = x^2 - 6x + 9$

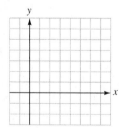

**23.** $y = x^2 - 4x + 6$

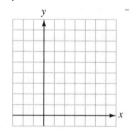

**24.** $y = x^2 + 2x - 3$

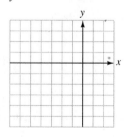

*In Exercises 25–36, find the vertex of the graph of each equation.* **Do not draw the graph.**

**25.** $y = -3(x - 2)^2 + 4$

**26.** $y = 4(x - 3)^2 + 2$

**27.** $y = 5(x + 1)^2 - 5$

**28.** $y = 4(x + 3)^2 + 1$        **29.** $y = (x - 1)^2$        **30.** $y = -(x - 2)^2$

**31.** $y = -7x^2 + 4$        **32.** $y = 5x^2 - 2$        **33.** $y = x^2 + 2x + 5$

**34.** $y = x^2 + 4x + 1$        **35.** $y = x^2 - 6x - 12$        **36.** $y = x^2 - 8x - 20$

*In Exercises 37–48, complete the square, if necessary, to determine the vertex of the graph of each equation. Then graph the equation. Check your work with a graphing calculator.*

**37.** $y = x^2 - 4x + 4$     **38.** $y = x^2 + 6x + 9$     **39.** $y = -x^2 - 2x - 1$     **40.** $y = -x^2 + 2x - 1$

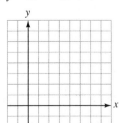 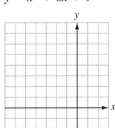 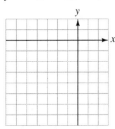 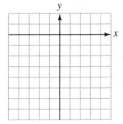

**41.** $y = x^2 + 2x - 3$     **42.** $y = x^2 + 6x + 5$     **43.** $y = -x^2 - 6x - 7$     **44.** $y = -x^2 + 8x - 14$

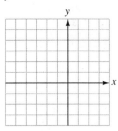 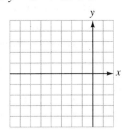 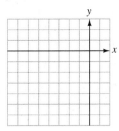 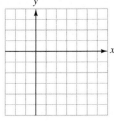

**45.** $y = 2x^2 + 8x + 6$     **46.** $y = 3x^2 - 12x + 9$     **47.** $y = -3x^2 + 6x - 2$     **48.** $y = -2x^2 - 4x + 2$

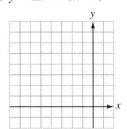 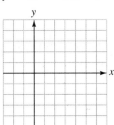 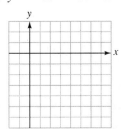 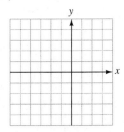

*In Exercises 49–54, find the vertex and the x- and y-intercepts of the graph of each equation. Then graph the equation.*

**49.** $y = x^2 - x - 2$        **50.** $y = x^2 - 6x + 8$        **51.** $y = -x^2 + 2x + 3$

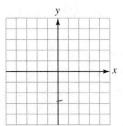 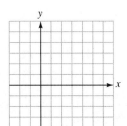 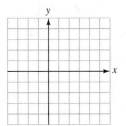

**52.** $y = -x^2 + 5x - 4$

**53.** $y = 2x^2 + 3x - 2$

**54.** $y = 3x^2 - 7x + 2$

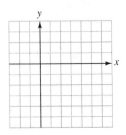

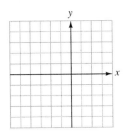

## APPLICATIONS

**55. Selling TVs**   A company has found that it can sell $x$ TVs at a price of $\$\left(450 - \frac{1}{6}x\right)$. How many TVs must the company sell to maximize its revenue?

**56. Finding maximum revenue**   In Exercise 55, find the maximum revenue.

**57. Selling CD players**   A wholesaler sells CD players for $150 each. However, he gives volume discounts on purchases of between 500 and 1,000 units according to the formula $\left(150 - \frac{1}{10}n\right)$, where $n$ represents the number of units purchased. How many units would a retailer have to buy for the wholesaler to obtain maximum revenue?

**58. Finding maximum revenue**   In Exercise 57, find the maximum revenue.

*In Exercises 59–62, use a graphing calculator to solve each equation.*

**59.** $x^2 - 5x + 6 = 0$

**60.** $6x^2 + 5x - 6 = 0$

**61.** $x^2 + 5x + 2 = 0$

**62.** $2x^2 - 5x + 1 = 0$

## WRITING

**63.** Explain why the $y$-intercept of the graph of $y = ax^2 + bx + c$ $(a \neq 0)$ is $(0, c)$.

**64.** Define the vertex of a parabola and explain how to find its coordinates.

## SOMETHING TO THINK ABOUT

**65.** The graph of $x = y^2$ is a parabola, but the equation does not define a function. Explain.

**66.** The graph of $x = -y^2$ is a parabola, but the equation does not define a function. Explain.

■ ■ ■ ■ ■ ■ ■ ■ ■

**MATHEMATICS IN SALES**   American Appliance (first discussed at the beginning of the chapter) can sell $x$ refrigerators for a price of $\$\left(1,800 - \frac{3}{2}x\right)$. If the store sells

0 refrigerators, the price will be $1,800, for a total revenue of $0;

50 refrigerators, the price will be $1,725, for a total revenue of $86,250;

100 refrigerators, the price will be $1,650, for a total revenue of $165,000;

and so on. In general, the total revenue $R$ is the product of the price and the number of refrigerators sold.

$$R = x\left(1,800 - \frac{3}{2}x\right)$$

**1.**  $R = -\frac{3}{2}x^2 + 1,800x$      Use the distributive property to remove parentheses.

Since the graph of Equation 1 is a parabola that opens downward, the maximum value of $R$ will be the $R$-value that corresponds to the $x$-coordinate of the vertex:

$$x = \frac{-b}{2a} = \frac{-1,800}{2\left(-\frac{3}{2}\right)} = \frac{-1,800}{-3} = 600 \qquad \text{The } x\text{-coordinate of the vertex is } x = \frac{-b}{2a}.$$

The store will maximize revenue by selling 600 refrigerators at a price of

$$1,800 - \frac{3}{2}x = 1,800 - \frac{3}{2}(600) = \$900$$

The store should order 600 refrigerators from the distributor each month to keep up with demand.

## ■ ■ ■ ■ ■ ■ ■ ■ ■ ■  PROJECTS

**PROJECT 1**  Solve the equation $x^2 + 5x + 1 = 0$. Are the solutions real numbers? Solve the equation $x^2 + x + 1 = 0$. Are the solutions real numbers? How can you tell without actually solving an equation whether the solutions are real or nonreal? Develop both an algebraic method and a graphing-calculator method.

   Test your method on the following quadratic equations.

   **a.** $x^2 - 7x + 2 = 0$        **b.** $x^2 + x - 2 = 0$

   **c.** $x^2 - 3x + 4 = 0$       **d.** $x^2 + 5x + 7 = 0$

   **e.** $3x^2 - 5x + 9 = 0$      **f.** $2x^2 + 5x + 2 = 0$

**PROJECT 2**  One of the world's greatest scientific geniuses, Sir Isaac Newton (1642–1727) contributed to every major area of science and mathematics known in his time. Because of his early interest in science and mathematics, Newton enrolled at Trinity College, Cambridge, where he studied the mathematics of René Descartes and the astronomy of Galileo but did not show any great talent. In 1665, the plague closed the school, and Newton went home. There, his genius appeared. Within a few years, he had made major discoveries in mathematics, physics, and astronomy.

   One of Newton's many contributions to mathematics is known as **Newton's method,** a way of finding better and better estimates of the solutions of certain

*(continued)*

■ ■ ■ ■ ■ ■ ■ ■ ■ ■ **PROJECTS** *(continued)*

equations. The process begins with a guess of the solution, and transforms that guess into a better estimate of the solution. When Newton's method is applied to this better answer, a third solution results, one that is more accurate than any of the previous ones. The method is applied again and again, producing better and better approximations of the solution.

To solve the quadratic equation $x^2 + x - 3 = 0$, for example, Newton's method uses the fraction $\dfrac{x^2 + 3}{2x + 1}$ to generate better solutions. We begin with a guess: 3, for example. We substitute 3 for $x$ in the fraction.

$$\frac{x^2 + 3}{2x + 1} = \frac{3^2 + 3}{2(3) + 1} \approx 1.714286$$

The number 1.714286 is a better estimate of the solution than the original guess, 3. We apply Newton's method again:

$$\frac{x^2 + 3}{2x + 1} = \frac{1.714286^2 + 3}{2(1.714286) + 1} \approx 1.341014$$

The number 1.341014 is the best estimate yet of the solution. However, more passes through Newton's method produce even better estimates: 1.303173, and then 1.302776. This final answer is accurate to six decimal places.

- The quadratic equation $x^2 + x - 3 = 0$ has two solutions. Find the other solution by using Newton's method and another initial guess. (Try a negative number.)
- To solve the quadratic equation $ax^2 + bx + c = 0$, Newton's method uses the fraction $\dfrac{ax^2 - c}{2ax + b}$ to generate solutions. Solve $2x^2 - 2x - 1 = 0$ using Newton's method. Find two solutions accurate to six decimal places.

# CHAPTER SUMMARY

## CONCEPTS

## REVIEW EXERCISES

### SECTION 9.1 — *Quadratic Equations*

**The zero-factor property:**
If $ab = 0$, then $a = 0$ or $b = 0$.

Many quadratic equations can be solved by factoring.

**1.** Solve each quadratic equation by factoring.
   **a.** $x^2 + 2x = 0$
   **b.** $2x^2 - 6x = 0$
   **c.** $x^2 - 9 = 0$
   **d.** $p^2 - 25 = 0$
   **e.** $a^2 - 7a + 12 = 0$
   **f.** $t^2 - 2t - 15 = 0$
   **g.** $2x - x^2 + 24 = 0$
   **h.** $2x^2 + x - 3 = 0$
   **i.** $x^2 - 7x = -12$

The two solutions of $x^2 = c$ are $x = \sqrt{c}$ and $x = -\sqrt{c}$.

**2.** Use the square root method to solve each quadratic equation.

   **a.** $x^2 = 25$        **b.** $x^2 = 36$        **c.** $2x^2 = 18$

   **d.** $4x^2 = 9$        **e.** $x^2 = 8$        **f.** $x^2 = 75$

**3.** Use the square root method to solve each equation.

   **a.** $(x - 1)^2 = 25$        **b.** $(x + 3)^2 = 36$

   **c.** $2(x + 1)^2 = 18$        **d.** $4(x - 2)^2 = 9$

   **e.** $(x - 8)^2 = 8$        **f.** $(x + 5)^2 = 75$

**4. Construction**  The 45-square-foot base of the preformed concrete panel shown in Illustration 1 is 3 feet longer than twice the height of the panel. How long is the base?

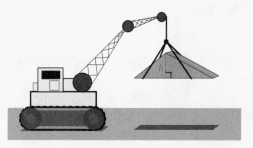

ILLUSTRATION 1

**5. Military application**  A pilot releases a bomb from an altitude of 3,000 feet. The bomb's height $h$ above the target $t$ seconds after its release is given by the formula

$$h = 3,000 + 40t - 16t^2$$

How long will it be until the bomb hits its target?

**6. Gardening**  A 27-square-foot rectangular flower bed is 3 feet longer than twice its width. Find its dimensions.

**7. Geometry**  A rectangle is 3 feet longer than it is wide. Its area is numerically equal to its perimeter. Find its dimensions.

---

**SECTION 9.2**       *Completing the Square*

---

To make $x^2 + 2bx$ a trinomial square, add the square of one-half of the coefficient of $x$:

$$x^2 + 2bx + b^2 = (x + b)^2$$

**8.** Complete the square to make each expression a trinomial square.

   **a.** $x^2 + 4x$        **b.** $y^2 + 8y$

   **c.** $z^2 - 10z$        **d.** $t^2 - 5t$

   **e.** $a^2 + \dfrac{3}{4}a$        **f.** $c^2 - \dfrac{7}{3}c$

To solve a quadratic equation by completing the square,

**1.** If necessary, divide both sides of the equation by the coefficient of $x^2$ to make that coefficient 1.

**2.** If necessary, get the constant on the right-hand side of the equation.

**3.** Complete the square.

**4.** Solve the resulting quadratic equation.

**5.** Check each solution.

**9.** Solve each quadratic equation by completing the square.
  **a.** $x^2 + 5x - 14 = 0$
  **b.** $x^2 - 8x + 15 = 0$
  **c.** $x^2 + 4x - 77 = 0$
  **d.** $x^2 - 2x - 1 = 0$
  **e.** $x^2 + 4x - 3 = 0$
  **f.** $x^2 - 6x + 4 = 0$
  **g.** $2x^2 + 5x - 3 = 0$
  **h.** $2x^2 - 2x - 1 = 0$

---

### SECTION 9.3     *The Quadratic Formula*

**The quadratic formula:**
$$x = \frac{-b \pm \sqrt{b^2 - 4ac}}{2a}$$

**10.** Use the quadratic formula to solve each quadratic equation.
  **a.** $x^2 - 2x - 15 = 0$
  **b.** $x^2 - 6x - 7 = 0$
  **c.** $x^2 - 15x + 26 = 0$
  **d.** $2x^2 - 7x + 3 = 0$
  **e.** $6x^2 - 7x - 3 = 0$
  **f.** $x^2 + 4x + 1 = 0$
  **g.** $x^2 - 6x + 7 = 0$
  **h.** $x^2 + 3x = 0$

**11. Geometry** The length of the rectangle in Illustration 2 is 14 centimeters greater than the width. Find the perimeter.

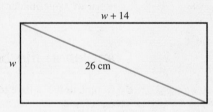

ILLUSTRATION 2

| **SECTION 9.4** | *Graphing Quadratic Functions* |
|---|---|

The graph of the equation $y = ax^2 + bx + c$ is a parabola. It opens upward when $a > 0$ and downward when $a < 0$.

The graph of an equation of the form $y = a(x - h)^2 + k$ is a parabola with vertex at $(h, k)$. It opens upward when $a > 0$ and downward when $a < 0$.

The $x$-coordinate of the vertex of the parabola $y = ax^2 + bx + c$ is $x = -\frac{b}{2a}$.

To find the $y$-coordinate of the vertex, substitute $-\frac{b}{2a}$ for $x$ in the equation of the parabola and find $y$.

**12.** Graph each equation.

**a.** $y = x^2 + 8x + 10$

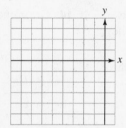

**b.** $y = -2x^2 - 4x - 6$

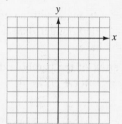

**13.** Find the vertex of the graph of each equation. **Do not draw the graph.**

**a.** $y = 5(x - 6)^2 + 7$

**b.** $y = 3(x + 3)^2 - 5$

**c.** $y = 2x^2 - 4x + 7$

**d.** $y = -3x^2 + 18x - 11$

## ■ Chapter Test

*In Problems 1–2, solve each equation by factoring.*

**1.** $6x^2 + x - 1 = 0$

**2.** $10x^2 + 43x = 9$

*In Problems 3–4, solve each equation by the square root method.*

**3.** $x^2 = 16$

**4.** $(x - 2)^2 = 3$

*In Problems 5–6, find the number required to complete the square.*

**5.** $x^2 + 14x$

**6.** $x^2 - 7x$

**7.** Use completing the square to solve $3a^2 + 6a - 12 = 0$.

**8.** Write the quadratic formula.

*In Problems 9–11, use the quadratic formula to solve each equation.*

**9.** $x^2 + 3x - 10 = 0$     **10.** $2x^2 - 5x = 12$     **11.** $2x^2 + 5x + 1 = 2$

**12.** The base of a triangle with an area of 40 square meters is 2 meters longer than it is high. Find the length of the base.

**13.** Find the vertex of the parabola determined by the equation $y = -4(x + 5)^2 - 4$.

**14.** Graph the equation $y = x^2 + 4x + 2$.

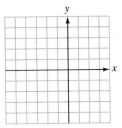

1. How many prime numbers are there between 20 and 30?
   **a.** 1   **b.** 2   **c.** 3   **d.** 4   **e.** none of the above

2. If $x = 3$, $y = -2$, and $z = -1$, find the value of $\dfrac{x + z}{y}$.

   **a.** 1   **b.** $-1$   **c.** $-2$   **d.** 2   **e.** none of the above

3. If $x = -2$, $y = -3$, and $z = -4$, find the value of $\dfrac{|x - z|}{|y|}$.

   **a.** $-2$   **b.** 2   **c.** $\dfrac{2}{3}$   **d.** $-\dfrac{2}{3}$   **e.** none of the above

4. The distributive property is written symbolically as
   **a.** $a + b = b + a$   **b.** $ab = ba$
   **c.** $a(b + c) = ab + ac$
   **d.** $(a + b) + c = a + (b + c)$   **e.** none of the above

5. Solve for $x$: $7x - 4 = 24$
   **a.** $x = 4$   **b.** $x = 5$   **c.** $x = -4$   **d.** $x = -5$
   **e.** none of the above

6. Solve for $z$: $6z - (9 - 3z) = -3(z + 2)$
   **a.** $\dfrac{4}{3}$   **b.** $\dfrac{3}{4}$   **c.** $-\dfrac{4}{3}$   **d.** $-\dfrac{1}{4}$   **e.** none of the above

7. Solve for $r$: $\dfrac{r}{5} - \dfrac{r - 3}{10} = 0$

   **a.** $-3$   **b.** $-2$   **c.** 3   **d.** 1   **e.** none of the above

8. Solve for $x$: $\dfrac{ax}{b} + c = 4$

   **a.** $\dfrac{b(4 - x)}{c}$   **b.** $\dfrac{4b - 4c}{b}$   **c.** $\dfrac{b(4 - c)}{a}$   **d.** none of the above

9. A man bought 25 pencils, some at 10 cents and some at 15 cents. The 25 pencils cost $3. How many 10-cent pencils did he buy?
   **a.** 5   **b.** 10   **c.** 15   **d.** 20   **e.** none of the above

10. Solve the inequality: $-3(x - 2) + 3 \geq 6$
    **a.** $x \geq -1$   **b.** $x \geq 1$   **c.** $x \leq -1$   **d.** $x \leq 1$
    **e.** none of the above

11. Simplify: $x^2 x^3 x^7$
    **a.** $12x$   **b.** $x^{12}$   **c.** $x^{42}$   **d.** $x^{35}$   **e.** none of the above

12. Simplify: $\dfrac{(x^2)^7}{x^3 x^4}$

    **a.** 0   **b.** 1   **c.** $x^7$   **d.** $x^2$   **e.** none of the above

13. Simplify: $\dfrac{x^{-2} y^3}{xy^{-1}}$

    **a.** $x^3 y^4$   **b.** $\dfrac{x^3}{y^4}$   **c.** $xy$   **d.** $\dfrac{y^4}{x^3}$   **e.** none of the above

14. Write 73,000,000 in scientific notation.
    **a.** $7.3 \times 10^7$   **b.** $7.3 \times 10^{-7}$   **c.** $73 \times 10^6$
    **d.** $0.73 \times 10^9$   **e.** none of the above

15. If $P(x) = 2x^2 + 3x - 4$, find $P(-2)$.
    **a.** 2   **b.** $-2$   **c.** $-18$   **d.** 10   **e.** none of the above

16. Simplify: $2(y + 3) - 3(y - 2)$
    **a.** $-y$   **b.** $5y$   **c.** $5y + 12$   **d.** $-y + 12$   **e.** none of the above

17. Multiply: $-3x^2 y^2 (2xy^3)$
    **a.** $6x^3 y^5$   **b.** $5x^2 y^5$   **c.** $-6x^3 y^5$   **d.** $-6xy^5$
    **e.** none of the above

**18.** Multiply: $2a^3b^2(3a^2b - 2ab^2)$
 **a.** $6a^6b^3 - 4a^4b^4$   **b.** $6a^5b^3 - 4a^4b^4$
 **c.** $6a^5b^3 - 4a^4b^2$   **d.** $6a^5b^4 - 4a^4b^4$   **e.** none of the above

**19.** Multiply: $(x + 7)(2x - 3)$
 **a.** $2x^2 - 11x - 21$   **b.** $2x^2 + 11x + 21$
 **c.** $2x^2 + 11x - 21$   **d.** $-2x^2 + 11x - 21$   **e.** none of the above

**20.** Divide: $(x^2 + 5x - 14) \div (x + 7)$
 **a.** $x + 2$   **b.** $x - 2$   **c.** $x + 1$   **d.** $x - 1$   **e.** none of the above

**21.** Factor completely: $r^2h + r^2a$
 **a.** $r(rh + ra)$   **b.** $r^2h(1 + a)$   **c.** $r^2a(h + 1)$
 **d.** $r^2(h + a)$   **e.** none of the above

**22.** Factor completely: $m^2n^4 - 49$
 **a.** $(mn^2 + 7)(mn^2 + 7)$   **b.** $(mn^2 - 7)(mn^2 - 7)$
 **c.** $(mn + 7)(mn - 7)$   **d.** $(mn^2 + 7)(mn^2 - 7)$
 **e.** none of the above

**23.** One of the factors of $x^2 - 5x + 6$ is
 **a.** $x + 3$   **b.** $x + 2$   **c.** $x - 6$   **d.** $x - 2$   **e.** none of the above

**24.** One of the factors of $36x^2 + 12x + 1$ is
 **a.** $6x - 1$   **b.** $6x + 1$   **c.** $x + 6$   **d.** $x - 6$
 **e.** none of the above

**25.** One of the factors of $2x^2 + 7xy + 6y^2$ is
 **a.** $2x + y$   **b.** $x - 3y$   **c.** $2x - y$   **d.** $x + 6y$
 **e.** none of the above

**26.** One of the factors of $8x^3 - 27$ is
 **a.** $2x + 3$   **b.** $4x^2 - 6x + 9$   **c.** $4x^2 + 12x + 9$
 **d.** $4x^2 + 6x + 9$   **e.** none of the above

**27.** One of the factors of $2x^2 + 2xy - 3x - 3y$ is
 **a.** $x - y$   **b.** $2x + 3$   **c.** $x - 3$   **d.** $2x - 3$   **e.** none of the above

**28.** Solve for $x$: $x^2 + x - 6 = 0$
 **a.** $x = -2, x = 3$   **b.** $x = 2, x = -3$   **c.** $x = 2, x = 3$   **d.** $x = -2, x = -3$   **e.** none of the above

**29.** Solve for $x$: $6x^2 - 7x - 3 = 0$
 **a.** $x = \dfrac{3}{2}, x = \dfrac{1}{3}$   **b.** $x = -\dfrac{3}{2}, x = \dfrac{1}{3}$
 **c.** $x = \dfrac{3}{2}, x = -\dfrac{1}{3}$   **d.** $x = -\dfrac{3}{2}, x = -\dfrac{1}{3}$   **e.** none of the above

**30.** Simplify the fraction: $\dfrac{x^2 - 16}{x^2 - 8x + 16}$
 **a.** $\dfrac{x + 4}{x - 4}$   **b.** $1$   **c.** $\dfrac{x - 4}{x + 4}$   **d.** $-\dfrac{1}{8x}$   **e.** none of the above

**31.** Multiply: $\dfrac{x^2 + 11x - 12}{x - 5} \cdot \dfrac{x^2 - 5x}{x - 1}$
 **a.** $-x(x + 12)$   **b.** $x(x + 12)$   **c.** $\dfrac{x^2 + 11x - 12}{(x - 5)(x - 1)}$
 **d.** $1$   **e.** none of the above

**32.** Divide: $\dfrac{t^2 + 7t}{t^2 + 5t} \div \dfrac{t^2 + 4t - 21}{t - 3}$
 **a.** $\dfrac{1}{t + 5}$   **b.** $t + 5$   **c.** $1$   **d.** $\dfrac{(t + 7)(t + 7)}{t + 5}$
 **e.** none of the above

**33.** Simplify: $\dfrac{3x}{2} - \dfrac{x}{4}$
 **a.** $-x$   **b.** $\dfrac{5}{4}$   **c.** $2x$   **d.** $\dfrac{5x}{4}$   **e.** none of the above

**34.** Simplify: $\dfrac{a + 3}{2a - 6} - \dfrac{2a + 3}{a^2 - 3a} + \dfrac{3}{4}$
 **a.** $\dfrac{5a + 4}{4a}$   **b.** $\dfrac{4a + 4}{5}$   **c.** $0$   **d.** $\dfrac{9 - a}{5a - a^2 - 2}$
 **e.** none of the above

**35.** Simplify: $\dfrac{x + \dfrac{1}{y}}{\dfrac{1}{x} + y}$
 **a.** $\dfrac{x^2y + x}{y + xy^2}$   **b.** $1$   **c.** $\dfrac{xy + x}{y + xy}$   **d.** $\dfrac{x}{y}$   **e.** none of the above

**36.** Solve for $x$: $\dfrac{1}{x} + \dfrac{1}{2x} = \dfrac{1}{4}$
 **a.** $4$   **b.** $5$   **c.** $6$   **d.** $7$   **e.** none of the above

**37.** Solve for $s$: $\dfrac{2}{s + 1} + \dfrac{1 - s}{s} = \dfrac{1}{s^2 + s}$
 **a.** $1$   **b.** $2$   **c.** $3$   **d.** $4$   **e.** none of the above

**38.** Find the $x$-intercept of the graph of $3x - 4y = 12$.
 **a.** $(3, 0)$   **b.** $(-3, 0)$   **c.** $(4, 0)$   **d.** $(-4, 0)$
 **e.** none of the above

**39.** The graph of $y = -3x + 12$ does not pass through
 **a.** quadrant I   **b.** quadrant II   **c.** quadrant III
 **d.** quadrant IV   **e.** none of the above

**40.** Find the slope of the line passing through $P(-2, 4)$ and $Q(8, -6)$.
 **a.** $-1$   **b.** $1$   **c.** $-\dfrac{1}{3}$   **d.** $3$   **e.** none of the above

**41.** The equation of the line passing through $P(-2, 4)$ and $Q(8, -6)$ is
   **a.** $y = -x + 2$   **b.** $y = -x + 6$   **c.** $y = -x - 6$
   **d.** $y = -x - 2$   **e.** none of the above

**42.** If $f(x) = x^2 - 2x$, find $f(a + 2)$.
   **a.** $a^2 - 2a$   **b.** $a^2 + 2a$   **c.** $a + 2$   **d.** $a + 2a^2$
   **e.** none of the above

**43.** Solve the system
$$\begin{cases} 2x - 5y = 5 \\ 3x - 2y = -16 \end{cases}$$
   for $x$.
   **a.** $x = 8$   **b.** $x = -8$   **c.** $x = 4$   **d.** $x = -4$
   **e.** none of the above

**44.** Solve the system
$$\begin{cases} 8x - y = 29 \\ 2x + y = 11 \end{cases}$$
   for $y$.
   **a.** $y = -4$   **b.** $y = 4$   **c.** $y = 3$   **d.** $y = -3$
   **e.** none of the above

**45.** Simplify: $\sqrt{12}$
   **a.** $2\sqrt{3}$   **b.** $4\sqrt{3}$   **c.** $6\sqrt{2}$   **d.** $4\sqrt{2}$   **e.** none of the above

**46.** Simplify: $\sqrt{\dfrac{3}{4}}$
   **a.** $\dfrac{\sqrt{3}}{4}$   **b.** $\dfrac{3}{2}$   **c.** $\dfrac{\sqrt{3}}{2}$   **d.** $\dfrac{9}{16}$   **e.** none of the above

**47.** Simplify: $\sqrt{75x^3}$
   **a.** $5\sqrt{x^3}$   **b.** $5x\sqrt{x}$   **c.** $x\sqrt{75x}$   **d.** $25x\sqrt{3x}$
   **e.** none of the above

**48.** Simplify: $3\sqrt{5} - \sqrt{20}$
   **a.** $3\sqrt{-15}$   **b.** $2\sqrt{5}$   **c.** $-\sqrt{5}$   **d.** $\sqrt{5}$   **e.** none of the above

**49.** Rationalize the denominator: $\dfrac{11}{\sqrt{11}}$
   **a.** $\dfrac{1}{11}$   **b.** $\sqrt{11}$   **c.** $\dfrac{\sqrt{11}}{11}$   **d.** $1$   **e.** none of the above

**50.** Rationalize the denominator: $\dfrac{7}{3 - \sqrt{2}}$
   **a.** $3 - \sqrt{2}$   **b.** $7(3 - \sqrt{2})$   **c.** $3 + \sqrt{2}$
   **d.** $7(3 + \sqrt{2})$   **e.** none of the above

**51.** Solve for $x$: $\sqrt{\dfrac{3x - 1}{5}} = 2$
   **a.** 7   **b.** 4   **c.** $-7$   **d.** $-4$   **e.** none of the above

**52.** Solve for $n$: $3\sqrt{n} - 1 = 1$
   **a.** $\dfrac{2}{3}$   **b.** $\sqrt{\dfrac{2}{3}}$   **c.** $\dfrac{4}{9}$   **d.** $n\sqrt{n + 1}$   **e.** none of the above

**53.** Simplify: $(a^6 b^4)^{1/2}$
   **a.** $(ab)^5$   **b.** $a^3 b^2$   **c.** $\dfrac{1}{a^3 b^2}$   **d.** $\dfrac{1}{a^6 b^4}$   **e.** none of the above

**54.** Simplify: $\left(\dfrac{8}{125}\right)^{2/3}$
   **a.** $\dfrac{4}{25}$   **b.** $\dfrac{25}{4}$   **c.** $\dfrac{2}{5}$   **d.** $\dfrac{5}{2}$   **e.** none of the above

**55.** What number must be added to $x^2 + 12x$ to make it a perfect trinomial square?
   **a.** 6   **b.** 12   **c.** 144   **d.** 36   **e.** none of the above

**56.** Write the quadratic formula.
   **a.** $x = \dfrac{b \pm \sqrt{b^2 - 4ac}}{2a}$   **b.** $x = \dfrac{-b \pm \sqrt{b^2 - 4ac}}{2a}$
   **c.** $x = \dfrac{-b \pm \sqrt{b^2 + 4ac}}{2a}$
   **d.** $x = \dfrac{-b \pm \sqrt{b^2 - 4ac}}{2b}$   **e.** none of the above

**57.** One solution of the equation $x^2 - 2x - 2 = 0$ is
   **a.** $2\sqrt{3}$   **b.** $2 + 2\sqrt{3}$   **c.** $1 - \sqrt{3}$   **d.** $2 - 2\sqrt{3}$
   **e.** none of the above

**58.** The vertex of the graph of the equation $y = x^2 - 2x + 1$ is
   **a.** $(1, 0)$   **b.** $(0, 1)$   **c.** $(-1, 0)$   **d.** $(0, -1)$
   **e.** none of the above

**59.** The graph of $y = -x^2 + 2x - 1$
   **a.** does not intersect the $y$-axis   **b.** does not intersect the $x$-axis   **c.** passes through the origin   **d.** passes through $(2, 2)$   **e.** none of the above

**60.** The formula that expresses the sentence "$x$ varies directly with the square of $y$ and inversely with $t$" is
   **a.** $x = ky^2 + t$   **b.** $x = \dfrac{ky^2}{t}$   **c.** $x = \dfrac{kt}{y^2}$
   **d.** $x = ky^2 t$   **e.** none of the above

# COMPLEX NUMBERS

■ COMPLEX NUMBERS ■ OPERATIONS WITH COMPLEX NUMBERS ■ COMPLEX CONJUGATES
■ RATIONALIZING DENOMINATORS ■ ABSOLUTE VALUE OF A COMPLEX NUMBER

**Getting Ready** *Do the following operations.*

**1.** $(2x + 4) + (3x - 5)$　　　　　**2.** $(3x - 4) - (2x + 3)$

**3.** $(x + 4)(x - 5)$　　　　　　　**4.** $(3x - 1)(2x - 1)$

We have seen that the solutions to some quadratic equations are not real numbers.

**EXAMPLE 1** Solve $x^2 + x + 1 = 0$.

*Solution* We can use the quadratic formula, with $a = 1$, $b = 1$, and $c = 1$.

$$x = \frac{-b \pm \sqrt{b^2 - 4ac}}{2a}$$

$$= \frac{-1 \pm \sqrt{1^2 - 4(1)(1)}}{2(1)}$$

$$= \frac{-1 \pm \sqrt{1 - 4}}{2}$$

$$= \frac{-1 \pm \sqrt{-3}}{2}$$

$$x = \frac{-1 + \sqrt{-3}}{2} \quad \text{or} \quad x = \frac{-1 - \sqrt{-3}}{2}$$ ■

Each solution in Example 1 involves $\sqrt{-3}$, which is not a real number, be-cause the square of no real number equals $-3$. For years, mathematicians believed that numbers like $\sqrt{-3}$, $\sqrt{-1}$, and $\sqrt{-9}$ were nonsense. Even the great English mathematician Sir Isaac Newton (1642–1727) called them impossible. These numbers were called **imaginary numbers** by René Descartes (1596–1650). Today they have important uses such as describing alternating current in electronics.

The imaginary number $\sqrt{-1}$ is usually denoted by the letter $i$. Since

$$i = \sqrt{-1}$$

it follows that

$$i^2 = -1$$

The powers of $i$ produce an interesting pattern:

$$i = \sqrt{-1} = i \qquad\qquad i^5 = i^4 \cdot i = 1 \cdot i = i$$
$$i^2 = (\sqrt{-1})^2 = -1 \qquad i^6 = i^4 \cdot i^2 = 1(-1) = -1$$
$$i^3 = i^2 \cdot i = -1 \cdot i = -i \qquad i^7 = i^4 \cdot i^3 = 1(-i) = -i$$
$$i^4 = i^2 \cdot i^2 = (-1)(-1) = 1 \qquad i^8 = i^4 \cdot i^4 = (1)(1) = 1$$

The pattern continues: $i, -1, -i, 1, \ldots$.

If we assume that multiplication of imaginary numbers is commutative and associative, then

$$(2i)^2 = 2^2 i^2$$
$$= 4(-1)$$
$$= -4$$

Since $(2i)^2 = -4$, it follows that $2i$ is a square root of $-4$, and we write

$$\sqrt{-4} = 2i$$

This result could have been obtained by the following process:

$$\sqrt{-4} = \sqrt{4(-1)}$$
$$= \sqrt{4}\sqrt{-1}$$
$$= 2i$$

Likewise, we have

$$\sqrt{-25} = \sqrt{25(-1)} = \sqrt{25}\sqrt{-1} = 5i$$
$$\sqrt{-\frac{1}{9}} = \sqrt{\frac{1}{9}(-1)} = \sqrt{\frac{1}{9}}\sqrt{-1} = \frac{1}{3}i$$
$$\sqrt{\frac{-100}{49}} = \sqrt{\frac{100}{49}(-1)} = \frac{\sqrt{100}}{\sqrt{49}}\sqrt{-1} = \frac{10}{7}i$$

In general, we have the following rules.

**Rules of Radicals**

If at least one of $a$ and $b$ is a nonnegative real number, then

$$\sqrt{ab} = \sqrt{a}\sqrt{b} \qquad \text{and} \qquad \sqrt{\frac{a}{b}} = \frac{\sqrt{a}}{\sqrt{b}} \quad (b \neq 0)$$

## ■ COMPLEX NUMBERS

Imaginary numbers such as $\sqrt{-3}$, $\sqrt{-1}$, and $\sqrt{-9}$ form a subset of a broader set of numbers called **complex numbers.**

> **Complex Number**
> A **complex number** is any number that can be written in the form $a + bi$, where $a$ and $b$ are real numbers, and $i = \sqrt{-1}$.
>     The number $a$ is called the **real part** and the number $b$ is called the **imaginary part** of the complex number $a + bi$.

If $b = 0$, the complex number $a + bi$ is the real number $a$. If $b \neq 0$ and $a = 0$, the complex number $0 + bi$ (or just $bi$) is an imaginary number.

Figure A-1 shows the relationship of the real numbers to the imaginary and complex numbers.

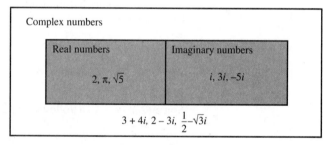

FIGURE A-1

> **Equality of Complex Numbers**
> The complex numbers $a + bi$ and $c + di$ are equal if and only if
> $$a = c \quad \text{and} \quad b = d$$

EXAMPLE 2

**a.** $2 + 3i = \sqrt{4} + \dfrac{6}{2}i$, because $2 = \sqrt{4}$ and $3 = \dfrac{6}{2}$.

**b.** $4 - 5i = \dfrac{12}{3} - \sqrt{25}i$, because $4 = \dfrac{12}{3}$ and $-5 = -\sqrt{25}$.

**c.** $x + yi = 4 + 7i$ if and only if $x = 4$ and $y = 7$.    ■

## ■ OPERATIONS WITH COMPLEX NUMBERS

### Addition and Subtraction of Complex Numbers
Complex numbers are added and subtracted as if they were binomials:
$$(a + bi) + (c + di) = (a + c) + (b + d)i$$

EXAMPLE 3

**a.** $(8 + 4i) + (12 + 8i) = 8 + 4i + 12 + 8i$
$$= 20 + 12i$$

**b.** $(7 - 4i) + (9 + 2i) = 7 - 4i + 9 + 2i$
$$= 16 - 2i$$

**c.** $(-6 + i) - (3 - 4i) = -6 + i - 3 + 4i$
$$= -9 + 5i$$

**d.** $(2 - 4i) - (-4 + 3i) = 2 - 4i + 4 - 3i$
$$= 6 - 7i$$    ■

To multiply a complex number by an imaginary number, we use the distributive property to remove parentheses and then simplify. For example,

$$-5i(4 - 8i) = -5i(4) - (-5i)(8i)$$
$$= -20i + 40i^2$$
$$= -40 - 20i \qquad \text{Remember that } i^2 = -1.$$

> **Multiplication of Complex Numbers**
> Complex numbers are multiplied as if they were binomials, with $i^2 = -1$:
> $$(a + bi)(c + di) = ac + adi + bci + bdi^2$$
> $$= (ac - bd) + (ad + bc)i$$

EXAMPLE 4

**a.** $(2 + 3i)(3 - 2i) = 6 - 4i + 9i - 6i^2$
$$= 6 + 5i + 6$$
$$= 12 + 5i$$

**b.** $(3 + i)(1 + 2i) = 3 + 6i + i + 2i^2$
$$= 3 + 7i - 2$$
$$= 1 + 7i$$

**c.** $(-4 + 2i)(2 + i) = -8 - 4i + 4i + 2i^2$
$$= -8 - 2$$
$$= -10$$

**d.** $(-1 - i)(4 - i) = -4 + i - 4i + i^2$
$$= -4 - 3i - 1$$
$$= -5 - 3i$$    ■

The next example shows how to write complex numbers in $a + bi$ form. When writing answers, it is acceptable to use $a - bi$ as a substitute for the form $a + (-b)i$.

EXAMPLE 5

**a.** $7 = 7 + 0i$

**b.** $3i = 0 + 3i$

**c.** $4 - \sqrt{-16} = 4 - \sqrt{-1(16)}$
$$= 4 - \sqrt{16}\sqrt{-1}$$
$$= 4 - 4i$$

**d.** $5 + \sqrt{-11} = 5 + \sqrt{-1(11)}$
$$= 5 + \sqrt{11}\sqrt{-1}$$
$$= 5 + \sqrt{11}\, i$$

**e.** $2i^2 + 4i^3 = 2(-1) + 4(-i)$
$$= -2 - 4i$$

**f.** $\dfrac{3}{2i} = \dfrac{3}{2i} \cdot \dfrac{i}{i}$
$$= \dfrac{3i}{2i^2}$$
$$= \dfrac{3i}{2(-1)}$$
$$= \dfrac{3i}{-2}$$
$$= 0 - \dfrac{3}{2}i$$

**g.** $-\dfrac{5}{i} = -\dfrac{5}{i} \cdot \dfrac{i^3}{i^3} = -\dfrac{5(-i)}{1} = 5i = 0 + 5i$    ∎

## ■ COMPLEX CONJUGATES

> **Complex Conjugates**
> The complex numbers $a + bi$ and $a - bi$ are called **complex conjugates** of each other.

For example,

- $3 + 4i$ and $3 - 4i$ are complex conjugates.
- $5 - 7i$ and $5 + 7i$ are complex conjugates.
- $8 + 17i$ and $8 - 17i$ are complex conjugates.

EXAMPLE 6

Find the product of $3 + i$ and its complex conjugate.

*Solution*    The complex conjugate of $3 + i$ is $3 - i$. We find their product as follows:

$$(3 + i)(3 - i) = 9 - 3i + 3i - i^2$$
$$= 9 - i^2 \qquad \text{Combine like terms.}$$
$$= 9 - (-1) \qquad \text{Because } i^2 = -1.$$
$$= 10$$

∎

In general, the product of the complex number $a + bi$ and its complex conjugate $a - bi$ is the real number $a^2 + b^2$.

$$(a + bi)(a - bi) = a^2 - abi + abi - b^2i^2$$
$$= a^2 - b^2(-1)$$
$$= a^2 + b^2$$

■ RATIONALIZING DENOMINATORS

To write complex numbers such as $\dfrac{1}{3+i}$, $\dfrac{3-i}{2+i}$, and $\dfrac{5+i}{5-i}$ in $a + bi$ form, we rationalize their denominators.

**EXAMPLE 7**   Write $\dfrac{1}{3+i}$ in $a + bi$ form.

*Solution*   Since the product of $3 + i$ and its conjugate is a real number, we can rationalize the denominator by multiplying both the numerator and the denominator of the fraction by the complex conjugate of the denominator and simplify.

$$\frac{1}{3+i} = \frac{1}{3+i} \cdot \frac{3-i}{3-i}$$
$$= \frac{3-i}{9 - 3i + 3i - i^2}$$
$$= \frac{3-i}{9-(-1)}$$
$$= \frac{3-i}{10}$$
$$= \frac{3}{10} - \frac{1}{10}i$$   ■

**EXAMPLE 8**   Write $\dfrac{3-i}{2+i}$ in $a + bi$ form.

*Solution*   We rationalize the denominator by multiplying the numerator and denominator by the complex conjugate of the denominator and simplify.

$$\frac{3-i}{2+i} = \frac{3-i}{2+i} \cdot \frac{2-i}{2-i}$$
$$= \frac{6 - 3i - 2i + i^2}{4 - 2i + 2i - i^2}$$
$$= \frac{5 - 5i}{4 - (-1)}$$
$$= \frac{5(1-i)}{5} \qquad \text{Factor out 5 in the numerator.}$$
$$= 1 - i \qquad \text{Divide out the common factor of 5.}$$   ■

EXAMPLE 9     Divide $5 + i$ by $5 - i$ and express the quotient in $a + bi$ form.

*Solution*     The quotient obtained when dividing $5 + i$ by $5 - i$ is expressed by the fraction $\frac{5+i}{5-i}$. To express this quotient in $a + bi$ form, we rationalize the denominator by multiplying both the numerator and the denominator by the complex conjugate of the denominator and simplify.

$$\frac{5+i}{5-i} = \frac{5+i}{5-i} \cdot \frac{5+i}{5+i}$$

$$= \frac{25 + 5i + 5i + i^2}{25 + 5i - 5i - i^2}$$

$$= \frac{25 + 10i - 1}{25 - (-1)}$$

$$= \frac{24 + 10i}{26}$$

$$= \frac{2(12 + 5i)}{26} \qquad \text{Factor out 2 in the numerator.}$$

$$= \frac{12 + 5i}{13} \qquad \text{Divide out a common factor of 2.}$$

$$= \frac{12}{13} + \frac{5}{13}i \qquad ■$$

 **WARNING!**   Complex numbers are not always written in $a + bi$ form. To avoid mistakes, always put complex numbers in $a + bi$ form before doing any arithmetic involving the numbers.

EXAMPLE 10     Write $\dfrac{4 + \sqrt{-16}}{2 + \sqrt{-4}}$ in $a + bi$ form.

*Solution*
$$\frac{4 + \sqrt{-16}}{2 + \sqrt{-4}} = \frac{4 + 4i}{2 + 2i} \qquad \text{Change to } a + bi \text{ form.}$$

$$= \frac{2(2 + 2i)}{2 + 2i} \qquad \text{Factor out 2 in the numerator.}$$

$$= 2 + 0i \qquad \text{Divide out } 2 + 2i. \qquad ■$$

## ■ ABSOLUTE VALUE OF A COMPLEX NUMBER

**Absolute Value of a Complex Number**
The **absolute value** of the complex number $a + bi$ is $\sqrt{a^2 + b^2}$. In symbols,
$$|a + bi| = \sqrt{a^2 + b^2}$$

**EXAMPLE 11**   **a.** $|3 + 4i| = \sqrt{3^2 + 4^2}$       **b.** $|5 - 12i| = \sqrt{5^2 + (-12)^2}$
$= \sqrt{9 + 16}$                  $= \sqrt{25 + 144}$
$= \sqrt{25}$                      $= \sqrt{169}$
$= 5$                              $= 13$   ∎

**EXAMPLE 12**   If $a$ and $b$ are both negative numbers, is the formula $\sqrt{a}\sqrt{b} = \sqrt{ab}$ still true?

*Solution*   We can let $a = -4$ and $b = -1$ and compute $\sqrt{a}\sqrt{b}$ and $\sqrt{ab}$ to see if their values are equal:

$$\sqrt{a}\sqrt{b} = \sqrt{-4}\sqrt{-1} \quad \text{and} \quad \sqrt{ab} = \sqrt{(-4)(-1)}$$
$$= 2i \cdot i \qquad\qquad\qquad = \sqrt{4}$$
$$= 2i^2 \qquad\qquad\qquad\quad = 2$$
$$= -2$$

Since $-2 \neq 2$, the formula $\sqrt{ab} = \sqrt{a}\sqrt{b}$ is false when both $a$ and $b$ are negative. ∎

*Orals*   *Do each operation.*

**1.** $(3 + 4i) + (2 + 3i)$          **2.** $(2 + 3i) - (1 - 2i)$
**3.** $(1 + i)(2 + i)$               **4.** $(2 - i)(1 - i)$
**5.** $(2 + i)(1 - i)$               **6.** $(1 + i)(2 - i)$

# EXERCISE II.1

***PRACTICE***   *In Exercises 1–10, solve each quadratic equation. Write all roots in bi or a + bi form.*

**1.** $x^2 + 9 = 0$        **2.** $x^2 + 16 = 0$       **3.** $3x^2 = -16$       **4.** $2x^2 = -25$

**5.** $x^2 + 2x + 2 = 0$              **6.** $x^2 + 3x + 3 = 0$

**7.** $2x^2 + x + 1 = 0$              **8.** $3x^2 + 2x + 1 = 0$

**9.** $3x^2 - 4x + 2 = 0$             **10.** $2x^2 - 3x + 2 = 0$

*In Exercises 11–18, simplify each expression.*

**11.** $i^{21}$        **12.** $i^{19}$        **13.** $i^{27}$        **14.** $i^{22}$
**15.** $i^{100}$       **16.** $i^{42}$        **17.** $i^{97}$        **18.** $i^{200}$

*In Exercises 19–60, express numbers in a + bi form, if necessary, and do the indicated operations. Give all answers in a + bi form.*

**19.** $(3 + 4i) + (5 - 6i)$

**20.** $(5 + 3i) - (6 - 9i)$

**21.** $(7 - 3i) - (4 + 2i)$

**22.** $(8 + 3i) + (-7 - 2i)$

**23.** $(8 + \sqrt{-25}) + (7 + \sqrt{-4})$

**24.** $(-7 + \sqrt{-81}) - (-2 - \sqrt{-64})$

**25.** $(-8 - \sqrt{-3}) - (7 - \sqrt{-27})$

**26.** $(2 + \sqrt{-8}) + (-3 - \sqrt{-2})$

**27.** $3i(2 - i)$

**28.** $-4i(3 + 4i)$

**29.** $(2 + 3i)(3 - i)$

**30.** $(4 - i)(2 + i)$

**31.** $(2 - 4i)(3 + 2i)$

**32.** $(3 - 2i)(4 - 3i)$

**33.** $(2 + \sqrt{-2})(3 - \sqrt{-2})$

**34.** $(5 + \sqrt{-3})(2 - \sqrt{-3})$

**35.** $(-2 - \sqrt{-16})(1 + \sqrt{-4})$

**36.** $(-3 - \sqrt{-81})(-2 + \sqrt{-9})$

**37.** $(2 + \sqrt{-3})(3 - \sqrt{-2})$

**38.** $(1 + \sqrt{-5})(2 - \sqrt{-3})$

**39.** $(8 - \sqrt{-5})(-2 - \sqrt{-7})$

**40.** $(-1 + \sqrt{-6})(2 - \sqrt{-3})$

**41.** $\dfrac{1}{i}$

**42.** $\dfrac{1}{i^3}$

**43.** $\dfrac{4}{5i^3}$

**44.** $\dfrac{3}{2i}$

**45.** $\dfrac{3i}{8\sqrt{-9}}$

**46.** $\dfrac{5i^3}{2\sqrt{-4}}$

**47.** $\dfrac{-3}{5i^5}$

**48.** $\dfrac{-4}{6i^7}$

**49.** $\dfrac{-6}{\sqrt{-32}}$

**50.** $\dfrac{5}{\sqrt{-125}}$

**51.** $\dfrac{3}{5 + i}$

**52.** $\dfrac{-2}{2 - i}$

**53.** $\dfrac{-12}{7 - \sqrt{-1}}$

**54.** $\dfrac{4}{3 + \sqrt{-1}}$

**55.** $\dfrac{5i}{6 + 2i}$

**56.** $\dfrac{-4i}{2 - 6i}$

**57.** $\dfrac{3 - 2i}{3 + 2i}$

**58.** $\dfrac{2 + 3i}{2 - 3i}$

**59.** $\dfrac{3 + \sqrt{-2}}{2 + \sqrt{-5}}$

**60.** $\dfrac{2 - \sqrt{-5}}{3 + \sqrt{-7}}$

*In Exercises 61–72, find each absolute value.*

**61.** $|6 + 8i|$

**62.** $|12 + 5i|$

**63.** $|12 - 5i|$

**64.** $|3 - 4i|$

**65.** $|5 + 7i|$

**66.** $|6 - 5i|$

**67.** $|4 + \sqrt{-2}|$

**68.** $|3 + \sqrt{-3}|$

**69.** $|8 + \sqrt{-5}|$

**70.** $|7 - \sqrt{-6}|$

**71.** $|5 - 0i|$

**72.** $|0 - 5i|$

**WRITING**   *Write a paragraph using your own words.*

**73.** Explain how to add or subtract two complex numbers.

**74.** Explain how to find the absolute value of $3 + 2i$.

# TABLE OF POWERS AND ROOTS

| $n$ | $n^2$ | $\sqrt{n}$ | $n^3$ | $\sqrt[3]{n}$ | $n$ | $n^2$ | $\sqrt{n}$ | $n^3$ | $\sqrt[3]{n}$ |
|---|---|---|---|---|---|---|---|---|---|
| 1 | 1 | 1.000 | 1 | 1.000 | 51 | 2,601 | 7.141 | 132,651 | 3.708 |
| 2 | 4 | 1.414 | 8 | 1.260 | 52 | 2,704 | 7.211 | 140,608 | 3.733 |
| 3 | 9 | 1.732 | 27 | 1.442 | 53 | 2,809 | 7.280 | 148,877 | 3.756 |
| 4 | 16 | 2.000 | 64 | 1.587 | 54 | 2,916 | 7.348 | 157,464 | 3.780 |
| 5 | 25 | 2.236 | 125 | 1.710 | 55 | 3,025 | 7.416 | 166,375 | 3.803 |
| 6 | 36 | 2.449 | 216 | 1.817 | 56 | 3,136 | 7.483 | 175,616 | 3.826 |
| 7 | 49 | 2.646 | 343 | 1.913 | 57 | 3,249 | 7.550 | 185,193 | 3.849 |
| 8 | 64 | 2.828 | 512 | 2.000 | 58 | 3,364 | 7.616 | 195,112 | 3.871 |
| 9 | 81 | 3.000 | 729 | 2.080 | 59 | 3,481 | 7.681 | 205,379 | 3.893 |
| 10 | 100 | 3.162 | 1,000 | 2.154 | 60 | 3,600 | 7.746 | 216,000 | 3.915 |
| 11 | 121 | 3.317 | 1,331 | 2.224 | 61 | 3,721 | 7.810 | 226,981 | 3.936 |
| 12 | 144 | 3.464 | 1,728 | 2.289 | 62 | 3,844 | 7.874 | 238,328 | 3.958 |
| 13 | 169 | 3.606 | 2,197 | 2.351 | 63 | 3,969 | 7.937 | 250,047 | 3.979 |
| 14 | 196 | 3.742 | 2,744 | 2.410 | 64 | 4,096 | 8.000 | 262,144 | 4.000 |
| 15 | 225 | 3.873 | 3,375 | 2.466 | 65 | 4,225 | 8.062 | 274,625 | 4.021 |
| 16 | 256 | 4.000 | 4,096 | 2.520 | 66 | 4,356 | 8.124 | 287,496 | 4.041 |
| 17 | 289 | 4.123 | 4,913 | 2.571 | 67 | 4,489 | 8.185 | 300,763 | 4.062 |
| 18 | 324 | 4.243 | 5,832 | 2.621 | 68 | 4,624 | 8.246 | 314,432 | 4.082 |
| 19 | 361 | 4.359 | 6,859 | 2.668 | 69 | 4,761 | 8.307 | 328,509 | 4.102 |
| 20 | 400 | 4.472 | 8,000 | 2.714 | 70 | 4,900 | 8.367 | 343,000 | 4.121 |
| 21 | 441 | 4.583 | 9,261 | 2.759 | 71 | 5,041 | 8.426 | 357,911 | 4.141 |
| 22 | 484 | 4.690 | 10,648 | 2.802 | 72 | 5,184 | 8.485 | 373,248 | 4.160 |
| 23 | 529 | 4.796 | 12,167 | 2.844 | 73 | 5,329 | 8.544 | 389,017 | 4.179 |
| 24 | 576 | 4.899 | 13,824 | 2.884 | 74 | 5,476 | 8.602 | 405,224 | 4.198 |
| 25 | 625 | 5.000 | 15,625 | 2.924 | 75 | 5,625 | 8.660 | 421,875 | 4.217 |
| 26 | 676 | 5.099 | 17,576 | 2.962 | 76 | 5,776 | 8.718 | 438,976 | 4.236 |
| 27 | 729 | 5.196 | 19,683 | 3.000 | 77 | 5,929 | 8.775 | 456,533 | 4.254 |
| 28 | 784 | 5.292 | 21,952 | 3.037 | 78 | 6,084 | 8.832 | 474,552 | 4.273 |
| 29 | 841 | 5.385 | 24,389 | 3.072 | 79 | 6,241 | 8.888 | 493,039 | 4.291 |
| 30 | 900 | 5.477 | 27,000 | 3.107 | 80 | 6,400 | 8.944 | 512,000 | 4.309 |
| 31 | 961 | 5.568 | 29,791 | 3.141 | 81 | 6,561 | 9.000 | 531,441 | 4.327 |
| 32 | 1,024 | 5.657 | 32,768 | 3.175 | 82 | 6,724 | 9.055 | 551,368 | 4.344 |
| 33 | 1,089 | 5.745 | 35,937 | 3.208 | 83 | 6,889 | 9.110 | 571,787 | 4.362 |
| 34 | 1,156 | 5.831 | 39,304 | 3.240 | 84 | 7,056 | 9.165 | 592,704 | 4.380 |
| 35 | 1,225 | 5.916 | 42,875 | 3.271 | 85 | 7,225 | 9.220 | 614,125 | 4.397 |
| 36 | 1,296 | 6.000 | 46,656 | 3.302 | 86 | 7,396 | 9.274 | 636,056 | 4.414 |
| 37 | 1,369 | 6.083 | 50,653 | 3.332 | 87 | 7,569 | 9.327 | 658,503 | 4.431 |
| 38 | 1,444 | 6.164 | 54,872 | 3.362 | 88 | 7,744 | 9.381 | 681,472 | 4.448 |
| 39 | 1,521 | 6.245 | 59,319 | 3.391 | 89 | 7,921 | 9.434 | 704,969 | 4.465 |
| 40 | 1,600 | 6.325 | 64,000 | 3.420 | 90 | 8,100 | 9.487 | 729,000 | 4.481 |
| 41 | 1,681 | 6.403 | 68,921 | 3.448 | 91 | 8,281 | 9.539 | 753,571 | 4.498 |
| 42 | 1,764 | 6.481 | 74,088 | 3.476 | 92 | 8,464 | 9.592 | 778,688 | 4.514 |
| 43 | 1,849 | 6.557 | 79,507 | 3.503 | 93 | 8,649 | 9.644 | 804,357 | 4.531 |
| 44 | 1,936 | 6.633 | 85,184 | 3.530 | 94 | 8,836 | 9.695 | 830,584 | 4.547 |
| 45 | 2,025 | 6.708 | 91,125 | 3.557 | 95 | 9,025 | 9.747 | 857,375 | 4.563 |
| 46 | 2,116 | 6.782 | 97,336 | 3.583 | 96 | 9,216 | 9.798 | 884,736 | 4.579 |
| 47 | 2,209 | 6.856 | 103,823 | 3.609 | 97 | 9,409 | 9.849 | 912,673 | 4.595 |
| 48 | 2,304 | 6.928 | 110,592 | 3.634 | 98 | 9,604 | 9.899 | 941,192 | 4.610 |
| 49 | 2,401 | 7.000 | 117,649 | 3.659 | 99 | 9,801 | 9.950 | 970,299 | 4.626 |
| 50 | 2,500 | 7.071 | 125,000 | 3.684 | 100 | 10,000 | 10.000 | 1,000,000 | 4.642 |

## Getting Ready (page 2)

**1.** 1, 2, 3, etc.   **2.** $\frac{1}{2}, \frac{2}{3}$, etc.   **3.** $-3, -21$, etc.

## Orals (page 11)

**11.** $-15$   **12.** 25

## Exercise 1.1 (page 11)

**1.** set   **3.** whole   **5.** subset   **7.** prime   **9.** is not equal to   **11.** is greater than or equal to   **13.** number   **15.** 1, 2, 6, 9
**17.** 1, 2, 6, 9   **19.** $-3, -1, 0, 1, 2, 6, 9$   **21.** $-3, -\frac{1}{2}, -1, 0, 1, 2, \frac{5}{3}, \sqrt{7}, 3.25, 6, 9$   **23.** $-3, -1, 1, 9$   **25.** 6, 9
**27.** 9; natural, odd, composite, and whole number   **29.** 0; even integer, whole number   **31.** 24; natural, even, composite, and
whole number   **33.** 3; natural, odd, prime, and whole number   **35.** =   **37.** <   **39.** >   **41.** =   **43.** =   **45.** <   **47.** =
**49.** $7 > 3$   **51.** $8 \le 8$   **53.** $3 + 4 = 7$   **55.** $7 \ge 3$   **57.** $0 < 6$   **59.** $8 < 3 + 8$   **61.** $10 - 4 > 6 - 2$   **63.** $3 \cdot 4 > 2 \cdot 3$
**65.** $\frac{24}{6} > \frac{12}{4}$   **67.** ; 6, 6   **69.** ; 11, 11   **71.** ; 2, 2
**73.** ; 8, 8   **75.**   **77.**
**79.**   **81.**   **83.**
**85.**   **87.** 36   **89.** 0   **91.** 230   **93.** 8

## Getting Ready (page 14)

**1.** 250   **2.** 148   **3.** 16,606   **4.** 105

## Orals (page 27)

**1.** $\frac{1}{2}$   **2.** $\frac{1}{2}$   **3.** $\frac{1}{2}$   **4.** $\frac{1}{3}$   **5.** $\frac{5}{12}$   **6.** $\frac{9}{20}$   **7.** $\frac{4}{9}$   **8.** $\frac{6}{25}$   **9.** $\frac{11}{9}$   **10.** $\frac{3}{7}$   **11.** $\frac{1}{6}$   **12.** $\frac{5}{4}$   **13.** 2.86   **14.** 1.24   **15.** 0.5   **16.** 3.9   **17.** 3.24
**18.** 3.25

## Exercise 1.2 (page 27)

**1.** true   **3.** false   **5.** true   **7.** true   **9.** =   **11.** =   **13.** numerator   **15.** simplify   **17.** proper   **19.** 1   **21.** multiply
**23.** numerators, denominator   **25.** plus   **27.** repeating   **29.** $\frac{1}{2}$   **31.** $\frac{3}{4}$   **33.** $\frac{4}{3}$   **35.** $\frac{9}{8}$   **37.** $\frac{3}{10}$   **39.** $\frac{8}{5}$   **41.** $\frac{3}{2}$   **43.** $\frac{1}{4}$   **45.** 10

**47.** $\frac{20}{3}$  **49.** $\frac{9}{10}$  **51.** $\frac{5}{8}$  **53.** $\frac{1}{4}$  **55.** $\frac{14}{5}$  **57.** 28  **59.** $\frac{1}{5}$  **61.** $\frac{6}{5}$  **63.** $\frac{1}{13}$  **65.** $\frac{5}{24}$  **67.** $\frac{19}{15}$  **69.** $\frac{17}{12}$  **71.** $\frac{22}{35}$  **73.** $\frac{9}{4}$  **75.** $\frac{29}{3}$  **77.** $5\frac{1}{5}$
**79.** $1\frac{2}{3}$  **81.** $1\frac{1}{4}$  **83.** $\frac{5}{9}$  **85.** 158.65  **87.** 44.785  **89.** 44.88  **91.** 4.55  **93.** 350.49  **95.** 55.21  **97.** 3,337.52  **99.** 10.02
**101.** $121\frac{3}{5}$ m  **103.** $53\frac{1}{6}$ ft  **105.** 2,514,820  **107.** 270 lb  **109.** 43.13 sec  **111.** $18,151.15  **113.** $2,143.23
**115.** the high-capacity boards  **117.** 205,200 lb  **119.** the high-efficiency furnace

## Getting Ready  (page 31)

**1.** 4  **2.** 9  **3.** 27  **4.** 8  **5.** $\frac{1}{4}$  **6.** $\frac{1}{27}$  **7.** $\frac{8}{125}$  **8.** $\frac{27}{1,000}$

## Orals  (page 40)

**1.** 32  **2.** 81  **3.** 64  **4.** 125  **5.** 24  **6.** 36  **7.** 11  **8.** 1  **9.** 16  **10.** 24

## Exercise 1.3  (page 40)

**1.**  **3.** prime number  **5.** exponent  **7.** multiplication  **9.** $P = 4s$
**11.** $P = 2l + 2w$  **13.** $P = a + b + c$  **15.** $P = a + b + c + d$  **17.** $C = \pi D$  **19.** $V = lwh$  **21.** $V = \frac{1}{3}Bh$  **23.** $V = \frac{4}{3}\pi r^3$
**25.** 16  **27.** 36  **29.** $\frac{1}{10,000}$  **31.** 493.039  **33.** 640.09  **35.** $x \cdot x$  **37.** $3 \cdot z \cdot z \cdot z \cdot z$  **39.** $5t \cdot 5t$  **41.** $5 \cdot 2x \cdot 2x \cdot 2x$  **43.** 36
**45.** 1,000  **47.** 18  **49.** 216  **51.** 11  **53.** 3  **55.** 28  **57.** 64  **59.** 13  **61.** 16  **63.** 2  **65.** 16  **67.** 21  **69.** 17  **71.** 9
**73.** 8  **75.** 8  **77.** $\frac{1}{144}$  **79.** 11  **81.** 1  **83.** $\frac{8}{9}$  **85.** 1  **87.** 4  **89.** 4  **91.** 12  **93.** 4  **95.** 11  **97.** 24  **99.** 12  **101.** 25
**103.** 1  **105.** 28  **107.** 35  **109.** 1  **111.** $(3 \cdot 8) + (5 \cdot 3)$  **113.** $(3 \cdot 8 + 5) \cdot 3$  **115.** 16 in.  **117.** 15 m  **119.** 25 m²
**121.** 60 ft²  **123.** 88 m  **125.** 1,386 ft²  **127.** 6 cm³  **129.** 905 m³  **131.** 1,056 cm³  **133.** 40,764.51 ft³  **135.** 480 ft³  **137.** 8
**141.** bigger

## Getting Ready  (page 44)

**1.** 17.52  **2.** 2.94  **3.** 2  **4.** 1  **5.** 96  **6.** 382

## Orals  (page 52)

**1.** 5  **2.** −3  **3.** 3  **4.** −11  **5.** 4  **6.** −12  **7.** 2  **8.** 16  **9.** −6  **10.** −6

## Exercise 1.4  (page 52)

**1.** 20  **3.** 24  **5.** arrows  **7.** subtract, greater  **9.** add, opposite  **11.** 12  **13.** −10  **15.** 2  **17.** −2  **19.** 0.5  **21.** $\frac{12}{35}$  **23.** 1
**25.** 2.2  **27.** 7  **29.** −1  **31.** −7  **33.** −8  **35.** 3  **37.** 1.3  **39.** −1  **41.** 3  **43.** 10  **45.** −3  **47.** −1  **49.** 9  **51.** 1
**53.** 4  **55.** 12  **57.** 5  **59.** $\frac{1}{2}$  **61.** $-8\frac{3}{4}$  **63.** −4.2  **65.** 4  **67.** −7  **69.** 10  **71.** 0  **73.** 8  **75.** 3  **77.** 2.45  **79.** 9  **81.** −3
**83.** −15  **85.** 1  **87.** 3  **89.** −1  **91.** 9.9  **93.** −7.1  **95.** $175  **97.** +9  **99.** −4°  **101.** 2,000 yr  **103.** 1,325 m
**105.** 4,000 ft  **107.** 5°  **109.** 9,187  **111.** 700  **113.** $422.66  **115.** $83,425.57

## Getting Ready  (page 56)

**1.** 56  **2.** 54  **3.** 72  **4.** 63  **5.** 9  **6.** 6  **7.** 8  **8.** 8

## Orals  (page 62)

**1.** −3  **2.** 10  **3.** 18  **4.** −24  **5.** 24  **6.** −24  **7.** −2  **8.** 2  **9.** −9  **10.** −1

## Exercise 1.5  (page 62)

**1.** 1,125 lb  **3.** 53  **5.** positive  **7.** positive  **9.** positive  **11.** $a$  **13.** 0  **15.** 48  **17.** 56  **19.** −144  **21.** −16  **23.** 2  **25.** 1
**27.** 72  **29.** −24  **31.** −420  **33.** −96  **35.** 4  **37.** −9  **39.** −2  **41.** 5  **43.** −3  **45.** −8  **47.** −8  **49.** 6  **51.** 5  **53.** 7
**55.** −4  **57.** 2  **59.** −4  **61.** −20  **63.** 2  **65.** 1  **67.** −6  **69.** −30  **71.** 7  **73.** −10  **75.** −10  **77.** 14  **79.** −81  **81.** 88

**83.** $-\frac{1}{6}$  **85.** $-\frac{11}{12}$  **87.** $-\frac{7}{36}$  **89.** $-\frac{11}{48}$  **91.** $(+2)(+3) = +6$  **93.** $(-30)(15) = -450$  **95.** $(+23)(-120) = -2,760$
**97.** $\frac{-18}{-3} = +6$  **99.** 2-point loss per day  **101.** yes

## Getting Ready (page 65)

**1.** sum  **2.** product  **3.** quotient  **4.** difference  **5.** quotient  **6.** difference  **7.** product  **8.** sum

## Orals (page 70)

**1.** 1  **2.** $-14$  **3.** $-11$  **4.** 7  **5.** 16  **6.** 64  **7.** $-12$  **8.** 36

## Exercise 1.6 (page 70)

**1.** 532  **3.** $\frac{1}{2}$  **5.** sum  **7.** multiplication  **9.** algebraic  **11.** variables  **13.** $x + y$  **15.** $x(2y)$  **17.** $y - x$  **19.** $\frac{y}{x}$  **21.** $z + \frac{x}{y}$
**23.** $z - xy$  **25.** $3xy$  **27.** $\frac{x+y}{y+z}$  **29.** $xy + \frac{y}{z}$  **31.** the sum of $x$ and 3  **33.** the quotient obtained when $x$ is divided by $y$
**35.** the product of 2, $x$, and $y$  **37.** the quotient obtained when 5 is divided by the sum of $x$ and $y$  **39.** the quotient obtained when
the sum of 3 and $x$ is divided by $y$  **41.** the product of $x$, $y$, and the sum of $x$ and $y$  **43.** $x + z$; 10  **45.** $y - z$; 2  **47.** $yz - 3$; 5
**49.** $\frac{xy}{z}$; 16  **51.** 1; 6  **53.** 3; $-1$  **55.** 4; 3  **57.** 3; $-4$  **59.** 4; 3  **61.** 19 and $x$  **63.** 29, $x$, $y$, and $z$  **65.** 3, $x$, $y$, and $z$
**67.** 17, $x$, and $z$  **69.** 5, 1, and 8  **71.** $x$ and $y$  **73.** 75  **75.** $x$ and $y$  **77.** $c + 4$  **79.** \$9,987$t$  **81.** $\frac{x}{5}$  **83.** \$(3d + 5)$

## Getting Ready (page 73)

**1.** 17  **2.** 17  **3.** 38.6  **4.** 38.6  **5.** 56  **6.** 56  **7.** 0  **8.** 1  **9.** 777  **10.** 777

## Exercise 1.7 (page 79)

**1.** $x + y^2 \geq z$  **3.** 0  **5.** positive  **7.** real  **9.** $a$  **11.** $(b + c)$  **13.** $ac$  **15.** $a$  **17.** element, multiplication  **19.** $\frac{1}{a}$  **21.** 10
**23.** $-24$  **25.** 144  **27.** 3  **29.** Both are 12.  **31.** Both are 29.  **33.** Both are 60.  **35.** Both are 0.  **37.** Both are $-6$.
**39.** Both are $-12$.  **41.** $3x + 3y$  **43.** $x^2 + 3x$  **45.** $-xa - xb$  **47.** $4x^2 + 4x$  **49.** $-5t - 10$  **51.** $-2ax - 2a^2$  **53.** $-2, \frac{1}{2}$
**55.** $-\frac{1}{3}, 3$  **57.** 0, none  **59.** $\frac{5}{2}, -\frac{2}{5}$  **61.** 0.2, $-5$  **63.** $-\frac{4}{3}, \frac{3}{4}$  **65.** comm. prop. of add.  **67.** comm. prop. of mult.
**69.** distrib. prop.  **71.** comm. prop. of add.  **73.** identity for mult.  **75.** add. inverse  **77.** $3x + 3 \cdot 2$  **79.** $xy^2$  **81.** $(y + x)z$
**83.** $x(yz)$  **85.** $x$

## Chapter 1 Summary (page 82)

**1. a.** 1, 2, 3, 4, 5  **b.** 2, 3, 5  **c.** 1, 3, 5  **d.** 4  **2. a.** $-6, 0, 5$  **b.** $-6, -\frac{2}{3}, 0, 2.6, 5$  **c.** 5  **d.** all of them  **e.** $-6, 0$  **f.** 5
**g.** $\sqrt{2}, \pi$  **3. a.** $<$  **b.** $<$  **c.** $=$  **d.** $>$  **4. a.** 8  **b.** $-8$  **5. a.** ⟨number line 14–20⟩
**b.** ⟨number line 19–25⟩  **c.** ⟨number line $-3$ to 2⟩  **d.** ⟨number line $-4$ to 3⟩  **6. a.** 11  **b.** 31  **7. a.** $\frac{5}{3}$  **b.** 11
**8. a.** $\frac{1}{3}$  **b.** $\frac{1}{3}$  **c.** 1  **d.** $\frac{5}{2}$  **e.** $\frac{4}{3}$  **f.** $\frac{1}{3}$  **g.** $\frac{10}{21}$  **h.** $\frac{73}{63}$  **i.** $\frac{11}{21}$  **j.** $\frac{2}{15}$  **k.** $8\frac{11}{12}$  **l.** $2\frac{11}{12}$  **9. a.** 48.61  **b.** 12.99  **c.** 18.55  **d.** 3.7
**10. a.** 4.70  **b.** 26.36  **c.** 3.57  **d.** 3.75  **11.** 6.85 hr  **12.** 57  **13.** 40.2 ft  **14. a.** 81  **b.** $\frac{4}{9}$  **c.** 0.25  **d.** 33  **15. a.** 81
**b.** 8  **16.** 15,133.6 ft$^3$  **17. a.** 32  **b.** 7  **c.** 6  **d.** 3  **e.** 98  **f.** 38  **g.** 3  **h.** 15  **18. a.** 58  **b.** 4  **c.** 7  **d.** 3  **19. a.** 22
**b.** 1  **20. a.** 15  **b.** $-57$  **c.** $-6.5$  **d.** $\frac{1}{2}$  **e.** $-12$  **f.** 16  **g.** 1.2  **h.** $-3.54$  **i.** 19  **j.** 1  **k.** $-5$  **l.** $-7$  **m.** $\frac{3}{2}$  **n.** 1  **o.** 1
**p.** $-\frac{1}{7}$  **21. a.** $-4$  **b.** $-1$  **c.** $-2$  **d.** 5  **e.** 4  **f.** 6  **22. a.** 12  **b.** 60  **c.** $\frac{1}{4}$  **d.** 1.3875  **e.** $-35$  **f.** $-105$  **g.** $-\frac{2}{3}$
**h.** $-45.14$  **i.** 5  **j.** 7  **k.** $\frac{7}{2}$  **l.** 6  **m.** $-5$  **n.** $-2$  **o.** 26  **p.** 7  **q.** 6  **r.** $\frac{3}{2}$  **23. a.** $-6$  **b.** 3  **c.** 2  **d.** 6  **e.** $-7$  **f.** 39
**g.** 6  **h.** $-2$  **24. a.** $xz$  **b.** $x + 2y$  **c.** $2(x + y)$  **d.** $x - yz$  **25. a.** the product of 3, $x$, and $y$  **b.** 5 decreased by the product
of $y$ and $z$  **c.** 5 less than the product of $y$ and $z$  **d.** the sum of $x$, $y$, and $z$, divided by twice their product  **26.** 3  **27.** 7  **28.** 1
**29.** 9  **30. a.** closure prop.  **b.** comm. prop. of mult.  **c.** assoc. prop. of add.  **d.** distrib. prop.  **e.** comm. prop. of add.
**f.** assoc. prop. of mult.  **g.** comm. prop. of add.  **h.** identity for mult.  **i.** add. inverse  **j.** identity for add.

## Chapter 1 Test  (page 88)

**1.** 31, 37, 41, 43, 47   **2.** 2   **3.**   **4.**   **5.** $-23$   **6.** 0
**7.** =   **8.** <   **9.** >   **10.** =   **11.** $\frac{13}{20}$   **12.** 1   **13.** $\frac{4}{5}$   **14.** $\frac{9}{2} = 4\frac{1}{2}$   **15.** $-1$   **16.** $-\frac{1}{13}$   **17.** 77.7   **18.** 301.57 ft$^2$   **19.** 64 cm$^2$
**20.** 1,539 in.$^3$   **21.** $-2$   **22.** $-14$   **23.** $-4$   **24.** 12   **25.** 5   **26.** $-23$   **27.** $\frac{xy}{x+y}$   **28.** $5y - (x + y)$   **29.** $24x + 14y$
**30.** $(12a + 8b)$   **31.** 3   **32.** 4   **33.** 0   **34.** 5   **35.** comm. prop. of mult.   **36.** distrib. prop.   **37.** comm. prop. of add.
**38.** mult. inverse prop.

## Getting Ready  (page 92)

**1.** $-3$   **2.** 7   **3.** 4   **4.** 7   **5.** 17   **6.** $x$

## Orals  (page 99)

**1.** 20   **2.** 16   **3.** 2   **4.** 4   **5.** 0   **6.** 0   **7.** $\frac{3}{5}$   **8.** 1   **9.** 80°   **10.** 100°

## Exercise 2.1  (page 100)

**1.** 15; integer, composite   **3.** $-1$; integer   **5.** closure prop. of add.   **7.** comm. prop. of add.   **9.** 64   **11.** 27   **13.** equation
**15.** root   **17.** equivalent   **19.** $x$   **21.** equal   **23.** markup   **25.** supplementary   **27.** yes   **29.** no   **31.** yes   **33.** yes   **35.** yes
**37.** no   **39.** yes   **41.** yes   **43.** yes   **45.** yes   **47.** 6   **49.** 19   **51.** 6   **53.** 519   **55.** 74   **57.** $-28$   **59.** 2   **61.** $\frac{5}{6}$   **63.** $-\frac{1}{5}$
**65.** $\frac{1}{2}$   **67.** $9,345   **69.** $90   **71.** $260   **73.** $53,000   **75.** $195   **77.** $145,149   **79.** 10°   **81.** 159°   **83.** 27°   **85.** 53°
**87.** 130°

## Getting Ready  (page 103)

**1.** 1   **2.** $\frac{1}{5}$   **3.** 1   **4.** 4   **5.** 4   **6.** 3   **7.** 63   **8.** 72

## Orals  (page 109)

**1.** 1   **2.** 1   **3.** $-2$   **4.** 0   **5.** 10   **6.** $-20$   **7.** $-12$   **8.** $-24$   **9.** 0.30   **10.** 8%

## Exercise 2.2  (page 109)

**1.** $\frac{22}{15}$   **3.** $\frac{25}{27}$   **5.** 14   **7.** $-317$   **9.** 25.2 ft$^2$   **11.** equal   **13.** $bc$   **15.** 100   **17.** 3   **19.** $-9$   **21.** 27   **23.** $-11$   **25.** 25
**27.** $-64$   **29.** 15   **31.** $-33$   **33.** $-2$   **35.** $-\frac{1}{2}$   **37.** 98   **39.** 5   **41.** $\frac{5}{2}$   **43.** 4,912   **45.** 1   **47.** 85   **49.** $-\frac{3}{2}$   **51.** 2.4   **53.** 80
**55.** 19   **57.** 320   **59.** 380   **61.** 150   **63.** 20%   **65.** 8%   **67.** 200%   **69.** 117   **71.** 1,519   **73.** 55%   **75.** $270   **77.** 5,600
**81.** about 3.16

## Getting Ready  (page 111)

**1.** 22   **2.** 36   **3.** 5   **4.** $\frac{13}{2}$   **5.** $-1$   **6.** $-1$   **7.** $\frac{7}{9}$   **8.** $-\frac{19}{3}$

## Orals  (page 117)

**1.** add 7   **2.** subtract 3   **3.** add 3   **4.** multiply by 7   **5.** add 5   **6.** subtract 5   **7.** multiply by 3   **8.** subtract 2   **9.** 3   **10.** 13

## Exercise 2.3  (page 118)

**1.** 50 cm   **3.** 80.325 in.$^2$   **5.** cost   **7.** percent   **9.** 1   **11.** $-1$   **13.** 3   **15.** $-2$   **17.** 2   **19.** $-5$   **21.** $\frac{3}{2}$   **23.** 2   **25.** 3
**27.** $-54$   **29.** $-9$   **31.** $-33$   **33.** 10   **35.** $-4$   **37.** 28   **39.** 5   **41.** 7   **43.** $-8$   **45.** 10   **47.** 4   **49.** 10   **51.** $\frac{17}{5}$   **53.** $-\frac{2}{3}$
**55.** 0   **57.** 6   **59.** $\frac{3}{5}$   **61.** 5   **63.** $250   **65.** 7 days   **67.** 29 min   **69.** $7,400   **71.** no chance; he needs 112   **73.** $50
**75.** 15% to 6%   **79.** $\frac{7x-3}{22} = \frac{1}{2}$

## Getting Ready  (page 120)

**1.** $3x + 4x$   **2.** $7x + 2x$   **3.** $8w - 3w$   **4.** $10y - 4y$   **5.** $7x$   **6.** $9x$   **7.** $5w$   **8.** $6y$

## Orals  (page 126)

**1.** $8x$   **2.** $y$   **3.** $0$   **4.** $-2y$   **5.** $12$   **6.** $6x$   **7.** $3$   **8.** impossible   **9.** $2$   **10.** $\frac{1}{3}$

## Exercise 2.4  (page 126)

**1.** $0$   **3.** $2$   **5.** $\frac{13}{56}$   **7.** $\frac{48}{35}$   **9.** variables, like   **11.** identity   **13.** $20x$   **15.** $3x^2$   **17.** $9x + 3y$   **19.** $7x + 6$   **21.** $7z - 15$
**23.** $12x + 121$   **25.** $6y + 62$   **27.** $-2x + 7y$   **29.** $2 + y$   **31.** $5x + 7$   **33.** $5x^2 + 24x$   **35.** $-2$   **37.** $3$   **39.** $1$   **41.** $1$   **43.** $\frac{1}{3}$
**45.** $2$   **47.** $6$   **49.** $35$   **51.** $-9$   **53.** $0$   **55.** $-20$   **57.** $-41$   **59.** $9$   **61.** $-1$   **63.** $8$   **65.** $5$   **67.** $4$   **69.** $-3$   **71.** $1$
**73.** identity   **75.** impossible equation   **77.** $16$   **79.** impossible equation   **81.** identity   **83.** identity   **89.** $0$

## Getting Ready  (page 128)

**1.** $(2x + 2)$ ft   **2.** $4x$ ft   **3.** $P = 2l + 2w$   **5.** $\$840$   **6.** $385$ mi   **7.** $5.6$ gal   **8.** $9.5$ lb

## Orals  (page 135)

**1.** $\$7d$   **2.** $\$18,000r$   **3.** $\frac{4}{6}$ ft   **4.** $\left(\frac{P}{2} - 9\right)$ ft

## Exercise 2.5  (page 136)

**1.** $200$ cm$^3$   **3.** $7x - 6$   **5.** $-\frac{3}{2}$   **7.** $\$1,488$   **9.** $2l + 2w$   **11.** vertex   **13.** $d = rt$   **15.** $4$ ft and $8$ ft   **17.** $19$ ft   **19.** $29$ m by $18$ m
**21.** $17$ in. by $39$ in.   **23.** $60°$   **25.** $\$4,500$ at 9% and $\$19,500$ at 8%   **27.** $\$3,750$   **29.** $\$5,000$   **31.** 6% and 7%   **33.** $3$ hr
**35.** $6.5$ hr   **37.** $7.5$ hr   **39.** $500$ mph   **41.** $20$ gal   **43.** $50$ gal   **45.** $7.5$ oz   **47.** $40$ lb lemon drops and $60$ lb jelly beans
**49.** $\$1.20$   **51.** $80$ lb

## Getting Ready  (page 140)

**1.** $3$   **2.** $-5$   **3.** $r$   **4.** $-a$   **5.** $7$   **6.** $12$   **7.** $d$   **8.** $s$

## Orals  (page 144)

**1.** $a = \frac{d-c}{b}$   **2.** $b = \frac{d-c}{a}$   **3.** $c = d - ab$   **4.** $d = ab + c$   **5.** $a = \frac{c}{d} - b$   **6.** $b = \frac{c}{d} - a$   **7.** $c = d(a + b)$   **8.** $d = \frac{c}{a+b}$

## Exercise 2.6  (page 145)

**1.** $5x - 5y$   **3.** $-x - 13$   **5.** literal   **7.** isolate   **9.** subtract   **11.** $I = E/R$   **13.** $w = V/(lh)$   **15.** $b = P - a - c$
**17.** $w = (P - 2l)/2$   **19.** $t = (A - P)/(Pr)$   **21.** $r = C/(2\pi)$   **23.** $w = 2gK/v^2$   **25.** $R = P/I^2$   **27.** $g = wv^2/(2K)$
**29.** $M = Fd^2/(Gm)$   **31.** $d^2 = GMm/F$   **33.** $r = G/(2b) + 1$ or $r = (G + 2b)/(2b)$   **35.** $t = \frac{d}{r}$; $t = 3$   **37.** $t = \frac{i}{pr}$; $t = 2$
**39.** $c = P - a - b$; $c = 3$   **41.** $h = \frac{2K}{a+b}$; $h = 8$   **43.** $I = E/R$; $I = 4$ amp   **45.** $r = C/(2\pi)$; $r = 2.28$ ft   **47.** $R = P/I^2$;
$R = 13.78$ ohms   **49.** $m = Fd^2/(GM)$   **51.** $D = (L - 3.25r - 3.25R)/2$; $D = 6$ ft   **55.** $90,000,000,000$ joules

## Getting Ready  (page 147)

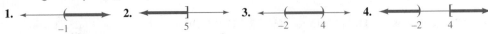

## Orals  (page 153)

**1.** $x < 2$   **2.** $x \geq 1$   **3.** $x \geq 2$   **4.** $x < -2$   **5.** $x < 6$   **6.** $x > -1$

## Exercise 2.7 (page 153)

**1.** $5x^2 - 2y^2$   **3.** $-x + 14$   **5.** is less than   **7.** $\geq$   **9.** inequality   **11.** $x > 3$;   **13.** $x \geq -10$;

**15.** $x < -1$;   **17.** $x \leq 4$;   **19.** $x < -2$;

**21.** $x < -4$;   **23.** $x \leq -1$;   **25.** $x \geq -13$;

**27.** $x > -3$;   **29.** $x < -2$;   **31.** $x \geq 2$;   **33.** $x > 3$;

**35.** $x > -15$;   **37.** $x \leq 20$;   **39.** $x \geq 3$;

**41.** $x > -7$;   **43.** $x \geq 4$;   **45.** $7 < x < 10$;

**47.** $-9 < x \leq 3$;   **49.** $-10 \leq x \leq 0$;   **51.** $-5 < x < -2$;

**53.** $-6 \leq x \leq 10$;   **55.** $2 \leq x < 3$;   **57.** $-1 \leq x < 2$;

**59.** $-4 < x < 1$;   **61.** $-2 < x < 2$;   **63.** $98\% \leq s \leq 100\%$   **65.** $r \geq 27$ mpg

**67.** $0$ ft $< s \leq 19$ ft   **69.** $0.1$ mi $\leq x \leq 2.5$ mi   **71.** $3.3$ mi $< x < 4.1$ mi   **73.** $66.2° < F < 71.6°$   **75.** $37.052$ in. $< C < 38.308$ in.
**77.** $68.18$ kg $< w < 86.36$ kg   **79.** $5$ ft $< w < 9$ ft

## Chapter 2 Summary (page 158)

**1. a.** yes   **b.** no   **c.** no   **d.** yes   **e.** yes   **f.** no   **2. a.** 1   **b.** 9   **c.** 16   **d.** 0   **e.** 4   **f.** $-2$   **3.** \$105.40   **4.** \$97.70   **5.** $21°$
**6.** $111°$   **7. a.** 5   **b.** $-2$   **c.** $\frac{1}{2}$   **d.** $\frac{3}{2}$   **e.** 18   **f.** $-35$   **g.** $-\frac{1}{2}$   **h.** 6   **8. a.** 245   **b.** 1,300   **c.** 37%   **d.** 12.5%   **9.** about 81%
**10. a.** 3   **b.** 2   **c.** 1   **d.** 1   **e.** 1   **f.** $-2$   **g.** 2   **h.** 7   **i.** $-2$   **j.** $-1$   **k.** 5   **l.** 3   **m.** 13   **n.** $-12$   **o.** 5   **p.** 7   **q.** 8
**r.** 30   **s.** $\frac{15}{2}$   **t.** 44   **11.** \$320   **12.** 6.5%   **13.** 96.4%   **14.** 53.8%   **15. a.** $14x$   **b.** $19a$   **c.** $5b$   **d.** $-2x$   **e.** $-2y$
**f.** not like terms   **g.** $9x$   **h.** $6 - 7x$   **i.** $4y^2 - 6$   **j.** 4   **16. a.** 7   **b.** 13   **c.** $-3$   **d.** $-41$   **e.** 9   **f.** $-7$   **g.** 7   **h.** 4   **i.** $-8$
**j.** $-18$   **17. a.** identity   **b.** contradiction   **c.** identity   **18.** 5 ft from one end   **19.** 13 in.   **20.** \$16,000 at 7%, \$11,000 at 9%
**21.** 20 min   **22.** 24 liters   **23.** 10 lb of each   **24.** 147 kwh   **25.** 85 ft   **26. a.** $R = \frac{E}{I}$   **b.** $t = \frac{i}{pr}$   **c.** $R = \frac{P}{I^2}$   **d.** $r = \frac{d}{t}$
**e.** $h = \frac{V}{lw}$   **f.** $m = \frac{y-b}{x}$   **g.** $h = \frac{V}{\pi r^2}$   **h.** $r = \frac{a}{2\pi h}$   **i.** $G = \frac{Fd^2}{Mm}$   **j.** $m = \frac{RT}{PV}$   **27. a.**   **b.**

**c.**   **d.**   **e.**   **f.**   **g.**

**h.**

## Chapter 2 Test (page 163)

**1.** solution   **2.** solution   **3.** not a solution   **4.** solution   **5.** $-36$   **6.** 47   **7.** $-12$   **8.** $-7$   **9.** $-2$   **10.** 1   **11.** 7   **12.** $-2$
**13.** $-3$   **14.** 0   **15.** $6x - 15$   **16.** $8x - 10$   **17.** $-18x$   **18.** $-36x^2 + 13x$   **19.** $\frac{3}{5}$ hr   **20.** $7\frac{1}{2}$ liters   **21.** $t = \frac{d}{r}$   **22.** $l = \frac{P-2w}{2}$
**23.** $h = \frac{A}{2\pi r}$   **24.** $r = \frac{A-P}{Pt}$   **25.**   **26.**   **27.**   **28.**

## Cumulative Review Exercises (page 163)

**1.** integer, rational, real, positive   **2.** rational, real, negative   **3.**   **4.**   **5.** 0

**6.** $\frac{10}{3}$   **7.** $8\frac{1}{10}$   **8.** 35.65   **9.** 0   **10.** $-2$   **11.** 16   **12.** 0   **13.** 24.75   **14.** 5,275   **15.** 5   **16.** 37, $y$   **17.** $-2x + 2y$   **18.** $x - 5$
**19.** $x^2y^3$   **20.** $4x^2$   **21.** 13   **22.** 41   **23.** $\frac{7}{4}$   **24.** $-11$   **25.** \$22,814.56   **26.** \$900   **27.** \$12,650   **28.** 125 lb   **29.** no
**30.** 7.3 and 10.7 ft   **31.** $h = \frac{2A}{b+B}$   **32.** $x = \frac{y-b}{m}$   **33.** $-9$   **34.** 1   **35.** 280   **36.** $-564$   **37.**
**38.**

## Getting Ready   (page 167)

**1.**  **2.** **3.** **4.**

## Orals   (page 176)

**2.** the origin   **3.** IV   **4.** *y*-axis

## Exercise 3.1   (page 176)

**1.** 12   **3.** 8   **5.** 7   **7.** −49   **9.** ordered pair   **11.** origin   **13.** rectangular coordinate   **15.** no   **17.** origin, left, up   **19.** II
**21.** 3 or −3, 5 or −5, 4 or −4, 5 or −5, 3 or −3, 5 or −5, 4 or −4   **23.** 10 minutes before the workout, her heart rate was 60
beats per min.   **25.** 150 beats per min   **27.** approximately 5 min and 50 min after starting   **29.** 10 beats per min faster after
cooldown   **31.**

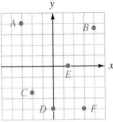

**33.** Carbondale (3, J), Champaign (4, D), Chicago (5, B), Peoria (3, C), Rockford (3, A),
Springfield (2, E), St. Louis (2, H)   **35. a.** 60°; 4 ft   **b.** 30°; 4 ft   **37. a.** $2   **b.** $4
**c.** $7   **d.** $9   **39.**

**a.** 35 mi
**b.** 4 gal
**c.** 32.5 mi

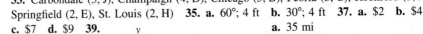

**41.**

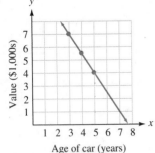

**a.** A 3-yr-old car is worth $7,000.   **b.** $1,000   **c.** 6 yr

## Getting Ready   (page 181)

**1.** 1   **2.** 5   **3.** −3   **4.** 2

## Orals   (page 192)

**1.** 3

## Exercise 3.2   (page 192)

**1.** −96   **3.** an expression   **5.** 1.25   **7.** 0.1   **9.** two   **11.** independent, dependent   **13.** linear   **15.** *y*-intercept   **17.** yes   **19.** no
**21.** −3, −2, −5   **23.** 0, −2, −6, 2, 4

**25.**

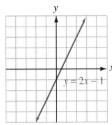

$y = 2x - 1$

**27.**

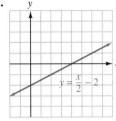

$y = \frac{x}{2} - 2$

**29.**

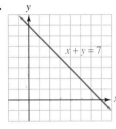

$x + y = 7$

**31.**

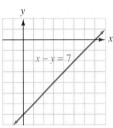

$x - y = 7$

**33.**

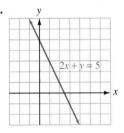

$2x + y = 5$

**35.**

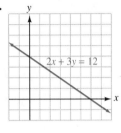

$2x + 3y = 12$

**37.**

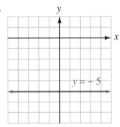

$y = -5$

**39.**

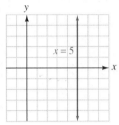

$x = 5$

**41.**

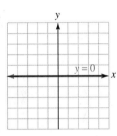

$y = 0$

**43.**

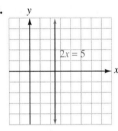

$2x = 5$

**45.**

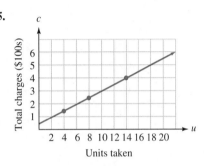

Total charges ($100s) vs. Units taken

**a.** $c = 50 + 25u$
**b.** 150, 250, 400
**c.** The service fee is $50.
**d.** $850

**47.**

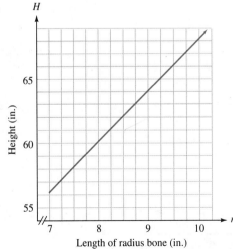

Height (in.) vs. Length of radius bone (in.)

**a.** 56.2, 62.1, 64.0   **b.** taller the woman is   **c.** 58 in.
**55.** (6, 6)   **57.** $\left(-\frac{1}{2}, \frac{5}{2}\right)$   **59.** (7, 6)

Getting Ready   (page 196)

**1.** $-3$   **2.** $-2$   **3.** 1   **4.** 6

Orals (page 205)

**1.** yes   **2.** yes   **3.** no   **4.** no

Exercise 3.3 (page 205)

**1.** 16   **3.** $-18$   **5.** system   **7.** independent   **9.** inconsistent   **11.** yes   **13.** yes   **15.** no   **17.** yes   **19.** no   **21.** no

**23.** 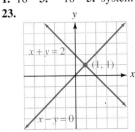   **25.** 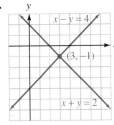   **27.** 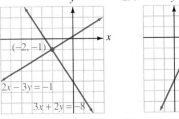   **29.**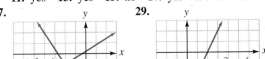

**31.** 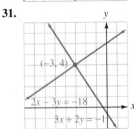   **33.** 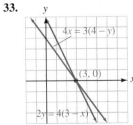   **35.** 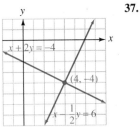   **37.**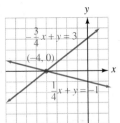

**39.** 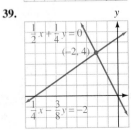   **41.**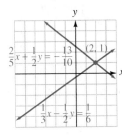

**43.** $(1, 3)$   **45.** $(0.67, -0.33)$   **47. a.** Donors outnumbered patients.
**b.** 1994; 4,100   **c.** The patients outnumber the donors.
**49. a.** Houston, New Orleans, St Augustine   **b.** St. Louis, Memphis, New Orleans   **c.** New Orleans

Getting Ready (page 209)

**1.** $6x + 4$   **2.** $-25 - 10x$   **3.** $2x - 4$   **4.** $3x - 12$

Orals (page 214)

**1.** $2z + 1$   **2.** $z + 2$   **3.** $3t + 3$   **4.** $\frac{t}{3} + 4$

Exercise 3.4 (page 214)

**1.** 5   **3.** 12   **5.** 10   **7.** $y$, terms   **9.** remove   **11.** infinitely many   **13.** $(2, 4)$   **15.** $(3, 0)$   **17.** $(-3, -1)$   **19.** inconsistent
system   **21.** $(-2, 3)$   **23.** $(3, 2)$   **25.** $(3, -2)$   **27.** $(-1, 2)$   **29.** $(-1, -1)$   **31.** dependent equations   **33.** $(1, 1)$   **35.** $(4, -2)$
**37.** $(-3, -1)$   **39.** $(-1, -3)$   **41.** $\left(\frac{1}{2}, \frac{1}{3}\right)$   **43.** $(1, 4)$   **45.** $(4, 2)$   **47.** $(-5, -5)$   **49.** $(-6, 4)$   **51.** $\left(\frac{1}{5}, 4\right)$   **53.** $(5, 5)$

Getting Ready (page 215)

**1.** $5x = 10$   **2.** $y = 6$   **3.** $2x = 33$   **4.** $18y = 28$

Orals  (page 221)

**1.** 1   **2.** 2   **3.** 3   **4.** 5

Exercise 3.5  (page 221)

**1.** 4   **3.** 4   **5.** $(-\infty, 2]$, 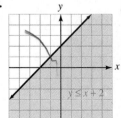 ⟵————⊢———⟶ 
2
  **7.** coefficient   **9.** general   **11.** 15   **13.** $(1, 4)$   **15.** $(-2, 3)$   **17.** $(-1, 1)$

**19.** $(2, 5)$   **21.** $(-3, 4)$   **23.** $(0, 8)$   **25.** $(2, 3)$   **27.** $(3, -2)$   **29.** $(2, 7)$   **31.** inconsistent system   **33.** dependent equations
**35.** $(4, 0)$   **37.** $\left(\frac{10}{3}, \frac{10}{3}\right)$   **39.** $(5, -6)$   **41.** $(-5, 0)$   **43.** $(-1, 2)$   **45.** $\left(1, -\frac{5}{2}\right)$   **47.** $(-1, 2)$   **49.** $(0, 1)$   **51.** $(-2, 3)$   **53.** $(2, 2)$
**57.** $(1, 4)$

Getting Ready  (page 223)

**1.** $x + y$   **2.** $x - y$   **3.** $xy$   **4.** $\frac{x}{y}$   **5.** $A = lw$   **6.** $P = 2l + 2w$

Orals  (page 232)

**1.** $2x$   **2.** $y + 1$   **3.** $2x + 3y$   **4.** $\$(3x + 2y)$   **5.** $\$(4x + 5y)$

Exercise 3.6  (page 232)

**1.** ⟵———————⟶ 
4
   **3.** ⟵——⊢———⊣——⟶ 
-1   2
   **5.** $8^3 c$   **7.** $a^2 b^2$   **9.** variable   **11.** system   **13.** 32, 64   **15.** 5, 8   **17.** 140

**19.** $15, $5   **21.** $5.40, $6.20   **23.** 10 ft, 15 ft   **25.** $100,000   **27.** causes: 24 min; outcome: 6 min   **29.** 90,000 accidents;
540,000 cancer   **31.** 25 ft by 30 ft   **33.** 60 ft$^2$   **35.** 9.9 yr   **37.** 80+   **39.** $2,000   **41.** 250   **43.** 10 mph   **45.** 50 mph
**47.** 5 L of 40% solution, 10 L of 55% solution   **49.** 32 lb peanuts, 16 lb cashews   **51.** 15   **53.** 9%   **57.** 2

Getting Ready  (page 236)

**1.** below   **2.** above   **3.** below   **4.** on   **5.** on   **6.** above   **7.** below   **8.** above

Orals  (page 245)

**1.** no   **2.** no   **3.** yes   **4.** yes   **5.** no   **6.** yes   **7.** no   **8.** yes

Exercise 3.7  (page 245)

**1.** 3   **3.** $t = \frac{A - P}{Pr}$   **5.** $7a - 15$   **7.** $-2a + 7b$   **9.** inequality   **11.** boundary   **13.** inequalities   **15.** doubly shaded   **17. a.** yes
**b.** no   **c.** yes   **d.** no   **19. a.** no   **b.** yes
**21.**

**23.**
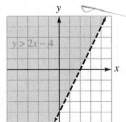
**25.**
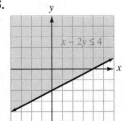
**27.**
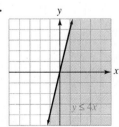

**29.**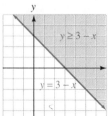

**31.**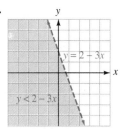

**33.**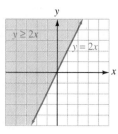

**35.**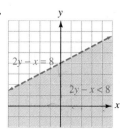

**37.**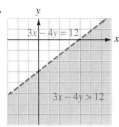

**39.**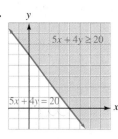

**41.**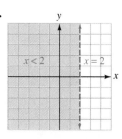

**43.**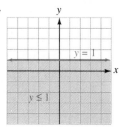

**45.** $(10, 10)$, $(20, 10)$, $(10, 20)$

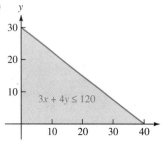

**47.** $(50, 50)$, $(30, 40)$, $(40, 40)$

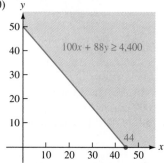

**49.** $(80, 40)$, $(80, 80)$, $(120, 40)$

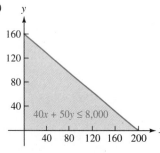

**51.**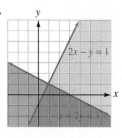

**53.**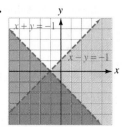

**55.**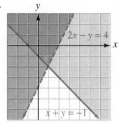

**57.**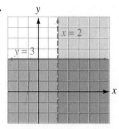

**59.**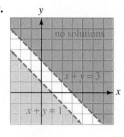

**61.**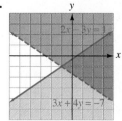

**63.**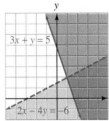

**65.**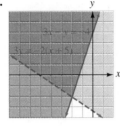

**67.**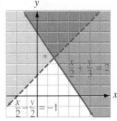

**69.** 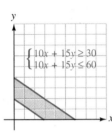 1 $10 CD and 2 $15 CDs; **71.** 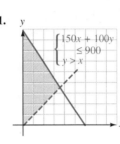 2 desk chairs and 4 side chairs;
4 $10 CDs and 1 $15 CD   1 desk chair and 5 side chairs

## Chapter 3 Summary  (page 253)

**1.**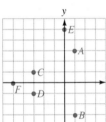

**2. a.** $(3, 1)$ **b.** $(-4, 5)$ **c.** $(-3, -4)$ **d.** $(2, -3)$ **e.** $(0, 0)$ **f.** $(0, 4)$ **g.** $(-5, 0)$ **h.** $(0, -3)$
**3. a.** no **b.** yes

**4. a.** 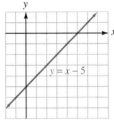 **b.** 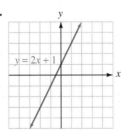 **c.** 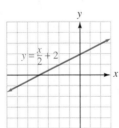 **d.**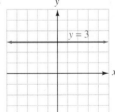

**e.** 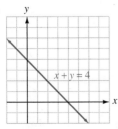 **f.** 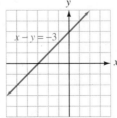 **g.** 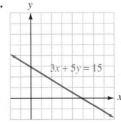 **h.**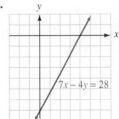

**5. a.** yes   **6. a.**
**b.** no
**c.** yes
**d.** yes

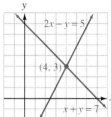

**b.**

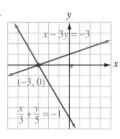

**c.**

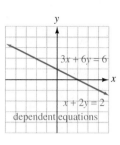

**d.**

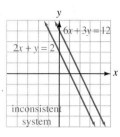

**7. a.** $(-1, -2)$   **b.** $(-2, 5)$   **c.** $(1, -1)$   **d.** $(-2, 1)$   **8. a.** $(3, -5)$   **b.** $\left(3, \frac{1}{2}\right)$   **c.** $(-1, 7)$   **d.** $\left(-\frac{1}{2}, \frac{7}{2}\right)$   **e.** $(0, 9)$
**f.** inconsistent system   **g.** dependent equations   **h.** $(0, 0)$   **9.** 3, 15   **10.** 3 ft by 9 ft   **11.** 50¢   **12.** $66   **13.** $1.69   **14.** $750

**15. a.**

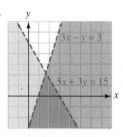

**b.**

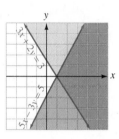

**16. a.**

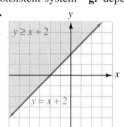

**b.**

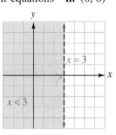

**c.**

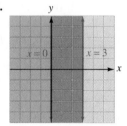

**d.**

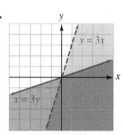

## Chapter 3 Test   (page 258)

**1.**

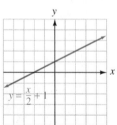

**2.**

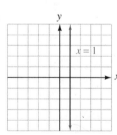

**3.**

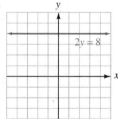

**4.**

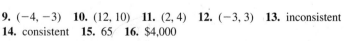

**5.** yes   **6.** no

**7.**

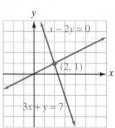

**8.**

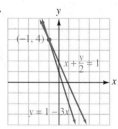

**9.** $(-4, -3)$   **10.** $(12, 10)$   **11.** $(2, 4)$   **12.** $(-3, 3)$   **13.** inconsistent
**14.** consistent   **15.** 65   **16.** $4,000

**17.**
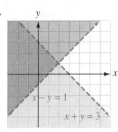

**18.**
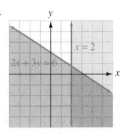

## Getting Ready (page 261)

**1.** 8   **2.** 9   **3.** 6   **4.** 6   **5.** 12   **6.** 32   **7.** 18   **8.** 3

## Orals (page 267)

**1.** base $x$, exponent 3   **2.** base 3, exponent $x$   **3.** base $b$, exponent $c$   **4.** base $ab$, exponent $c$   **5.** 36   **6.** 36   **7.** 9   **8.** 27

## Exercise 4.1 (page 268)

**1.**   **3.** the product of 3 and the sum of $x$ and $y$   **5.** $|2x| + 3$   **7.** $-5, 3$
**9.** $(3x)(3x)(3x)(3x)$   **11.** $y \cdot y \cdot y \cdot y \cdot y$   **13.** $x^n y^n$   **15.** $a^{b \cdot c}$   **19.** base 4, exponent 3   **21.** base $x$, exponent 5   **23.** base $2y$, exponent 3   **25.** base $x$, exponent 4   **27.** base $x$, exponent 1   **29.** base $x$, exponent 3   **31.** $5 \cdot 5 \cdot 5$   **33.** $x \cdot x \cdot x \cdot x \cdot x \cdot x \cdot x$
**35.** $-4 \cdot x \cdot x \cdot x \cdot x \cdot x$   **37.** $(3t)(3t)(3t)(3t)(3t)$   **39.** $2^3$   **41.** $x^4$   **43.** $(2x)^3$   **45.** $-4t^4$   **47.** 625   **49.** 13   **51.** 561   **53.** $-725$
**55.** $x^7$   **57.** $x^{10}$   **59.** $t^3$   **61.** $a^{12}$   **63.** $y^9$   **65.** $12x^7$   **67.** $-4y^5$   **69.** $6x^9$   **71.** $3^8$   **73.** $y^{15}$   **75.** $a^{21}$   **77.** $x^{25}$   **79.** $243z^{30}$
**81.** $x^{31}$   **83.** $r^{36}$   **85.** $s^{33}$   **87.** $x^3 y^3$   **89.** $r^6 s^4$   **91.** $16a^2 b^4$   **93.** $-8r^6 s^9 t^3$   **95.** $\dfrac{a^3}{b^3}$   **97.** $\dfrac{x^{10}}{y^{15}}$   **99.** $\dfrac{-32a^5}{b^5}$   **101.** $\dfrac{b^6}{27a^3}$
**103.** $x^2$   **105.** $y^4$   **107.** $3a$   **109.** $ab^4$   **111.** $\dfrac{10r^{13} s^3}{3}$   **113.** $\dfrac{x^{12} y^{16}}{2}$   **115.** $\dfrac{y^3}{8}$   **117.** $-\dfrac{8r^3}{27}$   **119.** 2 ft   **121.** \$16,000

## Getting Ready (page 270)

**1.** $\frac{1}{3}$   **2.** $\frac{1}{y}$   **3.** 1   **4.** $\frac{1}{xy}$

## Orals (page 275)

**1.** $\frac{1}{2}$   **2.** $\frac{1}{4}$   **3.** 2   **4.** 1   **5.** $x$   **6.** $\frac{1}{y^7}$   **7.** 1   **8.** $\frac{y}{x}$

## Exercise 4.2 (page 275)

**1.** 2   **3.** $\frac{6}{5}$   **5.** $s = \dfrac{f(P-L)}{i}$ or $s = \dfrac{fP-fL}{i}$   **7.** 1   **9.** 1   **11.** 8   **13.** 1   **15.** 1   **17.** 512   **19.** 2   **21.** 1   **23.** 1   **25.** $-2$
**27.** $\dfrac{1}{x^2}$   **29.** $\dfrac{1}{b^5}$   **31.** $\dfrac{1}{16y^4}$   **33.** $\dfrac{1}{a^3 b^6}$   **35.** $\dfrac{1}{y}$   **37.** $\dfrac{1}{r^6}$   **39.** $y^5$   **41.** 1   **43.** $\dfrac{1}{a^2 b^4}$   **45.** $\dfrac{1}{x^6 y^3}$   **47.** $\dfrac{1}{x^3}$   **49.** $\dfrac{1}{y^2}$   **51.** $a^8 b^{12}$
**53.** $-\dfrac{y^{10}}{32x^{15}}$   **55.** $a^{14}$   **57.** $\dfrac{1}{b^{14}}$   **59.** $\dfrac{256x^{28}}{81}$   **61.** $\dfrac{16y^{14}}{z^{10}}$   **63.** $\dfrac{x^{14}}{128y^{28}}$   **65.** $\dfrac{16u^4 v^8}{81}$   **67.** $\dfrac{1}{9a^2 b^2}$   **69.** $\dfrac{c^{15}}{216a^9 b^3}$   **71.** $\dfrac{1}{512}$
**73.** $\dfrac{17y^{27} z^5}{x^{35}}$   **75.** $x^{3m}$   **77.** $u^{5m}$   **79.** $y^{2m+2}$   **81.** $y^m$   **83.** $\dfrac{1}{x^{3n}}$   **85.** $x^{2m+2}$   **87.** $x^{8n-12}$   **89.** $y^{4n-8}$   **91.** \$6,678.04
**93.** \$3,183.76

## Getting Ready (page 277)

**1.** 100   **2.** 1,000   **3.** 10   **4.** $\frac{1}{100}$   **5.** 500   **6.** 8,000   **7.** 30   **8.** $\frac{7}{100}$

## Orals (page 282)

**1.** $3.72 \times 10^2$ **2.** 37.2 **3.** $4.72 \times 10^3$ **4.** $3.72 \times 10^3$ **5.** $3.72 \times 10^{-1}$ **6.** $2.72 \times 10^{-2}$

## Exercise 4.3 (page 282)

**1.** 5 **3.** comm. prop. of add. **5.** 6 **7.** scientific notation **9.** $2.3 \times 10^4$ **11.** $1.7 \times 10^6$ **13.** $6.2 \times 10^{-2}$ **15.** $5.1 \times 10^{-6}$
**17.** $4.25 \times 10^3$ **19.** $2.5 \times 10^{-3}$ **21.** 230 **23.** 812,000 **25.** 0.00115 **27.** 0.000976 **29.** 25,000,000 **31.** 0.00051
**33.** 714,000 **35.** 30,000 **37.** 200,000 **39.** $2.57 \times 10^{13}$ mi **41.** 114,000,000 mi **43.** $6.22 \times 10^{-3}$ mi **45.** $1.9008 \times 10^{11}$ ft
**47.** $3.3 \times 10^{-1}$ km/sec

## Getting Ready (page 284)

**1.** $2x^2y^3$ **2.** $3xy^3$ **3.** $2x^2 + 3y^2$ **4.** $x^3 + y^3$ **5.** $6x^3y^3$ **6.** $5x^2y^2z^4$ **7.** $5x^2y^2$ **8.** $x^3y^3z^3$

## Exercise 4.4 (page 294)

**1.** 8 **3.** ◄——|——► **5.** $x^{18}$ **7.** $y^9$ **9.** algebraic **11.** polynomial **13.** trinomial **15.** degree **17.** function
 $\quad\quad$ $-3$

**19.** domain **21.** yes **23.** yes **25.** binomial **27.** trinomial **29.** monomial **31.** binomial **33.** trinomial **35.** none of these
**37.** 4th **39.** 3rd **41.** 8th **43.** 6th **45.** 12th **47.** 0th **49.** 7 **51.** $-8$ **53.** $-4$ **55.** $-5$ **57.** 3 **59.** 11
**61.** $1, -2, -3, -2, 1$ **63.** $-6, 1, 2, 3, 10$ **65.** **67.** **69.** 64 ft **71.** 63 ft

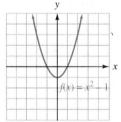

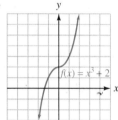

## Getting Ready (page 297)

**1.** $5x$ **2.** $2y$ **3.** $25x$ **4.** $5z$ **5.** $12r$ **6.** not possible **7.** 0 **8.** not possible

## Orals (page 301)

**1.** $4x^3$ **2.** $4xy$ **3.** $2y$ **4.** $2 - 2x$ **5.** $-2y^2$ **6.** $6x^2 + 4y$ **7.** $4x^2 + y$ **8.** $2y$

## Exercise 4.5 (page 302)

**1.** $-8$ **3.** $-9$ **5.** ◄——|——► **7.** monomial **9.** coefficients, variables **11.** like terms **13.** like terms, $7y$
 $\quad\quad\quad\quad\quad$ 3

**15.** unlike terms **17.** like terms, $13x^3$ **19.** like terms, $8x^3y^2$ **21.** like terms, $65t^6$ **23.** unlike terms **25.** $9y$ **27.** $-12t^2$
**29.** $16u^3$ **31.** $7x^5y^2$ **33.** $14rst$ **35.** $-6a^2bc$ **37.** $15x^2$ **39.** $4x^2y^2$ **41.** $95x^8y^4$ **43.** $7x + 4$ **45.** $2a + 7$ **47.** $7x - 7y$
**49.** $-19x - 4y$ **51.** $6x^2 + x - 5$ **53.** $7b + 4$ **55.** $3x + 1$ **57.** $5x + 15$ **59.** $3x - 3y$ **61.** $5x^2 - 25x - 20$
**63.** $5x^2 + x + 11$ **65.** $-7x^3 - 7x^2 - x - 1$ **67.** $2x^2y + xy + 13y^2$ **69.** $5x^2 + 6x - 8$ **71.** $-x^3 + 6x^2 + x + 14$
**73.** $-12x^2y^2 - 13xy + 36y^2$ **75.** $6x^2 - 2x - 1$ **77.** $t^3 + 3t^2 + 6t - 5$ **79.** $-3x^2 + 5x - 7$ **81.** $6x - 2$
**83.** $-5x^2 - 8x - 19$ **85.** $4y^3 - 12y^2 + 8y + 8$ **87.** $3a^2b^2 - 6ab + b^2 - 6ab^2$ **89.** $-6x^2y^2 + 4xy^2z - 20xy^3 + 2y$
**91.** \$114,000 **93.** \$132,000 **95. a.** \$263,000 **b.** \$263,000 **97.** $y = -1,100x + 6,600$ **99.** $y = -2,800x + 15,800$
**103.** $6x + 3h - 10$ **105.** 49

## Getting Ready (page 305)

**1.** $6x$ **2.** $3x^4$ **3.** $5x^3$ **4.** $8x^5$ **5.** $3x + 15$ **6.** $x^2 + 5x$ **7.** $4y - 12$ **8.** $2y^2 - 6y$

## Orals (page 313)

**1.** $6x^3 - 2x^2$ **2.** $10y^3 - 15y$ **3.** $7x^2y + 7xy^2$ **4.** $-4xy + 6y^2$ **5.** $x^2 + 5x + 6$ **6.** $x^2 - x - 6$ **7.** $2x^2 + 7x + 6$ **8.** $9x^2 - 1$
**9.** $x^2 + 6x + 9$ **10.** $x^2 - 10x + 25$

## Exercise 4.6 (page 313)

**1.** distrib. prop. **3.** comm. prop. of mult. **5.** 0 **7.** monomial **9.** trinomial **11.** $6x^2$ **13.** $15x$ **15.** $12x^5$ **17.** $-24b^6$
**19.** $6x^5y^5$ **21.** $-3x^4y^7z^8$ **23.** $x^{10}y^{15}$ **25.** $a^5b^4c^7$ **27.** $3x + 12$ **29.** $-4t - 28$ **31.** $3x^2 - 6x$ **33.** $-6x^4 + 2x^3$
**35.** $3x^2y + 3xy^2$ **37.** $6x^4 + 8x^3 - 14x^2$ **39.** $2x^7 - x^2$ **41.** $-6r^3t^2 + 2r^2t^3$ **43.** $-6x^4y^4 - 6x^3y^5$ **45.** $a^2 + 9a + 20$
**47.** $3x^2 + 10x - 8$ **49.** $6a^2 + 2a - 20$ **51.** $6x^2 - 7x - 5$ **53.** $2x^2 + 3x - 9$ **55.** $6s^2 + 7st - 3t^2$ **57.** $x^2 + xz + xy + yz$
**59.** $u^2 + 2tu + uv + 2tv$ **61.** $-4r^2 - 20rs - 21s^2$ **63.** $4x^2 + 11x + 6$ **65.** $12x^2 + 14xy - 10y^2$ **67.** $x^3 - 1$
**69.** $2x^3 + 7x^2 + x - 1$ **71.** $x^2 + 8x + 16$ **73.** $t^2 - 6t + 9$ **75.** $r^2 - 16$ **77.** $x^2 + 10x + 25$ **79.** $4s^2 + 4s + 1$
**81.** $16x^2 - 25$ **83.** $x^2 - 4xy + 4y^2$ **85.** $4a^2 - 12ab + 9b^2$ **87.** $16x^2 - 25y^2$ **89.** $2x^2 - 6x - 8$ **91.** $3a^3 - 3ab^2$
**93.** $4t^3 + 11t^2 + 18t + 9$ **95.** $-3x^3 + 25x^2y - 56xy^2 + 16y^3$ **97.** $x^3 - 8y^3$ **99.** $5t^2 - 11t$ **101.** $x^2y + 3xy^2 + 2x^2$
**103.** $2x^2 + xy - y^2$ **105.** $8x$ **107.** $5s^2 - 7s - 9$ **109.** $-3$ **111.** $-8$ **113.** $-1$ **115.** 0 **117.** 1 **119.** 4 m **121.** 90 ft

## Getting Ready (page 316)

**1.** $2xy^2$ **2.** $y$ **3.** $\frac{3xy}{2}$ **4.** $\frac{x}{y}$ **5.** $xy$ **6.** 3

## Orals (page 321)

**1.** $2x^2$ **2.** $2y$ **3.** $5bc^2$ **4.** $-2pq$ **5.** 1 **6.** $3x$

## Exercise 4.7 (page 321)

**1.** binomial **3.** none of these **5.** 2 **7.** polynomial **9.** two **11.** $\frac{a}{b}$ **13.** $\frac{1}{3}$ **15.** $-\frac{5}{3}$ **17.** $\frac{3}{4}$ **19.** 1 **21.** $-\frac{1}{4}$ **23.** $\frac{42}{19}$ **25.** $\frac{x}{z}$
**27.** $\frac{r^2}{s}$ **29.** $\frac{2x^2}{y}$ **31.** $-\frac{3u^3}{v^2}$ **33.** $\frac{4r}{y^2}$ **35.** $-\frac{13}{3rs}$ **37.** $\frac{x^4}{y^6}$ **39.** $a^8b^8$ **41.** $-\frac{3r}{s^9}$ **43.** $-\frac{x^3}{4y^3}$ **45.** $\frac{125}{8b^3}$ **47.** $\frac{xy^2}{3}$ **49.** $a^8$
**51.** $z^3$ **53.** $\frac{2}{y} + \frac{3}{x}$ **55.** $\frac{1}{5y} - \frac{2}{5x}$ **57.** $\frac{1}{y^2} + \frac{2y}{x^2}$ **59.** $3a - 2b$ **61.** $\frac{1}{y} - \frac{1}{2x} + \frac{2z}{xy}$ **63.** $3x^2y - 2x - \frac{1}{y}$ **65.** $5x - 6y + 1$
**67.** $\frac{10x^2}{y} - 5x$ **69.** $-\frac{4x}{3} + \frac{3x^2}{2}$ **71.** $xy - 1$ **73.** $\frac{x}{y} - \frac{11}{6} + \frac{y}{2x}$ **75.** 2 **77.** yes **79.** yes

## Getting Ready (page 323)

**1.** 13 **2.** 21 **3.** 19 **4.** 13

## Orals (page 328)

**1.** $2 + \frac{3}{x}$ **2.** $3 - \frac{5}{x}$ **3.** $2 + \frac{1}{x+1}$ **4.** $3 + \frac{2}{x+1}$ **5.** $x$ **6.** $x$

## Exercise 4.8 (page 328)

**1.** 21, 22, 24, 25, 26, 27, 28 **3.** 5 **5.** $-5$ **7.** $8x^2 - 6x + 1$ **9.** divisor, dividend **11.** remainder **13.** $4x^3 - 2x^2 + 7x + 6$
**15.** $6x^4 - x^3 + 2x^2 + 9x$ **17.** $0x^3$ and $0x$ **19.** $x + 2$ **21.** $y + 12$ **23.** $a + b$ **25.** $3a - 2$ **27.** $b + 3$ **29.** $x - 3y$
**31.** $2x + 1$ **33.** $x - 7$ **35.** $3x + 2y$ **37.** $2x - y$ **39.** $x + 5y$ **41.** $x - 5y$ **43.** $x^2 + 2x - 1$ **45.** $2x^2 + 2x + 1$

**47.** $x^2 + xy + y^2$   **49.** $x + 1 + \frac{-1}{2x+3}$   **51.** $2x + 2 + \frac{-3}{2x+1}$   **53.** $x^2 + 2x + 1$   **55.** $x^2 + 2x - 1 + \frac{6}{2x+3}$   **57.** $2x^2 + 8x + 14 + \frac{31}{x-2}$
**59.** $x + 1$   **61.** $2x - 3$   **63.** $x^2 - x + 1$   **65.** $a^2 - 3a + 10 + \frac{-30}{a+3}$   **67.** $5x^2 - x + 4 + \frac{16}{3x-4}$

## Chapter Summary  (page 331)

**1. a.** $(-3x)(-3x)(-3x)(-3x)$   **b.** $\left(\frac{1}{2}pq\right)\left(\frac{1}{2}pq\right)\left(\frac{1}{2}pq\right)$   **2. a.** 125   **b.** 243   **c.** 64   **d.** $-64$   **e.** 13   **f.** 25   **3. a.** $x^5$   **b.** $x^9$   **c.** $y^{21}$
**d.** $x^{42}$   **e.** $a^3b^3$   **f.** $81x^4$   **g.** $b^{12}$   **h.** $-y^2z^5$   **i.** $256s^3$   **j.** $-3y^6$   **k.** $x^{15}$   **l.** $4x^4y^2$   **m.** $x^4$   **n.** $\frac{x^2}{y^2}$   **o.** $\frac{2y^2}{x^2}$   **p.** $5yz^4$

**4. a.** 1   **b.** 1   **c.** 9   **d.** $9x^4$   **e.** $\frac{1}{x^3}$   **f.** $x$   **g.** $y$   **h.** $x^{10}$   **i.** $\frac{1}{x^2}$   **j.** $\frac{a^6}{b^3}$   **k.** $\frac{1}{x^5}$   **l.** $\frac{1}{9z^2}$   **5. a.** $7.28 \times 10^2$   **b.** $9.37 \times 10^3$
**c.** $1.36 \times 10^{-2}$   **d.** $9.42 \times 10^{-3}$   **e.** $7.73 \times 10^0$   **f.** $7.53 \times 10^5$   **g.** $1.8 \times 10^{-4}$   **h.** $6 \times 10^4$   **6. a.** 726,000   **b.** 0.000391
**c.** 2.68   **d.** 57.6   **e.** 7.39   **f.** 0.000437   **g.** 0.03   **h.** 160   **7. a.** 7th, monomial   **b.** 2nd, binomial   **c.** 5th, trinomial
**d.** 5th, binomial   **8. a.** 11   **b.** 2   **c.** $-4$   **d.** 4   **9. a.** 402   **b.** 0   **c.** 82   **d.** 0.3405   **10. a.** $-4$   **b.** 21   **c.** 0   **d.** $-\frac{15}{4}$
**11. a.**                        **b.**                        **12. a.** $7x$   **b.** in simplest terms   **c.** $4x^2y^2$   **d.** $x^2yz$   **e.** $8x^2 - 6x$

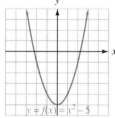

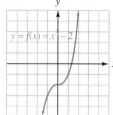

**f.** $4a^2 + 4a - 6$   **g.** $5x^2 + 19x + 3$   **h.** $6x^3 + 8x^2 + 3x - 72$   **13. a.** $10x^3y^5$   **b.** $x^7yz^5$   **14. a.** $5x + 15$   **b.** $6x + 12$
**c.** $3x^4 - 5x^2$   **d.** $2y^4 + 10y^3$   **e.** $-x^2y^3 + x^3y^2$   **f.** $-3x^2y^2 + 3x^2y$   **15. a.** $x^2 + 5x + 6$   **b.** $2x^2 - x - 1$   **c.** $6a^2 - 6$
**d.** $6a^2 - 6$   **e.** $2a^2 - ab - b^2$   **f.** $6x^2 + xy - y^2$   **16. a.** $x^2 + 6x + 9$   **b.** $x^2 - 25$   **c.** $y^2 - 4$   **d.** $x^2 + 8x + 16$
**e.** $x^2 - 6x + 9$   **f.** $y^2 - 2y + 1$   **g.** $4y^2 + 4y + 1$   **h.** $y^4 - 1$   **17. a.** $3x^3 + 7x^2 + 5x + 1$   **b.** $8a^3 - 27$   **18. a.** 1   **b.** $-1$
**c.** 7   **d.** 5   **e.** 1   **f.** 0   **19. a.** $\frac{3}{2y} + \frac{3}{x}$   **b.** $2 - \frac{3}{y}$   **c.** $-3a - 4b + 5c$   **d.** $-\frac{x}{y} - \frac{y}{x}$   **20. a.** $x + 1 + \frac{3}{x+2}$   **b.** $x - 5$   **c.** $2x + 1$
**d.** $x + 5 + \frac{3}{3x-1}$   **e.** $3x^2 + 2x + 1 + \frac{2}{2x-1}$   **f.** $3x^2 - x - 4$

## Chapter 4 Test  (page 334)

**1.** $2x^3y^4$   **2.** 134   **3.** $y^6$   **4.** $6b^7$   **5.** $32x^{21}$   **6.** $8r^{18}$   **7.** 3   **8.** $\frac{2}{y^3}$   **9.** $y^3$   **10.** $\frac{64a^3}{b^3}$   **11.** $2.8 \times 10^4$   **12.** $2.5 \times 10^{-3}$
**13.** 7,400   **14.** 0.000093   **15.** binomial   **16.** 10th degree   **17.** 0   **18.**                        **19.** $-7x + 2y$   **20.** $-3x + 6$

**21.** $5x^3 + 2x^2 + 2x - 5$   **22.** $-x^2 - 5x + 4$   **23.** $-4x^5y$   **24.** $3y^4 - 6y^3 + 9y^2$   **25.** $6x^2 - 7x - 20$   **26.** $2x^3 - 7x^2 + 14x - 12$
**27.** $\frac{1}{2}$   **28.** $\frac{y}{2x}$   **29.** $\frac{a}{4b} - \frac{b}{2a}$   **30.** $x - 2$

## Cumulative Review Exercises  (page 335)

**1.** 11   **2.** 71   **3.** $-\frac{11}{10}$   **4.** 7   **5.** 15   **6.** 4   **7.** $-10$   **8.** $-6$   **9.**                        **10.**

**11.**                    **12.**                    **13.** $r = \frac{A-P}{pt}$   **14.** $h = \frac{2A}{b}$

**15.**

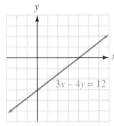

**16.**

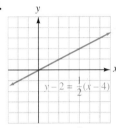

**17.**

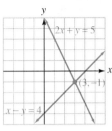

**18.**

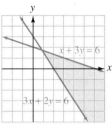

**19.** $(4, -3)$   **20.** $\left(\frac{1}{2}, \frac{2}{3}\right)$   **21.** $y^{14}$   **22.** $xy$   **23.** $\dfrac{a^7}{b^6}$   **24.** $x^2y^2$   **25.** $x^2 + 4x - 14$   **26.** $6x^2 + 10x - 56$   **27.** $x^3 - 8$   **28.** $2x + 1$

**29.** $4.8 \times 10^{18}$ m   **30.** 4 in.   **31.** 879.6 in.$^2$   **32.** \$512

## Getting Ready (page 339)

**1.** $5x + 15$   **2.** $7y - 56$   **3.** $3x^2 - 2x$   **4.** $5y^2 + 9y$   **5.** $ab + 9a$   **6.** $3x + x^2 + xy$   **7.** $x^2y - 4xy$   **8.** $2x^2y^2 - 5xy^3$

## Orals (page 345)

**1.** $2^2 \cdot 3^2$   **2.** $3^3$   **3.** $3^4$   **4.** $3^2 \cdot 5$   **5.** 3   **6.** $3ab$   **7.** $5(3xy + 2)$   **8.** $5xy(3 + 2y)$

## Exercise 5.1 (page 345)

**1.** 7   **3.** 11   **5.** prime   **7.** largest   **9.** 0, 0   **11.** $2^2 \cdot 3$   **13.** $3 \cdot 5$   **15.** $2^3 \cdot 5$   **17.** $2 \cdot 7^2$   **19.** $3^2 \cdot 5^2$   **21.** $2^5 \cdot 3^2$   **23.** 4
**25.** $r^2$   **27.** 4, $x$   **29.** $3(x + 2)$   **31.** $x(y - z)$   **33.** $t^2(t + 2)$   **35.** $r^2(r^2 - 1)$   **37.** $a^2b^3z^2(az - 1)$   **39.** $8xy^2z^3(3xyz + 1)$
**41.** $6uvw^2(2w - 3v)$   **43.** $3(x + y - 2z)$   **45.** $a(b + c - d)$   **47.** $2y(2y + 4 - x)$   **49.** $3r(4r - s + 3rs^2)$   **51.** $abx(1 - b + x)$
**53.** $2xyz^2(2xy - 3y + 6)$   **55.** $7a^2b^2c^2(10a + 7bc - 3)$   **57.** $-(a + b)$   **59.** $-(2x - 5y)$   **61.** $-(2a - 3b)$   **63.** $-(3m + 4n - 1)$
**65.** $-(3xy - 2z - 5w)$   **67.** $-(3ab + 5ac - 9bc)$   **69.** $-3xy(x + 2y)$   **71.** $-4a^2b^2(b - 3a)$   **73.** $-2ab^2c(2ac - 7a + 5c)$
**75.** $-7ab(2a^5b^5 - 7ab^2 + 3)$   **77.** $-5a^2b^3c(1 - 3abc + 5a^2)$   **79.** 2, $-3$   **81.** 4, $-1$   **83.** $\frac{5}{2}$, $-2$   **85.** 1, $-2$, 3   **87.** 0, 3
**89.** 0, $\frac{5}{2}$   **91.** 0, 7   **93.** 0, $-\frac{8}{3}$   **95.** 0, 2   **97.** 0, $-\frac{1}{5}$

## Getting Ready (page 348)

**1.** $3x + 3y + ax + ay$   **2.** $xy + x + 5y + 5$   **3.** $5x + 5 - yx - y$   **4.** $x^2 + 2x - yx - 2y$   **5.** $3x^2 + 2xy - y^2$   **6.** $-y^2 + 12y - 35$

## Orals (page 351)

**1.** $x + 3$   **2.** $a - 1$   **3.** $x - 2$   **4.** $y + 5$   **5.** $x - 7$   **6.** $2y + 9$

## Exercise 5.2 (page 351)

**1.** $u^9$   **3.** $\frac{a}{b}$   **5.** $a + b$   **7.** $(a + b)$   **9.** $(p - q)$   **11.** $(x + y)(2 + b)$   **13.** $(x + y)(3 - a)$   **15.** $(r - 2s)(3 - x)$
**17.** $(x - 3)(x - 2)$   **19.** $2(a^2 + b)(x + y)$   **21.** $3(r + 3s)(x^2 - 2y^2)$   **23.** $(a + b + c)(3x - 2y)$   **25.** $7xy(r + 2s - t)(2x - 3)$
**27.** $(x + 1)(x + 3 - y)$   **29.** $(x^2 - 2)(3x - y + 1)$   **31.** $(x + y)(2 + a)$   **33.** $(r + s)(7 - k)$   **35.** $(r + s)(x + y)$
**37.** $(2x + 3)(a + b)$   **39.** $(b + c)(2a + 3)$   **41.** $(x + y)(2x - 3)$   **43.** $(v - 3w)(3t + u)$   **45.** $(3p + q)(3m - n)$
**47.** $(m - n)(p - 1)$   **49.** $(a - b)(x - y)$   **51.** $x^2(a + b)(x + 2y)$   **53.** $4a(b + 3)(a - 2)$   **55.** $(x^2 + 1)(x + 2)$
**57.** $y(x^2 - y)(x - 1)$   **59.** $(x + 2)(x + y + 1)$   **61.** $(m - n)(a + b + c)$   **63.** $(d + 3)(a - b - c)$   **65.** $(a + b + c)(x^2 - y)$
**67.** $(r - s)(2 + b)$   **69.** $(x + y)(a + b)$   **71.** $(a - b)(c - d)$   **73.** $r(r + s)(a - b)$   **75.** $(b + 1)(a + 3)$   **77.** $(r - s)(p - q)$

## Getting Ready (page 353)

**1.** $a^2 - b^2$   **2.** $4r^2 - s^2$   **3.** $9x^2 - 4y^2$   **4.** $16x^4 - 9$

## Orals  (page 357)

**1.** $(x + 3)(x - 3)$   **2.** $(y + 6)(y - 6)$   **3.** $(z + 2)(z - 2)$   **4.** $(p + q)(p - q)$   **5.** $(5 + t)(5 - t)$   **6.** $(6 + r)(6 - r)$
**7.** $(10 + y)(10 - y)$   **8.** $(10 + y^2)(10 - y^2)$

## Exercise 5.3  (page 357)

**1.** $p = w\left(k - h - \dfrac{v^2}{2g}\right)$   **3.** difference of two squares   **5.** $(p - q)$   **7.** $(x - 3)$   **9.** $(2m - 3n)$   **11.** $(x + 4)(x - 4)$
**13.** $(y + 7)(y - 7)$   **15.** $(2y + 7)(2y - 7)$   **17.** $(3x + y)(3x - y)$   **19.** $(5t + 6u)(5t - 6u)$   **21.** $(4a + 5b)(4a - 5b)$   **23.** prime
**25.** $(a^2 + 2b)(a^2 - 2b)$   **27.** $(7y + 15z^2)(7y - 15z^2)$   **29.** $(14x^2 + 13y)(14x^2 - 13y)$   **31.** $8(x + 2y)(x - 2y)$
**33.** $2(a + 2y)(a - 2y)$   **35.** $3(r + 2s)(r - 2s)$   **37.** $x(x + y)(x - y)$   **39.** $x(2a + 3b)(2a - 3b)$   **41.** $3m(m + n)(m - n)$
**43.** $x^2(2x + y)(2x - y)$   **45.** $2ab(a + 11b)(a - 11b)$   **47.** $(x^2 + 9)(x + 3)(x - 3)$   **49.** $(a^2 + 4)(a + 2)(a - 2)$
**51.** $(a^2 + b^2)(a + b)(a - b)$   **53.** $(9r^2 + 16s^2)(3r + 4s)(3r - 4s)$   **55.** $(a^2 + b^4)(a + b^2)(a - b^2)$
**57.** $(x^4 + y^4)(x^2 + y^2)(x + y)(x - y)$   **59.** $2(x^2 + y^2)(x + y)(x - y)$   **61.** $b(a^2 + b^2)(a + b)(a - b)$
**63.** $3n(4m^2 + 9n^2)(2m + 3n)(2m - 3n)$   **65.** $3ay(a^4 + 2y^4)$   **67.** $3a^2(a^4 + b^2)(a^2 + b)(a^2 - b)$
**69.** $2y^2(x^4 + 4y^2)(x^2 + 2y)(x^2 - 2y)$   **71.** $a^2b^2(a^2 + b^2c^2)(a + bc)(a - bc)$   **73.** $a^2b^3(b^2 + 25)(b + 5)(b - 5)$
**75.** $3rs(9r^2 + 4s^2)(3r + 2s)(3r - 2s)$   **77.** $(4x - 4y + 3)(4x - 4y - 3)$   **79.** $(a + 3)(a + 3)(a - 3)$   **81.** $(y + 4)(y - 4)(y - 3)$
**83.** $3(x + 2)(x - 2)(x + 1)$   **85.** $3(m + n)(m - n)(m + a)$   **87.** $2(m + 4)(m - 4)(mn^2 + 4)$   **89.** $5, -5$   **91.** $7, -7$   **93.** $\frac{1}{2}, -\frac{1}{2}$
**95.** $\frac{2}{3}, -\frac{2}{3}$   **97.** $7, -7$   **99.** $\frac{9}{2}, -\frac{9}{2}$

## Getting Ready  (page 359)

**1.** $x^2 + 12x + 36$   **2.** $y^2 - 14y + 49$   **3.** $a^2 - 6a + 9$   **4.** $x^2 + 9x + 20$   **5.** $r^2 - 7r + 10$   **6.** $m^2 - 4m - 21$
**7.** $a^2 + ab - 12b^2$   **8.** $u^2 - 8uv + 15v^2$   **9.** $x^2 - 2xy - 24y^2$

## Orals  (page 367)

**1.** 4   **2.** $-, -$   **3.** $-, 3$   **4.** $-, 2$   **5.** 6, 1   **6.** 6, 1

## Exercise 5.4  (page 367)

**1.**   **3.**   **5.**   **7.**   **9.** $(x + y)^2$   **11.** 4, 2
**13.** $y, 2y$   **15.** $(x + 2)(x + 1)$   **17.** $(z + 11)(z + 1)$   **19.** $(a - 5)(a + 1)$   **21.** $(t - 7)(t - 2)$   **23.** prime   **25.** $(y - 6)(y + 5)$
**27.** $(a + 8)(a - 2)$   **29.** $(t - 10)(t + 5)$   **31.** prime   **33.** $(y + z)(y + z)$   **35.** $(x + 2y)(x + 2y)$   **37.** $(m + 5n)(m - 2n)$
**39.** $(a - 6b)(a + 2b)$   **41.** $(u + 5v)(u - 3v)$   **43.** $-(x + 5)(x + 2)$   **45.** $-(y + 5)(y - 3)$   **47.** $-(t + 17)(t - 2)$
**49.** $-(r - 10)(r - 4)$   **51.** $-(a + 3b)(a + b)$   **53.** $-(x - 7y)(x + y)$   **55.** $(x - 4)(x - 1)$   **57.** $(y + 9)(y + 1)$
**59.** $(c + 5)(c - 1)$   **61.** $-(r - 2s)(r + s)$   **63.** $(r + 3x)(r + x)$   **65.** $(a - 2b)(a - b)$   **67.** $2(x + 3)(x + 2)$   **69.** $3y(y + 1)(y + 1)$
**71.** $-5(a - 3)(a - 2)$   **73.** $3(z - 4t)(z - t)$   **75.** $4y(x + 6)(x - 3)$   **77.** $-4x(x + 3y)(x - 2y)$   **79.** $(x + 2)(ax + 2a + b)$
**81.** $(a + 5)(a + 3 + b)$   **83.** $(a + b + 2)(a + b - 2)$   **85.** $(b + y + 2)(b - y - 2)$   **87.** $(x + 3)(x + 3)$   **89.** $(y - 4)(y - 4)$
**91.** $(t + 10)(t + 10)$   **93.** $(u - 9)(u - 9)$   **95.** $(x + 2y)(x + 2y)$   **97.** $(r - 5s)(r - 5s)$   **99.** 12, 1   **101.** 5, -3   **103.** -3, 7
**105.** 8, 1   **107.** -3, -5   **109.** -4, 2   **111.** 0, -1, -2   **113.** 0, 9, -3   **115.** 1, -2, -3

## Getting Ready  (page 370)

**1.** $6x^2 + 7x + 2$   **2.** $6y^2 - 19y + 10$   **3.** $8t^2 + 6t - 9$   **4.** $4r^2 + 4r - 15$   **5.** $6m^2 - 13m + 6$   **6.** $16a^2 + 16a + 3$

## Orals  (page 376)

**1.** 2, 3   **2.** 2   **3.** $+, -$   **4.** $+, -$   **5.** 3, 1   **6.** 3, 1

## Exercise 5.5   (page 377)

**1.** $n = \frac{l - f + d}{d}$   **3.** descending   **5.** opposites   **7.** 2, 1   **9.** $y$, $y$   **11.** $(2x - 1)(x - 1)$   **13.** $(3a + 1)(a + 4)$   **15.** $(z + 3)(4z + 1)$
**17.** $(3y + 2)(2y + 1)$   **19.** $(3x - 2)(2x - 1)$   **21.** $(3a + 2)(a - 2)$   **23.** $(2x + 1)(x - 2)$   **25.** $(2m - 3)(m + 4)$
**27.** $(5y + 1)(2y - 1)$   **29.** $(3y - 2)(4y + 1)$   **31.** $(5t + 3)(t + 2)$   **33.** $(8m - 3)(2m - 1)$   **35.** $(3x - y)(x - y)$
**37.** $(2u + 3v)(u - v)$   **39.** $(2a - b)(2a - b)$   **41.** $(3r + 2s)(2r - s)$   **43.** $(2x + 3y)(2x + y)$   **45.** $(4a - 3b)(a - 3b)$
**47.** $(3x + 2)(x - 5)$   **49.** $(2a - 5)(4a - 3)$   **51.** $(4y - 3)(3y - 4)$   **53.** prime   **55.** $(2a + 3b)(a + b)$   **57.** $(3p - q)(2p + q)$
**59.** prime   **61.** $(4x - 5y)(3x - 2y)$   **63.** $2(2x - 1)(x + 3)$   **65.** $y(y + 12)(y + 1)$   **67.** $3x(2x + 1)(x - 3)$
**69.** $3r^3(5r - 2)(2r + 5)$   **71.** $4(a - 2b)(a + b)$   **73.** $4(2x + y)(x - 2y)$   **75.** $-2mn(4m + 3n)(2m + n)$
**77.** $-2uv^3(7u - 3v)(2u - v)$   **79.** $(2x + 3)^2$   **81.** $(3x + 2)^2$   **83.** $(4x - y)^2$   **85.** $(2x + y + 4)(2x + y - 4)$
**87.** $(3 + a + 2b)(3 - a - 2b)$   **89.** $(2x + y + a + b)(2x + y - a - b)$   **91.** $\frac{1}{2}$, 2   **93.** $\frac{1}{5}$, 1   **95.** $-\frac{1}{3}$, 3   **97.** $\frac{2}{3}$, $-\frac{1}{5}$   **99.** $-\frac{3}{2}$, $\frac{2}{3}$
**101.** $\frac{1}{8}$, 1   **103.** 0, $-3$, $-\frac{1}{3}$   **105.** 0, $-3$, $-3$

## Getting Ready   (page 379)

**1.** $x^3 - 27$   **2.** $x^3 + 8$   **3.** $y^3 + 64$   **4.** $r^3 - 125$   **5.** $a^3 - b^3$   **6.** $a^3 + b^3$

## Orals   (page 383)

**1.** $(x - y)(x^2 + xy + y^2)$   **2.** $(x + y)(x^2 - xy + y^2)$   **3.** $(a + 2)(a^2 - 2a + 4)$   **4.** $(b - 3)(b^2 + 3b + 9)$
**5.** $(1 + 2x)(1 - 2x + 4x^2)$   **6.** $(2 - r)(4 + 2r + r^2)$   **7.** $(xy + 1)(x^2y^2 - xy + 1)$   **8.** $(5 - 2t)(25 + 10t + 4t^2)$

## Exercise 5.6   (page 383)

**1.** 0.0000000000001 cm   **3.** $(x^2 - xy + y^2)$   **5.** $(y + 1)(y^2 - y + 1)$   **7.** $(a - 3)(a^2 + 3a + 9)$   **9.** $(2 + x)(4 - 2x + x^2)$
**11.** $(s - t)(s^2 + st + t^2)$   **13.** $(3x + y)(9x^2 - 3xy + y^2)$   **15.** $(a + 2b)(a^2 - 2ab + 4b^2)$   **17.** $(4x - 3)(16x^2 + 12x + 9)$
**19.** $(3x - 5y)(9x^2 + 15xy + 25y^2)$   **21.** $(a^2 - b)(a^4 + a^2b + b^2)$   **23.** $(x^3 + y^2)(x^6 - x^3y^2 + y^4)$   **25.** $2(x + 3)(x^2 - 3x + 9)$
**27.** $-(x - 6)(x^2 + 6x + 36)$   **29.** $8x(2m - n)(4m^2 + 2mn + n^2)$   **31.** $xy(x + 6y)(x^2 - 6xy + 36y^2)$
**33.** $3rs^2(3r - 2s)(9r^2 + 6rs + 4s^2)$   **35.** $a^3b^2(5a + 4b)(25a^2 - 20ab + 16b^2)$   **37.** $yz(y^2 - z)(y^4 + y^2z + z^2)$
**39.** $2mp(p + 2q)(p^2 - 2pq + 4q^2)$   **41.** $(x + 1)(x^2 - x + 1)(x - 1)(x^2 + x + 1)$
**43.** $(x^2 + y)(x^4 - x^2y + y^2)(x^2 - y)(x^4 + x^2y + y^2)$   **45.** $(x + y)(x^2 - xy + y^2)(3 - z)$   **47.** $(m + 2n)(m^2 - 2mn + 4n^2)(1 + x)$
**49.** $(a + 3)(a^2 - 3a + 9)(a - b)$   **51.** $(y + 1)(y - 1)(y - 3)(y^2 + 3y + 9)$

## Getting Ready   (page 385)

**1.** $3ax(x + a)$   **2.** $(x + 3y)(x - 3y)$   **3.** $(x - 2)(x^2 + 2x + 4)$   **4.** $2(x + 2)(x - 2)$   **5.** $(x - 5)(x + 2)$   **6.** $(2x - 3)(3x - 2)$
**7.** $2(3x - 1)(x - 2)$   **8.** $(a + b)(x + y)(x - y)$

## Orals   (page 387)

**1.** common factor   **2.** difference of two squares   **3.** sum of cubes   **4.** grouping   **5.** none, prime   **6.** common factor
**7.** difference of two squares   **8.** difference of cubes

## Exercise 5.7   (page 387)

**1.** $\frac{8}{3}$   **3.** 0, 7   **5.** factors   **7.** binomials   **9.** $3(2x + 1)$   **11.** $(x - 7)(x + 1)$   **13.** $(3t - 1)(2t + 3)$   **15.** $(2x + 5)(2x - 5)$
**17.** $(t - 1)(t - 1)$   **19.** $(a - 2)(a^2 + 2a + 4)$   **21.** $(y^2 - 2)(x + 1)(x - 1)$   **23.** $7p^4q^2(10q - 5 + 7p)$   **25.** $2a(b + 6)(b - 2)$
**27.** $-4p^2q^3(2pq^4 + 1)$   **29.** $(2a - b + 3)(2a - b - 3)$   **31.** prime   **33.** $-2x^2(x - 4)(x^2 + 4x + 16)$   **35.** $2t^2(3t - 5)(t + 4)$
**37.** $(x - a)(a + b)(a - b)$   **39.** $(2p^2 - 3q^2)(4p^4 + 6p^2q^2 + 9q^4)$   **41.** $(5p - 4y)(25p^2 + 20py + 16y^2)$
**43.** $-x^2y^2z(16x^2 - 24x^3yz^3 + 15yz^6)$   **45.** $(9p^2 + 4q^2)(3p + 2q)(3p - 2q)$   **47.** prime   **49.** $2(3a + 5y^2)(9x^2 - 15xy^2 + 25y^4)$
**51.** prime   **53.** $t(7t - 1)(3t - 1)$   **55.** $(x + y)(x - y)(x + y)(x^2 - xy + y^2)$   **57.** $2(a + b)(a - b)(c + 2d)$

## Getting Ready (page 389)

**1.** $s^2$   **2.** $2w + 4$   **3.** $x(x + 1)$   **4.** $w(w + 3)$

## Orals (page 392)

**1.** $A = lw$   **2.** $A = \frac{1}{2}bh$   **3.** $A = s^2$   **4.** $A = lwh$   **5.** $P = 2l + 2w$   **6.** $P = 4s$

## Exercise 5.8 (page 393)

**1.** $-10$   **3.** 675 cm$^2$   **5.** analyze   **7.** 5, 7   **9.** 9   **11.** 9 sec   **13.** $\frac{15}{4}$ sec and 10 sec   **15.** 2 sec   **17.** 4 m by 9 m   **19.** 48 ft
**21.** $b = 4$ in., $h = 18$ in.   **23.** 18 sq units   **25.** 1 m   **27.** 3 cm   **29.** 4 cm by 7 cm

## Chapter Summary (page 397)

**1. a.** $5 \cdot 7$   **b.** $3^2 \cdot 5$   **c.** $2^5 \cdot 3$   **d.** $2 \cdot 3 \cdot 17$   **e.** $3 \cdot 29$   **f.** $3^2 \cdot 11$   **g.** $2 \cdot 5^2 \cdot 41$   **h.** $2^{12}$   **2. a.** $3(x + 3y)$   **b.** $5a(x^2 + 3)$
**c.** $7x(x + 2)$   **d.** $3x(x - 1)$   **e.** $2x(x^2 + 2x - 4)$   **f.** $a(x + y - z)$   **g.** $a(x + y - 1)$   **h.** $xyz(x + y)$   **3. a.** $0, -2$   **b.** $0, 3$
**4. a.** $(x + y)(a + b)$   **b.** $(x + y)(x + y + 1)$   **c.** $2x(x + 2)(x + 3)$   **d.** $3x(y + z)(1 - 3y - 3z)$   **e.** $(p + 3q)(3 + a)$
**f.** $(r - 2s)(a + 7)$   **g.** $(x + a)(x + b)$   **h.** $(y + 2)(x - 2)$   **i.** $(x + y)(a + b)$   **5. a.** $(x + 3)(x - 3)$   **b.** $(xy + 4)(xy - 4)$
**c.** $(x + 2 + y)(x + 2 - y)$   **d.** $(z + x + y)(z - x - y)$   **e.** $6y(x + 2y)(x - 2y)$   **f.** $(x + y + z)(x + y - z)$   **6. a.** $3, -3$   **b.** $5, -5$
**7. a.** $(x + 3)(x + 7)$   **b.** $(x - 3)(x + 7)$   **c.** $(x + 6)(x - 4)$   **d.** $(x - 6)(x + 2)$   **8. a.** $(2x + 1)(x - 3)$   **b.** $(3x + 1)(x - 5)$
**c.** $(2x + 3)(3x - 1)$   **d.** $3(2x - 1)(x + 1)$   **e.** $x(x + 3)(6x - 1)$   **f.** $x(4x + 3)(x - 2)$   **9. a.** $3, 4$   **b.** $5, -3$   **c.** $-4, 6$   **d.** $2, 8$
**e.** $3, -\frac{1}{2}$   **f.** $1, -\frac{3}{2}$   **g.** $\frac{1}{2}, -\frac{1}{2}$   **h.** $\frac{2}{3}, -\frac{2}{3}$   **i.** $0, 3, 4$   **j.** $0, -2, -3$   **k.** $0, \frac{1}{2}, -3$   **l.** $0, -\frac{2}{3}, 1$   **10. a.** $(c - 3)(c^2 + 3c + 9)$
**b.** $(d + 2)(d^2 - 2d + 4)$   **c.** $2(x + 3)(x^2 - 3x + 9)$   **d.** $2ab(b - 1)(b^2 + b + 1)$   **11. a.** $y(3x - y)(x - 2)$   **b.** $5(x + 2)(x - 3y)$
**c.** $a(a + b)(2x + a)$   **d.** $(x + a + y)(x + a - y)$   **e.** $(x + 1)(ax + 3a - b)$   **f.** $a(x + y)(x^2 - xy + y^2)(x - y)(x^2 + xy + y^2)$
**12.** 5 and 7   **13.** $\frac{1}{3}$   **14.** 6 ft by 8 ft   **15.** 3 ft by 9 ft   **16.** 3 ft by 6 ft

## Chapter 5 Test (page 401)

**1.** $2^2 \cdot 7^2$   **2.** $3 \cdot 37$   **3.** $5a(12b^2c^3 + 6a^2b^2c - 5)$   **4.** $3x(a + b)(x - 2y)$   **5.** $(x + y)(a + b)$   **6.** $(x + 5)(x - 5)$
**7.** $3(a + 3b)(a - 3b)$   **8.** $(4x^2 + 9y^2)(2x + 3y)(2x - 3y)$   **9.** $(x + 3)(x + 1)$   **10.** $(x - 11)(x + 2)$   **11.** $(x + 9y)(x + y)$
**12.** $6(x - 4y)(x - y)$   **13.** $(3x + 1)(x + 4)$   **14.** $(2a - 3)(a + 4)$   **15.** $(2x - y)(x + 2y)$   **16.** $(4x - 3)(3x - 4)$
**17.** $6(2a - 3b)(a + 2b)$   **18.** $(x - 4)(x^2 + 4x + 16)$   **19.** $8(3 + a)(9 - 3a + a^2)$   **20.** $z^3(x^3 - yz)(x^6 + x^3yz + y^2z^2)$   **21.** $0, -3$
**22.** $-1, -\frac{3}{2}$   **23.** $3, -3$   **24.** $3, -6$   **25.** $\frac{9}{5}, -\frac{1}{2}$   **26.** $-\frac{9}{10}, 1$   **27.** $\frac{1}{5}, -\frac{9}{2}$   **28.** $-\frac{1}{10}, 9$   **29.** 12 sec   **30.** 10 m

## Getting Ready (page 403)

**1.** $\frac{1}{2}$   **2.** $\frac{2}{3}$   **3.** $-\frac{4}{5}$   **4.** $-\frac{5}{9}$

## Orals (page 408)

**1.** $\frac{5}{7}$   **2.** $\frac{50}{1}$   **3.** $\frac{1}{3}$   **4.** $\frac{7}{10}$

## Exercise 6.1 (page 408)

**1.** 17   **3.** 6   **5.** $2(x + 3)$   **7.** $(2x + 3)(x - 2)$   **9.** comparison, quotient   **11.** equal   **15.** $\frac{5}{7}$   **17.** $\frac{1}{2}$   **19.** $\frac{2}{3}$   **21.** $\frac{2}{7}$   **23.** $\frac{1}{3}$   **25.** $\frac{1}{5}$
**27.** $\frac{3}{7}$   **29.** $\frac{3}{4}$   **31.** \$1,825   **33.** $\frac{22}{365}$   **35.** \$8,725   **37.** $\frac{336}{1,745}$   **39.** $\frac{1}{16}$   **41.** $\frac{\$21.59}{17 \text{ gal}}$; \$1.27/gal   **43.** 7¢/oz   **45.** the 6-oz can   **47.** the
first student   **49.** $\frac{11,880 \text{ gal}}{27 \text{ min}}$; 440 gal/min   **51.** 5%   **53.** 65 mph   **55.** the second car

## Getting Ready (page 410)

**1.** 10   **2.** $\frac{7}{3}$   **3.** $\frac{20}{7}$   **4.** 5   **5.** $\frac{14}{3}$   **6.** 5   **7.** 21   **8.** $\frac{3}{2}$

## Orals (page 419)

**1.** proportion   **2.** not a proportion   **3.** not a proportion   **4.** proportion

## Exercise 6.2   (page 420)

**1.** 90%   **3.** $\frac{1}{3}$   **5.** 480   **7.** $73.50   **9.** proportion, ratios   **11.** means   **13.** shape   **15.** $ad$, $bc$   **17.** triangle   **19.** no   **21.** yes
**23.** no   **25.** yes   **27.** 4   **29.** 6   **31.** $-3$   **33.** 9   **35.** 0   **37.** $-17$   **39.** $-\frac{3}{2}$   **41.** $\frac{83}{2}$   **43.** $17   **45.** $6.50   **47.** 24
**49.** about $4\frac{1}{4}$   **51.** 47   **53.** $7\frac{1}{2}$ gal   **55.** $309   **57.** 49 ft $3\frac{1}{2}$ in.   **59.** 162   **61.** not exactly, but close   **63.** 39 ft   **65.** $46\frac{7}{8}$ ft
**67.** 6,750 ft   **69.** 15,840 ft

## Getting Ready   (page 423)

**1.** $\frac{3}{4}$   **2.** 2   **3.** $\frac{5}{11}$   **4.** $\frac{1}{2}$

## Orals   (page 429)

**1.** $\frac{2}{3}$   **2.** 2   **3.** $\frac{z}{w}$   **4.** $2x$   **5.** $\frac{x}{y}$   **6.** $\frac{1}{y}$   **7.** 1   **8.** $-1$

## Exercise 6.3   (page 429)

**1.** $(a+b)+c = a+(b+c)$   **3.** 0   **5.** $\frac{5}{3}$   **7.** numerator   **9.** 0   **11.** negatives   **13.** $\frac{a}{b}$   **15.** factor, common   **17.** $\frac{4}{5}$   **19.** $\frac{4}{5}$
**21.** $\frac{2}{13}$   **23.** $\frac{2}{9}$   **25.** $-\frac{1}{3}$   **27.** $2x$   **29.** $-\frac{x}{3}$   **31.** $\frac{5}{a}$   **33.** $\frac{2}{z}$   **35.** $\frac{a}{3}$   **37.** $\frac{2}{3}$   **39.** $\frac{3}{2}$   **41.** in lowest terms   **43.** $\frac{3x}{y}$   **45.** $\frac{7x}{8y}$   **47.** $\frac{1}{3}$   **49.** 5
**51.** $\frac{x}{2}$   **53.** $\frac{3x}{5y}$   **55.** $\frac{2}{3}$   **57.** $-1$   **59.** $-1$   **61.** $-1$   **63.** $\frac{x+1}{x-1}$   **65.** $\frac{x-5}{x+2}$   **67.** $\frac{-2x}{x-2}$   **69.** $\frac{x}{y}$   **71.** $\frac{x+2}{x^2}$   **73.** $\frac{x-4}{x+4}$   **75.** $\frac{2(x+2)}{x-1}$
**77.** in lowest terms   **79.** $\frac{3-x}{3+x}$ or $-\frac{x-3}{x+3}$   **81.** $\frac{4}{3}$   **83.** $x+3$

## Getting Ready   (page 431)

**1.** $\frac{2}{3}$   **2.** $\frac{14}{3}$   **3.** 3   **4.** 6   **5.** $\frac{5}{2}$   **6.** 1   **7.** $\frac{3}{4}$   **8.** 2

## Orals   (page 438)

**1.** $\frac{3}{2}$   **2.** $\frac{7}{5}$   **3.** 5   **4.** 1   **5.** $\frac{1}{4}$   **6.** $x$

## Exercise 6.4   (page 438)

**1.** $-6x^5y^6z$   **3.** $\dfrac{1}{81y^4}$   **5.** $\dfrac{1}{x^m}$   **7.** $4y^3 + 4y^2 - 8y + 32$   **9.** numerator   **11.** numerators, denominators   **13.** 1

**15.** divisor, multiply   **17.** $\frac{45}{91}$   **19.** $-\frac{3}{11}$   **21.** $\frac{5}{7}$   **23.** $\frac{3x}{2}$   **25.** $\frac{yx}{z}$   **27.** $\frac{14}{9}$   **29.** $x^2y^2$   **31.** $2xy^2$   **33.** $-3y^2$   **35.** $\dfrac{b^3c}{a^4}$   **37.** $\dfrac{r^3t^4}{s}$

**39.** $\dfrac{(z+7)(z+2)}{7z}$   **41.** $x$   **43.** $\frac{x}{5}$   **45.** $x+2$   **47.** $\frac{3}{2x}$   **49.** $x-2$   **51.** $x$   **53.** $\dfrac{(x-2)^2}{x}$   **55.** $\frac{(m-2)(m-3)}{2(m+2)}$   **57.** 1   **59.** $\dfrac{c^2}{ab}$

**61.** $\dfrac{x+1}{2(x-2)}$   **63.** $\frac{2}{3}$   **65.** $\frac{3}{5}$   **67.** $\dfrac{3}{2y}$   **69.** 3   **71.** $\dfrac{6}{y}$   **73.** 6   **75.** $\dfrac{2x}{3}$   **77.** $\dfrac{2}{y}$   **79.** $\dfrac{2}{3x}$   **81.** $\dfrac{2(z-2)}{z}$   **83.** $\dfrac{5z(z-7)}{z+2}$

**85.** $\frac{x+2}{3}$   **87.** 1   **89.** $\frac{x-2}{x-3}$   **91.** $x+5$   **93.** $\frac{9}{2x}$   **95.** $\frac{x}{36}$   **97.** $\frac{(x+1)(x-1)}{5(x-3)}$   **99.** 2   **101.** $\dfrac{2x(1-x)}{5(x-2)}$   **103.** $\dfrac{y^2}{3}$   **105.** $\dfrac{x+2}{x-2}$

## Getting Ready   (page 441)

**1.** $\frac{4}{5}$   **2.** 1   **3.** $\frac{7}{8}$   **4.** 2   **5.** $\frac{1}{9}$   **6.** $\frac{1}{2}$   **7.** $-\frac{2}{13}$   **8.** $\frac{13}{10}$

## Orals   (page 451)

**1.** equal   **2.** equal   **3.** not equal   **4.** equal   **5.** equal   **6.** not equal   **7.** equal   **8.** equal

## Exercise 6.5   (page 451)

**1.** $7^2$   **3.** $2^3 \cdot 17$   **5.** $2 \cdot 3 \cdot 17$   **7.** $2^4 \cdot 3^2$   **9.** LCD   **11.** numerators, common denominator   **13.** $\frac{2}{3}$   **15.** $\frac{1}{3}$   **17.** $\frac{4x}{y}$   **19.** $\frac{2}{y}$
**21.** $\dfrac{2y+6}{5z}$   **23.** 9   **25.** $\frac{1}{7}$   **27.** $-\frac{1}{8}$   **29.** $\frac{x}{y}$   **31.** $\frac{y}{x}$   **33.** $\frac{1}{y}$   **35.** 1   **37.** $\frac{4x}{3}$   **39.** $\frac{2x}{3y}$   **41.** $\dfrac{2(2x-y)}{y+2}$   **43.** $\dfrac{2(x+5)}{x-2}$   **45.** $\dfrac{125}{20}$   **47.** $\dfrac{8xy}{x^2y}$

**49.** $\dfrac{3x(x+1)}{(x+1)^2}$ **51.** $\dfrac{2y(x+1)}{x^2+x}$ **53.** $\dfrac{z(z+1)}{z^2-1}$ **55.** $\dfrac{2(x+2)}{x^2+3x+2}$ **57.** $6x$ **59.** $18xy$ **61.** $x^2-1$ **63.** $x^2+6x$

**65.** $(x+1)(x+5)(x-5)$ **67.** $\dfrac{7}{6}$ **69.** $\dfrac{5y}{9}$ **71.** $\dfrac{53x}{42}$ **73.** $\dfrac{4xy+6x}{3y}$ **75.** $\dfrac{2-3x^2}{x}$ **77.** $\dfrac{4y+10}{15y}$ **79.** $\dfrac{x^2+4x+1}{x^2y}$

**81.** $\dfrac{2x^2-1}{x(x+1)}$ **83.** $\dfrac{2xy+x-y}{xy}$ **85.** $\dfrac{x+2}{x-2}$ **87.** $\dfrac{2x^2+2}{(x-1)(x+1)}$ **89.** $\dfrac{2(2x+1)}{x-2}$ **91.** $\dfrac{x}{x-2}$ **93.** $\dfrac{5x+3}{x+1}$ **95.** $-\dfrac{1}{2(x-2)}$

Getting Ready (page 454)

**1.** 4 **2.** $-18$ **3.** 7 **4.** $-8$ **5.** $3+3x$ **6.** $2-y$ **7.** $12x-2$ **8.** $3y+2x$

Orals (4age 459)

**1.** $\frac{4}{3}$ **2.** 4 **3.** $\frac{1}{4}$ **4.** 3

Exercise 6.6 (page 459)

**1.** $t^9$ **3.** $-2r^7$ **5.** $\dfrac{81}{256r^8}$ **7.** $\dfrac{r^{10}}{9}$ **9.** complex fraction **11.** single, divide **13.** $\dfrac{8}{9}$ **15.** $\dfrac{3}{8}$ **17.** $\dfrac{5}{4}$ **19.** $\dfrac{5}{7}$ **21.** $\dfrac{x^2}{y}$ **23.** $\dfrac{5t^2}{27}$

**25.** $\dfrac{1-3x}{5+2x}$ **27.** $\dfrac{1+x}{2+x}$ **29.** $\dfrac{3-x}{x-1}$ **31.** $\dfrac{1}{x+2}$ **33.** $\dfrac{1}{x+3}$ **35.** $\dfrac{xy}{y+x}$ **37.** $\dfrac{y}{x-2y}$ **39.** $\dfrac{x^2}{(x-1)^2}$ **41.** $\dfrac{7x+3}{-x-3}$ **43.** $\dfrac{x-2}{x+3}$

**45.** $-1$ **47.** $\dfrac{y}{x^2}$ **49.** $\dfrac{x+1}{1-x}$ **51.** $\dfrac{a^2-a+1}{a^2}$ **53.** 2 **55.** $\dfrac{y-5}{y+5}$ **59.** $\frac{1}{2}, \frac{2}{3}, \frac{3}{5}, \frac{5}{8}$

Getting Ready (page 461)

**1.** $3x+1$ **2.** $8x-1$ **3.** $3+2x$ **4.** $y-6$ **5.** 19 **6.** $7x+6$ **7.** $y$ **8.** $3x+5$

Orals (page 466)

**1.** Multiply by 10. **2.** Multiply by $x(x-1)$. **3.** Multiply by 9. **4.** Multiply by 15.

Exercise 6.7 (page 466)

**1.** $x(x+4)$ **3.** $(2x+3)(x-1)$ **5.** $(x^2+4)(x+2)(x-2)$ **7.** extraneous **9.** LCD **11.** $xy$ **13.** 4 **15.** $-20$ **17.** 6 **19.** 60 **21.** $-12$ **23.** 0 **25.** $-7$ **27.** $-1$ **29.** 12 **31.** 0 **33.** $-3$ **35.** 3 **37.** no solution; 0 is extraneous **39.** 1 **41.** 5 **43.** no solution; $-2$ is extraneous **45.** no solution; 5 is extraneous **47.** $-1$ **49.** 6 **51.** 2 **53.** $-3$ **55.** 1 **57.** no solution; $-2$ is extraneous **59.** 1, 2 **61.** 3; $-3$ is extraneous **63.** 3, $-4$ **65.** 1 **67.** 0 **69.** $-2, 1$ **71.** $a=\frac{b}{b-1}$ **73.** $f=\dfrac{d_1 d_2}{d_1+d_2}$ **77.** 1

Getting Ready (page 469)

**1.** $\frac{1}{5}$ **2.** $\$(.05x)$ **3.** $\$\left(\frac{y}{.05}\right)$ **4.** $\frac{y}{52}$ hr

Orals (page 472)

**1.** $i=pr$ **2.** $d=rt$ **3.** $C=qd$

Exercise 6.8 (page 473)

**1.** $-1, 6$ **3.** $-2, -3, -4$ **5.** 0, 0, 1 **7.** 1, $-1$, 2, $-2$ **11.** 2 **13.** 5 **15.** $\frac{2}{3}, \frac{3}{2}$ **17.** $2\frac{2}{9}$ hr **19.** $2\frac{6}{11}$ days **21.** $7\frac{1}{2}$ hr **23.** 4 mph **25.** 7% and 8% **27.** 5 **29.** 30 **31.** 25 mph

## Chapter Summary   (page 477)

**1. a.** $\frac{1}{2}$ **b.** $\frac{4}{5}$ **c.** $\frac{2}{3}$ **d.** $\frac{5}{6}$ **2.** \$2.93 **3.** 568.75 kwh per week **4. a.** no **b.** yes **5. a.** $\frac{9}{2}$ **b.** 0 **c.** 7 **d.** 1 **6.** 20 ft **7. a.** $\frac{2}{5}$
**b.** $-\frac{2}{3}$ **c.** $-\frac{1}{3}$ **d.** $\frac{7}{3}$ **e.** $\frac{1}{2x}$ **f.** $\frac{5}{2x}$ **g.** $\frac{x}{x+1}$ **h.** $\frac{1}{x}$ **i.** 2 **j.** 1 **k.** $-1$ **l.** $\frac{x+7}{x+3}$ **m.** $\frac{x}{x-1}$ **n.** in lowest terms **8. a.** $\frac{3x}{y}$ **b.** $\frac{6}{x^2}$
**c.** 1 **d.** $\frac{2x}{x+1}$ **9. a.** $\frac{3y}{2}$ **b.** $\frac{1}{x}$ **c.** $x+2$ **d.** 1 **e.** $x+2$ **10. a.** 1 **b.** $\dfrac{2(x+1)}{x-7}$ **c.** $\dfrac{x^2+x-1}{x(x-1)}$ **d.** $\dfrac{x-7}{7x}$ **e.** $\dfrac{x-2}{x(x+1)}$
**f.** $\dfrac{x^2+4x-4}{2x^2}$ **g.** $\dfrac{x+1}{x}$ **h.** 0 **11. a.** $\frac{9}{4}$ **b.** $\frac{3}{2}$ **c.** $\dfrac{1+x}{1-x}$ **d.** $\dfrac{x(x+3)}{2x^2-1}$ **e.** $x^2+3$ **f.** $\dfrac{a(a+bc)}{b(b+ac)}$ **12. a.** 3 **b.** 1 **c.** 3
**d.** 4, $-\frac{3}{2}$ **e.** $-2$ **f.** 0 **13.** $r_1=\dfrac{rr_2}{r_2-r}$ **14.** $T_1=\dfrac{T_2}{1-E}$ or $T_1=\dfrac{-T_2}{E-1}$ **15.** $R=\dfrac{HB}{B-H}$ or $R=\dfrac{-HB}{H-B}$ **16.** $9\frac{9}{19}$ hr
**17.** $5\frac{5}{6}$ days **18.** 5 mph **19.** 40 mph

## Chapter 6 Test   (page 481)

**1.** $\frac{2}{3}$ **2.** yes **3.** $\frac{2}{3}$ **4.** 45 ft **5.** $\dfrac{8x}{9y}$ **6.** $\dfrac{x+1}{2x+3}$ **7.** 3 **8.** $\dfrac{5y^2}{4t}$ **9.** $\dfrac{x+1}{3(x-2)}$ **10.** $\dfrac{3t^2}{5y}$ **11.** $\dfrac{x^2}{3}$ **12.** $x+2$ **13.** $\dfrac{10x-1}{x-1}$
**14.** $\dfrac{13}{2y+3}$ **15.** $\dfrac{2x^2+x+1}{x(x+1)}$ **16.** $\dfrac{2x+6}{x-2}$ **17.** $\dfrac{2x^3}{y^3}$ **18.** $\dfrac{x+y}{y-x}$ **19.** $-5$ **20.** 6 **21.** 4 **22.** $B=\dfrac{RH}{R-H}$ **23.** $3\frac{15}{16}$ hr
**24.** 5 mph **25.** 8,050 ft

## Cumulative Review Exercises   (page 482)

**1.** $x^7$ **2.** $x^{10}$ **3.** $x^3$ **4.** 1 **5.** $6x^3-2x-1$ **6.** $2x^3+2x^2+x-1$ **7.** $13x^2-8x+1$ **8.** $16x^2-24x+2$ **9.** $-12x^5y^5$
**10.** $-35x^5+10x^4+10x^2$ **11.** $6x^2+14x+4$ **12.** $15x^2-2xy-8y^2$ **13.** $x+4$ **14.** $x^2+x+1$ **15.** $3xy(x-2y)$
**16.** $(a+b)(3+x)$ **17.** $(a+b)(2+b)$ **18.** $(5p^2+4q)(5p^2-4q)$ **19.** $(x-12)(x+1)$ **20.** $(x-3y)(x+2y)$
**21.** $(3a+4)(2a-5)$ **22.** $(4m+n)(2m-3n)$ **23.** $(p-3q)(p^2+3pq+9q^2)$ **24.** $8(r+2s)(r^2-2rs+4s^2)$
**25.** 15 **26.** 4 **27.** $\frac{2}{3}, -\frac{1}{2}$ **28.** 0, 2 **29.** $-1, -2$ **30.** $\frac{3}{2}, -4$ **31.** 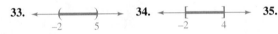 **32.**

**33.** 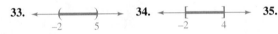 **34.** 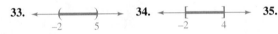 **35.** 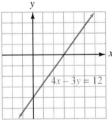 **36.** 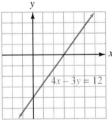 **37.** $(4,-3)$

**38.** $\left(\frac{1}{2}, \frac{2}{3}\right)$ **39.** 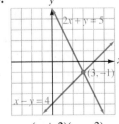 **40.** 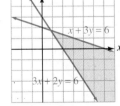 **41.** $-3$ **42.** 15 **43.** 5 **44.** $8x^2-3$ **45.** $\frac{x+1}{x-1}$ **46.** $\frac{x-3}{x-2}$

**47.** $\dfrac{(x-2)^2}{x-1}$ **48.** $\dfrac{(p+2)(p-3)}{3(p+3)}$ **49.** 1 **50.** $\dfrac{2(x^2+1)}{(x+1)(x-1)}$ **51.** $\dfrac{-1}{2(a-2)}$ **52.** $\dfrac{y+x}{y-x}$

## Getting Ready  (page 486)

**1.** 9  **2.** 16  **3.** 8  **4.** 125  **5.** 81  **6.** 256  **7.** −27  **8.** −32

## Orals  (page 494)

**1.** 5  **2.** 2  **3.** $\frac{1}{3}$  **4.** $\frac{5}{7}$  **5.** $2x$  **6.** $6x^2$  **7.** $9y^3$  **8.** $10z^2$

## Exercise 7.1  (page 494)

**1.**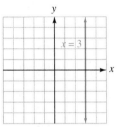

**3.**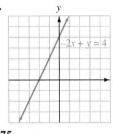

**5.** $b^2 = a$  **7.** positive  **9.** right  **11.** two, 5, −5  **13.** square
**15.** hypotenuse, $a^2 + b^2$  **17.** 3  **19.** 7  **21.** 6  **23.** $\frac{1}{9}$  **25.** −5
**27.** −9  **29.** 14  **31.** $\frac{3}{16}$  **33.** −17  **35.** 100  **37.** 18  **39.** −60
**41.** 1.414  **43.** 2.236  **45.** 2.449  **47.** 3.317  **49.** 4.796  **51.** 9.747
**53.** 80.175  **55.** −99.378  **57.** 4.621  **59.** 0.599  **61.** 0.996
**63.** −0.915  **65.** rational  **67.** rational  **69.** irrational  **71.** imaginary

**73.**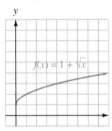

**75.**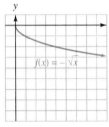

**77.** 5  **79.** 13  **81.** 20  **83.** 28  **85.** 12 ft  **87.** 30 ft  **89.** 127.3 ft
**91.** 5.8 mi  **93.** 2.4 in.  **95.** no  **97.** 4.2 ft  **99.** 24 in.

## Getting Ready  (page 498)

**1.** 8  **2.** 64  **3.** −125  **4.** $-\frac{1}{8}$  **5.** 81  **6.** $\frac{1}{16}$  **7.** 32  **8.** 64

## Orals  (page 504)

**1.** 4  **2.** $\frac{1}{4}$  **3.** 2  **4.** $\frac{1}{2}$  **5.** 1  **6.** 4  **7.** 2  **8.** $\frac{1}{2}$

## Exercise 7.2  (page 504)

**1.** −1  **3.** 9  **5.** $(x + 4y)(x − 4y)$  **7.** $(a + b)(x + y)$  **9.** cube root  **11.** index, radicand  **13.** $V = s^3$  **15.** $(−6)^3 = −216$
**17.** 1  **19.** 3  **21.** −2  **23.** −4  **25.** 5  **27.** 1  **29.** −4  **31.** 9  **33.** 31.78  **35.** −0.48
**37.** 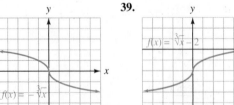 **39.**

**41.** 2  **43.** −2  **45.** 1  **47.** −2  **49.** 3.34  **51.** −5.70  **53.** $xy$
**55.** $x^2z^2$  **57.** $−x^2y$  **59.** $2z$  **61.** $−3x^2y$  **63.** $xyz$  **65.** $−xyz^2$
**67.** $−5x^2z^6$  **69.** $6z^{18}$  **71.** $−4z$  **73.** $3yz^2$  **75.** $−2p^2q$  **77.** 1.26 ft
**79.** 27.14 mph  **81.** 4

Orals  (page 513)

**1.** 16  **2.** 5  **3.** 3  **4.** $\frac{1}{2}$  **5.** 2  **6.** 1

Exercise 7.3  (page 513)

**1.** (2, 3)  **3.** (3, −2)  **5.** extraneous  **7.** $b^2$  **11.** 9  **13.** 49  **15.** none  **17.** 1  **19.** 30  **21.** none  **23.** 5  **25.** 6  **27.** 3
**29.** −1  **31.** 46  **33.** −3  **35.** 0  **37.** none  **39.** 3  **41.** 0, −1  **43.** 2  **45.** 2, 1  **47.** 3  **49.** 2  **51.** −1  **53.** 10  **55.** 65
**57.** 22  **59.** 5  **61.** 5  **63.** 13  **65.** 10  **67.** 256 ft  **69.** about 1.8 ft  **71.** 980 w  **73.** about 299 ft  **75.** about 88 ft
**77.** $2 \times 10^6$ m  **79.** $v^2 = c^2 - f^2c^2$  **81.** 540 w  **83.** about 16.9 mi  **87.** 207

Getting Ready  (page 516)

**1.** 10  **2.** 2  **3.** 5  **4.** 12  **5.** $3x$  **6.** $4x^2$  **7.** $3xy^2$  **8.** $-2x^2y^3$

Orals  (page 522)

**1.** $2\sqrt{2}$  **2.** $2\sqrt{3}$  **3.** $\dfrac{\sqrt{5}}{3}$  **4.** $\dfrac{\sqrt{7}}{5}$  **5.** $2x$  **6.** $3a^2$

Exercise 7.4  (page 522)

**1.** $\frac{1}{2xz}$  **3.** $\frac{a+1}{a+3}$  **5.** perfect  **7.** $\sqrt{a}\sqrt{b}$  **11.** $2\sqrt{5}$  **13.** $5\sqrt{2}$  **15.** $3\sqrt{5}$  **17.** $7\sqrt{2}$  **19.** $4\sqrt{3}$  **21.** $10\sqrt{2}$  **23.** $8\sqrt{3}$
**25.** $2\sqrt{22}$  **27.** 18  **29.** $7\sqrt{3}$  **31.** $6\sqrt{5}$  **33.** $12\sqrt{3}$  **35.** $48\sqrt{2}$  **37.** $-70\sqrt{10}$  **39.** $14\sqrt{5}$  **41.** $-45\sqrt{2}$  **43.** $5\sqrt{x}$
**45.** $a\sqrt{b}$  **47.** $3x\sqrt{y}$  **49.** $x^3y^2\sqrt{2}$  **51.** $48x^2y\sqrt{y}$  **53.** $-9x^2y^3z\sqrt{2xy}$  **55.** $6ab^2\sqrt{3ab}$  **57.** $-\dfrac{8n\sqrt{5m}}{5}$  **59.** $\dfrac{5}{3}$  **61.** $\dfrac{9}{8}$
**63.** $\dfrac{\sqrt{26}}{5}$  **65.** $\dfrac{2\sqrt{5}}{7}$  **67.** $\dfrac{4\sqrt{3}}{9}$  **69.** $\dfrac{4\sqrt{2}}{5}$  **71.** $\dfrac{5\sqrt{5}}{11}$  **73.** $\dfrac{7\sqrt{5}}{6}$  **75.** $\dfrac{6x\sqrt{2x}}{y}$  **77.** $\dfrac{5mn^2\sqrt{5}}{8}$  **79.** $\dfrac{4m\sqrt{2}}{3n}$  **81.** $\dfrac{2rs^2\sqrt{3}}{9}$
**83.** $2x$  **85.** $-4x\sqrt[3]{x^2}$  **87.** $3xyz^2\sqrt[3]{2y}$  **89.** $-3yz\sqrt[3]{3x^2z}$  **91.** $\dfrac{3m}{2n^2}$  **93.** $\dfrac{rs\sqrt[3]{2rs^2}}{5t}$  **95.** $\dfrac{5ab}{2}$

Getting Ready  (page 524)

**1.** $7x$  **2.** $3y$  **3.** $5xy$  **4.** $9t^2$  **5.** $2\sqrt{5}$  **6.** $3\sqrt{5}$  **7.** $2x\sqrt{2y}$  **8.** $3x\sqrt{2y}$

Orals  (page 528)

**1.** $5\sqrt{5}$  **2.** $5\sqrt{7}$  **3.** $2\sqrt{xy}$  **4.** $4\sqrt{5yz}$  **5.** $3\sqrt{2}$  **6.** $-\sqrt{3}$

Exercise 7.5  (page 528)

**1.** 8  **3.** −9  **5.** like terms  **7.** no  **9.** yes  **13.** $5\sqrt{3}$  **15.** $9\sqrt{3}$  **17.** $7\sqrt{5}$  **19.** $12\sqrt{5}$  **21.** $8\sqrt{5}$  **23.** $10\sqrt{10}$  **25.** $37\sqrt{5}$
**27.** $12\sqrt{7}$  **29.** $38\sqrt{2}$  **31.** $85\sqrt{2}$  **33.** $9\sqrt{5}$  **35.** $7\sqrt{6} + 4\sqrt{15}$  **37.** $\sqrt{2}$  **39.** $3 - 5\sqrt{2}$  **41.** $2\sqrt{2}$  **43.** $-2\sqrt{3}$  **45.** $\sqrt{3}$
**47.** $4\sqrt{10}$  **49.** $-7\sqrt{5}$  **51.** $22\sqrt{6}$  **53.** $-18\sqrt{2}$  **55.** $13\sqrt{10}$  **57.** $3\sqrt{2} - \sqrt{3}$  **59.** $10\sqrt{2} - \sqrt{3}$  **61.** $-6\sqrt{6}$
**63.** $10\sqrt{2} + 5\sqrt{3}$  **65.** $7\sqrt{3} - 6\sqrt{2}$  **67.** $6\sqrt{6} - 2\sqrt{3}$  **69.** $3x\sqrt{2}$  **71.** $3x\sqrt{2x}$  **73.** $3x\sqrt{2y} - 3x\sqrt{3y}$  **75.** $x^2\sqrt{2x}$
**77.** $19x\sqrt{6}$  **79.** $-5y\sqrt{10y}$  **81.** $2x\sqrt{5xy} + 3x^2y\sqrt{5xy} - 4x^3y^2\sqrt{5xy}$  **83.** $5\sqrt[3]{2}$  **85.** $\sqrt[3]{3}$  **87.** $2\sqrt[3]{5} + 5$  **89.** $x\sqrt[3]{x} - x^2\sqrt[3]{x}$
**91.** $2xy\sqrt[3]{3xy^2}$  **93.** $x^2y\sqrt[3]{5xy}$

Getting Ready  (page 530)

**1.** $x^5$  **2.** $y^7$  **3.** $x^3$  **4.** $y^3$  **5.** $x^2 + 2x$  **6.** $6y^5 - 8y^4$  **7.** $x^2 - x - 6$  **8.** $6x^2 + 13xy + 6y^2$

## Orals (page 537)

**1.** 5 **2.** 10 **3.** $x + 2\sqrt{x}$ **4.** $4\sqrt{y} - y$ **5.** $x - 1$ **6.** $1 - z$ **7.** $\dfrac{\sqrt{5}}{5}$ **8.** $\sqrt{x}$

## Exercise 7.6 (page 537)

**1.** $(x - 7)(x + 3)$ **3.** $3xy(2x - 5)$ **5.** radical **7.** $\sqrt{11}$ **9.** $\sqrt{7}$ **11.** 3 **13.** 4 **15.** 8 **17.** 4 **19.** $x^3$ **21.** $b^7$ **23.** $4\sqrt{15}$
**25.** $-60\sqrt{2}$ **27.** 24 **29.** $-8x$ **31.** $700x\sqrt{10}$ **33.** $4x^2\sqrt{y}$ **35.** $2 + \sqrt{2}$ **37.** $9 - \sqrt{3}$ **39.** $7 - 3\sqrt{7}$ **41.** $3\sqrt{5} - 5$
**43.** $3\sqrt{2} + \sqrt{3}$ **45.** $7 - 2\sqrt[3]{7}$ **47.** $x\sqrt{3} - 2\sqrt{x}$ **49.** $6x + 6\sqrt{x}$ **51.** $6\sqrt{x} + 3x$ **53.** 1 **55.** 1 **57.** $\sqrt[3]{4} + 2\sqrt[3]{2} + 1$
**59.** $7 - x^2$ **61.** $6x - 7$ **63.** $4x - 18$ **65.** $8xy + 4\sqrt{2xy} + 1$ **67.** $\dfrac{2x}{3}$ **69.** $\dfrac{3y\sqrt{2}}{5}$ **71.** $\dfrac{2y}{x}$ **73.** $\dfrac{2x}{3}$ **75.** $\dfrac{y\sqrt{xy}}{3}$ **77.** $\dfrac{5\sqrt{2y}}{x}$
**79.** $\dfrac{\sqrt{3}}{3}$ **81.** $\dfrac{2\sqrt{7}}{7}$ **83.** $\sqrt[3]{25}$ **85.** $\sqrt{3}$ **87.** $\dfrac{3\sqrt{2}}{8}$ **89.** $2\sqrt[3]{2}$ **91.** $\dfrac{\sqrt{15}}{3}$ **93.** $\dfrac{10\sqrt{x}}{x}$ **95.** $\dfrac{3\sqrt{2x}}{2x}$ **97.** $\dfrac{\sqrt{2xy}}{3y}$ **99.** $\dfrac{\sqrt[3]{20}}{2}$
**101.** $\sqrt[3]{x}$ **103.** $\dfrac{\sqrt[3]{2xy^2}}{xy}$ **105.** $-\dfrac{\sqrt[3]{5ab}}{ab}$ **107.** $\dfrac{3(\sqrt{3} + 1)}{2}$ **109.** $\sqrt{7} - 2$ **111.** $6 + 2\sqrt{3}$ **113.** $2 - \sqrt{2}$ **115.** $\dfrac{\sqrt{3} - 3}{2}$
**117.** $5\sqrt{3} - 5\sqrt{2}$ **119.** $\dfrac{x + 4\sqrt{x} + 4}{x - 4}$ **121.** $\dfrac{3x - 4\sqrt{3x} + 4}{3x - 4}$ **123.** $\dfrac{3x - 2\sqrt{3x} + 1}{3x - 1}$ **125.** $\dfrac{3y + 5\sqrt{3y} + 6}{3y - 4}$

## Getting Ready (page 540)

**1.** $x^5$ **2.** $x^{12}$ **3.** $x^5$ **4.** 1 **5.** $\frac{1}{ab}$ **6.** $a^{12}b^8$ **7.** $\dfrac{a^9}{b^6}$ **8.** $a^{24}$

## Orals (page 544)

**1.** 4 **2.** 5 **3.** 3 **4.** 3 **5.** 4 **6.** 8 **7.** $\frac{1}{3}$ **8.** $\frac{1}{4}$

## Exercise 7.7 (page 544)

**1.** $3(z - 4t)(z - t)$ **3.** 5 **5.** rational **7.** $x^{m+n}$ **9.** $\dfrac{x^n}{y^n}$ **11.** $\dfrac{1}{x}$ **13.** 9 **15.** $-12$ **17.** $\frac{1}{2}$ **19.** $\frac{2}{7}$ **21.** 3 **23.** $-5$ **25.** $-2$
**27.** $\frac{1}{4}$ **29.** $\frac{3}{4}$ **31.** 2 **33.** 2 **35.** $-3$ **37.** 729 **39.** 125 **41.** 25 **43.** 100 **45.** 4 **47.** 8 **49.** 27 **51.** $-8$ **53.** $\frac{4}{9}$ **55.** $\frac{8}{125}$
**57.** 6 **59.** 25 **61.** 7 **63.** 25 **65.** 8 **67.** 125 **69.** 6 **71.** 192 **73.** $\frac{1}{2}$ **75.** $\frac{1}{9}$ **77.** $\frac{1}{64}$ **79.** $\frac{1}{81}$ **81.** $x$ **83.** $x^2$ **85.** $x^4$
**87.** $x^2$ **89.** $y^2$ **91.** $x^{2/5}$ **93.** $x^{2/7}$ **95.** $x$ **97.** $y^7$ **99.** $y^2$ **101.** $x^{17/12}$ **103.** $b^{3/10}$ **105.** $t^{4/15}$ **107.** $x$ **109.** $a^{14/15}$ **113.** $2^x$

## Chapter Summary (page 548)

**1. a.** 5 **b.** 8 **c.** $-12$ **d.** $-17$ **e.** 16 **f.** $-8$ **g.** 13 **h.** $-15$ **2. a.** 4.583 **b.** $-3.873$ **c.** $-7.570$ **d.** 27.421
**3. a.** b. **4. a.** 35 **b.** 60 **c.** 1 **d.** $2\sqrt{6}$ **5.** 68 in. **6.** 45 ft **7. a.** $-3$

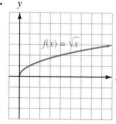

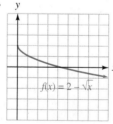

**b.** $-5$ **c.** 3 **d.** 2 **8. a.** 3.787 **b.** 0.144 **c.** $-0.380$ **d.** $-39.953$ **9. a.** $x^3$ **b.** $4x^2y$ **c.** $3x$ **d.** $10a^2b$ **10. a.** 6 **b.** $-3$
**c.** no solution **d.** 4 **e.** 4; $-1$ is extraneous **f.** 8; 0 is extraneous **g.** 2; $-2$ is extraneous **h.** 9; $-1$ is extraneous
**11. a.** 5 **b.** 13 **c.** 2 **d.** 29 **12. a.** $4\sqrt{2}$ **b.** $5\sqrt{2}$ **c.** $10\sqrt{5}$ **d.** $4\sqrt{7}$ **e.** $4x\sqrt{5}$ **f.** $3y\sqrt{7}$ **g.** $-5t\sqrt{10t}$ **h.** $-10z^2\sqrt{7z}$
**i.** $10x\sqrt{2y}$ **j.** $5y\sqrt{3z}$ **k.** $2y\sqrt[3]{x^2}$ **l.** $5xy\sqrt[3]{2x}$ **13. a.** $\dfrac{4}{5}$ **b.** $\dfrac{10}{7}$ **c.** $\dfrac{10}{3}$ **d.** $\dfrac{\sqrt[3]{2}}{2}$ **e.** $\dfrac{2\sqrt{15}}{7}$ **f.** $\dfrac{4\sqrt{5}}{15}$ **g.** $\dfrac{11x\sqrt{2}}{13}$

**h.** $\dfrac{15a^2\sqrt{2}}{14}$  **14. a.** 0  **b.** $2\sqrt{3}$  **c.** $18\sqrt{5}$  **d.** $\sqrt{7}$  **e.** $5x\sqrt{2y}$  **f.** $y^2\sqrt{5xy}$  **g.** $5\sqrt[3]{2}$  **h.** $6x\sqrt[3]{2}$  **15. a.** $-6\sqrt{6}$  **b.** $10x$
**c.** $36x\sqrt{2}$  **d.** $-6y^3\sqrt{6}$  **e.** $4\sqrt[3]{2}$  **f.** $-24x\sqrt[3]{x}$  **g.** $-2$  **h.** $2y\sqrt{3}+15\sqrt{2y}$  **i.** $-2$  **j.** $15+6x\sqrt{15}+9x^2$
**k.** $\sqrt[3]{9}+\sqrt[3]{3}-2$  **l.** $\sqrt[3]{25}-1$  **16. a.** $\dfrac{\sqrt{7}}{7}$  **b.** $\dfrac{\sqrt{2}}{2}$  **c.** $2\sqrt[3]{4}$  **d.** $\dfrac{5\sqrt[3]{2}}{2}$  **17. a.** $7(\sqrt{2}-1)$  **b.** $\dfrac{3\sqrt{3}+3}{2}$  **c.** $5-\sqrt{15}$
**d.** $\dfrac{4+\sqrt{7}}{3}$  **18. a.** 7  **b.** $-10$  **c.** 216  **d.** $\dfrac{32}{243}$  **e.** 64  **f.** 25  **g.** $x^{1/5}$  **h.** $\dfrac{1}{r}$  **i.** 6  **j.** 5  **k.** $x$  **l.** $\dfrac{1}{a^2b^4}$  **m.** $x^{11/15}$
**n.** $t^{1/12}$  **o.** $\dfrac{1}{x^{4/5}}$  **p.** $\dfrac{1}{r^{1/4}}$

## Chapter 7 Test  (page 552)

**1.** 10  **2.** $-20$  **3.** $-3$  **4.** $9x$  **5.** 13 in.  **6.** 10 ft  **7.** $2x\sqrt{2}$  **8.** $3x\sqrt{6xy}$  **9.** $4\sqrt{2}$  **10.** $3y\sqrt{x}$  **11.** $x^2y^2$  **12.** $\dfrac{2x^2}{y}$  **13.** 36
**14.** 66  **15.** 5  **16.** 5  **17.** 5; 0 is extraneous  **18.** 29  **19.** 10 units  **20.** 5 units  **21.** $5\sqrt{3}$  **22.** $-x\sqrt{2x}$  **23.** $-24x\sqrt{6}$
**24.** $2\sqrt{6}+3\sqrt{2}$  **25.** $-1$  **26.** $2x-4\sqrt{x}-6$  **27.** $\sqrt{2}$  **28.** $\dfrac{y\sqrt{xy}}{4x}$  **29.** $2\sqrt{5}+4$  **30.** $\dfrac{x\sqrt{3}-2\sqrt{3x}}{x-4}$  **31.** 11  **32.** $\frac{1}{81}$
**33.** $y^6$  **34.** $a^{10}$  **35.** $p^{17/12}$  **36.** $\dfrac{1}{x^{4/15}}$

## Getting Ready  (page 555)
**1.** 1  **2.** 3  **3.** $-3$  **4.** $-\frac{1}{2}$

## Orals  (page 564)
**1.** 3  **2.** 2  **3.** yes  **4.** 5  **5.** yes

## Exercise 8.1  (page 564)
**1.** $x^9y^6$  **3.** $\dfrac{x^{12}}{y^8}$  **5.** 1  **7.** $y, x$  **9.** run  **11.** undefined  **13.** perpendicular  **15.** 3  **17.** $-1$  **19.** $-\frac{1}{3}$  **21.** 0  **23.** undefined
**25.** $-1$  **27.** $-\frac{3}{2}$  **29.** $\frac{3}{4}$  **31.** $\frac{1}{2}$  **33.** 0  **35.** negative  **37.** positive  **39.** undefined  **41.** perpendicular  **43.** neither
**45.** parallel  **47.** parallel  **49.** perpendicular  **51.** neither  **53.** not the same line  **55.** not the same line  **57.** same line
**59.** $y=0, m=0$  **61.** $\frac{1}{220}$  **63.** $\frac{1}{3}$  **65.** 3.5 students per yr  **67.** \$642.86 per year  **71.** 4

## Getting Ready  (page 567)
**1.** $y=-2x+12$  **2.** $y=\frac{2x-9}{4}$

## Orals  (page 577)
**1.** $y-3=2(x-2)$  **2.** $y-8=2(x+3)$  **3.** $y=-3x+5$  **4.** $y=-3x-7$  **5.** parallel  **6.** perpendicular

## Exercise 8.2  (page 577)
**1.** 6  **3.** $-1$  **5.** 20 oz  **7.** $y-y_1=m(x-x_1)$  **9.** $-3, 7$  **11.** $Ax+By=C$  **13.** horizontal  **15.** $8x-y=-7$
**17.** $3x+y=12$  **19.** $2x-3y=-11$  **21.** $y=x$  **23.** $y=-\frac{1}{3}x+5$  **25.** $y=\frac{5}{4}x-\frac{1}{2}$  **27.** $y=5x+19$  **29.** $y=-9x+68$
**31.** $y=8$  **33.** $y=-\frac{1}{2}x+11$

**35.** 1, $(0, -1)$ 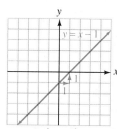   **37.** $\frac{2}{3}$, $(0, 2)$ 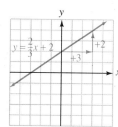   **39.** $-\frac{2}{3}$, $(0, 6)$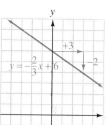

**41.** $\frac{3}{2}$, $(0, -4)$   **43.** $-\frac{1}{3}$, $\left(0, -\frac{5}{6}\right)$   **45.** $\frac{7}{2}$, $(0, 2)$   **47.** parallel   **49.** perpendicular   **51.** parallel   **53.** perpendicular
**55.** perpendicular   **57.** perpendicular   **59.** $y = 4x$   **61.** $y = 4x - 3$   **63.** $y = \frac{4}{5}x - \frac{26}{5}$   **65.** $y = -\frac{1}{4}x$   **67.** $y = -\frac{1}{4}x + \frac{11}{2}$
**69.** $y = -\frac{5}{4}x + 3$   **71.** perpendicular   **73.** parallel   **75.** $x = -2$   **77.** $x = 5$   **79.** $y = -\frac{A}{B}x + \frac{C}{B}$   **81.** $y = -3{,}200x + 24{,}300$
**83.** $y = 47{,}500x + 475{,}000$   **85.** $y = -\frac{710}{3}x + 1{,}900$   **87.** $90   **89.** $890   **91.** $37,200   **93.** $230   **95.** about 838
**99.** $a < 0, b > 0$

## Getting Ready  (page 582)

**1.** $-2$   **2.** 2   **3.** $-1$   **4.** 2   **5.** 1   **6.** $-7$   **7.** 2   **8.** $\frac{9}{8}$

## Orals  (page 589)

**1.** 1   **2.** 3   **3.** $-1$   **4.** $-3$

## Exercise 8.3  (page 589)

**1.** $-2$   **3.** 6   **5.** input, function   **7.** range   **9.** independent   **11.** cannot   **13.** 9, 0, $-3$   **15.** 3, $-3$, $-5$   **17.** 22, 7, 2
**19.** 3, 9, 11   **21.** 1, 4, 9   **23.** 0, $-9$, 26   **25.** 4, 1, 16   **27.** 1, 10, 15   **29.** 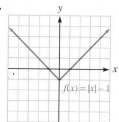   **31.**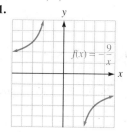
**33.** D = all reals; R = all reals $\le 2$   **35.** D = all reals;
R = all reals $\ge -2$   **37.** yes   **39.** no   **41.** $C = 0.10n + 12$
**43.** $14   **45.** $C = 0.30n + 15$   **47.** $28.50

## Getting Ready  (page 591)

**1.** 4   **2.** 16   **3.** $k = \frac{A}{bh}$   **4.** $k = \frac{PV}{T}$

## Orals  (page 597)

**1.** inverse   **2.** direct   **3.** combined   **4.** joint   **5.** direct   **6.** combined   **7.** joint   **8.** direct

## Exercise 8.4  (page 597)

**1.** $8x + 5$   **3.** $-3x^3 + 6x^2 - 6x$   **5.** direct   **7.** linear   **9.** joint   **11.** $d = kn$   **13.** $l = \frac{k}{w}$   **15.** $A = kr^2$   **17.** $d = kst$   **19.** $I = \frac{kV}{R}$
**21.** 35   **23.** 42   **25.** 675   **27.** 1   **29.** $\frac{80}{3}$   **31.** 24   **33.** 4   **35.** 80   **37.** $\frac{5}{2}$   **39.** 576 ft   **41.** $2,450   **43.** $2\frac{1}{2}$ hr   **45.** $53\frac{1}{3}$ m$^3$
**47.** $270   **49.** $3\frac{9}{11}$ amp

## Chapter Summary  (page 601)

**1. a.** $-1$   **b.** $-\frac{5}{4}$   **c.** $-\frac{1}{2}$   **d.** 0   **2. a.** $\frac{4}{3}$   **b.** $\frac{1}{4}$   **3. a.** positive   **b.** 0   **c.** undefined   **d.** negative   **4. a.** perpendicular
**b.** parallel   **c.** neither   **d.** parallel   **e.** perpendicular   **f.** parallel   **5.** $\frac{1}{4}$   **6.** $20,500 per year   **7. a.** $3x - y = 0$   **b.** $x + 3y = 3$

**c.** $x - 9y = -9$   **d.** $3x + 5y = 2$   **8. a.** $5, (0, 2)$   **b.** $-\frac{1}{2}, (0, 4)$   **c.** $0, (0, -3)$   **d.** $-\frac{1}{3}, \left(0, \frac{1}{3}\right)$   **9. a.** $3x + y = 2$   **b.** $y = -7$
**c.** $7x - y = 0$   **d.** $x - 2y = 3$

**10. a.** 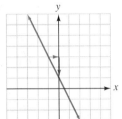   **b.** 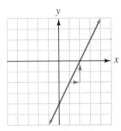   **c.** 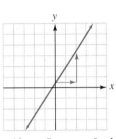   **d.**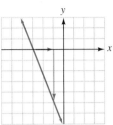

**11. a.** neither   **b.** perpendicular   **c.** parallel   **d.** perpendicular   **12. a.** $7x - y = 9$   **b.** $3x + 2y = 1$   **c.** $5x + 2y = 0$
**d.** $3x + y = -4$   **13. a.** $-\frac{4}{5}$   **b.** $\frac{3}{5}$   **14.** $1,200   **15. a.** $1$   **b.** $7$   **c.** $7$   **d.** $0.75$

**16. a.** 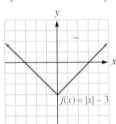   **b.**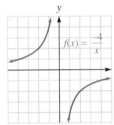

**17. a.** D = all reals; R = all reals $\geq -3$   **b.** D = all reals but 0;
R = all reals but 0   **18. a.** no   **b.** yes   **19. a.** $C = 5x + 50$
**b.** $70   **c.** $\bar{c} = \frac{5x + 50}{x}$   **d.** $11.25   **20. a.** $400$   **b.** $600$   **c.** $81$
**d.** $\frac{5}{7}$

## Chapter 8 Test   (page 604)

**1.** $\frac{4}{3}$   **2.** $-1$   **3.** $-2$   **4.** $(0, 21)$   **5.** $0$   **6.** undefined   **7.** equal   **8.** $-1$   **9.** perpendicular   **10.** parallel   **11.** $\frac{1}{4}$   **12.** $16,666.67
**13.** $2$   **14.** $-\frac{1}{2}$   **15.** $7x - y = -19$   **16.** $x - 2y = -6$   **17.** $x = -3$   **18.** $3x + y = 4$   **19.** $4$   **20.** $-11$
**21.** 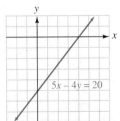 D = all reals; R = all reals $\leq 4$   **22.** 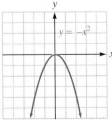 D = all reals but 0; R = all reals but 0
**23.** no   **24.** yes   **25.** $1$   **26.** $400$

## Cumulative Review Exercises   (page 606)

**1.** $6x^3 + 2x + 1$   **2.** $3x^3 + 3x^2 + x + 1$   **3.** $16x^2 - 5x - 1$   **4.** $19x^2 - 28x - 11$   **5.** $-15x^6y^6$   **6.** $15x^5 + 9x^4 - 6x^2$
**7.** $6x^2 + 17x + 12$   **8.** $12x^2 - 17xy + 6y^2$   **9.** $x - 4$   **10.** $x^2 + x + \frac{-3}{2x - 1}$   **11.** $2xy(x - 2y)$   **12.** $(x + y)(5 + a)$
**13.** $(a + b)(3 + b)$   **14.** $(7p^2 + 4q)(7p^2 - 4q)$   **15.** $(x - 12)(x + 3)$   **16.** $(x - 5y)(x + 2y)$   **17.** $(3a + 4)(4a - 5)$
**18.** $(5m + n)(2m - 3n)$   **19.** $(p - 4q)(p^2 + 4pq + 16q^2)$   **20.** $2(r + 3s)(r^2 - 3rs + 9s^2)$   **21.** $-1, -2$   **22.** $\frac{3}{2}, -4$   **23.** $2$   **24.** $2$
**25.** $-\frac{3}{2}$   **26.** $\frac{8}{11}$   **27.**   **28.**   **29.** $\frac{1}{2}$   **30.** $-\frac{1}{2}$   **31.** $y = \frac{2}{3}x + 5$   **32.** $3x - 4y = -22$

**33.** perpendicular   **34.** parallel   **35.**

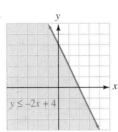

**36.**

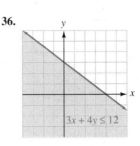

**37.** $-3$   **38.** 15   **39.** 5
**40.** $8x^2 - 3$   **41.** 12   **42.** 2

## Getting Ready (page 609)

**1.** $2x(2x - 1)$   **2.** $(x + 3)(x - 3)$   **3.** $(x + 3)(x - 2)$   **4.** $(2x - 3)(x + 3)$   **5.** 5   **6.** $-6$   **7.** $2\sqrt{5}$   **8.** $-5\sqrt{2}$

## Orals (page 617)

**1.** 5, $-5$   **2.** 2, 3   **3.** $\pm 10$   **4.** $\pm 7$

## Exercise 9.1 (page 617)

**1.** $y^2 - 2y + 1$   **3.** $x^2 + 2xy + y^2$   **5.** $4r^2 - 4rs + s^2$   **7.** quadratic   **9.** 0, 0   **11.** two   **13.** 2, $-3$   **15.** 4, $-1$   **17.** $\frac{5}{2}$, $-2$
**19.** 1, $-2$, 3   **21.** $\pm 3$   **23.** 0, $-3$   **25.** 2, 3   **27.** $-1$, $\frac{2}{3}$   **29.** $-\frac{1}{3}$, $-\frac{3}{2}$   **31.** $\frac{2}{5}$, $-\frac{1}{2}$   **33.** $\pm 1$   **35.** $\pm 3$   **37.** $\pm 2\sqrt{5}$ ($\pm 4.5$)
**39.** $\pm 3$   **41.** $\pm 2$   **43.** $\pm\sqrt{a}$   **45.** $-6$, 4   **47.** 7, $-11$   **49.** $2 \pm 2\sqrt{2}$ (4.8 and $-0.8$)   **51.** $3a$, $-a$   **53.** $-b \pm 4c$   **55.** 0, $-4$
**57.** 2, $-\frac{2}{3}$   **59.** $\dfrac{-1 \pm 2\sqrt{5}}{2}$ (1.7 and $-2.7$)   **61.** $\pm 3$   **63.** $\pm 5$   **65.** $-1 \pm 2\sqrt{2}$ (1.8 and $-3.8$)   **67.** 6, 8   **69.** $-6$, $-4$
**71.** 4 ft by 8 ft   **73.** 6 in.   **75.** 10 ft by 18 ft   **77.** about 9.5 sec   **79.** 288 ft   **81.** 324 ft   **83.** 68 sec

## Getting Ready (page 620)

**1.** 9   **2.** 25   **3.** 1   **4.** $\frac{25}{4}$   **5.** 16   **6.** 36   **7.** $\frac{1}{16}$   **8.** $\frac{1}{9}$

## Orals (page 624)

**1.** 4   **2.** 9   **3.** 16   **4.** 25   **5.** $\frac{25}{4}$   **6.** $\frac{9}{4}$

## Exercise 9.2 (page 624)

**1.** $-4$   **3.** 3   **5.** completing   **7.** $(x + b)^2$   **9.** square, 8   **11.** $x^2 + 2x + 1$   **13.** $x^2 - 4x + 4$   **15.** $x^2 + 7x + \frac{49}{4}$   **17.** $a^2 - 3a + \frac{9}{4}$
**19.** $b^2 + \frac{2}{3}b + \frac{1}{9}$   **21.** $c^2 - \frac{5}{2}c + \frac{25}{16}$   **23.** $-2$, $-4$   **25.** 2, 6   **27.** 5, $-3$   **29.** 3, 4   **31.** 1, $-6$   **33.** 1, $-2$   **35.** $-4$, $-4$
**37.** 2, $-\frac{1}{2}$   **39.** $-2$, $\frac{1}{4}$   **41.** $-2 \pm \sqrt{3}$ ($-0.3$ and $-3.7$)   **43.** $1 \pm \sqrt{5}$ (3.2 and $-1.2$)   **45.** $2 \pm \sqrt{7}$ (4.6 and $-0.6$)
**47.** $-1 \pm \sqrt{2}$ (0.4 and $-2.4$)   **49.** 1, $-4$   **51.** $\frac{3}{2}$, $-\frac{2}{3}$

## Getting Ready (page 625)

**1.** $-8$   **2.** $-7$   **3.** 8   **4.** 0

## Orals (page 632)

**1.** $a = 3$, $b = 4$, $c = -12$   **2.** $a = -2$, $b = -4$, $c = 10$   **3.** $a = 5$, $b = -1$, $c = -1$   **4.** $a = 1$, $b = 3$, $c = -9$

## Exercise 9.3   (page 632)

**1.** $r = \dfrac{A - p}{pt}$   **3.** $3x - 5y = -60$   **5.** $4\sqrt{5}$   **7.** $\dfrac{\sqrt{7x}}{7}$   **9.** $ax^2 + bx + c = 0$   **11.** 0   **13.** $-4, 8, 0$   **15.** $a = 1, b = 4, c = 3$

**17.** $a = 3, b = -2, c = 7$   **19.** $a = 4, b = -2, c = 1$   **21.** $a = 3, b = -5, c = -2$   **23.** $a = 7, b = 14, c = 21$   **25.** $a = 1,$

$b = -1, c = -5$   **27.** 2, 3   **29.** $-3, -4$   **31.** $1, -\dfrac{1}{2}$   **33.** $-1, -\dfrac{2}{3}$   **35.** $\dfrac{1}{2}, -\dfrac{3}{2}$   **37.** $2, -\dfrac{2}{5}$   **39.** $\dfrac{-3 \pm \sqrt{5}}{2}$ $(-0.4, -2.6)$

**41.** $\dfrac{-5 \pm \sqrt{37}}{2}$ $(0.5, -5.5)$   **43.** no real solutions   **45.** $\dfrac{-1 \pm \sqrt{41}}{4}$ $(1.4, -1.9)$   **47.** $-2 \pm \sqrt{3}$ $(-0.3, -3.7)$   **49.** no real

solutions   **51.** $-1 \pm \sqrt{2}$ $(0.4, -2.4)$   **53.** $\dfrac{3 \pm \sqrt{15}}{3}$ $(2.3, -0.3)$   **55.** 6 in. by 8 in.   **57.** 30 and 40 nautical miles   **59.** 7%

**61.** \$210,000   **63.** 2 in.   **65.** 4 hr

## Getting Ready   (page 635)

**1.** 2   **2.** 3   **3.** 5   **4.** 12   **5.** $-2$   **6.** 2

## Orals   (page 645)

**1.** up   **2.** up   **3.** down   **4.** down   **5.** down   **6.** up

## Exercise 9.4   (page 645)

**1.** $5\sqrt{3}$   **3.** 2   **5.** quadratic   **7.** $y$-intercept   **9.** $a > 0, a < 0$   **11.** $(0, c)$

**13.** 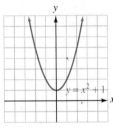   **15.** 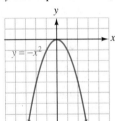   **17.** 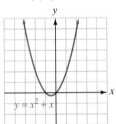   **19.**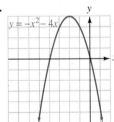

**21.** 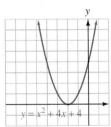   **23.** 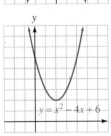   **25.** $(2, 4)$   **27.** $(-1, -5)$   **29.** $(1, 0)$   **31.** $(0, 4)$   **33.** $(-1, 4)$

**35.** $(3, -21)$   **37.** 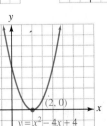   **39.** 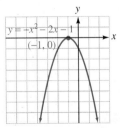   **41.**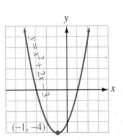

**43.**

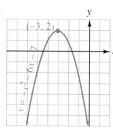

**45.**

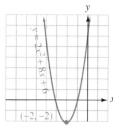

**47.**

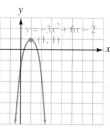

**49.** $\left(\frac{1}{2}, -\frac{9}{4}\right)$, (2, 0), (−1, 0), (0, −2)

**51.** (1, 4), (3, 0), (−1, 0), (0, 3)

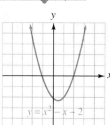

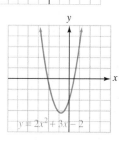

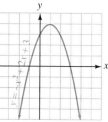

**53.** $\left(-\frac{3}{4}, -\frac{25}{8}\right)$, (−2, 0), $\left(\frac{1}{2}, 0\right)$, (0, −2)

**55.** 1,350   **57.** 750   **59.** 2, 3   **61.** −0.44, −4.56

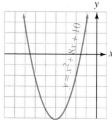

## Chapter Summary   (page 650)

**1. a.** 0, −2   **b.** 0, 3   **c.** 3, −3   **d.** 5, −5   **e.** 3, 4   **f.** 5, −3   **g.** 6, −4   **h.** 1, −$\frac{3}{2}$   **i.** 3, 4   **2. a.** ±5   **b.** ±6   **c.** ±3
**d.** ±$\frac{3}{2}$   **e.** ±2√2 (±2.8)   **f.** ±5√3 (±8.7)   **3. a.** −4, 6   **b.** 3, −9   **c.** 2, −4   **d.** $\frac{7}{2}, \frac{1}{2}$   **e.** 8 ± 2√2 (10.8 and 5.2)
**f.** −5 ± 5√3 (3.7 and −13.7)   **4.** 15 ft   **5.** 15 sec   **6.** 3 ft by 9 ft   **7.** 3 ft by 6 ft   **8. a.** $x^2 + 4x + 4$   **b.** $y^2 + 8y + 16$
**c.** $z^2 − 10z + 25$   **d.** $t^2 − 5t + \frac{25}{4}$   **e.** $a^2 + \frac{3}{4}a + \frac{9}{64}$   **f.** $c^2 − \frac{7}{3}c + \frac{49}{36}$   **9. a.** 2, −7   **b.** 3, 5   **c.** 7, −11

**d.** 1 ± √2 (2.4 and −0.4)   **e.** −2 ± √7 (0.6 and −4.6)   **f.** 3 ± √5 (5.2 and 0.8)   **g.** $\frac{1}{2}$, −3   **h.** $\dfrac{1 ± \sqrt{3}}{2}$ (1.4 and −0.4)

**10. a.** 5, −3   **b.** 7, −1   **c.** 13, 2   **d.** $\frac{1}{2}$, 3   **e.** $\frac{3}{2}$, −$\frac{1}{3}$   **f.** −2 ± √3 (−0.3 and −3.7)   **g.** 3 ± √2 (4.4 and 1.6)   **h.** 0, −3
**11.** 68 cm   **12. a.**   **b.**   **13. a.** (6, 7)   **b.** (−3, −5)   **c.** (1, 5)   **d.** (3, 16)

## Chapter 9 Test   (page 653)

**1.** $\frac{1}{3}$, −$\frac{1}{2}$   **2.** $\frac{1}{5}$, −$\frac{9}{2}$   **3.** 4, −4   **4.** 2 ± √3 (3.7 and 0.3)   **5.** 49   **6.** $\frac{49}{4}$   **7.** −1 ± √5 (1.2 and −3.2)

**8.** $x = \dfrac{-b \pm \sqrt{b^2 - 4ac}}{2a}$   **9.** $2, -5$   **10.** $-\frac{3}{2}, 4$   **11.** $\dfrac{-5 \pm \sqrt{33}}{4}$ (0.2 and $-2.7$)   **12.** 10 m   **13.** $(-5, -4)$

**14.**

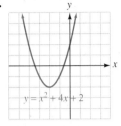

$y = x^2 + 4x + 2$

## Getting Ready   (page A-4)

**1.** $5x - 1$   **2.** $x - 7$   **3.** $x^2 - x - 20$   **4.** $6x^2 - 5x + 1$

## Orals   (page A-11)

**1.** $5 + 7i$   **2.** $1 + 5i$   **3.** $1 + 3i$   **4.** $1 - 3i$   **5.** $3 - i$   **6.** $3 + i$

## Exercise A.1   (page A-11)

**1.** $\pm 3i$   **3.** $\pm \dfrac{4\sqrt{3}}{3}i$   **5.** $-1 \pm i$   **7.** $\dfrac{-1}{4} \pm \dfrac{\sqrt{7}}{4}i$   **9.** $\dfrac{2}{3} \pm \dfrac{\sqrt{2}}{3}i$   **11.** $i$   **13.** $-i$   **15.** $1$   **17.** $i$   **19.** $8 - 2i$   **21.** $3 - 5i$

**23.** $15 + 7i$   **25.** $-15 + 2\sqrt{3}\,i$   **27.** $3 + 6i$   **29.** $9 + 7i$   **31.** $14 - 8i$   **33.** $8 + \sqrt{2}\,i$   **35.** $6 - 8i$

**37.** $6 + \sqrt{6} + \left(3\sqrt{3} - 2\sqrt{2}\right)i$   **39.** $-16 - \sqrt{35} + \left(2\sqrt{5} - 8\sqrt{7}\right)i$   **41.** $0 - i$   **43.** $0 + \dfrac{4}{5}i$   **45.** $\dfrac{1}{8} + 0i$   **47.** $0 + \dfrac{3}{5}i$

**49.** $0 + \dfrac{3\sqrt{2}}{4}i$   **51.** $\dfrac{15}{26} - \dfrac{3}{26}i$   **53.** $-\dfrac{42}{25} - \dfrac{6}{25}i$   **55.** $\dfrac{1}{4} + \dfrac{3}{4}i$   **57.** $\dfrac{5}{13} - \dfrac{12}{13}i$   **59.** $\dfrac{6 + \sqrt{10}}{9} + \dfrac{2\sqrt{2} - 3\sqrt{5}}{9}i$   **61.** $10$

**63.** $13$   **65.** $\sqrt{74}$   **67.** $3\sqrt{2}$   **69.** $\sqrt{69}$   **71.** $5$